W9-COJ-997

MEET THE TEAM BEHIND THE BEST TEST PREP FOR THE SAT

The best minds to help you get the best SAT scores

In this book, you'll find our commitment to excellence, an enthusiasm for the subject matter, and an unrivaled ability to help you master the SAT. REA's dedication to excellence and our passion for education make this book the very best test prep for the SAT.

Mel H. Friedman, M.S., has a diversified background in mathematics and has developed test items for Educational Testing Service. His teaching experience is at both the high school and college levels. He has also authored and co-authored books on SAT preparation.

Sally Wood, M.S., has been a certified exam reader for Educational Testing Service and a test-item writer for ACT, Inc. She has taught Advanced Placement and dual-credit courses, and has counseled college-bound students on such topics as academic preparation, test-taking, and college admissions.

Adel Arshaghi, M.S., has written and contributed to a number of books on high school and college math on topics ranging from geometry to statistics to the peculiarities of math items on the SAT. He has taught mathematics at both the secondary and college levels.

Michael Lee, M.A.T., is an SAT tutor and teaches high school literature and expository writing. He began his teaching career in 1996 as an adjunct professor at New Jersey's Mercer County Community College, where he still teaches composition on a part-time basis.

Michael McIrvin, M.A., taught university writing and literature courses for more than 12 years and has been a professional editor for more than a decade.

Penny Luczak, M.A., teaches math as a faculty member at Camden County College in New Jersey.

Drew D. Johnson, B.A., has written and contributed to a number of books on test preparation for grades K–12. He has wide-ranging experience as an educational writer, editor, and consultant.

Lina Miceli, M.A., worked for many years as a high school guidance counselor in New Jersey.

We also gratefully acknowledge the following for their editorial contributions:

Robert A. Bell, Ph.D.; Suzanne Coffield, M.A.; Anita Price Davis, Ed.D.; George DeLuca, J.D.; Joseph D. Fili, M.A.T.; Marilyn B. Gilbert, M.A.; Bernice E. Goldberg, Ph.D.; Leonard A. Kenner; Gary Lemco, Ph.D.; Marcia Mungenast; Sandra B. Newman, M.A.; and Richard C. Schmidt, Ph.D.

Research & Education Association

The Very Best Coaching & Study Course for the

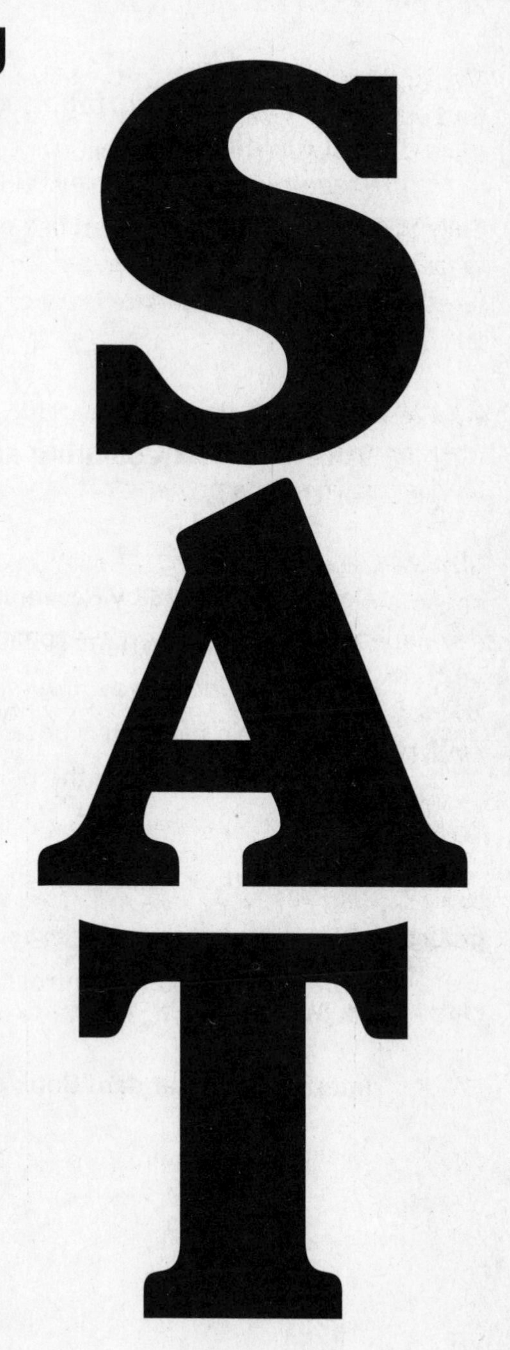

Visit our website at:

www.REA.com

Research & Education Association
61 Ethel Road West
Piscataway, New Jersey 08854
E-mail: info@rea.com

The Very Best Coaching and Study Course for the SAT

Printed in the United States of America

Library of Congress Control Number 2006923663

International Standard Book Number 0-7386-0056-3

D06

CONTENTS

ABOUT RESEARCH & EDUCATION ASSOCIATION

Founded in 1959, Research & Education Association is dedicated to publishing the finest and most effective educational materials—including software, study guides, and test preps—for students in middle school, high school, college, graduate school, and beyond.

REA's Test Preparation series includes books and software for all academic levels in almost all disciplines. Research & Education Association publishes test preps for students who have not yet completed high school, as well as high school students preparing to enter college. Students from countries around the world seeking to attend college in the United States will find the assistance they need in REA's publications. For college students seeking advanced degrees, REA publishes test preps for many major graduate school admission examinations in a wide variety of disciplines, including engineering, law, and medicine. Students at every level, in every field, with every ambition can find what they are looking for among REA's publications.

REA's practice tests are always based upon the most recently administered exams, and include every type of question that you can expect on the actual exams.

REA's publications and educational materials are highly regarded and continually receive an unprecedented amount of praise from professionals, instructors, librarians, parents, and students. Our authors are as diverse as the fields represented in the books we publish. They are well-known in their respective disciplines and serve on the faculties of prestigious high schools, colleges, and universities throughout the United States and Canada.

Today, REA's wide-ranging catalog is a leading resource for teachers, students, and professionals.

We invite you to visit us at *www.rea.com* to find out how "REA is making the world smarter."

STAFF ACKNOWLEDGMENTS

We would like to thank REA's Larry B. Kling, Vice President, Editorial, for supervising development; Pam Weston, Vice President, Publishing, for setting the quality standards for production integrity and managing the publication to completion; Anne Winthrop Esposito, Senior Editor, for coordinating revisions; Diane Goldschmidt, Associate Editor, for post-production quality assurance; Jeremy Rech, Graphic Artist, for interior page design; Christine Saul, Senior Graphic Designer, for cover design; and Jeff LoBalbo, Senior Graphic Designer, and Wende Solano, for post-production file mapping.

We also gratefully acknowledge the team at Publication Services for editing, proofreading, and page composition.

SAT STUDY SCHEDULE

The following study schedule will allow you enough time to master the SAT's content and format. Although it is designed for ten weeks, it can be condensed into a five-week course by combining each two-week period into one. If you are not enrolled in a structured course, be sure to budget enough time to study—about one hour each day for the regular schedule, or two hours if you're on the five-week schedule. If possible, set aside a four-hour block of time in order to take a Practice Test in one sitting.

Vocabulary is best learned by repetition, so plan to spend at least 15 minutes each day studying it in whatever manner works for you.

No matter which study schedule works best for you, the more time you spend preparing, the more relaxed and ready you will feel on the day of the exam.

Week	Activities
1	• Read and study Chapter 1, which will introduce you to the SAT. (30 minutes) • Take the Diagnostic Test, found in Chapter 2, to identify your strengths and weaknesses. (3 hours, 45 minutes) • Score each section using the Answer Key at the end of the test. The procedure for calculating your score is found in Chapter 1. The third column in the Answer Key lists the study topic associated with each question. Identify your weaknesses by highlighting the concepts tested by questions you missed. (You can concentrate on these concepts in later weeks.) (30 minutes) • Review the explanations for questions you missed (or guessed on) in the Diagnostic Test. (30 minutes)
2	• Begin the Mastering Vocabulary section in Chapter 3. This week, study the words and complete the drills for Groups 1–3. (1 hour) • Read and complete the drills in Chapter 3, Part II: Mastering Word Parts. (1 hour, 30 minutes) • Read and complete the drills in Chapter 4: Mastering Sentence Completion Questions. (2 hours) • Check your drill answers and revisit the review section for any type of problem that gave you trouble. (30 minutes) • Read the section corresponding to your weakest area, as identified by the Diagnostic Test. (1 hour)
3	• Read and complete the drills in Chapter 5: Mastering Reading Comprehension Questions. (3 hours, 30 minutes) • Read and complete the drills in Chapter 6, Part I: Writing. (1 hour, 30 minutes) • Check your drill answers and revisit the review section for any type of problem that gave you trouble. (30 minutes) • Review the vocabulary from Groups 1–3. (30 minutes) • This week, master the words and complete the drills for Groups 4–6. (1 hour)

Week	Activities
4	• Read and complete the drills in Chapter 7: Mastering Basic Math Skills. (6 hours) • Check your drill answers and revisit the review section for any type of problem that gave you trouble. (1 hour) • Review the vocabulary from Groups 1–6. (15 minutes) • No new vocabulary this week!
5	• Go back to Chapter 7 and reread the review sections for any type of problem that gave you trouble. (30 minutes) • Read and complete the drill in Chapter 6, Part II: Mastering the Essay. (2 hours) • Score and analyze your essay by comparing it to the sample essays provided at the end of the chapter. Revisit the review sections for any problem areas you identify. (30 minutes) • Read and complete the drills in Chapter 8: Mastering Regular Math Questions. (2 hours) • Review the vocabulary from Groups 1–6. (15 minutes) • This week, master the words and complete the drills for Groups 7–9. (1 hour)
6	• Read and complete the drills in Chapter 9: Mastering Student-Produced Response Questions. (3 hours, 30 minutes) • Check your drill answers and revisit the review section for any type of problem that gave you trouble. (30 minutes) • Review the vocabulary from Groups 1–9. (30 minutes) • This week, master the words and complete the drills for Groups 10–12. (1 hour)
7	• Take Practice Test 1, found on page 449. (3 hours, 45 minutes) • Score your exam, then carefully review the answer explanations for any questions you missed. (45 minutes) • If there are any types of questions that are particularly difficult for you, go over the corresponding topical reviews again. (30 minutes) • Review the vocabulary from Groups 1–12. (30 minutes) • Begin to study the Additional Vocabulary from the list that begins on page 129. This week, master the words on the first three pages, *abandon–flinch*. (1 hour, 30 minutes)
8	• Take Practice Test 2, found on page 541. (3 hours, 45 minutes) • Score your exam, then carefully review the answer explanations for any questions you missed. (45 minutes) • If there are any types of questions that are particularly difficult for you, go over the corresponding topical reviews again. (30 minutes) • Review the vocabulary from all previous weeks. (30 minutes) • Master the Additional Vocabulary on the next three pages, *fluency–quarantine*. (1 hour, 30 minutes)

Week	Activities
9	• Take Practice Test 3, found on page 625. (3 hours, 45 minutes) • Score your exam, then carefully review the answer explanations for any questions you missed. (45 minutes) • If there are any types of questions that are particularly difficult for you, go over the corresponding topical reviews again. (30 minutes) • Review the vocabulary from all previous weeks. (30 minutes) • Master the Additional Vocabulary on the next three pages, *quiescent–vindicate*. (1 hour, 30 minutes)
10	• Take Practice Test 4, found on page 721. (3 hours, 45 minutes) • Score your exam, then carefully review the answer explanations for any questions you missed. (45 minutes) • If there are any types of questions that are particularly difficult for you, go over the corresponding topical reviews again. (30 minutes) • Review the vocabulary from all previous weeks. (1 hour, 30 minutes) • Master the Additional Vocabulary on the last page, *vivacious–zephyr*. (15 minutes)

CHAPTER 1

MASTERING THE SAT

MASTERING THE SAT

THE BEST TEST PREP

You chose well!

The Very Best Coaching and Study Course for the SAT has all you need to master the SAT.

This book brings to bear the decades' worth of wisdom, hard-earned academic credentials, and—perhaps of greater interest to you—the kind of street smarts that comes from years of hands-on experience in preparing students for standardized tests.

Let's look at what's inside:

- The entire SAT Reasoning Test, from Critical Reading to Writing to Math, is covered within REA's comprehensive yet concise review. All the best test-taking tips, strategies, and preparation techniques are also provided.
- There's a diagnostic test to help you pinpoint your strengths and weaknesses—a real confidence booster to get you into the swing of things!
- Our four full-length practice exams, painstakingly crafted to depict the very exam administered by ETS (Educational Testing Service), will familiarize you with the SAT's format and level of difficulty. We take nothing for granted: Each practice-exam answer is thoroughly explained so you won't be left in the lurch.
- Because a strong vocabulary is bound to pillar your performance on two-thirds of the test, we not only isolate the types of words you're most likely to find on the SAT but also ensure you won't come up short, by providing the most frequently occurring Greek and Latin roots (see Appendix A). This way, even if you don't know the exact meaning of a word, you'll still have the basis for hunches that could be worth serious points.

Think of this book as a toolbox. We'll give you all the tools and their uses. By building a strong foundation in the elements that make up the SAT, you will be able to master the SAT itself.

THE SAT AT A GLANCE

Total testing time: 3 hours, 45 minutes

MATHEMATICS	CRITICAL READING*	WRITING	SCORING
• 70 minutes • Features two 25-minute sections and one 20-minute section. Includes Algebra II. • Includes multiple-choice questions and student-produced responses (which you may know as grid-ins).	• 70 minutes • Features two 25-minute sections and one 20-minute section. • Includes short reading passages added to the SAT's traditional paragraph-length passages. * *Formerly called Verbal*	• 60 minutes • 25 minutes allotted to analyze and respond to an assigned topic in essay form. • Your Essay will be graded on a six-point scale. • 35 minutes allotted for multiple-choice items testing grammar and usage.	• Top score is 2400. Stick with REA and we'll show you how to rake in the points by mastering the SAT's content and format.

TODAY'S SAT: WHAT'S IT ALL ABOUT?

In March 2005, the SAT underwent historic changes. The introduction of the new SAT stands as one of the most highly publicized events in college-admission testing. It featured substantial revisions in the test's two main sections (Critical Reading and Mathematics), as well as the addition of an all-new Writing section. In adding the Writing section, the College Board really did you a big favor! Whether or not you consider yourself a good writer, there's no time like the present to learn to write better. Effective communication by writing is vitally important—to your SAT score, to your college academics, and to your future.

It's also important to understand how SAT scoring works. Each of the three test sections (Critical Reading, Math, Writing) is scored on a 200–800 scale, making 2400 the highest score possible. You will also receive subscores on the multiple-choice and Essay portions of the Writing section. The total testing time is 3 hours, 45 minutes.

The Writing section contains an essay portion that asks you to take a position on an issue and support it with examples from your studies and experience. The question is designed to be open-ended, so you can successfully write your essay in many different ways. You are not required to have any prior specific knowledge about the topic to write your essay.

The Writing section also includes multiple-choice questions that test your ability to identify errors in sentences, improve sentences, and improve paragraphs. You are allotted 60 minutes to complete the Writing section.

The Critical Reading section of the SAT focuses on vocabulary skills. Quite simply, a solid vocabulary is a bedrock requirement for good across-the-board performance on all portions of Critical Reading. Other questions test your ability to read at a quick pace while grasping a solid understanding of the material. The time allotted for this section is 70 minutes.

The Mathematics section of today's SAT includes algebra I, algebra II, arithmetic, and geometry problems. The Student-Produced Response format remains the same as before and still offers the same degree of flexibility in the answering of questions. The time allotted for the Math section is 70 minutes.

ABOUT THE TEST

Who takes the SAT? What is it used for?

Juniors and seniors in high school are the ones most likely to take the SAT. College admissions personnel use your test results as a way to decide if you can be accepted to their school. Because high schools across the nation have a variety of grading systems, the SAT score is designed to put all students on an equal footing. Your SAT score, along with your grades and other school information, helps colleges predict how well you will do at the college level.

If you score poorly on the SAT, it does not mean you should change your plans about going to college. Nor does it mean you will not do well in college. It just means you scored low. Should this happen, remember that you have options:

First, you can register to take the SAT again. Use the time before the next SAT administration to prepare as best you can.

Second, a poor score does not automatically shut the door to all colleges. College admissions officers use several criteria when reviewing applicants including your high school grades, your extracurricular activities, and the levels of your courses in high school.

Who administers the test?

ETS, a client of the College Board, which owns the SAT, develops and scores the test and currently administers it with the assistance of educators across the United States.

When is it best to take the SAT?

You should take the test as a junior or senior in high school. We recommend taking the SAT early in the school year. This allows you more time to retake the test if you are not satisfied with your first set of scores.

When and where do I take the SAT?

The SAT is normally offered seven times a year nationwide. The test can be taken at hundreds of locations throughout the country, including high schools. The standard test day is normally on Saturday, but alternate days are permitted if a conflict—such as a religious obligation—exists.

For information on upcoming SAT testing dates, see your guidance counselor for an SAT Registration Bulletin or request a registration bulletin from ETS as follows:

Educational Testing Service
Rosedale Road
Princeton, NJ 08541
phone: (609) 921–9000 | e-mail: etsinfo@ets.org | www.ets.org

What about the registration fee?

You must pay a fee to register for the SAT. Some students may qualify to have this fee waived. To find out if you qualify for a fee waiver, contact your guidance counselor.

What is the Student Search Service?

The Student Search Service provides your SAT scores to colleges. Colleges enrolled in this service receive information about you, especially if you express interest in their school. On your SAT answer sheet, you can indicate that you want enrollment in this service.

HOW TO USE THIS BOOK

When should I start studying?

Make the best use of your time by making good use of our study schedule (located in the front of this book). The schedule is based on a ten-week program, but it can be shortened to five weeks if you are starting late.

It is never too early to start preparing for the SAT. Time is your ally here. The sooner you begin, the more time you can commit to performing better. It takes time to learn the test materials. It takes time to learn the SAT's format. It takes time to familiarize yourself with the test. Make the most of your time to master the essentials necessary to achieve a higher score.

Where do I begin?

Start with our subject reviews and the test-taking tips and strategies. Then take the diagnostic test to determine your weaknesses. Go back and focus your study on those areas. Reviewing the areas you did well on can help you improve your success on those types of questions. Repeated study of the review material helps build and reinforce your basic skills. After you have reviewed sufficiently, take the practice tests to familiarize yourself with the SAT's format.

I started too late. What do I do now?

You know that last-minute cramming is not the best way to prepare for the SAT but for whatever reason, you're just starting now and the test is right around the corner. Maybe you simply forgot . . . maybe you just procrastinate when it comes to these things. This book can still help you pass.

You won't master the SAT this way, but you may get a good enough score to get into your college of choice (see Appendix B for advice on getting into the right college for you). First, take the Diagnostic Test to pinpoint your strengths and weaknesses. Review the areas where you are weak. If you have time after this, review those areas you are strong in. Review the test-taking tips in this chapter; then take a practice exam. Grade yourself and see where you did well and where you did poorly. Review your weak areas again. Then take another practice exam. Repeat if time permits. With skill and effort (and a little luck), this crash course may help. While we find this last method to be the least desirable, we realize that not everyone can give themselves enough time to do it the right way. Of course, if you score poorly, you can always register for the next exam. Just remember to plan so you do that one right.

What should I expect?

Remember that your results may vary, but with enough time and preparation with this study guide, you are well on your way to mastering the SAT.

FORMAT OF THE SAT

Critical Reading Sections: 70 minutes

There are two types of Critical Reading questions on the SAT:

- Reading Comprehension & Sentence Completion: 67 multiple-choice questions. Reading Comprehension questions follow four reading passages that test your reading comprehension and analysis skills. Sentence Completion questions require you to choose the word or words that best fit the meaning of each sentence provided.

Mathematics Sections: 70 minutes

In the Mathematics sections, you will encounter the following question types that test your algebra, arithmetic, and geometry skills:

- Regular Math: 45 multiple-choice questions that test your general math knowledge.
- Student-Produced Response: 9 questions requiring you to solve problems and then enter your answers onto the provided grid. There are no multiple-choice answers in this section.

Writing Sections: 60 minutes

In the Writing sections, you will answer multiple-choice questions that test your grammar and reasoning skills as well as write an essay similar to the type required on in-class college essay exams.

- Writing (35 minutes): 49 multiple-choice questions to measure your ability to identify sentence errors, improve sentences, and improve paragraphs.
- Student Essay (25 minutes): Write an essay that effectively communicates your viewpoint as well as defines and supports your position.

On the use of calculators

Although solutions can be found to every math problem without them, calculators are permitted during the SAT. You may use a programmable or nonprogrammable four-function, scientific, or graphing calculator. Pocket organizers, hand-held mini PCs, PDAs, paper tape, noisy calculators, and calculators requiring an external power source are not allowed. Sharing calculators is not permitted.

PREPPING WITH OUR STUDY REVIEWS

Use the reviews in this book to help sharpen your basic skills. During your studying, you will acquire strategies for mastering each type of question. You will also find drills to reinforce what you have learned. Using the reviews along with the practice exams will better prepare you for the actual test.

Basic Verbal Skills Review

In this review, you will learn how to build your verbal skills, including an extensive vocabulary enhancer containing many of the words commonly found on the SAT. Separate reviews are provided for both types of verbal question: Critical Reading and Sentence Completion. You will also get a feel for the type of questions you can expect on the SAT, along with step-by-step strategies for mastering them.

Mathematics Review

The math drills in the Basic Math Skills Review will help to reinforce the algebra, arithmetic, and geometry concepts needed to master the SAT. Separate reviews are provided for Regular Math multiple-choice questions and Student-Produced Response questions. These reviews present the types of questions you can encounter on the actual test, along with key strategies and tips on solving math problems.

SCORING THE PRACTICE SAT EXAMS

How do I score my practice exam?

Math, Critical Reading, and Writing

To score these multiple-choice sections, count the number of correct responses. Enter the numbers into the corresponding blanks on the Scoring Worksheet that follows this section. Next, count up the number of incorrect responses and multiply this number by one-fourth; this is the penalty for answering incorrectly. Enter this number for each section on the Scoring Worksheet and subtract it from the number of correct responses. Total the scores for each subsection to get your total unrounded raw scores for Critical Reading and Math. (Wait to total the Writing subscore until you have scored your Essay.) Now, fractions should be rounded off. Round up for one-half or more, and round down for less than one-half.

Essay

Your SAT Essay will be evaluated by two scorers, so make sure to have two people read your practice essays. Because it is difficult to judge your own writing objectively, ask for a second opinion from a friend, parent, or teacher. The following scoring rubric summarizes the language and critical thinking skills that SAT Essay readers (who are all high school or college teachers) look for when evaluating essays. Comparing your essay according to the sample Essay in the Detailed Answers and Explanations section can also help you analyze your writing.

Add your two Essay scores together and enter this number in corresponding space on the Scoring Worksheet that follows. Your Essay score counts for about one-third of your Writing subscore.

Score	Criteria
6	*Clear and Consistent Competence* An essay with a score of 6 is extremely well-organized. The argument in the thesis is well-developed and effectively supported with examples. The essay is concise and easy to understand, displays an impressive range of accurately used vocabulary, and contains almost no language usage errors.
5	*Reasonably Consistent Competence* This essay is thoughtfully organized and presents adequate support for the main idea. Some language usage errors may appear, but there is still variety in sentence structure and a good range of vocabulary.
4	*Adequate Competence* An essay with a score of 4 shows some lapses in quality, but still addresses the topic and uses appropriate examples to support the thesis. The use of language is adequate, but there may be problems in grammar and usage.
3	*Developing Competence* An essay in this category shows weaknesses in organization and development, as well as significant errors in grammar and usage.
2	*Some Incompetence* This essay shows flaws in organization, has thin development, and provides few details to support the thesis. In addition, there are frequent errors in grammar and diction.
1	*Incompetence* An essay with a score of 1 is seriously flawed in every category. In this essay, there is virtually no structure or development of ideas, and numerous grammar and usage errors make the essay difficult to understand.
0	*Off-Topic* Essays that do not relate to the given topic (or are not written at all) receive a score of 0.

How do I convert my raw score?

Scores for each subsection on the SAT range from 200 to 800. For the Math and Critical Reading subsections, take the total number of points for each section (your raw score) and find the corresponding scaled score in the Score Conversion Chart at the end of this section.

The Writing subscore is a composite of your Essay score and your score on the multiple-choice Writing sections. First, score your Essay and find your multiple-choice raw score using the Scoring Worksheet. Then use the Writing Score Conversion Chart to find the subscore that corresponds to your total Essay score and your converted Writing multiple-choice score.

Add all three subscores together to find your total score, which is out of a maximum 2400.

When will I receive my SAT score report? What does the report contain?

Your score report will arrive approximately four weeks after you take the test. Your high school, any colleges you indicated on your answer sheet, and select scholarship services will also receive your scores.

Your SAT score report contains your total score for all three test sections and a list of the colleges you designated to receive your scores. Your total score will also be broken down into raw scores and scaled scores for the Critical Reading, Writing, and Math sections.

SCORING WORKSHEET

Math

Section 2 ________ – (1/4 × ________) = ________
no. correct / no. incorrect

Section 5 ________ – (1/4 × ________) = ________
no. correct / no. incorrect

Section 7 ________ – (1/4 × ________) = ________
no. correct / no. incorrect

Math Raw Score ________
(add Sections 2, 5, and 7)

Math Converted Subscore ________
(see the Score Conversion Chart: Math)

Critical Reading

Section 3 ________ – (1/4 × ________) = ________
no. correct / no. incorrect

Section 6 ________ – (1/4 × ________) = ________
no. correct / no. incorrect

Section 8 ________ – (1/4 × ________) = ________
no. correct / no. incorrect

Critical Reading Raw Score ________
(add Sections 3, 6, and 8)

Critical Reading Converted Subscore ________
(see the Score Conversion Chart: Critical Reading)

Writing and Essay

Section 4 ________ – (1/4 × ________) = ________
no. correct / no. incorrect

Section 9 ________ – (1/4 × ________) = ________
no. correct / no. incorrect

Writing Multiple-Choice Raw Score ________
(add Sections 4 and 9)

Essay ________ + ________ = ________
score 1 / score 2

Total Writing Scaled Subscore ________
(see the Writing Score Conversion Chart)

Total Converted Score

________ + ________ + ________ = ________
Math Subscore / Critical Reading Subscore / Writing Subscore / Total Score

SCORE CONVERSION CHART

Math

Raw Score	Scaled Score
54	800
53	800
52	790
51	780
50	770
49	760
48	740
47	730
46	720
45	710
44	700
43	690
42	680
41	660

Raw Score	Scaled Score
40	650
39	640
38	630
37	620
36	610
35	600
34	590
33	570
32	560
31	550
30	540
29	530
28	520
27	510
26	500

Raw Score	Scaled Score
25	480
24	470
23	460
22	450
21	440
20	430
19	420
18	410
17	390
16	380
15	370
14	360
13	350
12	340
11	330

Raw Score	Scaled Score
10	310
9	300
8	290
7	280
6	270
5	260
4	250
3	240
2	220
1	210
0 and below	200

Critical Reading

Raw Score	Scaled Score
67	800
66	790
65	780
64	770
63	760
62	760
61	750
60	740
59	730
58	720
57	710
56	700
55	690
54	680
53	670
52	670
51	660
50	650
49	640

Raw Score	Scaled Score
48	630
47	620
46	610
45	600
44	590
43	580
42	580
41	570
40	560
39	550
38	540
37	530
36	520
35	510
34	500
33	490
32	490

Raw Score	Scaled Score
31	480
30	470
29	460
28	450
27	440
26	430
25	420
24	410
23	400
22	400
21	390
20	380
19	370
18	360
17	350
16	340
15	330

Raw Score	Scaled Score
14	320
13	310
12	310
11	300
10	290
9	280
8	270
7	260
6	250
5	240
4	230
3	220
2	220
1	210
0 and below	200

WRITING SCORE CONVERSION CHART

Writing Multiple-Choice Raw Score	Essay Score						
	0	1–2	3–4	5–6	7–8	9–10	11–12
49	670	700	720	740	780	790	800
48	660	680	700	735	760	780	790
47	650	670	690	720	750	770	780
46	640	660	680	710	740	760	770
45	630	650	670	700	740	750	770
44	620	640	660	690	730	750	760
43	600	630	650	680	710	740	750
42	600	620	640	670	700	730	750
41	590	610	630	660	690	730	740
40	580	600	620	650	690	720	740
39	570	590	610	640	680	710	740
38	560	590	610	630	670	700	730
37	540	580	600	630	660	690	720
36	540	570	590	620	650	680	710
35	540	560	580	610	640	680	710
34	530	550	570	600	640	670	700
33	520	540	560	590	630	660	690
32	510	540	560	580	620	650	680
31	500	530	550	580	610	640	670
30	490	520	540	570	600	630	660
29	490	510	530	560	590	630	650
28	480	500	520	550	590	620	640
27	470	490	510	540	580	610	640
26	460	480	500	530	570	600	630
25	450	470	500	520	560	590	620
24	440	460	490	510	550	580	610
23	430	450	480	510	540	570	600
22	430	450	470	500	530	570	590
21	430	440	470	500	530	570	590
20	420	430	460	490	520	560	580
19	410	420	450	480	520	550	570
18	400	420	440	470	510	540	570
17	390	410	430	460	500	530	560
16	380	400	430	450	490	520	550
15	370	390	420	450	480	510	540
14	360	380	410	440	470	500	530
13	360	370	400	430	460	500	520

(cont.)

WRITING SCORE CONVERSION CHART (cont.)

Writing Multiple-Choice Raw Score	Essay Score						
	0	1–2	3–4	5–6	7–8	9–10	11–12
12	350	360	390	420	450	490	510
11	340	360	380	410	450	480	510
10	330	350	370	400	440	470	500
9	320	340	360	390	430	460	490
8	310	330	360	390	420	450	480
7	300	320	350	380	410	440	470
6	290	310	340	370	400	430	460
5	290	310	330	360	390	430	450
4	280	300	320	350	390	420	450
3	270	290	310	340	380	410	440
2	260	270	300	330	370	400	430
1	250	260	290	320	350	390	410
0	250	260	280	310	340	380	400
-1	240	250	270	290	320	360	380
-2	230	240	260	270	310	350	370
-3	220	230	250	260	300	330	360
-4	220	220	240	250	290	320	350
-5	200	210	230	240	280	310	340
-6	200	210	220	240	280	310	340
-7	200	210	220	230	270	300	330
-8	200	210	220	230	270	300	330
-9	200	210	220	230	270	300	330
-10	200	210	220	230	270	300	330
-11	200	210	220	230	270	300	330
-12	200	210	220	230	270	300	330

STUDYING FOR THE SAT

A key area to optimize your study time is to choose the time and location that works best for you. You may decide to study a certain number of hours each morning or each evening. You may decide to dedicate all your free time to studying whether waiting in line, eating lunch, or riding on the bus. Once you have decided how and when to study, we advise that you stick to your decision. Be consistent and use your time wisely. Studying works best when you have a routine and adhere to it faithfully.

When you take the practice exams, create a testing environment as close to the actual testing conditions as possible. Turn off your TV, stereo, and other media machines that serve as distractions. Sit in a chair at a table in a quiet room. Taking the test in bed, in a car, or any other similar location only makes you that much less prepared to take the actual test. Time yourself when taking a practice exam. If possible, use a timer in order to measure out the time allotted for each exam section.

After you finish a practice exam, score your test and thoroughly review the provided explanations to the questions you answered incorrectly. Avoid overdoing it by reviewing so many questions that you cannot remember what the correct answer was. It is better to concentrate on one problem area at a time, giving yourself the time necessary to digest the review material until you are confident you understand what you are reading.

You are allowed to write in the actual SAT test booklets during the exam. To practice this, consider writing in the margins of this book during practice exams.

Keep track of your practice exam scores and mark them on the Scoring Worksheet. By doing this, you will be better able to track your progress and identify general strengths and weaknesses in particular sections. You should carefully study the reviews that cover your weak areas as this will help build skills in those critical areas.

PROVEN TEST-TAKING TIPS & STRATEGIES

There are many ways to familiarize yourself with the format of the SAT. Familiarization is a great way to help alleviate test day anxieties. The following list has ways to help you become accustomed to the SAT.

Become comfortable with the SAT's format: When practicing, simulate the conditions of the actual test. Pace yourself. Stay calm. After repeating this process a few times, you are boosting your chances of performing well on test day. Success breeds confidence, and your successes here will give you much more confidence on test day.

Read all of the possible answers: If you believe you know the correct answer to a question, it is best not to assume that your answer is automatically the correct answer. Read through all answer choices to ensure that you are not making an error by jumping to conclusions.

Use the process of elimination: Examine each answer to a question. Eliminate as many of the answer choices as possible. By eliminating just two answer choices, you have given you're a better chance of the getting the correct answer out of the remaining three answer choices. Guess only if you can eliminate at least two answers. Remember that wrong answers are always penalized.

You don't have to answer every question: You are not penalized for not answering every question. Questions left blank are not counted. Maximize this advantage by using a smart guessing strategy to questions where you are unsure of the right answer. Always keep in mind that if you are truly stumped, you do not have to answer, especially because ¼ point is deducted for every wrong answer on multiple-choice questions.

Work quickly and steadily: You will only have 20 to 30 minutes to work on each section. Working quickly and steadily helps you avoid focusing too much attention and time on any one problem. Use the practice exams in this book to help you manage your timing.

Learn the directions and format for each section of the test: Familiarize yourself with the directions and form of each of the different test sections. This will help you avoid "direction shock" later on during the test, when you might read directions that were better read at the start of the test. Shocks like these cause nervousness. Nervousness causes mistakes. And these kinds of mistakes are completely avoidable.

Work on the easier questions first: The questions for each section of the SAT are arranged in ascending order of difficulty. The easier questions are at the beginning of each section, while the more difficult ones are at the end of the section. If you find yourself working too long on a single question, make a mark next to it in the test booklet and continue with the next question. After you have answered the remaining questions, return to the ones you skipped.

Mark answers carefully: Be sure that the answer sheet oval corresponds to the question and answer of your test booklet: Because the multiple-choice sections are graded mechanically, marking one wrong answer in this way can throw off your answer key and thus ruin your score. Be extremely careful.

Eliminating obviously wrong answers: Sometimes an SAT question has one or two answer choices that appear odd or out of place. These answers may be obviously wrong for one or more reasons:

- Impossible to achieve given the problem's conditions
- Violation of mathematical rules or principles
- Simply illogical

Being able to spot obviously wrong answers before you finish a problem gives you an advantage because you are able to make a better educated guess from the remaining choices. This works best when you find yourself unable to fully solve a problem.

Working from answer choices: Turn the multiple-choice format to your advantage by working backwards from the answer choices to solve a problem. This strategy is not applicable to all questions, but it is helpful when you can plug choices into a given formula or equation. The answer choices often narrow the scope of responses allowing you to make an educated guess based on eliminating choices that you know do not fit the problem.

YOUR TEST-DAY CHECKLIST

✓ Get a good night's sleep. Tired test-takers consistently perform poorly.

✓ Wake up early.

✓ Dress comfortably. Keep your clothing temperature appropriate. You'll be sitting in your test clothes for hours. Clothes that are itchy, tight, too warm, or too cold take away from your comfort level.

✓ Eat a good breakfast.

✓ Take these with you to the test center:
  - Several sharpened No. 2 pencils. Pencils are not provided at the test center
  - Admission ticket
  - A valid photo ID that contains your name and signature. Good examples of these are your driver's license, student ID card or a current alien registration card.

✓ Optional but helpful items to bring to the test center:
  - Noiseless wristwatch
  - Noiseless calculator

✓ Arrive at your test center early. Remember, no one is allowed into a test session after the test has begun.

✓ Compose your thoughts and try to relax before the test.

Remember that eating, drinking, and smoking are prohibited. Dictionaries, textbooks, notebooks, briefcases, and packages are also prohibited.

DURING THE TEST

Once in the test center, follow all rules and instructions given by the instructor and in the SAT booklet. If you fail to do this, you risk being dismissed from the test and having your SAT scores cancelled.

After all test materials are distributed, the test supervisor will provide directions on completing the answer sheet. As with all other sections of the test, fill in this section carefully. Fill in your name exactly as it appears on your admission ticket and identification document (unless otherwise instructed).

Remember that you can write in your test booklet. No scratch paper is provided or allowed during the test. Mark your answers in the appropriate spaces on the answer sheet. Each numbered row contains five ovals that correspond to each answer choice for that question. Fill in the oval that corresponds to your answer completely, darkly, and neatly. You can change an answer, but you must completely erase your original response. Only one answer should be marked. This is crucial to accurate score reporting, as your answer sheet will be scored by a machine. Stray lines and unnecessary marks will cause the machine to score answers incorrectly.

Because your essay will be scored by a person, not a machine, remember to write neatly. No points will be deducted for poor handwriting, but illegible handwriting impairs communication.

AFTER THE TEST

Once your test materials have been collected, you will be dismissed. Then your day is free. Go home and relax. Or reward yourself with some shopping. Or play a video game. Or hang with friends. The good news is that the hard part is over. Now you just have to wait for the results.

CHAPTER 2

A DIAGNOSTIC TEST

Answer sheets for this test start on page 839.

A DIAGNOSTIC TEST

Now that you have some background information about the SAT, you are ready to take the diagnostic test. This test is designed to help you identify your strengths and weaknesses. You will want to use this information to help you master the SAT. It is a complete test, so take it in the same way you would take the actual SAT. Situate yourself in a quiet room with no distractions. Write in pencil, and use the answer sheets provided in the back of the book. Keep track of the time allotted for each section.

When you are finished with the Diagnostic Test, refer to the charts that follow to evaluate your performance. The entries in the chart show you where to look in the book for a discussion of material covered in that type of problem.

TIME: 25 Minutes

ESSAY

DIRECTIONS: You have 25 minutes to plan and write an essay on the topic below. You may write on only the assigned topic.

Make sure to give examples to support your thesis. Proofread your essay carefully and take care to express your ideas clearly and effectively.

ESSAY TOPIC

According to author Anna Quindlen, "If your success is not on your own terms, if it looks good to the world but does not feel good in your heart, it is not success at all." In a similar vein, Bob Dylan, the famous songwriter, once said, "What's money? A man is a success if he gets up in the morning and goes to bed at night and in between does what he wants to do." However, in American culture, success is usually associated with precisely money and prestige.

ASSIGNMENT: Do you agree or disagree with the assertion that success is definable in many ways, and what is your personal definition of success? Plan and write an essay in which you defend your point of view on this issue. Support your reasoning with examples taken from your reading, studies, experience, and observations.

TIME: 25 Minutes
20 Questions

MATHEMATICS

DIRECTIONS: In this section solve each problem, using any available space on the page for scratchwork. Then decide which is the best of the choices given and fill in the corresponding oval on the answer sheet.

NOTES

(1) The use of a calculator is permitted.

(2) All numbers used are real numbers.

(3) Figures that accompany problems in this test are intended to provide information useful in solving the problems. They are drawn as accurately as possible EXCEPT when it is stated in a specific problem that the figure is not drawn to scale. All figures lie in a plane unless otherwise indicated.

(4) Unless otherwise specified, the domain of any function f is assumed to be the set of all real numbers x for which $f(x)$ is a real number.

REFERENCE INFORMATION

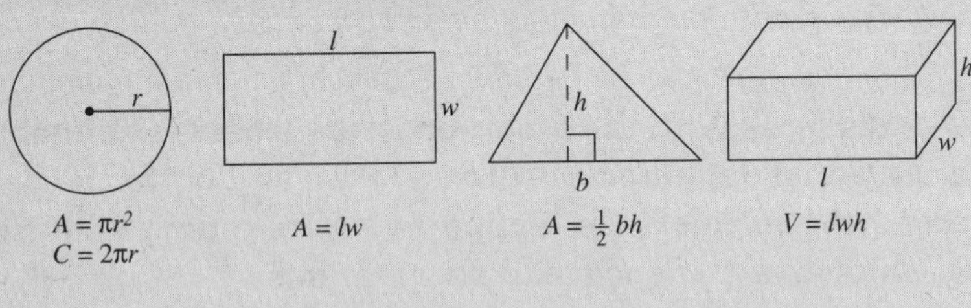

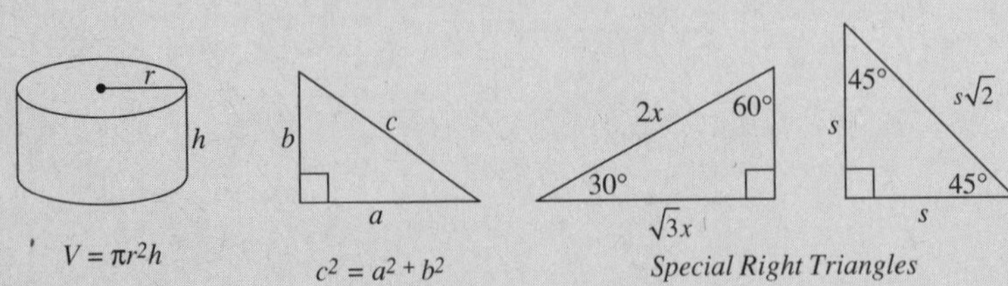

The number of degrees of arc in a circle is 360.

The sum of the measures in degrees of the angles of a triangle is 180.

1. If $f(x) = 5x^2 - 3^x$, what is the value of $f(3)$?

(A) 72

(B) 39

(C) 21

(D) 18

(E) 3

2. Casey works in a department store. One of his benefits is a 20% discount on all items he buys for himself. Casey paid $54 for a new coat. What was the price, in dollars, before the discount?

(A) $10.80

(B) $43.20

(C) $64.80

(D) $67.50

(E) $97.20

3. What is the value of x in the equation $\sqrt{2x-1}+9=14$?

(A) 9

(B) 12

(C) 13

(D) 264

(E) 265

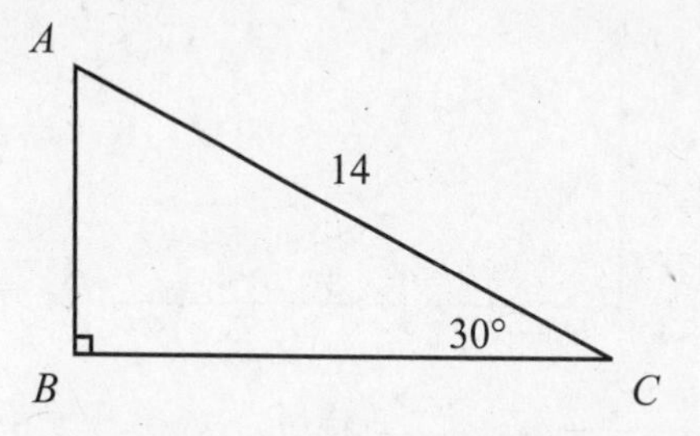

Note: Figure not drawn to scale.

4. In right triangle ABC shown above, what is the length of side BC?

(A) $7\sqrt{3}$

(B) $7\sqrt{2}$

(C) $14\sqrt{3}$

(D) 7

(E) $\sqrt{14}$

5. What are the values of x in the equation $|6x+7|=11$?

(A) $-\frac{2}{3}$ and $\frac{2}{3}$

(B) $-\frac{2}{3}$ and 3

(C) $\frac{3}{2}$ and -3

(D) $\frac{2}{3}$ and 3

(E) $\frac{2}{3}$ and -3

6. If the graph of $f(x)$ is undefined at $x = 7$, then the graph of $f(x - 3)$ must be undefined at which value of x?

(A) -10

(B) -4

(C) 4

(D) 10

(E) 21

GO ON TO THE NEXT PAGE

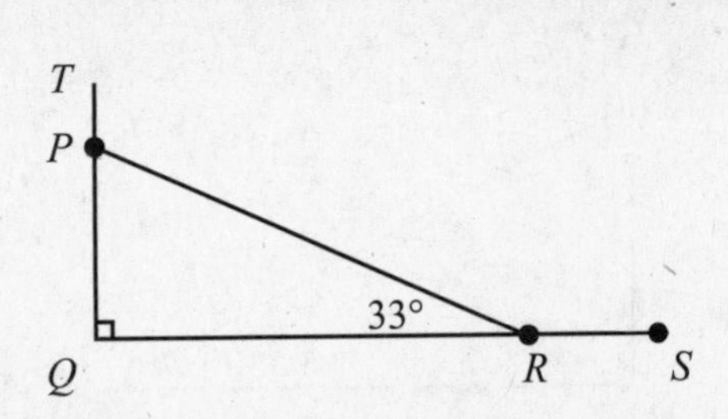

Note: Figure not drawn to scale.

7. In the right triangle PQR shown above, Q, R, and S are collinear points. Also, T, P, and Q are collinear points. How much larger is $\angle PRS$ than $\angle RPT$?

(A) 24°
(B) 33°
(C) 57°
(D) 90°
(E) 123°

8. A printing press used 28.8 gallons of ink over a three-day period. If $\frac{1}{9}$ of the ink was used on the first day, and twice that amount was used on the second day, how many gallons of ink were used on the third day?

(A) 6.4
(B) 9.6
(C) 12.8
(D) 14.4
(E) 19.2

9. One year ago, Pat was three times his sister's age. Next year, he will be only twice her age. How old will Pat be five years from now?

(A) 8
(B) 11
(C) 12
(D) 13
(E) 15

10. The population of insects that begins at 60 and quadruples every six years is given by the formula $P = (60)(4^{\frac{t}{6}})$, where P represents the population and t represents the number of years. Suppose the population is 60 in the year 2004. What will be the population in the year 2037?

(A) 1,320
(B) 122,880
(C) 309,640
(D) 1,228,800
(E) 39,008,730

11. In an apartment complex of 144 apartments, 9 out of every 16 apartments have terraces. If the landlord promised to increase the number of apartments with terraces by 25%, how many apartments would still not have terraces?

(A) 38
(B) 43
(C) 45
(D) 60
(E) 78

GO ON TO THE NEXT PAGE

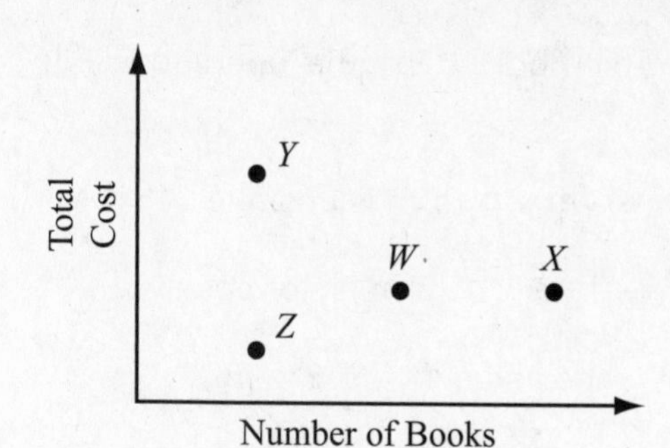

12. A math class bought trigonometry books from four different companies, identified as *W*, *X*, *Y*, and *Z*, as shown in the scatter plot above. For which two companies was the cost per book exactly the same?

(A) *Y* and *Z*

(B) *X* and *Z*

(C) *W* and *X*

(D) *X* and *Y*

(E) *W* and *Z*

13. What is the area, in square meters, of a trapezoid if the height is 17 meters and the parallel sides are 25 and 37 meters?

(A) 527

(B) 786

(C) 1,054

(D) 15,725

(E) 31,450

14. What is the domain of the function $f(x) = \sqrt{8 - x}$?

(A) $-8 \leq x \leq 8$

(B) $x \geq -8$

(C) $x \leq -8$

(D) $x \leq 8$

(E) $x \geq 8$

15. Which of the following is equivalent to $\left(\frac{x^{-6} \bullet x^{4}}{x^{-5}}\right)^{-2}$?

(A) x^{-6}

(B) x^{-1}

(C) x^{4}

(D) x^{5}

(E) x^{10}

16. The volume of a cube changed from 216 cubic centimeters to 64 cubic centimeters. What is the best approximation of the percent change in length for a side of the cube?

(A) 70%

(B) 67%

(C) 50%

(D) 33%

(E) 30%

17. It took a jet $4\frac{1}{2}$ hours to complete a trip of 2,400 miles. For the first part of the trip, it flew at 530 mph. For the second part of the trip, the tailwind picked up and the jet flew at 540 mph. How long did the jet fly at 540 mph?

(A) 45 minutes

(B) 60 minutes

(C) 75 minutes

(D) 90 minutes

(E) 105 minutes

GO ON TO THE NEXT PAGE

18. Which interval of x values represents the solution to $-x^2 + 9 < 0$?

(A) (0,3)

(B) (–3,0)

(C) (–3,∞)

(D) (–∞,–3) and (3,∞)

(E) (–∞,0) and (3,∞)

19. Sam works at an electronics store and earns a commission for each stereo he sells. His monthly commission in dollars is given by the equation $C = 270s - 3s^2$, where C is the monthly commission and s is the number of stereos he sells. How many stereos should he sell each month to earn the greatest commission possible?

(A) 30

(B) 45

(C) 90

(D) 180

(E) 270

20. Define symbol ∞ by the operations below:

$m \infty n = \sqrt{m} + n^2$ if m is positive

$m \infty n = m^2 + \sqrt{n}$ if m is not positive

What is the value of $-2\sqrt{2} \infty 16$?

(A) –4

(B) 8

(C) 12

(D) 254

(E) 257.68

STOP

If time remains, you may go back and check your work. When the time allotted is up, you may go on to the next section.

TIME: 25 Minutes
24 Questions

CRITICAL READING

DIRECTIONS: Each sentence below has one or two blanks, each blank indicating that something has been omitted. Beneath the sentence are five lettered words or sets of words. Choose the word or set of words that BEST fits the meaning of the sentence as a whole.

EXAMPLE

Although critics found the book _______ , many readers found it rather _______ .

(A) obnoxious . . . perfect
(B) spectacular . . . interesting
(C) boring . . . intriguing
(D) comical . . . persuasive
(E) popular . . . rare

EXAMPLE ANSWER

1. Despite the _______ of their tactics and habits with those of the White Huns, the Magyars, the Mongols, and the Turks, their connection with those peoples is either _______ or—in the case of the Magyars and the Turks—unfounded.

(A) variability . . . courageous
(B) redundancy . . . vigorous
(C) perplexity . . . perplexing
(D) similarity . . . tenuous
(E) conviviality . . . paramount

2. Robbe-Grillet is considered the originator of the French *nouveau roman*, or new novel, in which story is _______ to structure and the significance of objects is stressed above that of human _______ or action.

(A) subordinated . . . motivation
(B) perpetrator . . . organization
(C) facilitated . . . forgery
(D) contrasted . . . frivolity
(E) appropriate . . . convolution

3. By 300 B.C., and until Columbian times, Yucatan was _______ by the Maya.

(A) populated
(B) figured
(C) delivered
(D) stipulated
(E) prefabricated

4. Belief in vampires has existed from the earliest times and has resulted in an _______ of legends and superstitions, ultimately giving rise to the convoluted story that survives into the present.

(A) enunciation
(B) inaccuracy
(C) ordinance
(D) inauguration
(E) amalgam

5. Enigmatic and _______, Bob Dylan shuns interviews and other trappings of success, but he has become a cult figure nevertheless, continuing to tour and produce new albums.

(A) voyeuristic
(B) discerning
(C) reclusive
(D) redemptive
(E) indigent

6. A momentous volcanic explosion on August 23, 1883, blew up most of the island of Krakatau and altered the _______ of the strait; the accompanying tsunami caused great destruction and loss of life along the nearby coasts of Java and Sumatra.

(A) distribution
(B) configuration
(C) assimilation
(D) reformulation
(E) incongruity

GO ON TO THE NEXT PAGE

7. Down each of the four faces of the obelisk, in most cases, ran a line of deeply _______ hieroglyphs and representations, setting forth the names and titles of the Pharaoh. The cap was sometimes _______ with copper or other metal.

   (A) reiterated . . . received
   (B) distraught . . . calamitous
   (C) valued . . . absorbed
   (D) incised . . . sheathed
   (E) deteriorated . . . evolved

8. Many species of *Anopheles* mosquitoes, _______ by their tilted resting position, carry the protozoan parasites that cause malaria. Species of the genus *Aedes* _______ the viruses responsible for yellow fever, jungle yellow fever, and dengue fever.

   (A) recognizable . . . transmit
   (B) regurgitated . . . hold
   (C) perpetuated . . . desire
   (D) confabulated . . . organize
   (E) flown . . . treated

Questions 9–10 are based on the following passage.

This passage is the beginning of a short story.

His father used to tell him that there are only two kinds of people. "There are winners and there are losers, and you don't get to pick which one you are." Those words had haunted him his entire life, and now that his marriage was falling apart and he had lost his job, he had no doubt that his father's reductive definition of the world was absolutely correct, and he knew exactly which category was his.

As he stood in a phone booth wondering how to give his wife the news of his most recent failure, it occurred to him to call his dad and tell him just that—that the old man was right all along. He did not have enough change or a credit card on him, and so he would have to call collect, which he thought perfect symbolism. But when his dad answered the phone, he was really concerned that something was wrong because his son had never called him collect before. And when the son repeated that long-ago axiom, his father asked if he was crazy to live by such a negative creed, even if he himself had said it so many years ago. The older man said that he must have just been depressed, and then he apologized to his son, something he had never done before either.

GO ON TO THE NEXT PAGE

9. The main idea in this passage is

(A) a son believes his father's negative philosophy of life is the reason his own luck is so bad.

(B) a son believes that what his father told him about life is true.

(C) a son believes that his father is to blame for his bad luck.

(D) a son learns that his father's negative philosophy of life is absolutely accurate.

(E) a son learns that what he thought was his father's negative philosophy of life was just an offhand comment the older man made once.

10. The passage says that the father's definition of the world was "reductive," which implies

(A) the man's worldview was simplistic.

(B) the man's worldview was horrific.

(C) the man's worldview was demonic.

(D) the man's worldview was mathematical.

(E) the man's worldview was organic.

Questions 11–12 are based on the following passage.

A winter fly is a miraculous thing. He buzzes about the house while snow flies outside, oblivious to the fact that his days are numbered. I imagine that he wonders, however, where all of the others of his kind could be. "Certainly, they are not *all* dead," he must be thinking even as I swat him away from my sandwich and chips.

But then it occurs to me that there are equivalent creatures among humans. My great-grandfather is a very old man, in his nineties I guess, and he is the only one he knows who is that old. He has told me so. As I take a bite of my peanut-butter-and-jelly, I imagine him wandering around the halls of the nursing home where he will spend his last days, wondering where all of the others of his kind have gone: those who remember the First World War, even if he was a child then, those who were in the second and have not seen a just war since, those who have voted in perhaps 18 presidential elections, those who know a level of loneliness the rest of us can't even imagine.

11. Which of the following best describes the purpose the first sentence serves?

(A) A fly in winter is akin to the miracles described in holy texts.

(B) This fly is emblematic of survival.

(C) Flies usually die when it gets cold, and this one is being compared to the speaker's great-grandfather, who is also living beyond when men usually do.

(D) Flies are obnoxious creatures and so is the speaker's great-grandfather.

(E) The speaker likes flies and describes this one in a positive way in this sentence.

12. Which of the following best explains the last clause in this passage?

    (A) The speaker ponders visiting his great-grandfather.

    (B) The reader is expected to feel sorry for the speaker's great-grandfather.

    (C) The old man will die soon, and this clause foreshadows that fact.

    (D) The old man realizes that he is the last one of his generation alive.

    (E) The speaker realizes that his great-grandfather's loneliness is beyond comprehension for anyone who is not the last of his generation alive.

GO ON TO THE NEXT PAGE

Questions 13–24 are based on the following passage.

(1) Older Americans face a variety of stereotypes, almost all of them negative. Many people perceive the elderly as feeble in mind and body and as economic burdens on society. Even though Americans have an average lifespan of 76.5 years, which means that a sizable percentage of those who hold these negative views will themselves join the ranks of the aged, many believe that their fellow citizens have little to contribute once they reach their sixties.

(2) At one time, American attitudes toward the elderly were more positive. In the seventeenth and eighteenth centuries, for example, the aged were respected and venerated because they helped transmit wisdom and tradition to the younger generations. They were given the best seats in church, and Puritan teachings instructed youth on how to behave toward their elders. One reason that the aged garnered this respect was because there were so few of them in colonial society. According to social historian David Hackett Fischer, only two percent of the population at that time was over sixty-five years old.

(3) By the nineteenth century, however, American society had changed significantly. Ironically, the elderly suffered as America progressed. The rise of an urban and industrialized nation meant that the skills and education of many of the aged were no longer useful. Because younger, healthier employees were more desirable for factory work, mandatory retirement laws were passed as early as 1777, forcing aging workers to leave their jobs and frequently plunging them into poverty. Old-age homes were established for those elderly who were poor and had no family to look after them, and although such homes gave these people shelter, they further isolated the elderly from society.

(4) Negative attitudes toward the elderly have continued into the twentieth and twenty-first centuries. Because of the increased mobility of the population in the last several decades, many of the elderly are isolated from their families. The majority live with their spouse or alone, and in fact, one recent study indicates that 60 percent of women over the age of eighty-five live alone due to the difference in mortality rates between the sexes.

(5) Commercials and jokes frequently rely on stereotypes of the elderly, such as the belief that they are desperate to appear young and virile, and some people believe that the Social Security system allows senior citizens to drain money away from tax-paying workers. However, aging Americans who want to continue to work or return to the workplace face age discrimination. As one researcher put it: "Since occupation and work are the principal criteria of social prestige in America, the old, by being excluded from work, are therefore devalued."

(6) The United States is not unique in its attitudes toward the elderly, however. In other cultures and nations, the rise of urban industrialization has led to similar results. When African and Asian nations were largely rural, older relatives used to live with their children and grandchildren. However, the limited space of urban housing makes such intergenerational homes less practical and less desirable. As a consequence, the elderly in these countries have lost their status in the family and society as a whole. In these nations in the past, the elderly were involved in the rearing of children, helped make important decisions for the family, and were supported as they aged. But as society became more urban, the aged began to lose their economic security as well as their favored status within the family.

GO ON TO THE NEXT PAGE

(7) Japanese society has changed in a similar fashion. Although 55 percent of elderly parents live with their children and grandchildren, a rate nearly three times that in other industrialized nations, the percentage has dropped sharply since 1970. Then, 80 percent of homes housed multiple generations.

(8) Despite the devalued status of the elderly in many cultures, aging is not a universally negative experience, however. For example, recent research has refuted many of the stereotypes associated with the aging process, such as the beliefs that the elderly are in poor health or unable to learn new skills. These researchers also note that many elderly Americans contribute significantly to society and the economy, but because much of the work done by the elderly is unpaid, they do not receive their rightful recognition. In fact, almost all older men and women are productive in this larger sense—one-third work for pay and one-third work as volunteers in churches, hospitals, and other organizations.

13. The writer points out that the average lifespan in America is 76.5 years, which implies

(A) the overall population of the country is getting older.

(B) what constitutes "old" is constantly changing.

(C) the aged population is growing.

(D) the people now thinking negatively of the aged are themselves likely to live to be elderly.

(E) Americans are living longer than they used to live.

14. The writer points out the following in paragraph 2: "One reason that the aged garnered this respect was because there were so few of them in colonial society." What might this observation imply about the situation for the elderly in our age?

(A) They are not held in high regard because the elderly are not a small population group; they are not rare.

(B) The young feel that the old will soon outnumber them.

(C) The young feel burdened by the old.

(D) The aged population in America is growing.

(E) The elderly deserve better treatment because they represent a valuable demographic to politicians.

GO ON TO THE NEXT PAGE

15. Which of the following best delineates the arrangement of this passage?

(A) discussion of stereotypes of the aged in America, examples of how the old are treated in other countries, conclusion that the elderly should be treated better

(B) discussion of stereotypes of the aged in America, examples of how the elderly were treated in Colonial America, call to arms for the aged suggesting they vote for older candidates

(C) discussion of stereotypes of the aged in America, examples of how the aged should be treated from other cultures

(D) discussion of stereotypes of the aged in America, examples of how the elderly were treated better in previous centuries, examples of the changing attitudes toward the elderly in other countries, research that refutes some of the stereotypes

(E) discussion of stereotypes in America, indictment of stereotypes generally and of the aged in general, examples of how the old are treated in other countries

16. Why did the lot of the elderly change during the industrial revolution?

(A) Machines replaced many workers and the old were the first to go.

(B) Machines killed many unskilled workers in accidents.

(C) The skills and education of many of the aged were no longer useful, and healthier employees were needed for factory work.

(D) Harsher conditions killed many elderly workers.

(E) Management decided that younger employees were more productive.

17. In paragraph 5, the author points out the following: "Commercials and jokes frequently rely on stereotypes of the elderly, such as the belief that they are desperate to appear young and virile, . . ." If this stereotype is false, why might the stereotype exist?

(A) the young are the ones making commercials and telling the jokes, and youth is overvalued in our culture

(B) those who tell jokes and make commercials are trying to appeal to the young

(C) to make some people feel better about themselves

(D) because it is really true

(E) to answer criticism of the young with criticism of the old, whether true or not

GO ON TO THE NEXT PAGE

18. In paragraph 5, the writer suggests that the elderly are in an impossible situation regarding work. Which of the following best describes this situation?

    (A) The old want to work but can't because they are retired.

    (B) The elderly are viewed by some as unproductive if they do not work, but finding a job is difficult because of the bias against the aged.

    (C) The young expect the old to work but do not give them jobs.

    (D) Working is difficult for the aged but they need to work to make ends meet.

    (E) The working elderly do not want to give up their jobs but are forced to do so.

19. The writer quotes a researcher who declares that "occupation and work are the principal criteria of social prestige in America . . . . " Which of the following best defines this statement relative to the overall passage?

    (A) Value is ascribed to people by what they do for a living, and so the retired are not held to be valuable in America.

    (B) The young have more value in America because they work.

    (C) Professionals are more important in America than nonprofessionals.

    (D) We are not valued for who we are in America.

    (E) Value is not assigned to nonwork tasks.

20. What reason does the author give for the loss of prestige and value for elderly people in Asian countries?

    (A) The American model has been adopted in these countries.

    (B) The young no longer associate with the elderly.

    (C) The young now use daycare for their children.

    (D) The elderly used to be involved in the rearing of children and helped make important family decisions, but now there is no room for them in urban households.

    (E) Urban dwellings are smaller, so the elderly do not live with their children and grandchildren anymore.

21. Although 55 percent of elderly parents live with their children and grandchildren in Japan, why is this indicative of a decline in how the elderly are treated in that country?

    (A) The elderly are not living as long as they did 30 years ago.

    (B) The elderly are not allowed to take part in raising the children.

    (C) The elderly are no longer satisfied with this living situation.

    (D) A far greater percentage of the elderly lived with their families 30 years ago.

    (E) The young no longer have sufficient space for their elders to live with them.

GO ON TO THE NEXT PAGE

22. What stereotypes about the elderly have researchers refuted?

(A) The elderly wish to live with their children.

(B) The elderly wish to work after retirement and wish to help raise their grandchildren.

(C) The elderly are in poor health and unable to learn new skills.

(D) The aged do not care if they are stereotyped or not.

(E) The aged are unhealthy and accident prone.

23. Given the overall argument being made in the passage, which of the following seems the most logical assumption about what the writer would suggest regarding our treatment of the aged?

(A) They should be put in nursing homes and left alone.

(B) They should be treated with respect for what they know and be allowed to work or retire as it suits them.

(C) They should be the center of our cultural attention.

(D) No jokes or commercials filled with clichés about them should be allowed.

(E) They should work longer so that the Social Security system remains solvent.

24. What reasons does the writer give for the elderly not being appreciated in America?

(A) More of the elderly work for pay than many Americans realize, and those who volunteer are not recognized for what they contribute because the work is unpaid.

(B) The work of the elderly is largely invisible to most Americans.

(C) The elderly are really hard-working and productive employees.

(D) Volunteering is work too.

(E) The elderly can be proud of their contributions to America.

TIME: 25 Minutes
35 Questions

WRITING

DIRECTIONS: In each of the following sentences, some portion of the sentence is underlined. Under each sentence are five choices. The first choice has the same wording as the original. The other four choices are reworded. Sometimes the first choice containing the original wording is the best; sometimes one of the other choices is the best. Choose the letter of the best choice. Your choice should produce a sentence that is not ambiguous or awkward and that is correct, clear, and precise.

This is a test of correct and effective English expression. Keep in mind the standards of English usage, punctuation, grammar, word choice, and construction.

EXAMPLE

When you listen to opera, a person may not appreciate it.

(A) a person may not appreciate it.

(B) it may not be appreciated by a person.

(C) you may not appreciate it.

(D) which may not be appreciated by you.

(E) appreciating it may be a problem for you.

EXAMPLE ANSWER

1. The typical amphitheater was elliptical in shape, with seats, supported on vaults of masonry, rising in many tiers around an arena at the center; corridors and stairs facilitated the circulation of great throngs.

   (A) the center; corridors and stairs facilitated the circulation of great throngs.
   (B) the center corridors and stairs facilitated the circulation of great throngs.
   (C) the center, corridors and stairs facilitated the circulation of great throngs.
   (D) the center as corridors and stairs facilitated the circulation of great throngs.
   (E) the center facilitated corridors and stairs and the circulation of great throngs.

2. The existence of cartels are in opposition to classic theories of economic competition and the free market, and they are forbidden by law in many nations.

   (A) The existence of cartels are in opposition to classic theories of economic competition
   (B) The existence of cartels is in opposition to classic theories of economic competition
   (C) The existence of cartels in opposition to classic theories of economic competition
   (D) The existence of cartels, forbidden by law in many countries in opposition to classic theories of economic competition
   (E) The existence of cartels are classic theories of economic competition

3. The concept of sovereignty has had a long history of development and it may be said that every political theorist since Plato has dealt with the notion in some manner although not always explicitly.

   (A) long history of development and it may be said that every political theorist since Plato has dealt with the notion in some manner although not always explicitly.
   (B) long history of development and may be said that every political theorist since Plato has dealt with the notion in some manner although not always explicitly.
   (C) long history of development and every political theorist since Plato has dealt with the notion in some manner although not always explicitly.
   (D) long history of development every political theorist since Plato has dealt with the notion in some manner, although not always explicitly.
   (E) long history of development, and it may be said that every political theorist since Plato has dealt with the notion in some manner, although not always explicitly.

GO ON TO THE NEXT PAGE

4. In the *Leviathan*, Hobbes argued that life is simply the motions of the organism and that man is by nature a selfishly individualistic animal at constant war with all <u>other men in a state of nature, men are equal in their self-seeking and live out lives that are "nasty brutish and short."</u>

(A) other men in a state of nature, men are equal in their self-seeking and live out lives that are "nasty brutish and short."

(B) other men. In a state of nature, men are equal in their self-seeking and live out lives that are "nasty, brutish, and short."

(C) other men in a state of nature. Men are equal in their self-seeking and live out lives that are "nasty brutish and short."

(D) other men. In a state of nature, men are equal in their self-seeking and live out lives that are "nasty brutish and short."

(E) other men. In a state of nature men are equal in their self-seeking and live out lives that are "nasty brutish and short."

5. Because absolute motion cannot be confirmed <u>by objective measurement Einstein suggested that it be discarded from physical reasoning he explained</u> the results of the Michelson-Morley experiment by means of the special relativity theory, which he enunciated in 1905.

(A) by objective measurement Einstein suggested that it be discarded from physical reasoning he explained

(B) by objective measurement, Einstein suggested that it be discarded from physical reasoning he explained

(C) by objective measurement, Einstein suggested that it be discarded from physical reasoning; he explained

(D) by objective measurement, Einstein suggested that it be discarded from physical reasoning, he explained

(E) by objective measurement Einstein suggested that it be discarded, from physical reasoning, he explained

GO ON TO THE NEXT PAGE

6. From the end of the 11th to the early 13th century, classical Georgian poetry, secular in nature and strongly influenced by the Persian epic, enjoyed its greatest flowering.

   (A) century, classical Georgian poetry, secular in nature and strongly influenced by the Persian epic, enjoyed its greatest flowering.
   (B) century classical Georgian poetry, secular in nature and strongly influenced by the Persian epic, enjoyed its greatest flowering.
   (C) century, classical Georgian poetry secular in nature and strongly influenced by the Persian epic, enjoyed its greatest flowering.
   (D) century, classical Georgian poetry secular in nature and strongly influenced by the Persian epic enjoyed its greatest flowering.
   (E) century, classical Georgian poetry was secular in nature and strongly influenced by the Persian epic, enjoyed its greatest flowering.

7. An estimated 90% of peptic ulcers are caused by infection with a bacterium, *Helicobacter pylori* which promotes the formation of ulcers by weakening the defenses of the stomach wall, making it more susceptible to the hydrochloric acid secreted by the stomach.

   (A) are caused by infection with a bacterium, *Helicobacter pylori* which promotes the formation of ulcers by weakening the defenses of the stomach wall, making
   (B) are caused by infection with a bacterium, *Helicobacter pylori,* which promotes the formation of ulcers by weakening the defenses of the stomach wall, made up of
   (C) are caused by infection with a bacterium *Helicobacter pylori* which promotes the formation of ulcers by weakening the defenses of the stomach wall, making
   (D) are caused by infection with a bacterium, *Helicobacter pylori,* which promotes the formation of ulcers by weakening the defenses of the stomach wall, making
   (E) are caused by infection with a bacterium, *Helicobacter pylori* which promotes the formation of ulcers by weakening the defenses of the stomach wall making

GO ON TO THE NEXT PAGE

8. Molinism, or quietism, developed within the Roman Catholic Church in Spain and spread especially to France, where its most influential exponent was Madame Guyon.

(A) Molinism, or quietism, developed within the Roman Catholic Church in Spain
(B) Molinism or quietism developed within the Roman Catholic Church in Spain
(C) Molinism developed within the Roman Catholic Church in Spain
(D) Molinism, or quietism, developed within the Roman Catholic Church, in Spain
(E) Molinism or quietism developed within the Roman Catholic Church, in Spain

9. Although most phloxes are perennial the common garden phloxes are annual hybrids of the Texas species *Phlox drummondii.*

(A) most phloxes are perennial the common garden phloxes are
(B) most phloxes are perennial, the common garden phloxes are
(C) most phloxes is perennial, the common garden phloxes is
(D) most phloxes aren't perennial, the common garden phloxes are
(E) the common garden phloxes are

10. Each ecosystem consists of a community of plants and animals in an environment that supplies them with raw materials for life, that is, chemical elements and water. The ecosystem delimited by the climate, altitude, water and soil characteristics, and other physical conditions of the environment.

(A) delimited by the climate, altitude, water and soil characteristics, and other physical conditions of the environment.
(B) delimited by the climate, altitude, water and soil characteristics, and other physical conditions of the environment.
(C) delimited by the climate, altitude, water and soil characteristics, and other physical conditions of the environment.
(D) delimited by the climate altitude water and soil characteristics, and other physical conditions of the environment.
(E) delimited by the climate, altitude, water and soil characteristics, and other physical conditions of the environment.

GO ON TO THE NEXT PAGE

11. Melanin is one of two pigments found in human skin and hair <u>and adds brown to skin color, the other pigment is carotene, which contributes yellow coloring.</u>

(A) and adds brown to skin color, the other pigment is carotene, which contributes yellow coloring.

(B) and adds brown to skin color. The other pigment is carotene, which contributes yellow coloring.

(C) and adds brown to skin color the other pigment is carotene, which contributes yellow coloring.

(D) and adds brown to skin color: the other pigment is carotene, which contributes yellow coloring.

(E) and adds brown to skin color, the other pigment, carotene, contributes yellow coloring.

GO ON TO THE NEXT PAGE

DIRECTIONS: Each of the following sentences may contain an error in diction, usage, idiom, or grammar. Some sentences are correct. Some sentences contain one error. No sentence contains more than one error.

If there is an error, it will appear in one of the underlined portions labeled A, B, C, or D. If there is no error, choose the portion labeled E. If there is an error, select the letter of the portion that must be changed in order to correct the sentence.

EXAMPLE

He drives slowly (A) and cautiously (B) in order to hopefully (C) avoid having an accident (D). No error (E)

EXAMPLE ANSWER

12. *The Platform Sutra*, attributed to Hui-neng, (A) defines (B) enlightenment as the direct seeing of one's "original Mind" or "original Nature," which (C) is Buddha, and this teaching (D) has remained characteristic of Zen. No error (E)

13. The Yalta conferees (A) decide (B) to ask China and France to join them in sponsoring the founding conference of the United Nations to be convened in San Francisco on April 25, 1945; agreement (C) was reached on using the veto system of voting in the projected Security Council. Future meetings of the foreign ministers of the "Big Three" were (D) planned. No error (E)

GO ON TO THE NEXT PAGE

14. Bounded by fierce mountains and deserts (A) the high plateau of Iran has seen (B) the flow of many migrations and the development of many cultures, all (C) of which have added distinctive features to the many styles of Persian art and architecture (D). No error (E)

15. A surrealist, Neruda revitalized (A) everyday expressions and employed (B) bold metaphors in free verse, his (C) evocative poems are filled with grief and despair and bespeak (D) a quest for simplicity. No error (E)

16. A 1994 survey found (A) that 12 million Americans had experienced (B) homelessness at some point in their lives (C). The vast majority men and families with children (D). No error (E)

17. In England, (A) the clergy keep records of christenings, marriages, and burials (B) as early as the 16th century; and (C) during the 17th century, the clergy in France, Italy, and Spain (D) began to keep similar records. No error (E)

18. During the Industrial Revolution, advancements in technology sanitation and the means of food distribution (A) made possible a drop (B) in the death rate so significant that (C) between (D) 1650 and 1900 the population of Europe almost quadrupled (from about 100 million to about 400 million) in spite of considerable emigration. No error (E)

19. Mt. Rainier is (A) snow-crowned and has 26 glaciers which offered (B) a challenge to serious climbers; (C) its heavily forested lower slopes are popular with hikers (D). No error (E)

GO ON TO THE NEXT PAGE

20. Acute arterial thrombosis often results
A
from the deposition of atherosclerotic
B
material in the wall of an artery, that grad-
C
ually narrows the channel, precipitating
D
clot formation. No error
E

21. The disaster at Love Canal led to the
A
creation of the Environmental Protection
B
Agency's "Superfund," which makes re-
C
sponsible parties liable for the cleanup
of environmental hazards, more than
D
$20,000,000 in settlement damages was
paid by the chemical company and the
city of Niagara Falls to a group of former
residents. No error
E

22. Nucleotides consisting of either a purine
A B
or a pyrimidine base, a ribose or deoxyri-
C D
bose, and a phosphate group. No error
E

23. Tammany Hall's decline were accelerated
A
by women's suffrage, immigration restric-
B
tion, and the social programs of the New
Deal, which weakened voters' depen-
C D
dence on the political machine. No error
E

24. The Kelvin degree is the same size as the
A
Celsius degree; hence the two reference
B
temperatures for Celsius, the freezing
point of water (0°C), and the boiling point
C
of water (100°C), correspond to 273.15°K
and 373.15°K, respectively. No error
D E

25. Beyond the specifically Jewish notions
contained within the kabbalah, some
A B
scholars believe that it reflects a strong
Neoplatonic influence, especially in its
C
doctrines of emanation and the transmi-
D
gration of souls. No error
E

GO ON TO THE NEXT PAGE

26. <u>The mystery of childbearing,</u> (A) <u>the sorrow of a tragic love,</u> (B) and a burning desire for justice <u>are</u> (C) recurrent themes of Gabriela Mistral's <u>fluent and lyric</u> (D) verse. <u>No error</u> (E)

27. <u>Intaglio design cut</u> (A) into stone or other material or etched or engraved in a <u>metal plate, producing</u> (B) a concave instead of a convex <u>effect</u> (C). <u>It is the reverse of a relief or cameo</u> (D). <u>No error</u> (E)

28. The term "alternative medicine" <u>can encompass</u> (A) a wide range of <u>therapies,</u> (B) including <u>chiropractic, homeopathy, acupuncture, herbal medicine, meditation, biofeedback, massage therapy,</u> (C) and various "new age" therapies such as guided <u>imagery and naturopathy</u> (D). <u>No error</u> (E)

29. After the atomic bomb <u>was used</u> (A) against <u>Japan Robert J. Oppenheimer</u> (B) <u>became</u> (C) one of the foremost proponents of <u>civilian and international</u> (D) control of atomic energy. <u>No error</u> (E)

GO ON TO THE NEXT PAGE

<u>DIRECTIONS:</u> The following passages are considered early draft efforts of a student. Some sentences need to be rewritten to make the ideas clearer and more precise.

Read each passage carefully and answer the questions that follow. Some of the questions are about particular sentences or parts of sentences and ask you to make decisions about sentence structure, diction, and usage. Some of the questions refer to the entire essay or parts of the essay and ask you to make decisions about organization, development, appropriateness of language, audience, and logic. Choose the answer that most effectively makes the intended meaning clear and follows the requirements of standard written English. After you have chosen your answer, fill in the corresponding oval on your answer sheet.

EXAMPLE

(1) On the one hand, I think television is bad, But it also does some good things for all of us. (2) For instance, my little sister thought she wanted to be a policemen until she saw police shows on television.

Which of the following is the best revision of the underlined portion of sentence (1) below?

One the one hand, I think television <u>is bad, But it also</u> does some good things for all of us.

(A) is bad; But it also

(B) is bad. but it also

(C) is bad, but it also

(D) is bad, and it also

(E) is bad because it also

EXAMPLE ANSWER

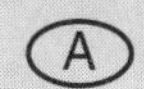    

GO ON TO THE NEXT PAGE

Questions 30–35 refer to following passage.

(1) Bilingual education uses the students' native language as a tool to learn a second language. (2) In the initial stages of education, the students' native language helps keep them from falling behind their fellow students while learning English. (3) The first language serves as a bridge for learning, but more importantly, the knowledge acquired in one language transfers to the other language. (4) This means that a child who is not fluent in English but fluent in Spanish will learn English easily because he or she has already learned the foundational processes in the first language. (5) The "knowledge-transfer" hypothesis rests on the premise that the process of reading is similar across languages, even though the languages and writing systems are different.

(6) One researcher explains that when schools provide children quality education in their primary language, they give them two things: knowledge and literacy. (7) The *knowledge* that children get through their first language helps make the English they hear and read more comprehensible. (8) *Literacy* developed in the primary language transfers to the second language. (9) The reason is simple: Because we learn by reading, by making sense of what is on the page, it is easier to learn to read in a language we understand. (10) Once we can read in one language, we can read in general.

(11) Note that the most effective bilingual education programs are *two-way* bilingual programs. (12) The aim is to teach both native speakers of Spanish and native speakers of English attending the same classes academic subjects in both languages. (13) The non-native students initially receive 90 percent of their instruction in Spanish and 10 percent in English, and then the amount of English increases with each grade. (14) One recent study concludes that English language learners do better academically over the long term if English is introduced slowly rather than submerging learners in intensive English instruction in a regular classroom.

30. What is the best way to deal with sentence 11 (reproduced below)?

Note that the most effective bilingual education programs are two-way bilingual programs.

(A) Leave it as is.

(B) Change "effective" to "affective."

(C) Remove the sentence from the paragraph.

(D) Insert a comma after "education programs."

(E) Insert a hyphen between "education" and "programs."

GO ON TO THE NEXT PAGE

31. Which revision is the best choice for this passage from paragraph 2 (reproduced below)?

Because we learn by reading, by making sense of what is on the page, it is easier to learn to read in a language we understand. Once we can read in one language, we can read in general.

(A) We learn by reading, by making sense of what is on the page, and it is obviously easier to learn to read in a language we understand. Once we can read in one language, we can read in general, and we continue to learn in the process by virtue of taking in new information all the while.

(B) Because we learn by reading, by making sense of what is on the page, it is easier to learn to read in a language we understand, and once we can read in one language, we can read in general.

(C) Because we learn by reading, making sense, it is easier to learn to read. Once we can read in one language, we can read in general.

(D) Because we learn by making sense of what is on the page, it is easier to learn to read in a language we understand. Once we can read in one language, we can read in general.

(E) Once we can read in one language, we can read in general because we learn by reading, by making sense of what is on the page. So it is easier to learn to read in a language we understand.

32. What word(s) in sentence 12 can be eliminated to improve the sentence?

(A) both languages

(B) the aim is

(C) native

(D) attending the same classes

(E) languages

33. Which of the following phrases should be inserted at the beginning of sentence 14 to link it to sentence 13?

(A) Somewhere done the road,

(B) To the contrary,

(C) In fact,

(D) If this is possible,

(E) In study after study,

34. Which of the following best characterizes the meaning of the last sentence?

(A) learning English is easier if the students are allowed to use only that language

(B) learning any language takes time

(C) learning math in one's native language is easier than learning it in a new language

(D) teaching language in bits and pieces is preferable to teaching the larger concepts all at once

(E) teaching students a language gradually is more effective in the long run than presenting them with all information in the new language only

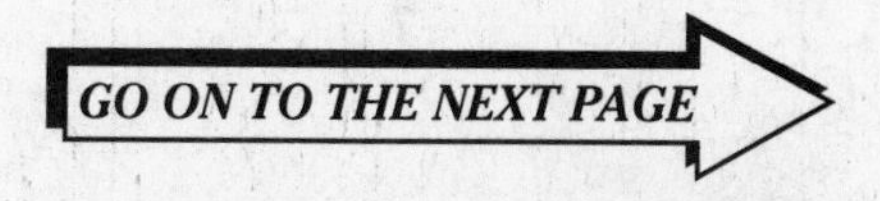

35. Which of the following is most likely to follow the final paragraph in this passage?

(A) a definition of other language learning models

(B) more discussion of the effects of gradually learning a new language on academic performance overall

(C) examples from classrooms where the languages are something other than Spanish and English

(D) a discussion of the present state of bilingual education in the United States

(E) a discussion of a particular classroom

TIME: 25 Minutes
18 Questions

MATHEMATICS

DIRECTIONS: In this section solve each problem, using any available space on the page for scratchwork. Then decide which is the best of the choices given and fill in the corresponding oval on the answer sheet.

NOTES

(1) The use of a calculator is permitted.

(2) All numbers used are real numbers.

(3) Figures that accompany problems in this test are intended to provide information useful in solving the problems. They are drawn as accurately as possible EXCEPT when it is stated in a specific problem that the figure is not drawn to scale. All figures lie in a plane unless otherwise indicated.

(4) Unless otherwise specified, the domain of any function f is assumed to be the set of all real numbers x for which $f(x)$ is a real number.

REFERENCE INFORMATION

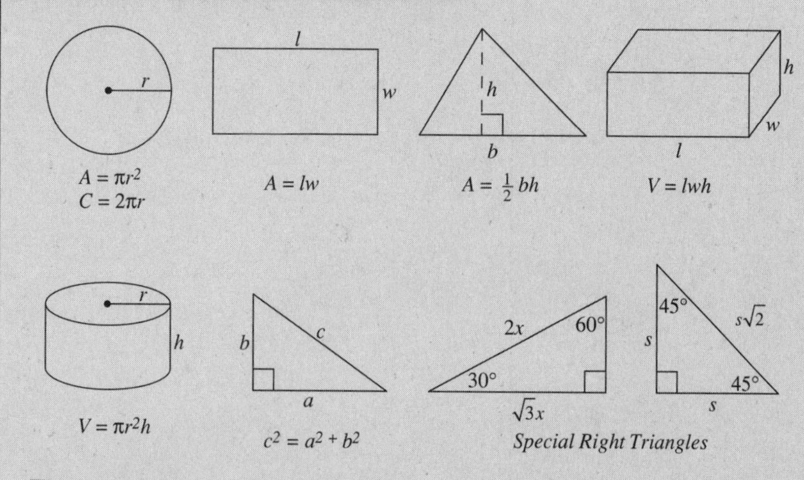

The number of degrees of arc in a circle is 360.

The sum of the measures in degrees of the angles of a triangle is 180.

1. If $m^2 = 625$, and $m > 0$, what is the value of $3m^{-\frac{1}{2}}$?

(A) 75

(B) 37.5

(C) 15

(D) 0.6

(E) 0.2

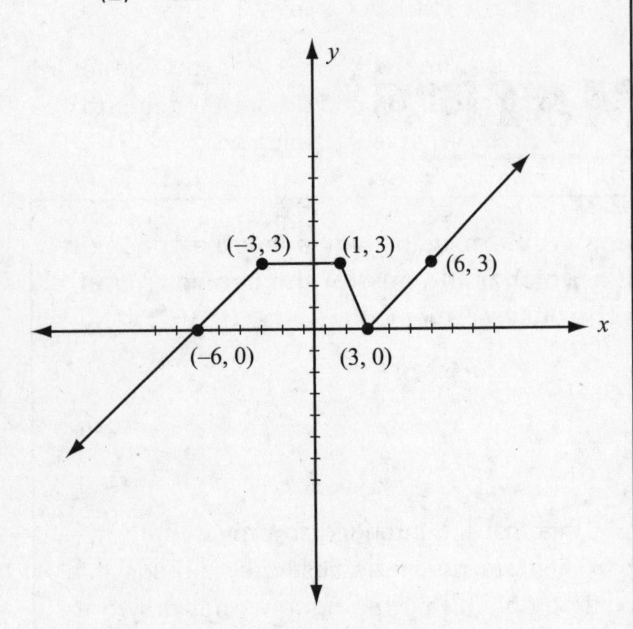

2. In the graph shown above, for what complete list of values of x are the y values less than 3?

(A) $1 < x < 6$

(B) $x < -3$; $1 < x < 6$

(C) $x < 1$

(D) $x < -3$

(E) $x < -3$; $x > 1$

3. Given {13, 7, 9, 19, 9, 27}, what is the value of the median of this set?

(A) 18

(B) 14

(C) 13

(D) 11

(E) 9

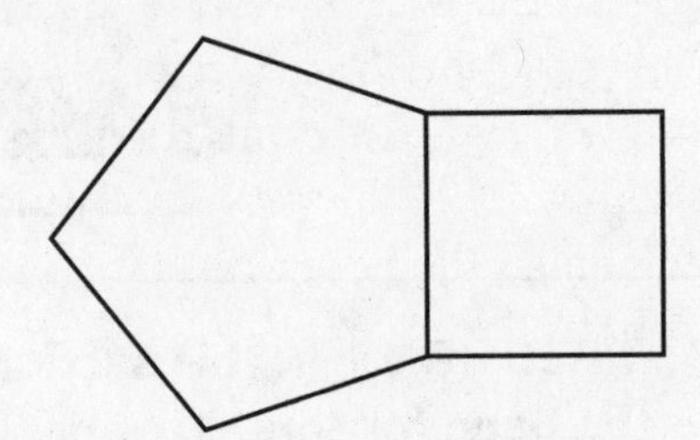

4. In the figure above, a regular pentagon and a square share a common side. If the perimeter of the pentagon is 12.5, what is the area of the square?

(A) 6.25

(B) 10

(C) 18

(D) 25

(E) 50

GO ON TO THE NEXT PAGE

	Men	Women	Children
Small	$15	$12	$8
Medium	$20	$16	$10
Large	$25	$18	$14

5. The table above shows the price per shirt for men, women, and children. The three different sizes are also shown. A family has $200 to spend and must buy 3 children's shirts in a medium size, 4 women's shirts in a small size, and some men's shirts. How much money will be left if the family buys the maximum number of men's shirts they can afford?

(A) $1.00

(B) $1.50

(C) $2.00

(D) $5.00

(E) $8.00

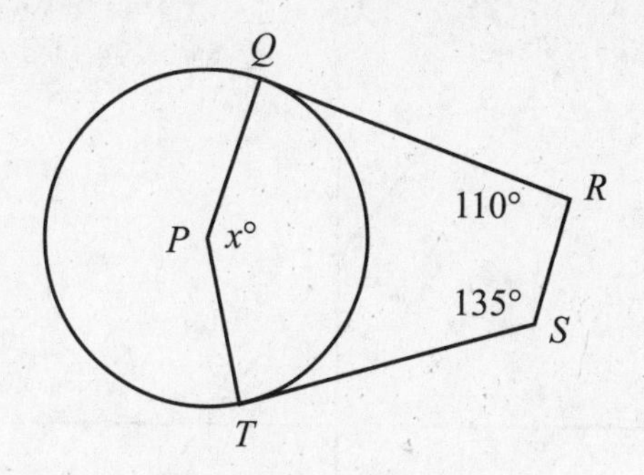

Note: Figure not drawn to scale.

6. In the figure above, P is the center of the circle. $\overline{QR}$ and $\overline{ST}$ are tangents to the circle. What is the value of x?

(A) 115

(B) 120

(C) 125

(D) 140

(E) 145

7. If $\frac{-4x+5}{3} > -6$, which of the following describes all possible values of x?

(A) $x < \frac{23}{4}$

(B) $x > -\frac{23}{4}$

(C) $x < -\frac{13}{4}$

(D) $-\frac{23}{4} < x < \frac{13}{4}$

(E) $-\frac{23}{4} < x < -\frac{13}{4}$

GO ON TO THE NEXT PAGE

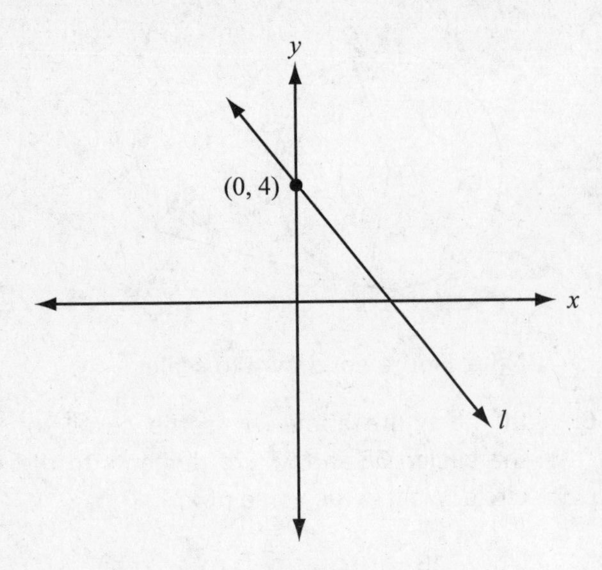

Note: Figure not drawn to scale.

8. In the figure above, if the slope of line l is $-\frac{8}{7}$, what is the x-intercept of line l?

(A) 3

(B) $\frac{13}{4}$

(C) $\frac{7}{2}$

(D) $\frac{15}{4}$

(E) 4

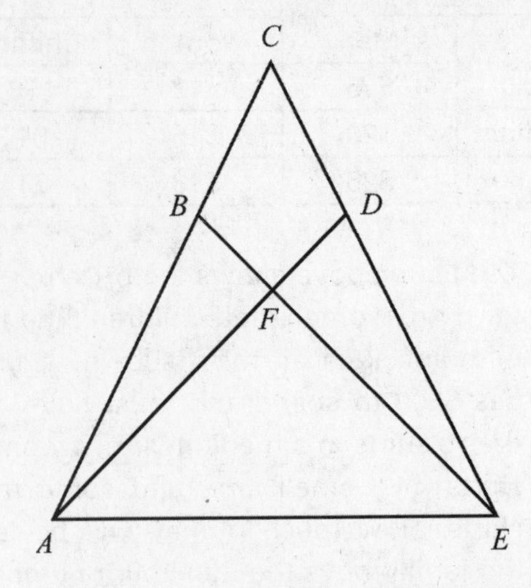

9. In the figure above, $\overline{AC} = \overline{CE}$ and $\overline{AB} = \overline{DE}$. Which of the following statements must be true?

I. $\overline{AB} = \overline{FE}$

II. $\angle ABE \cong \angle EDA$

III. $\triangle ACD \cong \triangle ABE$

(A) I only

(B) II only

(C) II and III only

(D) I, II, and III

(E) I and II only

10. Lines ℓ_1 and ℓ_2 are perpendicular and intersect at (1, 4). If ℓ_1 contains the point (3, 9), which of the following points lies on ℓ_2?

(A) (3, –1)

(B) (–9, 8)

(C) (–1, –1)

(D) (6, 6)

(E) (4, 13)

GO ON TO THE NEXT PAGE

11. Triangle XYZ is an isosceles right triangle, with $\angle Y$ as the right angle. The sum of sides $\overline{XY}$ and $\overline{YZ}$ is best approximated by what percent of the perimeter of $\triangle XYZ$?

(A) 70%

(B) 68%

(C) 65%

(D) 63%

(E) 59%

12. Which of the following functions has its highest point at (–1, 5)?

(A) $f(x) = -2(x - 1)^2 - 5$

(B) $f(x) = 2(x + 1)^2 - 5$

(C) $f(x) = -2(x - 1)^2 + 5$

(D) $f(x) = -2(x + 1)^2 + 5$

(E) $f(x) = 2(x + 1)^2 + 5$

13. A geometric sequence for positive numbers is given by $x, 2, y, 72, \ldots$. What is the value of $x + y$?

(A) 3

(B) $12\frac{1}{3}$

(C) $18\frac{1}{2}$

(D) 36

(E) 37

14. Angela, Kim, and Tanya together require 2 hours to paint a kitchen. To complete the job working alone, Angela would need 12 hours and Kim would need 24 hours. How many hours would Tanya need if she were working alone?

(A) 10

(B) 8

(C) 6

(D) $5\frac{1}{3}$

(E) $2\frac{2}{3}$

15. Numbers w, x, and y are all positive. If $wx = 2$, $xy = 9$, and $wy = 8$, what is the value of wxy?

(A) 1

(B) $5\frac{1}{3}$

(C) 6

(D) 9

(E) 12

GO ON TO THE NEXT PAGE

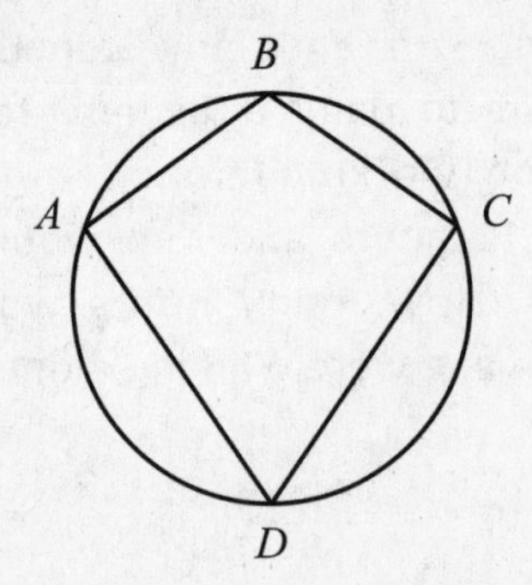

Note: Figure not drawn to scale.

16. In the figure above, quadrilateral *ABCD* is inscribed in the circle. If the measure of $\angle B$ is 105° and the measure of $\angle C$ is 95°, what is the measure, in degrees, of arc *ABC*?

(A) 75
(B) 80
(C) 90
(D) 150
(E) 160

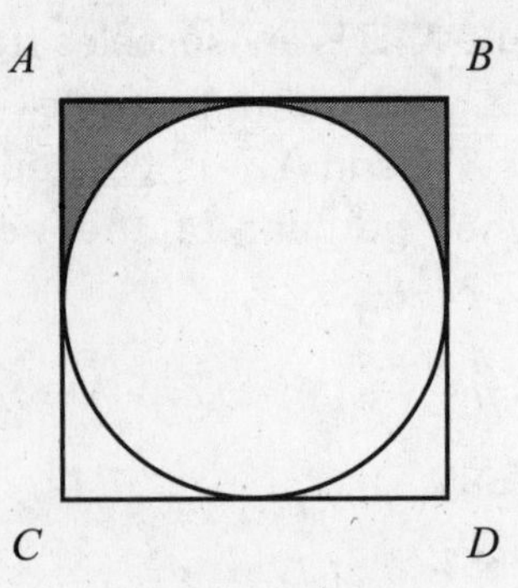

17. The figure above represents a dartboard in which a circle is inscribed in square *ABCD*. The radius of the circle is 4. If a dart is thrown and lands on the dartboard, what is the probability that it landed in a shaded region?

(A) $\frac{2-\pi}{\pi}$
(B) $\frac{4-\pi}{8}$
(C) $\frac{4-\pi}{\pi}$
(D) $\frac{\pi}{16}$
(E) $\frac{\pi}{8}$

18. Which of the following is equivalent to $\frac{\sqrt{75x^3y^5}}{\sqrt{625x^5y^3}}$?

(A) $\frac{y\sqrt{3xy}}{5x\sqrt{xy}}$

(B) $\frac{5x\sqrt{3xy}}{y\sqrt{x}}$

(C) $\frac{y\sqrt{3x}}{5x\sqrt{y}}$

(D) $\frac{5y\sqrt{3}}{x\sqrt{xy}}$

(E) $\frac{y\sqrt{3xy}}{5x\sqrt{5xy}}$

STOP

If time remains, you may go back and check your work. When the time allotted is up, you may go on to the next section.

TIME: 25 Minutes
24 Questions

CRITICAL READING

DIRECTIONS: Each sentence below has one or two blanks, each blank indicating that something has been omitted. Beneath the sentence are five lettered words or sets of words. Choose the word or set of words that BEST fits the meaning of the sentence as a whole.

EXAMPLE

Although critics found the book _______ , many readers found it rather _______ .

(A) obnoxious . . . perfect
(B) spectacular . . . interesting
(C) boring . . . intriguing
(D) comical . . . persuasive
(E) popular . . . rare

EXAMPLE ANSWER

   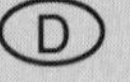

1. Just as the theory of relativity _______ importance in the special situation where very large speeds are involved, so the quantum theory is necessary for the special situation where very _______ quantities are involved (i.e., on the scale of molecules, atoms, and elementary particles).

   (A) assumes . . . small
   (B) purports . . . brilliant
   (C) contraindicates . . . violent
   (D) validates. . . maximal
   (E) recognizes . . . gargantuan

2. Nationalism is basically a collective state of mind or _______ in which people believe their primary duty and loyalty is to the nation-state. Often nationalism implies national superiority and _______ various national virtues.

   (A) consciousness . . . glorifies
   (B) villification . . . defrocks
   (C) a priori . . . purports
   (D) conscience . . . denies
   (E) purpose . . . decries

3. Orpheus _______ to Hades searching for Eurydice. He was granted the chance to regain her if he could refrain from looking at her until he had led her back to sunlight. Orpheus could not resist, and Eurydice vanished forever. Grieving _______ , he became a recluse and wandered for many years.

   (A) discouraged . . . invariably
   (B) descended . . . inconsolably
   (C) resolved . . . lugubriously
   (D) fled . . . conversely
   (E) decried . . . absolutely

4. The purpose of collective bargaining may be either a discussion of the terms and conditions of employment (wages, work hours, job safety, or job security) or a _______ of the collective relations between both sides (the right to organize workers, recognition of a union, or a guarantee of no _______ against the workers if a strike has occurred).

   (A) perpetuation . . . reprises
   (B) occasion . . . corruption
   (C) fomentation . . . configuration
   (D) trepidation . . . violence
   (E) consideration . . . reprisals

5. Piracy is _______ from privateering in that the pirate holds no commission from and receives the protection of no nation but usually attacks vessels of all countries.

   (A) distinguished
   (B) perpetuated
   (C) compared
   (D) eradicated
   (E) established

GO ON TO THE NEXT PAGE

Questions 6–9 are based on the following passages.

Passage 1

(1) When kids are tossed together every day, six hours a day for the entire school year, they seem to form friendship groupings naturally. Better known as cliques, such groupings are small, tightly knit, autonomous, and sometimes inclusive groups of people that share the same interests or characteristics. In short, members of cliques often hold the same values and exhibit the same behavior. Although they have been known to form in elementary school, cliques are commonly associated with middle and high school students. In a recent nationwide survey, 96.3 percent of the female respondents claimed that cliques existed in their schools and 84.2 percent of all respondents reported that most of their classmates belonged to cliques.

(2) Cliques can be based on appearance, athletic ability, academic achievement, social or economic status, talent, ability to attract the opposite sex, or some shared definition of sophistication, according to adolescent development experts. The prominent characteristic of a clique usually becomes the clique's label. For instance, a group of self-assured, varsity-jacket-wearing male students might be known as "jocks," whereas another group's unkempt appearance and calculated spacey demeanor could earn them the "stoner" or "druggie" label.

(3) There are strong incentives for adolescents to join cliques. For example, teenagers use cliques to simplify their lives, perhaps using their group as a sort of shortcut to develop friendships and romantic relationships. Teenagers also use cliques to categorize their peers, especially when they move on to middle or high school, where student populations can reach the thousands. Cliques and peer groups also help adolescents establish an identity, but most importantly, teenagers join cliques to gain a sense of belonging.

Passage 2

(1) The cliquishness among students in American high schools has been treated as a normal and relatively harmless phenomenon. However, the perception that domineering high school cliques can worsen many students' feelings of depression, alienation, and rage emerged strongly after the Columbine High School shootings in Littleton, Colorado.

(2) In fact, the Columbine tragedy and the recent spate of other school shootings have generated much criticism of high school cliques, especially if the group dynamics both within and between these groups is not monitored and controlled to the degree possible by adults. Some commentators suggest that cliques can be socially counterproductive because they create hierarchies that alienate some teenagers. For example, according to one violence prevention consultant, "the downside of cliques is that there are some groups that are valued more highly than others, and those who cannot latch onto groups are disenfranchised." Another commentator insists that cliques thrive at the emotional cost of the students outside the group, particularly the disenfranchised: "One of the ways cliques reinforce themselves is by putting down whoever isn't in the group by teasing, taunting, and in extreme cases even physically abusing them."

(3) Some critics even warn that adolescents who are persecuted or rejected by popular or mainstream cliques may react and form cliques that defy the entire school, such as the Trench Coat Mafia, the name of the clique to which the two young men belonged who killed their fellow students and a teacher at Columbine. As one psychologist asserts: "All kids need to belong, and if they can't belong in a positive way at school, they'll find a way to belong to a marginal group."

6. The writer of Passage 1 declares that "teenagers use cliques to simplify their lives." Which of the following best defines what this statement means?

   (A) Cliques insulate students from the harsh realities of growing up.

   (B) Cliques develop social skills.

   (C) Cliques help students make friends and develop relationships within an otherwise overwhelmingly large population.

   (D) Cliques allow students to easily identify their friends and their enemies

   (E) Cliques will teach students how to live in a hierarchically arranged world.

7. In paragraph 2 writer of Passage 2 uses the term "disenfranchised." Which of the following best defines this term?

   (A) to lose rights accorded to others and any sense of belonging

   (B) to be beaten down verbally

   (C) to feel physically threatened or otherwise unsafe in an environment

   (D) to be continually frightened and isolated

   (E) to be ostracized for one's appearance

8. Which of the following best defines the most dire consequences of clique behavior mentioned in Passage 2?

   (A) The disenfranchised are more likely to drop out of school.

   (B) Those who are excluded from all cliques may look outside the school for their sense of belonging.

   (C) The members of cliques will get a false sense of how the world outside of high school actually operates.

   (D) The members of cliques may lose touch with their sense of empathy for others not like themselves.

   (E) Students who are disenfranchised may form their own cliques in opposition to the entire school, and the result could be antisocial behavior.

9. Which of the following best describes the relationship of Passage 1 to Passage 2?

   (A) Passage 1 describes cliques and accepts them as positive, and Passage 2 describes the negative possibilities in this behavior.

   (B) Passage 2 refutes the primary claim of Passage 1 that cliques are good for the school.

   (C) Passage 2 refutes the argument in Passage 1 point-by-point.

   (D) Both passages describe clique behaviors.

   (E) Passage 1 argues for accepting clique behavior as part of being young, and Passage 2 argues that cliques should be closely monitored.

Questions 10–15 are based on the following passage.

(1) Parents, educators, and policymakers have responded in myriad ways to dire reports regarding the state of American education. Many have pushed for reforms (such as state-funded tuition vouchers and charter schools) that would allow parents to choose which school their children attend. The proponents of such reforms argue that these reforms would enable poor parents to use state funds to send their children to high-quality private schools and allow nongovernmental groups to use public money to operate their own schools.

(2) However, the "education crisis" thus envisioned is largely a myth concocted by a few ideologues who have used distorted data on student achievement with the intent of winning approval and taxpayer support for private schools. Such misguided schemes for education reform could drain needed funds away from public institutions and seriously damage American schools overall.

(3) That said, however, whether one agrees that American education has reached a crisis state, it remains true that the public school system has room for improvement. One reform measure that has received wide support is the push for national academic standards. Currently, because public schools are administered at the state and local levels, admittedly academic standards vary widely from region to region. National standards would clearly identify what concepts and skills all U.S. students should master at certain grade levels. For example, the signatories of "A Nation Still at Risk" maintain that "America needs solid national academic standards and . . . standards-based assessments, shielded from government control, and independent of partisan politics, interest groups, and fads." Supposedly, these standards would give educators distinct guidelines for tracking student progress and deciding on curricula and teaching methods. Proponents contend that national standards would, in the end, increase academic achievement and ensure that all U.S. students receive roughly the same education.

(4) However, there is a real risk that the implementation of national standards could lead to increased governmental intrusion in local school board decisions. Despite policymakers' assurances to the contrary, national standards will create politically motivated controversies over what kind of information should be taught in schools. Conservative groups such as the Christian Coalition, for example, contend that national standards could eventually impose an overly liberal and secular curriculum on those who hold traditional Christian beliefs, whereas liberal groups argue the opposite, that a conservative administration will push for reforms antithetical to the separation of church and state. Furthermore, it seems inevitable that the new mission of the public schools would be to coach children on how to pass the test rather than on real education.

(5) Such ironclad standards would also prove to be devastating for low-income school districts that cannot afford to hire better teachers, revise curricula, or renovate crumbling classrooms. In short, holding all children to the same standards without providing all children with equal educational resources would damage morale and raise dropout rates when these students fail to meet the standards. Policymakers must first address the problem of educational funding inequities before setting tough national standards.

(6) The bottom line is that the debate over improving education in America remains politically charged and politically motivated. Since all independent commissions charged with studying the issue seem to be less than completely objective, it remains to be seen if government initiatives will drive real reform. Perhaps the only way that education will be reformed is from within, via a revolution perpetrated by a unique source of energy and creativity: front line education professionals and parents.

GO ON TO THE NEXT PAGE

10. The overall purpose of this passage is

    (A) to argue for a particular strategy for public school reform.

    (B) to argue against the need for education reform.

    (C) to argue that the present system is flawed but the best we can do.

    (D) to dispute the prominent suggestions for school reform and to suggest that real change will have to come from within the public school system rather than from government.

    (E) to argue for more resources, and a better allocation of those resources, for students and teachers before any discussion of reform so that the playing field is level for all schools.

11. The writer argues against the idea that there is an "education crisis," but the author admits that there is room for improvement in the education system. Which of the following best explains this strategy?

    (A) The argument ceases to be about education reform.

    (B) The tone of the argument changes from blame and controversy to reasoned inquiry into how best to improve education.

    (C) The author can now refute other suggestions for reform.

    (D) The author can now concentrate on just a few options rather than taking on all suggestions for education reform.

    (E) The tone becomes dismissive of all opposition.

12. Which of the following best describes what takes place in paragraphs 3 and 4?

    (A) description of the argument for national standards in public education and proponents' reasoning followed by a refutation of this strategy

    (B) explanation of why the author is for the standards followed by why others are against this strategy

    (C) description of the "A Nation Still at Risk" findings followed by a refutation of those findings

    (D) explanation of yet one more misguided strategy for improving schools and a declaration that this will not work

    (E) reiteration of the writer's stance that there is not really a crisis in education followed by an explanation of proponents' motives

13. In paragraph 5, the author declares, "Policymakers must first address the problem of educational funding inequities before setting tough national standards." Which of the following best explains this sentence?

    (A) Poor school districts can't meet tougher standards.

    (B) Rich school districts need to give up some of their resources to poor districts.

    (C) Establishing standards is useless until poorer school districts have the resources to improve.

    (D) Establishing standards will drive school funding reform.

    (E) Schools should all be privatized so that they can achieve such standards.

GO ON TO THE NEXT PAGE

14. Which of the following best characterizes the various aspects of this passage in order?

(A) description of reform suggestions, dismissal of these suggestions as politically motivated, admission that there is room to improve, description of national standards argument, caveat that this suggestion requires funding reform, call for change from within

(B) discussion of reform suggestions, caveats regarding these changes, description of national standards, call for change from within

(C) plea for education reform, dismissal of opponents' ideas, conclusion that new strategies are called for other than those presently being considered

(D) dismissal of vouchers and charter schools in favor of national standards, call for changes initiated from within the school system

(E) demand for greater teacher accountability, dismissal of calls for national standards, plea for teachers and parents to cooperate in achieving funding reforms

15. Which of the following best explains the author's claim that "Perhaps the only way that education will be reformed is from within, via a revolution perpetrated by a unique source of energy and creativity: front line education professionals and parents."

(A) Government-initiated reforms will inevitably fail.

(B) Education reform is impossible in the current political climate.

(C) External reform suggestions are always politically motivated.

(D) Our education system will never be improved in the absence of a commitment from teachers and parents.

(E) Only teachers, administrators, and parents are motivated by what is best for students rather than by a political agenda.

GO ON TO THE NEXT PAGE

Questions 16–24 are based on the following passage.

This passage is the beginning of a short story.

(1) Both sisters were silent, feeling delight shudder through them as they listened to their young neighbor sing. Miss Abigail stood at the window, leaning against the frame as the notes sailed in on the breeze. She was tall and straight as an arrow, in spite of the fact that she was nearly 70 years old. Her snow-white hair was brushed straight up from her high forehead, and her blue eyes were keen and bright as a sharpened blade. She wore a black dress and a white apron, and her hands showed the wear and tear of years of serving, of hard work of all kinds.

(2) No one would have thought that Abigail and Darnel were sisters, unless they were a witness to the loving looks that sometimes passed between the siblings when they were alone together. The face that lay on the pillow was white and withered, like a crumpled flower of some indeterminate variety, and the dark eyes had a pleading, wistful look. Darnel had white hair too, but hers had a warm yellowish tinge, very different from the brilliant white of Abigail's tresses. Darnel's hair also curled in little ringlets around her beautiful old face. In short, Darnel was nothing short of gorgeous in spite of her advanced age and deterioring state, while no one would think of calling her sister anything but merely lovely.

(3) The younger sister was confined in perpetuity to her bed. It had been many years since she met with the accident that changed her from a vivacious, laughing girl into a helpless but still beautiful woman. There had been a party, all participants mindless of anything except the joy of the moment, and she had gone riding alone in the moonlight. Her horse had leaped sideways suddenly, frightened by a startled bird flying from the tall grass, and she had fallen, striking her head on a sharp stone. This was Darnel's little story.

(4) People in the neighborhood had forgotten that there was any story at all. Even her lifelong friends had almost forgotten that Darnel had ever been other than she was now. But Abigail never forgot. She had left her job in the neighboring town and came home to take care of her sister — they had not been apart for more than a few hours since. Once, when Darnel's bitterness first began to diminish, which took more than a year, and she realized that she was still a living creature and not a condemned person suffering for the sins of humankind, she asked her sister about her former life, about leaving it behind. Darnel was feeling guilty that her sister had deserted a perfectly admirable existence to wait upon her every need, but Abigail assured her that she was miserable before and happy to be of service besides.

(5) Since these only questions about Abigail's past, the sisters had lived their lives together without another thought that there might be room for anyone else, until the young girl moved in next door the previous year, until she came into their world and filled it with sunshine and music.

GO ON TO THE NEXT PAGE

16. The first sentence foreshadows what?

    (A) The story will be about music.

    (B) The story will be about sisters.

    (C) The story will take place in spring and have two women in it.

    (D) This story will be be about sisters who like music.

    (E) This story will be about a young woman and her music and the joy it gives to two women who are sisters.

17. There are two similes used to describe Abigail in paragraph 1. What do these two similes imply about this character?

    (A) She seems younger than her age, both physically and mentally.

    (B) She is athletic.

    (C) She is prone to violence, since both similes include weapons.

    (D) She will be hard upon her sister because of the sacrifices she has made.

    (E) She is a rigid and harsh person.

18. In paragraph 2, the narrator tells the reader, "Darnel was nothing short of gorgeous in spite of her advanced age and deterioring state, while no one would think of calling her sister anything but merely lovely." This indicates

    (A) These sisters have always competed with each other.

    (B) These sisters are both quite vain about their appearance.

    (C) The sisters compliment each other's looks often.

    (D) Both women are pretty, but Darnel is prettier in spite of being bedridden for many years.

    (E) Both women have been declared beautiful for their age by those who know them.

19. What is the effect of referring to the tragic event that left her bedridden as "Darnel's little story" (paragraph 3)?

    (A) It makes the reader wonder if she is faking it.

    (B) It universalizes tragedy as common to all people, but also suggests that others can't understand another person's tragic tale.

    (C) It categorizes the disabled as not important.

    (D) It suggests that no one takes her accident very seriously and diminishes her condition.

    (E) It suggests that she tells this story over and over again.

20. What does this reprinted passage from paragraph 4 imply about people in general and about Abigail?

People in the neighborhood had forgotten that there was any story at all. Even her lifelong friends had almost forgotten that Darnel had ever been other than she was now. But Abigail never forgot.

(A) People in general lack the capacity for empathy, but Abigail does not.

(B) People have short memories and assume that a person's condition is simply that person, but Abigail remembers that this was a tragic event for her sister and changed her life in an extraordinary way.

(C) People are too caught up in their own lives to pay much heed to anyone else, but Abigail is selfless.

(D) Other people do not care what happens to Darnel, but her sister does.

(E) The disabled are often forgotten by their neighbors and must rely on their families for support.

21. How did Darnel first react to being bedridden?

(A) She was thankful for her sister's willingness to help her.

(B) She was in denial that this injury would permanently disable her.

(C) She wished she were dead.

(D) She thought that she had been stupid to go riding at night.

(E) She was bitter and felt like she had been singled out unjustly for this fate.

22. Given how this passage ends, which of the following is the best extrapolation of what will follow?

(A) further explanation of the sisters' relationship

(B) further description of both sisters

(C) a description of the little girl and an explanation for why the sisters are so fond of her

(D) a story about the sisters when they were young, which will connect to the little girl somehow

(E) the arrival of the girl and a conversation with the sisters in which it is revealed that Darnel is not really disabled but chooses not to get out of bed

23. In paragraph 3, the phrase "in perpetuity" is used, which means

(A) for good.

(B) for two weeks.

(C) in the end.

(D) after all.

(E) frequently.

24. From the description in this passage, which of the following best describes the relationship between these sisters?

(A) complex

(B) redundant

(C) perplexing

(D) devoted

(E) variable

GO ON TO THE NEXT PAGE

TIME: 20 Minutes
16 Questions

MATHEMATICS

DIRECTIONS: **In this section solve each problem, using any available space on the page for scratchwork. Then decide which is the best of the choices given and fill in the corresponding oval on the answer sheet.**

NOTES

(1) The use of a calculator is permitted.

(2) All numbers used are real numbers.

(3) Figures that accompany problems in this test are intended to provide information useful in solving the problems. They are drawn as accurately as possible EXCEPT when it is stated in a specific problem that the figure is not drawn to scale. All figures lie in a plane unless otherwise indicated.

(4) Unless otherwise specified, the domain of any function f is assumed to be the set of all real numbers x for which $f(x)$ is a real number.

REFERENCE INFORMATION

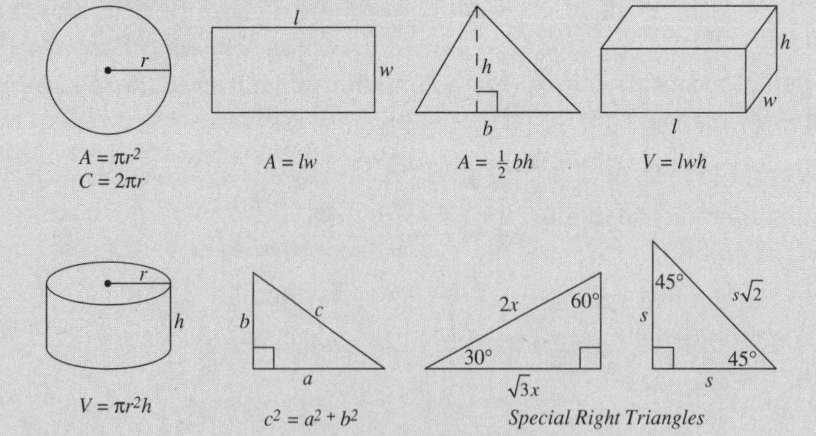

The number of degrees of arc in a circle is 360.

The sum of the measures in degrees of the angles of a triangle is 180.

Annual Sales at
Sal's Shoe Store

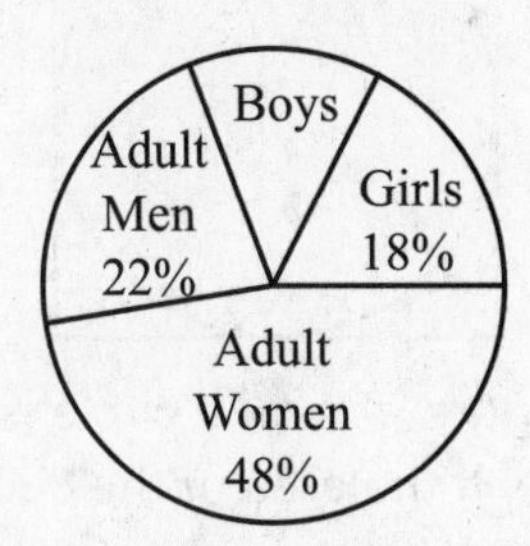

1. As shown in the circle graph above, Sal's Shoe Store sold $6,000 worth of shoes for boys. What were the sales for the amount of shoes sold for adult women?

(A) $8,800
(B) $12,500
(C) $24,000
(D) $26,400
(E) $42,000

2. If P = {prime numbers between 4 and 20} and $P \cup Q$ = {all positive integers less than 20}, what is the <u>least</u> number of elements in set Q?

(A) 3
(B) 6
(C) 10
(D) 13
(E) 17

3. In the graph of a linear function, the x-intercept is (–6, 0) and the slope is $\frac{7}{2}$. What is the y-intercept?

(A) (0, –21)
(B) (0, –7)
(C) (0, 6)
(D) (0, 7)
(E) (0, 21)

4. If $(7 - 3x) < (5x - 10)$ and x <u>must</u> be an integer, which of the following represents the solution set?

(A) $x \geq 0$
(B) $x \geq 2$
(C) $x \geq 3$
(D) $x \leq 2$
(E) $x \leq 3$

5. Which of the following shows that the length (L) varies inversely as the cube of the width (w)?

(A) $L = 5 + w^3$
(B) $L = 5w^3$
(C) $L = w^3/5$
(D) $L = 5 - w^3$
(E) $L = 5/w^3$

6. What is the range of the function $f(x) = |4 - x| + 3$?

(A) All numbers greater than or equal to 4
(B) All numbers less than or equal to 4
(C) All numbers between 3 and 4, inclusive
(D) All numbers greater than or equal to 3
(E) All numbers less than or equal to 3

GO ON TO THE NEXT PAGE

7. Which of the following represents the solution set for the equation $2x^2 = x + 21$?

(A) $\left(3, \frac{7}{2}\right)$

(B) $\left(3, \frac{2}{7}\right)$

(C) $\left(-3, \frac{7}{2}\right)$

(D) $\left(-3, -\frac{2}{7}\right)$

(E) $\left(-3, -\frac{7}{2}\right)$

8. If $x \neq 6, 8$, what is the equivalent reduced fraction for $\frac{3x^2 + 6x - 144}{2x^2 - 28x + 96}$?

(A) $\frac{3(x+8)}{2(x-8)}$

(B) $\frac{3(x+6)}{2(x-6)}$

(C) $\frac{3(x+8)}{2(x-6)}$

(D) $-\frac{3}{2}$

(E) $\frac{3}{2}$

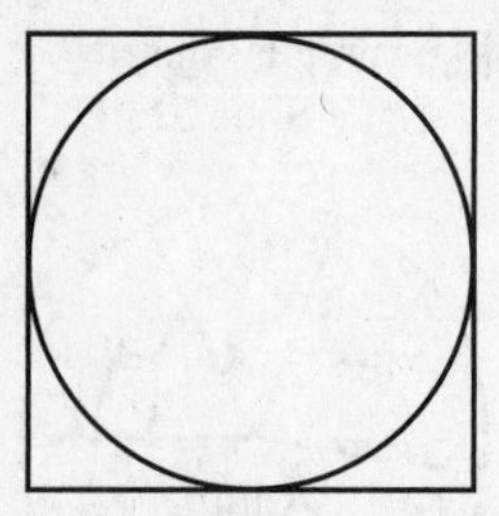

9. In the figure above, a circle is inscribed in a square. If the area of the circle is A, what is the length of the side of the square?

(A) $\frac{A}{\pi}$

(B) $\frac{2A}{\pi}$

(C) $\sqrt{\frac{A}{\pi}}$

(D) $2 \times \sqrt{\frac{A}{\pi}}$

(E) $\sqrt{\frac{2A}{\pi}}$

10. A child's piggy bank contains only nickels and dimes. In all, there are 42 coins with a total value of $3.85. How many nickels are in the piggy bank?

(A) 3
(B) 7
(C) 17
(D) 28
(E) 35

11. A line segment, L, has endpoints located at (2,1) and (6,7). Which of the following represents the equation of the perpendicular bisector of L?

(A) $7x + 3y = 17$
(B) $3x + 2y = 32$
(C) $2x + 3y = 20$
(D) $3x - 2y = 4$
(E) $2x - 3y = 1$

GO ON TO THE NEXT PAGE

DIRECTIONS FOR STUDENT-PRODUCED RESPONSE QUESTIONS

For each of the following questions, solve the problem and indicate your answer by marking the ovals in the special grid, as shown in the examples below.

Answer: $\frac{9}{5}$ **or 9/5 or 1.8**

Either position is correct.

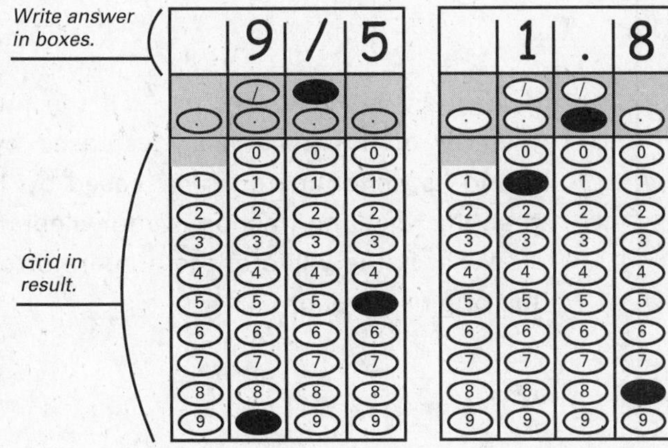

NOTE: You may start your anwers in any column, space permitting. Columns not needed should be left blank.

- Mark no more than one oval in any column.
- Because the answer sheet will be machine scored, you will receive credit only if the ovals are filled in correctly.
- Although not required, it is suggested that you write your answer in the boxes at the top of the columns to help you fill in the ovals accurately.
- Some problems may have more than one correct answer. In such cases, grid only one answer.
- No question has a negative answer.
- Mixed numbers such as $3\frac{1}{2}$ must be gridded as 3.5 or 7/2.

(If 3 1 / 2 is gridded, it will be interpreted as $\frac{31}{2}$, not $3\frac{1}{2}$.)

- **Decimal Accuracy:** If you obtain a decimal answer, enter the most accurate value the grid will accommodate. For example, if you obtain an answer such as 0.6666 ..., you should record the result as .666 or .667. Less accurate values such as .66 or .67 are not acceptable.

Acceptable ways to grid $\frac{2}{3}$ **= .6666...**

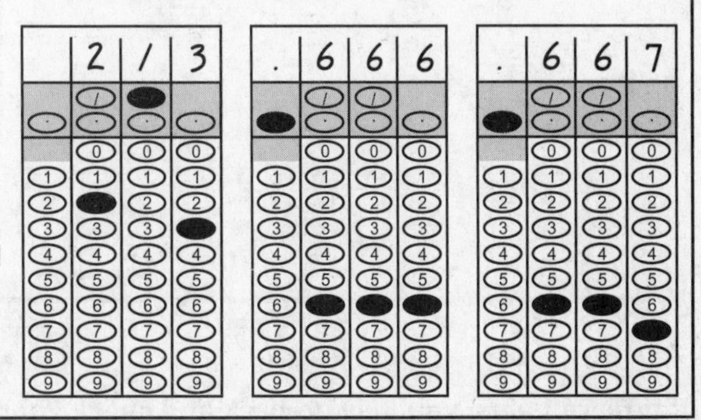

12. A bag contains a total of 20 balls. Six are white, three are red, and the rest are blue. If one ball is drawn from the bag, what is the probability that it is blue?

13. What is the smallest positive integer such that when it is divided by 4, by 5, and by 6, it will always leave a remainder of 3?

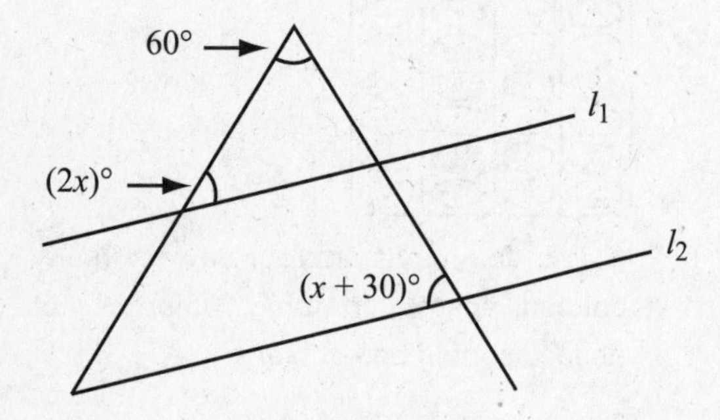

Note: Figure is not drawn to scale.

14. In the figure above, line l_1 is parallel to line l_2. What is the value of x?

15. The diameter of a cylinder is 6, equal to both the width and height of a rectangular prism. Also, the height of the cylinder is 10, the length of the rectangular prism. What is the ratio of the volume of the cylinder to the volume of the rectangular prism?

16. A fraction is equivalent to $\frac{2}{5}$. If the numerator of the fraction is decreased by 2 and its denominator is increased by 1, then the resulting fraction is equivalent to $\frac{1}{4}$. What is the value of the numerator of the original fraction?

STOP

If time remains, you may go back and check your work. When the time allotted is up, you may go on to the next section.

TIME: 20 Minutes
19 Questions

CRITICAL READING

DIRECTIONS: Each sentence below has one or two blanks, each blank indicating that something has been omitted. Beneath the sentence are five lettered words or sets of words. Choose the word or set of words that BEST fits the meaning of the sentence as a whole.

EXAMPLE

Although critics found the book _______ , many readers found it rather _______ .

(A) obnoxious . . . perfect

(B) spectacular . . . interesting

(C) boring . . . intriguing

(D) comical . . . persuasive

(E) popular . . . rare

EXAMPLE ANSWER

1. Potential energy is the _______ for doing work that a body possesses because of its position or condition. For example, a stone resting on the edge of a cliff has potential energy due to its position in the earth's gravitational field.

(A) voracity
(B) capacity
(C) incapacity
(D) terminus
(E) desire

2. Some parasites' larvae may be _______ by penetrating the skin of new hosts, but other parasite larvae live in intermediate hosts that are normally eaten by the final host within which the adult parasites _______ .

(A) dispersed . . . develop
(B) required . . . grow
(C) remediated . . . convolute
(D) retrained . . . gather
(E) reverberate . . . decay

3. Igneous rock originates from the cooling and _______ of molten matter from the earth's interior.

(A) reorganization
(B) fomentation
(C) fortification
(D) reunification
(E) solidification

4. Serfdom is _______ from slavery chiefly by the body of rights the serfs held by a custom generally recognized as inviolable, by the strict arrangement that made the peasants servile in a group rather than _______ , and by the fact that they could usually pass the right to work their land on to a son.

(A) distinguished . . . individually
(B) indelible . . . differentially
(C) credible . . . perpetually
(D) fortified . . . briefly
(E) designated . . . corporately

5. A coenzyme may either be attached by covalent bonds to a particular enzyme or exist freely in solution, but in either case it _______ intimately in the chemical reactions _______ by the enzyme.

(A) charges . . . instantiated
(B) responds . . . initiated
(C) participates . . . catalyzed
(D) perpetuates . . . perpetuated
(E) contraindicates . . . contradicted

6. Because its natural pollinating agents (certain bees and hummingbirds) are uniquely adapted for this _______ , commercial plants raised in greenhouses must be pollinated by hand.

(A) symptom
(B) exposure
(C) connection
(D) function
(E) outcome

GO ON TO THE NEXT PAGE

Questions 7–19 are based on the following passages.

Passage 1

(1) Many researchers into the effects of technology on our lives warn that heavy use of the Internet can have quite negative, and sometimes irreparable, consequences. In fact, several studies indicate that a segment of the Internet user population neglects their relationships and careers to the point that marriages have been ruined and they have lost jobs. "It's as addictive as alcohol or drugs," according to one 43-year-old woman. This woman told an interviewer that her Internet compulsion led to her divorce and estrangement from her children. "I believe it could be really bad and really dangerous for anyone," she said.

(2) Likewise, students can also find their education suffering because of too much time spent online. A graduate student says in an interview that one reason he joined a Webaholics group was because he'd witnessed his friends' academic careers suffering due to excessive Internet use, and a study of retention rates at another university found that nearly half the students who dropped out had been logging excessive amounts of time on the Internet.

(3) The mental health community was initially quite skeptical that Internet use could actually be addictive, but several recent studies by clinical psychologists now indicate that it is as real an addiction as alcoholism or drug abuse. According to one psychology researcher, "Our findings suggest that use of the Internet can definitely disrupt one's academic, social, financial, and occupational life the same way other well-documented addictions such as gambling, eating disorders, and alcoholism can. In fact, the respondents to our study who self-report spending many hours on the Internet each day also say that they display the classic symptoms of addictive behavior. They go online to escape real-life problems, are unable to control their Internet use, and feel restless and irritable when they try to cut back their use."

GO ON TO THE NEXT PAGE

Passage 2

(1) Once again, researchers are trumping up a disease that does not exist, creating an issue out of changing behavior among the populace. IAD, Internet addiction disorder, is the latest term for mostly normal behavior to be falsely deemed a crisis. After all, the average TV viewer spends more than 28 hours a week in front of the tube, while the average Internet user clocks in at a comparatively measly five and a half hours, and no one is declaring TV an addiction—in spite of the fact that there is no discernible use for this behavior when Internet surfers are either communicating with each other or seeking useful information.

(2) I conducted a very unscientific poll among my friends who might qualify for this so-called disease, and all asserted that they could easily reduce their Internet time if required. That is, none thought the activity more important than their job or their marriage. Other interviewees contended that logging onto the Internet is no more of a problem than spending time on the telephone, which points out the primary use for the Internet for most of us: as a communication tool just like the telephone and television and the post office. And once again, I've never heard any of these being classified as the cause of some disorder.

(3) The devoted Internet users I interviewed do contend that exploring the Internet is an extremely enriching experience, a daily ritual that they eagerly look forward to, however. After all, one benefit of the Internet is the development of online relationships among users. However, there is a treasure trove of educational and other useful information available on the World Wide Web with a click of a desktop mouse, and no one accuses someone who checks out every book in the library of being addicted to books.

(4) Perhaps anything can be deemed an addiction, but it seems silly to compare communicating and seeking information to things like drug use. However, assuming any activity can become an obsession, just as in the case of any other potentially overly time-consuming activities, whether Internet use becomes detrimental depends on an individual's capacity for self-control. For users who lack adequate self-control, I would suggest setting a daily online time limit and sticking to it. An alcoholic can't control the number of drinks he has, but I doubt anyone would have a problem getting back to the rest of his life with a little planning and will power.

7. What evidence does the author of Passage 1 use in the first paragraph to set up the argument to follow?

   (A) assertions by researchers about the possible negative effects of Internet use followed by a quote from a self-proclaimed addict about the results of her addiction
   (B) discussion of the ramifications of Internet addiction
   (C) statistical analysis of Internet usage growth
   (D) proof that Internet addiction exists
   (E) a history of the Internet

8. What other forms of evidence are presented in Passage 1?

   (A) statistics
   (B) quotations from Internet users and researchers, research results
   (C) eyewitness accounts
   (D) interviews with friends who use the Internet
   (E) historical precedence

GO ON TO THE NEXT PAGE

9. By contrast, the writer of Passage 2 uses what forms of evidence to support the argument the author makes?

(A) expert witnesses
(B) statistical analysis
(C) research results
(D) personal experience with the addiction
(E) statistics for comparison, interviews with his friends who use the Internet a lot

10. In paragraph 3, the writer of Passage 1 admits the following: "The mental health community was initially quite skeptical that Internet use could actually be addictive, . . ." What effect does this assertion have on the reader?

(A) do not trust the writer because the admission comes late in the essay
(B) trust the writer more for this admission and realize that this is a recent phenomenon
(C) recognize that this whole argument is built on little evidence
(D) realize the writer may not have sufficient evidence to support the argument
(E) conclude that the writer is honest

11. The author of Passage 2 compares the average number of hours Americans watch television to the average number of hours an Internet user is online, which serves to

(A) ground his argument in history.
(B) inform the reader of the primary source of information for the passage.
(C) contextualize the argument within the ongoing debate over the effects of television viewing.
(D) illustrate the author's claim that Internet use is not at epidemic levels and that not every activity that takes up our time is deemed a disorder.
(E) provide evidence for the author's claim that too much television watching is worse than too much Internet use.

12. From Passage 2: "I conducted a very unscientific poll among my friends who might qualify for this so-called disease, and all asserted that they could easily reduce their Internet time if required." What might the opposition say about this sentence?

(A) This study lacks sound statistical proof.
(B) This poll is not scientific research and therefore not very valid, and many addicts of any variety deny their inability to control their addictive behavior.
(C) The writer has no data to back up these assertions, merely comments from friends.
(D) These assertions are misleading because made by avid Internet users.
(E) These people may well be lying about their Internet use and their ability to control it.

GO ON TO THE NEXT PAGE

13. Which of the following best outlines Passage 2?

    (A) denial that IAD is a disorder, comparison of television viewing to Internet use, unscientific poll results, discussion of what is good about the Internet, suggestion that those who lack self-control should set a time limit for being online
    (B) discussion of IAD as a disorder, advocacy for television viewing as well as Internet use, poll results, discussion of good and bad Internet uses, comparison of Internet use to other addictions
    (C) refutation of IAD as a disorder by arguing that TV is worse, comments from friends, discussion of the Internet as a valid tool, demand that everyone in America exercise more self-control
    (D) history of addiction as proof that IAD is not a real disorder, comments from friends, discussion of the Internet as a means of communication
    (E) discussion of the need for increased control of the Internet, comparison of the Internet to television as a tool for communication, call for self-control that contradicts the author's original premise

14. The first sentence of Passage 2 contains the word "however." Which of the following best explains why this word is used in this sentence?

    (A) to serve as a transition between this paragraph and the next
    (B) to admit that there is some correspondence between the authors' friends' behavior and addiction
    (C) to indicate that the writer recognizes that the description of Internet users in this sentence could be construed as representative of addictive behavior
    (D) to signal that the writer now believes that this behavior is addictive
    (E) to prepare the reader for the next issue to be discussed in the author's argument

15. The author of Passage 1 refers to potential damage from Internet addiction as "irreparable," which means

    (A) deviant.
    (B) resultant.
    (C) unimportant.
    (D) important.
    (E) permanent.

16. Which choice best defines the tone of the first sentence of Passage 2?

    (A) ironic
    (B) perceptive
    (C) angry
    (D) debatable
    (E) sarcastic

GO ON TO THE NEXT PAGE

17. Which of the following best characterizes and then explains the first sentence of paragraph 4, Passage 2?

    (A) a call to reason: real addictions are diminished by calling every obsessive behavior an addiction
    (B) complete disregard for those who think this disorder is real: they are merely "silly"
    (C) an appeal to reason: those who believe this is a real disorder are irrational
    (D) despair that he and his friends might have this disorder after all: he feels "silly"
    (E) a denial of any credibility on the part of those who believe this disorder is real: they are just following the latest media trend

18. Passage 2 ends this way: "For users who lack adequate self-control, I would suggest setting a daily online time limit and sticking to it. An alcoholic can't control the number of drinks he has, but I doubt anyone would have a problem getting back to the rest of his life with a little planning and will power." This maneuver begs the question of whether or not the disorder is real because it

    (A) assumes that using the Internet is not addictive but merely an activity about which some users are slightly obsessive.
    (B) assumes that all people possess will power.
    (C) does not prove that will power is sufficient.
    (D) does not prove that users are not in fact addicted.
    (E) creates a "straw man" argument.

19. Which passage presents a more convincing argument and why?

    (A) Passage 2 because the writer is more passionate than the author of Passage 1
    (B) Passage 2 because the evidence is stronger
    (C) Passage 1 because a person with the disorder is quoted
    (D) Passages 2 because I do not know anyone with this problem
    (E) Passage 1 because research is cited whereas the writer of Passage 2 merely interviewed his or her friends and offers mostly personal opinion

TIME: 10 Minutes
14 Questions

WRITING

DIRECTIONS: In each of the following sentences, some portion of the sentence is underlined. Under each sentence are five choices. The first choice has the same wording as the original. The other four choices are reworded. Sometimes the first choice containing the original wording is the best; sometimes one of the other choices is the best. Choose the letter of the best choice. Your choice should produce a sentence that is not ambiguous or awkward and that is correct, clear, and precise.

This is a test of correct and effective English expression. Keep in mind the standards of English usage, punctuation, grammar, word choice, and construction.

EXAMPLE

When you listen to opera, a person may not appreciate it.

(A) a person may not appreciate it.

(B) it may not be appreciated by a person.

(C) you may not appreciate it.

(D) which may not be appreciated by you.

(E) appreciating it may be a problem for you.

EXAMPLE ANSWER

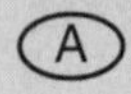 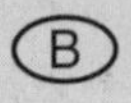  

1. The Renaissance was a golden age of translations, especially into English, renewed interest in the Latin classics created a demand for renderings of Ovid's *Metamorphoses*, Virgil's *Aeneid*, and Plutarch's *Lives*.

(A) into English, renewed interest in the Latin classics created a demand for renderings of Ovid's *Metamorphoses*, Virgil's *Aeneid*, and Plutarch's *Lives*.

(B) into English. Renewed interest in the Latin classics created a demand for renderings of Ovid's *Metamorphoses*, Virgil's *Aeneid*, and Plutarch's *Lives*.

(C) into English, renewed interest in the Latin classics created a demand for renderings of Ovid's *Metamorphoses*, Virgil's *Aeneid*, and Plutarch's *Lives*.

(D) into English, renewed interest in the Latin classics created a demand for renderings of Ovid's *Metamorphoses* Virgil's *Aeneid* and Plutarch's *Lives*.

(E) into English. Renewed interest in the Latin classics created a demand for renderings of Ovid's *Metamorphoses* Virgil's *Aeneid*, and Plutarch's *Lives*.

2. Moreover the resourcefulness and daring of the fleeing slaves themselves who were usually helped only after the most dangerous part of their journey (i.e., the Southern part) was over, were probably more important factors in the success of their escape than many conductors along the Underground Railroad readily admitted.

(A) Moreover the resourcefulness and daring of the fleeing slaves themselves who

(B) Moreover, the resourcefulness and daring of the fleeing slaves themselves who,

(C) Moreover, the resourcefulness and daring of the fleeing slaves themselves who

(D) Moreover the resourcefulness and daring of the fleeing slaves themselves, who

(E) Moreover, the resourcefulness and daring of the fleeing slaves themselves, who

GO ON TO THE NEXT PAGE

3. Because their legs are set far back on their bodies, penguins waddle awkwardly on land, and often travel by tobogganing on their bellies over the ice as they migrate, sometimes great distances—each fall to their nesting sites.

(A) on their bodies, penguins waddle awkwardly on land, and often travel by tobogganing on their bellies over the ice as they migrate, sometimes great distances—each fall to their nesting sites.

(B) on their bodies, penguins waddle awkwardly on land and often travel by tobogganing on their bellies over the ice as they migrate, sometimes great distances—each fall to their nesting sites.

(C) on their bodies, penguins waddle awkwardly on land and often travel by tobogganing on their bellies over the ice as they migrate to their nesting sites each fall, sometimes over great distances.

(D) on their bodies penguins waddle awkwardly on land and often travel by tobogganing on their bellies over the ice as they migrate, sometimes great distances—each fall to their nesting sites.

(E) on their bodies, penguins waddle awkwardly on land, and often travel by tobogganing on their bellies over the ice as they migrate sometimes great distances each fall to their nesting sites.

4. His early poems show the influence of the Greek Archilochus but his later verse displays complete and individualized adaptation of Greek meters to Latin. As his genius matured, Horace's themes turned from personal vilification to more generalized satire and to literary criticism.

(A) the Greek Archilochus but his later verse displays complete and individualized adaptation of Greek meters to Latin. As his genius matured,

(B) the Greek Archilochus, but his later verse displays complete and individualized adaptation of Greek meters to Latin. As his genius matured,

(C) the Greek Archilochus, but his later verse displays complete and individualized adaptation of Greek meters to Latin as his genius matured,

(D) the Greek Archilochus, but his later verse displays complete and individualized adaptation of Greek meters to Latin, as his genius matured,

(E) the Greek Archilochus but his later verse displays complete and individualized adaptation of Greek meters to Latin, as his genius matured,

GO ON TO THE NEXT PAGE

5. The Quetzal is strikingly beautiful, with a crested head, bronze-green back, and crimson and white underparts. Quetzals nest in holes, and lay from two to four eggs per clutch. The male shares incubation duties with the female.

   (A) The Quetzal is strikingly beautiful, with a crested head, bronze-green back, and crimson and white underparts.
   (B) The Quetzal is strikingly beautiful with a crested head, bronze-green back, and crimson and white underparts.
   (C) The Quetzal is strikingly beautiful with a crested head bronze-green back, and crimson and white underparts.
   (D) The Quetzal is strikingly beautiful, with a crested head bronze-green back, and crimson and white underparts.
   (E) The Quetzal is strikingly beautiful with a crested head bronze-green back and crimson and white underparts.

6. In the Channel Islands a tall fodder variety, known as Jersey kale, Jersey cabbage, or cow cabbage, grows to more than 7 feet.

   (A) In the Channel Islands a tall fodder variety, known as Jersey kale, Jersey cabbage, or cow cabbage, grows to more than 7 feet.
   (B) In the Channel Islands a tall fodder variety known as Jersey kale, Jersey cabbage, or cow cabbage, grows to more than 7 feet.
   (C) In the Channel Islands, a tall fodder variety, known as Jersey kale, Jersey cabbage, or cow cabbage, grows to more than 7 feet.
   (D) In the Channel Islands, a tall fodder variety known as Jersey kale Jersey cabbage, or cow cabbage, grows to more than 7 feet.
   (E) In the Channel Islands a tall fodder variety, known as Jersey kale, Jersey cabbage, or cow cabbage grows to more than 7 feet.

GO ON TO THE NEXT PAGE

7. For jelly, only those fruits may successfully be used that contain a sufficient amount of pectin (the chief gelling substance) and acid. Among these are plums, apples, grapes, and quinces and such berries as currants, gooseberries, raspberries, blackberries, and cranberries.

(A) Among these are plums, apples, grapes, and quinces and such berries as currants, gooseberries, raspberries, blackberries, and cranberries.
(B) Among these are plums, apples, grapes, and quinces; and such berries as currants, gooseberries, raspberries, blackberries, and cranberries.
(C) Among these are plums, apples, grapes, and quinces. And such berries as currants, gooseberries, raspberries, blackberries, and cranberries.
(D) Among these are plums, apples, grapes, and quinces, and currants, gooseberries, raspberries, blackberries, and cranberries.
(E) Among these are plums, and apples, and grapes, and quinces and such berries as currants, and gooseberries, and raspberries, and blackberries, and cranberries.

8. Honolulu the state capital and the economic center of Hawaii, is on the highly urbanized southern coast of Oahu.

(A) Honolulu the state capital and the economic center of Hawaii, is on
(B) Honolulu, the state capital and the economic center of Hawaii, is on
(C) Honolulu, the state capital, and the economic center of Hawaii, is on
(D) Honolulu the state capital and the economic center of Hawaii is on
(E) Honolulu, the state capital and the economic center of Hawaii is on

9. Wombats are native to savanna forests and grasslands they are solitary, nocturnal animals that feed chiefly on grass, roots, and bark and have been known to gnaw down large trees.

(A) forests and grasslands they are solitary, nocturnal animals that feed chiefly on grass, roots, and bark
(B) forests and grasslands and they are solitary, nocturnal animals that feed chiefly on grass, roots, and bark
(C) forests and grasslands. They are solitary, nocturnal animals that feed chiefly on grass, roots, and bark
(D) forests and grasslands they are solitary, nocturnal animals that feed chiefly on grass roots and bark
(E) forests and grasslands, are solitary, nocturnal animals who feed chiefly on grass, roots, and bark

10. The National Woman Suffrage Association, <u>led by Susan B. Anthony and Elizabeth Cady Stanton, were formed in 1869 to agitate for an amendment</u> to the U.S. Constitution.

    (A) led by Susan B. Anthony and Elizabeth Cady Stanton, were formed in 1869 to agitate for an amendment
    (B) led by Susan B. Anthony and Elizabeth Cady Stanton and formed in 1869 to agitate for an amendment
    (C) led by Susan B. Anthony and Elizabeth Cady Stanton, were formed in 1869 and agitated for an amendment
    (D) led by Susan B. Anthony and Elizabeth Cady Stanton were formed in 1869 to agitate for an amendment
    (E) led by Susan B. Anthony and Elizabeth Cady Stanton, was formed in 1869 to agitate for an amendment

11. Best known and most controversial <u>of Beckett's dramas are *Waiting for Godot* (1952) and *Endgame* (1957), which have been performed throughout the world, Beckett was awarded</u> the 1969 Nobel Prize in Literature.

    (A) of Beckett's dramas are *Waiting for Godot* (1952) and *Endgame* (1957), which have been performed throughout the world, Beckett was awarded
    (B) of Beckett's dramas are *Waiting for Godot* (1952) and *Endgame* (1957), which have been performed throughout the world. Beckett was awarded
    (C) of Beckett's dramas are *Waiting for Godot* (1952) and *Endgame* (1957), which have been performed throughout the world, and Beckett was awarded
    (D) of Beckett's dramas are *Waiting for Godot* (1952) and *Endgame* (1957), which have been performed throughout the world, but Beckett was awarded
    (E) of Beckett's dramas are *Waiting for Godot* (1952) and *Endgame* (1957) which have been performed throughout the world. Beckett was awarded

12. Reliquaries were often designed <u>in shapes that reflected the nature of their contents, such as hands, shoes, buildings, and heads. They are richly decorated</u> with gold, silver, enamel, and jewels.

    (A) shapes that reflected the nature of their contents, such as hands, shoes, buildings, and heads. They are richly decorated
    (B) shapes that reflected the nature of their contents, such as hands, shoes, buildings, and heads, and they are richly decorated
    (C) shapes that reflected the nature of their contents, such as hands, shoes, buildings, and heads richly decorated
    (D) shapes that reflected the nature of their contents, such as hands, shoes, buildings, and heads. They were richly decorated
    (E) shapes that reflected the nature of their contents, such as hands, shoes, buildings, and heads — richly decorated

13. Coins and medals preserve old <u>forms of writing, portraits of eminent persons, and reproductions of lost works of art; they also assist in the study of early customs, in ascertaining dates, in clarifying economic status and trade relations, and in</u> tracing changes in political attitudes.

    (A) forms of writing, portraits of eminent persons, and reproductions of lost works of art; they also assist in the study of early customs, in ascertaining dates, in clarifying economic status and trade relations, and in
    (B) forms of writing, portraits of eminent persons, and reproductions of lost works of art, they also assist in the study of early customs, in ascertaining dates, in clarifying economic status and trade relations, and in
    (C) forms of writing; portraits of eminent persons; and reproductions of lost works of art. They also assist in the study of early customs in ascertaining dates; in clarifying economic status and trade relations; and in
    (D) forms of writing, portraits of eminent persons, and reproductions of lost works of art; they also assist in the study of early customs; in ascertaining dates; in clarifying economic status and trade relations; and in
    (E) forms of writing, portraits of eminent persons, and reproductions of lost works of art also assist in the study of early customs, in ascertaining dates, in clarifying economic status and trade relations, and in

GO ON TO THE NEXT PAGE

14. In 1908, the Young Turk movement, a reformist and strongly nationalist group with many adherents in the army, forced the restoration of the constitution of 1876, and in 1909 the parliament deposed the sultan and put Muhammad V on the throne.

    (A) In 1908, the Young Turk movement, a reformist and strongly nationalist group
    (B) In 1908 the Young Turk movement, a reformist and strongly nationalist group
    (C) In 1908, the Young Turk movement a reformist and strongly nationalist group
    (D) In 1908, the Young Turk movement a reformist, and strongly nationalist group,
    (E) In 1908, the Young Turk movement a reformist, and strongly nationalist group

DIAGNOSTIC TEST

ANSWER KEY

Section 1—Essay

Mastering the Essay, pages 222–228

Refer to the Detailed Explanation for essay analysis

Section 2—Math

Mastering Basic Math Skills, pages 233–394

Mastering Regular Math Questions, pages 395–416

Question Number	Correct Answer	If you answered this question incorrectly, study . . .
1	D	Algebra; substitution and simplifying expressions (exponents)
2	D	Arithmetic; word problem (percentages)
3	C	Algebra; rational and radical equations
4	A	Geometry; Pythagorean Theorem and special triangles
5	E	Algebra; absolute value
6	D	Algebra, Geometry; transformations, coordinate geometry
7	A	Geometry; angle relationships (complements)
8	E	Algebra; word problem (fractions)
9	C	Algebra; word problem (ages)
10	B	Algebra; sequences and series (geometric)
11	B	Arithmetic; percent increase and percent decrease
12	E	Regular Math; data interpretation (slope)
13	A	Geometry; area of a polygon
14	D	Algebra; concepts of functions
15	A	Algebra; properties of exponents
16	D	Geometry; volume of a cube
17	D	Algebra; average speed
18	D	Algebra; quadratic equations, factoring
19	B	Algebra; word problem, parabolas
20	C	Algebra; newly defined symbols

Section 3—Critical Reading

Mastering Sentence Completion Questions, pages 153–170

Mastering Reading Comprehension Questions, pages 171–208

Question Number	Correct Answer	If you answered this question incorrectly, study . . .
1	D	Two-word completions; mixed value words
2	A	Two-word completions; mixed value words
3	A	One-word completions; context clues
4	E	One-word completions; context clues
5	C	One-word completions; negative value words
6	B	One-word completions; context clues
7	D	Two-word completions; etymology
8	A	Two-word completions; etymology
9	E	Narrative passages; synthesis/analysis questions
10	A	Narrative passages; evaluation questions
11	C	Narrative passages; synthesis/analysis questions
12	E	Narrative passages; interpretation questions
13	D	Social sciences passages; evaluation questions
14	A	Social sciences passages; interpretation questions
15	D	Social sciences passages; synthesis/analysis questions
16	C	Social sciences passages; synthesis/analysis questions
17	A	Social sciences passages; interpretation questions
18	B	Social sciences passages; evaluation questions
19	A	Social sciences passages; synthesis/analysis questions
20	D	Social sciences passages; synthesis/analysis questions
21	D	Social sciences passages; synthesis/analysis questions
22	C	Social sciences passages; synthesis/analysis questions
23	B	Social sciences passages; evaluation questions
24	A	Social sciences passages; synthesis/analysis questions

Section 4—Writing

Mastering the SAT Writing Section, pages 210–221

Question Number	Correct Answer	If you answered this question incorrectly, study . . .
1	A	Usage questions; run-ons, comma splices
2	B	Usage questions; subject/verb agreement
3	E	Usage questions; independent/dependent clauses, effective sentence structure
4	B	Usage questions; run-ons, comma splices, capitalization, punctuation

5	C	Usage questions; independent/dependent clauses, effective sentence structure/capitalization, punctuation
6	A	Usage questions; independent/dependent clauses, effective sentence structure/capitalization, punctuation
7	D	Usage questions; capitalization, punctuation
8	A	Usage questions; capitalization, punctuation
9	B	Usage questions; capitalization, punctuation
10	C	Usage questions; sentence fragments
11	B	Usage questions; run-ons, comma splices
12	E	Sentence Correction questions; capitalization, punctuation
13	B	Sentence Correction questions; verbs
14	A	Sentence Correction questions; capitalization, punctuation
15	C	Sentence Correction questions; run-ons, comma splices
16	D	Sentence Correction questions; sentence fragments
17	E	Sentence Correction questions; capitalization, punctuation
18	A	Sentence Correction questions; capitalization, punctuation
19	B	Sentence Correction questions; verbs/capitalization, punctuation
20	C	Sentence Correction questions; independent/dependent clauses, effective sentence structure
21	D	Sentence Correction questions; run-ons, comma splices
22	A	Sentence Correction questions; verbs/sentence fragments
23	A	Sentence Correction questions; subject/verb agreement
24	C	Sentence Correction questions; independent/dependent clauses, effective sentence structure
25	C	Sentence Correction questions; independent/dependent clauses, effective sentence structure
26	E	Sentence Correction questions; capitalization, punctuation
27	A	Sentence Correction questions; sentence fragments/nouns
28	E	Sentence Correction questions; independent/dependent clauses, effective sentence structure/capitalization, punctuation
29	B	Sentence Correction questions; capitalization, punctuation
30	A	Paragraph Improvement questions
31	A	Paragraph Improvement questions
32	D	Paragraph Improvement questions
33	C	Paragraph Improvement questions
34	E	Paragraph Improvement questions
35	B	Paragraph Improvement questions

Section 5—Math

Mastering Basic Math Skills, pages 233–394

Mastering Regular Math Questions, pages 395–416

Question Number	Correct Answer	If you answered this question incorrectly, study . . .
1	D	Algebra; substitution and simplifying expressions (exponents)
2	B	Algebra; solutions of linear equations and inequalities
3	D	Arithmetic; statistics (median)
4	A	Geometry; area and perimeter of a polygon
5	C	Algebra; word problems
6	A	Geometry; angle relationships (tangents)
7	A	Algebra; solutions of linear equations and inequalities
8	C	Algebra; equations of lines
9	B	Geometry; similarity
10	B	Algebra; properties of parallel and perpendicular lines
11	E	Geometry; Pythagorean Theorem and special triangles
12	D	Algebra; parabolas
13	B	Algebra; sequences and series (geometric)
14	E	Algebra; word problems (work)
15	E	Algebra; substitution and simplifying expressions
16	D	Geometry; angle relationships
17	B	Geometry; area of a circle
18	A	Algebra; rational and radical equations

Section 6—Critical Reading

Mastering Sentence Completion Questions, pages 153–170

Mastering Reading Comprehension Questions, pages 171–208

Question Number	Correct Answer	If you answered this question incorrectly, study . . .
1	A	Two-word completions; context clues
2	A	Two-word completions; positive value words
3	B	Two-word completions; negative value words
4	E	Two-word completions; mixed value words
5	A	One-word completions; negative value words
6	C	Social science passages; two passages; interpretation questions
7	A	Social science passages; two passages; vocabulary-in-context questions
8	E	Social science passages; two passages; synthesis/analysis questions
9	A	Social science passages; two passages; synthesis/analysis questions
10	D	Social science passages; evaluation questions

11	B	Social science passages; evaluation questions
12	A	Social science passages; synthesis/analysis questions
13	C	Social science passages; interpretation questions
14	A	Social science passages; synthesis/analysis questions
15	E	Social science passages; interpretation questions
16	A	Narrative passages; evaluation questions
17	A	Narrative passages; interpretation questions
18	D	Narrative passages; interpretation questions
19	B	Narrative passages; evaluation questions
20	B	Narrative passages; evaluation questions
21	E	Narrative passages; synthesis/analysis questions
22	C	Narrative passages; evaluation questions
23	A	Narrative passages; vocabulary-in-context questions
24	D	Narrative passages; evaluation questions

Section 7—Math

Mastering Basic Math Skills, pages 233–394

Mastering Regular Math Questions, pages 395–416

Mastering Student-Produced Response Questions, pages 417–448

Question Number	Correct Answer	If you answered this question incorrectly, study . . .
1	C	Arithmetic; percent increase and percent decrease
2	D	Arithmetic; sets
3	E	Algebra; equations of lines
4	C	Algebra; solutions of linear equations and inequalities
5	E	Algebra; variation (inverse)
6	D	Algebra; concepts of functions
7	C	Algebra; factoring
8	A	Algebra; factoring
9	D	Geometry; area of a circle
10	B	Algebra; substitution and simplifying expressions
11	C	Algebra; properties of parallel and perpendicular lines
12	11/20 or 0.55	Student-Produced Response questions; Arithmetic; probability
13	63	Student-Produced Response questions; Arithmetic; word problem (integers)
14	30	Student-Produced Response questions; Geometry; angle relationships
15	0.79	Student-Produced Response questions; Geometry; volume of a cylinder
16	6	Student-Produced Response questions; Arithmetic; word problems (fractions)

Section 8—Critical Reading

Mastering Sentence Completion Questions, pages 153–170

Mastering Reading Comprehension Questions, pages 171–208

Question Number	Correct Answer	If you answered this question incorrectly, study . . .
1	B	One-word completions; context clues
2	A	Two-word completions; context clues
3	E	One-word completions; context clues
4	A	Two-word completions; context clues
5	C	Two-word completions; positive value words
6	D	One-word completions; context clues
7	A	Social science passages; two passages; evaluation questions
8	B	Social science passages; two passages; evaluation questions
9	E	Social science passages; two passages; evaluation questions
10	B	Social science passages; two passages; evaluation questions
11	D	Social science passages; two passages; evaluation questions
12	B	Social science passages; two passages; interpretation questions
13	A	Social science passages; two passages; synthesis/analysis questions
14	C	Social science passages; two passages; synthesis/analysis questions
15	E	Social science passages; two passages; vocabulary-in-context questions
16	E	Social science passages; two passages; evaluation questions
17	A	Social science passages; two passages; evaluation questions
18	A	Social science passages; two passages; interpretation questions
19	E	Social science passages; two passages; evaluation questions

Section 9—Writing

Mastering the New SAT Writing Section, pages 210–221

Question Number	Correct Answer	If you answered this question incorrectly, study . . .
1	B	Usage questions; run-ons, comma splices
2	E	Usage questions; independent/dependent clauses, effective sentence structure/capitalization, punctuation
3	C	Usage questions; independent/dependent clauses, effective sentence structure/capitalization, punctuation
4	B	Usage questions; independent/dependent clauses, effective sentence structure
5	A	Usage questions; independent/dependent clauses, effective sentence structure/capitalization, punctuation
6	C	Usage questions; capitalization, punctuation
7	A	Usage questions; capitalization, punctuation

8	B	Usage questions; independent/dependent clauses, effective sentence structure
9	C	Usage questions; independent/dependent clauses, effective sentence structure
10	E	Usage questions; subject/verb Agreement, independent/dependent clauses
11	B	Usage questions; subject/verb agreement
12	D	Usage questions; verbs
13	A	Usage questions; independent/dependent clauses, effective sentence structure/capitalization, punctuation
14	A	Usage questions; independent/dependent clauses, effective sentence structure/capitalization, punctuation

DETAILED EXPLANATIONS

SECTION 1—ESSAY

<u>ASSIGNMENT:</u> **Do you agree or disagree with the assertion that success is definable in many ways, and what is your personal definition of success? Plan and write an essay in which you defend your point of view on this issue. Support your reasoning with examples taken from your reading, studies, experience, and observations.**

<u>SAMPLE ESSAY:</u>

The definition of success mentioned in the prompt, success as measured by wealth and prestige, grows out of the American Dream. This is the foremost capitalist nation on earth, and therefore, it stands to reason that the values that many Americans live by would be dictated by economic standards. We also tend to nearly worship famous people, and consequently, Americans also strive for the same kind of attention and adoration that such success brings. However, over the last several decades, many people have opted not to chase that dream quite so hard, opting instead to find a way to make a living that is fulfilling and to define what success is within their own lives.

This issue comes down to time: we are only allotted so much of it, and the majority of our time is spent making a living. Consequently, some people have decided to find a way to make a living that they enjoy as opposed to one that provides them with wealth. From my own observations, this is not an easy thing to achieve (many among us are obviously miserable going about their jobs), but it seems like those who choose to provide some service for others almost universally enjoy how they spend their days, perhaps because what they accomplish is easily measured. For example, teachers witness the "lights going on" in their students' eyes, and healthcare workers can see their patients' health improve. Likewise, a researcher can witness the good that their investigations bring about, and a firefighter knows his day has been successful if lives have been saved and fires extinguished.

Many people choose to define success outside of work altogether. We can't all be nurses or teachers, perhaps because of responsibilities such as raising children that disallow furthering our education. However, bringing up children to be happy healthy adults is no small feat in our age, and being the best gardener in town does not pay but it could certainly be definable as success. Perhaps the people I admire most are those who defy the common definition of success altogether in favor of taking their own path in life. I know a nun who makes no money whatsoever, but her service to the poor is invaluable. I also know a writer who struggles to pay his rent on his waiter's paycheck, but he writes every day of his life and plans to continue doing so whether he ever becomes rich and famous. Both of these people have chosen a path over their families' objections and for the joy it brings them, which is truly admirable.

Increasingly, I agree with Bob Dylan's comment mentioned in the prompt. What counts is doing what you enjoy between getting up and going to bed. My own choice will be to teach children how to read, which is why I am going to college, but any path that brings a person happiness and does not do harm to others would seem a valid definition for personal success. Given how success is defined by the society as a whole, this might require courage and even some defiance; but time is short and so not to be wasted. As Anna Quindlen says, what good is the praise of others if your success does not feel like success in your own heart?

ANALYSIS:

The Sample Essay has a score range of 5–6. It is well-organized, presenting a thesis in the introduction. The essay develops that idea throughout the essay, and expands upon it yet further in the conclusion. The author offers examples to prove the main idea. The writing demonstrates a variety of sentence structure and length, and the grammar and punctuation are correct. The vocabulary is also fairly sophisticated.

DETAILED EXPLANATIONS

SECTION 2—MATH

1. **(D)** Substitute $x = 3$ to get $5 \times 3^2 - 3^3 = (5)(9) - 27 = 18$.

2. **(D)** Let x = price before the discount. Then $x - 0.20x = 54$. This simplifies to $0.80x = 54$, so $x = \frac{54}{0.80} = 67.5$.

3. **(C)** Subtract 9 from both sides to get $\sqrt{2x-1} = 5$, then square both sides to get $2x - 1 = 25$. So $2x = 26$, thus $x = 13$.

4. **(A)** In a 30°–60°–90° right triangle, the ratio of the sides is $1:\sqrt{3}:2$. These numbers correspond to the sides opposite 30°, 60°, and 90°, respectively. Because BC is opposite 60° and AC is opposite 90°, use the proportion $BC/AC = \frac{\sqrt{3}}{2}$. Then $\frac{BC}{14} = \frac{\sqrt{3}}{2}$. Solving, $BC = 7\sqrt{3}$.

5. **(E)** Rewrite this absolute value equation as two separate linear equations, namely, $6x + 7 = 11$ and $6x + 7 = -11$. For the first of these equations, $6x = 4$, so $x = \frac{4}{6} = \frac{2}{3}$. For the second of these equations, $6x = -18$, so $x = -3$.

6. **(D)** The graph of $f(x)$ will be shifted 3 units to the right, so the new graph $f(x - 3)$ will not be defined for $x = 7 + 3 = 10$.

7. **(A)** $\angle PRS = 180° - 33° = 147°$. Also, $\angle QPR = 180° - 90° - 33° = 57°$, because the sum of the angles of any triangle is 180°. Then $\angle RPT = 180° - 57° = 123°$. Finally, $147° - 123° = 24°$.

8. **(E)** The amount of ink used on the first day was $28.8 \times \frac{1}{9} = 3.2$ gallons. The amount of ink used on the second day was twice that amount, 6.4 gallons. Because a total of 9.6 gallons was used on the first two days, the amount used on the third day was $28.8 - 9.6 = 19.2$ gallons.

9. **(C)** Let x = Pat's present age and let y = Pat's sister's present age. Then, one year ago, Pat was $x - 1$ years old and his sister was $y - 1$ years old, so $x - 1 = 3(y - 1)$. Next year, Pat will be $x + 1$ years old and his sister will be $y + 1$ years old, so $x + 1 = 2(y + 1)$. The first equation simplifies to $x - 3y = -2$ and the second equation simplifies to $x - 2y = 1$. Subtracting equation 1 from equation 2 yields $y = 3$. Substituting $y = 3$ into one of the equations we find that $x = 7$. So, Pat's age five years from now is $7 + 5 = 12$.

10. **(B)** The number of years, t, is equal to $2037 - 2004 = 33$. Substitute 33 for t in the formula $P = (60)(4^{\frac{t}{6}})$. Then, $P = 60 \times 4^{\frac{33}{6}} = 60 \times 4^{5.5} = (60)(2{,}048) = 122{,}880$.

11. **(B)** Let x = the number of apartments with terraces. Use the proportion $\frac{9}{16} = \frac{x}{144}$. Then $16x = 1{,}296$, so $x = 81$. To increase the number of apartments with terraces by 25%, take $81 \times 1.25 = 101.25$. Therefore, 101 apartments will have terraces and $144 - 101 = 43$ apartments will not.

12. **(E)** Connect each point to (0,0). The cost per book is represented by the slope of each line. In order for the cost per book to be the same, the points that represent the companies must lie on the same line. Only points *W* and *Z* satisfy this requirement.

13. **(A)** The area = one half the height times the sum of the bases = $(\frac{1}{2})(17)(25 + 37) = (\frac{1}{2})(17)(62) = 527$.

14. **(D)** To find the domain, recognize that any quantity inside a square root must be at least equal to 0. Thus, $8 - x \geq 0$. Add x to each side to get $8 \geq x$, which is equivalent to $x \leq 8$.

15. **(A)** Work with the variables in parentheses first (PEMDAS rule). Use the rules of exponents to simplify the numerator to $x^{-6} \times x^{4} = x^{-2}$ then divide to get $\frac{x^{-2}}{x^{-5}} = x^{3}$. Next, $(x^{3})^{-2} = x^{-6}$.

16. **(D)** The respective sides of the cube prior to and after the change are $\sqrt[3]{216} = 6$ cm and $\sqrt[3]{64} = 4$ cm. The percent change from 6 to 4 is $(\frac{2}{6})(100\%)$, which is approximately 33%.

17. **(D)** To solve this type of problem, set up a grid that is based on the formula distance = rate × time, with lines for each speed:

	d	r	t
Speed 1	$2{,}400 - d$	530	$4.5 - t$
Speed 2	d	540	t
Total	2,400	—	4.5

Then, create equations for Speed 1 and Speed 2, combine, simplify, and solve for t by substituting for d.

$$2{,}400 - d = (530)(4.5 - t) \quad \& \quad d = (540)(t)$$
$$2{,}400 - 2{,}385 = -530t - d$$
$$15 = -530t - 540t$$
$$15 = 10t$$
$$1.5 = t$$

Now, convert hours to minutes: 1.5 hours × 60 = 90 minutes.

18. **(D)** Rewrite the inequality as $9 - x^2 < 0$, and then factor the left side to get $(3 - x)(3 + x) < 0$.

Case 1: $3 - x < 0$ and $3 + x > 0$. This implies that $x > 3$ and $x > -3$, which results in $x > 3$.

Case 2: $3 - x > 0$ and $3 + x < 0$. This implies that $x < 3$ and $x < -3$, which results in $x < -3$.

Thus, the solution is $x > 3$ or $x < -3$. The inequality $x > 3$ can be expressed as $(3, \infty)$, and the inequality $x < -3$ can be expressed as $(-\infty, -3)$.

Graphically, $y = -x^2 + 9$ is a parabola with a vertex at (0,9) and two x-intercepts: (–3,0) and (3,0). The solution in this problem represents values of x where the parabola lies below the x-axis.

19. **(B)** The equation is a parabola that opens "down," so Sam's maximum commission is at the top of the parabola. First, find the zeroes of the equation by factoring:

$270s - 3s^2$
$3s\,(90 - s)$
$s = 0, 90$

Therefore, the parabola crosses the x-axis at the points (0,0) and (90,0). The maximum point is halfway between these two points, at (45,0). When Sam sells 45 stereos, he maximizes his commission.

20. **(C)** Because m is negative, $m \infty n = m^2 + \sqrt{n} = (-2\sqrt{2})(-2\sqrt{2}) + \sqrt{16} = 8 + 4 = 12$.

DETAILED EXPLANATIONS

SECTION 3—CRITICAL READING

1. **(D)** The clues that this is the correct choice are in the words "despite," "connection," and "unfounded," as well as in the use of the preposition "with."
2. **(A)** The clues that this is the correct choice are in the context and in the word "action."
3. **(A)** The clues that this is the correct choice are in the context (Yucatan is a place) and the preposition.
4. **(E)** The clues that this is the correct choice are in the plural nouns and the word "convoluted."
5. **(C)** The clue that this is the correct choice is in the second clause.
6. **(B)** The clues that this is the correct choice are in the overall context.
7. **(D)** The clues that this is the correct choice are in the overall context and in the words "deeply" and "hieroglyphs."
8. **(A)** The clues that this is the correct choice are in the context.
9. **(E)** The clue that this is the correct choice is in the second to last sentence. All choices are true to some degree, but choice (E) is the most accurate.
10. **(A)** This choice means "to reduce," and the connotation is "an oversimplification."
11. **(C)** Other choices have merit, but choice (C) is the most complete and most accurate.
12. **(E)** Other choices have merit, but choice (E) is the most complete and most accurate.
13. **(D)** The writer interprets this statistic this way for the reader.
14. **(A)** All other choices have merit, but choice (A) is the most complete and most accurate.
15. **(D)** The passage first discusses stereotypes of the aged in America, then gives examples of the elderly being treated better in the past. Examples of changing attitudes toward the elderly in other countries follow, after which comes a discussion of research that refutes some of the stereotypes.
16. **(C)** This is discussed in paragraph 3.
17. **(A)** This question obviously requires speculation, but the other choices are not adequate.
18. **(B)** Other choices have merit, but choice (B) is the most complete and most accurate.
19. **(A)** Other choices have merit (i.e., choice E), but choice (A) is the most complete and most accurate.

20. **(D)** Other choices have merit, but choice (D) is the most complete and most accurate. See paragraph 6.

21. **(D)** See paragraph 7.

22. **(C)** See paragraph 8.

23. **(B)** Based on the passage, it seems that the writer advocates deep respect for the elderly. They should be able to work or retire as they please.

24. **(A)** All other choices have merit, but choice (A) is the most complete and most accurate. See the last paragraph.

DETAILED EXPLANATIONS

SECTION 4—WRITING

1. **(A)** The independent clause requires the semicolon, or a new sentence must begin at "corridors." (Run-ons, Comma Splices, Sentence Combining)
2. **(B)** The verb should be singular: "is." (Subject/Verb Agreement)
3. **(E)** The word "and" is a coordinating conjunction and therefore requires a comma. The word "although" begins a modifying clause that also requires a comma. (Independent/Dependent Clauses, Effective Sentence Structure)
4. **(B)** A new sentence must begin at "in." (Run-ons, Comma Splices, Sentence Combining/Capitalization, Punctuation)
5. **(C)** The introductory clause requires a comma, and the independent clause beginning with "reasoning" must either be a separate sentence or set off with a semicolon. (Independent/Dependent Clauses, Effective Sentence Structure/Capitalization, Punctuation)
6. **(A)** The introductory and modifying clauses both require commas. (Independent/Dependent Clauses, Effective Sentence Structure/Capitalization, Punctuation)
7. **(D)** The word "which" signals a nonrestrictive relative clause and requires a comma. (Capitalization, Punctuation)
8. **(A)** No error as written. The clause following the subject requires commas. Choice (C) is grammatically correct but less complete because the clause is merely deleted. (Capitalization, Punctuation)
9. **(B)** The introductory clause requires a comma, and where the comma is used correctly in other choices, noun/verb agreement is wrong or the meaning is altered. (Capitalization, Punctuation)
10. **(C)** The last construction is a fragment and requires a verb to be corrected. (Sentence Fragments)
11. **(B)** A new sentence needs to start at "the other." (Run-ons, Comma Splices, Sentence Combining)
12. **(E)** No error as written. All commas are used correctly to set off modifying clauses. (Capitalization, Punctuation)
13. **(B)** The first verb should be past tense: "decided." (Verbs)
14. **(A)** The introductory clause requires a comma. (Capitalization, Punctuation)
15. **(C)** A new sentence begins with "his." (Run-ons, Comma Splices, Sentence Combining)

16. **(D)** A new sentence begins with "the." (Sentence Fragments)

17. **(E)** No error as written. The commas are used correctly to divide a series. The semicolon is used correctly as well. (Capitalization, Punctuation)

18. **(A)** The series must be divided by commas. (Capitalization, Punctuation)

19. **(B)** The word "which" signals a nonrestrictive relative clause and requires a comma, and the verb tense is incorrect. (Verbs/Capitalization, Punctuation)

20. **(C)** The word "that" denotes a restrictive clause and does not require a comma. The correct word for the nonrestrictive clause is "which." (Independent/Dependent Clauses, Effective Sentence Structure)

21. **(D)** A new sentence begins with "more." (Run-ons, Comma Splices, Sentence Combining)

22. **(A)** The verb should be "consists." (Verbs/Sentence Fragments)

23. **(A)** The verb should be "was." (Subject/Verb Agreement)

24. **(C)** The word "and" is not a coordinating conjunction and thus does not require the comma. (Independent/Dependent Clauses, Effective Sentence Structure)

25. **(C)** The modifying clause requires a comma. (Independent/Dependent Clauses, Effective Sentence Structure)

26. **(E)** No error as written. Commas correctly separate the series. (Capitalization, Punctuation)

27. **(A)** The verb is missing and so is an article. (Sentence Fragments/Nouns)

28. **(E)** No error as written. Commas correctly separate the series and set off the dependent clause. (Independent/Dependent Clauses, Effective Sentence Structure/Capitalization, Punctuation)

29. **(B)** The introductory clause requires a comma. (Capitalization, Punctuation)

30. **(A)** The sentence is fine as written. None of the other options make the passage better.

31. **(A)** (A) corrects ambiguities and further develops ideas only glossed in the original. Other choices either do not improve readability or change the meaning of the original. (Independent/Dependent Clauses, Effective Sentence Structure)

32. **(D)** Including "attending the same classes" makes the sentence too wordy. In addition, the phrase is superfluous..

33. **(C)** "In fact," is the only answer choice that ties sentence 14 to sentence 13..

34. **(E)** The final sentence of the passage states that gradually intoducing a nonnative speaker to a language is more effective than intensive instruction in the language.

35. **(B)** The question requires speculation, and although other choices are not necessarily wrong, given what the last sentence says, (B) is the most complete and accurate.

DETAILED EXPLANATIONS

SECTION 5—MATH

1. **(D)** If $m^2 = 625$, then $m = \pm\sqrt{625} = \pm 25$. But because $m > 0$, use only the positive 25. Then $3m^{-\frac{1}{2}} = (3)(25^{-\frac{1}{2}}) = \frac{3}{\sqrt{25}} = \frac{3}{5} = 0.6$.

2. **(B)** When $x < -3$, the left-most arrow of the graph shows $y < 3$. Also, note that when $1 < x < 6$, the y value is always less than 3 (and at least equal to 0). Furthermore, when $-3 \cdot x \cdot 1$, the y value is exactly 3; when $x > 6$, the y value is greater than 3. So, the correct list of x values is $x < -3$; $1 < x < 6$.

3. **(D)** To find the median, first arrange the numbers in ascending order to read: 7, 9, 9, 13, 19, 27. By definition, the median is the average of the two middle numbers (3rd and 4th). So, the median is $\frac{(9+13)}{2} = 11$.

4. **(A)** Because the pentagon is regular, each of its sides must be $\frac{12.5}{5} = 2.5$, which also represents the side of the square. Then the area of the square is $(2.5)^2 = 6.25$.

5. **(C)** The cost of 3 children's shirts = (3)($10) = $30. The cost of 4 women's shirts = (4)($12) = $48. After these purchases, the family has $200 – $30 – $48 = $122 left. Of the men's shirts, the small size is the least expensive. The family can afford to buy $\frac{\$122}{\$15} = 8.13$ shirts. Buying a fraction of a shirt is impossible, so they will spend ($15)(8) = $120 on men's shirts. Therefore, the family will have $122 – $120 = $2 left.

6. **(A)** A tangent to a circle forms a right angle at the point of tangency, so $\angle PQR = \angle PTS = 90°$. The sum of the angles of any 5-sided figure is $(180°)(5 - 2) = 540°$. Thus, $\angle P = x° = 540° - 90° - 90° - 110° - 135° = 115°$.

7. **(A)** Multiply both sides of the inequality by 3 to get $-4x + 5 > -18$. Subtracting 5 from both sides, we get $-4x > -23$. The last step requires division by -4, so the order of inequality is reversed. Thus $x < \frac{23}{4}$.

8. **(C)** Let $y = mx + b$ represent the equation of line ℓ, where m = slope and b = y-intercept. In this example, $m = -8/7$ and $b = 4$. So $y = (-\frac{8}{7})x + 4$. To find the x-intercept, substitute $y = 0$. Then $0 = (-\frac{8}{7})x + 4$, subtracting 4 from both sides we have $(-\frac{8}{7})x = -4$, and so $x = \frac{4}{\left(-\frac{8}{7}\right)} = \frac{28}{8} = \frac{7}{2}$.

9. **(B)** All that is given is $\overline{AC} = \overline{CE}$ and $\overline{AB} = \overline{DE}$. It may be tempting to look at the figure and draw other conclusions (such as "ΔACE looks like an equilateral triangle" or "$\overline{BE}$ and $\overline{AD}$ look like angle bisectors"). By working with only the information given, we can see that ΔACD and ΔECB are congruent because they have a Side–Angle–Side relationship (because they share angle C). Likewise, ΔABE and ΔAED are congruent because they have a Side–Side–Angle relationship.

10. **(B)** The slope of line $\ell_1 = \frac{(9-4)}{(3-1)} = 5/2$. Because the slopes of perpendicular lines are negative reciprocals of each other, the slope of $\ell_2 = -\frac{2}{5}$. We know that ℓ_2 contains (1,4). The only choice among the five points given for which the slope would be $-\frac{2}{5}$ is (–9,8). Note that $\frac{(8-4)}{(-9-1)} = -\frac{2}{5}$.

11. **(E)** In an isosceles right triangle, we can use the ratio $1:1:\sqrt{2}$, corresponding to leg : leg : hypotenuse. In this right triangle, $\overline{XY}$ and $\overline{YZ}$ are both legs. The perimeter becomes $1 + 1 + \approx 3.414$. Then, $\frac{2}{3.414} \approx 59\%$.

12. **(D)** When a quadratic function is written in the form $f(x) = A(x-h)^2 + k$, the vertex is located at (h,k). This vertex is the highest point when A is negative, and the lowest point when A is positive. Since we seek a highest point, A must be negative. The values of h and k are –1 and 5 respectively. Thus, $f(x) = A(x - [-1])^2 + 5 = A(x + l)^2 + 5$. Only choice (D) satisfies all the requirements.

13. **(B)** The formula for a geometric sequence is $L_n = AR^{n-1}$, where L_n is the last term, A is the first term, and R is the common ratio. It is easier to find y by first assuming that 2 is actually the first term, y is the second term, and 72 is the third term. Then $y = 2 \times R^1$ and $72 = 2 \times R^2$. Then $R^2 = 36$, so $R = 6$. This means that $y = 12$. We now have x, 2, 12, 72, ... Because $R = 6$, x must be $2/6 = 1/3$. Now, $x + x = 1/3 + 12 = 12\frac{1}{3}$.

14. **(E)** The formula for a work problem involving 3 people is (Time together)/(Time alone for person 1) + (Time together)/(Time alone for person 2) + (Time together)/(Time alone for person 3) = 1. If Tanya is person 3, let x represent her time alone. Then, $\frac{2}{12} + \frac{2}{24} + \frac{2}{x} = 1$. Multiply the entire equation by $24x$ to get $4x + 2x + 48 = 24x$. This simplifies to $48 = 18x$, so $x = \frac{48}{18} = 2\frac{2}{3}$.

15. **(E)** Solve by substitution. Because $wx = 2$, $x = \frac{2}{w}$. Then $xy = (\frac{2}{w})(y) = (\frac{2y}{w}) = 9$, which means $2y = 9w$ or $w = \frac{2y}{9}$. Also, $wy = 8$ becomes $(\frac{2y}{9})(y) = 8$. This leads to $2y^2 = 72$, $y^2 = 36$, so $y = 6$. Because $xy = 9$, $x = \frac{9}{6} = \frac{3}{2}$. Then, because $wx = 2(w)(\frac{3}{2}) = 2$, so $w = \frac{2}{\left(\frac{3}{2}\right)}) = \frac{4}{3}$. Finally, $wxy = (\frac{4}{3})(\frac{3}{2})(6) = 12$.

16. **(D)** As inscribed angles, $\angle B = (\frac{1}{2})(ADC)$ and $\angle D = (\frac{1}{2})(ABC)$. If a quadrilateral is inscribed in a circle, then its opposite angles are supplementary. Therefore, $\angle B + \angle D = 180°$. The measure of $\angle D$ is $180° - 105° = 75°$. Finally, $75° = (\frac{1}{2})(ABC)$, so $ABC = 150°$.

17. **(B)** The area of the circle is $(\pi)(42) = 16\pi$. Because the radius is 4, the diameter is 8. Therefore, 8 is the side of the square, so the area of the square is $82 = 64$. The area of the region outside the circle but inside the square is $64 - 16\pi$. The area of the shaded region is half of this,

$$\frac{64-16\pi}{2} = 32 - 8\pi$$

Thus, the required probability is the ratio of the shaded region to the whole

$$\frac{32-8\pi}{64} = \frac{4-\pi}{8}$$

18. **(A)** To simplify a radical expression, factor everything under the radical sign. So

$$\frac{\sqrt{75x^3y^5}}{\sqrt{625x^5y^3}} \text{ becomes } \frac{\sqrt{5\cdot5\cdot3\cdot x\cdot x\cdot x\cdot y\cdot y\cdot y\cdot y\cdot y}}{\sqrt{5\cdot5\cdot5\cdot5\cdot x\cdot x\cdot x\cdot x\cdot x\cdot y\cdot y\cdot y}}$$

Move pairs of terms out from under the radical sign, because this is the equivalent of taking the square root of squared terms. This leaves

$$\frac{5xyy\sqrt{3xy}}{5\cdot5xxy\sqrt{xy}} = \frac{5xy^2\sqrt{3xy}}{25x^2y\sqrt{xy}}$$

Reducing the fraction further results in the final answer, $\frac{y\sqrt{3xy}}{5x\sqrt{xy}}$.

DETAILED EXPLANATIONS

SECTION 6—CRITICAL READING

1. **(A)** The clues that this is the correct choice are in the parallel structure and what is in the parentheses.
2. **(A)** The clues that this is the correct choice are in the words "state of mind" and "superiority."
3. **(B)** The clues that this is the correct choice are in the word "Hades" and in the final clause.
4. **(E)** The clues that this is the correct choice are in the words "discussion" and "against."
5. **(A)** The clues that this is the correct choice are in the obvious comparison and in the use of the preposition "from."
6. **(C)** This is noted in the last paragraph. Other options have merit, but choice (C) is the most accurate and complete.
7. **(A)** See paragraphs 2 and 3. Other options have merit, but choice (A) is the most accurate and complete.
8. **(E)** See the first and last paragraphs.
9. **(A)** Other options have merit, but choice (A) is the most accurate and complete.
10. **(D)** The passage's purpose is to dispute the need and effectiveness of school reform initiatives imposed by the federal government. Instead the passage suggests that practical reform must come from within the public school system, educators, and parents.
11. **(B)** Clues to the change of tone are in the reference to proponents of some strategies in paragraph 1 as ideologues and the text in paragraph 2 beginning "whether or not one agrees. . . ."
12. **(A)** The writer cites the "A Nation Still at Risk" report and then argues against the strategy.
13. **(C)** See paragraphs 4 and 5.
14. **(A)** Other options have merit, but choice (A) is the most accurate and complete relative to the overall argument.
15. **(E)** Other options have merit (i.e., A, C, and D), but choice (E) is the most accurate and complete relative to the overall argument.
16. **(A)** Foreshadowing is used in fiction to hint at something that is coming. Things that are stated outright—or even implied—are not good examples of foreshadowing, so choices (B), (D), and (E) are incorrect. There is no indication that the story will be set in spring, so choice (C) is not correct. Choice (A) is the correct answer.

17. **(A)** "Straight as an arrow" depicts her physical strength and the description of her eyes suggests her strength of mind and character.

18. **(D)** The clues that this is the correct answer are in the descriptors and the word "merely."

19. **(B)** This question calls for speculation, but the clues that this is the correct choice are in what follows this section.

20. **(B)** This question calls for speculation, and although other choices have merit, choice (B) is more accurate and complete.

21. **(E)** See paragraph 4.

22. **(C)** The clues that this is the correct choice are in the mention of the girl in the first and last paragraphs and in the assertion that there was no thought that there might be room in their lives for anyone else until the girl arrived.

23. **(A)** The term means "perpetually."

24. **(D)** See the reference to looks of love exchanged between them and the last paragraph for the assertion that there was no thought that there might be room in their lives for anyone else until the girl arrived.

DETAILED EXPLANATIONS

SECTION 7—MATH

1. **(C)** The percent of all sales for boys is 100 – 22 – 48 – 18 = 12%. Because the sales for adult women is 48%, this is 4 times as large as the sales for boys. Thus, the required amount is (4)($6,000) = $24,000.

2. **(D)** Set P contains six numbers: 5, 7, 11, 13, 17, and 19. Set $P \cup Q$ contains the nineteen numbers 1, 2, 3, . . . , 19. Because $P \cup Q$ contains elements in P, Q, or both P and Q, Q must contain <u>at least</u> 19 – 6 = 13 elements. These numbers are 1, 2, 3, 4, 6, 8, 9, 10, 12, 14, 15, 16, and 18.

3. **(E)** Let $(0,y)$ represent the y-intercept. Then by the definition of slope, $\frac{7}{2} = \frac{(y-0)}{(0-[-6])}$. So, $\frac{7}{2} = \frac{y}{6}$, then cross-multiply to get $2y = 42$. Then $y = 21$.

4. **(C)** Rewrite as $-3x < 5x - 17$. Then $-3x - 5x < -17$, which becomes $-8x < -17$. Dividing by –8 reverses the order of inequality, so that $x > \frac{17}{8}$. Because x must be an integer, its lowest possible value is 3. Thus, $x = 3$.

5. **(E)** When two quantities are inversely related, their product must be a constant. Let the constant be k, so that $L \times w^3 = k$. Replace k with 5 to get $L \times w^3 = 5$, which can be rewritten as $L = \frac{4}{w^3}$.

6. **(D)** The domain refers to the values of $f(x)$. The minimum value of $|4 - x|$ is 0, since absolute values must never be negative. Since $f(x) = |4 - x| + 3$, the minimum value must be $0 + 3 = 3$.

7. **(C)** Rewrite the equation as $2x^2 - x - 21 = 0$. By factoring the left side, we get $(x + 3)(2x - 7) = 0$. Then $x + 3 = 0$ and $2x - 7 = 0$. The solution set becomes $(-3, \frac{7}{2})$.

8. **(A)** Rewrite the expression as $\frac{[3(x^2+2x-48)]}{[2(x^2-14+48)]}$ which factors further as $\frac{[3(x+8)(x-6)]}{[2(x-8)(x-6)]}$. Cancel the common factor $x - 6$ to get $\frac{[3(x-8)]}{[2(x-8)]}$.

9. **(D)** Area of a circle $A = \pi \cdot R^2$, where R is the radius. Then $R^2 = \frac{A}{\pi}$, so $R = \frac{A}{\pi}$. Notice that any one side of the square is the same length as the diameter of the circle, which is twice the radius, $2 \times \sqrt{\frac{A}{\pi}}$.

10. **(B)** Let x = number of nickels and $42 - x$ = number of dimes. The value of the nickels is $0.05x$ and the value of the dimes is $0.10(42 - x)$. Now since the total value is $3.85, we get the equation $0.05x + 0.10(42 - x) = 3.85$. Simplifying, $0.05x - 4.20 - 10x = 3.85$. Then $-0.05x + 4.20 = 3.85$. So $-0.05x = -0.35$, and thus $x = 7$.

11. **(C)** L has its midpoint at $(\frac{[2+6]}{2}, \frac{[1+7]}{2}) = (4,4)$, and a slope of $\frac{(7-1)}{(6-2)} = \frac{6}{4} = \frac{3}{2}$. The perpendicular bisector of L must have a slope equal to the negative reciprocal of L, which is $-\frac{2}{3}$. The perpendicular bisector must also contain the point (4,4). Then $y - 4 = (-\frac{2}{3})(x - 4)$. This becomes $y - 4 = (-\frac{2}{3})x + \frac{8}{3}$. Multiply the equation by 3 to get $3y - 12 = -2x + 8$, which simplifies to $2x + 3y = 20$.

12. **$\frac{11}{20}$ or 0.55** Because 6 balls are white, and 3 are red, $20 - 6 - 3 = 11$ must be blue. Thus the probability of drawing a blue ball is $\frac{11}{20}$ or 0.55.

13. **63** First find the least common denominator for 4, 5, and 6, which is 60. The quickest way to find this number is to take multiples of 5 and 6, starting with 30 and continuing this process until we get a multiple of 4 also. In order to find the smallest number that will leave a remainder of 3, just add 3 to 60 to get 63.

14. **30** The third angle in the small triangle, which lies on ℓ_1, has a measure of $(x + 30)°$ because it represents a corresponding angle on ℓ_2. The sum of the angles of any triangle is 180°, so using the 3 angles of the small triangle, $(2x) + (x + 30) + 60 = 180$; $3x + 90 = 180$, $3x = 90$, so $x = 30$.

15. **0.79** Let 6 = diameter of the cylinder = width and height of the rectangular prism. Let 10 = height of the cylinder = length of the rectangular prism. Then the volume of the cylinder = $\pi \times 32 \times 10 = 90\pi$, and the volume of the rectangular prism = $(6)(6)(10) = 360$. The required ratio is $\frac{90\pi}{360} = \frac{\pi}{4}$. In decimal form, this is equivalent to 0.7853981634, which is rounded to 0.79 in order to fit in the grid.

16. **6** Let the original fraction be represented as $\frac{2x}{5x}$, because this always has a value of $\frac{2}{5}$, no matter what x is worth (except zero). Then, $\frac{(2x-2)}{(5x+1)} = \frac{1}{4}$. Cross-multiply to get $4(2x - 2) = 1(5x + 1)$. So, $8x - 8 = 5x + 1$. Simplifying, $3x = 9$, which reduces to $x = 3$. This means that the numerator of the original fraction is $(2)(3) = 6$.

DETAILED EXPLANATIONS

SECTION 8—CRITICAL READING

1. **(B)** The clues that this is the correct choice are in the words "potential" and "possesses."
2. **(A)** The clues that this is the correct choice are in the context and in the contrast between "larvae" and "adult."
3. **(E)** The clues that this is the correct choice are in the words "rock" and "molten."
4. **(A)** The clues that this is the correct choice are in the context and in the contrast between "individual" and "group."
5. **(C)** The clues that this is the correct choice are in the context (chemical reactions) and in the word "intimately." Other choices have merit, but (C) is common usage.
6. **(D)** The clues that this is the correct choice are in the first clause and in the word "greenhouses" and what this environment implies.
7. **(A)** Other choices have merit, but (A) is more accurate and more complete.
8. **(B)** The argument presented in Passage 1 is supported by quotations from Internet users, researchers, and research results.
9. **(E)** See paragraphs 1 and 2.
10. **(B)** Other choices have merit, but choice (B) is more accurate and complete.
11. **(D)** The author compares the average number of hours Americans spend watching television to the average number of hours we spend online to show that Internet is not at epidemic levels, not is every time-consuming activity deemed a disorder.
12. **(B)** Other choices have merit, but choice (B) is more accurate and complete.
13. **(A)** Passage 2 argues that IAD is not a disorder by comparing Internet use to television viewing, reporting unscientific poll results, discussing the benfits of Internet use, and suggesting a time limit for those who struggle to control their Internet use.
14. **(C)** The clues that this is the correct choice are in the words "devoted," "daily ritual," "extremely," and "eagerly." Other choices have merit (such as A and B), but (C) is more accurate and complete.
15. **(E)** The word means "not repairable."
16. **(E)** The clue that this is the best choice is in the word "silly."
17. **(A)** The first sentence of Passage 2's fourth paragraph is a call to reason.

18. **(A)** The opposition's primary assumption is that Internet use can be addictive. The suggestion in question does not prove otherwise and in fact merely assumes that the practice is not addictive. Choices (C) and (D) have merit, but (A) is more complete and accurate.

19. **(E)** Passage 1 cites scientific data, whereas Passage 2 cites unscientific data and personal opinion. For this reason, Passagge 1 presents a more convincing argument.

DETAILED EXPLANATIONS

SECTION 9—WRITING

1. **(B)** A new sentence is required beginning with "renewed." (Run-ons, Comma Splices, Sentence Combining)

2. **(E)** The introductory clause and the modifying clause that begins with "who" must be set off with commas. (Independent/Dependent Clauses, Effective Sentence Structure/Capitalization, Punctuation)

3. **(C)** Answer (B) is grammatical, but the location of the prepositional phrase beginning with "each fall" makes this choice less readable. Also, "and" is not a coordinating conjunction and thus does not require a comma, and the dash is misused. (Independent/Dependent Clauses, Effective Sentence Structure/Capitalization, Punctuation)

4. **(B)** The word "but" serves as a coordinating conjunction and thus requires a comma. (Independent/Dependent Clauses, Effective Sentence Structure)

5. **(A)** No errors as written. Commas are used correctly within a series and to set off a modifying clause. (Independent/Dependent Clauses, Effective Sentence Structure/Capitalization, Punctuation)

6. **(C)** The introductory clause requires a comma. (Capitalization, Punctuation)

7. **(A)** No error as written. Commas are used correctly. Answer (D) is grammatically correct, but this choice deletes important information and the multiple use of "and" is problematic for readability. (Capitalization, Punctuation)

8. **(B)** The modifying clause requires a comma. (Independent/Dependent Clauses, Effective Sentence Structure)

9. **(C)** A new sentence should begin with "they." (Independent/Dependent Clauses, Effective Sentence Structure)

10. **(E)** A new sentence needs to begin with "thereafter," and the introductory clause this creates requires a comma. Answer (D) is grammatically correct but changes the meaning of the passage.

11. **(B)** The verb form is wrong. It should be "was." (Subject/Verb Agreement)

12. **(D)** The second verb must be past tense. Answer (C) is grammatical but changes the meaning. (Verbs)

13. **(A)** No error as written. The commas to set off the constituents in the series and the semicolon are used correctly. Answer (E) is grammatical but changes the meaning. (Independent/Dependent Clauses, Effective Sentence Structure/Capitalization, Punctuation)

14. **(A)** No error as written. The introductory and modifying clauses require commas. (Independent/Dependent Clauses, Effective Sentence Structure/Capitalization, Punctuation)

CHAPTER 3

BASIC VERBAL SKILLS REVIEW

BASIC VERBAL SKILLS REVIEW

In order to perform successfully on the SAT, you must be as familiar as possible with the different aspects of the test. One particular characteristic that becomes obvious when practicing the SAT verbal questions is that without a strong, college level-vocabulary, it is virtually impossible to do well on the Sentence Completion section or the Critical Reading section.

However, don't give up! Not all students have strong vocabularies, and if you don't, there are still ways to build on it before taking the SAT. Even a game of Scrabble will help build your vocabulary. In addition, the most effective way to build your vocabulary is to read. Reading increases your familiarity with words and their uses in different contexts. By reading every day, you can increase your chances of recognizing a word and its meaning in a question.

As you read books, newspapers, and magazines, ask yourself these questions:

- What was the main idea of this material?
- What was the author's purpose in writing this material?
- How did the author make his or her arguments?
- What tone did the author use?

Doing so will help you comprehend what you are reading. This ultimately leads to a stronger vocabulary and an increase in your reading abilities.

Unfortunately, you may not remember all the words that you read. This is why we have provided you with a necessary tool to building your vocabulary. We know that learning new words requires much time and concentration. Therefore, rather than give an extensive list of thousands of words, we have narrowed down our vocabulary list so that you will be able to study all the words without becoming overwhelmed. Our list consists of 600 words. The first 300 are essential vocabulary, the most frequently tested words on the SAT. These words have appeared over and over again on the SAT. In addition, we have provided a second list of 300 words that are commonly tested on the SAT. It is very important to study all 600 words, because any of them could appear on the test. Although these are not the only words you may encounter, they will give you a strong indication as to the appropriate level of vocabulary that is present on the SAT and prevent you from wasting time studying words that will never appear on the test.

MASTERING VOCABULARY

The Mastering Vocabulary section lists the most frequently tested words on the SAT. The most effective way to study these words is one section at a time. The 10-week study schedule (found at the front of the book) can help you manage your vocabulary program. Identify the words you don't know or that are defined in unusual ways, and write them on index cards with the word on one side and the definition on the other. Test yourself on these words by completing the drills that follow each section. Doing the same for the list of additional words will also help you. In addition, studying our table of prefixes, roots, and suffixes will allow you to dissect words in order to determine the meanings of any unfamiliar words.

I. MASTERING VOCABULARY (The most frequently tested words on the SAT)

GROUP 1

abstract – *adj.* – not easy to understand; theoretical

acclaim – *n.* – loud approval; applause

acquiesce – *v.* – to agree or consent to an opinion

adamant – *adj.* – not yielding; firm

adversary – *n.* – an enemy; foe

advocate – 1. *v.* – to plead in favor of; 2. *n.* – supporter; defender

aesthetic – *adj.* – showing good taste; artistic

alleviate – *v.* – to lessen or make easier

aloof – *adj.* – distant in interest; reserved; cool

altercation – *n.* – controversy; dispute

altruistic – *adj.* – unselfish

amass – *v.* – to collect together; accumulate

ambiguous – *adj.* – not clear; uncertain; vague

ambivalent – *adj.* – undecided

ameliorate – *v.* – to make better; to improve

amiable – *adj.* – friendly

amorphous – *adj.* – having no determinate form

anarchist – *n.* – one who believes that a formal government is unnecessary

antagonism – *n.* – hostility; opposition

apathy – *n.* – lack of emotion or interest

appease – *v.* – to make quiet; to calm

apprehensive – *adj.* – fearful; aware; conscious

arbitrary – *adj.* – based on one's preference or judgment

arrogant – *adj.* – acting superior to others; conceited

articulate – 1. *v.* – to speak distinctly; 2. *adj.* – eloquent; fluent; 3. *adj.* – capable of speech; 4. *v.* – to hinge; to connect; 5. *v.* – to convey; to express effectively

DRILL: GROUP 1

<u>DIRECTIONS:</u> Match each word in the left column with the word in the right column that is most <u>opposite</u> in meaning.

Word		Match	
1. ___articulate	6. ___abstract	A. hostile	F. disperse
2. ___apathy	7. ___acquiesce	B. concrete	G. enthusiasm
3. ___amiable	8. ___arbitrary	C. selfish	H. certain
4. ___altruistic	9. ___amass	D. reasoned	I. resist
5. ___ambivalent	10. ___adversary	E. ally	J. incoherent

<u>DIRECTIONS:</u> Match each word in the left column with the word in the right column that is most <u>similar</u> in meaning.

Word		Match	
11. ___adamant	14. ___antagonism	A. afraid	D. insistent
12. ___aesthetic	15. ___altercation	B. disagreement	E. hostility
13. ___apprehensive		C. tasteful	

GROUP 2

assess – *v.* – to estimate the value of

astute – *adj.* – cunning; sly; crafty

atrophy – *v.* – to waste away through lack of nutrition

audacious – *adj.* – fearless; bold

augment – *v.* – to increase or add to; to make larger

austere – *adj.* – harsh; severe; strict

authentic – *adj.* – real; genuine; trustworthy

authoritarian – *n.* – acting as a dictator; demanding obedience banal – *adj.* – common; petty; ordinary

belittle – *v.* – to make small; to think lightly of

benefactor – *n.* – one who helps others; a donor

benevolent – *adj.* – kind; generous

benign – *adj.* – mild; harmless

biased – *adj.* – prejudiced; influenced; not neutral

blasphemous – *adj.* – irreligious; away from acceptable standards

blithe – *adj.* – happy; cheery; merry

brevity – *n.* – briefness; shortness

candid – *adj.* – honest; truthful; sincere

capricious – *adj.* – changeable; fickle

caustic – *adj.* – burning; sarcastic; harsh

censor – *v.* – to examine and delete objectionable material

censure – *v.* – to criticize or disapprove of

charlatan – *n.* – an imposter; fake

coalesce – *v.* – to combine; come together

collaborate – *v.* – to work together; cooperate

DRILL: GROUP 2

<u>DIRECTIONS:</u> Match each word in the left column with the word in the right column that is most <u>opposite</u> in meaning.

Word		Match	
1. ___augment	6. ___authentic	A. permit	F. malicious
2. ___biased	7. ___candid	B. heroine	G. neutral
3. ___banal	8. ___belittle	C. praise	H. mournful
4. ___benevolent	9. ___charlatan	D. diminish	I. unusual
5. ___censor	10. ___blithe	E. dishonest	J. fake

<u>DIRECTIONS:</u> Match each word in the left column with the word in the right column that is most <u>similar</u> in meaning.

Word		Match	
11. ___collaborate	14. ___censure	A. harmless	D. cooperate
12. ___benign	15. ___capricious	B. cunning	E. criticize
13. ___astute		C. changeable	

GROUP 3

compatible – *adj.* – in agreement with; harmonious

complacent – *adj.* – content; self-satisfied; smug

compliant – *adj.* – yielding; obedient

comprehensive – *adj.* – all-inclusive; complete; thorough

compromise – *v.* – to settle by mutual adjustment

concede – *v.* – 1. to acknowledge; admit; 2. to surrender; to abandon one's position

concise – *adj.* – in few words; brief; condensed

condescend – *v.* – to come down from one's position or dignity

condone – *v.* – to overlook; to forgive

conspicuous – *adj.* – easy to see; noticeable

consternation – *n.* – amazement or terror that causes confusion

consummation – *n.* – the completion; finish

contemporary – *adj.* – living or happening at the same time; modern

contempt – *n.* – scorn; disrespect

contrite – *adj.* – regretful; sorrowful

conventional – *adj.* – traditional; common; routine

cower – *v.* – to crouch down in fear or shame

defamation – *n.* – harming a name or reputation; slandering

deference – *adj.* – yielding to the opinion of another

deliberate – 1. *v.* – to consider carefully; to weigh in the mind; 2. *adj.* – intentional

denounce – *v.* – to speak out against; condemn

depict – *v.* – to portray in words; to present a visual image

deplete – *v.* – to reduce; to empty

depravity – *n.* – moral corruption; badness

deride – *v.* – to ridicule; to laugh at with scorn

DRILL: GROUP 3

DIRECTIONS: Match each word in the left column with the word in the right column that is most opposite in meaning.

Word		Match	
1. ___deplete	6. ___condone	A. unintentional	F. support
2. ___contemporary	7. ___conspicuous	B. disapprove	G. beginning
3. ___concise	8. ___consummation	C. invisible	H. ancient
4. ___deliberate	9. ___denounce	D. respect	I. virtue
5. ___depravity	10. ___contempt	E. fill	J. verbose

DIRECTIONS: Match each word in the left column with the word in the right column that is most similar in meaning.

Word		Match	
11. ___compatible	14. ___comprehensive	A. portray	D. thorough
12. ___depict	15. ___complacent	B. content	E. common
13. ___conventional		C. harmonious	

GROUP 4

desecrate – *v.* – to violate a holy place or sanctuary

detached – *adj.* – separated; not interested; standing alone

deter – *v.* – to prevent; to discourage; to hinder

didactic – *adj.* – 1. instructive; 2. dogmatic; preachy

digress – *v.* – to stray from the subject; to wander from topic

diligence – *n.* – hard work

discerning – *adj.* – distinguishing one thing from another

discord – *n.* – disagreement; lack of harmony

discriminating – 1. *v.* to distinguish one thing from another; 2. *v.* – to demonstrate bias; 3. *adj.* – able to distingush

disdain – 1. *n.* – intense dislike; 2. *v.* – to look down upon; scorn

disparage – *v.* – to belittle; to undervalue

disparity – *n.* – difference in form, character, or degree

dispassionate – *adj.* – lack of feeling; impartial

disperse – *v.* – to scatter; to separate

disseminate – *v.* – to circulate; to scatter

dissent – *v.* – to disagree; to differ in opinion

dissonance – *n.* – harsh contradiction

diverse – *adj.* – different; dissimilar

document – 1. *n.* – official paper containing information; 2. *v.* – to support; to substantiate; to verify

dogmatic – *adj.* – stubborn; biased; opinionated

dubious – *adj.* – doubtful; uncertain; skeptical; suspicious

eccentric – *adj.* – odd; peculiar; strange

efface – *v.* – to wipe out; to erase

effervescence – *n.* – 1. liveliness; spirit; enthusiasm; 2. bubbliness

egocentric – *adj.* – self-centered

DRILL: GROUP 4

DIRECTIONS: Match each word in the left column with the word in the right column that is most opposite in meaning.

Word		Match	
1. ___detached	6. ___dubious	A. agree	F. respect
2. ___deter	7. ___diligence	B. certain	G. compliment
3. ___dissent	8. ___disdain	C. lethargy	H. sanctify
4. ___discord	9. ___desecrate	D. connected	I. harmony
5. ___efface	10. ___disparage	E. assist	J. restore

DIRECTIONS: Match each word in the left column with the word in the right column that is most similar in meaning.

Word		Match	
11. ___effervescence	14. ___document	A. stubborn	D. liveliness
12. ___dogmatic	15. ___eccentric	B. distribute	E. odd
13. ___disseminate		C. substantiate	

GROUP 5

elaboration – *n.* – act of clarifying; adding details

eloquence – *n.* – the ability to speak well

elusive – *adj.* – hard to catch; difficult to understand

emulate – *v.* – to imitate; to copy; to mimic

endorse – *v.* – to support; to approve of; to recommend

engender – *v.* – to create; to bring about

enhance – *v.* – to improve; to compliment; to make more attractive

enigma – *n.* – mystery; secret; perplexity

ephemeral – *adj.* – temporary; brief; short-lived

equivocal – *adj.* – doubtful; uncertain

erratic – *adj.* – unpredictable; strange

erroneous – *adj.* – untrue; inaccurate; not correct

esoteric – *adj.* – incomprehensible; obscure

euphony – *n.* – pleasant sound

execute – *v.* – 1. to put to death; to kill; 2. to carry out; to fulfill

exemplary – *adj.* – serving as an example; outstanding

exhaustive – *adj.* – thorough; complete

expedient – *adj.* – helpful; practical; worthwhile

expedite – *v.* – to speed up

explicit – *adj.* – specific; definite

extol – *v.* – to praise; to commend

extraneous – *adj.* – irrelevant; not related; not essential

facilitate – *v.* – to make easier; to simplify

fallacious – *adj.* – misleading

fanatic – *n.* – enthusiast; extremist

DRILL: GROUP 5

<u>DIRECTIONS:</u> Match each word in the left column with the word in the right column that is most <u>opposite</u> in meaning.

Word		Match	
1. ___extraneous	6. ___erratic	A. incomplete	F. eternal
2. ___ephemeral	7. ___explicit	B. delay	G. condemn
3. ___exhaustive	8. ___euphony	C. dependable	H. relevant
4. ___expedite	9. ___elusive	D. comprehensible	I. indefinite
5. ___erroneous	10. ___extol	E. dissonance	J. accurate

DIRECTIONS: Match each word in the left column with the word in the right column that is most similar in meaning.

Word		Match	
11. ___endorse	14. ___fallacious	A. enable	D. worthwhile
12. ___expedient	15. ___engender	B. recommend	E. deceptive
13. ___facilitate		C. create	

GROUP 6

fastidious – *adj.* – fussy; hard to please

fervent – *adj.* – passionate; intense

fickle – *adj.* – changeable; unpredictable

fortuitous – *adj.* – accidental; happening by chance; lucky

frivolity – *adj.* – giddiness; lack of seriousness

fundamental – *adj.* – basic; necessary

furtive – *adj.* – secretive; sly

futile – *adj.* – worthless; unprofitable

glutton – *n.* – overeater

grandiose – *adj.* – extravagant; flamboyant

gravity – *n.* – seriousness

guile – *n.* – slyness; deceit

gullible – *adj.* – easily fooled

hackneyed – *adj.* – commonplace; trite

hamper – *v.* – to interfere with; to hinder

haphazard – *adj.* – disorganized; random

hedonistic – *adj.* – pleasure seeking

heed – *v.* – to obey; to yield to

heresy – *n.* – opinion contrary to popular belief

hindrance – *n.* – blockage; obstacle

humility – *n.* – lack of pride; modesty

hypocritical – *adj.* – two-faced; deceptive

hypothetical – *adj.* – assumed; uncertain

illuminate – *v.* – to make understandable

illusory – *adj.* – unreal; false; deceptive

DRILL: GROUP 6

DIRECTIONS: Match each word in the left column with the word in the right column that is most opposite in meaning.

Word		Match	
1. ___heresy	6. ___fervent	A. predictable	F. beneficial
2. ___fickle	7. ___fundamental	B. dispassionate	G. orthodoxy
3. ___illusory	8. ___furtive	C. simple	H. organized
4. ___frivolity	9. ___futile	D. extraneous	I. candid
5. ___grandiose	10. ___haphazard	E. real	J. seriousness

DIRECTIONS: Match each word in the left column with the word in the right column that is most similar in meaning.

Word		Match	
11. ___glutton	14. ___hackneyed	A. hinder	D. overeater
12. ___heed	15. ___hindrance	B. obstacle	E. obey
13. ___hamper		C. trite	

GROUP 7

immune – *adj.* – protected; unthreatened by

immutable – *adj.* – unchangeable; permanent

impartial – *adj.* – unbiased; fair

impetuous – *adj.* – 1. rash; impulsive; 2. forcible; violent

implication – *n.* – suggestion; inference

inadvertent – *adj.* – not on purpose; unintentional

incessant – *adj.* – constant; continual

incidental – *adj.* – extraneous; unexpected

inclined – *adj.* – 1. apt to; likely to; 2. angled

incoherent – *adj.* – illogical; rambling

incompatible – *adj.* – disagreeing; disharmonious

incredulous – *adj.* – unwilling to believe; skeptical

indifferent – *adj.* – unconcerned

indolent – *adj.* – lazy; inactive

indulgent – *adj.* – lenient; patient

inevitable – *adj.* – sure to happen; unavoidable

infamous – *adj.* – having a bad reputation; notorious

infer – *v.* – to form an opinion; to conclude

initiate – 1. *v.* – to begin; to admit into a group; 2. *n.* – a person who is in the process of being admitted into a group

innate – *adj.* – natural; inborn

innocuous – *adj.* – harmless; innocent

innovate – *v.* – to introduce a change; to depart from the old

insipid – *adj.* – uninteresting; bland

instigate – *v.* – to start; to provoke

intangible – *adj.* – incapable of being touched; immaterial

DRILL: GROUP 7

DIRECTIONS: Match each word in the left column with the word in the right column that is most opposite in meaning.

Word		Match	
1. ___immutable	6. ___innate	A. intentional	F. changeable
2. ___impartial	7. ___incredulous	B. articulate	G. avoidable
3. ___inadvertent	8. ___inevitable	C. gullible	H. harmonious
4. ___incoherent	9. ___intangible	D. material	I. learned
5. ___incompatible	10. ___indolent	E. biased	J. energetic

DIRECTIONS: Match each word in the left column with the word in the right column that is most similar in meaning.

Word		Match	
11. ___impetuous	14. ___instigate	A. lenient	D. conclude
12. ___incidental	15. ___indulgent	B. impulsive	E. extraneous
13. ___infer		C. provoke	

GROUP 8

ironic – *adj.* – contradictory; inconsistent; sarcastic

irrational – *adj.* – not logical

jeopardy – *n.* – danger

kindle – *v.* – to ignite; to arouse

languid – *adj.* – weak; fatigued

laud – *v.* – to praise

lax – *adj.* – careless; irresponsible

lethargic – *adj.* – lazy; passive

levity – *n.* – silliness; lack of seriousness

lucid – *adj.* – 1. shining; 2. easily understood

magnanimous – *adj.* – forgiving; unselfish

malicious – *adj.* – spiteful; vindictive

marred – *adj.* – damaged

meander – *v.* – to wind on a course; to go aimlessly

melancholy – *n.* – depression; gloom

meticulous – *adj.* – exacting; precise

minute – *adj.* – extremely small; tiny

miser – *n.* – penny pincher; stingy person

mitigate – *v.* – to alleviate; to lessen; to soothe

morose – *adj.* – moody; despondent

negligence – *n.* – carelessness

neutral – *adj.* – impartial; unbiased

nostalgic – *adj.* – longing for the past; filled with bittersweet memories

novel – *adj.* – new

DRILL: GROUP 8

DIRECTIONS: Match each word in the left column with the word in the right column that is most opposite in meaning.

Word		Match	
1. ___irrational	6. ___magnanimous	A. extinguish	F. ridicule
2. ___kindle	7. ___levity	B. jovial	G. kindly
3. ___meticulous	8. ___minute	C. selfish	H. sloppy
4. ___malicious	9. ___laud	D. logical	I. huge
5. ___morose	10. ___novel	E. seriousness	J. stale

DIRECTIONS: **Match each word in the left column with the word in the right column that is most similar in meaning.**

Word		Match	
11. ___ironic	14. ___jeopardy	A. lessen	D. carelessness
12. ___marred	15. ___negligence	B. damaged	E. danger
13. ___mitigate		C. sarcastic	

GROUP 9

nullify – *v.* – to cancel; to invalidate

objective – 1. *adj.* – open-minded; impartial; 2. *n.* – goal

obscure – *adj.* – not easily understood; dark

obsolete – *adj.* – out of date; passe

ominous – *adj.* – threatening

optimist – *n.* – person who hopes for the best; person who sees the good side

orthodox – *adj.* – traditional; accepted

pagan – 1. *n* – polytheist; 2. *adj.* – polytheistic

partisan – 1. *n* – supporter; follower; 2. *adj.* – biased; one-sided

perceptive – *adj.* – full of insight; aware

peripheral – *adj.* – marginal; outer

pernicious – *adj.* – dangerous; harmful

pessimism – *n.* – seeing only the gloomy side; hopelessness

phenomenon – *n.* – 1. miracle; 2. occurrence

philanthropy – *n.* – charity; unselfishness

pious – *adj.* – religious; devout; dedicated

placate – *v.* – to pacify

plausible – *adj.* – probable; feasible

pragmatic – *adj.* – matter-of-fact; practical

preclude – *v.* – to inhibit; to make impossible

predecessor – *n.* – one who has occupied an office before another

prodigal – *adj.* – wasteful; lavish

prodigious – *adj.* – exceptional; tremendous

profound – *adj.* – deep; knowledgeable; thorough

profusion – *n.* – great amount; abundance

DRILL: GROUP 9

DIRECTIONS: Match each word in the left column with the word in the right column that is most opposite in meaning.

Word		Match	
1. ___objective	6. ___plausible	A. scanty	F. minute
2. ___obsolete	7. ___preclude	B. assist	G. anger
3. ___placate	8. ___prodigious	C. superficial	H. pessimism
4. ___profusion	9. ___profound	D. biased	I. modern
5. ___peripheral	10. ___optimism	E. improbable	J. central

DIRECTIONS: Match each word in the left column with the word in the right column that is most similar in meaning.

Word		Match	
11. ___nullify	14. ___pernicious	A. invalidate	D. threatening
12. ___ominous	15. ___prodigal	B. follower	E. harmful
13. ___partisan		C. lavish	

GROUP 10

prosaic – *adj.* – tiresome; ordinary

provincial – *adj.* – regional; unsophisticated

provocative – *adj.* – 1. tempting; 2. irritating

prudent – *adj.* – wise; careful; prepared

qualified – *adj.* – experienced; indefinite

rectify – *v.* – to correct

redundant – *adj.* – repetitious; unnecessary

refute – *v.* – to challenge; to disprove

relegate – *v.* – to banish; to put to a lower position

relevant – *adj.* – of concern; significant

remorse – *n.* – guilt; sorrow

reprehensible – *adj.* – wicked; disgraceful

repudiate – *v.* – to reject; to cancel

rescind – *v.* – to retract; to discard

resignation – *n.* – 1. quitting; 2. submission

resolution – *n.* – proposal; promise; determination

respite – *n.* – recess; rest period

reticent – *adj.* – silent; reserved; shy

reverent – *adj.* – respectful

rhetorical – *adj.* – having to do with verbal communication

rigor – *n.* – severity

sagacious – *adj.* – wise; cunning

sanguine – *adj.* – 1. optimistic; cheerful; 2. red

saturate – *v.* – to soak thoroughly; to drench

scanty – *adj.* – inadequate; sparse

DRILL: GROUP 10

DIRECTIONS: Match each word in the left column with the word in the right column that is most opposite in meaning.

Word		Match	
1. ___provincial	6. ___remorse	A. inexperienced	F. affirm
2. ___reticent	7. ___repudiate	B. joy	G. extraordinary
3. ___prudent	8. ___sanguine	C. pessimistic	H. sophisticated
4. ___qualified	9. ___relevant	D. unrelated	I. forward
5. ___relegate	10. ___prosaic	E. careless	J. promote

DIRECTIONS: Match each word in the left column with the word in the right column that is most similar in meaning.

Word		Match	
11. ___provocative	14. ___rescind	A. drench	D. severity
12. ___rigor	15. ___reprehensible	B. tempting	E. disgraceful
13. ___saturate		C. retract	

GROUP 11

scrupulous – *adj.* – honorable; exact

scrutinize – *v.* – to examine closely; to study

servile – *adj.* – slavish; groveling

skeptic – *n.* – doubter

slander – *v.* – to defame; to maliciously misrepresent

solemnity – *n.* – seriousness

solicit – *v.* – to ask; to seek

stagnant – *adj.* – motionless; uncirculating

stanza – *n.* – group of lines in a poem having a definite pattern

static – *adj.* – inactive; changeless

stoic – *adj.* – detached; unruffled; calm

subtlety – *n.* – 1. understatement; 2. propensity for understatement; 3. sophistication; 4. cunning

superficial – *adj.* – on the surface; narrowminded; lacking depth

superfluous – *adj.* – unnecessary; extra

surpass – *v.* – to go beyond; to outdo

sycophant – *adj.* – flatterer

symmetry – *n.* – correspondence of parts; harmony

taciturn – *adj.* – reserved; quiet; secretive

tedious – *adj.* – time-consuming; burdensome; uninteresting

temper – *v.* – to soften; to pacify; to compose

tentative – *adj.* – not confirmed; indefinite

thrifty – *adj.* – economical; pennywise

tranquility – *n.* – peace; stillness; harmony

trepidation – *n.* – apprehension; uneasiness

trivial – *adj.* – unimportant; small; worthless

DRILL: GROUP 11

DIRECTIONS: Match each word in the left column with the word in the right column that is most opposite in meaning.

Word		Match	
1. ___scrutinize	6. ___tentative	A. frivolity	F. skim
2. ___skeptic	7. ___thrifty	B. enjoyable	G. turbulent
3. ___solemnity	8. ___tranquility	C. prodigal	H. active
4. ___static	9. ___solicit	D. chaos	I. believer
5. ___tedious	10. ___stagnant	E. give	J. confirmed

DIRECTIONS: **Match each word in the left column with the word in the right column that is most similar in meaning.**

Word		Match	
11. ___symmetry	14. ___subtle	A. understated	D. fear
12. ___superfluous	15. ___trepidation	B. unnecessary	E. flatterer
13. ___sycophant		C. balance	

GROUP 12

tumid – *adj.* – swollen; inflated

undermine – *v.* – to weaken; to ruin

uniform – *adj.* – consistent; unvaried; unchanging

universal – *adj.* – concerning everyone; existing everywhere

unobtrusive – *adj.* – inconspicuous; reserved

unprecedented – *adj.* – unheard of; exceptional

unpretentious – *adj.* – simple; plain; modest

vacillation – *n.* – fluctuation

valid – *adj.* – acceptable; legal

vehement – *adj.* – intense; excited; enthusiastic

venerate – *v.* – to revere

verbose – *adj.* – wordy; talkative

viable – *adj.* – 1. capable of maintaining life; 2. possible; attainable

vigor – *n.* – energy; forcefulness

vilify – *v* . – to slander

virtuoso – *n.* – highly skilled artist

virulent – *adj.* – deadly; harmful; malicious

vital – *adj.* – important; spirited

volatile – *adj.* – changeable; undependable

vulnerable – *adj.* – open to attack; unprotected

wane – *v.* – to grow gradually smaller

whimsical – *adj.* – fanciful; amusing

wither – *v.* – to wilt; to shrivel; to humiliate; to cut down

zealot – *n.* – believer; enthusiast; fan

zenith – *n.* – point directly overhead in the sky

DRILL: GROUP 12

DIRECTIONS: Match each word in the left column with the word in the right column that is most opposite in meaning.

Word		Match	
1. ___uniform	6. ___vigor	A. amateur	F. support
2. ___virtuoso	7. ___volatile	B. trivial	G. constancy
3. ___vital	8. ___vacillation	C. visible	H. lethargy
4. ___wane	9. ___undermine	D. placid	I. wax
5. ___unobtrusive	10. ___valid	E. unacceptable	J. varied

DIRECTIONS: Match each word in the left column with the word in the right column that is most similar in meaning.

Word		Match	
11. ___wither	14. ___vehement	A. intense	D. possible
12. ___whimsical	15. ___virulent	B. deadly	E. shrivel
13. ___viable		C. amusing	

ADDITIONAL VOCABULARY

The following additional vocabulary terms are commonly found on the SAT.

abandon – 1. *v.* – to leave behind; 2. *v.* – to give something up; 3. *n.* – freedom; enthusiasm; impetuosity

abase – *v.* – to degrade; to humiliate; to disgrace

abbreviate – *v.* – to shorten; to compress; to diminish

aberrant – *adj.* – abnormal

abhor – *v.* – to hate

abominate – *v.* – to loathe; to hate

abridge – *v.* – 1. to shorten; 2. to limit; to take away

absolve – *v.* – to forgive; to acquit

abstinence – *n.* – self-control; abstention; chastity

accede – *v.* – to comply with; to consent to

accomplice – *n.* – coconspirator; partner; partner-in-crime

accrue – *v.* – to collect; to build up

acrid – *adj.* – sharp; bitter; foul-smelling

adept – *adj.* – skilled; practiced

adverse – *adj.* – negative; hostile; antagonistic; inimical

affable – *adj.* – friendly; amiable; good-natured

aghast – *adj.* – 1. astonished; amazed; 2. horrified; terrified; appalled

alacrity – *n.* – 1. enthusiasm; fervor; 2. liveliness; sprightliness

allocate – *v.* – set aside; designate; assign

allure – 1. *v.* – to attract; to entice; 2. *n.* – attraction; temptation; glamour

amiss – 1. *adj.* – wrong; awry; 2. *adv.* – wrongly; mistakenly

analogy – *n.* – similarity; correlation; parallelism; simile; metaphor

anoint – *v.* – 1. to crown; to ordain; 2. to smear with oil

anonymous – *adj.* – nameless; unidentified

arduous – *adj.* – difficult; burdensome

awry – 1. *adv.* – crooked(ly); uneven(ly); 2. *adj.* wrong; askew

baleful – *adj.* – sinister; threatening; evil; deadly

baroque – *adj.* – extravagant; ornate

behoove – *v.* – to be advantageous; to be necessary

berate – *v.* – to scold; to reprove; to reproach; to criticize

bereft – *adj.* – hurt by someone's death

biennial – 1. *adj.* – happening every two years; 2. *n.* – a plant which blooms every two years

blatant – *adj.* – 1. obvious; unmistakable; 2. crude; vulgar

bombastic – *adj.* – pompous; wordy; turgid

burly – *adj.* – strong; bulky; stocky

cache – *n.* – 1. stockpile; store; heap; 2. hiding place for goods

calamity – *n.* – disaster

cascade – 1. *n.* – waterfall; 2. *v.* – to pour; to rush; to fall

catalyst – *n.* – anything that creates a situation in which change can occur

chagrin – *n.* – distress; shame

charisma – *n.* – appeal; magnetism; presence

chastise – *v.* – to punish; to discipline; to admonish; to rebuke

choleric – *adj.* – cranky; cantankerous

cohesion – *n.* – the act of holding together

colloquial – *adj.*– casual; common; conversational; idiomatic

conglomeration – *n.* – mixture; collection

connoisseur – *n.* – expert; authority (usually refers to a wine or food expert) consecrate – *n.* – sanctify; make sacred; immortalize

craven – *adj.* – cowardly; fearful

dearth – *n.* – scarcity; shortage

debilitate – *v.* – to deprive of strength

deign – *v.* – to condescend; to stoop

delineate – *v.* – to outline; to describe

demur – 1. *v.* – to object; 2. *n.* – objection; misgiving

derision – *n.* – ridicule; mockery

derogatory – *adj.* – belittling; uncomplimentary

destitute – *adj.* – poor; poverty-stricken

devoid – *adj.* – lacking; empty

dichotomy – *n.* – branching into two parts

disheartened – *adj.* – discouraged; depressed

diverge – *v.* – to separate; to split

docile – *adj.* – manageable; obedient

duress – *n.* – force; constraint

ebullient – *adj.* – showing excitement

educe – *v.* – to draw forth

effervescence – *n.* – bubbliness; enthusiasm; animation

emulate – *v.* – to follow the example of

ennui – *n.* – boredom; apathy

epitome – *n.* – model; typification; representation

errant – *adj.* – wandering

ethnic – *adj.* – native; racial; cultural

evoke – *v.* – to call forth; to provoke

exotic – *adj.* – unusual; striking

facade – *n.* – front view; false appearance

facsimile – *n.* – copy; reproduction; replica

fathom – *v.* – to comprehend; to uncover

ferret – *v.* – to drive or hunt out of hiding

figment – *n.* – product; creation

finite – *adj.* – measurable; limited; not everlasting

fledgling – *n.* – 1. inexperienced person; beginner; 2. young bird

flinch – *v.* – to wince; to draw back; to retreat

fluency – *n.* – smoothness of speech

flux – *n.* – current; continuous change

forbearance – *n.* – patience; self-restraint

foster – *v.* – to encourage; to nurture; to support

frivolity – *n.* – lightness; folly; fun

frugality – *n.* – thrift

garbled – *adj.* – mixed up

generic – *adj.* – common; general; universal

germane – *adj.* – pertinent; related; to the point

gibber – *v.* – to speak foolishly

gloat – *v.* – to brag; to glory over

guile – *n.* – slyness; fraud

haggard – *adj.* – tired-looking; fatigued

hiatus – *n.* – interval; break; period of rest

hierarchy – *n.* – body of people, things, or concepts divided into ranks

homage – *n.* – honor; respect

hubris – *n.* – arrogance

ideology – *n.* – set of beliefs; principles

ignoble – *adj.* – shameful; dishonorable

imbue – *v.* – to inspire; to arouse

impale – *v.* – to fix on a stake; to stick; to pierce

implement – *v.* – to begin; to enact

impromptu – *adj.* – without preparation

inarticulate – *adj.* – speechless; unable to speak clearly

incessant – *adj.* – uninterrupted

incognito – *adj.* – unidentified; disguised; concealed

indict – *v.* – to charge with a crime

inept – *adj.* – incompetent; unskilled

innuendo – *n.* – hint; insinuation

intermittent – *adj.* – periodic; occasional

invoke – *v.* – to ask for; to call upon

itinerary – *n.* – travel plan; schedule; course

jovial – *adj.* – cheery; jolly; playful

juncture – *n.* – critical point; meeting

juxtapose – *v.* – to place side-by-side

knavery – *n* – rascality; trickery

knead – *v.* – mix; massage

labyrinth – *n.* – maze

laggard – *n.* – a lazy person; one who lags behind

larceny – *n.* – theft; stealing

lascivious – *adj.* – indecent; immoral

lecherous – *adj.* – impure in thought and act

lethal – *adj.* – deadly

liaison – *n.* – connection; link

limber – *adj.* – flexible; pliant

livid – *adj.* – 1. black-and-blue; discolored; 2. enraged; irate

lucrative – *adj.* – profitable; gainful

lustrous – *adj.* – bright; radiant

malediction – *n.* – curse; evil spell

mandate – *n.* – order; charge

manifest – *adj.* – obvious; clear

mentor – *n.* – teacher

mesmerize – *v.* – to hypnotize

metamorphosis – *n.* – change of form

mimicry – *n.* – imitation

molten – *adj.* – melted

motif – *n.* – theme

mundane – *adj.* – ordinary; commonplace

myriad – *adj.* – innumerable; countless

narcissistic – *adj.* – egotistical; self-centered

nautical – *adj.* – of the sea

neophyte – *n.* – beginner; newcomer

nettle – *v.* – to annoy; irritate

notorious – *adj.*– infamous; renowned

obdurate – *adj.* – stubborn; inflexible

obligatory – *adj.* – mandatory; necessary

obliterate – *v.* – to destroy completely

obsequious – *adj.* – slavishly attentive; servile

obstinate – *adj.* – stubborn

occult – *adj.* – mystical; mysterious

opaque – *adj.* – dull; cloudy; nontransparent

opulence – *n.* – wealth; fortune

ornate – *adj.* – elaborate; lavish; decorated

oust – *v.* – to drive out; to eject

painstaking – *adj.* – thorough; careful; precise

pallid – *adj.* – sallow; colorless

palpable – *adj.* – tangible; apparent

paradigm – *n.* – model; example

paraphernalia – *n.* – equipment; accessories

parochial – *adj.* – religious; narrow-minded

passive – *adj.* – submissive; unassertive

pedestrian – *adj.* – mediocre; ordinary

pensive – *adj.* – reflective; contemplative

percussion – *adj.* – striking one object against another

perjury – *n.* – the practice of lying

permeable – *adj.* – porous; allowing to pass through

perpetual – *adj.* – enduring for all time

pertinent – *adj.* – related to the matter at hand

pervade – *v.* – to occupy the whole of

petty – *adj.* – unimportant; of subordinate standing

phlegmatic – *adj.* – without emotion or interest

phobia – *n.* – morbid fear

pittance – *n.* – small allowance

plethora – *n.* – condition of going beyond what is needed; excess; overabundance

potent – *adj.* – having great power or physical strength

privy – *adj.* – private; confidential

progeny – *n.* – children; offspring

provoke – *v.* – to stir action or feeling; arouse

pungent – *adj.* – sharp; stinging

quaint – *adj.* – old-fashioned; unusual; odd

quandary – *n.* – dilemma

quarantine – *n.* – isolation of a person to prevent spread of disease quiescent – *adj.* – inactive; at rest

quirk – *n.* – peculiar behavior; startling twist

rabid – *adj.* – furious; with extreme anger

rancid – *adj.* – having a bad odor

rant – *v.* – to speak in a loud, pompous manner; to rave

ratify – *v.* – to make valid; to confirm

rationalize – *v.* – to offer reasons for; to account for

raucous – *adj.* – disagreeable to the sense of hearing; harsh

realm – *n.* – an area; sphere of activity

rebuttal – *n.* – refutation

recession – *n.* – withdrawal; depression

reciprocal – *n.* – mutual; having the same relationship to each other

recluse – *n.* – solitary and shut off from society

refurbish – *v.* – to make new

regal – *adj.* – royal; grand

reiterate – *v.* – to repeat; to state again

relinquish – *v.* – to let go; to abandon

render – *v.* – to deliver; to provide; to give up a possession

replica – *n.* – copy; representation

resilient – *adj.* – flexible; capable of withstanding stress

retroaction – *n.* – an action elicited by a stimulus

reverie – *n.* – the condition of being unaware of one's surroundings; trance

rummage – *v.* – to search thoroughly

rustic – *adj.* – plain and unsophisticated; homely

saga – *n.* – a legend; story

salient – *adj.* – noticeable; prominent

salvage – *v.* – to rescue from loss

sarcasm – *n.* – ironic; bitter humor designed to wound

satire – *n.* – a novel or play that uses humor or irony to expose folly

saunter – *v.* – to walk at a leisurely pace; to stroll

savor – *v.* – to receive pleasure from; to enjoy

seethe – *v.* – to be in a state of emotional turmoil; to become angry

serrated – *adj.* – having a sawtoothed edge

shoddy – *adj.* – of inferior quality; cheap

skulk – *v.* – to move secretly

sojourn – *n.* – temporary stay; visit

solace – *n.* – hope; comfort during a time of grief

soliloquy – *n.* – a talk one has with oneself (esp. on stage)

somber – *adj.* – dark and depressing; gloomy

sordid – *adj.* – filthy; base; vile

sporadic – *adj.* – rarely occurring or appearing; intermittent

stamina – *n.* – endurance

steadfast – *adj.* – loyal

stigma – *n.* – a mark of disgrace

stipend – *n.* – payment for work done

stupor – *n.* – a stunned or bewildered condition

suave – *adj.* – effortlessly gracious

subsidiary – *adj.* – subordinate

succinct – *adj.* – consisting of few words; concise

succumb – *v.* – to give in; to yield; to collapse

sunder – *v.* – to break; to split in two

suppress – *v.* – to bring to an end; to hold back

surmise – *v.* – to draw an inference; to guess

susceptible – *adj.* – easily imposed; inclined

tacit – *adj.* – not voiced or expressed

tantalize – *v.* – to tempt; to torment

tarry – *v.* – to go or move slowly; to delay

taut – *adj.* – stretched tightly

tenacious – *adj.* – persistently holding to something

tepid – *adj.* – lacking warmth, interest, enthusiasm; lukewarm

terse – *adj.* – concise; abrupt

thwart – *v.* – to prevent from accomplishing a purpose; to frustrate

timorous – *adj.* – fearful

torpid – *adj.* – lacking alertness and activity; lethargic

toxic – *adj.* – poisonous

transpire – *v.* – to take place; to come about

traumatic – *adj.* – causing a violent injury

trek – *v.* – to make a journey

tribute – *n.* – expression of admiration

trite – *adj.* – commonplace; overused

truculent – *adj.* – aggressive; eager to fight

turbulence – *n.* – condition of being physically agitated; disturbance

turmoil – *n.* – unrest; agitation

tycoon – *n.* – wealthy leader

tyranny – *n.* – absolute power; autocracy

ubiquitous – *adj.* – ever present in all places; universal

ulterior – *adj.* – buried; concealed

uncanny – *adj.* – of a strange nature; weird

unequivocal – *adj.* – clear; definite

unique – *adj.* – without equal; incomparable

unruly – *adj.* – not submitting to discipline; disobedient

unwonted – *adj.* – not ordinary; unusual

urbane – *adj.* – cultured; suave

usurpation – *n.* – act of taking something for oneself; seizure

usury – *n.* – the act of lending money at illegal rates of interest

utopia – *n.* – imaginary land with perfect social and political systems

vacuous – *adj.* – containing nothing; empty

vagabond – *n.* – wanderer; one without a fixed place

vagrant – 1. *n.* – homeless person; 2. *adj.* – rambling; wandering; transient

valance – *n.* – short drapery hanging over the window frame

valor – *n.* – bravery

vantage – *n.* – position giving an advantage

vaunted – *adj.* – boasted of

velocity – *n.* – speed

vendetta – *n.* – feud

venue – *n.* – location

veracious – *adj.* – conforming to fact; accurate

verbatim – *adj.* – employing the same words as another; literal

versatile – *adj.* – having many uses; multifaceted

vertigo – *n.* – dizziness

vex – *v.* – to trouble the nerves; to annoy

vindicate – *v.* – to free from charge; to clear

vivacious – *adj.* – animated; gay

vogue – *n.* – modern fashion

voluble – *adj.* – fluent

waft – *v.* – to move gently by wind or breeze

waive – *v.* – to give up possession or right

wanton – *adj.* – unruly; excessive

warrant – *v.* – to justify; to authorize

wheedle – *v.* – to try to persuade; to coax

whet – *v.* – to sharpen

wrath – *n.* – violent or unrestrained anger; fury

wry – *adj.* – mocking; cynical

xenophobia – *n.* – fear of foreigners

yoke – *n.* – harness; collar; bond

yore – *n.* – former period of time

zephyr – *n.* – a gentle wind; breeze

II. MASTERING WORD PARTS

While taking the SAT, you will have nothing but your own knowledge to refer to when you come into contact with unfamiliar words. Even though we have provided you with the 600 most commonly tested SAT words, there is a very good chance that you will come across words that you still do not know. Therefore, you will need to review our list of the most common prefixes, roots, and suffixes in order to be prepared.

Learn the meanings of the prefixes, roots, and suffixes in the same way that you learned the vocabulary words and their meanings. Be sure to use index cards for the items you don't know or find unusual. Look over the examples given and then try to think of your own. Testing yourself in this way will allow you to see if you really do know the meaning of each item. Knowledge of prefixes, roots, and suffixes is essential to a strong vocabulary and, therefore, to a high score on the verbal SAT.

PREFIXES

PREFIX	MEANING	EXAMPLE
ab –, a –, abs –	away, without, from	absent – away, not present apathy – without interest abstain – keep from doing, refrain
ad –	to, toward	adjacent – next to address – to direct toward

PREFIX	MEANING	EXAMPLE
ante –	before	antecedent – going before in time anterior – occurring before
anti –	against	antidote – remedy to act against an evil antibiotic – substance that fights against bacteria
be –	over, thoroughly	bemoan – to mourn over belabor – to exert much labor upon
bi –	two	bisect – to divide biennial – happening every two years
cata –, cat –, cath –	down	catacombs – underground passageways catalogue – descriptive list catheter – tubular medical device
circum –	around	circumscribe – to draw a circle around circumspect – watchful on all sides
com –	with	combine – to join together communication – to have dealings with
contra –	against	contrary – opposed contrast – to stand in opposition
de –	down, from	decline – to bend downward decontrol – to release from government control
di –	two	dichotomy – cutting in two diarchy – system of government with two authorities
dis –, di–	apart, away	discern – to distinguish as separate dismiss – to send away digress – to turn aside
epi –, ep –, eph –	upon, among	epidemic – happening among many people epicycle – circle whose center moves around in the circumference of a greater circle epaulet – decoration worn to ornament or protect the shoulder ephedra – any of a large genus of desert shrubs
ex –, e –	from, out	exceed – go beyond the limit emit – to send forth
extra –	outside, beyond	extraordinary – beyond or out of the common method extrasensory – beyond the senses
hyper –	beyond, over	hyperactive – over the normal activity level hypercritic – one who is critical beyond measure
hypo –	beneath, lower	hypodermic – parts beneath the skin hypocrisy – to be under a pretense of goodness
in –, il –, im –, ir –	not	inactive – not active illogical – not logical imperfect – not perfect irreversible – not reversible

PREFIX	MEANING	EXAMPLE
in –, il –, im –, ir –	in, on, into	instill – to put in slowly illation – action of bringing in impose – to lay on irrupt – to break in
inter –	among, between	intercom – to exchange conversations between people interlude – performance given between parts in a play
intra –	within	intravenous – within a vein intramural – within a single college or its students
meta –	beyond, over, along with	metamorphosis – change over in form or nature metatarsus – part of foot beyond the flat of the foot
mis –	badly, wrongly	misconstrue – to interpret wrongly misappropriate – to use wrongly
mono –	one	monogamy – to be married to one person at a time monotone – a single, unvaried tone
multi –	many	multiple – of many parts multitude – a great number
non –	no, not	nonsense – lack of sense nonentity – not existing
ob –	against	obscene – offensive to modesty obstruct – to hinder the passage of
para –, par –	beside	parallel – continuously at equal distance apart parenthesis – sentence inserted within a passage
per –	through	persevere – to maintain an effort permeate – to pass through
poly –	many	polygon – a plane figure with many sides or angles polytheism – belief in the existence of many gods
post –	after	posterior – coming after postpone – to put off until a future time
pre –	before	premature – ready before the proper time premonition – a previous warning
pro –	in favor of, forward	prolific – bringing forth offspring project – throw or cast forward
re –	back, against	reimburse – to pay back retract – to draw back
semi –	half	semicircle – half a circle semiannual – half-yearly
sub –	under	subdue – to bring under one's power submarine – to travel under the surface of the sea
super –	above	supersonic – above the speed of sound superior – higher in place or position

PREFIX	MEANING	EXAMPLE
tele –, tel –	across	telecast – transmit across a distance telepathy – communication between mind and mind at a distance
trans –	across	transpose – to change the position of two things transmit – to send from one person to another
ultra –	beyond	ultraviolet – beyond the limit of visibility ultramarine – beyond the sea
un –	not	undeclared – not declared unbelievable – not believable
uni –	one	unity – state of oneness unison – sounding togethe
with –	away, against	withhold – to hold back withdraw – to take away

DRILL: PREFIXES

DIRECTIONS: Provide a definition for each prefix.

1. pro– __________
2. com– __________
3. epi– __________
4. ob– __________
5. ad– __________

DIRECTIONS: Identify the prefix in each word.

6. effące __________
7. hypothetical __________
8. permeate __________
9. contrast __________
10. inevitable __________

ROOTS

ROOT	MEANING	EXAMPLE
act, ag	do, act, drive	activate – to make active agile – having quick motion
alt	high	altitude – height alto – high singing voice

ROOT	MEANING	EXAMPLE
alter, altr	other, change	alternative – choice between two things altruism – living for the good of others
am, ami	love, friend	amiable – worthy of affection amity – friendship
anim	mind, spirit	animated – spirited animosity – violent hatred
annu, enni	year	annual – every year centennial – every hundred years
aqua	water	aquarium – tank for water animals and plants aquamarine – semiprecious stone of sea-green color
arch	first, ruler	archenemy – chief enemy archetype – original pattern from which things are copied
aud, audit	hear	audible – capable of being heard audience – assembly of hearers audition – the power or act of hearing
auto	self	automatic – self-acting autobiography – story about a person who also wrote it
bell	war	belligerent – a party taking part in a war bellicose – warlike
ben, bene	good	benign – kindly disposition beneficial – advantageous
bio	life	biotic – relating to life biology – the science of life
brev	short	abbreviate – make shorter brevity – shortness
cad, cas	fall	cadence – fall in voice casualty – loss caused by death
capit, cap	head	captain – the head or chief decapitate – to cut off the head
cede, ceed, cess	to go, to yield	recede – to move or fall back proceed – to move onward recessive – tending to go back
cent	hundred	century – hundred years centipede – insect with a hundred legs
chron	time	chronology – science dealing with historical dates chronicle – register of events in order of time
cide, cis	to kill, to cut	homicide – one who kills; planned killing of a person incision – a cut

ROOT	MEANING	EXAMPLE
clam, claim	to shout	acclaim – receive with applause proclamation – announce publicly
cogn	to know	recognize – to know again cognition – awareness
corp	body	incorporate – combine into one body corpse – dead body
cred	to trust, to believe	incredible – unbelievable credulous – too prone to believe
cur, curr, curs	to run	current – flowing body of air or water excursion – short trip
dem	people	democracy – government formed for the people epidemic – affecting all people
dic, dict	to say	dictate – to read aloud for another to transcribe verdict – decision of a jury
doc, doct	to teach	docile – easily instructed indoctrinate – to instruct
domin	to rule	dominate – to rule dominion – territory of rule
duc, duct	to lead	conduct – act of guiding induce – to overcome by persuasion
eu	well, good	eulogy – speech or writing in praise euphony – pleasantness or smoothness of sound
fac, fact, fect, fic	to do, to make	facilitate – to make easier factory – location of production confect – to put together fiction – something invented or imagined
fer	to bear, to carry	transfer – to move from one place to another refer – to direct to
fin	end, limit	infinity – unlimited finite – limited in quantity
flect, flex	to bend	flexible – easily bent reflect – to throw back
fort	luck	fortunate – lucky fortuitous – happening by chance
fort	strong	fortify – strengthen fortress – stronghold
frag, fract	break	fragile – easily broken fracture – break
fug	flee	fugitive – fleeing refugee – one who flees to a place of safety
gen	class, race	engender – to breed generic – of a general nature in regard to all members

ROOT	MEANING	EXAMPLE
grad, gress	to go, to step	regress – to go back graduate – to divide into regular steps
graph	writing	telegraph – message sent by telegraph autograph – person's own handwriting or signature
ject	to throw	projectile – capable of being thrown reject – to throw away
leg	law	legitimate – lawful legal – defined by law
leg, lig, lect	to choose, gather, read	illegible – incapable of being read ligature – something that binds election – the act of choosing
liber	free	liberal – favoring freedom of ideals liberty – freedom from restraint
log	study, speech	archaeology – study of human antiquities prologue – address spoken before a performance
luc, lum	light	translucent – slightly transparent illuminate – to light up
magn	large, great	magnify – to make larger magnificent – great
mal, male	bad, wrong	malfunction – to operate incorrectly malevolent – evil
mar	sea	marine – pertaining to the sea submarine – below the surface of the sea
mater, matr	mother	maternal – motherly matriarchy – government by mothers or women
mit, miss	to send	transmit – to send from one person or place to another mission – the act of sending
morph	shape	metamorphosis – a changing in shape anthropomorphic – having a human shape
mut	change	mutable – subject to change mutate – to change a
nat	born	innate – inborn native – a person born in a place
neg	deny	negative – expressing denial renege – to deny
nom	name	nominate – to put forward a name nomenclature – process of naming
nov	new	novel – new renovate – to make as good as new
omni	all	omnipotent – all powerful omnipresent – all present

ROOT	MEANING	EXAMPLE
oper	to work	operate – to work on something cooperate – to work with others
pass, path	to feel	pathetic – affecting the tender emotions passionate – moved by strong emotion
pater, patr	father	paternal – fatherly patriarchy – government by fathers or men
ped, pod	foot	pedestrian – one who travels on foot podiatrist – foot doctor
pel, puls	to drive, to push	impel – to drive forward compulsion – irresistible force
phil	love	philharmonic – loving harmony or music philanthropist – one who loves and seeks to do good for others
port	carry	export – to carry out of the country portable – able to be carried
psych	mind	psychology – study of the mind psychiatrist – specialist in mental disorders
quer, ques, quir, quis	to ask	querist – one who inquires inquiry – to ask about question – that which is asked inquisitive – inclined to ask questions
rid, ris	to laugh	ridiculous – laughable derision – to mock
rupt	to break	interrupt – to break in upon erupt – to break through
sci	to know	science – systematic knowledge of physical or natural phenomena conscious – having inward knowledge
scrib, script	to write	transcribe – to write over again script – text of words
sent, sens	to feel, to think	sentimental – feel great emotion sensitive – easily affected by changes
sequ, secut	to follow	sequence – connected series consecutive – following one another in unbroken order
solv, solu, solut	to loosen	dissolve – to break up absolute – without restraint
spect	to look at	spectator – one who watches inspect – to look at closely
spir	to breathe	inspire – to breathe in respiration – process of breathing
string, strict	to bind	stringent – binding strongly restrict – to restrain within bounds

ROOT	MEANING	EXAMPLE
stru, struct	to build	strut – a structural piece designed to resist pressure construct – to build
tang, ting, tact, tig	to touch	tactile – perceptible by touching tangent – touching, but not intersecting contact – touching contiguous – to touch along a boundary
ten, tent, tain	to hold	tenure – holding of office contain – to hold
term	to end	terminate – to end terminal – having an end
terr	earth	terrain – tract of land terrestrial – existing on earth
therm	heat	thermal – pertaining to heat thermometer – instrument for measuring temperature
tort, tors	to twist	contortionist – one who twists violently torsion – act of turning or twisting
tract	to pull, to draw	attract – draw toward distract – to draw away
vac	empty	vacant – empty evacuate – to empty out
ven, vent	to come	prevent – to stop from coming intervene – to come between
ver	true	verify – to prove to be true veracious – truthful
verb	word	verbose – use of excess words verbatim – word for word
vid, vis	to see	video – picture phase of television vision – act of seeing external objects
vinc, vict, vanq	to conquer	invincible – unconquerable victory – defeat of enemy vanquish – to defeat
viv, vit	life	vital – necessary to life vivacious – lively
voc	to call	vocation – a summons to a course of action vocal – uttered by voice
vol	to wish, to will	involuntary – outside the control of will volition – the act of willing or choosing

DRILL: ROOTS

DIRECTIONS: Provide a definition for each root.

1. cede ______________________
2. fact ______________________
3. path ______________________
4. ject ______________________
5. ver ______________________

DIRECTIONS: Identify the root in each word.

6. acclaim ______________________
7. verbatim ______________________
8. benefactor ______________________
9. relegate ______________________
10. tension ______________________

SUFFIXES

SUFFIX	MEANING	EXAMPLE
–able, –ble	capable of	believable – capable of believing legible – capable of being read vivacious – full of life
–acious, –icious, –ous	full of	delicious – full of pleasurable smell or taste wondrous – full of wonder
–ant, –ent	full of	eloquent – full of eloquence expectant – full of expectation
–ary	connected with	honorary – for the sake of honor disciplinary – relating to a field of study
–ate	to make	ventilate – to make public consecrate – to dedicate
–fy	to make	magnify – to make larger testify – to make witness
–ile	pertaining to, capable of	docile – capable of being managed easily infantile – pertaining to infancy
–ism	belief, ideal	conservationism – ideal of keeping safe sensationalism – matter, language designed to excite
–ist	doer	artist – one who creates art pianist – one who plays the piano

SUFFIX	MEANING	EXAMPLE
–ose	full of	verbose – full of words grandiose – striking, imposing
–osis	condition	neurosis – nervous condition psychosis – psychological condition
–tude	state	magnitude – state of greatness multitude – state of quantity

DRILL: SUFFIXES

<u>DIRECTIONS:</u> Provide a definition for each suffix.

1. –ant, –ent ____________________
2. –tude ____________________
3. –ile ____________________
4. –fy ____________________
5. –ary ____________________

<u>DIRECTIONS:</u> Identify the suffix in each word.

6. audacious ____________________
7. expedient ____________________
8. gullible ____________________
9. grandiose ____________________
10. antagonism ____________________

VERBAL DRILLS

ANSWER KEY

Drill: Group 1

1. (J)
2. (G)
3. (A)
4. (C)
5. (H)
6. (B)
7. (I)
8. (D)
9. (F)
10. (E)
11. (D)
12. (C)
13. (A)
14. (E)
15. (B)

Drill: Group 2

1. (D)
2. (G)
3. (I)
4. (F)
5. (A)
6. (J)
7. (E)
8. (C)
9. (B)
10. (H)
11. (D)
12. (A)
13. (B)
14. (E)
15. (C)

Drill: Group 3

1. (E)
2. (H)
3. (J)
4. (A)
5. (I)
6. (B)
7. (C)
8. (G)
9. (F)
10. (D)
11. (C)
12. (A)
13. (E)
14. (D)
15. (B)

Drill: Group 4

1. (D)
2. (E)
3. (A)
4. (I)
5. (J)
6. (B)
7. (C)
8. (F)
9. (H)
10. (G)
11. (D)
12. (A)
13. (B)
14. (C)
15. (E)

Drill: Group 5

1. (H)
2. (F)
3. (A)
4. (B)
5. (J)
6. (C)
7. (I)
8. (E)
9. (D)
10. (G)
11. (B)
12. (D)
13. (A)
14. (E)
15. (C)

Drill: Group 6

1. (G)
2. (A)
3. (E)
4. (J)
5. (C)
6. (B)
7. (D)
8. (I)
9. (F)
10. (H)
11. (D)
12. (E)
13. (A)
14. (C)
15. (B)

Drill: Group 7

1. (F)
2. (E)
3. (A)
4. (B)
5. (H)
6. (I)
7. (C)
8. (G)
9. (D)
10. (J)
11. (B)
12. (E)
13. (D)
14. (C)
15. (A)

Drill: Group 8

1. (D)
2. (A)
3. (H)
4. (G)
5. (B)
6. (C)
7. (E)
8. (I)
9. (F)
10. (J)
11. (C)
12. (B)
13. (A)
14. (E)
15. (D)

Drill: Group 9

1. (D)
2. (I)
3. (G)
4. (A)
5. (J)
6. (E)
7. (B)
8. (F)
9. (C)
10. (H)
11. (A)
12. (D)
13. (B)
14. (E)
15. (C)

Drill: Group 10

1. (H)
2. (I)
3. (E)
4. (A)
5. (J)
6. (B)
7. (F)
8. (C)
9. (D)
10. (G)
11. (B)
12. (D)
13. (A)
14. (C)
15. (E)

Drill: Group 11

1. (F)
2. (I)
3. (A)
4. (H)
5. (B)
6. (J)
7. (C)
8. (D)
9. (E)
10. (G)
11. (C)
12. (B)
13. (E)
14. (A)
15. (D)

Drill: Group 12

1.	(J)	5.	(C)	9.	(F)	13.	(D)
2.	(A)	6.	(H)	10.	(E)	14.	(A)
3.	(B)	7.	(D)	11.	(E)	15.	(B)
4.	(I)	8.	(G)	12.	(C)		

Drill: Prefixes

1.	forward	6.	ef–
2.	with	7.	hypo–
3.	upon, among	8.	per–
4.	against	9.	con–
5.	to, toward	10.	in–

Drill: Roots

1.	to go, to yield	6.	claim
2.	to do, to make	7.	verb
3.	to feel	8.	ben(e)
4.	to throw	9.	leg
5.	true	10.	ten

Drill: Suffixes

1.	full of	6.	(a)cious
2.	state	7.	ent
3.	pertaining to, capable of	8.	ible
4.	to make	9.	ose
5.	connected with	10.	ism

CHAPTER 4

MASTERING SENTENCE COMPLETION QUESTIONS

MASTERING SENTENCE COMPLETION QUESTIONS

Regardless of the Critical Reading or Writing SAT section in which one is working, all problem-solving techniques should be divided into two main categories: skills and strategies. This chapter will present skills and strategies that are effective in helping the test-taker successfully answer Sentence Completions. These techniques include the recognition of a context clue, a knowledge of the levels of difficulty in a Sentence Completion section, the application of deductive reasoning, and familiarity with the logical structure of sentence completions. You will encounter Sentence Completion questions in two different sections.

Success on the SAT begins with one fundamental insight: the underlying intent is to test your vocabulary. No matter what section you are working in, you will be expected to demonstrate a command of a wide array of vocabulary words drawn from a treasury of Greek and Latin roots and prefixes, which can be found in Appendix A of this book. Devote as much time as possible to strengthening your vocabulary, especially by studying the prefixes and roots of Greek- and Latin-derived words. The Mastering Vocabulary section of the Basic Verbal Skills Review should be studied thoroughly.

ABOUT THE DIRECTIONS

The directions for Sentence Completion questions are relatively straightforward.

DIRECTIONS: Each sentence below has one or two blanks, each blank indicating that something has been omitted. Beneath the sentence are five lettered words or sets of words. Choose the word or set of words that BEST fits the meaning of the sentence as a whole.

EXAMPLE

Although the critics found the book _____, many of the readers found it rather _____.

(A) obnoxious . . . perfect
(B) spectacular . . . interesting
(C) boring . . . intriguing
(D) comical . . . persuasive
(E) popular . . . rare

(A) (B) ● (D) (E)

ABOUT THE QUESTIONS

You will encounter two main types of questions in the Sentence Completion sections of the SAT. In addition, the questions will appear in varying difficulties, which we will call Level I (easy), Level II (average), and Level III (difficult). The following explains the structure of the questions.

Question Type 1: One-Word Completions

One-Word Completions will require you to fill in one blank. The one-word completion can appear as a Level I, II, or III question depending on the difficulty of the vocabulary included.

Question Type 2: Two-Word Completions

Two-Word Completions will require you to fill in two blanks. As with the one-word completion, this type may be a Level I, II, or III question. This will depend not only on the difficulty of the vocabulary, but also on the relationship between the words and between the words and the sentence.

The remainder of this review will provide explicit details on what you will encounter when dealing with Sentence Completion questions, in addition to strategies for correctly completing these sentences.

POINTS TO REMEMBER

- Like other verbal sections of the test, Sentence Completions can be divided into three basic levels of difficulty, and, as a general rule, SAT verbal exercises *increase* in difficulty as they progress through a section.
- Level I exercises allow you to rely on your instincts and common sense. You should not be obsessed with analysis or second-guessing in Level I problems.
- In Level II questions, the SAT often presents words that appear easy at first glance but that may have secondary meanings. Be wary of blindly following your gut reactions and common sense.
- All SAT word problems contain "magnet words," answer choices that look good but are designed to draw the student away from the correct answer. Magnet words can effectively mislead you in Level III questions. Always watch for them. Remember that Level III questions are intentionally designed to work against your common sense and natural inclinations.
- Deductive reasoning is a tool that will be of constant assistance to you as you work through SAT word problems. To deduce means to derive a truth (or answer) through a reasoning process.

- Sentence Completion questions are puzzles, and they are put together with a certain amount of predictability. One such predictable characteristic is the *structure* of an SAT word exercise. Since there are always five possible answers from which to choose, you must learn to see which answers are easy to eliminate first. Use the process of elimination.
- Most SAT word problems are designed around a "three-two" structure. This means that there are three easier answers to eliminate before you have to make the final decision between the remaining two.
- Use word roots, prefixes, and suffixes to find the meanings of words you do not know.

MASTERING SENTENCE COMPLETION QUESTIONS

Follow these steps as you attempt to answer each question.

Identifying context clues is one of the most successful ways for students to locate correct answers in Sentence Completions. Practicing constantly in this area will help you strengthen one of your main strategies in this type of word problem. The following Sentence Completion is an example of a Level I question.

Pamela played her championship chess game _____ , avoiding all traps and making no mistakes.

(A) hurriedly
(B) flawlessly
(C) prodigally
(D) imaginatively
(E) aggressively

The phrase "avoiding all traps and making no mistakes" is your context clue. Notice that the phrase both follows *and* modifies the word in question. Because you know that Sentence Completions are exercises seeking to test your vocabulary knowledge, attack these problems accordingly. For example, ask yourself what word means "avoiding all traps and making no mistakes." In so doing, you discover the answer flawlessly (B), which means perfectly or without mistakes. If Pamela played hurriedly (A), she might well make mistakes. Difficult words are seldom the answer in easier questions; therefore, prodigally (C) stands out as a suspicious word. This could be a magnet word. However, before you eliminate it, ask yourself whether you know its meaning. If so, does it surpass flawlessly (B) in defining the context clue, "making no mistakes"? It does not. Imaginatively (D) is a tempting answer, since one might associate a perfect game of chess as one played imaginatively; however, there is no connection between the imagination and the absence of mistakes. Aggressively (E) playing a game may, in fact, cause you to make mistakes.

Here is an example of a Level II Sentence Completion. Try to determine the context clue.

Although most people believe the boomerang is the product of a _____ design, that belief is deceptive; in fact, the boomerang is a(n) _____ example of the laws of aerodynamics.

(A) foreign . . . modern
(B) symbolic . . . complex
(C) practical . . . scientific
(D) primitive . . . sophisticated
(E) faulty . . . invalid

The most important context clue in this sentence is the opening word "although," which indicates that some kind of antonym relationship is present in the sentence. It tells us there is a reversal in meaning. Therefore, be on the lookout for words which will form an opposite relationship. The phrase "that belief is deceptive" makes certain the idea that there will be an opposite meaning between the missing words. Primitive... sophisticated (D) is the best answer, since the two are exact opposites. "Primitive" means crude and elementary, whereas "sophisticated" means refined and advanced. Foreign . . . modern (A) and symbolic . . . complex (B) have no real opposite relationship. Also, "complex" is a magnet word that sounds right in the context of scientific laws, but "symbolic" is not its counterpart. Practical . . . scientific (C) and faulty . . . invalid (E) are rejectable because they are generally synonymous pairs of relationships.

The following is an example of a Level III question:

The weekly program on public radio is the most _____ means of educating the public about pollution.

(A) proficient
(B) effusive
(C) effectual
(D) capable
(E) competent

The context clue in this sentence is "means of educating the public about pollution." Effectual (C) is the correct answer. Effectual means having the power to produce the exact effect or result. Proficient (A) is not correct as it implies competency above the average—radio programs are not described in this manner. Effusive (B) does not fit the sense of the sentence. Both capable (D) and competent (E) are incorrect because they refer to people, not things.

Because the Critical Reading SAT sections are fundamentally a vocabulary test, it must resort to principles and techniques necessary for testing your vocabulary. Therefore, certain dynamics like antonyms (word opposites) and synonyms (word similarities) become very useful in setting up a question or word problem. This idea can be taken one step further.

Another type of technique that utilizes the tension of opposites and the concurrence of similarities is *word values.* Word values begin with the recognition that most pivotal words in an SAT word problem can be assigned a positive or negative value. Marking a "+" or "–" next to choices may help you eliminate inappropriate choices. In turn, you will be able to more quickly identify possible correct answers.

Dealing with Positive Value Words

Positive value words are usually easy to recognize. They usually convey a meaning which can be equated with gain, advantage, liveliness, intelligence, virtue, and positive emotions, conditions, or actions.

The ability to recognize positive and negative word values, however, will not bring you very far if you do not understand how to apply it to your advantage in Sentence Completions. Below you will find examples of how to do this, first with a study of positive value Sentence Completions, then with a study of negative value Sentence Completions. The following is an example of a Level I question.

An expert skateboarder, Tom is truly _____ ; he smoothly blends timing with balance.

(A) coordinated
(B) erudite
(C) a novice
(D) supportive
(E) casual

As you know, the context clue is the clause after the word in question, which acts as a modifier. Naturally, anyone who "smoothly blends" is creating a *positive* situation. Look for the positive answer.

An expert skateboarder, Tom is truly __+__ ; he smoothly blends timing with balance.

+(A) coordinated
+(B) erudite
–(C) a novice
+(D) supportive
–(E) casual

Coordinated (A), a positive value word that means ordering two or more things, fits the sentence perfectly. Erudite (B) is positive, but it is too difficult to be a Level I answer. A novice (C) in this context is negative. Supportive (D) and casual (E) don't fulfill the definition of the context clue, and casual is negative, implying a lack of attention. Notice that eliminating negatives immediately reduces the number of options from which you have to choose. This raises the odds of selecting the correct answer. (One of the analytic skills you should develop for the SAT I is being able to see the hidden vocabulary question in any exercise.)

A Level II question may appear as follows:

Despite their supposedly primitive lifestyle, Australian aborigines developed the boomerang, a _____ and _____ hunting tool that maximizes gain with minimum effort.

(A) ponderous . . . expensive
(B) clean . . . dynamic
(C) dangerous . . . formidable
(D) sophisticated . . . efficient
(E) useful . . . attractive

In this case, the context clues begin and end the sentence (in italics below).

Despite their supposedly primitive lifestyle, Australian aborigines developed the boomerang, a __+__ and __+__ hunting tool that *maximizes gain* with *minimum effort.*

–(A) ponderous . . . expensive
+(B) clean . . . dynamic
–(C) dangerous . . . formidable
+(D) sophisticated . . . efficient
+(E) useful . . . attractive

The first context clue (*despite*) helps you determine that this exercise entails an antonym relationship with the word primitive, which means simple or crude. The second context clues offer a definition of the missing words. Since the meaning of primitive in this context is a negative word value, you can be fairly confident that the answer will be a pair of positive word values. Sophisticated . . . efficient (D) is positive *and* it satisfies the definition of the latter context clue. This is the best answer. Ponderous ... expensive (A) is not correct. Clean . . . dynamic (B) is positive, but does not meet the definition of the latter context clues. Dangerous ...formidable (C) is negative. Useful . . . attractive (E) is positive, but it does not work with the latter context clues.

Here is a Level III example.

When physicians describe illnesses to colleagues, they must speak an _____ language, using professional terms and concepts understood mostly by members of the profession.

(A) extrinsic
(B) inordinate
(C) ambulatory
(D) esoteric
(E) abbreviated

Looking at this question, we can see an important context clue. This appears in italics below.

When physicians describe illnesses to colleagues, they must speak an __+__ language, *using professional terms and concepts understood mostly by members of the profession.*

+(A) extrinsic
–(B) inordinate
+(C) ambulatory
+(D) esoteric
–(E) abbreviated

This clue gives us a definition of the missing word. Begin by eliminating the two obvious negatives, inordinate (B) and abbreviated (E). This leaves us with three positives. Because this is a Level III exercise, at first you may be intimidated by the level of vocabulary. In the section on etymology you will be given insights into how to handle difficult word problems. For now, note that esoteric (D) is the best answer, since it is an adjective that means *inside* or *part of a group.* Ambulatory (C) is positive, but it is a trap. It seems like an easy association with the world of medicine. In Level III there are no easy word associations. Extrinsic (A) is positive, but it means *outside of*, which would not satisfy the logic of the sentence.

Dealing with Negative Value Words

Here are examples of how to work with negative value Sentence Completion problems. The first example is Level I.

Although Steve loves to socialize, his fellow students find him _____ and strive to _____ his company.

(A) generous . . . enjoy
(B) boring . . . evade
(C) altruistic . . . accept
(D) sinister . . . delay
(E) weak . . . limit

The context clue (in italics) tells us that a reversal is being set up between what Steve thinks and what his fellow students think.

Although Steve loves to socialize, his fellow students find him __–__ and strive to __–__ his company.

+(A) generous . . . enjoy
–(B) boring . . . evade
+(C) altruistic . . . accept
–(D) sinister . . . delay
–(E) weak . . . limit

Boring . . . evade (B) is the best answer. The words appearing in Level 1 questions are not overly difficult, and they satisfy the logic of the sentence. Generous . . . enjoy (A) is positive. Altruistic . . . accept (C) is not only positive but contains a very difficult word (altruistic), and it would be unlikely that this would be a Level I answer. The same is true of sinister . . . delay (D), even though it is negative. Weak . . . limit (E) does not make sense in the context of the sentence.

This next example is Level II.

Because they reject _____ , conscientious objectors are given jobs in community work as a substitute for participation in the armed services.

(A) labor
(B) belligerence
(C) peace
(D) dictatorships
(E) poverty

Essentially, this example is a synonym exercise. The description of the alternative to participation in the military offered to conscientious objectors (in italics) acts as a strong context clue. Conscientious objectors avoid ("reject") militancy.

Because they reject __–__ , conscientious objectors are given jobs in *community work as a substitute for participation in the armed services.*

+(A) labor
–(B) belligerence
+(C) peace
–(D) dictatorships
–(E) poverty

Because we are looking for a negative word value (something to do with militancy), labor (A) is incorrect because it is positive. Belligerence (B) fits perfectly, because it is a negative value word having to do with war. Not only is peace (C) a positive value word, it is hardly something to be rejected by conscientious objectors. Dictatorships (D), although a negative word value, has no logical place in the context of this sentence. The same is true of poverty (E).

Here is a Level III example:

Dictators understand well how to centralize power, and that is why they combine a(n) _____ political process with military _____.

(A) foreign . . . victory
(B) electoral . . . escalation
(C) agrarian . . . strategies
(D) domestic . . . decreases
(E) totalitarian . . . coercion

Totalitarian . . . coercion (E) is the best answer. These are difficult words, and both have to do with techniques useful in the centralizing of power by a dictator. *Totalitarian* means centralized, and *coercion* means force.

Dictators understand well how to *centralize power,* and that is why they combine a(n) __–__ political process with military __–__.

+(A) foreign . . . victory
+(B) electoral . . . escalation
+(C) agrarian . . . strategies
+(D) domestic . . . decreases
–(E) totalitarian . . . coercion

Foreign . . . victory (A) are not only easy words, they do not appear to be strictly negative. Remember that easy word answers should be suspect in Level III. Agrarian . . . strategies (C) is positive. Domestic... decreases (D) is a positive combination. Since you are searching for two negatives, this answer is incorrect. There will be more about this in the next section.

Dealing with Mixed Value Words

In examples with two-word answers so far, you have searched for answers composed with identical word values, such as negative/negative and positive/positive. However, every SAT Sentence Completion section will have exercises in which two-word answers are found in combinations. Below you will find examples of how to work with these. Here is a Level I example:

Despite a healthy and growing environmental _____ in America, there are many people who prefer to remain _____.

(A) awareness . . . ignorant
(B) movement . . . enlightened
(C) bankruptcy . . . wealthy
(D) crisis . . . unencumbered
(E) industry . . . satisfied

The context clue *despite* sets up the predictable antonym warning. In this case, the sentence seems to call for a positive (+) and then a negative (–) value word answer.

Despite a healthy and growing environmental _____ in America, there are many people who prefer to remain _____.

+/–(A) awareness . . . ignorant
+/+(B) movement . . . enlightened
–/+(C) bankruptcy . . . wealthy
–/+(D) crisis . . . unencumbered
+/+(E) industry . . . satisfied

Awareness . . . ignorant (A) is the best answer. These are logical antonyms, and they fit the meaning of the sentence. Notice that the order of the missing words is positive, *then* negative. This should help you eliminate (C) and (D) immediately, as they are a reversal of the correct order. Furthermore, industry . . . satisfied (E) and movement . . . enlightened (B) are both identical values, and so are eliminated. Practice these techniques until you confidently can recognize word values and the order in which they appear in a sentence.

Here is a Level II example:

Prone to creating characters of _____ quality, novelist Ed Abbey cannot be accused of writing _____ stories.

(A) measly . . . drab
(B) romantic . . . imaginative
(C) mythic . . . mundane
(D) sinister . . . complete
(E) two-dimensional . . . flat

The best answer is mythic . . . mundane (C). Measly . . . drab (A) does not make sense when you consider the context clue cannot, which suggests the possibility of antonyms. The same is true for sinister . . . complete (D), romantic . . . imaginative (B), and two-dimensional . . . flat (E).

Prone to creating characters of __+__ quality, novelist Ed Abbey cannot be accused of writing __–__ stories.

–/–(A) measly . . . drab
+/+(B) romantic . . . imaginative
+/–(C) mythic . . . mundane
–/+(D) sinister . . . complete
–/–(E) two-dimensional . . . flat

Notice that the value combinations help you determine where to search for the correct answer.

Here is a Level III example:

Reminding his students that planning ahead would protect them from _____, Mr. McKenna proved to be a principal who understood the virtues of _____.

(A) exigency . . . foresight
(B) grades . . . examinations
(C) poverty . . . promotion
(D) deprivation . . . abstinence
(E) turbulence . . . amelioration

The best answer is exigency . . . foresight (A). The first context clue tells us that we are looking for a negative value word. The second context clue tells us the missing word is most likely positive. Furthermore, exigency . . . foresight is a well-suited antonym combination. Exigencies are emergencies, and foresight helps to lessen their severity, if not their occurrence.

Reminding his students that planning ahead would *protect them* from _____, Mr. McKenna proved to be a principal who understood the *virtues* of _____.

–/+(A) exigency . . . foresight

0/0(B) grades . . . examinations

–/+(C) poverty . . . promotion

–/–(D) deprivation . . . abstinence

–/+(E) turbulence . . . amelioration

Grades . . . examinations (B) are a trap, since they imply school matters. Furthermore, they are neutrals. There will be more on this below. Poverty . . . promotion (C) is an easy word answer and should be immediately suspect, especially if there are no difficult words in the sentence completion itself. Also, this answer does not satisfy the logic of the sentence. Turbulence . . . amelioration (E) is a negative/positive combination, but it does not make sense in this sentence. Even if you are forced to guess between this answer and exigency . . . foresight (A), you have narrowed the field to two. These are excellent odds for success.

Dealing with Neutral Value Words

There is another category of word values that will help you determine the correct answer in a Sentence Completion problem. These are neutral word values. Neutral words are words that convey neither loss nor gain, advantage nor disadvantage. Consider the previous example, once again:

Reminding his students that planning ahead would protect them from _____, Mr. McKenna proved to be a principal who understood the virtues of _____.

–/+(A) exigency . . . foresight

0/0(B) grades . . . examinations

–/+(C) poverty . . . promotion

–/–(D) deprivation . . . abstinence

–/+(E) turbulence . . . amelioration

Notice that grades . . . examinations (B) is rated as neutral. In fact, in this case, both words are considered of neutral value. This is because neither word conveys a usable value. Grades in and of themselves are not valued until a number is assigned. Examinations are not significant until a passing or failing value is implied or applied.

Neutral word values are significant because they are *never* the correct answer. Therefore, when you identify a neutral word or combination of words, you may eliminate that choice from your selection. You may eliminate a double-word answer even if only one of the words is obviously neutral.

Neutral words are rare, and you should be careful to measure their value before you make a choice. Here is another example from an exercise seen previously (Note: The answer choices have been altered.):

Dictators understand well how to centralize power, and that is why they combine a(n) _____ political process with military _____.

0/+(A) foreign . . . victory

0/+(B) electoral . . . escalation

0/+(C) agrarian . . . strategies

0/0(D) current . . . jobs

–/–(E) totalitarian . . . coercion

Here, current . . . jobs (D) is an obvious neutral word combination, conveying no positive or negative values. You may eliminate this choice immediately. There is no fixed list of words that may be considered neutral. Rather, you should determine *from the context* of a word problem whether you believe a word or word combination is of a neutral value. This ability will come with practice and a larger vocabulary. As before, the correct answer remains totalitarian . . . coercion (E).

Another way to determine the correct answer is by using etymology. Etymology is the study of the anatomy of words. The most important components of etymology on the SAT are prefixes and roots. SAT vocabulary is derived almost exclusively from the etymology of Greek and Latin word origins, and that is where you should concentrate your study. In this section, you will learn how to apply your knowledge of prefixes and roots to Sentence Completion problems.

Etymological skills will work well in conjunction with other techniques you have learned, including positive/negative word values. Furthermore, the technique of "scrolling" will help you understand how to expand your knowledge of etymology.

Scrolling is a process whereby you "scroll" through a list of known related words, roots, or prefixes to help you discover the meaning of a word. As an example, consider the common SAT word *apathy*. The prefix of apathy is *a*. This means *without*. To scroll this prefix, think of any other words that may begin with this prefix, such as *a*moral, *a*typical, *a*symmetrical. In each case, the meaning of the word is preceded by the meaning *without*.

At this point, you know that *apathy* means without something. Now try to scroll the root, *path*, which comes from the Greek word *pathos*. Words like pathetic, sympathy, antipathy, and empathy may come to mind. These words all have to do with feeling or sensing. In fact, that is what *pathos* means: feeling. So apathy means without feeling.

With this process you can often determine the fundamental meaning of a word or part of a word, and this may give you enough evidence with which to choose a correct answer. Consider the following familiar Level I example:

An expert skateboarder, Tom is truly _____; he smoothly blends timing with balance.

+(A) coordinated

+(B) erudite

–(C) a novice

+(D) supportive

–(E) casual

As you should remember, the correct answer is coordinated (A). The prefix of this word is co, meaning together, and the root is *order*. Something that is "ordered together" fits the context clue perfectly. Combining that with the knowledge that you are looking for a positive value word certifies coordinated (A) as the correct answer.

Here is a Level II example:

> Because they reject __–__ , conscientious objectors are given jobs in community work as a substitute for participation in the armed services.
>
> +(A) labor –(D) dictatorships
>
> –(B) belligerence –(E) poverty
>
> +(C) peace

From working with this example previously, you know that the correct answer is belligerence (B). The root of this word is *bellum*, Latin for war. Belligerence is an inclination toward war. Other words that may be scrolled from this are bellicose, belligerent, and antebellum, all of which have to do with war. Study your roots and prefixes well. A casual knowledge is not good enough. Another root, *bellis*, might be confused with *bellum*. *Bellis* means beauty. Is it logical that a conscientious objector would reject beauty? Know when to use which root and prefix. This ability will come with study and practice.

Here is a Level III example:

> When a physician describes an illness to a colleague, he must speak an __+__ language, using professional terms and concepts understood mostly by members of his profession.
>
> +(A) extrinsic +(D) esoteric
>
> –(B) inordinate –(E) abbreviated
>
> +(C) ambulatory

Recalling this example, you will remember that the context clue defines the missing word as one meaning language that involves a special group of people, that is, "inside information." The correct answer is esoteric (D). *Eso* is a prefix that means *inside*. The prefix of extrinsic (A) is *ex*, which means *out*, the opposite of the meaning you seek. Inordinate (B) means *not ordered*. In this case, the prefix *in* means *not*. This is Level III, so beware of easy assumptions! The root of ambulatory (C) is *ambulare*, which means *to walk*. Abbreviated (E) breaks down to *ab*, meaning *to*; and *brevis*, Latin for brief or short.

In many Level III words you may not be able to scroll or break down a word completely. However, often, as in the example above, a partial knowledge of the etymology may be enough to find the correct answer.

Now, take what you have learned and apply it to the questions appearing in the following drill. If you are unsure of an answer, refer back to the review material for help.

DRILL: SENTENCE COMPLETIONS

DIRECTIONS: Each sentence below has one or two blanks, each blank indicating that something has been omitted. Beneath the sentence are five lettered words or sets of words. Choose the word or set of words that BEST fits the meaning of the sentence as a whole.

EXAMPLE

Although the critics found the book _____, many of the readers found it rather _____.

(A) obnoxious . . . perfect
(B) spectacular . . . interesting
(C) boring . . . intriguing
(D) comical . . . persuasive
(E) popular . . . rare

Ⓐ Ⓑ ● Ⓓ Ⓔ

1. The problems of the homeless were so desperate that he felt a need to help _____ them.

(A) increase
(B) ameliorate
(C) authenticate
(D) collaborate
(E) justify

2. The activities of the business manager were so obviously unethical that the board had no choice but to _____ him.

(A) censure
(B) commend
(C) consecrate
(D) censor
(E) reiterate

3. _____ people often are taken in by _____ salespeople.

(A) Suave . . . futile
(B) Benevolent . . . inept
(C) Gullible . . . larcenous
(D) Erratic . . . passive
(E) Pious . . . obstinate

4. The speaker _____ the work of environmentalists as ineffective.

(A) dissented
(B) savored
(C) disparaged
(D) conceded
(E) tantalized

5. Her exceptionally well-written first novel was happily reviewed by the critics with _____.

(A) ennui
(B) pessimism
(C) remorse
(D) acclaim
(E) chagrin

6. That commentator never has anything good to say; every remark is _____.

(A) inept
(B) frivolous
(C) aberrant
(D) bombastic
(E) caustic

7. The principal's plan to gain students' and parents' cooperation by forming small work groups has worked well; it is both creative and _____.

(A) sagacious
(B) ignoble
(C) dissonant
(D) erroneous
(E) conventional

8. My boss is so arrogant that we're surprised when he _____ to speak to us in the cafeteria.

(A) forbears
(B) declines
(C) abhors
(D) delays
(E) deigns

9. Scientists and environmentalists are very concerned about the _____ of the ozone layer.

(A) depletion
(B) dissonance
(C) conglomeration
(D) defamation
(E) enhancement

10. Resolving racist attitudes seems to happen most successfully in communities where different ethnic groups _____ around issues of justice.

(A) educe
(B) collapse
(C) dissolve
(D) coalesce
(E) diverge

11. One obstacle to solving the mass transit problem is a _____ of funds to build and repair systems.

(A) euphony
(B) profusion
(C) periphery
(D) dearth
(E) vindication

12. Terrorists, who are usually _____, seldom can be dealt with _____.

(A) rabid . . . timorously
(B) unruly . . . fairly
(C) zealots . . . rationally
(D) blasphemous . . . tersely
(E) hedonistic . . . honestly

13. The candidate argued that it was _____ to _____ democracy and yet not vote.

(A) malicious . . . denounce
(B) lucrative . . . allocate
(C) commendable . . . delineate
(D) inarticulate . . . defend
(E) hypocritical . . . advocate

14. Of all the boring speeches I have ever heard, last night's address had to be the most _____ yet!

(A) arrogant
(B) insipid
(C) effervescent
(D) fervent
(E) indolent

15. Malcolm X was a _____ of Martin Luther King, Jr., yet he had a _____ different view of integration.

(A) disciple . . . reciprocally
(B) codependent . . . uniquely
(C) contemporary . . . radically
(D) fanatic . . . futilely
(E) biographer . . . unrealistically

16. In order to pass the hearing test, you have to be able to _____ high pitch tones from low pitch tones.

(A) surmise
(B) define
(C) discriminate
(D) document
(E) vindicate

17. The committee _____ carefully before making the final report; nevertheless, a minority report _____ its conclusions.

(A) deliberated . . . refuted
(B) analyzed . . . abased
(C) discerned . . . emulated
(D) gloated . . . implemented
(E) argued . . . accepted

18. In general, _____ behavior will bring rewards.

(A) languid
(B) rhetorical
(C) questionable
(D) disruptive
(E) exemplary

19. Colonial Americans, who had little extra money or leisure time, built simple and _____ homes.

(A) baroque
(B) unpretentious
(C) grandiose
(D) prodigious
(E) disreputable

20. Communist countries today are trying to _____ the ineffective economic policies of the past.

(A) ignore
(B) rectify
(C) reiterate
(D) condone
(E) provoke

21. Albert Einstein is the _____ of a genius.

(A) rebuttal
(B) digression
(C) antithesis
(D) mentor
(E) epitome

22. He was tempted to cheat but did not want to _____ his morals.

(A) obscure
(B) refute
(C) compromise
(D) concede
(E) succumb

23. Religious services have spiritual significance for those who are _____.

(A) fastidious
(B) pessimistic
(C) pragmatic
(D) pious
(E) phlegmatic

24. Her _____ remarks seemed innocent enough, but in reality they were _____.

(A) caustic . . . fallacious
(B) acrid . . . insipid
(C) magnanimous . . . affable
(D) innocuous . . . malicious
(E) ebullient . . . frivolous

25. The legend of Beowulf is a famous Norse _____ .

(A) saga
(B) reverie
(C) soliloquy
(D) utopia
(E) satire

26. An effective way to prevent the spread of infectious disease is to _____ the sick person.

(A) alleviate
(B) efface
(C) quarantine
(D) salvage
(E) absolve

27. In international business and politics, English is virtually a _____ language.

(A) mundane
(B) finite
(C) dead
(D) palpable
(E) universal

28. Spring break is a welcome _____ from _____ school work.

(A) dichotomy . . . garbled
(B) quandary . . . lax
(C) respite . . . arduous
(D) liaison . . . difficult
(E) zenith . . . phenomenal

29. If you can't find an original form, just prepare a reasonable _____.

(A) paradigm
(B) facsimile
(C) facade
(D) aberration
(E) equivocation

30. Their generous donation provided the _____ needed to raise the entire goal.

(A) catalyst
(B) alacrity
(C) duress
(D) ideology
(E) plethora

31. The protestors got a better response to their requests when they _____ their anger.

(A) invoked
(B) appeased
(C) condoned
(D) tempered
(E) censured

32. Carelessly dumping chemicals has created many _____ waste sites.

(A) choleric
(B) utopian
(C) torpid
(D) toxic
(E) vaunted

33. The investment counselor had a _____ reputation for purchasing companies and then stripping them of all the assets.

(A) potent
(B) commendable
(C) subtle
(D) notorious
(E) pervasive

34. Staying active is important for people of all ages, so that neither the brain nor the muscles _____.

(A) expand
(B) atrophy
(C) endure
(D) vacillate
(E) condescend

35. When a student consistently does not turn in homework, the teacher often _____ that the student is _____.

(A) implies . . . zealous
(B) concludes . . . ambitious
(C) assumes . . . depraved
(D) assures . . . taciturn
(E) infers . . . indolent

36. Looking through the photo album brought warm feelings of _____.

(A) remorse
(B) nostalgia
(C) complacence
(D) ambiguity
(E) deference

37. There is a marked _____ between the salaries of skilled and unskilled workers.

(A) disparity
(B) increase
(C) cohesion
(D) calamity
(E) amorphousness

38. Although the chairperson seemed to be neutral in her support of the plan, they suspected she had _____ motives.

(A) satirical
(B) palpable
(C) occult
(D) guileless
(E) ulterior

39. They were elated to learn that the salary increases were _____ to the beginning of the year.

(A) reciprocal
(B) recessive
(C) retroactive
(D) germane
(E) subsidiary

40. Scientists debate whether it is possible to even _____ exactly how life begins.

(A) fathom
(B) juxtapose
(C) gloat
(D) rebut
(E) suppress

SENTENCE COMPLETION DRILL

ANSWER KEY

Drill: Sentence Completions

1.	(B)	11.	(D)	21.	(E)	31.	(D)
2.	(A)	12.	(C)	22.	(C)	32.	(D)
3.	(C)	13.	(E)	23.	(D)	33.	(D)
4.	(C)	14.	(B)	24.	(D)	34.	(B)
5.	(D)	15.	(C)	25.	(A)	35.	(E)
6.	(E)	16.	(C)	26.	(C)	36.	(B)
7.	(A)	17.	(A)	27.	(E)	37.	(A)
8.	(E)	18.	(E)	28.	(C)	38.	(E)
9.	(A)	19.	(B)	29.	(B)	39.	(C)
10.	(D)	20.	(B)	30.	(A)	40.	(A)

CHAPTER 5

MASTERING READING COMPREHENSION QUESTIONS

MASTERING READING COMPREHENSION QUESTIONS

The role that Reading Comprehension plays in the SAT should be fully comprehended by anyone hoping to earn a top score. In the Critical Reading sections, Reading Comprehension questions outnumber Sentence Completions 5 to 2. Half of the more than one hundred verbal multiple-choice questions on the test measure your ability to extract meaning efficiently from prose passages. "Why," you must wonder, "would this much importance be attached to reading?" The reason is simple. Your ability to read at a strong pace while grasping a solid understanding of the material is a key factor in your high school performance and your potential college success. And your grasp of the SAT passages must go beyond simple comprehension. You'll need to be able to analyze and evaluate the passages and make inferences about the writers' meaning. Even your ability to understand vocabulary in context will come under scrutiny. "Can I meet the challenge?" you ask yourself. Yes, and preparation is the means!

MASTERING READING COMPREHENSION PASSAGES AND QUESTIONS

There are three Critical Reading sections in the test (two 25-minute sections and one 20-minute section). More than two-thirds of the questions in each of these sections are based on short and long passages (the rest are Sentence Completion questions). The short passages are about one hundred words long and are typically followed by two questions. The long passages fall into two categories: 400- to 550-word passages followed by 7 or 8 questions and 700- to 850-word passages followed by as many as 13 or 14 questions. Both the short and long passages may appear as paired passages—two passages addressing the same topic or theme from difference perspectives. Whatever the combination of short, long, or paired passages, they will form the basis for 17 to 20 Reading Comprehension questions per section.

The reading content of the passages will cover:

humanities (philosophy, the fine arts)

social sciences (psychology, archeology, anthropology, economics, political science, sociology, history)

natural sciences (biology, geology, astronomy, chemistry, physics)

narration (fiction, nonfiction).

Familiarize yourself with these basic departments of human expression and inquiry. As you read the sample passages, find out which interest you most. Because you can start anywhere in a given section, you may want to start with a reading passage that interests you and gives you the momentum needed to plow efficiently through the questions.

You will encounter four kinds of Reading Comprehension questions:

1. Synthesis/Analysis
2. Evaluation
3. Vocabulary-in-Context
4. Interpretation

Although you'll never be required to identify these question types, getting familiar with them as you review for the test and do practice questions will hone your reading skills and help make the test seem more manageable.

ABOUT THE DIRECTIONS

Make sure to study and learn the directions to save yourself time during the actual test. You should simply skim them when beginning the section. The directions will be similar to the following.

DIRECTIONS: Read each passage and answer the questions that follow. Each question will be based on the information stated or implied in the passage or its introduction.

A variation of these directions will be presented as follows for the double passage.

DIRECTIONS: Read the passages and answer the questions that follow. Each question will be based on the information stated or implied in the selections or their introductions, and may be based on the relationship between the passages.

IDENTIFYING PASSAGE AND QUESTION TYPES

Below are four short passages, each drawn from one of the content areas mentioned above (i.e., the humanities, social sciences, sciences, and narration). The questions that follow have been designed to illustrate the four kinds of Reading Comprehension questions. The explanations of correct answer choices will suggest strategies for identifying and approaching each question type.

An in-depth explanation of the four question types occurs after the sample short passages. Consult it as you try to identify the question types.

SHORT PASSAGE #1

1 A biologist has to look no further than cyanobacteria to find an ecological illustration of the dangers posed by too much of a good thing. Also known as blue-green algae, cyanobacteria helps keep the atmosphere
5 life-sustaining; through photosynthesis, it produces more oxygen than all land plants combined. Phytoplankton, a kind of cyanobacteria that floats on the surface of lakes and oceans, is the principle food source for many organisms. But too much blue-green algae can be toxic.
10 When an excess of nitrates allows it to grow unchecked, cyanobacteria can start to blanket lakes and ponds with a smelly blue-green film that kills off fish populations and even threatens human health.

The preceding passage is drawn from which of the following?

(A) Humanities, (B) Social Sciences, (C) Natural Sciences, or (D) Narration

1. Based on the passage, the writer would most likely characterize cyanobacteria as
   - (A) a necessary evil within the global environment.
   - (B) an important but potentially harmful organism.
   - (C) an insignificant contributor to the atmosphere.
   - (D) a powerful antidote to fresh and saltwater toxins.
   - (E) proof that too much of a good thing is not enough.

Question Type: Synthesis/Analysis

Explanation: Synthesis and analysis are complementary modes of thought. To analyze is to "break down" into parts; to synthesize is to "put together" to form a whole. Both the question stem and the answer choices indicate that the test taker is being asked to assess the author's general attitude toward cyanobacteria. We can only know how an author might be likely to "characterize" something (which would be an expression of attitude) if the passage provides enough clues about the author's attitude. These clues come in the form of specific word choices that can be identified (i.e., analyzed) within a text. Taken together—that is, synthesized—these word choices form an overall tone, which in turn can suggest a general attitude.

The author describes cyanobacteria as a "good thing" that "helps" make the air around us "life-sustaining" and that acts as the "principle food source" for many creatures. Together, the words in quotes suggest a positive attitude based on the conviction that blue-green algae has an important role to play. The author also describes the algae as potentially "toxic," "smelly," and capable of "killing" or "threatening." Again, the word choice establishes a tone—in this case, a tone that suggests the author's repugnance for the algae's "dark side." Choice (B) best captures the author's attitude toward cyanobacteria's positive and negative aspects. Choice (A) is incorrect because it disregards the author's positive attitude toward the algae. Choice (C) directly contradicts the author's assertion that the algae contributes more oxygen to the atmosphere than all other sources combined. Choice (D) is not correct because, although the notion of toxicity is introduced, cyanobateria is described as a potential toxin, not a neutralizer of toxins. Choice (E) is incorrect because it completely contradicts the passage's opening statement while seeming to rephrase it in simpler terms.

2. As used in the passage, the word "unchecked"(line 11) most nearly means
   (A) unobserved.
   (B) uncontrolled.
   (C) unexplained.
   (D) undisciplined.
   (E) unexpected.

Question Type: Vocabulary-in-Context

Explanation: All vocabulary-in-context questions have the same format. A word or phrase from the passage is isolated, and the test taker is asked to choose the best possible synonym based on how the word is used in the passage. The sentences immediately preceding and immediately following the sentence that contains the vocabulary word are usually all the "context" you'll need to determine the contextual meaning of the word.

In the passage, the author states that an "excess of nitrates" may cause the cyanobacteria to "blanket" the body of water it inhabits. The words "excess" and "blanket" both suggest an overabundance—a state of affairs in which too much of something is resulting in the water being nearly or totally covered by the algae. The implication is that the cyanobacteria had *not* been *controlled* by the usual environmental factors that keep the algae's growth *in check*. The correct answer is (B). Choice (A) is not correct because the algae's growth has clearly been observed by biologists. Choice (C) is incorrect because the passage explains why the algae grows—because of an "excess of nitrates." Choice (D) is not correct because "undisciplined" is a human characteristic, and nothing in the passage warrants personifying the bacteria. Choice (E) is not correct because the passage itself states the circumstances under which cyanobacteria is expected to grow. Indeed, the passage is all about scientific expectations concerning cyanobacteria.

SHORT PASSAGE #2

1 The ancient Egyptian word *ka* perfectly illustrates the hazards of interpreting another culture purely in terms of one's own. The ancient Egyptians used the word *ka* to refer to a fundamental part of the self that is distinct from
5 the physical body. Because the *ka* becomes a crucial factor in one's fate after death, it has often been translated as "soul" or "spirit." But this translation is misleading. After an individual's death, his or her *ka* required sustenance like a living person; it was not so much the essence
10 of the person who had died as it was a stand-in who could act on behalf of the deceased, whom death had incapacitated. The ka was more of a helper or guide who ministered to the real focus in the Egyptian conception of the afterlife: the mummified body.

The preceding passage is drawn from which of the following?

(A) Humanities, (B) Social Sciences, (C) Natural Sciences, or (D) Narration

3. The author assumes that the reader will recognize the "soul" as being each of the following EXCEPT

(A) a part of the self distinct from the body.

(B) a determining factor in life after death.

(C) an essential part of the self.

(D) an entity that does not require food.

(E) a substitute for the dead body.

Question Type: Evaluation

Explanation: Writers are human. Like all of us, they possess prejudices, make assumptions, and sometimes even argue in the face of reason. Sometimes a writer's assumptions are justified; other times they need to be critically evaluated. A good reader doesn't just get at the writer's meaning; he or she notes and passes judgment on the arguments and assumptions the writer makes in the process of getting that meaning across.

In the preceding passage, the writer attempts to convince us that "soul" does not accurately translate the Egyptian word "ka." The writer then proceeds to detail all the nuances of the word "ka." The writer assumes that we will recognize differences between the meaning of the word "ka" and the meaning of "soul." However, the writer never defines "soul." He or she assumes we already have a rough-and-ready definition in our minds. The evaluation question asks us to piece together the writer's implicit definition of "soul"—the one it is assumed we already know—and identify the one choice that does not constitute an assumption. Choice (A) is not correct because the writer does in fact imply that "soul" and "ka" both designate noncorporeal parts of the self. Choice (B) is not correct because the writer cites the importance of the "ka" in regards to the afterlife as the primary reason that the "ka" has been equated with "soul," implying that the "soul" is important in the same way. Choice (C) is incorrect because, in the portion of the passage where the writer is implicitly contrasting "ka" with "soul," the writer says that the "ka" was NOT the "essence of the person," implying that "soul" does designate "an essential part of the self." Choice (D) is not correct for the same reason that (C) is not correct. Again, in that portion of the passage where "ka" and "soul" are implicitly contrasted, the writer asserts that the "ka" required sustenance, implying that the soul does NOT need nourishment. The correct answer is (E). The writer describes the "ka" as a stand-in (i.e., substitute) when implicitly detailing differences between "ka" and "soul." The writer never assumes that the "soul" functions as a "substitute."

SHORT PASSAGE #3

1 In the past few years, the status of comic books has risen dramatically. Long dismissed as childish fare, comic books—or "graphic novels"—are now embraced as serious art. Poignant works such as Daniel Clowe's Ghost
5 world have begun to receive the critical attention once reserved for serious fiction and poetry. Of course, the newfound legitimacy of comic books is due in no small part to a sea of change in their content. In the world of comic

book art, the superhero battling evil has yielded to the
10 flawed everyman slogging through an existence strangely
like our own.

The preceding passage is drawn from which of the following?

(A) Humanities, (B) Social Sciences, (C) Natural Sciences, or (D) Narration

4. It can be inferred from the passage that serious comic book readers prefer
   (A) characters that resemble real people.
   (B) comic books that blend traditions.
   (C) stories in which the superhero loses.
   (D) works with a supernatural element.
   (E) graphic novels that are critically acclaimed.

Question Type: Interpretation

Explanation: In the section entitled "ABOUT THE QUESTIONS," it states that interpretation questions will ask you to "distinguish probable motivations and effects or actions not stated outright in the essay." In the preceding passage, we are told why comic books have begun to receive critical attention: it's because they've gone from being tales of superheroes to real-life narratives. But in this simple assertion of cause and effect, something crucial concerning the change in content is implied rather than stated: serious critics and readers prefer stories about real people. The correct answer is (A). Answer choice (B) is incorrect because the only implied comic book tradition in the passage is the depiction of superheroes, which has been superceded by, not blended with, the real-life content. Answer choice (C) is incorrect because no stories about superheroes losing are mentioned. The passage does state that the "superhero battling evil has yielded to the flawed everyman," which suggests a plot in which a superhero is overthrown. But the writer is being figurative; he or she is not describing the plot of a specific comic book but the evolution of comic books *as though* it were a comic book with a doomed hero. Answer choice (D) completely contradicts the main thrust of the passage, which asserts readers' preference for realism over supernatural content. Choice (E) ignores the basic cause-and-effect argument of the passage: the new content resulted in the new-found legitimacy. Of course, critical acclaim often does generate interest. But, whereas nothing in the passage rules out this possibility, it never implies it.

Now that we have illustrated the four questions types and the kinds of reasoning that might go into solving them, apply what you have learned to the following passage and two questions (most short passages are followed by two questions).

SHORT PASSAGE # 4

1 Up in her bedroom window Sally Happer rested her
nineteen-year-old chin on the sill and watched Clark Dar-
row's ancient Ford turn into the driveway. Clark laboriously
climbed the drive's gentle incline, the wheels squeaking
5 indignantly, and then with a terrifying expression he gave
the steering wheel a final wrench and deposited self and
car approximately in front of the Happer steps. There was
a plaintive heaving sound, a death-rattle, followed by a

short silence; then the air was rent by a startling whistle.
10 Sally gazed down sleepily. She started to yawn, but finding this quite impossible unless she raised her chin from the window-sill, changed her mind and continued silently to regard the car, whose owner sat at attention as he waited for an answer to his signal.

The preceding passage is drawn from which of the following?

(A) Humanities, (B) Social Sciences, (C) Natural Sciences, or (D) Narration

5. Which of the following reactions does Clark's arrival elicit from Sally?
   (A) Complete indifference
   (B) Mild disapproval
   (C) Feigned boredom
   (D) Sudden interest
   (E) Sleepy attention

6. The wheels of Clark's car "squeak indignantly" because
   (A) Sally's driveway is steep.
   (B) Clark is a poor driver.
   (C) Clark's car is decrepit.
   (D) Sally ignores the whistle.
   (E) the road is deeply rutted.

ABOUT THE QUESTIONS

As previously mentioned, there are four major question types that appear in the Critical Reading section. The following explains what these questions will cover.

Question Type 1: Synthesis/Analysis

Synthesis/analysis questions deal with the structure of the passage and how one part relates to another part or to the text as a whole. These questions may ask you to look at passage details and, from them, point out general themes or concepts. They might ask you to trace problems, causes, effects, and solutions or to understand the points of an argument or persuasive passage. They might ask you to compare or contrast different aspects of the passage. Synthesis/analysis questions may also involve inferences, asking you to decide what the details of the passage imply about the author's general tone or attitude. Key terms in synthesis/analysis questions are example, difference, general, compare, contrast, cause, effect, and result.

Question Type 2: Evaluation

Evaluation questions involve judgments about the worth of the essay as a whole. You may be asked to consider concepts the author assumes rather than factually proves and to judge whether or not the author presents a logically consistent case. Does he/she prove the points through generalization, citing an authority, use of example, implication, personal experience, or factual data? You'll need to be able to distinguish the supportive bases for the argumentative theme. Almost as a book reviewer, you'll also be asked to pinpoint the author's writing techniques. What is the style, the tone? Who is the intended audience? How might the author's points relate to information outside the essay itself? Key terms you'll often see in evaluation questions and answer choices are generalization, implication, and support.

Question Type 3: Vocabulary-in-Context

Vocabulary-in-context questions occur in several formats. You'll be given easy words with challenging choices or the reverse. You'll need to know multiple meanings of words. You'll encounter difficult words and difficult choices. In some cases, your knowledge of prefixes, roots, and suffixes will gain you clear advantage. In addition, connotations will be the means of deciding, in some cases, which answer is the best. Of course, how the term works in the textual context is the key to the issue.

Question Type 4: Interpretation

Interpretation questions ask you to decide on a valid explanation or clarification of the author's points. Based on the text, you'll be asked to distinguish probable motivations and effects or actions not stated outright in the essay. Furthermore, you'll need to be familiar with clichés, euphemisms, catch phrases, colloquialisms, metaphors, and similes and be able to explain them in straightforward language. Interpretation question stems usually have a word or phrase enclosed in quotation marks.

Keep in mind that being able to categorize accurately is not of prime importance. What is important, however, is that you are familiar with all the types of information you will be asked and that you have a set of basic strategies to use when answering questions. The remainder of this review will give you these skills.

MASTERING READING COMPREHENSION QUESTIONS

Of course, there is more to doing well on Reading Comprehension than identifying question and passage types. One should have a systematic plan for attacking the passages and questions in an efficient manner. You should follow these steps as you begin each critical reading passage. They will act as a guide when answering the questions.

Before you address reading comprehension, answer all sentence completions within the given verbal section. You can answer more questions per minute in these short sections than in the reading, and because all answers are credited equally, you'll get the most for your time here.

Now, find the Reading Comprehension passage(s). If more than one passage appears, give each a brief overview. Attack the easiest and most interesting passages first. Reading Comprehension passages are not automatically presented in the order of least-to-most difficult. The difficulty or ease of a reading selection is an individual matter, determined by the reader's own specific interests and past experience, so what you might consider easy, someone else might consider hard, and vice versa. Again, time is an issue, so you need to begin with something you can quickly understand in order to get to the questions, where the payoff lies.

First, read the question stems following the passage, making sure to block out the answer choices with your free hand. (You don't want to be misled by incorrect choices.)

In question stems, underline key words, phrases, and dates. For example:

In line 27, "stand" means:

From 1776 to 1812, King George did:

Lincoln was similar to Pericles in that:

The act of underlining takes little time and will force you to focus on the main ideas in the questions and then in the essays.

You will notice that questions often note a line number for reference. Place a small mark by the appropriate lines in the essay itself to remind yourself to read those parts very carefully. You'll still have to refer to these lines when answering the questions, but you'll be able to find them quickly.

STEP 3A *If you are addressing a short passage, read it with an eye toward formulating its main idea or concept in a concise sentence. A firm grasp of the passage's focus and purpose will make most questions about the passage seem straightforward.*

Short Passage #3 in the preceding "Identifying Passage and Question Types" could be summarized thus:

> Today's comic books are more respected, largely because they portray real people instead of superheroes.

See how readily you can provide one-sentence summaries for the other Short Passages in this review section.

STEP 3B *If you are addressing a long passage and it is not divided into paragraphs, read the first 10 lines. If the passage is divided into manageable paragraphs, read the first paragraph. Make sure to read at a moderate pace because fast skimming will not be sufficient for comprehension and slow, forced reading will take too much time and yield too little understanding of the overall passage.*

In the margin of your test booklet, using two or three words, note the main point of the paragraph/section. Don't labor long over the exact wording. Underline key terms, phrases, or ideas when you notice them. If a sentence is particularly difficult, don't spend too much time trying to figure it out. Bracket it, though, for easy reference in the remote instance that it might serve as the basis for a question.

You should proceed through each paragraph/section in a similar manner. Don't read the whole passage with the intention of going back and filling in the main points. Read carefully and consistently, annotating and underlining to keep your mind on the context.

Upon finishing the entire passage, quickly review your notes in the margin. They should give you main ideas and passage structure (chronological, cause and effect, process, comparison/contrast). Ask yourself what the author's attitude is toward his/her subject. What might you infer from the selection? What might the author say next? Some of these questions may appear, and you'll be immediately prepared to answer.

Start with the first question and work through to the last question. The order in which the questions are presented follows the order of the passage, so going for the "easy" questions first rather than answering the questions consecutively will cost you valuable time in searching and backtracking.

Be sure to block the answer choices for each question before you read the question itself. Again, you don't want to be misled.

If a line number is mentioned, quickly reread that section. In addition, circle your own answer to the question *before* viewing the choices. Then, carefully examine each answer choice, eliminating those that are obviously incorrect. If you find a close match to your own answer, don't assume that it is the best answer; an even better one may be among the last choices. Remember, in the SAT only one answer is correct, and it is the *best* one, not simply one that will work.

Once you've proceeded through all the choices, eliminating incorrect answers as you go, choose from among those remaining. If the choice is not clear, reread the question stem and the referenced passage lines to seek tone or content you might have missed. If the answer now is not readily obvious and you have reduced your choices by eliminating at least one, then simply choose one of the remaining and proceed to the next question. Place a small mark in your test booklet to remind you, should you have time at the end of this test section, to review the question and seek a more accurate answer.

POINTS TO REMEMBER

- Do not spend too much time answering any one question.
- Vocabulary plays a large part in successful critical reading. As a long-term approach to improving your ability and therefore your test scores, read as much as you can of any type of material. Your speed, comprehension, and vocabulary will grow.
- Be an engaged reader. Don't let your mind wander. Focus through annotation and key terms.
- Time is an important factor on the SAT. Therefore, the rate at which you are reading is very important. If you are concerned that you may be reading too slow, try to compete with yourself. For example, if you are reading at 120 words per minute, try to improve your speed to 250 words per minute (without decreasing your understanding). Remember that improving reading speed is not an end in itself. Improved comprehension with fewer regressions must accompany this speed increase. Make sure to read, read, read. The more you read, the more you will sharpen your skills.

APPLYING THE STEPS TO A LONG PASSAGE

Perhaps even more than the short passages, long passages (and the questions that follow them) need to be processed in an efficient, time-saving way. It's easy to get bogged down in a lengthy prose passage when you're under the gun. Apply the steps listed previously to the passage that follows. In addition, review the question types illustrated earlier and use your familiarity with them to help you answer the questions.

The following article was written by a physical chemist and recounts the conflict between volcanic matter in the atmosphere and airplane windows. It was published in a scientific periodical in 1989.

1 Several years ago the airlines discovered a new kind of problem—a window problem. The acrylic windows on some of their 747s were getting hazy and dirty-looking. Suspicious travelers thought the airlines might have stopped cleaning them, but the windows were not dirty; they were 5 inexplicably deteriorating within as little as 390 hours of flight time, even though they were supposed to last for five to ten years. Boeing looked into it.

At first the company thought the culprit might be one well known in modern technology, the component supplier who changes materials with- 10 out telling the customer. Boeing quickly learned this was not the case, so there followed an extensive investigation that eventually brought in the Air Transport Association, geologists, and specialists in upper-atmosphere chemistry, and the explanation turned out to be not nearly so mundane. Indeed, it began to look like a grand reenactment of an ancient Aztec 15 myth: the struggle between the eagle and the serpent, which is depicted on the Mexican flag.

The serpent in this case is an angry Mexican volcano, El Chichon. Like its reptilian counterpart, it knows how to spit venom at the eyes of its adversary. In March and April of 1982 the volcano, in an unusual eruption 20 pattern, ejected millions of tons of sulfur-rich material directly into the stratosphere. In less than a year, a stratospheric cloud had blanketed the entire Northern Hemisphere. Soon the photochemistry of the upper atmosphere converted much of the sulfur into tiny droplets of concentrated sulfuric acid.

25 The eagle in the story is the 747, poking occasionally into the lower part of the stratosphere in hundreds of passenger flights daily. Its two hundred windows are made from an acrylic polymer, which makes beautifully clear, strong windows but was never intended to withstand attack by strong acids.

30 The stratosphere is very different from our familiar troposphere environment. Down here the air is humid, with a lot of vertical convection to carry things up and down; the stratosphere is bone-dry, home to the continent-striding jet stream, with unceasing horizontal winds at an average of 120 miles per hour. A mist of acid droplets accumulated gradually near the 35 lower edge of the stratosphere, settling there at a thickness of about a mile a year, was able to wait for planes to come along.

As for sulfuric acid, most people know only the relatively benign liquid in a car battery: 80 percent water and 20 percent acid. The stratosphere dehydrated the sulfuric acid into a persistent, corrosive mist 75 40 percent pure acid, an extremely aggressive liquid. Every time the 747 poked into the stratosphere—on almost every long flight—acid droplets struck the windows and began to react with their outer surface, causing it

to swell. This built up stresses between the softened outer layer and the underlying material. Finally, parallel hairline cracks developed, creating
45 the hazy appearance. The hazing was sped up by the mechanical stresses always present in the windows of a pressurized cabin.

The airlines suffered through more than a year of window replacements before the acid cloud finally dissipated. Ultimately the drops reached the lower edge of the stratosphere, were carried away into the
50 lower atmosphere, and finally came down in the rain. In the meantime, more resistant window materials and coatings were developed. (As for the man-made sulfur dioxide that causes acid rain, it never gets concentrated enough to attack the window material. El Chichon was unusual in its ejection of sulfur directly into the stratosphere, and the 747 is unusual in
55 its frequent entrance into the stratosphere.)

As for the designers of those windows, it is hard to avoid the conclusion that a perfectly adequate engineering design was defeated by bad luck. After all, this was the only time since the invention of the airplane that there were acid droplets of this concentration in the upper atmosphere.
60 But reliability engineers, an eminently rational breed, are very uncomfortable when asked to talk about luck. In principle it should be possible to anticipate events, and the failure to do so somehow seems like a professional failure. The cosmos of the engineer has no room for poltergeists, demons, or other mystic elements. But might it accommodate the inexo-
65 rable scenario of an ancient Aztec myth?

1. Initially the hazy windows were thought by the company to be a result of
   (A) small particles of volcanic glass abrading their surfaces.
   (B) substandard window material substituted by the parts supplier.
   (C) ineffectual cleaning products used by the maintenance crew.
   (D) build-up of the man-made sulfur dioxide that also causes acid rain.
   (E) humidity.

2. When first seeking a reason for the abraded windows, both the passengers and Boeing management exhibited attitudes of
   (A) disbelief.
   (B) optimism.
   (C) cynicism.
   (D) pacifism.
   (E) disregard.

3. In line 13, "mundane" means
   (A) simple.
   (B) complicated.
   (C) far-reaching.
   (D) ordinary.
   (E) important.

4. In what ways is El Chichon like the serpent on the Mexican flag, knowing how to "spit venom at the eyes of its adversary" (lines 18–19)?

   (A) It seeks to poison its adversary with its bite.
   (B) It carefully plans its attack on an awaited intruder.
   (C) It ejects tons of destructive sulfuric acid to damage jet windows.
   (D) It angrily blankets the Northern Hemisphere with sulfuric acid.
   (E) It protects itself with the acid rain it produces.

5. The term "photochemistry" in line 22 refers to a chemical change caused by

   (A) the proximity of the sun.
   (B) the drop in temperature at stratospheric altitudes.
   (C) the jet stream's "unceasing horizontal winds."
   (D) the vertical convection of the troposphere.
   (E) the amount of sulfur present in the atmosphere.

6. Unlike the troposphere, the stratosphere

   (A) is extremely humid because it is home to the jet stream.
   (B) contains primarily vertical convections to cause air particles to rise and fall rapidly.
   (C) is approximately one mile thick.
   (D) contains powerful horizontal winds resulting in an excessively dry atmosphere.
   (E) contains very little wind activity.

7. In line 40, "aggressive" means

   (A) exasperating.
   (B) enterprising.
   (C) prone to attack.
   (D) assertive.
   (E) surprising.

8. As the eagle triumphed over the serpent in the Mexican flag,

   (A) El Chichon triumphed over the plane as the 747s had to change their flight altitudes.
   (B) The newly designed window material deflected the damaging acid droplets.
   (C) the 747 was able to fly unchallenged by acid droplets a year later as they drifted away to the lower atmosphere.
   (D) the reliability engineers are now prepared for any run of "bad luck" that may approach their aircraft.
   (E) the component supplier of the windows changed materials without telling the customers.

9. The reliability engineers are typified as people who

   (A) are uncomfortable considering natural disasters.
   (B) believe that all events are predictable through scientific methodology.
   (C) accept luck as an inevitable and unpredictable part of life.
   (D) easily accept their failure to predict and protect against nature's surprises.
   (E) are extremely irrational and are comfortable speaking about luck.

The questions following the passage that you just read are typical of those in the Critical Reading section. After carefully reading the passage, you can begin to answer these questions. Let's look again at the questions.

1. Initially the hazy windows were thought by the company to be a result of

    (A) small particles of volcanic glass abrading their surfaces.

    (B) substandard window material substituted by the parts supplier.

    (C) ineffectual cleaning products used by the maintenance crew.

    (D) a build-up of the man-made sulfur dioxide that also causes acid rain.

    (E) the humidity.

As you read the question stem, blocking the answer choices, you'll note the key term "result," which should alert you to the question category *synthesis/analysis*. Argument structure is the focus here. Ask yourself what part of the argument is being questioned: cause, problem, result, or solution. Careful reading of the stem and perhaps mental rewording to "_____ caused hazy windows" reveals cause is the issue. Once you're clear on the stem, proceed to the choices.

The word "initially" clues you in to the fact that the correct answer should be the first cause considered. Answer choice (B) is the correct response, as "substandard window material" was the *company's* first (initial) culprit, as explained in the first sentence of the second paragraph. They had no hint of (A) a volcanic eruption's ability to cause such damage. In addition, they were not concerned, as were the *passengers*, that (C) the windows were not properly cleaned. Answer (D) is not correct because scientists had yet to consider testing the atmosphere. Along the same lines, answer choice (E) is incorrect.

2. When first seeking a reason for the abraded windows, both the passengers and Boeing management exhibited attitudes of

    (A) disbelief.

    (B) optimism.

    (C) cynicism.

    (D) pacifism.

    (E) disregard.

As you read the stem before viewing the choices, you'll know you're being asked to judge or *evaluate* the tone of a passage. The tone is not stated outright, so you'll need to rely on your perception as you re-read that section, if necessary. Remember, questions follow the order of the passage, so you know to look after the initial company reaction to the windows, but not far after, as many more questions are to follow. Now, formulate your own word for the attitude of the passengers and employees. "Skepticism" or "criticism" work well. If you can't come up with a term, at least note if the tone is negative or positive. In this case, negative is clearly indicated as the passengers are distrustful of the maintenance crew and the company mistrusts the window supplier. Proceed to each choice, seeking the closest match to your term and/or eliminating words with positive connotations.

Choice (C) is correct because "cynicism" best describes the skepticism and distrust with which the passengers view the cleaning company and the parts suppliers. Choice (A) is not correct because both Boeing and the passengers believed the windows were hazy, they just didn't know why. Choice (B) is not correct because people were somewhat agitated that the windows were hazy—certainly not "optimistic." Choice (D), "pacifism," has a rather positive connotation, which the tone of the section does not. Choice (E) is incorrect because the people involved took notice of the situation and did not disregard it. In addition to the ability to discern tone, of course, your vocabulary knowledge is being tested. "Cynicism," should you be unsure of the term, can be viewed in its root, "cynic," which may trigger you to remember that it is negative, and therefore, appropriate in tone.

3. In line 13, "mundane" means
   (A) simple.
   (B) complicated.
   (C) far-reaching.
   (D) ordinary.
   (E) important.

This question obviously tests *vocabulary-in-context*. Your strategy here should be to quickly view line 13 to confirm usage, block answer choices while devising your own synonym for "mundane," perhaps "common," and then viewing each choice separately, looking for the closest match. Although you might not be familiar with "mundane," the choices are all relatively simple terms. Look for contextual clues in the passage if you can't define the term outright. Whereas the "component supplies" explanation is "mundane," the Aztec myth is not. Perhaps you could then look for an opposite of mythical; "real" or "down-to-earth" comes to mind.

Choice (D), "ordinary," fits best as it is clearly the opposite of the extraordinary Aztec myth of the serpent and the eagle, which is not as common as a supplier switching materials. Choice (A), "simple," works contextually but not as an accurate synonym for the word "mundane"; it does not deal with "mundane's" "down-to-earth" definition. Choice (B), "complicated," is inaccurate because the parts switch is anything but complicated. Choice (C), "far-reaching," is not better because it would apply to the myth rather than the common, everyday action of switching parts. Choice (E), "important," does not work either because the explanation was an integral part of solving the problem. If you eliminated (B), (C), and (E) because of contextual inappropriateness, you would be left with "ordinary" and "simple." A quick rereading of the section, then, should clarify the better choice. But, if the rereading did not clarify the better choice, your strategy would be to choose one answer, place a small mark in the booklet, and proceed to the next question. If time is left at the end of the test, you could then review your answer choice.

4. In what ways is El Chichon like the serpent on the Mexican flag, knowing how to "spit venom at the eyes of its adversary" (lines 18–19)?
   (A) It seeks to poison its adversary with its bite.
   (B) It carefully plans its attack on an awaited intruder.
   (C) It ejects tons of destructive sulfuric acid to damage jet windows.
   (D) It angrily blankets the Northern Hemisphere with sulfuric acid.
   (E) It protects itself with the acid rain it produces.

As you view the question, note the word "like" indicates a comparison is being made. The quoted simile forms the comparative basis of the question, and you must *interpret* that phrase with respect to the actual process. You must carefully seek to duplicate the tenor of the terms, coming close to the spitting action in which a harmful substance is expelled in the direction of an object similar to the eyes of an opponent. Look for key words when comparing images. "Spit," "venom," "eyes," and "adversary" are these keys.

In choice (C), the verb that is most similar to the serpent's "spitting" venom is the sulfuric acid "ejected" from the Mexican volcano, El Chichon. Also, the jet windows most closely resemble the "eyes of the adversary" that are struck by El Chichon. Being a volcano, El Chichon is certainly incapable of injecting poison into an adversary, as in choice (A), or planning an attack on an intruder, as in choice (B). In choice (D), although the volcano does indeed "blanket the Northern Hemisphere" with sulfuric acid, this image does not coincide with the "spitting" image of the serpent. Finally, in choice (E), although a volcano can indirectly cause acid rain, it cannot produce acid rain on its own and then spew it out into the atmosphere.

5. The term "photochemistry" in line 22 refers to a chemical change caused by

(A) the proximity of the sun.

(B) the drop in temperature at stratospheric altitudes.

(C) the jet stream's "unceasing horizontal winds."

(D) the vertical convection of the troposphere.

(E) the amount of sulfur present in the atmosphere.

Even if you are unfamiliar with the term "photochemistry," you probably know its root or its prefix. Clearly, this question fits in the *vocabulary-in-context* mode. Your first step may be a quick reference to line 22. If you don't know the term, context may provide you a clue. The conversion of sulfur-rich *upper* atmosphere into droplets may help. If context does not yield information, look at the term "photochemistry" itself. "Photo" has to do with light or sun, as in photosynthesis. Chemistry deals with substance composition and change. Knowing these two parts can take you a long way toward a correct answer.

Answer choice (A) is the correct response, as the light of the sun closely compares with the prefix "photo." Although choice (B), "the drop in temperature," might lead you to associate the droplet formation with condensation, light is not a factor here, nor is it in choice (C), "the jet stream's winds"; choice (D), "the vertical convection"; or choice (E), "the amount of sulfur present."

6. Unlike the troposphere, the stratosphere

(A) is extremely humid because it is home to the jet stream.

(B) contains primarily vertical convections to cause air particles to rise and fall rapidly.

(C) is approximately one mile thick.

(D) contains powerful horizontal winds resulting in an excessively dry atmosphere.

(E) contains very little wind activity.

"Unlike" should immediately alert you to a *synthesis/analysis* question asking you to contrast specific parts of the text. Your margin notes should take you right to the section contrasting the atmospheres. Quickly scan it before considering the answers. Usually you won't remember this broad type of comparison from your first passage reading. Don't spend much time, though, on the scan before beginning to answer because time is still a factor.

This question is tricky because all the answer choices contain key elements/phrases in the passage, but again, a quick, careful scan will yield results. Answer (D) proves best because the "horizontal winds" dry the air of the stratosphere. Choices (A), (B), and (E) are all characteristic of the troposphere. Choice (C) is incorrect because the acid droplets accumulate at the rate of one mile per year within the much larger stratosphere. As you answer such questions, remember to eliminate incorrect choices as you go; don't be misled by what seems familiar yet isn't accurate—read all the answer choices.

7. In line 40, "aggressive" means

(A) exasperating. (D) assertive.

(B) enterprising. (E) surprising.

(C) prone to attack.

Another *vocabulary-in-context* surfaces here; but this time, the word is probably familiar to you. Again, before forming a synonym, quickly refer to the line number, aware that perhaps a secondary meaning is appropriate as the term already is a familiar one. Upon reading the line, you'll note "persistent" and "corrosive," both strong terms, the latter being quite negative in its destruction. Now, form an appropriate synonym for aggressive, one that has a negative connotation. "Hostile" might come to mind. You are ready at this point to view all choices for a match.

Using your vocabulary knowledge, you can answer this question. "Hostile" most closely resembles choice (C), "prone to attack," and is therefore the correct response. Choice (A), "exasperating," or irritating, is too weak a term, whereas choices (B), "enterprising," and (D), "assertive," are too positive. Choice (E), "surprising," is not a synonym for "aggressive."

8. As the eagle triumphed over the serpent in the Mexican flag,

(A) El Chichon triumphed over the plane as the 747s had to change their flight altitudes.

(B) the newly designed window material deflected the damaging acid droplets.

(C) the 747 was able to fly unchallenged by acid droplets a year later as they drifted away to the lower atmosphere.

(D) the reliability engineers are now prepared for any run of "bad luck" that may approach their aircraft.

(E) the component supplier of the windows changed materials without telling the customer.

This question asks you to compare the eagle's triumph over the serpent to another part of the text. "As" often signals comparative relationships, so you are forewarned of the *synthesis/analysis* question. You are also dealing again with a simile, so, of course, the question can also be categorized as *interpretation.* The eagle-serpent issue is a major theme in the text. You are being asked, as you will soon discover in the answer choices, what this general theme is. Look at the stem keys: eagle, triumphed, and serpent. Ask yourself to what each corresponds. You'll arrive at the eagle and the 747, some sort of victory, and the volcano or its sulfur. Now that you've formed that corresponding image in your own mind, you're ready to view the choices.

Choice (C) is the correct choice because we know the statement "the 747 was able to fly unchallenged . . . " to be true. Not only do the remaining choices fail to reflect the eagle-triumphs-over-serpent image but also choice (A) is inaccurate because the 747 did not "change its flight altitudes." In choice (B), the windows did not deflect "the damaging acid droplets." Furthermore, in choice (D), "the reliability engineers" cannot be correct because they cannot possibly predict the future and, therefore, cannot anticipate what could go wrong in the future. Finally, we know that in (E), the window materials were never changed.

9. The reliability engineers are typified as people who

(A) are uncomfortable considering natural disasters.

(B) believe that all events are predictable through scientific methodology.

(C) accept luck as an inevitable and unpredictable part of life.

(D) easily accept their failure to predict and protect against nature's surprises.

(E) are extremely irrational and are comfortable speaking about luck.

When the question involves such terms as type, kind, example, or typified, be aware of possible *synthesis/analysis* or *interpretation* issues. Here the question deals with implications: what the author means but doesn't state outright. Types can also lead you to situations that ask you to make an unstated generalization based on specifically stated details. In fact, this question could even be categorized as *evaluation* because specific detail to generalization is a type of argument/essay structure. In any case, before viewing the answer choices, ask yourself what general traits the reliability engineers portray. You may need to check back in the text for typical characteristics. You'll find the engineers to be rational unbelievers in luck. These key characteristics will help you to make a step toward a correct answer.

Choice (B) is the correct answer because the passage specifically states that the reliability engineers "are very uncomfortable when asked to talk about luck" and believe "it should be possible to anticipate events" scientifically. The engineers might be uncomfortable, as in choice (A), but this is not a main concern in the passage. Choice (C) is obviously incorrect, because the engineers do not believe in luck at all, and choice (D) is not correct because "professional failure" is certainly unacceptable to these scientists. There is no indication in the passage that (E) the scientists are "irrational and are comfortable speaking about luck."

The following drill should be used to test what you have just learned. Read the passages and answer the questions. If you are unsure of an answer, refer back to the review for help.

DRILL: READING COMPREHENSION—SHORT PASSAGES

DIRECTIONS: Read each passage and answer the questions that follow. Each question will be based on information stated or implied in the passages.

Questions 1–2 are based on the following passage.

1 As an animal class, insects owe their considerable success to a number of factors. Compared to many other kinds of life, modern insects are small—between 1.5 to 50 millimeters in length. Being small helps
5 them to avoid predators; it also makes their food requirement modest, practically eliminating the threat of starvation. As one of the few classes of animal that can fly, insects can fly toward food sources or away from predators. Finally, insects are great adaptors, as
10 their presence in virtually all land and water environments suggests. From snow-capped mountains to deep-sea thermal vents, insects have managed to thrive wherever they find (or take) themselves.

1. The author would most likely attribute the wide distribution of human beings throughout the world to their

   (A) mastery of flight.

   (B) modest food needs.

   (C) relatively small size.

   (D) ability to adapt.

   (E) pursuit of knowledge.

2. As used in the passage, modest (line 6) most nearly means

   (A) shy.

   (B) gradual.

   (C) unpretentious.

   (D) moderate.

   (E) simple.

Questions 3–4 are based on the following passage.

1 From the hour Catherine came downstairs until the hour she went to bed, the family had not a minute's security that she would not be in mischief. Her spirits were always at high-water mark, her tongue always going—
5 singing, laughing, and plaguing everyone who would not do the same. A wild, wicked slip she was—but she had the bonniest eye, the sweetest smile, and lightest foot in the parish: and none could believe she meant any harm; for when once she made you cry in good
10 earnest, it seldom happened that she would not keep you company, and oblige you to be quiet that you might comfort her.

3. It can be inferred from the passage that Catherine's emotional state was.

   (A) characterized by highs and lows.

   (B) consistent throughout the day.

   (C) readily affected by criticism.

   (D) indifferent to others' suffering.

   (E) the result of emotional abuse.

4. The speaker would most likely describe others' attitude toward Cathy as

   (A) exasperated.

   (B) contemptuous.

   (C) resentful.

   (D) indebted.

   (E) ambivalent.

Questions 5–8 are based on the following passages.

Passage 1

1 The pseudo-science of astrology—not to be confused with the genuine science of astronomy—has been a bane of scientific thinking for untold generations. In the guise of a true science, complete with weighty
5 books and complex chartings of planetary positions, astrology claims to be able to explain individual personality. No scientific study has ever been able to validate this claim. Indeed, attempts to test astrology's hypothesis have revealed just how nebulous that hypothesis
10 is. The claim that being born under a particular sign makes an individual "creative" or "goal-oriented" can't really be tested, since the presence of such qualities in an individual is largely a matter of interpretation as well as degree.

Passage 2

1 In their understandable zeal to show that astrology is, indeed, a pseudo-science, critics of astrology have perhaps too vehemently dismissed the whole astrological project. Many early cultures independently developed
5 sophisticated astrological systems that became the precursors of astronomy—indeed, for scientific investigation itself. Enticed by the predictive claims of astrology, Sir Isaac Newton studied geometry to learn how to cast horoscopes. This mathematical study would eventually
10 lead to some of the greatest scientific discoveries of any age. Even psychology owes something to astrology and its emphasis on individual personality.

5. As used in Passage 1, "guise" (line 4) most nearly means
   - (A) appearance.
   - (B) fashion.
   - (C) attire.
   - (D) assistance.
   - (E) need.

6. The author of Passage 2 would most likely agree with what about Sir Isaac Newton?
   - (A) He was extremely skeptical about astrology's claims.
   - (B) He saw that astrology was a precursor of astronomy.
   - (C) He was motivated by astrology to further his studies.
   - (D) He sought to discredit claims of astrology.
   - (E) He didn't consider geometry a branch of mathematics.

7. Based on the passages, the authors of Passage 1 and 2 most likely agree that astrology
   (A) was a forerunner of astronomy.
   (B) has falsely claimed to be a science.
   (C) was advanced psychological thinking.
   (D) has been confused with astronomy.
   (E) has inspired scientific investigations.

8. The author of passage 2 would most likely react to the statement in passage 1 that "astrology has been the bane of scientific thinking" by pointing out that
   (A) not all of astrology's claims have been disproved.
   (B) astrology has been a source of inspiration for science.
   (C) even genuine science has made unclear hypotheses.
   (D) science has been equally detrimental to astrology.
   (E) even Newton recognized the limitations of science.

Questions 9–10 are based on the following passage.

1 Fred liked to speculate, with another's money as well as his own, and he enjoyed doing so not least because he wanted the money and hoped to win. But he was not a gambler; he had not that specific disease in which the
5 suspension of the whole nervous system on a chance or risk becomes as necessary as whisky to the drunkard; he had only the tendency toward that diffusive form of gambling which has no alcoholic intensity but is carried on with the healthiest chyle-fed blood, keeping up a joyous
10 imaginative activity which fashions events according to desire, and having no fears about its own weather, only sees the advantage there must be to others in going aboard with it.

9. As used in the passage, "speculate" (line 1) most nearly means
   (A) gamble.
   (B) consider.
   (C) meditate.
   (D) ponder.
   (E) inspect.

10. The author of the passage does not consider Fred to be a true gambler because Fred
   (A) genuinely enjoyed winning.
   (B) was not addicted to risk-taking.
   (C) gambled only to make a profit.
   (D) didn't risk other's money.
   (E) speculated with a level head.

Questions 11–12 are based on the following passage.

1 An underappreciated characteristic of ancient Greek culture was its emphasis on the *agon*, or competition. The pursuit of "the good," for which the Greeks are well known, is ultimately a search for "the best," which the
5 Greeks sought to identify and promote by introducing contests and competition in virtually every aspect of public life. The ancient Olympics epitomize this competitive spirit today, when popular competition is readily associated with sports. But the Greeks did not reserve vital public
10 contests for shows of physical excellence. The religious festivals that were celebrated with dramatic performances were also occasions for popular literary contests.

11. As used in the passage, "vital" (line 9) most nearly means
    (A) essential.
    (B) important.
    (C) life-giving.
    (D) indispensable.
    (E) animated.

12. It can be inferred from the passage that the plays commemorating religious festivals
    (A) competed for the distinction of "best play."
    (B) did not elicit as much interest as athletics.
    (C) depicted athletic victories from past competitions.
    (D) are still performed during the modern Olympics.
    (E) were more like sermons than popular entertainment.

Questions 13–14 are based on the following passage.

1 Rock samples taken from the seabed bolster the claims of plate tectonics theorists. Analysis of these rocks has revealed that they are relatively young—the oldest are no more than two hundred million years old. Plate tectonics
5 theory provides a startling explanation for the comparatively recent formation of the ocean floor: The rocks that form the seabed are, amazingly, continuously recycled on a grand scale. At divergent plate boundaries, molten rock wells up, hardens, and travels outward, pushed from
10 its source as new material is formed. At convergent plate boundaries, the expelled material is drawn back under the crust and returned to its molten source.

13. The author's attitude toward the recycling process described in the passage could best be described as
    (A) skeptical.
    (B) dispassionate.
    (C) intrigued.
    (D) baffled.
    (E) approving.

14. It can be inferred from the passage that the oldest rocks on the ocean floor
    (A) remain close to their point of formation.
    (B) are scattered across the entire seabed.
    (C) exist along the convergent plate boundaries.
    (D) somehow escape the recycling process.
    (E) exist in a molten rather than solid form.

DRILL: READING COMPREHENSION—LONG PASSAGES

DIRECTIONS: Read each passage and answer the questions that follow. Each question will be based on the information stated or implied in the passage or its introduction.

In this excerpt from Dickens's Oliver Twist, we read the early account of Oliver's birth and the beginning of his impoverished life.

1 Although I am not disposed to maintain that the being born in a workhouse, is in itself the most fortunate and enviable circumstance that can possibly befall a human being, I do mean to say that in this particular instance, it was the best thing for Oliver Twist that could by possibility
5 have occurred. The fact is, that there was considerable difficulty in inducing Oliver to take upon himself the office of respiration,—a troublesome practice, but one which custom has rendered necessary to our easy existence; and for some time he lay gasping on a little flock mattress, rather unequally poised between this world and the next: the balance being de-
10 cidedly in favour of the latter. Now, if, during this brief period, Oliver had been surrounded by careful grandmothers, anxious aunts, experienced nurses, and doctors of profound wisdom, he would most inevitably and indubitably have been killed in no time. There being nobody by, however, but a pauper old woman, who was rendered rather misty by an unwonted
15 allowance of beer; and a parish surgeon who did such matters by contract; Oliver and Nature fought out the point between them. The result was, that, after a few struggles, Oliver breathed, sneezed, and proceeded to advertise to the inmates of the workhouse the fact of a new burden having been imposed upon the parish, by setting up as loud a cry as could reasonably
20 have been expected from a male infant who had not been possessed of that very useful appendage, a voice, for a much longer space of time than three minutes and a quarter. . . .

For the next eight or ten months, Oliver was the victim of a systematic course of treachery and deception. He was brought up by hand. The
25 hungry and destitute situation of the infant orphan was duly reported by the workhouse authorities to the parish authorities. The parish authorities inquired with dignity of the workhouse authorities, whether there was no female then domiciled in 'the house' who was in a situation to impart to Oliver Twist, the consolation and nourishment of which he stood in need.
30 The workhouse authorities replied with humility, that there was not. Upon this, the parish authorities magnanimously and humanely resolved, that Oliver should be 'farmed,' or, in other words, that he should be despatched to a branchworkhouse some three miles off, where twenty or thirty other juvenile offenders against the poor-laws, rolled about the floor
35 all day, without the inconvenience of too much food or too much clothing, under the parental superintendence of an elderly female, who received the culprits at and for the consideration of sevenpence-halfpenny per small head per week. Sevenpence-halfpenny's worth per week is a good round diet for a child; a great deal may be got for sevenpence-halfpenny: quite
40 enough to overload its stomach, and make it uncomfortable. The elderly female was a woman of wisdom and experience; she knew what was good for children; and she had a very accurate perception of what was good for herself. So, she appropriated the greater part of the weekly stipend to her own use, and consigned the rising parochial generation to even a shorter
45 allowance than was originally provided for them. Thereby finding in the lowest depth a deeper still; and proving herself a very great experimental philosopher.

Everybody knows the story of another experimental philosopher, who had a great theory about a horse being able to live without eating, and who
50 demonstrated it so well, that he got his own horse down to a straw a day, and would most unquestionably have rendered him a very spirited and rampacious animal on nothing at all, if he had not died, just four-and-twenty hours before he was to have had his first comfortable bait of air. Unfortunately for the experimental philosophy of the female to whose
55 protecting care Oliver Twist was delivered over, a similar result usually attended the operation of *her* system . . .

It cannot be expected that this system of farming would produce any very extraordinary or luxuriant crop. Oliver Twist's ninth birth-day found him a pale thin child, somewhat diminutive in stature, and decidedly small
60 in circumference. But nature or inheritance had implanted a good sturdy spirit in Oliver's breast. It had had plenty of room to expand, thanks to the spare diet of the establishment; and perhaps to this circumstance may be attributed his having any ninth birth-day at all.

1. After Oliver was born, he had an immediate problem with his
   (A) heart rate.
   (B) breathing.
   (C) vision.
   (D) hearing.
   (E) memory.

2. What are the two worlds that Oliver stands "unequally poised between" in lines 8–10?
   (A) Poverty and riches
   (B) Infancy and childhood
   (C) Childhood and adolescence
   (D) Love and hatred
   (E) Life and death

3. What does the author imply about "careful grandmothers, anxious aunts, experienced nurses, and doctors of profound wisdom" in lines 11–12?
   (A) They can help nurse sick children back to health.
   (B) They are necessary for every being's survival.
   (C) They are the pride of the human race.
   (D) They tend to adversely affect the early years of children.
   (E) Their involvement in Oliver's birth would have had no outcome on his survival.

4. What is the outcome of Oliver's bout with Nature?
   (A) He is unable to overcome Nature's fierceness.
   (B) He loses, but gains some dignity from his will to fight.
   (C) It initially appears that Oliver has won, but moments later he cries out in crushing defeat.
   (D) Oliver cries out with the breath of life in his lungs.
   (E) There is no way of knowing who won the struggle.

5. What is the "systematic course of treachery and deception" that Oliver falls victim to in the early months of his life?
   (A) He is thrown out into the streets.
   (B) His inheritance is stolen by caretakers of the workhouse.
   (C) He is relocated by the uncaring authorities of the workhouse and the parish.
   (D) The records of his birth are either lost or destroyed.
   (E) He is publicly humiliated by the parish authorities.

6. What is meant when the residents of the workhouse are referred to by the phrase "juvenile offenders against the poor-laws" (line 34)?
   (A) They are children who have learned to steal early in life.
   (B) They are adolescents who work on probation.
   (C) They are infants who have no money to support them.
   (D) They are infants whose parents were law offenders.
   (E) They are adults who have continuously broken the law.

7. What is the author's tone when he writes that the elderly caretaker "knew what was good for children" (lines 41–42)?
   (A) Sarcastic
   (B) Complimentary
   (C) Impressed
   (D) Astonished
   (E) Outraged

8. What does the author imply when he further writes that the elderly caretaker "had a very accurate perception of what was good for herself" (lines 42–43)?
   (A) She knew how to keep herself groomed and clean.
   (B) She knew how to revenge herself on her enemies.
   (C) She had a sense of confidence that inspired others.
   (D) She really had no idea how to take care of herself.
   (E) She knew how to selfishly benefit herself despite the cost to others.

9. Why is the elderly caretaker considered "a very great experimental philosopher" (lines 46–47)?
   (A) She was scientifically weaning the children off of food trying to create stronger humans.
   (B) She experimented with the survival of the children in her care.
   (C) She thought children were the key to a meaningful life.
   (D) She made sure that the children received adequate training in philosophy.
   (E) She often engaged in parochial and philosophical discussions.

10. In line 53, "bait" most nearly means
   (A) worms.
   (B) a hook.
   (C) a breeze.
   (D) a trap.
   (E) a meal.

11. To what does the author attribute Oliver's survival to his ninth year?
   (A) A strong, healthy diet
   (B) Money from an anonymous donor
   (C) Sheer luck
   (D) His diminutive stature
   (E) His sturdy spirit

12. Based upon the passage, what is the author's overall attitude concerning the city where Oliver lives?
   (A) It is the best of all possible worlds.
   (B) It should be the prototype for future cities.
   (C) It is a dark place filled with greedy, selfish people.
   (D) Although impoverished, most of its citizens are kind.
   (E) It is a flawed place, but many good things often happen there.

The following passage analyzes the legal and political philosophy of John Marshall, a chief justice of the Supreme Court in the nineteenth century.

1 As chief justice of the Supreme Court from 1801 until his death in 1835, John Marshall was a staunch nationalist and upholder of property rights. He was not, however, as the folklore of American politics would have it, the lonely and embattled Federalist defending these values
5 against the hostile forces of Jeffersonian democracy. On the contrary, Marshall's opinions dealing with federalism, property rights, and national economic development were consistent with the policies of the Republi-

can Party in its mercantilist phase from 1815 to 1828. Never an extreme Federalist, Marshall opposed his party's reactionary wing in the crisis of
10 1798–1800. Like almost all Americans of his day, Marshall was a Lockean republican who valued property not as an economic end in itself, but rather as the foundation of civil liberty and a free society. Property was the source both of individual happiness and social stability and progress.

Marshall evinced strong centralizing tendencies in his theory of
15 federalism and completely rejected the compact theory of the Union expressed in the Virginia and Kentucky Resolutions. Yet his outlook was compatible with the Unionism that formed the basis of the post-1815 American System of the Republican Party. Not that Marshall shared the democratic sensibilities of the Republicans; like his fel-
20 low Federalists, he tended to distrust the common people and saw in legislative majoritarianism a force that was potentially hostile to constitutionalism and the rule of law. But aversion to democracy was not the hallmark of Marshall's constitutional jurisprudence. Its central features rather were a commitment to federal authority versus
25 states' rights and a socially productive and economically dynamic conception of property rights. Marshall's support of these principles placed him near the mainstream of American politics in the years between the War of 1812 and the conquest of Jacksonian democracy.

In the long run, the most important decisions of the Marshall Court
30 were those upholding the authority of the federal government against the states. *Marbury v. Madison* provided a jurisprudential basis for this undertaking, but the practical significance of judicial review in the Marshall era concerned the state legislatures rather than Congress. The most serious challenge to national authority resulted from state attempts to
35 administer their judicial systems independent of the Supreme Court's appellate supervisions as directed by the Judiciary Act of 1789. In successfully resisting this challenge, the Marshall Court not only averted a practical disruption of the federal system but also evolved doctrines of national supremacy that helped preserve the Union during the Civil War.

13. The primary purpose of this passage is to
    (A) describe Marshall's political jurisprudence.
    (B) discuss the importance of centralization to the preservation of the Union.
    (C) criticize Marshall for being disloyal to his party.
    (D) examine the role of the Supreme Court in national politics.
    (E) chronicle Marshall's tenure on the Supreme Court.

14. According to the author, Marshall viewed property as
    (A) an investment.
    (B) irrelevant to constitutional liberties.
    (C) the basis of a stable society.
    (D) inherent to the upper class.
    (E) an important centralizing incentive.

15. In line 15, the "compact theory" was most likely a theory
    (A) supporting states' rights.
    (B) of the extreme Federalists.
    (C) of the Marshall Court's approach to the Civil War.
    (D) supporting centralization.
    (E) advocating jurisprudential activism.

16. According to the author, Marshall's attitude toward mass democratic politics can best be described as
    (A) hostile.
    (B) supportive.
    (C) indifferent.
    (D) nurturing.
    (E) distrustful.

17. In line 22, the word "aversion" means
    (A) loathing.
    (B) acceptance.
    (C) fondness.
    (D) forbidding.
    (E) misdirection.

18. The author argues the Marshall Court
    (A) failed to achieve its centralizing policies.
    (B) failed to achieve its decentralizing policies.
    (C) helped to bring on the Civil War.
    (D) supported federalism via judicial review.
    (E) had its greatest impact on Congress.

19. According to the author, Marshall's politics were
    (A) extremist.
    (B) right-wing.
    (C) democratic.
    (D) moderate.
    (E) majoritarian.

In this passage, the author discusses the properties and uses of selenium cells, which convert sunlight to energy, creating solar power.

1 The physical phenomenon responsible for converting light to electricity—the photovoltaic effect—was first observed in 1839 by the renowned French physicist, Edmund Becquerel. Becquerel noted that a voltage appeared when one of two identical electrodes in a weak conducting solution
5 was illuminated. The PV effect was first studied in solids, such as selenium, in the 1870s. In the 1880s, selenium photovoltaic cells were built that exhibited 1–2 percent efficiency in converting light to electricity. Selenium converts light in the visible part of the sun's spectrum; for this reason, it was quickly adopted by the then emerging field of photography
10 for photometric (light-measuring) devices. Even today, the light-sensitive cells on cameras used for adjusting shutter speed to match illumination are made of selenium.

Selenium cells have never become practical as energy converters because their cost is too high relative to the tiny amount of power they 15 produce (at 1 percent efficiency). Meanwhile, work on the physics of PV phenomena has expanded. In the 1920s and 1930s, quantum mechanics laid the theoretical foundation for our present understanding of PV. A major step forward in solar-cell technology came in the 1940s and early 1950s when a new method (called the Czochralski method) was developed for 20 producing highly pure crystalline silicon. In 1954, work at Bell Telephone Laboratories resulted in a silicon photovoltaic cell with a 4 percent efficiency. Bell Labs soon bettered this to a 6 percent and then 11 percent efficiency, heralding an entirely new era of power-producing cells.

A few schemes were tried in the 1950s to use silicon PV cells com- 25 mercially. Most were for cells in regions geographically isolated from electric utility lines. But an unexpected boom in PV technology came from a different quarter. In 1958, the U.S. Vanguard space satellite used a small (less than one-watt) array of cells to power its radio. The cells worked so well that space scientists soon realized the PV could be an effective power 30 source for many space missions. Technology development of the solar cell has been a part of the space program ever since.

Today, photovoltaic systems are capable of transforming one kilowatt of solar energy falling on one square meter into about a hundred watts of electricity. One hundred watts can power most household appliances: a 35 television, a stereo, an electric typewriter, or a lamp. In fact, standard solar cells covering the sun-facing roof space of a typical home can provide about 8,500 kilowatt-hours of electricity annually, which is about the average household's yearly electric consumption. By comparison, a modern, 200-ton electric-arc steel furnace, demanding 50,000 kilowatts of electric- 40 ity, would require about a square kilometer of land for a PV power supply.

Certain factors make capturing solar energy difficult. Besides the sun's low illuminating power per square meter, sunlight is intermittent, affected by time of day, climate, pollution, and season. Power sources based on photovoltaics require either back-up from other sources or stor- 45 age for times when the sun is obscured.

In addition, the cost of a photovoltaic system is far from negligible (electricity from PV systems in 1980 cost about 20 times more than that from conventional fossil-fuel-powered systems).

Thus, solar energy for photovoltaic conversion into electricity is 50 abundant, inexhaustible, and clean; yet, it also requires special techniques to gather enough of it effectively.

20. To the author, Edmund Becquerel's research was
    - (A) unimportant.
    - (B) of some significance.
    - (C) not recognized in its time.
    - (D) weak.
    - (E) an important breakthrough.

21. In the first paragraph, it can be concluded that the photovoltaic effect is the result of
    (A) two identical negative electrodes.
    (B) one weak solution and two negative electrodes.
    (C) two positive electrodes of different qualities.
    (D) positive electrodes interacting in a weak environment.
    (E) one negative electrode and one weak solution.

22. The author establishes that selenium was used for photometric devices because
    (A) selenium was the first solid to be observed to have the PV effect.
    (B) selenium is inexpensive.
    (C) selenium converts the visible part of the sun's spectrum.
    (D) selenium can adjust shutter speeds on cameras.
    (E) selenium is abundant.

23. Which of the following can be concluded from the passage?
    (A) Solar energy is still limited by problems of technological efficiency.
    (B) Solar energy is the most efficient source of heat for most families.
    (C) Solar energy represents the PV effect in its most complicated form.
    (D) Solar energy is 20 percent cheaper than fossil-fuel-powered systems.
    (E) Solar energy is 40 percent more expensive than fossil-fuel-powered systems.

24. In line 22, the word "heralding" most nearly means
    (A) celebrating.
    (B) observing.
    (C) commemorating.
    (D) anticipating.
    (E) introducing.

25. According to the passage, commercially used PV cells have powered
    (A) car radios.
    (B) space satellite radios.
    (C) telephones.
    (D) electric utility lines.
    (E) space stations.

26. Through the information in lines 32–34, it can be inferred that two kilowatts of solar energy transformed by a PV system equal
    (A) 200 watts of electricity.
    (B) 100 watts of electricity.
    (C) no electricity.
    (D) two square meters.
    (E) 2,000 watts of electricity.

27. Sunlight is difficult to procure for transformation into solar energy. Which of the following statements most accurately supports this belief derived from the passage?
(A) Sunlight is erratic and subject to variables.
(B) Sunlight is steady but never available.
(C) Sunlight is not visible because of pollution.
(D) Sunlight would have to be artificially produced.
(E) Sunlight is never erratic.

28. The author's concluding paragraph would be best supported with additional information regarding
(A) specific benefits of solar energy for photovoltaic conversion into electricity.
(B) the negative effects of solar energy for photovoltaic conversion into electricity.
(C) the negative effects of photovoltaic conversion.
(D) why solar energy is clean.
(E) why solar energy is abundant.

DIRECTIONS: Read the passages and answer the questions that follow. Each question will be based on the information stated or implied in the selections or their introductions, and may be based on the relationship between the passages.

In Passage 1, the author writes a general summary about the nature of comedy. In Passage 2, the author sums up the essentials of tragedy.

Passage 1

1 The primary aim of comedy is to amuse us with a happy ending, although comedies can vary according to the attitudes they project, which can be broadly identified as either high or low, terms having nothing to do with an evaluation of the play's merit. Generally, the amusement found in
5 comedy comes from an eventual victory over threats or ill fortune. Much of the dialogue and plot development might be laughable, yet a play need not be funny to be comic. In fact, some critics in the Renaissance era thought that the highest form of comedy should elicit no laughter at all from its audience. A comedy that forced its audience into laughter failed
10 in the highest comic endeavor, whose purpose was to amuse as subtly as possible. Note that Shakespeare's comedies themselves were often under attack for their appeal to laughter.

"Farce" is low comedy intended to make us laugh by means of a series of exaggerated, unlikely situations that depend less on plot and character
15 than on gross absurdities, sight gags, and coarse dialogue. The "higher" a comedy goes, the more natural the characters seem and the less boisterous their behavior. The plots become more sustained, and the dialogue shows more weighty thought. As with all dramas, comedies are about things that go wrong. Accordingly, comedies create deviations from accepted nor-
20 malcy, presenting problems which we might or might not see as harmless. If these problems make us judgmental about the involved characters and

events, the play takes on the features of satire, a rather high comic form implying that humanity and human institutions are in need of reform. If the action triggers our sympathy for the characters, we feel even less protected from the incongruities as the play tilts more in the direction of tragicomedy. In other words, the action determines a figurative distance between the audience and the play. Such factors as characters' personalities and the plot's predictability influence this distance. The farther away we sit, the more protected we feel and usually the funnier the play becomes. Closer proximity to believability in the script draws us nearer to the conflict, making us feel more involved in the action and less safe in its presence.

Passage 2

The term "tragedy" when used to define a play has historically meant something very precise, not simply a drama which ends with unfortunate consequences. This definition originated with Aristotle, who insisted that the play be an imitation of complex actions which should arouse an emotional response combining fear and pity. Aristotle believed that only a certain kind of plot could generate such a powerful reaction. Comedy shows us a progression from adversity to prosperity. Tragedy must show the reverse; moreover, this progression must be experienced by a certain kind of character, says Aristotle, someone whom we can designate as the tragic hero. This central figure must be basically good and noble: "good" because we will not be aroused to fear and pity over the misfortunes of a villain, and "noble" both by social position and moral stature because the fall to misfortune would not otherwise be great enough for tragic impact. These virtues do not make the tragic hero perfect, however, for he must also possess hamartia—a tragic flaw—the weakness which leads him to make an error in judgment which initiates the reversal in his fortunes, causing his death or the death of others or both. These dire consequences become the hero's catastrophe. The most common tragic flaw is hubris; an excessive pride that adversely influences the protagonist's judgment.

Often the catastrophic consequences involve an entire nation because the tragic hero's social rank carries great responsibilities. Witnessing these events produces the emotional reaction Aristotle believed the audience should experience, the catharsis. Although tragedy must arouse our pity for the tragic hero as he endures his catastrophe and must frighten us as we witness the consequences of a flawed behavior which anyone could exhibit, there must also be a purgation, "a cleansing," of these emotions which should leave the audience feeling not depressed but relieved and almost elated. The assumption is that while the tragic hero endures a crushing reversal, somehow he is not thoroughly defeated as he gains new stature through suffering and the knowledge that comes with suffering. Classical tragedy insists that the universe is ordered. If truth or universal law is ignored, the results are devastating, causing the audience to react emotionally; simultaneously, the tragic results prove the existence of truth, thereby reassuring our faith that existence is sensible.

29. In Passage 1, the term "laughable" (line 6) suggests that on occasion comic dialogue and plot development can be
   (A) senselessly ridiculous.
   (B) foolishly stupid.
   (C) amusingly droll.
   (D) theoretically depressing.
   (E) critically unsavory.

30. The author of Passage 1 makes an example of Shakespeare (lines 11–12) in order to
   (A) make the playwright look much poorer in our eyes.
   (B) emphasize that he wrote the highest form of comedy.
   (C) degrade higher forms of comedy.
   (D) suggest the foolishness of Renaissance critics.
   (E) show that even great authors do not always use high comedy.

31. The protagonist in a play discovers he has won the lottery, only to misplace the winning ticket. According to the author's definition, this situation would be an example of which type of comedy?
   (A) Satire
   (B) Farce
   (C) Tragicomedy
   (D) Sarcasm
   (E) Slapstick

32. In line 26, the phrase "figurative distance" suggests
   (A) the distance between the seats in the theater and the stage.
   (B) the lengths the comedy will go to elicit laughter.
   (C) the years separating the composition of the play and the time of its performance.
   (D) the degree to which an audience relates with the play's action.
   (E) that the play's matter is too high for the audience to grasp.

33. What is the author trying to espouse in lines 28–32?
   (A) He warns us not to get too involved with the action of the drama.
   (B) He wants the audience to immerse itself in the world of the drama.
   (C) He wants us to feel safe in the presence of the drama.
   (D) He wants us to be critical of the drama's integrity.
   (E) He feels that we should not enjoy the drama overly much.

34. In Passage 2, the author introduces Aristotle as a leading source for the definition of tragedy. He does this

(A) to emphasize how outdated the tragedy is for the modern audience.

(B) because Greek philosophy is the only way to truly understand the world of the theater.

(C) because Aristotle was one of Greece's greatest actors.

(D) because Aristotle instituted the definition of tragedy still used widely today.

(E) in order to prove that Aristotle's sense of tragedy was based on false conclusions.

35. In line 42, "noble" most nearly means

(A) of high degree and superior virtue.

(B) of great wealth and self-esteem.

(C) of quick wit and high intelligence.

(D) of manly courage and great strength.

(E) of handsome features and social charm.

36. Which of the following is an example of hamartia (line 47)?

(A) Courtesy to the lower class

(B) The ability to communicate freely with others

(C) A refusal to acknowledge the power of the gods

(D) A weak, miserly peasant

(E) A desire to do penance for one's crimes

37. Which of the following best summarizes the idea of catharsis explained in lines 52–55?

(A) All of the tragic consequences are reversed at the last moment; the hero is rescued from certain doom and is allowed to live happily for the rest of his life.

(B) The audience gains a perverse pleasure from watching another's suffering.

(C) The play's action ends immediately, unresolved, and the audience is left in a state of blissful confusion.

(D) When the play ends, the audience is happy to escape the drudgery of the tragedy's depressing conclusion.

(E) The audience lifts itself from a state of fear and pity for the tragic hero to a sense of renewal and absolution for the hero's endurance of great suffering.

38. The authors of both passages make an attempt to

(A) ridicule their subject matter.

(B) outline the general terms and guidelines of a particular aspect of drama.

(C) thrill their readers with sensational information.

(D) draw upon Shakespeare as an authority to back up their work.

(E) persuade their readers to study only one or the other type of drama (that is, comedy or tragedy).

39. Which of the following best describes the differences between the structure of both passages?

(A) Passage 1 is concerned primarily with the Renaissance era. Passage 2 is concerned primarily with Classical Greece.

(B) Passage 1 is concerned with dividing its subject into subcategories. Passage 2 is concerned with extracting its subject's individual elements.

(C) Passage 1 makes fun of its subject matter. Passage 2 treats its subject matter very solemnly.

(D) Passage 1 draws upon a series of plays that serve as examples. Passage 2 draws upon no outside sources.

(E) Passage 1 introduces special vocabulary to illuminate the subject matter. Passage 2 fails to do this.

40. What assumption do both passages seem to draw upon?

(A) Tragedy is a higher form of drama than comedy.

(B) Tragedy is on the decline in modern society; comedy, however, is on the rise.

(C) Catharsis is an integral part of both comedy and tragedy.

(D) An audience's role in the performance of either comedy or tragedy is a vital one.

(E) The tragicomedy is a form that is considered greater than drama that is merely comic or tragic.

READING COMPREHENSION DRILLS

ANSWER KEY

Drill: Reading Comprehension—Short Passages

1.	(D)	6.	(C)	11.	(B)
2.	(D)	7.	(B)	12.	(A)
3.	(B)	8.	(B)	13.	(C)
4.	(E)	9.	(A)	14.	(E)
5.	(A)	10.	(B)		

Drill: Reading Comprehension—Long Passages

1.	(B)	11.	(E)	21.	(D)	31.	(C)
2.	(E)	12.	(C)	22.	(C)	32.	(D)
3.	(D)	13.	(A)	23.	(A)	33.	(B)
4.	(D)	14.	(C)	24.	(E)	34.	(D)
5.	(C)	15.	(A)	25.	(B)	35.	(A)
6.	(C)	16.	(E)	26.	(A)	36.	(C)
7.	(A)	17.	(A)	27.	(A)	37.	(E)
8.	(E)	18.	(D)	28.	(A)	38.	(B)
9.	(B)	19.	(D)	29.	(C)	39.	(B)
10.	(E)	20.	(E)	30.	(E)	40.	(D)

CHAPTER 6

MASTERING THE SAT WRITING SECTIONS

MASTERING THE SAT WRITING SECTIONS

The SAT begins with a 25-minute Essay section. Two sections, one 25 minutes and the other 10 minutes, will consist of multiple-choice questions on Usage, Sentence Correction, and Paragraph Improvement.

I. WRITING

MASTERING USAGE QUESTIONS

In all likelihood, you'll find that the Usage questions cover familiar territory. They'll test your ability to identify typical writing errors, the kind that your English teachers have been warning you about for years. The Basic Verbal Skills Review included in this book offers a concise but thorough review of standard written English usage. If you have the time, work systematically through the review and complete the drill exercises.

But if you're pressed for time, don't despair. The Usage questions, for the most part, focus on the relatively small number of error types that many writers—even proficient ones—are prone to committing. In this chapter, we will cover the most common types of errors you'll encounter in the Writing section.

STEPS FOR MASTERING USAGE QUESTIONS

Listen for an error. Even if you lack confidence in your knowledge of grammar, you probably know enough to "hear" most errors as you encounter them.

Having identified the part of the sentence that seems incorrect, try to determine what makes that part wrong, and then replace the underlined error with the correct word or phrase. Doing so will help you to test your intuition.

If you do not "hear" anything wrong with the sentence, it may be error-free. Nevertheless, mentally review the kinds of errors you are told to expect and apply them to the underlined parts of the sentence. If you see evidence of a given type of error, apply STEP 2 to that part of the sentence.

USAGE QUESTION FORMAT

Each usage question will consist of a sentence that may or may not contain an error. Choose the underlined part that contains the error or select "No error."

Be sure to read the sentence exactly as it is written:

Some people fail to realize that regular dental cleaning, accompanied by a thorough
A **B**
exam, are essential to avoid more serious complications later. No error.
C **D** **E**

As you read the preceding sample questions, you may have "heard" something wrong with "are" (choice C). When a verb such as "are" doesn't sound right, the cause is almost always noun-verb disagreement. In other words, the verb doesn't agree with the subject in number.

The reason "are" sounds wrong is that the subject of the sentence, "dental cleaning," is singular, not plural. To agree with this singular subject, the form of the verb "to be" must be present singular, "is."

If you didn't hear the error in the sample question, it's probably because you didn't "hear" that the subject of the sentence is singular. After all, the phrase "accompanied by a thorough exam" follows the subject, creating the impression that both "dental cleaning and exam are the subject—in other words, a compound (and therefore plural) subject. But the phrase mentioning "exam" only modifies the subject; it is not a part of it.

KINDS OF USAGE ERRORS

The following mini-review is not exhaustive, but it does cover the major kinds of errors likely to appear on the Writing exam.

Disagreement Errors—Noun-Verb

Each of the following sentences illustrates a classic way in which noun-verb disagreement manages to slip past our "grammar radar."

1. Attendance at <u>poetry festivals</u> over the past few years <u>have</u> increased dramatically as a <u>result of</u> poetry clubs and <u>writers'</u> workshops.
   A: poetry festivals; B: have; C: result of; D: writers'

2. There <u>is</u>, when you consider the issue, several reasons <u>for supporting</u> the current legislation now <u>before</u> the <u>Appropriations Committee</u>.
   A: is; B: for supporting; C: before; D: Appropriations Committee

3. In the <u>depths</u> of the woods <u>live</u> a species of bird that <u>hasn't</u> been thoroughly studied <u>by</u> wild-life experts.
   A: depths; B: live; C: hasn't; D: by

Answers

1. (B) The subject of the sentence is "attendance," which is singular. The main verb should be "has," not "have." Because the plural nouns "festivals" and "years" come between the subject and the verb, the writer failed to hear the disagreement.

2. (A) The subject of a clause beginning with "There is" or "There are" comes after the verb. The subject of sentence 2 is "reasons," a plural noun, so the main verb should be "are," not "is."

3. (B) Atypical word order can occur in contexts other than the one mentioned above. The subject of sentence 3 is " species," a singular noun. Therefore the verb should be "lives," not "live." Plural "woods" and "depths," which occur in prepositional phrases, probably encouraged the writer to use the plural verb form.

Disagreement Errors—Noun-Pronoun

Disagreement can also occur between nouns and pronouns:

4. Both the geography book I used <u>as</u> a high school student and <u>the one</u> I used as a college student <u>have</u> the same vivid photograph of Mount Everest on <u>its cover</u>.
   A: as; B: the one; C: have; D: its cover

5. In <u>today's</u> competitive world, even a student <u>who</u> earns high grades and top test scores may <u>have</u> a tough time getting into the college of <u>their</u> choice.
   A: today's; B: who; C: have; D: their

Answers

4. (D) The subject of sentence 4 is a compound noun—in other words, it's plural. Consequently, the same photograph appears on two separate "covers." Choice D should be "their covers."

5. (D) "Student" in sentence 5 is, of course, singular. But because the gender of student is not specified, the writer mistakenly uses the gender-free plural possessive pronoun "their."

Shift Errors

There are two kinds of shift errors: tense shift and pronoun shift.

6. When I saw her again after almost twenty years, I can't help but approach her and say "How's life
   A B C
   been treating you?"
   D

7. Experience teaches us that we have to work hard and play hard if you want to have a genuinely
   A B C D
   fulfilling life.

Answers

6. (B) The action of the sentence is taking place in the past—"I saw her" "Can't" is a present tense modal. The modal helping to describe the nature and conditions of the encounter needs to be in the past tense: "I couldn't help but approach her"

7. (C) In conversation, speakers often use the pronoun "you" to refer to people in general. Sometimes writers lapse into this conversational tendency in their writing, even when it violates pronoun consistency. The writer of sentence 7 errs by going from "we" and "us"—both first person plural pronouns—to second person "you."

Comparison Errors

Two types of comparison errors may occur in the Writing section: comparative/superlative errors and unequal comparison errors.

8. Mons Olympus, a volcanic mountain on Mars, is the taller of all the other known mountains in
   A B C D
   the Solar System.

9. From a distance, Betty seems rather short to me, but, when I stand next to her, she's clearly
   A B C
   the tallest.
   D

10. My high school's track team is much better at long distance running than the high school that
    A B C D
    won the state championship last year.

Answers

8. (B) The writer discusses Mons Olympus in superlative terms: no mountain is taller than it. Hence, it is the "tallest" of all the other mountains, not simply "taller." Also, the comparative "taller" should appear with the conjunction "than," which does not appear in the sentence.

9. (D) The writer compares his or her height to that of one other person. If two things or people are being compared, the adjective needs to be in the comparative form, not the superlative.

10. (D) The writer clearly intended to compare two track teams; grammatically, the writer is comparing a track team to a high school. Although it's acceptable to substitute the whole for the part—for example, high school = track team—this substitution should be applied consistently, a rule the writer fails to apply in this case.

Word Choice Errors

In a sense, all grammatical errors are the result of choosing the wrong word or phrase. But in the grammatical mistakes we've considered so far, the errors in word choice have been logical errors. Treating a plural noun as though it's singular, or grammatically comparing unequal things, is simply illogical. The category "Word Choice Errors" concerns violations of accepted idioms and definitions. Nonnative speakers of English have particular difficulty identifying these kinds of errors.

Try to "hear" improperly used words and expressions in the following:

11. After three hours of (A) negotiations, the Trade Union insisted that management agree on (B) its (C) demands for a reduced work week and improved working (D) conditions.

12. Despite of (A) the bad weather, Mark (B) and Heather's garage sale went well (C), earning the couple a tidy (D) sum.

13. Even (A) at an early age, Picasso displayed signs of (B) imaginary (C) genius that would fuel ground-breaking (D) artwork.

Answers

11. (B) Like many verbs, "agree" combines with several prepositions to form new concepts. To "agree with" someone is to share his or her opinion; to "agree to" something is to concede to a demand or request. A trade union usually doesn't care if management *agrees with* its philosophy, so long as it *agrees to* (concedes to) to the union's demands.

12. (A) "In spite of" and "despite" have virtually the same meaning and sound alike. Consequently, characteristics of each are sometimes erroneously blended, as in sentence 12.

13. (C) Imaginary and imaginative are adjectives derived from the same root word. Nevertheless, they have very different meanings. "Imaginative" means "gifted with imagination"—which Picasso was; "imaginary" means "illusory."

Faulty Parallelism

Consider the following sentence:

Jim likes hiking, skiing, and to snowboard.

Clearly, the third element in the series of things Jim likes should be "snowboarding," not "to snowboard." Whenever possible, elements catalogued or linked with conjunctions should be grammatically equal. This rule of "parallelism"—think "equal"—is more typically violated when the sentence is more complex than the preceding one. See if you can identify the faulty parallelisms below.

14. A low-calorie (A) diet, most nutritionists agree (B), is the best way (C) to reduce cholesterol and the achievement (D) of weight loss.

15. When one takes a high-stakes test, it's perfectly natural to be a little nervous, irritable, sweaty palms.
A B C D

Answers

14. (D) In naming the two things that a low-calorie diet helps the dieter to achieve, the writer begins with an infinitival phrase: "to reduce" Disobeying the rule of parallelism, the writer describes the second benefit of dieting as "the achievement of weight loss"—a noun phrase. Clearly, following the first infinitival phrase with another to create "to lower cholesterol and to achieve weight loss" sounds better and improves clarity.

15. (D) "Nervous" and "irritable" are adjectives. "Sweaty" is also an adjective, but, because it is followed by "palms," the third element in this catalogue is a noun phrase modified by an adjective. Parallelism could be achieved simply by removing "palms."

Additional Error Types

The preceding review is a representative sample of the kinds of errors you will encounter; it doesn't cover all error types. Below is a "mixed bag" of other error types.

Double Negative:
I can't hardly *see the bridge from here due to the fog.*

Explanation: The writer's view of the bridge is obscured. He or she either *can't* see the bridge or *can hardly* see it. Joined together, "can't " and "hardly" cancel each other out.

Who/Which/That Substitution:
The students that *protested the tuition increase actually had lots of disposable income.*

Explanation: "Who" is an animate relative pronoun (i.e., it refers to people and sometimes to pets/animals). "Which" and "that" are inanimate relative pronouns (i.e., they refer to things).

Correlative Conjunctions (Misused):
The new movie was seen both by my friend Mary as well as *her brother.*

Explanation: Correlative conjunctions function as pairs. If you see one half of a correlative conjunction pair in a sentence, the second half should soon follow. "And," not "as well as," should appear soon after "both." Here are some other common corrective conjunctions:

either . . . or neither . . . nor not only . . . but also

MASTERING SENTENCE CORRECTION QUESTIONS

These questions test your ability to make appropriate revisions in accordance with the rules of standard written English. In the following sentences some part of the sentence, or all of the sentence, is underlined. Below each you will find five ways of phrasing the underlined part. Choose the answer that most effectively expresses the meaning of the original sentence. If you think the original sentence needs no revision, choose (A), which is always the same as the underlined part.

Although familiarity with the error types described in the Usage Questions review may help you to answer some Sentence Correction questions, be advised that this section focuses particularly on the following error types:

1. Ambiguous (unclear) sentences
2. Awkward sentences
3. Fragments (incomplete sentences)
4. Run-on sentences

Read the sentence carefully:

Intelligence is determined contrary to public opinion not by the size of the brain but by the number and complexity of the dendrites in the brain.

(A) Intelligence is determined contrary to public opinion not by
(B) Intelligence is contrary to public opinion and determined not by
(C) Contrary is public opinion, intelligence is determined not by
(D) Not by public opinion is intelligence determined but by
(E) Contrary to public opinion, intelligence determines not by

If you determine that the sentence contains an error, proceed as follows:

Eliminate choice (A) because it restates exactly what is underlined in the sentence.

Eliminate the obviously incorrect choices.
(B) ELIMINATE—changes the meaning of the sentence
(C) Maybe
(D) Maybe
(E) ELIMINATE—awkward and unclear

Choose between the most likely answers
(C) is clearer than (D). (C) directly links the phrase "Intelligence is determined not by" with the part of the sentence that logically follows this statement.

Note that the preceding sample question falls under the "awkward sentences" category; it is best revised by placing part of the underlined portion at the beginning of the sentence.

Answer the following questions and try to identify the kind of error contained in each.

1. When I arrived at the stadium, which was jam-packed with spectators, the halftime show underway.

   (A) the halftime show underway.
   (B) the halftime show being underway.
   (C) underway, the half-time show was nearly over.
   (D) the halftime show was underway.
   (E) was underway.

2. Leaving behind a middle-class existence in France, Paul Gaugin's sojourn in Tahiti brought the artist much-needed inspiration.

   (A) France, Paul Gaugin's sojourn in Tahiti brought the artist much-needed inspiration.
   (B) France, Paul Gaugin sojourned in Tahiti and gained much-needed inspiration.
   (C) A sojourn in Tahiti brought Paul Gaugin much-needed inspiration.
   (D) Paul Gaugin was inspired to sojourn in Tahiti.
   (E) Much-needed inspiration was sought by Gaugin in Tahiti.

3. Today's poetry lovers don't really read much poetry, they prefer books *about* poetry and poets.

   (A) poetry, they prefer books about poetry and poets.
   (B) poetry; they prefer books about poetry and poets.
   (C) poetry because of books about poetry and poets themselves.
   (D) poetry; books about poetry are largely read by poets themselves.
   (E) poetry, poetry itself being less preferable to books about poetry.

4. America's literacy rates have increased far from being in decline, as book sales suggest.

   (A) America's literacy rates have increased far from being in decline,
   (B) Far from being in decline, America's literacy rates have increased
   (C) Far from the literacy rates of Americans being in decline, they have actually increased
   (D) Unlike the decline of literacy rates among most Americans, there is an increase
   (E) American's literacy rates far from being in decline have actually increased

Answers

1. (D) The underlined portion is the main clause of the sentence, yet it lacks a verb. Choice (D) provides the appropriate verb (i.e., "was"). The remaining choices still result in a sentence fragment, either by not supplying an appropriate verb (choices (B) and (C)), or by eliminating the main clause's subject (choice (E)).

2. (B) The original sentence is both confusing and ungrammatical. Paul Gaugin, not his "sojourn," in "leaving behind a middle-class existence in France." Choices (C) and (E) similarly confuse the reader by failing to mention Gaugin directly after the participial phrase. Choice (D) doesn't commit this grammatical error—known as a dangling modifier—but it does not reproduce the original sentence's intended meaning. Choice (B) reproduces the original sentence's intended meaning and eliminates the dangling modifier. The correct answer is (B).

3. (B) The sentence in question 3 contains two clauses. Separate clauses should either be joined with conjunctions or separated by a period or semicolon. If the sentences are short and structurally similar, a semicolon is preferable to a period. The correct answer is (B). The remaining choices either alter the original meaning or result in a grammatical error.

4. (B) The underlined portion of sentence 4 is awkward; its elements need to be rearranged. Choice (B) moves a participial phrase to the beginning of the sentence and separates it from the main clause with a comma. Now it is clear that what is "far from being in decline" is, in fact, "America's literacy rates." Choice (C) is ambiguous because it's unclear

whether "they" refers to literacy rates or to Americans. In choice (D), it's not clear what "the decline in literacy rates" is being contrasted with. Choice (E) would be acceptable if the phrase "far from being in decline" was offset by commas. Also, "American's literacy rates" means the literacy rates of just one American, a nonsensical notion.

MASTERING PARAGRAPH IMPROVEMENT QUESTIONS

As its name implies, Paragraph Improvement questions are more about strengthening paragraph cohesion and essay unity than about correcting out-and-out errors. A paragraph's elements may be grammatically correct but stylistically and rhetorically inadequate.

Problem Types

Below is a chart listing some of the stylistic deficiencies typically encountered in the Improving Paragraphs questions:

Problem	Solution
wordy, redundant	eliminate unnecessary words
choppy sentences	combine sentences
unclear relationships between sentences	clarify relationships (usually with conjuctions)
vague language	replace with specific words
awkward passive voice	use active voice
off-topic sentence	find and eliminate sentence
lacking in support/details	select appropriate details

Steps for Mastering Paragraph Improvement Questions

The following is an introduction excerpted from a student's essay. The essay was written in response to a writing prompt asking the student to explain how his or her use of language differs from proper English when speaking with a friend.

(1) The language I use in talking to a friend would certainly differ from Standard English. (2) In talking with a friend I would be prone to phrases drawn from popular culture that we are both aware of. (3) I would know that by employing certain phrases that my friends and I were members of a peer group, and thus friends.

1. Which of the following revisions most clearly states the meaning of sentence 3?
   (A) No change
   (B) I would know, by employing certain phrases, that my friends and I shared certain interests.
   (C) My friends and I would be peers by employing certain phrases that show we are friends.
   (D) I would know that my friends and I had things in common by our similar uses of language.
   (E) I would know, that by employing certain phrases and words, that my friends and I shared common interests and were, because of this knowledge, peers.

If you feel that the sentence is not in need of revision, choose answer choice (A). If you determine that the sentence would benefit from revision, proceed as follows:

STEP 1 *Eliminate choice (A) because it makes no revision.*

STEP 2 *Eliminate the obviously incorrect choices. Choice (D) says something completely different from the idea implied by the original sentence, while (E) is a verbose and repetitive run-on sentence.*

STEP 3 *Choose between the most likely answers. Choice (B) is clearer than (C), and (B) better expresses the thought that sentence 3 attempts to convey.*

Now read the extended passage below and apply the steps to the questions that follow. Also, keep in mind the typical error types and their solutions as you both read the passage and answer the questions.

(1) Some high schools are making community service a requirement for graduating from high school. (2) School officials assert that students benefit from this policy. (3) Officials believe community service has both educational value as well as making students civic-minded. (4) They learn good work habits and assistance is provided where they are needed.

(5) Almost everyone has positive feelings about the new graduating requirement. (6) Some students feel the community service requirement, which involves volunteering one's time and assistance for forty hours or more, is too demanding. (7) Many students already have very full schedules. (8) They participate in sports. (9) They work part-time jobs. (10) Some even belong to clubs. (11) Many of which already provide valuable services to the community. (12) Also, teenagers just need those relatively care-free years to enjoy themselves, which won't last forever.

(13) More students would be in favor of the community service requirement if changes were made. (14) The number of community service hours required to graduate should be reduced. (15) Participation in community-centered clubs should be counted toward a student's community service requirement.

1. Which is the best way to revise the underlined portion of sentence 3 (reproduced below)?

   Officials believe community service <u>has both educational value as well as making students civic-minded.</u>

   (A) has both educational value as well as making students civic-minded.
   (B) teaches students valuable lessons and makes students civic-minded.
   (C) has both educational and civic value.
   (D) both educates and civilizes students.
   (E) is educationally as well as civically valuable.

2. In context, which of the following best replaces the word "They" in sentence 4?

   (A) They
   (B) Students
   (C) Educators
   (D) Officials
   (E) Aid recipients

3. In context, which is the best way to revise the underlined portion of sentence 4 (reproduced below)?

   They learn good work habits and assistance is provided where they are needed.

   (A) and assistance is provided where they are needed.
   (B) and provide assistance where it is needed.
   (C) and assist the needy and provide other kinds of assistance.
   (D) and, where needed, assist in providing for others.
   (E) and assistance skills provided where they are needed.

4. In context, which is the best way to revise the underlined portion of sentence 5 (reproduced below)?

   Almost everyone has positive feelings about the new graduating requirement.

   (A) Almost everyone
   (B) Although hardly anyone
   (C) Not everyone
   (D) Because everyone
   (E) No one

5. Which of the following is the best way to revise the underlined portion of sentence 6 (reproduced below)?

   Some students feel the community service requirement, which involves volunteering one's time and assistance for forty hours or more, is too demanding.

   (A) which involves volunteering one's time and assistance for forty hours or more
   (B) which involves volunteering forty or more hours of one's time
   (C) which involves volunteering forty or more hours
   (D) which involves volunteering one's time and assistance
   (E) a forty-hour or more volunteer commitment

6. In context, which is the best way to revise and combine the underlined portions of sentences 8, 9, and 10 (reproduced below)?

   They participate in sports. They work part-time jobs. Some even belong to clubs.

   (A) They participate in sports. They work part-time jobs. Some even belong to clubs.
   (B) They participate in sports, work part-time jobs, and belong to clubs.
   (C) They already devote time to sports, to working part-time, and to belong to clubs.
   (D) Their participation in sports, part-time work, and clubs is a kind of community service.
   (E) Sports, working part-time jobs, and clubs take up lots of a student's time.

7. Which of the following would be the most appropriate sentence to insert after sentence 15 ?

(A) A community service requirement should recognize the reality of students' lives.

(B) Community service should be a volunteer program for busy students.

(C) The benefits of a community service program should be weighed against its potential liabilities.

(D) Community service isn't the only way students contribute their time and effort.

(E) As long as students' concerns are addressed, community service programs will remain counterproductive.

Answers

1. (B) The underlined portion is wrong for two reasons. First, it uses the correlative conjunction "both" but not its partner, "and." Second, the elements linked by the phrase "as well as" are not grammatically equal. Choice (B) correctly joins two verb phrases with the conjunction "and." Choice (B) is the right answer. Choice (C) properly uses the correlative conjunctions "both" and "and"; it also links two grammatically equal phrases. But it does not reproduce the meaning of the original. Similarly, (D) and (E) are grammatically correct but alter what the writer is saying.

2. (B) The "they" is ambiguous, if only grammatically. The context strongly suggests that "they" refers to the students, making choice (B) correct. But because two plural nouns are mentioned in the previous sentence, the reader may initially be unclear about who "they" are. Choice (C) is clearly wrong because "educators" are previously mentioned. Choice (E) is wrong for the same reason.

3. (B) The underlined portion contains two passive voice constructions. Although the passive voice is acceptable, it should be used sparingly. Using the passive voice can also unnecessarily increase your chances of writing an awkward and grammatically ambiguous sentence. Does the "they" in "they are needed" refer to "habits" or "students"? The correct answer, choice (B), puts the first passive construction into the active voice and replaces the ambiguous "they" with "it," which unambiguously refers to "assistance." Choice (C), with its overuse of "and," lacks concision. Similarly, choice (D) needlessly complicates and expands the sentence. Choice (E) tries, but fails, to eliminate correctly one of the passive voice constructions.

4. (C) The relationship between sentences 5 and 6 needs to be clarified. The underlined part of sentence 5—"almost everyone"—downplays the existence of the minority that has negative feelings. But it soon becomes clear that the writer wants to focus on, not dismiss, the minority. Choice (C) properly signals the discussion of a dissatisfied minority. Choices (B) and (D) both contain subordinating conjunctions and would turn sentence 5 into a dependent clause fragment. Choice (E) would create a contradiction in the paragraph, which discusses students' negative feelings.

5. (C) The underlined portion in question 5 is too wordy and redundant. In the context of the essay, "volunteering" means devoting one's "time and assistance." Choice (C) eliminates this redundant phrase. In choice (B), "of one's time" is redundant because "hours" clearly establishes that time is being volunteered. (D) eliminates valuable information while retaining the original redundancy. Choice (E) introduces a new redundancy. "Volunteer" needlessly modifies "commitment," which we already know is of a volunteer nature.

6. (B) Choppy sentences should be combined. Be sure to choose the option that combines the sentences grammatically without changing the original meaning or creating an awkward sentence. Choice (B) combines the sentences by forming a parallel structure that consists of verb phrases. The sentence is clear and grammatical. (B) is the correct answer. Choice (C) is a faulty parallelism consisting of two noun phases and one infinitival phrase. Both choices (D) and (E) alter the original meaning of the sentences. They are also awkward—notice how late the main verb occurs in each sentence.

7. (A) The final question of a given Paragraph Improvement set may ask you to choose an appropriate final sentence for the passage. This sentence should be a good concluding sentence for the passage as a whole. The passage focuses on why many students are unhappy with the community service requirement, which doesn't really seem to take into account just how busy most students are. The best answer is (A). Choice (B) is incorrect because the writer never suggests totally eliminating the requirement for one group of students. Choice (C) looks like an attractive choice; the writer does begin by mentioning the supposed benefits of community service. But the writer never really specifies how the community service requirement has impacted students (e.g., the jobs they've had to forego, the activities they've given up). Choice (D) does follow from sentence 15, but it does not account for the passage's overall meaning or purpose. (E) contradicts the passage, which implies that students' concerns have yet to (and very much need to) be taken into account.

II. MASTERING THE ESSAY

The SAT Essay contains one writing exercise. You will have 25 minutes to plan and write an essay on a given topic. You must write on only that topic. Because you will have only 25 minutes to complete the essay, efficient use of your time is essential.

Writing under pressure can be frustrating, but if you study this review, practice and polish your essay skills, and have a realistic sense of what to expect, you can turn problems into possibilities. The following review will show you how to plan and write a logical, coherent, and interesting essay.

PREWRITING/PLANNING

Before you actually begin to write, there are certain preliminary steps you need to take. A few minutes spent planning pays off—your final essay will be more focused, well-developed, and clear. For a 25-minute essay, you should spend about five minutes on the prewriting process.

Understand the Question

Read the essay question very carefully and ask yourself the following questions:

- What is the meaning of the topic statement?
- Is the question asking me to persuade the reader of the validity of a certain opinion?
- Do I agree or disagree with the statement? What will be my thesis (main idea)?
- What kinds of examples can I use to support my thesis? Explore personal experiences, historical evidence, current events, and literary subjects.

Consider Your Audience

Essays would be pointless without an audience. Why write an essay if no one wants or needs to read it? Why add evidence, organize your ideas, or correct bad grammar? The reason to do any of these things is because someone out there needs to understand what you mean or say.

What does the audience need to know to believe you or to come over to your position? Imagine someone you know listening to you declare your position or opinion and then saying, "Oh, yeah? Prove it!" This is your audience—write to them. Ask yourself the following questions so that you will not be confronted with a person who says, "Prove it!"

- What evidence do I need to prove my idea to this audience?
- What would the audience disagree with me about?
- What does the audience share with me as common knowledge? What do I need to tell them?

WRITING YOUR ESSAY

Once you have considered your position on the topic and thought of several examples to support it, you are ready to begin writing.

Organizing Your Essay

Decide how many paragraphs you will write. In a 25-minute exercise, you will probably have time for no more than four or five paragraphs. In such a format, the first paragraph will be the introduction, the next two or three will develop your thesis with specific examples, and the final paragraph should be a strong conclusion.

The Introduction

The focus of your introduction should be the thesis statement. This statement allows your reader to understand the point and direction of your essay. The statement identifies the central idea of your essay and should clearly state your attitude about the subject. It will also dictate the basic content and organization of your essay. If you do not state your thesis clearly, your essay will suffer.

The thesis is the heart of the essay. Without it, readers won't know what your major message or central idea is in the essay.

The thesis must be something that can be argued or needs to be proven, not just an accepted fact. For example, "Animals are used every day in cosmetic and medical testing," is a fact—it needs no proof. But if the writer says, "Using animals for cosmetic and medical testing is cruel and should be stopped," we have a point that must be supported and defended by the writer.

The thesis can be placed in any paragraph of the essay, but in a short essay, especially one written for evaluative exam purposes, the thesis is most effective when placed in the last sentence of the opening paragraph.

Consider the following sample question:

ESSAY TOPIC

"That government is best which governs least."

ASSIGNMENT: Do you agree or disagree with this statement? Choose a specific example from current events, personal experience, or your reading to support your position.

After reading the topic statement, decide if you agree or disagree. If you agree with this statement, your thesis statement could be the following:

"Government has the right to protect individuals from interference, but it has no right to extend its powers and activities beyond this function."

This statement clearly states the writer's opinion in a direct manner. It also serves as a blueprint for the essay. The remainder of the introduction should give two or three brief examples that support your thesis.

Supporting Paragraphs

The next two or three paragraphs of your essay will elaborate on the supporting examples you gave in your introduction. Each paragraph should discuss only one idea. Like the introduction, each paragraph should be coherently organized with a topic sentence and supporting details.

The topic sentence is to each paragraph what the thesis statement is to the essay as a whole. It tells the reader what you plan to discuss in that paragraph. It has a specific subject and is neither too broad nor too narrow. It also establishes the author's attitude and gives the reader a sense of the direction in which the writer is going. An effective topic sentence also arouses the reader's interest.

Although it may occur in the middle or at the end of the paragraph, the topic sentence usually appears at the beginning of the paragraph. Placing the topic sentence at the beginning is advantageous because it helps you stay focused on the main idea.

The remainder of each paragraph should support the topic sentence with examples and illustrations. Each sentence should progress logically from the previous one and be centrally connected to your topic sentence. Do not include any extraneous material that does not serve to develop your thesis.

Conclusion

Your conclusion should briefly restate your thesis and explain how you have shown it to be true. Since you want to end your essay on a strong note, your conclusion should be concise and effective.

Do not introduce any new topics that you cannot support. If you were watching a movie that suddenly shifted plot and characters at the end, you would be disappointed or even angry. Similarly, conclusions must not drift away from the major focus and message of the essay. Make sure your conclusion is clearly on the topic and represents your perspective without any confusion about what you really mean and believe. The reader will respect you for staying true to your intentions.

The conclusion is your last chance to grab and impress the reader. You can even use humor, if appropriate, but a dramatic close will remind the reader you are serious, even passionate, about what you believe.

EFFECTIVE USE OF LANGUAGE

Clear organization, although vitally important, is not the only factor the graders of your essay consider. You must also demonstrate that you can express your ideas clearly, using correct grammar, diction, usage, spelling, and punctuation. For rules on grammar, usage, and mechanics, consult the Basic Verbal Skills Review in this book.

Point of View

Depending on the audience, essays may be written from one of three points of view:

1. *Subjective/Personal Point of View:*

   "I think . . . "

   "I believe cars are more trouble than they are worth."

   "I feel . . . "

2. *Second Person Point of View* (We...You; I...You):

   "If you own a car, you will soon find out that it is more trouble than it is worth."

3. *Third Person Point of View* (focuses on the idea, not what "I" think of it):

   "Cars are more trouble than they are worth."

It is very important to maintain a consistent point of view throughout your essay. If you begin writing in the first-person ("I"), do not shift to the second- or third-person in the middle of the essay. Such inconsistency is confusing to your reader and will be penalized by the graders of your essay.

Tone

A writer's tone results from his or her attitude toward the subject and the reader. If the essay question requires you to take a strong stand, the tone of your essay should reflect this.

Your tone should also be appropriate for the subject matter. A serious topic demands a serious tone. For a more light-hearted topic, you may wish to inject some humor into your essay.

Whatever tone you choose, be consistent. Do not make any abrupt shifts in tone in the middle of your essay.

Verb Tense

Make sure to remain in the same verb tense in which you began your essay. If you start in the past, make sure all verbs are past tense. Staying in the same verb tense improves the continuity and flow of ideas. Avoid phrases such as "now was," a confusing blend of present and past. Consistency of time is essential to the reader's understanding.

Transitions

Transitions are like the links of a bracelet, holding the beads or major points of your essay together. They help the reader follow the smooth flow of your ideas and show a connection between major and minor ideas. Transitions are used either at the beginning of a paragraph, or to show the connections among ideas within a single paragraph. Without transitions, you will jar the reader and distract him from your true ideas.

Here are some typical transitional words and phrases:

Linking similar ideas

again	for example	likewise
also	for instance	moreover
and	further	nor
another	furthermore	of course
besides	in addition	similarly
equally important	in like manner	too

Linking dissimilar/contradictory ideas

although	however	otherwise
and yet	in spite of	provided that
as if	instead	still
but	nevertheless	yet
conversely	on the contrary	on the other hand

Indicating cause, purpose, or result

as	for	so
as a result	for this reason	then
because	hence	therefore
consequently	since	thus

Indicating time or position

above	before	meanwhile
across	beyond	next
afterwards	eventually	presently
around	finally	second
at once	first	thereafter
at the present time	here	thereupon

Indicating an example or summary

as a result	in any event	in short
as I have said	in brief	on the whole
for example	in conclusion	to sum up
for instance	in fact	in other words

Common Writing Errors

The four writing errors most often made by beginning writers are run-ons (also known as fused sentences), fragments, lack of subject-verb agreement, and incorrect use of object:

1. **Run-ons:** "She swept the floor it was dirty" is a run-on, because the pronoun "it" stands as a noun subject and starts a new sentence. A period or semicolon is needed after "floor."
2. **Fragments:** "Before Jimmy learned how to play baseball" is a fragment, even though it has a subject and verb (Jimmy learned). The word "before" fragmentizes the clause, and the reader needs to know what happened before Jimmy learned how to play baseball.
3. **Problems with subject-verb agreement:** "Either Maria or Robert are going to the game" is incorrect because either Maria is going or Robert is going, but not both. The sentence should say, "Either Maria or Robert is going to the game."
4. **Incorrect object:** Probably the most common offender in this area is saying "between you and I," which sounds correct, but isn't. "Between" is a preposition that takes the objective case "me." The correct usage is "between you and me."

SAT Essay test graders also cite lack of thought and development, misspellings, and incorrect pronouns or antecedents, and lack of development as frequently occurring problems. Finally, keep in mind that clear, coherent handwriting always works to your advantage. Readers will appreciate an essay they can read with ease.

Five Words Weak Writers Overuse

Weak and beginning writers overuse the vague pronouns "you, we, they, this, and it" often without telling exactly who or what is represented by the pronoun.

1. Beginning writers often shift to second person **"you,"** when the writer means, "a person." This shift confuses readers and weakens the flow of the essay. Although "you" is commonly accepted in creative writing, journalism, and other arenas, in a short, formal essay, it is best to avoid "you" altogether.
2. **"We"** is another pronoun that should be avoided. If by "we" the writer means "Americans," "society," or some other group, then he or she should say so.
3. **"They"** is often misused in essay writing, because it is overused in conversation: "I went to the doctor, and they told me to take some medicine." Tell the reader who "they" are.
4. **"This"** is usually used incorrectly without a referent: "She told me she received a present. This sounded good to me." This what? This idea? This news? This present? Be clear—don't make your readers guess what you mean. The word "this" should be followed by a noun or referent.
5. **"It"** is a common problem among weak writers. To what does "it" refer? Your readers don't appreciate vagueness, so take the time to be clear and complete in your expression of ideas.

Use Your Own Vocabulary

Is it a good idea to use big words that sound good in the dictionary or thesaurus, but that you don't really use or understand? No. So whose vocabulary should you use? Your own. You will be most comfortable with your own level of vocabulary.

This "comfort zone" doesn't give you license to be informal in a formal setting or to violate the rules of standard written English, but if you try to write in a style that is not yours, your writing will be awkward and lack a true voice.

You should certainly improve and build your vocabulary at every opportunity, but remember: you should not attempt to change your vocabulary level at this point.

Avoid the Passive Voice

In writing, the active voice is preferable because it is emphatic and direct. A weak passive verb leaves the doer unknown or seemingly unimportant. However, the passive voice is essential when the action of the verb is more important than the doer, when the doer is unknown, or when the writer wishes to place the emphasis on the receiver of the action rather than on the doer.

PROOFREADING

Make sure to leave yourself enough time at the end to read over your essay for errors such as misspellings, omitted words, or incorrect punctuation. You will not have enough time to make large-scale revisions, but take this chance to make any small changes that will make your essay stronger. Consider the following when proofreading your work:

- Are all your sentences really sentences? Have you written any fragments or run-on sentences?
- Are you using vocabulary correctly?
- Did you leave out any punctuation? Did you capitalize correctly?
- Are there any misspellings, especially of difficult words?

If you have time, read your essay backwards from end to beginning. By doing so, you may catch errors that you missed reading forward only.

DRILL: ESSAY WRITING

DIRECTIONS: You have 25 minutes to plan and write an essay on the topic below. You may write only on the assigned topic.

Make sure to give specific examples to support your thesis. Proofread your essay carefully and take care to express your ideas clearly and effectively.

Write your essay on the lined pages at the back of the book.

ESSAY TOPIC

In the last 40 years, the deterioration of the environment has become a growing concern among both scientists and ordinary citizens.

ASSIGNMENT: Choose one pressing environmental problem, explain its negative impact, and discuss possible solutions.

DETAILED EXPLANATIONS OF ANSWERS

Drill: Essay Writing

This Answer Key provides three sample essays that represent possible responses to the essay topic. Compare your own response to those given on the next few pages. Allow the strengths and weaknesses of the sample essays help you critique your own essay and improve your writing skills.

ESSAY I (Score: 5–6)

There are many pressing environmental problems facing both this country and the world today. Pollution, the misuse and squandering of resources, and the cavalier attitude many people express all contribute to the problem. But one of the most pressing problems this country faces is the apathetic attitude many Americans have toward recycling.

Why is recycling so imperative? There are two major reasons. First, recycling previously used materials conserves precious natural resources. Many people never stop to think that reserves of metal ores are not unlimited. There is only so much gold, silver, tin, and other metals in the ground. Once it has all been mined, there will never be any more unless we recycle what has already been used.

Second, the United States daily generates more solid waste than any other country on earth. Our disposable consumer culture consumes fast food meals in paper or styrofoam containers, uses disposable diapers with plastic liners that do not biodegrade, receives pounds, if not tons, of unsolicited junk mail every year, and relies more and more on prepackaged rather than fresh food.

No matter how it is accomplished, increased recycling is essential. We have to stop covering our land with garbage, and the best ways to do this are to reduce our dependence on prepackaged goods and to minimize the amount of solid waste disposed of in landfills. The best way to reduce solid waste is to recycle it. Americans need to band together to recycle, to preserve our irreplaceable natural resources, reduce pollution, and preserve our precious environment.

Analysis

This essay presents a clearly defined thesis, and the writer elaborates on this thesis in a thoughtful and sophisticated manner. Various aspects of the problem under consideration are presented and explored, along with possible solutions. The support provided for the writer's argument is convincing and logical. There are few usage or mechanical errors to interfere with the writer's ability to communicate effectively. This writer demonstrates a comprehensive understanding of the rules of written English.

ESSAY II (Score: 3–4)

A pressing environmental problem today is the way we are cutting down too many trees and not planting any replacements for them. Trees are beneficial in many ways, and without them, many environmental problems would be much worse.

One of the ways trees are beneficial is that, like all plants, they take in carbon dioxide and produce oxygen. They can actually help clean the air this way. When too many trees are cut down in a small area, the air in that area is not as good and can be unhealthy to breath.

Another way trees are beneficial is that they provide homes for many types of birds, insects, and animals. When all the trees in an area are cut down, these animals lose their homes and sometimes they can die out and become extinct that way. Like the spotted owls in Oregon, that the loggers wanted to cut down the trees they lived in. If the loggers did cut down all the old timber stands that the spotted owls lived in, the owls would have become extinct.

But the loggers say that if they can't cut the trees down then they will be out of work, and that peoples' jobs are more important than birds. The loggers can do two things—they can either get training so they can do other jobs, or they can do what they should have done all along, and start replanting trees. For every mature tree they cut down, they should have to plant at least one tree seedling.

Cutting down the trees that we need for life, and that lots of other species depend on, is a big environmental problem that has a lot of long term consaquences. Trees are too important for all of us to cut them down without thinking about the future.

Analysis

This essay has a clear thesis, which the author does support with good examples. But the writer shifts between the chosen topic, which is that indiscriminate tree-cutting is a pressing environmental problem, and a list of the ways in which trees are beneficial and a discussion about the logging profession. Also, although there are few mistakes in usage and mechanics, the writer does have some problems with sentence structure. The writing is pedestrian and the writer does not elaborate on the topic as much as he or she could have. The writer failed to provide the kind of critical analysis that the topic required.

ESSAY III (Score: 1–2)

The most pressing environmental problem today is that lots of people and companies don't care about the environment, and they do lots of things that hurt the environment.

People throw littur out car windows and don't use trash cans, even if their all over a park, soda cans and fast food wrappers are all over the place. Cigarette butts are the worst cause the filters never rot. Newspapers and junk mail get left to blow all over the neighborhood, and beer bottles too.

Companies pollute the air and the water. Sometimes the ground around a company has lots of tocsins in it. Now companies can buy credits from other companies that let them pollute the air even more. They dump all kinds of chemacals into lakes and rivers that kills off the fish and causes acid rain and kills off more fish and some trees and small animuls and insects and then noone can go swimming or fishing in the lake.

People need to respect the environment because we only have one planet, and if we keep polluting it pretty soon nothing will grow and then even the people will die.

Analysis

The writer of this essay does not define his or her thesis for this essay. Because of this lack of a clear thesis, the reader is left to infer the topic from the body of the essay. It is possible to perceive the writer's intended thesis; however, the support for this thesis is very superficial. The writer presents a list of common complaints about polluters, without any critical discussion of the problems and possible solutions. Many sentences are run-ons and the writer has made several spelling errors. Although the author manages to communicate his or her position on the issue, he or she does so on such a superficial level and with so many errors in usage and mechanics that the writer fails to demonstrate an ability to effectively communicate.

CHAPTER 7

MASTERING BASIC MATH SKILLS

MASTERING BASIC MATH SKILLS

I. ARITHMETIC

II. ALGEBRA

III. GEOMETRY

IV. WORD PROBLEMS

Are you ready to master the math sections of the SAT? Well, the chances are that you will be, but only after some reviewing of basic concepts in arithmetic, algebra, and geometry. The more familiar you are with these fundamental principles, the better you will do on the math sections of the SAT. Our math review represents the various mathematical topics that will appear on the SAT. You will not find any trigonometry, calculus, or imaginary numbers in our math review. Why? Because these concepts are not tested on the math sections of the SAT. The mathematical concepts presented on the SAT are ones with which you are already familiar and simply need to review in order to score well.

Along with a knowledge of these topics, how quickly and accurately you can answer the math questions will have an effect upon your success. Therefore, memorize the directions in order to save time and decrease your chances of making careless mistakes. Then, complete the practice drills that are provided for you in our review. Even if you are sure you know your fundamental math concepts, the drills will help to warm you up so that you can go into the math sections of the SAT with quick, sharp math skills.

REFERENCE TABLE

SYMBOLS AND THEIR MEANINGS

$=$	is equal to	$\leq$	is less than or equal to
$\neq$	is unequal to	$\geq$	is greater than or equal to
$<$	is less than	$\parallel$	is parallel to
$>$	is greater than	$\perp$	is perpendicular to

FORMULAS

DESCRIPTION	FORMULA
Area (A) of a:	
square	$A = s^2$; where s = side
rectangle	$A = lw$; where l = length, w = width
parallelogram	$A = bh$; where b = base, h = height
triangle	$A = \frac{1}{2}bh$; where b = base, h = height
circle	$A = \pi r^2$; where $\pi = 3.14$, r = radius
Perimeter (P) of a:	
square	$P = 4s$; where s = side
rectangle	$P = 2l + 2w$; where l = length, w = width
triangle	$P = a + b + c$; where a, b, and c are the sides
circumference (C) of a circle	$C = \pi d$; where $\pi = 3.14$, d = diameter = $2r$
Volume (V) of a:	
cube	$V = s^3$; where s = side
rectangular container	$V = lwh$; where l = length, w = width, h = height
Pythagorean Theorem	$c^2 = a^2 + b^2$; where c = hypotenuse, a and b are legs of a right triangle
Distance (d):	
between two points in a plane	$d = \sqrt{(x_2 - x_1)^2 + (y_2 - y_1)^2}$ where (x_1, y_1) and (x_2, y_2) are two points in a plane
as a function of rate and time	$d = rt$; where r = rate, t = time
Mean	$\text{mean} = \frac{x_1 + x_2 + \ldots + x_n}{n}$ where the xs are the values for which a mean is desired, and n = number of values in the series
Median	**median** = the point in an ordered set of numbers at which half of the numbers are above and half of the numbers are below this value
Simple Interest (i)	$i = prt$; where p = principal, r = rate, t = time
Total Cost (c)	$c = nr$; where n = number of units, r = cost per unit

I. ARITHMETIC

MASTERING INTEGERS AND REAL NUMBERS

Most of the numbers used in algebra belong to a set called the **real numbers** or **reals**. This set can be represented graphically by the real number line.

Given the number line that follows, we arbitrarily fix a point and label it with the number 0. In a similar manner, we can label any point on the line with one of the real numbers, depending on its position relative to 0. Numbers to the right of 0 are positive, whereas those to the left are negative. Value increases from left to right, so that if *a* is to the right of *b*, it is said to be greater than *b*.

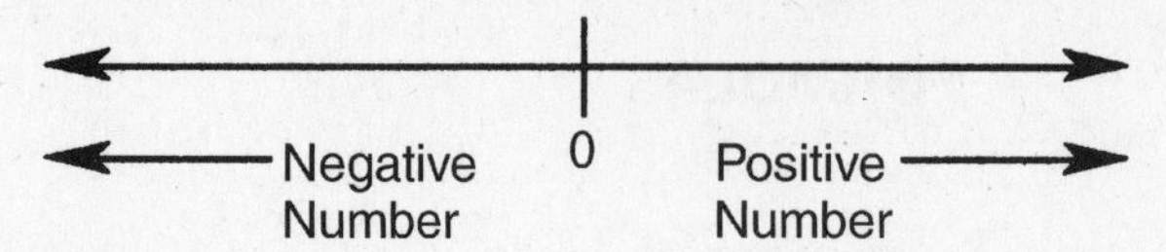

If we now divide the number line into equal segments, we can label the points on this line with real numbers. For example, the point 2 lengths to the left of 0 is –2, whereas the point 3 lengths to the right of 0 is +3 (the + sign is usually assumed, so +3 is written simply as 3). The number line now looks like this:

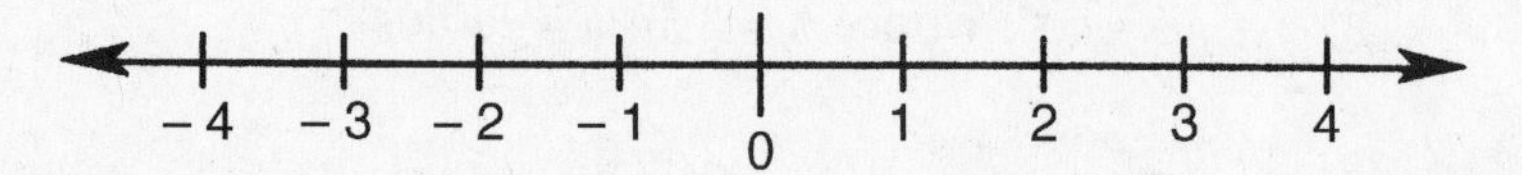

These boundary points represent the subset of the reals known as the **integers**. The set of integers is made up of both the positive and negative whole numbers:

$\{\ldots, -4, -3, -2, -1, 0, 1, 2, 3, 4, \ldots\}$.

Some subsets of integers are as follows:

Natural Numbers or Positive Numbers—the set of integers starting with 1 and increasing:

$N = \{1, 2, 3, 4, \ldots\}$.

Whole Numbers—the set of integers starting with 0 and increasing:

$W = \{0, 1, 2, 3, \ldots\}$.

Negative Numbers—the set of integers starting with –1 and decreasing:

$Z = \{-1, -2, -3, \ldots\}$.

Prime Numbers—the set of positive integers greater than 1 that are divisible only by 1 and themselves:

$\{2, 3, 5, 7, 11, \ldots\}$.

Even Integers—the set of integers divisible by 2:

$\{\ldots, -4, -2, 0, 2, 4, 6, \ldots\}$.

Odd Integers—the set of integers not divisible by 2:

$\{\ldots, -3, -1, 1, 3, 5, 7, \ldots\}$.

Consecutive Integers—the set of integers that differ by 1:

$\{n, n + 1, n + 2, \ldots\}$ (n = an integer).

PROBLEM

Classify each of the following numbers into as many different sets as possible. Example: real, integer . . .

(1) 0	*(3)* $\sqrt{6}$	*(5)* $\frac{2}{3}$
(2) 9	*(4)* $\frac{1}{2}$	*(6)* 1.5

SOLUTION

(1) 0 is a real number, an integer, and a whole number.

(2) 9 is a real number, an odd number, and a natural number.

(3) $\sqrt{6}$ is a real number.

(4) $\frac{1}{2}$ is a real number.

(5) $\frac{2}{3}$ is a real number.

(6) 1.5 is a real number and a decimal.

SETS

A **set** is a group of items. Each item in a set is called an **element** of that set. For example, let set X represent the set of positive odd integers between 0 and 10. This set has five elements: the numbers 1, 3, 5, 7, and 9. The elements of a set are often written within brackets, as shown next.

$$X = \{1, 3, 5, 7, 9\}$$

The **union** of two sets includes all of the elements in the first set and all of the elements in the second set. If set $Y = \{6, 7, 8, 9\}$, then the union of sets X and Y is the set $\{1, 3, 5, 6, 7, 8, 9\}$.

The **intersection** of two sets includes only the elements that the two sets have in common. The intersection of sets X and Y is the set $\{7, 9\}$.

One way to show the relationship between two sets is with a Venn diagram. In a Venn diagram, each set is represented by a circle. The intersection of the sets is represented by the area where the circles overlap. The following Venn diagram shows sets x and y. The shaded area shows their intersection.

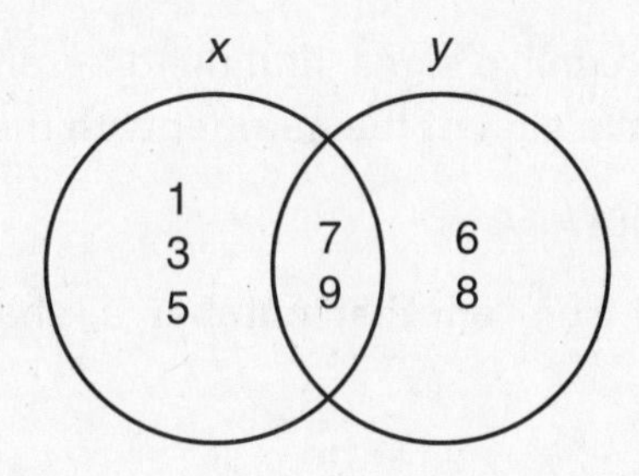

PROBLEM

A = {factors of 20}

B = {factors of 16}

C = {whole numbers}

Find the union and intersection of each pair of sets.

(1) set A and set B *(2) set A and set C*

SOLUTION

(1) A factor of a given number is a whole number that divides evenly into the given number. List the elements of sets A and B.

$A = \{1, 2, 4, 5, 10, 20\}$

$B = \{1, 2, 4, 8, 16\}$

The union of set A and set B is $\{1, 2, 4, 5, 8, 10, 16, 20\}$. The intersection of set A and set B is $\{1, 2, 4\}$.

(2) All of the elements in set A are whole numbers. Because set C is the set of whole numbers, each element in set A is also in set C. Therefore, the union of sets A and C is equal to set C, the set of whole numbers.

The elements that set A and set C have in common are all the elements in set A. Therefore, the intersection of sets A and C is equal to $\{1, 2, 4, 5, 10, 20\}$.

POSITIVE AND NEGATIVE NUMBERS

The **absolute value** of a number is represented by two vertical lines around the number and is equal to the given number, regardless of sign.

The absolute value of a real number A is defined as follows:

$$|A| = \begin{cases} A \text{ if } A \geq 0 \\ -A \text{ if } A < 0 \end{cases}$$

A) **To add two numbers with like signs,** add their absolute values and write the sum with the common sign. So,

$6 + 2 = 8, (-6) + (-2) = -8$

B) **To add two numbers with unlike signs,** find the difference between their absolute values, and write the result with the sign of the number with the greater absolute value. So,

$(-4) + 6 = 2, 15 + (-19) = -4$

C) **To subtract a number** b **from another number** a, change the sign of b and add to a. Examples:

$10 - (3) = 10 + (-3) = 7$ (1)

$2 - (-6) = 2 + 6 = 8$ (2)

$(-5) - (-2) = -5 + 2 = -3$ (3)

D) **To multiply (or divide) two numbers having like signs,** multiply (or divide) their absolute values and write the result with a positive sign. Examples:

$$(5)\,(3) = 15 \quad (1)$$

$$(-6) \div (-3) = 2 \quad (2)$$

E) **To multiply (or divide) two numbers having unlike signs,** multiply (or divide) their absolute values and write the result with a negative sign. Examples:

$$(-2)\,(8) = -16 \quad (1)$$

$$9 \div (-3) = -3 \quad (2)$$

According to the law of signs for real numbers, the square of a positive or negative number is always positive. This means that it is impossible to take the square root of a negative number in the real number system.

PROBLEM

Calculate the value of each of the following expressions:

(1) $(2-5)+6-14$ *(2)* $-8\times 2+\frac{-12}{4}$

SOLUTION

Before solving this problem, one must use the rules for the **order of operations.** Always work within the parentheses or with absolute values first while keeping in mind that multiplication and division are carried out before addition and subtraction.

(1)
$$\begin{aligned}(2-5)+6-14 &= -3+6-14\\ &= 3-14\\ &= -11\end{aligned}$$

(2)
$$\begin{aligned}-8\times 2+\frac{-12}{4} &= -16+\frac{-12}{4}\\ &= -16+(-3)\\ &= -19\end{aligned}$$

ODD AND EVEN NUMBERS

When dealing with odd and even numbers keep in mind the following:

Adding:

even + even = even

odd + odd = even

even + odd = odd

Multiplying:

even × even = even

even × odd = even

odd × odd = odd

DRILL: INTEGERS AND REAL NUMBERS

Sets

$X = \{-3, -1, 0, 4, 5\}$

$Y = \{-1, 3, 4, 6\}$

1. What is the intersection of sets X and Y?

   (A) $\{-1, 4\}$ (B) $\{-1, 3, 4\}$ (C) $\{-3, -1, 4\}$
   (D) $\{-3, -1, 0, 4\}$ (E) $\{-3, -1, 0, 3, 4, 5, 6\}$

2. How many different elements are in the union of sets X and Y?

   (A) 2 (B) 4 (C) 5 (D) 7 (E) 9

3. Set J is the set of factors of 12. Set K is the set of factors of 8. What is the union of sets J and K?

   (A) $\{1, 2, 4\}$ (B) $\{1, 2, 4, 8\}$ (C) $\{1, 2, 4, 6, 8\}$
   (D) $\{1, 2, 3, 4, 6, 12\}$ (E) $\{1, 2, 3, 4, 6, 8, 12\}$

4. What is the intersection of the set of positive integers and the set of even numbers?

   (A) the set of positive integers
   (B) the set of even positive integers
   (C) the set of even integers
   (D) the set of odd positive integers
   (E) the set of even negative integers

5. What is the union of the set of multiples of 3 and the set of multiples of 6?

   (A) the set of multiples of 3
   (B) the set of multiples of 6
   (C) the set of multiples of 18
   (D) the set of whole numbers
   (E) the set of positive integers

Addition

6. Simplify $4 + (-7) + 2 + (-5)$.

   (A) -6 (B) -4 (C) 0 (D) 6 (E) 18

7. Simplify $144 + (-317) + 213$.

   (A) -357 (B) -40 (C) 40 (D) 357 (E) 674

8. Simplify $4 + (-3) + (-2)$.

   (A) -2 (B) -1 (C) 1 (D) 3 (E) 9

9. What integer makes the equation $-13 + 12 + 7 + ? = 10$ a true statement?

   (A) -22 (B) -10 (C) 4 (D) 6 (E) 10

10. Simplify $4 + 17 + (-29) + 13 + (-22) + (-3)$.

    (A) -44 (B) -20 (C) 23 (D) 34 (E) 78

Subtraction

11. Simplify $319 - 428$.

    (A) -111 (B) -109 (C) -99 (D) 109 (E) 747

12. Simplify 91,203 – 37,904 + 1,073.

(A) 54,372 (B) 64,701 (C) 128,034 (D) 129,107 (E) 130,180

13. Simplify 43 – 62 – (–17 – 3).

(A) –39 (B) –19 (C) –1 (D) 1 (E) 39

14. Simplify –(–4 – 7) + (–2).

(A) –22 (B) –13 (C) –9 (D) 7 (E) 9

15. In the St. Elias Mountains, Mt. Logan rises from 1,292 meters above sea level to 7,243 meters above sea level. How tall is Mt. Logan?

(A) 4,009 m (B) 5,951 m (C) 5,699 m (D) 6,464 m (E) 7,885 m

Multiplication

16. Simplify (–3) × (–18) × (–1).

(A) –108 (B) –54 (C) –48 (D) 48 (E) 54

17. Simplify (–42) × 7.

(A) –294 (B) –49 (C) –35 (D) 284 (E) 294

18. Simplify (–6) × 5 × (–10) × (–4) × 0 × 2.

(A) –2,400 (B) –240 (C) 0 (D) 280 (E) 2,700

19. Simplify –(–6 × 8).

(A) –48 (B) –42 (C) 2 (D) 42 (E) 48

20. A city in Georgia had a record low temperature of –3°F one winter. During the same year, a city in Michigan experienced a record low that was nine times the record low set in Georgia. What was the record low in Michigan that year?

(A) –31°F (B) –27°F (C) –21°F (D) –12°F (E) –6°F

Division

21. Simplify (–24) ÷ 8.

(A) –4 (B) –3 (C) –2 (D) 3 (E) 4

22. Simplify (–180) ÷ (–12).

(A) –30 (B) –15 (C) 1.5 (D) 15 (E) 216

23. Simplify 76 ÷ (–4).

(A) –21 (B) –19 (C) 13 (D) 19 (E) 21.5

24. Simplify –(–216 ÷ 6) .

(A) –36 (B) –12 (C) 36 (D) 38 (E) 43

25. At the end of the year, a small firm has $2,996 in its account for bonuses. If the entire amount is equally divided among the 14 employees, how much does each one receive?

(A) $107 (B) $114 (C) $170 (D) $210 (E) $214

Order of Operations

26. Simplify $\frac{4+8\times2}{5-1}$

(A) 4 (B) 5 (C) 6 (D) 8 (E) 12

27. $96 \div 3 \div 4 \div 2 =$

(A) 65 (B) 64 (C) 16 (D) 8 (E) 4

28. $3 + 4 \times 2 - 6 \div 3 =$

(A) -1 (B) $\frac{5}{3}$ (C) $\frac{8}{3}$ (D) 9 (E) 12

29. $[(4 + 8) \times 3] \div 9 =$

(A) 4 (B) 8 (C) 12 (D) 24 (E) 36

30. $18 + 3 \times 4 \div 3 =$

(A) 3 (B) 5 (C) 10 (D) 22 (E) 28

31. $(29 - 17 + 4) \div 4 + 2 =$

(A) 22/3 (B) 4 (C) $4\frac{2}{3}$ (D) 6 (E) 15

32. $(-3) \times 5 - 20 \div 4 =$

(A) -75 (B) -20 (C) -10 (D) $8\frac{3}{4}$ (E) 20

33. $\frac{11\times2+2}{16-2\times2} =$

(A) 11/16 (B) 1 (C) 2 (D) $3\frac{2}{3}$ (E) 4

34. (–8 – 4) ÷ (–3) #6 + 2 =

(A) 20 (B) 26 (C) 32 (D) 62 (E) 212

35. $32 \div 2 + 4 - 15 \div 3 =$

(A) 0 (B) 7 (C) 15 (D) 23 (E) 63

MASTERING FRACTIONS

The fraction $\frac{a}{b}$, where the **numerator** is *a* and the **denominator** is *b*, implies that *a* is being divided by *b*. The denominator of a fraction can never be zero since a number divided by zero is not defined. If the numerator is greater than the denominator, the fraction is called an **improper fraction**. A **mixed number** is the sum of a whole number and a fraction:

$$4\frac{3}{8} = 4 + \frac{3}{8}.$$

Operations with Fractions

A) **To change a mixed number to an improper fraction,** simply multiply the whole number by the denominator of the fraction and add the numerator. This product becomes the numerator of the result and the denominator remains the same:

$$5\frac{2}{3} = \frac{(5\times3)+2}{3} = \frac{15+2}{3} = \frac{17}{3}$$

To change an improper fraction to a mixed number, simply divide the numerator by the denominator. The remainder becomes the numerator of the fractional part of the mixed number, and the denominator remains the same:

$$\frac{35}{4} = 35 \div 4 = 8\frac{3}{4}$$

To check your work, change your result back to an improper fraction to see if it matches the original fraction.

B) **To find the sum of fractions having a common denominator,** simply add together the numerators of the given fractions and put this sum over the common denominator.

$$\frac{11}{3} + \frac{5}{3} = \frac{11+5}{3} = \frac{16}{3}$$

Similarly for subtraction,

$$\frac{11}{3} - \frac{5}{3} = \frac{11-5}{3} = \frac{6}{3} = 2$$

C) **To find the sum of two fractions having different denominators,** it is necessary to find the **lowest common denominator (LCD)** of the different denominators using a process called **factoring.**

To **factor** a number means to find two numbers that when multiplied together have a product equal to the original number. These two numbers are then said to be **factors** of the original number. For example, the factors of 6 are

(1) 1 and 6 because $1 \times 6 = 6$.

(2) 2 and 3 because $2 \times 3 = 6$.

Every number is the product of itself and 1. A **prime factor** is a number that does not have any factors besides itself and 1. This is important when finding the LCD of two fractions having different denominators.

To find the LCD of $\frac{11}{16}$ and $\frac{5}{16}$, we must first find the prime factors of each of the two denominators.

$$6 = 2 \times 3$$

$$16 = 2 \times 2 \times 2 \times 2$$

$$\text{LCD} = 2 \times 2 \times 2 \times 2 \times 3 = 48$$

Note that we do not need to repeat the 2 that appears in both the factors of 6 and 16.

Once we have determined the LCD of the denominators, each of the fractions must be converted into equivalent fractions having the LCD as a denominator.

Rewrite $\frac{11}{16}$ and $\frac{5}{16}$ to have 48 as their denominators.

$6 \times ? = 48$ $\quad$ $16 \times ? = 48$

$6 \times 8 = 48$ $\quad$ $16 \times 3 = 48$

If the numerator and denominator of each fraction is multiplied (or divided) by the same number, the value of the fraction will not change. This is because a fraction $\frac{b}{b}$, b being any number, is equal to the multiplicative identity, 1.

Therefore

$$\frac{11}{6}\times\frac{8}{8}=\frac{88}{48} \quad \frac{5}{16}\times\frac{3}{3}=\frac{15}{48}$$

We may now find

$$\frac{11}{6}+\frac{5}{16}=\frac{88}{48}+\frac{15}{48}=\frac{103}{48}$$

Similarly for subtraction

$$\frac{11}{6}-\frac{5}{16}=\frac{88}{48}-\frac{15}{48}=\frac{73}{48}$$

D) **To find the product of two or more fractions,** simply multiply the numerators of the given fractions to find the numerator of the product and multiply the denominators of the given fractions to find the denominator of the product:

$$\frac{2}{3}\times\frac{1}{5}\times\frac{4}{7}=\frac{2\times1\times4}{3\times5\times7}=\frac{8}{105}$$

E) **To find the quotient of two fractions,** simply invert (or flip over) the divisor and multiply:

$$\frac{8}{9}\div\frac{1}{3}=\frac{8}{9}\times\frac{3}{1}=\frac{24}{9}=\frac{8}{3}$$

F) **To simplify a fraction** is to convert it into a form in which the numerator and denominator have no common factor other than 1:

$$\frac{12}{18}=\frac{12\div6}{18\div6}=\frac{2}{3}$$

G) A **complex fraction** is a fraction whose numerator and/or denominator is made up of fractions. To simplify the fraction, find the LCD of all the fractions. Multiply both the numerator and denominator by this number and simplify.

PROBLEM

If a = 4 and b = 7, find the value of $\dfrac{a+\frac{a}{b}}{a-\frac{a}{b}}$.

SOLUTION

By substitution,

$$\frac{a+\frac{a}{b}}{a-\frac{a}{b}}=\frac{4+\frac{4}{7}}{4-\frac{4}{7}}$$

In order to combine the terms, we must find the LCD of 1 and 7. Since both are prime factors, the LCD = 1 × 7 = 7.

Multiplying both the numerator and denominator by 7, we get

$$\frac{7\left(4+\frac{4}{7}\right)}{7\left(4-\frac{4}{7}\right)}=\frac{28+4}{28-4}=\frac{32}{24}$$

By dividing both the numerator and denominator by 8, $\frac{32}{24}$ can be reduced to $\frac{4}{3}$.

DRILL: FRACTIONS

Changing an Improper Fraction to a Mixed Number

DIRECTIONS: Write each improper fraction as a mixed number in simplest form.

1. $\frac{50}{4}$

   (A) $10\frac{1}{4}$ (B) $11\frac{1}{2}$ (C) $12\frac{1}{4}$ (D) $12\frac{1}{2}$ (E) 25

2. $\frac{17}{5}$

   (A) $3\frac{2}{5}$ (B) $3\frac{3}{5}$ (C) $3\frac{4}{5}$ (D) $4\frac{1}{5}$ (E) $4\frac{2}{5}$

3. $\frac{42}{3}$

   (A) $10\frac{2}{3}$ (B) 12 (C) $13\frac{1}{3}$ (D) 14 (E) $21\frac{1}{3}$

4. $\frac{85}{6}$

   (A) $9\frac{1}{6}$ (B) $10\frac{5}{6}$ (C) $11\frac{1}{2}$ (D) 12 (E) $14\frac{1}{6}$

5. $\frac{151}{7}$

   (A) $19\frac{6}{7}$ (B) $20\frac{1}{7}$ (C) $21\frac{4}{7}$ (D) $31\frac{2}{7}$ (E) $31\frac{4}{7}$

Changing a Mixed Number to an Improper Fraction

DIRECTIONS: Change each mixed number to an improper fraction in simplest form.

6. $2\frac{3}{5}$

   (A) $\frac{4}{5}$ (B) $\frac{6}{5}$ (C) $\frac{11}{5}$ (D) $\frac{13}{5}$ (E) $\frac{17}{5}$

7. $4\frac{3}{4}$

   (A) $\frac{7}{4}$ (B) $\frac{13}{4}$ (C) $\frac{16}{3}$ (D) $\frac{19}{4}$ (E) $\frac{21}{4}$

8. $6\frac{7}{6}$

   (A) $\frac{13}{6}$ (B) $\frac{43}{6}$ (C) $\frac{19}{36}$ (D) $\frac{42}{36}$ (E) $\frac{48}{6}$

9. $12\frac{3}{7}$

(A) $\frac{87}{7}$ (B) $\frac{164}{14}$ (C) $\frac{34}{3}$ (D) $\frac{187}{21}$ (E) $\frac{252}{7}$

10. $21\frac{1}{2}$

(A) $\frac{11}{2}$ (B) $\frac{22}{2}$ (C) $\frac{24}{2}$ (D) $\frac{42}{2}$ (E) $\frac{43}{2}$

Adding Fractions with the Same Denominator

DIRECTIONS: Add and write the answer in simplest form.

11. $\frac{5}{12}+\frac{3}{12}=$

(A) $\frac{5}{24}$ (B) $\frac{1}{3}$ (C) $\frac{8}{12}$ (D) $\frac{2}{3}$ (E) $\frac{2}{3}$

12. $\frac{5}{8}+\frac{7}{8}+\frac{3}{8}=$

(A) $\frac{5}{24}$ (B) $\frac{3}{4}$ (C) $\frac{5}{6}$ (D) $\frac{7}{8}$ (E) $1\frac{7}{8}$

13. $131\frac{2}{15}+28\frac{3}{15}=$

(A) $159\frac{1}{6}$ (B) $159\frac{1}{5}$ (C) $159\frac{1}{3}$ (D) $159\frac{1}{2}$ (E) $159\frac{3}{5}$

14. $3\frac{5}{18}+2\frac{1}{18}+8\frac{7}{18}=$

(A) $13\frac{13}{18}$ (B) $13\frac{3}{4}$ (C) $13\frac{7}{9}$ (D) $14\frac{1}{6}$ (E) $14\frac{2}{9}$

15. $17\frac{9}{20}+4\frac{3}{20}+8\frac{11}{20}=$

(A) $29\frac{23}{60}$ (B) $29\frac{23}{20}$ (C) $30\frac{3}{20}$ (D) $30\frac{1}{5}$ (E) $30\frac{3}{5}$

Subtracting Fractions with the Same Denominator

DIRECTIONS: Subtract and write the answer in simplest form.

16. $4\frac{7}{8}-3\frac{1}{8}=$

(A) $1\frac{1}{4}$ (B) $1\frac{3}{4}$ (C) $1\frac{12}{16}$ (D) $1\frac{7}{8}$ (E) 2

17. $132\frac{5}{12} - 37\frac{3}{12} =$

(A) $94\frac{1}{6}$ (B) $95\frac{1}{12}$ (C) $95\frac{1}{6}$ (D) $105\frac{1}{6}$ (E) $169\frac{2}{3}$

18. $19\frac{1}{3} - 2\frac{2}{3} =$

(A) $16\frac{2}{3}$ (B) $16\frac{5}{6}$ (C) $17\frac{1}{3}$ (D) $17\frac{2}{3}$ (E) $17\frac{5}{6}$

19. $\frac{8}{21} - \frac{5}{21} =$

(A) $\frac{1}{21}$ (B) $\frac{1}{7}$ (C) $\frac{3}{21}$ (D) $\frac{2}{7}$ (E) $\frac{3}{7}$

20. $82\frac{7}{10} - 38\frac{9}{10} =$

(A) $43\frac{4}{5}$ (B) $44\frac{1}{5}$ (C) $44\frac{2}{5}$ (D) $44\frac{1}{5}$ (E) $45\frac{2}{10}$

Finding the LCD

DIRECTIONS: Find the lowest common denominator of each group of fractions.

21. $\frac{2}{3}, \frac{5}{9}$, and $\frac{1}{6}$

(A) 9 (B) 18 (C) 27 (D) 54 (E) 162

22. $\frac{1}{2}, \frac{5}{6}$, and $\frac{3}{4}$

(A) 2 (B) 4 (C) 6 (D) 12 (E) 48

23. $\frac{7}{16}, \frac{5}{6}$, and $\frac{2}{3}$

(A) 3 (B) 6 (C) 12 (D) 24 (E) 48

24. $\frac{8}{15}, \frac{2}{5}$, and $\frac{12}{25}$

(A) 5 (B) 15 (C) 25 (D) 75 (E) 375

25. $\frac{2}{3}, \frac{1}{5}$, and $\frac{5}{6}$

(A) 15 (B) 30 (C) 48 (D) 90 (E) 120

26. $\frac{1}{3}, \frac{9}{42}$, and $\frac{4}{21}$

(A) 21 (B) 42 (C) 126 (D) 378 (E) 4,000

27. $\frac{4}{9}, \frac{2}{5}$, and $\frac{1}{3}$

(A) 15 (B) 17 (C) 27 (D) 45 (E) 135

28. $\frac{7}{12}, \frac{11}{36}$, and $\frac{1}{9}$

(A) 12 (B) 36 (C) 108 (D) 324 (E) 432

29. $\frac{3}{7}, \frac{5}{21}$, and $\frac{2}{3}$

(A) 21 (B) 42 (C) 31 (D) 63 (E) 441

30. $\frac{13}{16}, \frac{5}{8}$, and $\frac{1}{4}$

(A) 4 (B) 8 (C) 16 (D) 32 (E) 64

Adding Fractions with Different Denominators

DIRECTIONS: Add and write the answer in simplest form.

31. $\frac{1}{3}+\frac{5}{12}=$

(A) $\frac{2}{5}$ (B) $\frac{1}{2}$ (C) $\frac{9}{12}$ (D) $\frac{3}{4}$ (E) $1\frac{1}{3}$

32. $3\frac{5}{9}+2\frac{1}{3}=$

(A) $5\frac{1}{2}$ (B) $5\frac{2}{3}$ (C) $5\frac{8}{9}$ (D) $6\frac{1}{9}$ (E) $6\frac{2}{3}$

33. $12\frac{9}{16}+17\frac{3}{4}+8\frac{1}{8}=$

(A) $37\frac{7}{16}$ (B) $38\frac{7}{16}$ (C) $38\frac{1}{2}$ (D) $38\frac{2}{3}$ (E) $39\frac{3}{16}$

34. $28\frac{4}{5}+11\frac{16}{25}=$

(A) $39\frac{2}{3}$ (B) $39\frac{4}{5}$ (C) $40\frac{9}{25}$ (D) $40\frac{2}{5}$ (E) $40\frac{11}{25}$

35. $2\frac{1}{8}+1\frac{3}{16}+\frac{5}{12}=$

(A) $3\frac{35}{48}$ (B) $3\frac{3}{4}$ (C) $3\frac{19}{24}$ (D) $3\frac{13}{16}$ (E) $4\frac{1}{12}$

Subtracting Fractions with Different Denominators

DIRECTIONS: Subtract and write the answer in simplest form.

36. $8\frac{9}{12}-2\frac{2}{3}=$

(A) $6\frac{1}{12}$ (B) $6\frac{1}{6}$ (C) $6\frac{1}{3}$ (D) $6\frac{7}{12}$ (E) $6\frac{2}{3}$

37. $185\frac{11}{15}-107\frac{2}{5}=$

(A) $77\frac{2}{15}$ (B) $78\frac{1}{5}$ (C) $78\frac{3}{10}$ (D) $78\frac{1}{3}$ (E) $78\frac{1}{3}$

38. $34\frac{2}{3}-16\frac{5}{6}$

(A) 16 (B) $16\frac{1}{3}$ (C) $17\frac{1}{2}$ (D) 17 (E) $17\frac{5}{6}$

39. $3\frac{11}{48}-2\frac{3}{16}=$

(A) $\frac{47}{48}$ (B) $1\frac{1}{48}$ (C) $1\frac{1}{24}$ (D) $1\frac{8}{48}$ (E) $1\frac{7}{24}$

40. $81\frac{4}{21}-31\frac{1}{3}=$

(A) $47\frac{3}{7}$ (B) $49\frac{6}{7}$ (C) $49\frac{1}{6}$ (D) $49\frac{5}{7}$ (E) $49\frac{13}{21}$

Multiplying Fractions

DIRECTIONS: Multiply and reduce the answer.

41. $\frac{2}{3}\times\frac{4}{5}=$

(A) $\frac{6}{8}$ (B) $\frac{3}{4}$ (C) $\frac{8}{15}$ (D) $\frac{10}{12}$ (E) $\frac{6}{5}$

42. $\frac{7}{10}\times\frac{4}{21}=$

(A) $\frac{2}{15}$ (B) $\frac{11}{31}$ (C) $\frac{28}{210}$ (D) $\frac{1}{6}$ (E) $\frac{4}{15}$

43. $5\frac{1}{3} \times \frac{3}{8} =$

(A) $\frac{4}{11}$ (B) 2 (C) $\frac{8}{5}$ (D) $5\frac{1}{8}$ (E) $5\frac{17}{24}$

44. $6\frac{1}{2} \times 3 =$

(A) $9\frac{1}{2}$ (B) $18\frac{1}{2}$ (C) $19\frac{1}{2}$ (D) 20 (E) $12\frac{1}{2}$

45. $3\frac{1}{4} \times 2\frac{1}{3} =$

(A) $5\frac{7}{12}$ (B) $6\frac{2}{7}$ (C) $6\frac{5}{7}$ (D) $7\frac{7}{12}$ (E) $7\frac{11}{12}$

Dividing Fractions

DIRECTIONS: Divide and reduce the answer.

46. $\frac{3}{16} \div \frac{3}{4} =$

(A) $\frac{9}{64}$ (B) $\frac{1}{4}$ (C) $\frac{6}{16}$ (D) $\frac{9}{16}$ (E) $\frac{3}{4}$

47. $\frac{4}{9} \div \frac{2}{3} =$

(A) $\frac{1}{3}$ (B) $\frac{1}{2}$ (C) $\frac{2}{3}$ (D) $\frac{7}{11}$ (E) $\frac{8}{9}$

48. $5\frac{1}{4} \div \frac{7}{10} =$

(A) $2\frac{4}{7}$ (B) $3\frac{27}{40}$ (C) $5\frac{19}{20}$ (D) $7\frac{1}{2}$ (E) $8\frac{1}{4}$

49. $4\frac{2}{3} \div \frac{7}{9} =$

(A) $2\frac{24}{27}$ (B) $3\frac{2}{9}$ (C) $4\frac{14}{27}$ (D) $5\frac{12}{27}$ (E) 6

50. $3\frac{2}{5} \div 1\frac{7}{10} =$

(A) 2 (B) $3\frac{4}{7}$ (C) $4\frac{7}{25}$ (D) $5\frac{1}{10}$ (E) $5\frac{2}{7}$

MASTERING DECIMALS

When we divide the denominator of a fraction into its numerator, the result is a **decimal**. The decimal is based on a fraction with a denominator of 10, 100, 1,000, . . . and is written with a **decimal point**. Whole numbers are placed to the left of the decimal point, where the first place to the left is the units place, the second to the left is the tens, the third to the left is the hundreds, and so on. The fractions are placed on the right where the first place to the right is the tenths, the second to the right is the hundredths, and so on.

EXAMPLES

$12\frac{3}{10} = 12.3 \qquad 4\frac{17}{100} = 4.17 \qquad \frac{3}{100} = 0.03$

Since a **rational number** is of the form $\frac{a}{b}$, $b \neq 0$, then all rational numbers can be expressed as decimals by dividing b into a. The result is either a **terminating decimal**, meaning that b divides a with a remainder of 0 after a certain point, or **repeating decimal**, meaning that b continues to divide a so that the decimal has a repeating pattern of integers.

EXAMPLES

(A) $\frac{1}{2} = 0.5$

(B) $\frac{1}{3} = 0.333\ldots$

(C) $\frac{11}{16} = 0.285714285714\ldots$

(D) $\frac{2}{7} = 0.285714285714\ldots$

(A) and (C) are terminating decimals; (B) and (D) are repeating decimals. This explanation allows us to define **irrational numbers** as numbers whose decimal form is nonterminating and nonrepeating:

$\sqrt{2} = 1.414\ldots$

$\sqrt{3} = 1.732\ldots$

PROBLEM

Express $-\frac{10}{20}$ *as a decimal.*

SOLUTION

$-\frac{10}{20} = -\frac{50}{100} = -0.5$

PROBLEM

Write $\frac{2}{7}$ as a repeating decimal.

SOLUTION

To write a fraction as a repeating decimal divide the numerator by the denominator until a pattern of repeated digits appears.

$$2 \div 7 = 0.285714285714\ldots$$

Identify the entire portion of the decimal that is repeated. The repeating decimal can then be written in the shortened form:

$$\frac{2}{7} = 0.\overline{285714}$$

Operations with Decimals

A) **To add numbers containing decimals,** write the numbers in a column making sure the decimal points are lined up, one beneath the other. Add the numbers as usual, placing the decimal point in the sum so that it is still in line with the others. It is important not to mix the digits in the tenths place with the digits in the hundredths place, and so on.

EXAMPLES

2.558 + 6.391

$$\begin{array}{r} 2.558 \\ +\ 6.391 \\ \hline 8.949 \end{array}$$

57.51 + 6.2

$$\begin{array}{r} 57.51 \\ +\ 6.20 \\ \hline 63.71 \end{array}$$

Similarly with subtraction,

78.54 – 21.33

$$\begin{array}{r} 78.54 \\ -\ 21.33 \\ \hline 57.21 \end{array}$$

7.11 – 4.2

$$\begin{array}{r} 7.11 \\ -\ 4.20 \\ \hline 2.91 \end{array}$$

Note that if two numbers differ according to the number of digits to the right of the decimal point, zeros must be added.

0.63 – 0.214

$$\begin{array}{r} 0.630 \\ -\ 0.214 \\ \hline 0.416 \end{array}$$

15.224 – 3.6891

$$\begin{array}{r} 15.2240 \\ -\ 3.6891 \\ \hline 11.5349 \end{array}$$

B) **To multiply numbers with decimals,** simply multiply as usual. Then, to figure out the number of decimal places that belong in the product, find the total number of decimal places in the numbers being multiplied.

EXAMPLES

```
   6.555  (3 decimal places)          5.32  (2 decimal places)
×    4.5  (1 decimal place)         × 0.04  (2 decimal places)
   32775                              2128
  26220                               000
  294975                              2128

 29.4975  (4 decimal places)        0.2128  (4 decimal places)
```

C) **To divide numbers with decimals,** you must first make the divisor a whole number by moving the decimal point the appropriate number of places to the right. The decimal point of the dividend should also be moved the same number of places. Place a decimal point in the quotient, directly in line with the decimal point in the dividend.

EXAMPLES

```
12.92 ÷ 3.4                      40.376 ÷ 7.21

        3.8                               5.6
3.4. )12.9.2                     7.21. )40.37.6
     -102                              -3605
       272                               4326
      -272                              -4326
         0                                  0
```

If the question asks you to find the correct answer to two decimal places, simply divide until you have three decimal places and then round off. If the third decimal place is a 5 or larger, the number in the second decimal place is increased by 1. If the third decimal place is less than 5, that number is simply dropped.

PROBLEM

Find the answer to the following to two decimal places:

(1) 44.3 ÷ 3 *(2) 56.99 ÷ 6*

SOLUTION

```
(1)     14.766                       9.498
     3 )44.300                    6 )56.990
        -3                           -54
         14                            29
        -12                           -24
          23                            59
         -21                           -54
           20                            50
          -18                           -48
            20                            2
           -18
             2
```

14.766 can be rounded off to 14.77

9.498 can be rounded off to 9.50

D) When comparing two numbers with decimals to see which is the larger, first look at the tenths place. The larger digit in this place represents the larger number. If the two digits are the same, however, take a look at the digits in the hundredths place, and so on.

EXAMPLES

0.518 and 0.216

5 is larger than 2, therefore 0.518 is larger than 0.216

0.723 and 0.726

6 is larger than 3, therefore 0.726 is larger than 0.723

DRILL: DECIMALS

Addition

DIRECTIONS: Solve the following equations.

1. 1.032 + 0.987 + 3.07 =
   (A) 4.089 (B) 5.089 (C) 5.189 (D) 6.189 (E) 13.972

2. 132.03 + 97.1483 =
   (A) 98.4686 (B) 110.3513 (C) 209.1783
   (D) 229.1486 (E) 229.1783

3. 7.1 + 0.62 + 4.03827 + 5.183 =
   (A) 0.2315127 (B) 16.45433 (C) 16.94127
   (D) 18.561 (E) 40.4543

4. 8 + 17.43 + 9.2 =
   (A) 34.63 (B) 34.86 (C) 35.63 (D) 176.63 (E) 189.43

5. 1,036.173 + 289.04 =
   (A) 382.6573 (B) 392.6573 (C) 1,065.077
   (D) 1,325.213 (E) 3,926.573

Subtraction

DIRECTIONS: Solve the following equations.

6. 3.972 – 2.04 =
   (A) 1.932 (B) 1.942 (C) 1.976 (D) 2.013 (E) 2.113

7. 16.047 – 13.06 =
   (A) 2.887 (B) 2.987 (C) 3.041 (D) 3.141 (E) 4.741

8. 87.4 – 56.27 =
   (A) 30.27 (B) 30.67 (C) 31.1 (D) 31.13 (E) 31.27

9. 1,046.8 – 639.14 =
(A) 303.84 (B) 313.74 (C) 407.66 (D) 489.74 (E) 535.54

10. 10,000 – 842.91 =
(A) 157.09 (B) 942.91 (C) 5,236.09
(D) 9,057.91 (E) 9,157.09

Multiplication

DIRECTIONS: Solve the following equations.

11. 1.03 × 2.6 =
(A) 2.18 (B) 2.678 (C) 2.78 (D) 3.38 (E) 3.63

12. 93 × 4.2 =
(A) 39.06 (B) 97.2 (C) 223.2 (D) 390.6 (E) 3,906

13. 0.04 × 0.23 =
(A) 0.0092 (B) 0.092 (C) 0.27 (D) 0.87 (E) 0.920

14. 0.0186 × 0.03 =
(A) 0.000348 (B) 0.000558 (C) 0.0548 (D) 0.0848 (E) 0.558

15. 51.2 × 0.17 =
(A) 5.29 (B) 8.534 (C) 8.704 (D) 36.352 (E) 36.991

Division

DIRECTIONS: Solve the following equations.

16. 123.39 ÷ 3 =
(A) 31.12 (B) 41.13 (C) 401.13 (D) 411.3 (E) 4,113

17. 1,428.6 ÷ 6 =
(A) 0.2381 (B) 2.381 (C) 23.81 (D) 238.1 (E) 2,381

18. 25.2 ÷ 0.3 =
(A) 0.84 (B) 8.04 (C) 8.4 (D) 84 (E) 840

19. 14.95 ÷ 6.5 =
(A) 2.3 (B) 20.3 (C) 23 (D) 230 (E) 2,300

20. 46.33 ÷ 1.13 =
(A) 0.41 (B) 4.1 (C) 41 (D) 410 (E) 4,100

Comparing

DIRECTIONS: Solve the following equations.

21. Which is the **largest** number in this set: {0.8, 0.823, 0.089, 0.807, 0.852}?
(A) 0.8 (B) 0.823 (C) 0.089 (D) 0.807 (E) 0.852

22. Which is the **smallest** number in this set: {32.98, 32.099, 32.047, 32.5, 32.304}?

(A) 32.98 (B) 32.099 (C) 32.047 (D) 32.5 (E) 32.304

23. In which set below are the numbers arranged correctly from smallest to largest?

(A) {0.98, 0.9, 0.993} (B) {0.113, 0.3, 0.31}
(C) {7.04, 7.26, 7.2} (D) {0.006, 0.061, 0.06}
(E) {12.84, 12.801, 12.6}

24. In which set below are the numbers arranged correctly from largest to smallest?

(A) {1.018, 1.63, 1.368} (B) {4.219, 4.29, 4.9}
(C) {0.62, 0.6043, 0.643} (D) {16.34, 16.304, 16.3}
(E) {12.98, 12.601, 12.86}

25. Which is the **largest** number in this set: {0.87, 0.89, 0.889, 0.8, 0.987}?

(A) 0.87 (B) 0.89 (C) 0.889 (D) 0.8 (E) 0.987

Changing a Fraction to a Decimal

DIRECTIONS: Solve the following equations.

26. What is $\frac{1}{4}$ written as a decimal?

(A) 1.4 (B) 0.14 (C) 0.2 (D) 0.25 (E) 0.3

27. What is $\frac{3}{5}$ written as a decimal?

(A) 0.3 (B) 0.35 (C) 0.6 (D) 0.65 (E) 0.8

28. What is $\frac{7}{20}$ written as a decimal?

(A) 0.35 (B) 0.4 (C) 0.72 (D) 0.75 (E) 0.9

29. What is $\frac{2}{3}$ written as a decimal?

(A) 0.23 (B) 0.33 (C) 0.5 (D) 0.6 (E) $0.\overline{6}$

30. What is $\frac{11}{25}$ written as a decimal?

(A) 0.1125 (B) 0.25 (C) 0.4 (D) 0.44 (E) 0.5

MASTERING PERCENTAGES

A **percent** is a way of expressing the relationship between part and whole, where whole is defined as 100%. A percent can be defined by a fraction with a denominator of 100. Decimals can also represent a percent. For instance

$$56\% = 0.56 = \frac{56}{100}$$

PROBLEM

Compute the value of

(1) 90% of 400 *(3) 50% of 500*

(2) 180% of 400 *(4) 200% of 4*

SOLUTION

The symbol % means per hundred, therefore 5% = 5/100

(1) $90\% \text{ of } 400 = 90 \div 100 \times 400 = 90 \times 4 = 360$

(2) $180\% \text{ of } 400 = 180 \div 100 \times 400 = 180 \times 4 = 720$

(3) $50\% \text{ of } 500 = 50 \div 100 \times 500 = 50 \times 5 = 250$

(4) $200\% \text{ of } 4 = 200 \div 100 \times 4 = 2 \times 4 = 8$

PROBLEM

What percent of

(1) 100 is 99.5 *(2) 200 is 4*

SOLUTION

(1) $99.5 = x \times 100$

$99.5 = 100x$

$0.995 = x$; but this is the value of x per hundred. Therefore,

$99.5\% = x$

(2) $4 = x \times 200$

$4 = 200x$

$0.02 = x$. Again this must be changed to percent, so

$2\% = x$

Equivalent Forms of a Number

Some problems may call for converting numbers into an equivalent or simplified form in order to make the solution more convenient.

A) Converting a fraction to a decimal:

$$\frac{1}{2} = 0.50$$

Divide the numerator by the denominator:

$$\begin{array}{r} 0.50 \\ 2\overline{)1.00} \\ \underline{-10} \\ 00 \end{array}$$

B) Converting a number to a percent:

$$0.50 = 50\%$$

Multiply by 100:

$$0.50 = (0.50 \times 100)\% = 50\%$$

C) Converting a percent to a decimal:

$$30\% = 0.30$$

Divide by 100:

$$30\% = 30 \div 100 = 0.30$$

D) Converting a decimal to a fraction:

$$0.500 = \frac{1}{2}$$

Convert 0.500 to $\frac{500}{1000}$ and then simplify the fraction by dividing the numerator and denominator by common factors:

$$\frac{\cancel{2} \times \cancel{2} \times \cancel{5} \times \cancel{5} \times \cancel{5}}{\cancel{2} \times \cancel{2} \times \cancel{2} \times \cancel{5} \times \cancel{5} \times \cancel{5}}$$

and then cancel out the common numbers to get $\frac{1}{2}$.

PROBLEM

Express

(1) 1.65 as a percent *(2) 0.7 as a fraction*

(3) $-\frac{10}{20}$ *as a decimal* *(4)* $\frac{4}{2}$ *as an integer*

SOLUTION

(1) $1.65 \times 100 = 165\%$

(2) $0.7 = \frac{7}{10}$

(3) $-\frac{10}{20} = -0.5$

(4) $\frac{4}{2} = 2$

DRILL: PERCENTAGES

Finding Percents

DIRECTIONS: Solve to find the correct percentages.

1. Find 3% of 80.

   (A) 0.24 (B) 2.4 (C) 24 (D) 240 (E) 2,400

2. Find 50% of 182.

(A) 9 (B) 90 (C) 91 (D) 910 (E) 9,100

3. Find 83% of 166.

(A) 0.137 (B) 1.377 (C) 13.778 (D) 137 (E) 137.78

4. Find 125% of 400.

(A) 425 (B) 500 (C) 525 (D) 600 (E) 825

5. Find 300% of 4.

(A) 12 (B) 120 (C) 1,200 (D) 12,000 (E) 120,000

6. Forty-eight percent of the 1,200 students at Central High are males. How many male students are there at Central High?

(A) 57 (B) 576 (C) 580 (D) 600 (E) 648

7. For 35% of the last 40 days, there has been measurable rainfall. How many days out of the last 40 days have had measurable rainfall?

(A) 14 (B) 20 (C) 25 (D) 35 (E) 40

8. Of every 1,000 people who take a certain medicine, 0.2% develop severe side effects. How many people out of every 1,000 who take the medicine develop the side effects?

(A) 0.2 (B) 2 (C) 20 (D) 22 (E) 200

9. Of 220 applicants for a job, 75% were offered an initial interview. How many people were offered an initial interview?

(A) 75 (B) 110 (C) 120 (D) 155 (E) 165

10. Find 0.05% of 4,000.

(A) 0.05 (B) 0.5 (C) 2 (D) 20 (E) 400

Changing Percents to Fractions

DIRECTIONS: Solve to find the correct fractions.

11. What is 25% written as a fraction?

(A) $\frac{1}{25}$ (B) $\frac{1}{5}$ (C) $\frac{1}{4}$ (D) $\frac{1}{3}$ (E) $\frac{1}{2}$

12. What is $33\frac{1}{3}\%$ written as a fraction?

(A) $\frac{1}{4}$ (B) $\frac{1}{3}$ (C) $\frac{1}{2}$ (D) $\frac{2}{3}$ (E) $\frac{5}{9}$

13. What is 200% written as a fraction?

(A) $\frac{1}{2}$ (B) $\frac{2}{1}$ (C) $\frac{20}{1}$ (D) $\frac{200}{1}$ (E) $\frac{2000}{1}$

14. What is 84% written as a fraction?

(A) $\frac{1}{84}$ (B) $\frac{4}{8}$ (C) $\frac{17}{25}$ (D) $\frac{21}{25}$ (E) $\frac{44}{50}$

15. What is 2% written as a fraction?

(A) $\frac{1}{50}$ (B) $\frac{1}{25}$ (C) $\frac{1}{10}$ (D) $\frac{1}{4}$ (E) $\frac{1}{2}$

Changing Fractions to Percents

DIRECTIONS: Solve to find the following percentages.

16. What is $\frac{2}{3}$ written as a percent?

(A) 23% (B) 32% (C) $33\frac{1}{3}\%$ (D) $57\frac{1}{3}\%$ (E) $66\frac{2}{3}\%$

17. What is $\frac{3}{5}$ written as a percent?

(A) 30% (B) 35% (C) 53% (D) 60% (E) 65%

18. What is $\frac{17}{20}$ written as a percent?

(A) 17% (B) 70% (C) 75% (D) 80% (E) 85%

19. What is $\frac{45}{50}$ written as a percent?

(A) 45% (B) 50% (C) 90% (D) 95% (E) 97%

20. What is $1\frac{1}{4}$ written as a percent?

(A) 114% (B) 120% (C) 125% (D) 127% (E) 133%

Changing Percents to Decimals

DIRECTIONS: Convert the percentages to decimals.

21. What is 42% written as a decimal?

(A) 0.42 (B) 4.2 (C) 42 (D) 420 (E) 422

22. What is 0.3% written as a decimal?

(A) 0.0003 (B) 0.003 (C) 0.03 (D) 0.3 (E) 3

23. What is 8% written as a decimal?

(A) 0.0008 (B) 0.008 (C) 0.08 (D) 0.80 (E) 8

24. What is 175% written as a decimal?

(A) 0.175 (B) 1.75 (C) 17.5 (D) 175 (E) 17,500

25. What is 34% written as a decimal?

(A) 0.00034 (B) 0.0034 (C) 0.034 (D) 0.34 (E) 3.4

Changing Decimals to Percents

DIRECTIONS: Convert the following decimals to percents.

26. What is 0.43 written as a percent?

(A) 0.0043% (B) 0.043% (C) 4.3% (D) 43% (E) 430%

27. What is 1 written as a percent?

(A) 1% (B) 10% (C) 100% (D) 111% (E) 150%

28. What is 0.08 written as a percent?

(A) 0.08% (B) 8% (C) 8.8% (D) 80% (E) 800%

29. What is 3.4 written as a percent?

(A) 0.0034% (B) 3.4% (C) 34% (D) 304% (E) 340%

30. What is 0.645 written as a percent?

(A) 64.5% (B) 65% (C) 69% (D) 70% (E) 645%

MASTERING RADICALS

The **square root** of a number is a number that when multiplied by itself results in the original number. Thus, the square root of 81 is 9 since $9 \times 9 = 81$. However, –9 is also a root of 81 because $(-9)(-9) = 81$. Every positive number will have two roots. The principal root is the positive one. Zero has only one square root, whereas negative numbers do not have real numbers as their roots.

A **radical sign** indicates that the root of a number or expression will be taken. The **radicand** is the number of which the root will be taken. The **index** tells how many times the root needs to be multiplied by itself to equal the radicand:

(1) $\sqrt[3]{64}$;

3 is the index and 64 is the radicand. Since $4 \times 4 \times 4 = 64$, then $\sqrt[3]{64} = 4$.

(2) $\sqrt[5]{32}$;

5 is the index and 32 is the radicand. Since $2 \times 2 \times 2 \times 2 \times 2 = 32$, then $\sqrt[5]{32} = 2$.

Operations with Radicals

A) **To multiply two or more radicals,** we utilize the law that states

$$\sqrt{a} \times \sqrt{b} = \sqrt{ab}$$

Simply multiply the whole numbers as usual. Then, multiply the radicands and put the product under the radical sign and simplify:

(1) $\sqrt{12} + \sqrt{5} = \sqrt{60} = 2\sqrt{15}$

(2) $3\sqrt{2} \times 4\sqrt{8} = 12\sqrt{16} = 48$

(3) $2\sqrt{10} + 6\sqrt{5} = 12\sqrt{50} = 60\sqrt{2}$

B) **To divide radicals,** simplify both the numerator and the denominator. By multiplying the radical in the denominator by itself, you can make the denominator a rational number. The numerator, however, must also be multiplied by this radical so that the value of the expression does not change. You must choose as many factors as necessary to rationalize the denominator:

(1) $\dfrac{\sqrt{128}}{\sqrt{2}} = \dfrac{\sqrt{64} \times \sqrt{2}}{\sqrt{2}} = \dfrac{8\sqrt{2}}{\sqrt{2}} = 8$

(2) $\dfrac{\sqrt{10}}{\sqrt{3}} = \dfrac{\sqrt{10} \times \sqrt{3}}{\sqrt{3} \times \sqrt{3}} = \dfrac{\sqrt{30}}{3}$

(3) $\dfrac{\sqrt{8}}{2\sqrt{3}} = \dfrac{\sqrt{8} \times \sqrt{3}}{2\sqrt{3} \times \sqrt{3}} = \dfrac{\sqrt{24}}{2 \times 3} = \dfrac{2\sqrt{6}}{6} = \dfrac{\sqrt{6}}{3}$

C) **To add two or more radicals,** the radicals must have the same index and the same radicand. Only where the radicals are simplified can these similarities be determined.

EXAMPLES

(1) $6\sqrt{2} + 2\sqrt{2} = (6+2)\sqrt{2} = 8\sqrt{2}$

(2) $\sqrt{27} + 5\sqrt{3} = \sqrt{9}\sqrt{3} + 5\sqrt{3} = 3\sqrt{3} + 5\sqrt{3} = 8\sqrt{3}$

(3) $7\sqrt{3} + 8\sqrt{2} + 5\sqrt{3} = 12\sqrt{3} + 8\sqrt{2}$

Similarly, to subtract

(1) $12\sqrt{3} - 7\sqrt{3} = (12-7)\sqrt{3} = 5\sqrt{3}$

(2) $\sqrt{80} - \sqrt{20} = \sqrt{16}\sqrt{5} - \sqrt{4}\sqrt{5} = 4\sqrt{5} - 2\sqrt{5} = 2\sqrt{5}$

(3) $\sqrt{50} - \sqrt{3} = 5\sqrt{2} - \sqrt{3}$

DRILL: RADICALS

Multiplication

DIRECTIONS: Multiply and simplify each answer.

1. $\sqrt{6} \times \sqrt{5} =$

(A) $\sqrt{11}$ (B) $\sqrt{30}$ (C) $2\sqrt{5}$ (D) $3\sqrt{10}$ (E) $2\sqrt{3}$

2. $\sqrt{3} \times \sqrt{12} =$

(A) 3 (B) $\sqrt{15}$ (C) $\sqrt{36}$ (D) 6 (E) 8

3. $\sqrt{7} \times \sqrt{7} =$

(A) 7 (B) 49 (C) $\sqrt{14}$ (D) $2\sqrt{7}$ (E) $2\sqrt{14}$

4. $3\sqrt{5} \times 2\sqrt{5} =$

(A) $5\sqrt{5}$ (B) 25 (C) 30 (D) $5\sqrt{25}$ (E) $6\sqrt{5}$

5. $4\sqrt{6} \times \sqrt{2} =$

(A) $4\sqrt{8}$ (B) $8\sqrt{2}$ (C) $5\sqrt{8}$ (D) $4\sqrt{12}$ (E) $8\sqrt{3}$

Division

DIRECTIONS: **Divide and simplify the answer.**

6. $\sqrt{10} \div \sqrt{2} =$

(A) $\sqrt{8}$ (B) $2\sqrt{2}$ (C) $\sqrt{5}$ (D) $2\sqrt{5}$ (E) $2\sqrt{3}$

7. $\sqrt{30} \div \sqrt{15} =$

(A) $\sqrt{2}$ (B) $\sqrt{45}$ (C) $3\sqrt{5}$ (D) $\sqrt{15}$ (E) $5\sqrt{3}$

8. $\sqrt{100} \div \sqrt{25} =$

(A) $\sqrt{4}$ (B) $5\sqrt{5}$ (C) $5\sqrt{3}$ (D) 2 (E) 4

9. $\sqrt{48} \div \sqrt{8} =$

(A) $4\sqrt{3}$ (B) $3\sqrt{2}$ (C) $\sqrt{6}$ (D) 6 (E) 12

10. $3\sqrt{12} \div \sqrt{3} =$

(A) $3\sqrt{15}$ (B) 6 (C) 9 (D) 12 (E) $3\sqrt{36}$

Addition

DIRECTIONS: **Simplify each radical and add.**

11. $\sqrt{7} + 3\sqrt{7} =$

(A) $3\sqrt{7}$ (B) $4\sqrt{7}$ (C) $3\sqrt{14}$ (D) $4\sqrt{14}$ (E) $3\sqrt{21}$

12. $\sqrt{5} + 6\sqrt{5} + 3\sqrt{5} =$

(A) $9\sqrt{5}$ (B) $9\sqrt{15}$ (C) $5\sqrt{10}$ (D) $10\sqrt{5}$ (E) $18\sqrt{15}$

13. $3\sqrt{32} + 2\sqrt{2} =$

(A) $5\sqrt{2}$ (B) $\sqrt{34}$ (C) $14\sqrt{2}$ (D) $5\sqrt{34}$ (E) $6\sqrt{64}$

14. $6\sqrt{15} + 8\sqrt{15} + 16\sqrt{15} =$

(A) $15\sqrt{30}$ (B) $30\sqrt{45}$ (C) $30\sqrt{30}$ (D) $15\sqrt{45}$ (E) $30\sqrt{15}$

15. $6\sqrt{5} + 2\sqrt{45} =$

(A) $12\sqrt{5}$ (B) $8\sqrt{50}$ (C) $40\sqrt{2}$ (D) $12\sqrt{50}$ (E) $8\sqrt{5}$

Subtraction

DIRECTIONS: Simplify each radical and subtract.

16. $8\sqrt{5} - 6\sqrt{5} =$

    (A) $2\sqrt{5}$ (B) $3\sqrt{5}$ (C) $4\sqrt{5}$ (D) $14\sqrt{5}$ (E) $48\sqrt{5}$

17. $16\sqrt{33} - 5\sqrt{33} =$

    (A) $3\sqrt{33}$ (B) $33\sqrt{11}$ (C) $1\sqrt{33}$ (D) $11\sqrt{0}$ (E) $\sqrt{33}$

18. $14\sqrt{2} - 19\sqrt{2} =$

    (A) $5\sqrt{2}$ (B) $-5\sqrt{2}$ (C) $-33\sqrt{2}$ (D) $33\sqrt{2}$ (E) $-4\sqrt{2}$

19. $10\sqrt{2} - 3\sqrt{8} =$

    (A) $6\sqrt{6}$ (B) $-2\sqrt{2}$ (C) $7\sqrt{6}$ (D) $4\sqrt{2}$ (E) $-6\sqrt{6}$

20. $4\sqrt{3} - 2\sqrt{12} =$

    (A) $-2\sqrt{9}$ (B) $-6\sqrt{15}$ (C) 0 (D) $6\sqrt{15}$ (E) $2\sqrt{12}$

MASTERING MEAN, MEDIAN, MODE

Mean

The mean is the arithmetic average. It is the sum of the variables divided by the total number of variables. For example, the mean of 4, 3, and 8 is

$$\frac{4+3+8}{3} = \frac{15}{3} = 5.$$

PROBLEM

Find the mean salary for four company employees who make $5/hr., $8/hr., $12/hr., and $15/hr.

SOLUTION

The mean salary is the average.

$$\frac{\$5+\$8+\$12+\$15}{4} = \frac{\$40}{10} = \$10/\text{hr}.$$

PROBLEM

Find the mean length of five fish with lengths of 7.5 in., 7.75 in., 8.5 in., 8.5 in., and 8.25 in.

SOLUTION

The mean length is the average length.

$$\frac{7.5+7.75+8.5+8.5+8.25}{5} = \frac{40.5}{5} = 8.1\text{ in}.$$

Median

The median is the middle value in a set when there is an odd number of values. There is an equal number of values larger and smaller than the median. When the set is an even number of values, the average of the two middle values is the median. For example:

The median of (2, 3, 5, 8, 9) is 5.

The median of (2, 3, 5, 9, 10, 11) is $\frac{5+9}{2}=7$.

MODE

The mode is the most frequently occurring value in the set of values. For example, the mode of 4, 5, 8, 3, 8, and 2 would be 8 because it occurs twice, whereas the other values occur only once.

PROBLEM

For this series of observations find the mean, median, and mode.

500, 600, 800, 800, 900, 900, 900, 900, 900, 1,000, 1,100

SOLUTION

The mean is the value obtained by adding all the measurements and dividing by the number of measurements.

$$\frac{500+600+800+800+900+900+900+900+900+1{,}000+1{,}100}{11}=\frac{9{,}300}{11}=845.45$$

The median is the value appearing in the middle. We have 11 values, so here the sixth, 900, is the median.

The mode is the value that appears most frequently. That is also 900, which has five appearances.

All three of these numbers are measures of central tendency. They describe the "middle" or "center" of the data.

PROBLEM

Nine rats run through a maze. The time in minutes each rat took to traverse the maze is recorded and these times (in minutes) are listed below.

1, 2.5, 3, 1.5, 2, 1.25, 1, 0.9, 30

Which of the three measures of central tendency would be the most appropriate in this case?

SOLUTION

We will calculate the three measures of central tendency and then compare them to determine which would be the most appropriate in describing these data.

The mean is the sum of the values listed divided by the number of values. In this case

$$\frac{1+2.5+3+1.5+2+1.25+1+0.9+30}{9}=\frac{43.15}{9}=4.79$$

The median is the "middle number" in an array of the values from the lowest to the highest.

0.9, 1.0, 1.0, 1.25, 1.5, 2.0, 2.5, 3.0, 30.0

The median is the fifth value in this ordered array or 1.5. There are four values larger than 1.5 and four values smaller than 1.5.

The mode is the most frequently occurring value in the sample. In this data set the mode is 1.0.

mean = 4.79

median = 1.5

mode = 1.0

The mean is not appropriate here. Only one rat took more than 4.79 minutes to run the maze and this rat took 30 minutes. We see that the mean has been distorted by this one large value.

The median or mode seems to describe this data set better and would be more appropriate to use.

DRILL: AVERAGES

Mean

DIRECTIONS: Find the mean of each set of numbers.

1. 18, 25, and 32

(A) 3 (B) 25 (C) 50 (D) 75 (E) 150

2. $\frac{4}{9}, \frac{2}{3}$, and $\frac{5}{6}$

(A) $\frac{11}{18}$ (B) $\frac{35}{54}$ (C) $\frac{41}{54}$ (D) $\frac{35}{18}$ (E) $\frac{54}{18}$

3. 97, 102, 116, and 137

(A) 40 (B) 102 (C) 109 (D) 113 (E) 116

4. 12, 15, 18, 24, and 31

(A) 18 (B) 19.3 (C) 20 (D) 25 (E) 100

5. 7, 4, 6, 3, 11, and 14

(A) 5 (B) 6.5 (C) 7 (D) 7.5 (E) 8

Median

DIRECTIONS: Find the median value of each set of numbers.

6. 3, 8, and 6

(A) 3 (B) 6 (C) 8 (D) 17 (E) 20

7. 19, 15, 21, 27, and 12

(A) 19 (B) 15 (C) 21 (D) 27 (E) 94

8. $1\frac{2}{3}, 1\frac{7}{8}, 1\frac{3}{4}$, and $1\frac{5}{6}$

(A) $1\frac{30}{48}$ (B) $1\frac{2}{3}$ (C) $1\frac{3}{4}$ (D) $1\frac{19}{24}$ (E) $1\frac{21}{24}$

9. 29, 18, 21, and 35

(A) 29 (B) 18 (C) 21 (D) 35 (E) 25

10. 8, 15, 7, 12, 31, 3, and 28

(A) 7 (B) 11.6 (C) 12 (D) 14.9 (E) 104

Mode

DIRECTIONS: Find the mode(s) of each set of numbers.

11. 1, 3, 7, 4, 3, and 8

(A) 1 (B) 3 (C) 7 (D) 4 (E) None

12. 12, 19, 25, and 42

(A) 12 (B) 19 (C) 25 (D) 42 (E) None

13. 16, 14, 12, 16, 30, and 28

(A) 6 (B) 14 (C) 16 (D) $19.\overline{3}$ (E) None

14. 4, 3, 9, 2, 4, 5, and 2

(A) 3 and 9 (B) 5 and 9 (C) 4 and 5 (D) 2 and 4 (E) None

15. 87, 42, 111, 116, 39, 111, 140, 116, 97, and 111

(A) 111 (B) 116 (C) 39 (D) 140 (E) None

MASTERING SEQUENCES

A **sequence** is a list of terms in a particular order. The terms of a sequence often follow a pattern that can be described by an algebraic expression. For example, look at the following sequence.

2, 4, 6, 8, 10, . . .

If n represents the term number, the preceding sequence can be represented by the expression $2n$. Evaluating this expression for $n = 1$ gives the first term of the sequence: $2n = 2(1) = 2$. Evaluating the expression for $n = 2$ gives the second term of the sequence: $2n = 2(2) = 4$, and so on.

A **geometric sequence** is a special type of sequence in which there is a constant ratio between consecutive terms. The following sequence is geometric because the ratio of each term to the previous term is equal to –2.

–2, 4, –8, 16, –32, . . .

$\frac{4}{2} = -2 \quad \frac{-8}{4} = -2 \quad \frac{16}{-8} = -2 \quad \frac{-32}{16} = -2$

A geometric sequence can be described by using an algebraic expression having the form $a(r^{n-1})$. In this expression, a is a constant equal to the first term of the series, r is the constant ratio, and n is the term number.

PROBLEM

Write an expression for the nth term of the following geometric sequence.

6, 18, 54, 162, . . .

SOLUTION

The sequence is geometric, so it can be described by the expression $a(r^{n-1})$. The first term of the sequence is 6, so the value of a is 6. Divide the second term of the sequence by the first term to find the constant ratio, r.

$$r = \frac{18}{6} = 3$$

The algebraic expression that describes the sequence is $6(3^{n-1})$. Check that this expression correctly gives the terms of the sequence, as shown in the table below.

n	$6(3^{n-1})$
1	$6(3^{1-1}) = 6(3^0) = 6(1) = 6$
2	$6(3^{2-1}) = 6(3^1) = 6(3) = 18$
3	$6(3^{3-1}) = 6(3^2) = 6(9) = 54$
4	$6(3^{4-1}) = 6(3^3) = 6(27) = 162$

One real-world application of geometric sequences involves population growth. The expression $P_0\left(2^{\frac{t}{t_D}}\right)$ can be used to determine a population t years after it begins to grow. In this formula, P_0 is the initial population, and t_D is the number of years it takes the population to double in size.

PROBLEM

A population that initially has 200 members is doubling every 10 years. What will the population be after 100 years?

SOLUTION

Use the expression $P_0\left(2^{\frac{t}{t_D}}\right)$ to determine the population. The initial population is 200, so P_0 = 200. The population is doubling every 10 years, so t_D = 10. We want to know the population after 100 years, so t = 100. Substitute these values into the expression and then simplify.

$$P_0\left(2^{\frac{t}{t_D}}\right) = 200\left(2^{\frac{100}{10}}\right) = 200\left(2^{10}\right) = 200(1{,}024) = 204{,}800$$

After 100 years, the population will be 204,800.

DRILL: SEQUENCES

Geometric Sequences

1. Which of the following is a geometric sequence?

   (A) 1, 4, 9, 16, . . . (B) 4, 8, 12, 16, . . . (C) 5, 10, 20, 40, . . .
   (D) 4, 5, 7, 10, . . . (E) 2, 4, 12, 48, . . .

2. What is the 6th term in the following geometric sequence?
   3, 9, 27, 81, . . .

   (A) 189 (B) 243 (C) 486 (D) 729 (E) 810

3. What is the 9th term in the following geometric sequence?
   6, 12, 24, 48, . . .

   (A) 288 (B) 432 (C) 512 (D) 1,024 (E) 1,536

4. Which expression can be used to determine the nth term in the following geometric sequence?
   20, 100, 500, 2,500, . . .

   (A) $2(10^{n-1})$ (B) $5(20^{n-1})$ (C) $20(n-1)$
   (D) 20^{n-1} (E) $20(5^{n-1})$

5. A population that initially has 60 members is doubling every 15 years. What will the population be after 75 years?

   (A) 300 (B) 600 (C) 1,920 (D) 2,250 (E) 7,200

ARITHMETIC DRILLS

ANSWER KEY

Drill: Integers and Real Numbers

1. (A)
2. (D)
3. (E)
4. (B)
5. (A)
6. (A)
7. (C)
8. (B)
9. (C)
10. (B)
11. (B)
12. (A)
13. (D)
14. (E)
15. (B)
16. (B)
17. (A)
18. (C)
19. (E)
20. (B)
21. (B)
22. (D)
23. (B)
24. (C)
25. (E)
26. (B)
27. (E)
28. (D)
29. (A)
30. (D)
31. (D)
32. (B)
33. (C)
34. (B)
35. (C)

Drill: Fractions

1. (D)
2. (A)
3. (D)
4. (E)
5. (C)
6. (D)
7. (D)
8. (B)
9. (A)
10. (E)
11. (D)
12. (E)
13. (C)
14. (A)
15. (C)
16. (B)
17. (C)
18. (A)
19. (B)
20. (A)
21. (B)
22. (D)
23. (E)
24. (D)
25. (B)
26. (B)
27. (D)
28. (B)
29. (A)
30. (C)
31. (D)
32. (C)
33. (B)
34. (E)
35. (A)
36. (A)
37. (D)
38. (E)
39. (C)
40. (B)
41. (C)
42. (A)
43. (B)
44. (C)
45. (D)
46. (B)
47. (C)
48. (D)
49. (E)
50. (A)

Drill: Decimals

1.	(B)	9.	(C)	17.	(D)	25.	(E)
2.	(E)	10.	(E)	18.	(D)	26.	(D)
3.	(C)	11.	(B)	19.	(A)	27.	(C)
4.	(A)	12.	(D)	20.	(C)	28.	(A)
5.	(D)	13.	(A)	21.	(E)	29.	(E)
6.	(A)	14.	(B)	22.	(C)	30.	(D)
7.	(B)	15.	(C)	23.	(B)		
8.	(D)	16.	(B)	24.	(D)		

Drill: Percentages

1.	(B)	9.	(E)	17.	(D)	25.	(D)
2.	(C)	10.	(C)	18.	(E)	26.	(D)
3.	(E)	11.	(C)	19.	(C)	27.	(C)
4.	(B)	12.	(B)	20.	(C)	28.	(B)
5.	(A)	13.	(B)	21.	(A)	29.	(E)
6.	(B)	14.	(D)	22.	(B)	30.	(A)
7.	(A)	15.	(A)	23.	(C)		
8.	(B)	16.	(E)	24.	(B)		

Drill: Radicals

1.	(B)	6.	(C)	11.	(B)	16.	(A)
2.	(D)	7.	(A)	12.	(D)	17.	(C)
3.	(A)	8.	(D)	13.	(C)	18.	(B)
4.	(C)	9.	(C)	14.	(E)	19.	(D)
5.	(E)	10.	(B)	15.	(A)	20.	(C)

Drill: Exponents

1.	(B)	9.	(B)
2.	(A)	10.	(D)
3.	(C)	11.	(C)
4.	(D)	12.	(E)
5.	(B)	13.	(B)
6.	(E)	14.	(A)
7.	(B)	15.	(D)
8.	(C)		

Drill: Averages

1.	(B)	9.	(E)
2.	(B)	10.	(C)
3.	(D)	11.	(B)
4.	(C)	12.	(E)
5.	(D)	13.	(C)
6.	(B)	14.	(D)
7.	(A)	15.	(A)
8.	(D)		

Drill: Sequences

1.	(G)	2.	(D)	3.	(E)	4.	(E)	5.	(C)

II. ALGEBRA

In algebra, letters or variables are used to represent numbers. A **variable** is defined as a placeholder, which can take on any of several values at a given time. A **constant**, on the other hand, is a symbol that takes on only one value at a given time. A **term** is a constant, a variable, or a combination of constants and variables. For example: 7.76, $3x$, xyz, $\frac{5z}{x}$, $(0.99)x^2$ are terms. If a term is a combination of constants and variables, the constant part of the term is referred to as the **coefficient** of the variable. If a variable is written without a coefficient, the coefficient is assumed to be 1.

EXAMPLES

$3x^2$	y^3
coefficient: 3	coefficient: 1
variable: x	variable: y

An **expression** is a collection of one or more terms. If the number of terms is greater than 1, the expression is said to be the sum of the terms.

EXAMPLES

$$9,\ 9xy,\ 6x + \frac{x}{3},\ 8yz - 2x$$

An algebraic expression consisting of only one term is called a **monomial**; two terms is called a **binomial**; three terms is called a **trinomial**. In general, an algebraic expression consisting of two or more terms is called a **polynomial**.

Before we examine polynomials in detail, we will first review the laws of exponents.

MASTERING EXPONENTS

When a number is multiplied by itself a specific number of times, it is said to be **raised to a power**. The way this is written is $a^n = b$, where a is the number or **base**, n is the **exponent** or **power** that indicates the number of times the base is to be multiplied by itself, and b is the product of this multiplication.

In the expression 3^2, 3 is the base and 2 is the exponent. This means that 3 is multiplied by itself 2 times and the product is 9.

An exponent can be either positive or negative. A negative exponent implies a fraction such that if n is an integer

$$a^{-n} = \frac{1}{a^n},\ a \neq 0. \text{ So, } 2^{-4} = \frac{1}{2^4} = \frac{1}{16}$$

An exponent that is 0 gives a result of 1, assuming that the base is not equal to 0.

$$a^0 = 1,\ a \neq 0.$$

An exponent can also be a fraction. If m and n are positive integers,

$$a^{\frac{m}{n}} = \sqrt[n]{a^m}$$

The numerator remains the exponent of a, but the denominator tells what root to take. For example,

$$(1) \quad 4^{\frac{3}{2}} = \sqrt[2]{4^3} = \sqrt{64} = 8 \qquad (2) \quad 3^{\frac{4}{2}} = \sqrt[2]{3^4} = \sqrt{81} = 9$$

If a fractional exponent were negative, the same operation would take place, but the result would be a fraction. For example,

$$(1) \quad 27^{-\frac{2}{3}} = \frac{1}{27^{\frac{2}{3}}} = \frac{1}{\sqrt[3]{27^2}} = \frac{1}{\sqrt[3]{729}} = \frac{1}{9}$$

PROBLEM

Simplify the following expressions:

(1) -3^{-2}

(2) $(-3)^{-2}$

(3) $\frac{-3}{4^{-1}}$

SOLUTION

(1) Here the exponent applies only to 3. Since

$$x^{-y} = \frac{1}{x^y},\ -3^{-2} = (3)^{-2} = \left(\frac{1}{3^2}\right) = \frac{1}{9}$$

(2) In this case the exponent applies to the negative base. Thus,

$$(-3)^{-2} = \frac{1}{(3)^{-2}} = \frac{1}{(-3)(-3)} = \frac{1}{9}$$

$$\frac{-3}{4^{-1}} = \frac{-3}{\left(\frac{1}{4}\right)^1} = \frac{-3}{\frac{1^1}{4^1}} = \frac{-3}{\frac{1}{4}}$$

(3) Division by a fraction is equivalent to multiplication by that fraction's reciprocal, thus

$$\frac{-3}{\frac{1}{4}} = -3 \times \frac{4}{1} = -12 \quad \text{and} \quad \frac{-3}{4^{-1}} = -12$$

General Laws of Exponents

A) $a^p a^q = a^{p+q}$

$4^2 4^3 = 4^2 + 3 = 1{,}024$

B) $(a^p)^q = a^{pq}$

$(2^3)^2 = 2^6 = 64$

C) $\frac{a^p}{a^q} = a^{p-q}$

$\frac{3^6}{3^2} = 3^4 = 81$

D) $(ab)^p = a^p b^p$

$(3 \times 2)^2 = 3^2 \times 2^2 = (9)\ (4) = 36$

E) $\left(\frac{a}{b}\right)^p = \frac{a^p}{b^p},\ b \neq 0$

$\left(\frac{4}{5}\right)^2 = \frac{4^2}{5^2} = \frac{16}{25}$

DRILL: EXPONENTS

Multiplication

DIRECTIONS: Simplify.

1. $4^6 \times 4^2 =$

   (A) 4^4 (B) 4^8 (C) 4^{12} (D) 16^8 (E) 16^{12}

2. $2^2 \times 2^5 \times 2^3 =$

   (A) 2^{10} (B) 4^{10} (C) 8^{10} (D) 2^{30} (E) 8^{30}

3. $6^6 \times 6^2 \times 6^4 =$

   (A) 18^8 (B) 18^{12} (C) 6^{12} (D) 6^{48} (E) 18^{48}

4. $a^4b^2 \times a^3b =$

   (A) ab (B) $2a^7b^2$ (C) $2a^{12}b$ (D) a^7b^3 (E) a^7b^2

5. $m^8n^3 \times m^2n \times m^4n^2 =$

   (A) $3m^{16}n^6$ (B) $m^{14}n^6$ (D) $3m^{14}n^5$ (D) $3m^{14}n^5$ (E) m^2

Division

DIRECTIONS: Simplify.

6. $6^5 \div 6^3 =$

   (A) 0 (B) 1 (C) 6 (D) 12 (E) 36

7. $11^8 \div 11^5 =$

   (A) 1^3 (B) 11^3 (C) 11^{13} (D) 11^{40} (E) 88^5

8. $x^{10}y^8 \div x^7y^3 =$

   (A) x^2y^5 (B) x^3y^4 (C) x^3y^5 (D) x^2y^4 (E) x^5y^3

9. $a^{14} \div a^9 =$

   (A) 1^5 (B) a^5 (C) $2a^5$ (D) a^{23} (E) $2a^{23}$

10. $c^{17}d^{12}e^4 \div c^{12}d^8e =$

   (A) $c^4d^5e^3$ (B) $c^4d^4e^3$ (C) $c^5d^8e^4$ (D) $c^5d^4e^3$ (E) $c^5d^4e^4$

Power to a Power

DIRECTIONS: Simplify.

11. $(3^6)^2 =$
    (A) 3^4 (B) 3^8 (C) 3^{12} (D) 9^6 (E) 9^8

12. $(4^3)^5 =$
    (A) 4^2 (B) 2^{15} (C) 4^8 (D) 20^3 (E) 4^{15}

13. $(a^4b^3)^2 =$
    (A) $(ab)^9$ (B) a^8b^6 (C) $(ab)^{24}$ (D) a^6b^5 (E) $2a^4b^3$

14. $(r^3p^6)^3 =$
    (A) r^9p^{18} (B) $(rp)^{12}$ (C) r^6p^9 (D) $3r^3p^6$ (E) $3r^9p^{18}$

15. $(m^6n^5q^3)^2 =$
    (A) $2m^6n^5q^3$ (B) m^4n^3q (C) $m^8n^7q^5$
    (D) $m^{12}n^{10}q^6$ (E) $2m^{12}n^{10}q^6$

Negative and Rational Exponents

DIRECTIONS: Identify the expression equivalent to the given expression.

16. $\dfrac{x^4}{x^{-2}} =$
    (A) x^{-6} (B) x^{-2} (C) x^2 (D) x^6 (E) x^8

17. $8^{\frac{2}{3}} =$
    (A) $\dfrac{1}{8}$ (B) 2 (C) 4 (D) $5\dfrac{1}{3}$ (E) $21\dfrac{1}{3}$

18. $\dfrac{3}{\sqrt{b}} =$
    (A) $b\dfrac{3}{2}$ (B) $\dfrac{3}{2}b^{-1}$ (C) $3b^{-2}$ (D) $3b^{-\frac{1}{2}}$ (E) $3b^{\frac{1}{2}}$

19. $(3p)^{-3} =$
    (A) $-27p^3$ (B) $-9p$ (C) $\dfrac{1}{27p^3}$ (D) $\dfrac{3}{p^3}$ (E) $\dfrac{27}{p^3}$

20. $\left(\dfrac{1}{4}\right)^{-2} =$
    (A) $\dfrac{1}{16}$ (B) $\dfrac{1}{8}$ (C) 2 (D) 8 (E) 16

MASTERING OPERATIONS WITH POLYNOMIALS

A) **Addition of polynomials** is achieved by combining like terms, terms which differ only in their numerical coefficients, e.g.,

$$P(x) = (x^2 - 3x + 5) + (4x^2 + 6x - 3)$$

Note that the parentheses are used to distinguish the polynomials.

By using the commutative and associative laws, we can rewrite P(x) as:

$$P(x) = (x^2 + 4x^2) + (6x - 3x) + (5 - 3)$$

Using the distributive law, $ab + ac = a(b + c)$, yields:

$$(1 + 4)x^2 + (6 - 3)x + (5 - 3)$$

$$= 5x^2 + 3x + 2$$

B) **Subtraction of two polynomials** is achieved by first changing the sign of all terms in the expression that are being subtracted and then adding this result to the other expression, e.g.,

$$(5x^2 + 4y^2 + 3z^2) - (4xy + 7y^2 - 3z^2 + 1)$$

$$= 5x^2 + 4y^2 + 3z^2 - 4xy - 7y^2 + 3z^2 - 1$$

$$= 5x^2 + (4y^2 - 7y^2) + (3z^2 + 3z^2) - 4xy - 1$$

$$= 5x^2 + (-3y^2) + 6z^2 - 4xy - 1$$

C) **Multiplication of two or more polynomials** is achieved by using the laws of exponents, the rules of signs, and the commutative and associative laws of multiplication. Begin by multiplying the coefficients and then multiply the variables according to the laws of exponents, e.g.,

$$(y^2)\ (5)\ (6y^2)\ (yz)\ (2z^2)$$

$$= (1)\ (5)\ (6)\ (1)\ (2)\ (y^2)\ (y^2)\ (yz)\ (z^2)$$

$$= 60[(y^2)\ (y^2)\ (y)]\ [(z)\ (z^2)]$$

$$= 60(y^5)\ (z^3)$$

$$= 60y^5z^3$$

D) **Multiplication of a polynomial by a monomial** is achieved by multiplying each term of the polynomial by the monomial and combining the results, e.g.,

$$(4x^2 + 3y)\ (6xz^2)$$

$$= (4x^2)\ (6xz^2) + (3y)\ (6xz^2)$$

$$= 24x^3z^2 + 18xyz^2$$

E) **Multiplication of a polynomial by a polynomial** is achieved by multiplying each of the terms of one polynomial by each of the terms of the other polynomial and combining the result, e.g.,

$$(5y + z + 1)\,(y^2 + 2y)$$

$$[(5y)\,(y^2) + (5y)\,(2y)] + [(z)\,(y^2) + (z)\,(2y)] + [(1)\,(y^2) + (1)\,(2y)]$$

$$= (5y^3 + 10y^2) + (y^2z + 2yz) + (y^2 + 2y)$$

$$= (5y^3) + (10y^2 + y^2) + (y^2z) + (2yz) + (2y)$$

$$= 5y^3 + 11y^2 + y^2z + 2yz + 2y$$

F) **Division of a monomial by a monomial** is achieved by first dividing the constant coefficients and the variable factors separately and then multiplying these quotients, e.g.,

$$6xyz^2 \div 2y^2z$$

$$= \left(\frac{6}{2}\right)\left(\frac{x}{1}\right)\left(\frac{y}{y^2}\right)\left(\frac{z^2}{z}\right)$$

$$= 3xy^{-1}z$$

$$= \frac{3xz}{y}$$

G) **Division of a polynomial by a polynomial** is achieved by following the given procedure, called long division.

STEP 1: *The terms of both the polynomials are arranged in order of ascending or descending powers of one variable.*

STEP 2: *The first term of the dividend is divided by the first term of the divisor, which gives the first term of the quotient.*

STEP 3: *This first term of the quotient is multiplied by the entire divisor and the result is subtracted from the dividend.*

STEP 4: *Using the remainder obtained from Step 3 as the new dividend, Steps 2 and 3 are repeated until the remainder is zero or the degree of the remainder is less than the degree of the divisor.*

STEP 5: *The result is written as follows:*

$$\frac{\text{dividend}}{\text{divisor}} = \text{quotient} + \frac{\text{remainder}}{\text{divisor}}$$

$$\text{divisor} \neq 0$$

e.g., $(2x^2 + x + 6) \div (x + 1)$

The result is $\left(2x^2 + x + 6\right) \div \left(x + 1\right) = 2x - 1 + \dfrac{7}{x+1}$

DRILL: OPERATIONS WITH POLYNOMIALS

Addition

DIRECTIONS: Add the following polynomials.

1. $9a^2b + 3c + 2a^2b + 5c =$

(A) $19a^2bc$ (B) $11a^2b + 8c$ (C) $11a^4b^2 + 8c^2$
(D) $19a^4b^2c^2$ (E) $12a^2b + 8c^2$

2. $14m^2n^3 + 6m^2n^3 + 3m^2n^3 =$

(A) $20m^2n^3$ (B) $23m^6n^9$ (C) $23m^2n^3$
(D) $32m^6n^9$ (E) $23m^8n^{27}$

3. $3x + 2y + 16x + 3z + 6y =$

(A) $19x + 8y$ (B) $19x + 11yz$ (C) $19x + 8y + 3z$
(D) $11xy + 19xz$ (E) $30xyz$

4. $(4d^2 + 7e^3 + 12f) + (3d^2 + 6e^3 + 2f) =$

(A) $23d^2e^3f$ (B) $33d^2e^2f$ (C) $33d^4e^6f^2$
(D) $7d^2 + 13e^3 + 14f$ (E) $23d^2 + 11e^3f$

5. $3ac^2 + 2b^2c + 7ac^2 + 2ac^2 + b^2c =$

(A) $12ac^2 + 3b^2c$ (B) $14ab^2c^2$ (C) $11ac^2 + 4ab^2c$
(D) $15ab^2c^2$ (E) $15a^2b^4c^4$

Subtraction

DIRECTIONS: Subtract the following polynomials.

6. $14m^2n - 6m^2n =$

(A) $20m^2n$ (B) $8m^2n$ (C) $8m$ (D) 8 (E) $8m^4n^2$

7. $3x^3y^2 - 4xz - 6x^3y^2 =$

(A) $-7x^2y^2z$ (B) $3x3y^2 - 10x4y^2z$ (C) $-3x^3y^2 - 4xz$
(D) $-x^2y^2z - 6x3y^2$ (E) $-7xyz$

8. $9g^2 + 6h - 2g^2 - 5h =$

(A) $15g^2h - 7g^2h$ (B) $7g^4h^2$ (C) $11g^2 + 7h$
(D) $11g^2 - 7h^2$ (E) $7g^2 + h$

9. $7b^3 - 4c^2 - 6b^3 + 3c^2 =$

(A) $b^3 - c^2$ (B) $-11b^2 - 3c^2$ (C) $13b^3 - c$
(D) $7b - c$ (E) 0

10. $11q^2r - 4q^2r - 8q^2r =$

(A) $22q^2r$ (B) q^2r (C) $-2q^2r$
(D) $-q^2r$ (E) $2q^2r$

Multiplication

DIRECTIONS: Multiply the following polynomials.

11. $5p^2t \times 3p^2t =$
    (A) $15p^2t$ (B) $15p^4t$ (C) $15p^4t^2$
    (D) $8p^2t$ (E) $8p^4t^2$

12. $(2r + s)14r =$
    (A) $28rs$ (B) $28r^2 + 14sr$ (C) $16r^2 + 14rs$
    (D) $28r + 14sr$ (E) $17r^2s$

13. $(4m + p)(3m - 2p) =$
    (A) $12m^2 + 5mp + 2p^2$ (B) $12m^2 - 2mp + 2p^2$ (C) $7m - p$
    (D) $12m - 2p$ (E) $12m^2 - 5mp - 2p^2$

14. $(2a + b)(3a^2 + ab + b^2) =$
    (A) $6a^3 + 5a^2b + 3ab^2 + b^3$ (B) $5a^3 + 3ab + b^3$
    (C) $6a^3 + 2a^2b + 2ab^2$ (D) $3a^2 + 2a + ab + b + b^2$
    (E) $6a^3 + 3a^2b + 5ab^2 + b^3$

15. $(6t^2 + 2t + 1)3t =$
    (A) $9t^2 + 5t + 3$ (B) $18t^2 + 6t + 3$ (C) $9t^3 + 6t^2 + 3t$
    (D) $18t^3 + 6t^2 + 3t$ (E) $12t^3 + 6t^2 + 3t$

Division

DIRECTIONS: Divide the following polynomials.

16. $(x^2 + x - 6) \div (x - 2) =$
    (A) $x - 3$ (B) $x + 2$ (C) $x + 3$ (D) $x - 2$ (E) $2x + 2$

17. $24b^4c^3 \div 6b^2c =$
    (A) $3b^2c^2$ (B) $4b^4c^3$ (C) $4b^3c^2$ (D) $4b^2c^2$ (E) $3b^4c^3$

18. $(3p^2 + pq - 2q^2) \div (p + q) =$
    (A) $3p + 2q$ (B) $2q - 3p$ (C) $3p - q$
    (D) $2q + 3p$ (E) $3p - 2q$

19. $(y^3 - 2y^2 - y + 2) \div (y - 2) =$
    (A) $(y - 1)^2$ (B) $y^2 - 1$ (C) $(y + 2)(y - 1)$
    (D) $(y + 1)^2$ (E) $(y + 1)(y - 2)$

20. $(m^2 + m - 14) \div (m + 4) =$
    (A) $m - 2$ (B) $m - 3 + \frac{-2}{m+4}$ (C) $m - 3 + \frac{4}{m+4}$
    (D) $m - 3$ (E) $m - 2 + \frac{-3}{m+4}$

MASTERING FACTORING ALGEBRAIC EXPRESSIONS

To factor a polynomial completely is to find the prime factors of the polynomial with respect to a specified set of numbers.

The following concepts are important when factoring or simplifying expressions.

A) The factors of an algebraic expression consist of two or more algebraic expressions that, when multiplied together, produce the given algebraic expression.

B) A **prime factor** is a polynomial with no factors other than itself and 1. The **least common multiple (LCM)** for a set of numbers is the smallest quantity divisible by every number of the set. For algebraic expressions the least common numerical coefficients for each of the given expressions will be a factor.

C) The **greatest common factor (GCF)** for a set of numbers is the largest factor that is common to all members of the set.

D) For algebraic expressions, the greatest common factor is the polynomial of highest degree and the largest numerical coefficient that is a factor of all the given expressions.

Some important formulas, useful for the factoring of polynomials, are listed below.

$$a(c + d) = ac + ad$$

$$(a + b)(a - b) = a^2 - b^2$$

$$(a + b)(a + b) = (a + b)^2 = a^2 + 2ab + b^2$$

$$(a - b)(a - b) = (a - b)^2 = a^2 - 2ab + b^2$$

$$(x + a)(x + b) = x^2 + (a + b)x + ab$$

$$(ax + b)(cx + d) = acx^2 + (ad + bc)x + bd$$

$$(a + b)(c + d) = ac + bc + ad + bd$$

$$(a + b)(a + b)(a + b) = (a + b)^3 = a^3 + 3a^2b + 3ab^2 + b^3$$

$$(a - b)(a - b)(a - b) = (a - b)^3 = a^3 - 3a^2b + 3ab^2 - b^3$$

$$(a - b)(a^2 + ab + b^2) = a^3 - b^3$$

$$(a + b)(a^2 - ab + b^2) = a^3 + b^3$$

$$(a + b + c)^2 = a^2 + b^2 + c^2 + 2ab + 2ac + 2bc$$

$$(a - b)(a^3 + a^2b + ab^2 + b^3) = a^4 - b^4$$

$$(a - b)(a^4 + a^3b + a^2b^2 + ab^3 + b^4) = a^5 - b^5$$

$$(a - b)(a^5 + a^4b + a^3b^2 + a^2b^3 + ab^4 + b^5) = a^6 - b^6$$

$$(a - b)(a^{n-1} + a^{n-2}b + a^{n-3}b^2 + \ldots + ab^{n-2} + b^{n-1}) = a^n - b^n$$

where n is any positive integer (1, 2, 3, 4, . . .).

$$(a + b)(a^{n-1} - a^{n-2}b + a^{n-3}b^2 - \ldots - ab^{n-2} + b^{n-1}) = a^n + b^n$$

where n is any positive odd integer (1, 3, 5, 7, . . .).

The procedure for factoring an algebraic expression completely is as follows:

STEP 1: *First find the greatest common factor if there is any. Then examine each factor remaining for greatest common factors.*

STEP 2: *Continue factoring the factors obtained in Step 1 until all factors other than monomial factors are prime.*

EXAMPLE

Factoring $4 - 16x^2$,

$$4 - 16x^2 = 4(1 - 4x^2) = 4(1 + 2x)(1 - 2x)$$

PROBLEM

Express each of the following as a single term.

(1) $3x^2 + 2x^2 - 4x^2$ *(2)* $5axy^2 - 7axy^2 - 3xy^2$

SOLUTION

(1) Factor x^2 in the expression.

$$3x^2 + 2x^2 - 4x^2 = (3 + 2 - 4)x^2 = 1x^2 = x^2$$

(2) Factor xy^2 in the expression and then factor a.

$$\begin{aligned} 5axy^2 - 7axy^2 - 3xy^2 &= (5a - 7a - 3)xy^2 \\ &= [(5 - 7)a - 3]xy^2 \\ &= (-2a - 3)xy^2 \end{aligned}$$

PROBLEM

Simplify $\dfrac{\dfrac{1}{x-1} - \dfrac{1}{x-2}}{\dfrac{1}{x-2} - \dfrac{1}{x-3}}$.

SOLUTION

Simplify the expression in the numerator by using the addition rule:

$$\frac{a}{b} + \frac{c}{d} = \frac{ad + bc}{bd}$$

Notice bd is the Lowest Common Denominator, LCD. We obtain

$$\frac{x - 2 - (x - 1)}{(x - 1)(x - 2)} = \frac{-1}{(x - 1)(x - 2)}$$

in the numerator.

Repeat this procedure for the expression in the denominator:

$$\frac{x-3-(x-2)}{(x-2)(x-3)}=\frac{-1}{(x-2)(x-3)}$$

We now have

$$\frac{\frac{-1}{(x-1)(x-2)}}{\frac{-1}{(x-2)(x-3)}}$$

which is simplified by inverting the fraction in the denominator and multiplying it by the numerator and cancelling like terms

$$\frac{-1}{(x-1)(x-2)}\times\frac{(x-2)(x-3)}{-1}=\frac{x-3}{x-1}.$$

DRILL: SIMPLIFYING ALGEBRAIC EXPRESSIONS

DIRECTIONS: Simplify the following expressions.

1. $16b^2 - 25z^2 =$
   (A) $(4b - 5z)^2$ (B) $(4b + 5z)^2$ (C) $(4b - 5z)(4b + 5z)$
   (D) $(16b - 25z)^2$ (E) $(5z - 4b)(5z + 4b)$

2. $x^2 - 2x - 8 =$
   (A) $(x - 4)^2$ (B) $(x - 6)(x - 2)$ (C) $(x + 4)(x - 2)$
   (D) $(x - 4)(x + 2)$ (E) $(x - 4)(x - 2)$

3. $2c^2 + 5cd - 3d^2 =$
   (A) $(c - 3d)(c + 2d)$ (B) $(2c - d)(c + 3d)$ (C) $(c - d)(2c + 3d)$
   (D) $(2c + d)(c + 3d)$ (E) $(2d + c)(c + 3d)$

4. $4t^3 - 20t =$
   (A) $4t(t^2 - 5)$ (B) $4t^2(t - 20)$ (C) $4t(t + 4)(t - 5)$
   (D) $2t(2t^2 - 10)$ (E) $12t(t - 20)$

5. $x^2 + xy - 2y^2 =$
   (A) $(x - 2y)(x + y)$ (B) $(x - 2y)(x - y)$ (C) $(x + 2y)(x + y)$
   (D) $(x + 2y)(x - y)$ (E) $(x - y)(2y - x)$

6. $5b^2 + 17bd + 6d^2 =$
   (A) $(5b + d)(b + 6d)$ (B) $(5b + 2d)(b + 3d)$ (C) $(5b - 2d)(b - 3d)$
   (D) $(5b - 2d)(b + 3d)$ (E) $(b + 3d)(5b - 2d)$

7. $x^2 + 2x + 1 =$
   (A) $(x + 1)^2$ (B) $(x + 2)(x - 1)$ (C) $(x - 2)(x + 1)$
   (D) $(x + 1)(x - 1)$ (E) $(x - 1)(x + 1)$

8. $3z^3 + 6z^2 =$

(A) $3(z^3 + 2z^2)$ (B) $3z^2(z + 2)$ (C) $3z(z^2 + 2z)$

(D) $z^2(3z + 6)$ (E) $3z^2(1 + 2z)$

9. $m^2p^2 + mpq - 6q^2 =$

(A) $(mp - 2q)(mp + 3q)$ (B) $mp(mp - 2q)(mp + 3q)$

(C) $mpq(1 - 6q)$ (D) $(mp + 2q)(mp + 3q)$

(E) $(mp + 2q)(1 - 6q)$

10. $2h^3 + 2h^2t - 4ht^2 =$

(A) $2(h^3 - t)(h + t)$ (B) $2h(h - 2t)^2$ (C) $4h(ht - t^2)$

(D) $2h(h + t) - 4ht^2$ (E) $2h(h + 2t)(h - t)$

MASTERING EQUATIONS

An **equation** is defined as a statement that two separate expressions are equal.

A **solution** to an equation containing a single variable is a number that makes the equation true when it is substituted for the variable. For example, in the equation $3x = 18$, 6 is the solution since $3(6) = 18$. Depending on the equation, there can be more than one solution. Equations with the same solutions are said to be **equivalent equations**. An equation without a solution is said to have a solution set that is the **empty** or **null set** and is represented by $\varnothing$.

Replacing an expression within an equation by an equivalent expression will result in a new equation with solutions equivalent to the original equation. Suppose we are given the equation

$$3x + y + x + 2y = 15$$

By combining like terms we get

$$3x + y + x + 2y = 4x + 3y$$

Since these two expressions are equivalent, we can substitute the simpler form into the equation to get

$$4x + 3y = 15$$

Performing the same operation to both sides of an equation by the same expression will result in a new equation that is equivalent to the original equation.

A) **Addition or subtraction**

$$y + 6 = 10$$

We can add (–6) to both sides

$$y + 6 + (-6) = 10 + (-6)$$

to get $y + 0 = 10 - 6$ $y = 4$

B) **Multiplication or division**

$$3x = 6$$
$$\frac{3x}{3} = \frac{6}{3}$$
$$x = 2$$

$3x = 6$ is equivalent to $x = 2$.

C) **Raising to a power**

$$a = x^2y$$
$$a^2 = (x^2y)^2$$
$$a^2 = x^4y^2$$

This can be applied to negative and fractional powers as well, e.g.,

$$x^2 = 3y^4$$

If we raise both sides to the –2 power, we get

$$(x^2)^{-2} = (3y^4)^{-2}$$
$$\frac{1}{(x^2)^2} = \frac{1}{\left(3y^4\right)^2}$$
$$\frac{1}{x^4} = \frac{1}{9y^8}$$

If we raise both sides to the $\frac{1}{2}$ power, which is the same as taking the square root, we get

$$\left(x^2\right)^{1/2} = \left(3y^4\right)^{1/2}$$
$$x = \pm\sqrt{3}y^2$$

D) The **reciprocal** of both sides of an equation are equivalent to the original equation. Note: The reciprocal of zero is undefined.

$$\frac{2x+y}{z} = \frac{5}{2} \quad \frac{z}{2x+y} = \frac{2}{5}$$

PROBLEM

Solve for x, justifying each step.

$3x - 8 = 7x + 8$

SOLUTION

	$3x - 8 = 7x + 8$
Add 8 to both sides:	$3x - 8 + 8 = 7x + 8 + 8$
Additive inverse property:	$3x + 0 = 7x + 16$
Additive identity property:	$3x = 7x + 16$
Add $(-7x)$ to both sides:	$3x - 7x = 7x + 16 - 7x$

Commute: $-4x = 7x - 7x + 16$

Additive inverse property: $-4x = 0 + 16$

Additive identity property: $-4x = 16$

Divide both sides by –4: $x = \frac{16}{-4}$

$x = -4$

Check: Replacing x with –4 in the original equation:

$$3x - 8 = 7x + 8$$

$$3(-4) - 8 = 7(-4) + 8$$

$$-12 - 8 = -28 + 8$$

$$-20 = -20$$

Linear Equations

A linear equation with one unknown is one that can be put into the form $ax + b = 0$, where a and b are constants, $a \neq 0$.

To solve a linear equation means to transform it in the form $x = -\frac{b}{a}$.

A) If the equation has unknowns on both sides of the equality, it is convenient to put similar terms on the same sides. Refer to the following example.

$$4x + 3 = 2x + 9$$

$$4x + 3 - 2x = 2x + 9 - 2x$$

$$(4x - 2x) + 3 = (2x - 2x) + 9$$

$$2x + 3 = 0 + 9$$

$$2x + 3 - 3 = 0 + 9 - 3$$

$$2x = 6$$

$$\frac{2x}{2} = \frac{6}{2}$$

$$x = 3$$

B) If the equation appears in fractional form, it is necessary to transform it, using cross-multiplication, and then repeating the same procedure as in A), we obtain:

$$\frac{3x+4}{3} \times \frac{7x+2}{5}$$

By using cross-multiplication we would obtain:

$$3(7x + 2) = 5(3x + 4).$$

This is equivalent to:

$$21x + 6 = 15x + 20,$$

which can be solved as in A).

$$21x + 6 = 15x + 20$$

$$21x - 15x + 6 = 15x - 15x + 20$$

$$6x + 6 - 6 = 20 - 6$$

$$6x = 14$$

$$x = \frac{14}{6}$$

$$x = \frac{7}{3}$$

PROBLEM

Solve the equation $2(x + 3) = (3x + 5) - (x - 5)$.

SOLUTION

We transform the given equation to an equivalent equation in which we can easily recognize the solution set.

	$2(x + 3) = 3x + 5 - (x - 5)$
Distribute:	$2x + 6 = 3x + 5 - x + 5$
Combine terms:	$2x + 6 = 2x + 10$
Subtract $2x$ from both sides:	$6 = 10$

Since $6 = 10$ is not a true statement, there is no real number x which will make the original equation true. The equation is inconsistent and the solution set is $\varnothing$, the empty set.

PROBLEM

Solve the equation $2(\frac{2}{3}y + 5) + 2(y + 5) = 130$.

SOLUTION

The procedure for solving this equation is as follows:

Distribute:	$\frac{4}{3}y + 10 + 2y + 10 = 130$
Combine like terms:	$\frac{4}{3}y + 2y + 20 = 130$
Subtract 20 from both sides:	$\frac{4}{3}y + 2y = 110$
Convert $2y$ into a fraction with denominator 3:	$\frac{4}{3}y + \frac{6}{3}y = 110$
Combine like terms:	$\frac{10}{3}y = 110$
Divide by $\frac{10}{3}$:	$y = 110 \times \frac{3}{10} = 33$

Check: Replace y with 33 in the original equation.

$$2\left(\frac{2}{3}(33)+5\right)+2(33+5)=130$$

$$2(22+5)+2(38)=130$$

$$2(27)+76=130$$

$$54+76=130$$

$$130=130$$

Therefore, the solution to the given equation is $y = 33$.

Radical Equations

A **radical equation** includes at least one radical expression. To solve an equation with a radical expression, first isolate the radical on one side of the equation. Then raise each side of the equation to a power to eliminate the radical. Finally, use inverse operations to isolate the variable.

PROBLEM

Solve the equation $\sqrt{3x+1}=5$.

SOLUTION

The radical is already isolated on one side of the equation. Square both sides to eliminate the radical.

$\sqrt{3x+1}=5$	
$\left(\sqrt{3x+1}\right)^2=5^2$	Square both sides of the equation.
$3x+1=25$	Simplify.
$3x=24$	Subtract 1 from both sides.
$x=8$	Divide both sides by 3.

The solution of the equation is $x = 8$.

PROBLEM

Solve the equation $3\sqrt[3]{x}=\sqrt[3]{x}+6$.

SOLUTION

First, isolate a radical expression on one side of the equation. Then raise each side to a power.

$3\sqrt[3]{x}=\sqrt[3]{x}+6$	
$2\sqrt[3]{x}=6$	Subtract $\sqrt[3]{x}$ from both sides of the equation.
$\sqrt[3]{x}=3$	Divide both sides by 2.
$(\sqrt[3]{x})^3=3^3$	Cube both sides of the equation.
$x=27$	Simplify.

The solution of the equation is $x = 27$.

DRILL: EQUATIONS

Linear Equations

DIRECTIONS: Solve for *x*.

1. $4x - 2 = 10$

(A) -1 (B) 2 (C) 3 (D) 4 (E) 6

2. $7z + 1 - z = 2z - 7$

(A) -2 (B) 0 (C) 1 (D) 2 (E) 3

3. $\frac{1}{3}b + 3 = \frac{1}{2}b$

(A) $\frac{1}{2}$ (B) 2 (C) $3\frac{3}{5}$ (D) 6 (E) 18

4. $0.4p + 1 = 0.7p - 2$

(A) 0.1 (B) 2 (C) 5 (D) 10 (E) 12

5. $4(3x + 2) - 11 = 3(3x - 2)$

(A) -3 (B) -1 (C) 2 (D) 3 (E) 7

Radical Equations

DIRECTIONS: Solve each equation for *x*.

6. $\sqrt{4x - 4} = 6$

(A) 2.5 (B) 4 (C) 8 (D) 10 (E) 25

7. $3\sqrt{x} + 12 = 42$

(A) 25 (B) 81 (C) 100 (D) 324 (E) 900

8. $\frac{\sqrt[3]{x+5}}{3} = 2$

(A) 13 (B) 19 (C) 49 (D) 211 (E) 221

MASTERING LINEAR EQUATIONS IN TWO VARIABLES

Equations of the form $ax + by = c$, where *a*, *b*, *c* are constants and $a, b \neq 0$ are called **linear equations** with two unknown variables.

There are several ways to solve systems of linear equations with two variables.

METHOD 1: *Addition or subtraction—if necessary, multiply the equations by numbers that will make the coefficients of one unknown in the resulting equations numerically equal. If the signs of equal coefficients are the same, subtract the equation, otherwise add.*

The result is one equation with one unknown; we solve it and substitute the value into the other equations to find the unknown that we first eliminated.

METHOD 2: *Substitution—find the value of one unknown in terms of the other. Substitute this value in the other equation and solve.*

METHOD 3: *Graph—graph both equations. The point of intersection of the drawn lines is a simultaneous solution for the equations and its coordinates correspond to the answer that would be found analytically.*

If the lines are parallel they have no simultaneous solution.

Dependent equations are equations that represent the same line; therefore, every point on the line of a dependent equation represents a solution. Since there is an infinite number of points on a line there is an infinite number of simultaneous solutions, for example,

$$\begin{cases} 2x + y = 8 \\ 4x + 2y = 16 \end{cases}$$

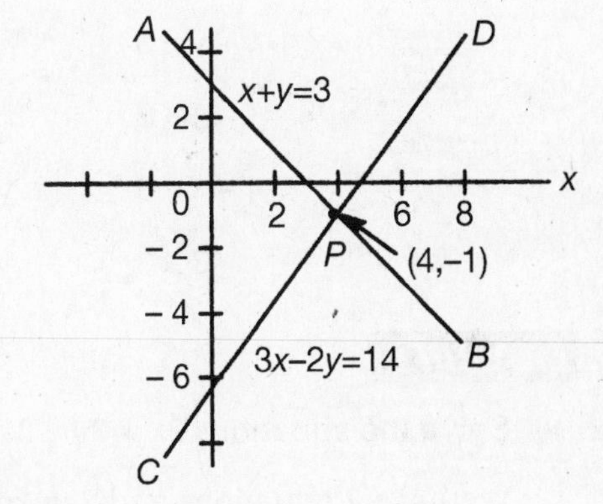

The equations on the previous page are dependent. Since they represent the same line, all points that satisfy either of the equations are solutions of the system.

A system of linear equations is consistent if there is only one solution for the system.

A system of linear equations is inconsistent if it does not have any solutions.

EXAMPLE

Find the point of intersection of the graphs of the equations as shown in the previous figure.

$x + y = 3$

$3x - 2y = 14$

To solve these linear equations, solve for y in terms of x. The equations will be in the form $y = mx + b$, where m is the slope and b is the intercept on the y-axis.

$x + y = 3$

Subtract x from both sides: $y = 3 - x$

Subtract $3x$ from both sides: $3x - 2y = 14$

Divide by –2: $-2y = 14 - 3x$

$y = -7 + \frac{3}{2}x$

The graphs of the linear functions, $y = 3 - x$ and $y = 7 + \frac{3}{2}x$ can be determined by plotting only two points. For example, for $y = 3 - x$, let $x = 0$, then $y = 3$. Let $x = 1$, then $y = 2$. The two points on this first line are (0,3) and (1,2). For $y = -7 + \frac{3}{2}x$, let $x = 0$, then $y = -7$. Let $x = 1$, then $y = -5\frac{1}{2}$. The two points on this second line are (0,–7) and $(1, -5\frac{1}{2})$.

To find the point of intersection P of

$$x + y = 3 \quad \text{and} \quad 3x - 2y = 14$$

solve them algebraically. Multiply the first equation by 2. Add these two equations to eliminate the variable y.

$$2x + 2y = 6$$

$$3x - 2y = 14$$

$$5x = 20$$

Solve for x to obtain $x = 4$. Substitute this into $y = 3 - x$ to get $y = 3 - 4 = -1$. P is (4,–1). AB is the graph of the first equation, and CD is the graph of the second equation. The point of intersection P of the two graphs is the only point on both lines. The coordinates of P satisfy both equations and represent the desired solution of the problem. From the graph, P seems to be the point (4,–1). These coordinates satisfy both equations, and hence are the exact coordinates of the point of intersection of the two lines.

To show that (4,–1) satisfies both equations, substitute this point into both equations.

$x + y = 3$	$3x - 2y = 14$
$4 + (-1) = 3$	$3(4) - 2(-1) = 14$
$4 - 1 = 3$	$12 + 2 = 14$
$3 = 3$	$14 = 14$

EXAMPLE

Solve the equations $2x + 3y = 6$ and $4x + 6y = 7$ simultaneously.

We have 2 equations and 2 unknowns,

$$2x + 3y = 6 \qquad (1)$$

and

$$4x + 6y = 7 \qquad (2)$$

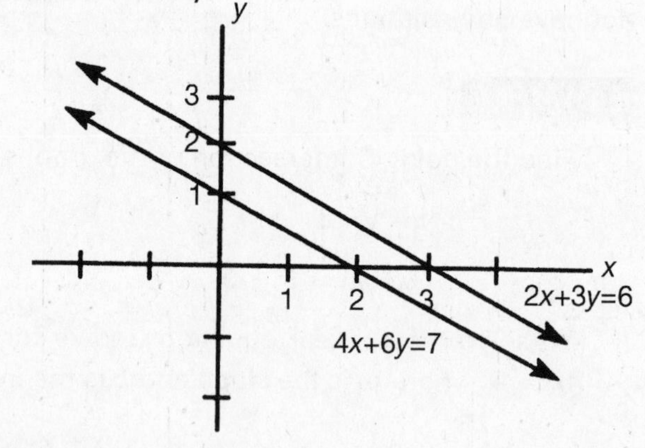

There are several methods to solve this problem. We have chosen to multiply each equation by a different number so that when the two equations are added, one of the variables drops out. Thus,

Multiply equation (1) by 2:	$4x + 6y = 12$	(3)
Multiply equation (2) by –1:	$-4x - 6y = -7$	(4)
Add equations (3) and (4):	$0 = 5$	

We obtain a peculiar result!

Actually, what we have shown in this case is that if there were a simultaneous solution to the given equations, then 0 would equal 5. But the conclusion is impossible; therefore there can be no simultaneous solution to these two equations, hence no point satisfying both.

The straight lines that are the graphs of these equations must be parallel if they never intersect, but not identical, which can be seen from the graph of these equations (see the accompanying diagram).

EXAMPLE

Solve the equations $2x + 3y = 6$ and $y = -(\frac{2x}{3}) + 2$ simultaneously.

We have 2 equations and 2 unknowns.

$$2x + 3y = 6 \qquad (1)$$

and

$$y = -\left(\frac{2x}{3}\right) + 2 \qquad (2)$$

There are several methods of solution for this problem. Since equation (2) already gives us an expression for y, we use the method of substitution. Substitute: $-(\frac{2x}{3}) + 2$ for y in the first equation:

$$2x + 3(-\frac{2x}{3} + 2) = 6$$

Distribute:

$$2x - 2x + 6 = 6$$

$$6 = 6$$

Apparently we have gotten nowhere! The result $6 = 6$ is true, but indicates no solution. Actually, our work shows that no matter what real number x is, if y is determined by the second equation, then the first equation will always be satisfied.

The reason for this peculiarity may be seen if we take a closer look at the equation $y = -(\frac{2x}{3}) + 2$. It is equivalent to $3y = -2x + 6$, or $2x + 3y = 6$.

In other words, the two equations are equivalent. Any pair of values of x and y that satisfies one satisfies the other.

It is hardly necessary to verify that in this case the graphs of the given equations are identical lines and that there are an infinite number of simultaneous solutions of these equations.

A system of three linear equations in three unknowns is solved by eliminating one unknown from any two of the three equations and solving them. After finding two unknowns substitute them in any of the equations to find the third unknown.

PROBLEM

Solve the system

$$2x + 3y - 4z = -8 \quad (1)$$

$$x + y - 2z = -5 \quad (2)$$

$$7x - 2y + 5z = 4 \quad (3)$$

SOLUTION

We cannot eliminate any variable from two pairs of equations by a single multiplication. However, both x and z may be eliminated from equations 1 and 2 by multiplying equation (2) by –2. Then

$$2x + 3y - 4z = -8 \quad (1)$$

$$-2x - 2y + 4z = 10 \quad (4)$$

By addition, we have $y = 2$. Although we may now eliminate either x or z from another pair of equations, we can more conveniently substitute $y = 2$ in equations (2) and (3) to get two equations in two variables. Thus, making the substitution $y = 2$ in equations (2) and (3), we have

$$x - 2z = -7 \quad (5)$$

$$7x + 5z = 8 \quad (6)$$

Multiply equation (5) by 5 and multiply (6) by 2. Then add the two new equations. Then $x = -1$. Substitute x in either equation (5) or (6) to find z.

The solution of the system is $x = -1$, $y = 2$, and $z = 3$. Check by substitution.

A system of equations, as shown below, that has all constant terms $b_1, b_2, \ldots, b_n$ equal to zero is said to be a homogeneous system.

$$\begin{cases} a_{11}x_1 + a_{12}x_2 + \ldots + a_{1n}x_m = b_1 \\ a_{21}x_1 + a_{22}x_2 + \ldots + a_{2n}x_m = b_2 \\ \vdots \quad \vdots \quad \vdots \quad \vdots \\ a_{n1}x_1 + a_{n2}x_2 + \ldots + a_{nn}x_m = b_n \end{cases}$$

A homogeneous system (one in which each variable can be replaced by a constant and the constant can be factored out) always has at least one solution, which is called the trivial solution, that is $x_1 = 0$, $x_2 = 0, \ldots, x_m = 0$.

For any given homogeneous system of equations, in which the number of variables is greater than or equal to the number of equations, there are nontrivial solutions.

Two systems of linear equations are said to be equivalent if and only if they have the same solution set.

PROBLEM

Solve for x and y.

$x + 2y = 8$ (1)

$3x + 4y = 20$ (2)

SOLUTION

Solve equation (1) for x in terms of y: $x = 8 - 2y$ (3)

Substitute $(8 - 2y)$ for x in (2): $3(8 - 2y) + 4y = 20$ (4)

Solve (4) for y as follows:

Distribute: $24 - 6y + 4y = 20$

Combine like terms and then subtract 24 from both sides:

$$24 - 2y = 20$$

$$24 - 24 - 2y = 20 - 24$$

$$-2y = -4$$

Divide both sides by -2: $y = 2$

Substitute 2 for y in equation (1): $x + 2(2) = 8$

$$x = 4$$

Thus, our solution is $x = 4$, $y = 2$.

Check: Substitute $x = 4$, $y = 2$ in equations (1) and (2):

$$4 + 2(2) = 8$$

$$8 = 8$$

$$3(4) + 4(2) = 20$$

$$20 = 20$$

PROBLEM

Solve algebraically.

$4x + 2y = -1$ (1)

$5x - 3y = 7$ (2)

SOLUTION

We arbitrarily choose to eliminate x first.

Multiply (1) by 5: $20x + 10y = -5$ (3)

Multiply (2) by 4: $20x - 12y = 28$ (4)

Subtract (3) from (4): $22y = -33$ (5)

Divide (5) by 22: $y = -\frac{33}{22} = -\frac{3}{2}$

To find x, substitute $y = -\frac{3}{2}$ in either of the original equations. If we use equation (1), we obtain $4x + 2(-\frac{3}{2}) = -1$, $4x - 3 = -1$, $4x = 2$, $x = \frac{1}{2}$.

The solution $(\frac{1}{2}, -\frac{3}{2})$ should be checked in both equations of the given system.

Replacing $(\frac{1}{2}, -\frac{3}{2})$ in equation (1):

$$4x + 2y = -1$$

$$4(\frac{1}{2}) + 2(-\frac{3}{2}) = -1$$

$$\frac{4}{2} - 3 = -1$$

$$2 - 3 = -1$$

$$-1 = -1$$

Replacing $(\frac{1}{2}, -\frac{3}{2})$ in equation (2):

$$5x - 3y = 7$$

$$5\left(\frac{1}{2}\right) - 3\left(-\frac{3}{2}\right) = 7$$

$$\frac{5}{2} + \frac{9}{2} = 7$$

$$\frac{14}{2} = 7$$

$$7 = 7$$

(Instead of eliminating x from the two given equations, we could have eliminated y by multiplying equation (1) by 3, multiplying equation (2) by 2, and then adding the two derived equations.)

DRILL: LINEAR EQUATIONS IN TWO VARIABLES

DIRECTIONS: Find the solution set for each pair of equations.

1. $3x + 4y = -2$

   $x - 6y = -8$

   (A) (2, –1) (B) (1, –2) (C) (–2, –1) (D) (1, 2) (E) (–2, 1)

2. $2x + y = -10$

   $-2x - 4y = 4$

   (A) (6, –2) (B) (–6, 2) (C) (–2, 6) (D) (2, 6) (E) (–6, –2)

3. $6x + 5y = -4$

   $3x - 3y = 9$

   (A) (1, –2) (B) (1, 2) (C) (2, –1) (D) (–2, 1) (E) (–1, 2)

4. $4x + 3y = 9$

$2x - 2y = 8$

(A) (–3, 1) (B) (1, –3) (C) (3, 1) (D) (3, –1) (E) (–1, 3)

5. $x + y = 7$

$x = y - 3$

(A) (5, 2) (B) (–5, 2) (C) (2, 5) (D) (–2, 5) (E) (2, –5)

6. $5x + 6y = 4$

$3x - 2y = 1$

(A) (3, 6) (B) $\left(\frac{1}{2}, \frac{1}{4}\right)$ (C) (–3, 6) (D) (2, 4) (E) $\left(\frac{1}{3}, \frac{3}{2}\right)$

7. $x - 2y = 7$

$x + y = -2$

(A) (–2, 7) (B) (3, –1) (C) (–7, 2) (D) (1, –3) (E) (1, –2)

8. $4x + 3y = 3$

$-2x + 6y = 3$

(A) $\left(\frac{1}{2}, \frac{2}{3}\right)$ (B) (–0.3, 0.6) (C) $\left(\frac{2}{3}, -1\right)$

(D) (–0.2, 0.5) (E) (0.3, 0.6)

9. $4x - 2y = -14$

$8x + y = 7$

(A) (0, 7) (B) (2, –7) (C) (7, 0) (D) (–7, 2) (E) (0, 2)

10. $6x - 3y = 1$

$-9x + 5y = -1$

(A) (1, –1) (B) $\left(\frac{2}{3}, 1\right)$ (C) $\left(1, \frac{2}{3}\right)$

(D) (–1, 1) (E) $\left(\frac{2}{3}, -1\right)$

MASTERING QUADRATIC EQUATIONS

A second degree equation in x of the type $ax^2 + bx + c = 0$, $a \neq 0$, a, b, and c are real numbers, is called a **quadratic equation.**

To solve a quadratic equation is to find values of x that satisfy $ax^2 + bx + c = 0$. These values of x are called **solutions**, or **roots**, of the equation.

A quadratic equation has a maximum of two roots. Methods of solving quadratic equations are

A) **Direct solution**: Given $x^2 - 9 = 0$.

We can solve directly by isolating the variable x.

$x^2 = 9$

$x = \pm 3$

B) **Factoring**: Given a quadratic equation $ax^2 + bx + c = 0$, $a, b, c \neq 0$, to factor means to express it as the product $a(x - r_1)(x - r_2) = 0$, where r_1 and r_2 are the two roots.

Some helpful hints to remember are

a) $r_1 + r_2 = -\frac{b}{a}$.

b) $r_1 r_2 = \frac{c}{a}$.

Given $x^2 - 5x + 4 = 0$.

Since

$$r_1 + r_2 = -\frac{b}{a} = -\frac{(-5)}{1} = 1$$

the possible solutions are (3, 2), (4, 1), and (5, 0). Also

$$r_1 r_2 = \frac{c}{a} = \frac{4}{1} = 4$$

This equation is satisfied only by the second pair, so $r_1 = 4$, $r_2 = 1$, and the factored form is $(x - 4)(x - 1) = 0$.

If the coefficient of x^2 is not 1, it is necessary to divide the equation by this coefficient and then factor.

Given $2x^2 - 12x + 16 = 0$

Dividing by 2, we obtain

$x^2 - 6x + 8 = 0$

Since

$$r_1 + r_2 = -\frac{b}{a} = 6$$

the possible solutions are (6, 0), (5, 1), (4, 2), and (3, 3). Also $r_1 r_2 = 8$, so the only possible answer is (4, 2) and the expression $x^2 - 6x + 8 = 0$ can be factored as $(x - 4)(x - 2)$.

C) **Completing the Squares**: If it is difficult to factor the quadratic equation using the previous method, we can complete the squares.

Given $x^2 - 12x + 8 = 0$.

We know that the two roots added up should be 12 because

$$r_1 + r_2 = -\frac{b}{a} = \frac{-(-12)}{1} = 12$$

The possible roots are (12, 0), (11, 1), (10, 2), (9, 3), (8, 4), (7, 5), #and (6, 6).

But none of these satisfy $r_1 r_2 = 8$, so we cannot use (B).

To complete the square, it is necessary to isolate the constant term,

$$x^2 - 12x = -8$$

Then take $\frac{1}{2}$ coefficient of x, square it and add to both sides.

$$x^2 - 12x + \left(\frac{-12}{2}\right)^2 = -8 + \left(\frac{-12}{2}\right)^2$$

$$x^2 - 12x + 36 = -8 + 36 = 28$$

Now we can use the previous method to factor the left side.

$$r_1 + r_2 = 12, \; r_1 r_2 = 36$$

is satisfied by the pair (6, 6), so we have

$$(x - 6)^2 = 28.$$

Now extract the root of both sides and solve for x.

$$(x - 6) = \pm\sqrt{28} = \pm 2\sqrt{7}$$

$$x = \pm 2\sqrt{7} + 6$$

So the roots are

$$x = 2\sqrt{7} + 6, \; x = -2\sqrt{7} + 6$$

PROBLEM

Solve the equation $x^2 + 8x + 15 = 0$.

SOLUTION

Since

$$(x + a)(x + b) = x^2 + bx + ax + ab$$

$$= x^2 + (a + b)x + ab$$

we may factor the given equation,

$$0 = x^2 + 8x + 15$$

replacing $a + b$ by 8 and ab by 15. Thus,

$$a + b = 8, \quad \text{and} \quad ab = 15$$

We want the two numbers a and b whose sum is 8 and whose product is 15. We check all pairs of numbers whose product is 15.

(a) $1 \times 15 = 15$; thus, $a = 1$, $b = 15$, and $ab = 15$.

$1 + 15 = 16$; therefore, we reject these values because $a + b \neq 8$.

(b) $3 \times 5 = 15$; thus, $a = 3$, $b = 5$, and $ab = 15$.

$3 + 5 = 8$; therefore, $a + b = 8$, and we accept these values.

Hence, $x^2 + 8x + 15 = 0$ is equivalent to

$$0 = x^2 + (3 + 5)x + 3 \times 5 = (x + 3)(x + 5)$$

Hence,

$$x + 5 = 0 \text{ or } x + 3 = 0$$

since the product of these two numbers is zero, one of the numbers must be zero. Hence, $x = -5$, or $x = -3$, and the solution set is $x = \{-5, -3\}$.

The student should note that $x = -5$ or $x = -3$. We are certainly not making the statement that $x = -5$ and $x = -3$. Also, the student should check that both these numbers do actually satisfy the given equations and hence are solutions.

Check: Replacing x by (-5) in the original equation:

$$x^2 + 8x + 15 = 0$$
$$(-5)^2 + 8(-5) + 15 = 0$$
$$25 - 40 + 15 = 0$$
$$-15 + 15 = 0$$
$$0 = 0$$

Replacing x by (-3) in the original equation:

$$x^2 + 8x + 15 = 0$$
$$(-3)^2 + 8(-3) + 15 = 0$$
$$9 - 24 + 15 = 0$$
$$-15 + 15 = 0$$
$$0 = 0$$

PROBLEM

Solve the following equations by factoring.

(1) $2x^2 + 3x = 0$ *(2)* $y^2 - 2y - 3 = y - 3$

(3) $z^2 - 2z - 3 = 0$ *(4)* $2m^2 - 11m - 6 = 0$

SOLUTION

(1) $2x^2 + 3x = 0$. Factor out the common factor of x from the left side of the given equation.

$$x(2x + 3) = 0$$

Whenever a product $ab = 0$, where a and b are any two numbers, either $a = 0$ or $b = 0$. Then, either

$$x = 0 \quad \text{or} \quad 2x + 3 = 0$$
$$2x = -3$$
$$x = -\frac{3}{2}$$

Hence, the solution set to the original equation $2x^2 + 3x = 0$ is: $\{-\frac{3}{2}, 0\}$.

(2) $y^2 - 2y - 3 = y - 3$. Subtract $(y - 3)$ from both sides of the given equation.

$$y^2 - 2y - 3 - (y - 3) = y - 3 - (y - 3)$$

$$y^2 - 2y - 3 - y + 3 = y - 3 - y + 3$$

$$y^2 - 2y - \cancel{3} - y + \cancel{3} = \cancel{y} - \cancel{3} - \cancel{y} + \cancel{3}$$

$$y^2 - 3y = 0$$

Factor out a common factor of y from the left side of this equation:

$$y(y - 3) = 0$$

Thus, $y = 0$ or $y - 3 = 0$, $y = 3$.

Therefore, the solution set to the original equation $y^2 - 2y - 3 = y - 3$ is $\{0, 3\}$.

(3) $z^2 - 2z - 3 = 0$. Factor the original equation into a product of two polynomials.

$$z^2 - 2z - 3 = (z - 3)(z + 1) = 0$$

Hence,

$(z - 3)(z + 1) = 0$; and $z - 3 = 0$ or $z + 1 = 0$

$$z = 3 \qquad z = -1$$

Therefore, the solution set to the original equation $z^2 - 2z - 3 = 0$ is $\{-1, 3\}$.

(4) $2m^2 - 11m - 6 = 0$. Factor the original equation into a product of two polynomials.

$$2m^2 - 11m - 6 = (2m + 1)(m - 6) = 0$$

Thus,

$$2m + 1 = 0 \quad \text{or} \quad m - 6 = 0$$

$$2m = -1 \qquad m = 6$$

$$m = -\frac{1}{2}$$

Therefore, the solution set to the original equation $2m^2 - 11m - 6 = 0$ is $\{-\frac{1}{2}, 6\}$.

DRILL: QUADRATIC EQUATIONS

DIRECTIONS: Solve for all values of x.

1. $x^2 - 2x - 8 = 0$

(A) 4 and –2 (B) 4 and 8 (C) 4 (D) –2 and 8 (E) –2

2. $x^2 + 2x - 3 = 0$

(A) –3 and 2 (B) 2 and 1 (C) 3 and 1 (D) –3 and 1 (E) –3

3. $x^2 - 7x = -10$

(A) –3 and 5 (B) 2 and 5 (C) 2 (D) –2 and –5 (E) 5

4. $x^2 - 8x + 16 = 0$

(A) 8 and 2 (B) 1 and 16 (C) 4 (D) –2 and 4 (E) 4 and –4

5. $3x^2 + 3x = 6$

(A) 3 and –6 (B) 2 and 3 (C) –3 and 2 (D) 1 and –3 (E) 1 and –2

6. $x^2 + 7x = 0$

(A) 7 (B) 0 and –7 (C) –7 (D) 0 and 7 (E) 0

7. $x^2 - 25 = 0$

(A) 5 (B) 5 and –5 (C) 15 and 10 (D) –5 and 10 (E) –5

8. $2x^2 + 4x = 16$

(A) 2 and –2 (B) 8 and –2 (C) 4 and 8 (D) 2 and –4 (E) 2 and 4

9. $6x^2 - x - 2 = 0$

(A) 2 and 3 (B) $\frac{1}{2}$ and $\frac{1}{3}$ (C) $-\frac{1}{2}$ and $\frac{2}{3}$ (D) $\frac{2}{3}$ and 3 (E) 2 and $-\frac{1}{3}$

10. $12x^2 + 5x = 3$

(A) $\frac{1}{3}$ and $-\frac{1}{4}$ (B) 4 and –3 (C) 4 and $\frac{1}{6}$ (D) $\frac{1}{3}$ and –4 (E) $-\frac{3}{4}$ and $\frac{1}{3}$

MASTERING ABSOLUTE VALUE EQUATIONS

The **absolute value** of a, $|a|$, is defined as

$$|a| = a \text{ when } a > 0,$$

$$|a| = -a \text{ when } a < 0,$$

$$|a| = 0 \text{ when } a = 0.$$

EXAMPLES

$|5| = 5$, $|-8| = -(-8) = 8$

Absolute values follow the given rules:

(A) $|-a| = |a|$

(B) $|a| \geq 0$, equality holding only if $a = 0$

(C) $\left|\frac{a}{b}\right| = \frac{|a|}{|b|}$, $b \neq 0$

(D) $|ab| = |a| \times |b|$

(E) $|a|^2 = a^2$

Absolute value can also be expressed on the real number line as the distance of the point represented by the real number from the point labeled 0.

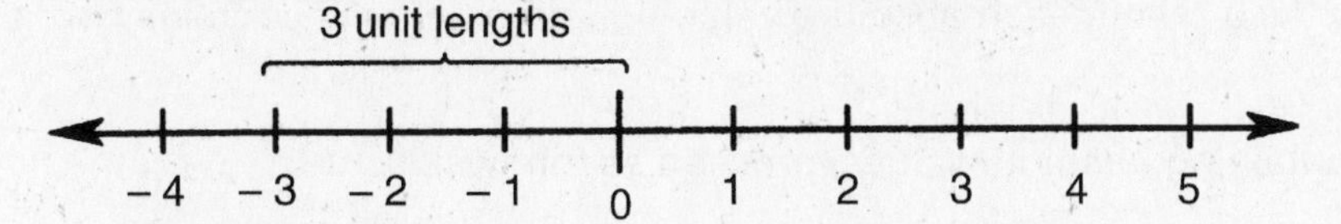

So $|-3| = 3$ because -3 is 3 units to the left of 0.

PROBLEM

Classify each of the following statements as true or false. If it is false, explain why.

(1) $|-120| > 1$

(2) $|4 - 12| = |4| - |12|$

(3) $|4 - 9| = 9 - 4$

(4) $|12 - 3| = 12 - 3$

(5) $|-12a| = 12|a|$

SOLUTION

(1) True

(2) False, $|4 - 12| = |4| - |12|$

$$|-8| = 4 - 12$$

$$8 \neq 8$$

In general, $|a + b| \neq |a| + |b|$

(3) True

(4) True

(5) True

PROBLEM

Find the absolute value for each of the following:

(1) 0

(2) 4

(3) $-\pi$

(4) *a, where a is a real number*

SOLUTION

(1) $|0| = 0$

(2) $|4| = 4$

(3) $|-\pi| = \pi$

(4) for $a > 0$, $|a| = a$

for $a = 0$, $|a| = 0$

for $a < 0$, $|a| = -a$

$$\text{i.e., } |a| = \begin{cases} a \text{ if } a > 0 \\ 0 \text{ if } a = 0 \\ -a \text{ if } a < 0 \end{cases}$$

When the definition of absolute value is applied to an equation, the quantity within the absolute value symbol is considered to have two values. This value can be either positive or negative before the absolute value is taken. As a result, each absolute value equation actually contains two separate equations.

When evaluating equations containing absolute values, proceed as follows:

EXAMPLE

$|5 - 3x| = 7$ is valid if either

$$5 - 3x = 7 \quad \text{or} \quad 5 - 3x = -7$$

$$-3x = 2 \qquad\qquad -3x = -12$$

$$x = -\frac{2}{3} \qquad\qquad x = 4$$

The solution set is therefore $x = (-\frac{2}{3}, 4)$.

Remember, the absolute value of a number cannot be negative. So, for the equation $|5x + 4| = -3$, there would be no solution.

EXAMPLE

Solve for x in $|2x - 6| = |4 - 5x|$.

There are four possibilities here. $2x - 6$ and $4 - 5x$ can be either positive or negative. Therefore,

$$2x - 6 = 4 - 5x \quad (1)$$

$$-(2x - 6) = 4 - 5x \quad (2)$$

$$2x - 6 = -(4 - 5x) \quad (3)$$

$$-(2x - 6) = -(4 - 5x) \quad (4)$$

Equations (2) and (3) result in the same solution, as do equations (1) and (4). Therefore, it is necessary to solve only for equations (1) and (2). This gives

$$2x - 6 = 4 - 5x \quad \text{or} \quad -(2x - 6) = 4 - 5x$$

$$7x = 10 \qquad\qquad -2x + 6 = 4 - 5x$$

$$x = \frac{10}{7} \qquad\qquad x = -\frac{2}{3}$$

The solution set is $(\frac{10}{7}, -\frac{2}{3})$.

DRILL: ABSOLUTE VALUE EQUATIONS

Absolute Value

DIRECTIONS: Simplify each expression.

1. $|-3| - |-17|$

   (A) -20 (B) -14 (C) 4 (D) 14 (E) 20

2. $-2 + |-8 - 8 \div 4|$

(A) -2 (B) 2 (C) 6 (D) 8 (E) 12

3. $\dfrac{-|8-20|}{|-4|}$

(A) -13 (B) -7 (C) -3 (D) 3 (E) 7

4. $|7b|^2$

(A) $-49b^2$ (B) $-7b^2$ (C) $7b^2$ (D) $49b$ (E) $49b^2$

5. $|6n + 4|$, if $n < -1$

(A) $-6n + 4$ (B) $6n - 4$ (C) $6n + 4$ (D) $-(6n + 4)$ (E) $-(6n - 4)$

Equations

<u>DIRECTIONS</u>: Find the appropriate solutions.

6. $|4x - 2| = 6$

(A) -2 and -1 (B) -1 and 2 (C) 2

(D) $\frac{1}{2}$ and -2 (E) No solution

7. $\left|3 - \frac{1}{2}y\right| = -7$

(A) -8 and 20 (B) 8 and -20 (C) 2 and -5

(D) 4 and -2 (E) No solution

8. $2|x + 7| = 12$

(A) -13 and -1 (B) -6 and 6 (C) -1 and 13

(D) 6 and -13 (E) No solution

9. $|5x| - 7 = 3$

(A) 2 and 4 (B) $\frac{4}{5}$ and 3 (C) -2 and 2

(D) 2 (E) No solution

10. $\left|\frac{3}{4}m\right| = 9$

(A) 24 and -16 (B) $\frac{4}{27}$ and $-\frac{4}{3}$ (C) $\frac{4}{3}$ and 12

(D) -12 and 12 (E) No solution

MASTERING INEQUALITIES

An inequality is a statement where the value of one quantity or expression is greater than (>), less than (<), greater than or equal to ($\geq$), less than or equal to ($\leq$), or not equal to ($\neq$) that of another.

EXAMPLE

$5 > 4$

The expression above means that the value of 5 is greater than the value of 4.

A **conditional inequality** is an inequality whose validity depends on the values of the variables in the sentence. That is, certain values of the variables will make the sentence true, and others will make it false.

$3 - y > 3 + y$

is a conditional inequality for the set of real numbers, since it is true for any replacement less than zero and false for all others.

$x + 5 > x + 2$

is an **absolute inequality** for the set of real numbers, meaning that for any real value x, the expression on the left is greater than the expression on the right.

$5y < 2y + y$

is inconsistent for the set of non-negative real numbers. For any y greater than 0 the sentence is always false. A sentence is inconsistent if it is always false when its variables assume allowable values.

The solution of a given inequality in one variable x consists of all values of x for which the inequality is true.

The graph of an inequality in one variable is represented by either a ray or a line segment on the real number line.

The endpoint is not a solution if the variable is strictly less than or greater than a particular value.

EXAMPLE

$x > 2$

−4 −3 −2 −1 0 1 2 3 4 5

2 is not a solution and should be represented as shown.

The endpoint is a solution if the variable is either (1) less than or equal to or (2) greater than or equal to, a particular value.

EXAMPLE

$5 > x \geq 2$

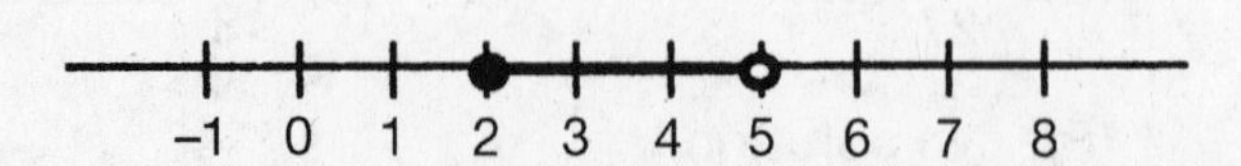

In this case 2 is the solution and should be represented as shown.

Properties of Inequalities

If x and y are real numbers, then one and only one of the following statements is true.

$x > y$, $x = y$, or $x < y$.

This is the order property of real numbers.

If a, b, and c are real numbers, the following are true:

A) If $a < b$ and $b < c$ then $a < c$.

B) If $a > b$ and $b > c$ then $a > c$.

This is the transitive property of inequalities.

If a, b, and c are real numbers and $a > b$, then $a + c > b + c$ and $a - c > b - c$. This is the **addition property of inequality**.

Two inequalities are said to have the same **sense** if their signs of inequality point in the same direction.

The sense of an inequality remains the same if both sides are multiplied or divided by the same positive real number.

EXAMPLE

$4 > 3$

If we multiply both sides by 5, we will obtain

$4 \times 5 > 3 \times 5$

$20 > 15$

The sense of the inequality does not change.

The sense of an inequality becomes opposite if each side is multiplied or divided by the same negative real number.

EXAMPLE

$4 > 3$

If we multiply both sides by –5, we would obtain

$4 \times -5 < 3 \times -5$

$-20 < -15$

The sense of the inequality becomes opposite.

If $a > b$ and a, b, and n are positive real numbers, then

$a^n > b^n$ and $a^{-n} < b^{-n}$

If $x > y$ and $q > p$, then $x + q > y + p$.

If $x > y > 0$ and $q > p > 0$, then $xq > yp$.

Inequalities that have the same solution set are called **equivalent inequalities**.

PROBLEM

Solve the inequality $2x + 5 > 9$.

SOLUTION

Add –5 to both sides:	$2x + 5 + (-5) > 9 + (-5)$
Additive inverse property:	$2x + 0 > 9 + (-5)$
Additive identity property:	$2x > 9 + (-5)$
Combine terms:	$2x > 4$
Multiply both sides by $\frac{1}{2}$:	$\frac{1}{2}(2x) > \frac{1}{2} \times 4$
	$x > 2$

The solution set is

$$X = \{x \mid 2x + 5 > 9\}$$
$$= \{x \mid x > 2\}$$

(that is all x, such that x is greater than 2)

PROBLEM

Solve the inequality $4x + 3 < 6x + 8$.

SOLUTION

In order to solve the inequality $4x + 3 < 6x + 8$, we must find all values of x that make it true. Thus, we wish to obtain x alone on one side of the inequality.

Add –3 to both sides:

$$\begin{array}{r} 4x + 3 < 6x + 8 \\ \underline{\quad -3 \qquad\quad -3} \\ 4x < 6x + 5 \end{array}$$

Add $-6x$ to both sides:

$$\begin{array}{l} \;\;4x < 6x + 5 \\ \underline{-6x \quad -6x \quad\;} \\ -2x < 5 \end{array}$$

In order to obtain x alone we must divide both sides by (–2). Recall that dividing an inequality by a negative number reverses the inequality sign, hence

$$\frac{-2x}{-2} > \frac{5}{-2}$$

Cancelling $\frac{-2}{-2}$ we obtain, $x > -\frac{5}{2}$.

Thus, our solution is $\{x : x > -\frac{5}{2}\}$ (the set of all x such that x is greater than $-\frac{5}{2}$).

DRILL: INEQUALITIES

DIRECTIONS: Find the solution set for each inequality.

1. $3m + 2 < 7$

(A) $m \geq \frac{5}{3}$ (B) $m \leq 2$ (C) $m < 2$

(D) $m > 2$ (E) $m < \frac{5}{3}$

2. $\frac{1}{2}x - 3 \leq 1$

(A) $-4 \leq x \leq 8$ (B) $x \geq -8$ (C) $x \leq 8$

(D) $2 \leq x \leq 8$ (E) $x \geq 8$

3. $-3p + 1 \geq 16$

(A) $p \geq -5$ (B) $p \geq \frac{-17}{3}$ (C) $p \leq \frac{-17}{3}$

(D) $p \leq -5$ (E) $p \leq 5$

4. $-6 < \frac{2}{3}r + 6 \leq 2$

(A) $-6 < r \leq -3$ (B) $-18 < r \leq -6$ (C) $r \geq -6$

(D) $-2 < r \leq -\frac{4}{3}$ (E) $r \leq -6$

5. $0 < 2 - y < 6$

(A) $-4 < y < 2$ (B) $-4 < y < 0$ (C) $-4 < y < -2$

(D) $-2 < y < 4$ (E) $0 < y < 4$

MASTERING RATIOS AND PROPORTIONS

The ratio of two numbers x and y written $x : y$ is the fraction $\frac{x}{y}$ where $y \neq 0$. A ratio compares x to y by dividing one by the other. Therefore, in order to compare ratios, simply compare the fractions.

A proportion is an equality of two ratios. The laws of proportion are listed below.

If $\frac{a}{b} = \frac{c}{d}$, then

(A) $ad = bd$

(B) $\frac{b}{a} = \frac{d}{c}$

(C) $\frac{a}{c} = \frac{b}{d}$

(D) $\frac{a+b}{b} = \frac{c+d}{d}$

(E) $\frac{a-b}{b} = \frac{c-d}{d}$

Given a proportion $a : b = c : d$, then a and d are called extremes, b and c are called the means, and d is called the fourth proportion to a, b, and c.

PROBLEM

Solve the proportion $\frac{x+1}{4} = \frac{15}{12}$.

SOLUTION

Cross-multiply to determine x; that is, multiply the numerator of the first fraction by the denominator of the second, and equate this to the product of the numerator of the second and the denominator of the first.

$$(x + 1)\ 12 = 4 \times 15$$

$$12x + 12 = 60$$

$$x = 4$$

PROBLEM

Find the ratios of $x : y : z$ *from the equations*

$7x = 4y + 8z, \ \ 3z = 12x + 11y.$

SOLUTION

By transposition we have

$$7x - 4y - 8z = 0$$

$$12x + 11y - 3z = 0$$

To obtain the ratio of $x : y$, we convert the given system into an equation in terms of just x and y. We may eliminate z as follows: Multiply each term of the first equation by 3 and each term of the second equation by 8 (because they are both z variables which we wish to eliminate), and then subtract the second equation from the first. We thus obtain

$$21x - 12y - 24z = 0$$

$$\underline{-(96x + 88y - 24z = 0)}$$

$$-75x - 100y \quad = 0$$

Dividing each term of the last equation by 25, we obtain

$$-3x - 4y = 0$$

or,

$$-3x = 4y$$

Dividing both sides of this equation by 4 and by –3, we have the proportion

$$\frac{x}{4} = \frac{y}{-3}$$

We are now interested in obtaining the ratio of $y : z$. To do this we convert the given system of equations into an equation in terms of just y and z, by eliminating x as follows: Multiply each term of the first equation by 12, and each term of the second equation by 7, and then subtract the second equation from the first. We thus obtain

$$\begin{array}{r} 84x - 48y - 96z = 0 \\ \underline{-(84x + 77y - 21z = 0)} \\ -125y - 75z = 0 \end{array}$$

Dividing each term of the last equation by 25, we obtain

$$-5y - 3z = 0$$

or,

$$-3z = 5y$$

Dividing both sides of this equation by 5 and by –3, we have the proportion

$$\frac{z}{5} = \frac{y}{-3}.$$

From this result and our previous result we obtain

$$\frac{x}{4} = \frac{y}{-3} = \frac{z}{5}$$

as the desired ratios.

DRILL: RATIOS AND PROPORTIONS

DIRECTIONS: Find the appropriate solutions.

1. Solve for n: $\frac{4}{n} = \frac{8}{5}$.

   (A) 10 (B) 8 (C) 6 (D) 2.5 (E) 2

2. Solve for n: $\frac{2}{3} = \frac{n}{72}$.

   (A) 12 (B) 48 (C) 64 (D) 56 (E) 24

3. Solve for n: $n : 12 = 3 : 4$.

   (A) 8 (B) 1 (C) 9 (D) 4 (E) 10

4. Four out of every five students at West High take a mathematics course. If the enrollment at West is 785, how many students take mathematics?

   (A) 628 (B) 157 (C) 705 (D) 655 (E) 247

5. At a factory, three out of every 1,000 parts produced are defective. In a day, the factory can produce 25,000 parts. How many of these parts would be defective?

   (A) 7 (B) 75 (C) 750 (D) 7,500 (E) 75,000

6. A summer league softball team won 28 out of the 32 games they played. What is the ratio of games won to games played?

   (A) 4 : 5 (B) 3 : 4 (C) 7 : 8 (D) 2 : 3 (E) 1 : 8

7. A class of 24 students contains 16 males. What is the ratio of females to males?

(A) 1 : 2 (B) 2 : 1 (C) 2 : 3 (D) 3 : 1 (E) 3 : 2

8. A family has a monthly income of $1,250, but they spend $450 a month on rent. What is the ratio of the amount of income to the amount paid for rent?

(A) 16 : 25 (B) 25 : 9 (C) 25 : 16 (D) 9 : 25 (E) 36 : 100

9. A student attends classes 7.5 hours a day and works a part-time job for 3.5 hours a day. She knows she must get 7 hours of sleep a night. Write the ratio of the number of free hours in this student's day to the total number of hours in a day.

(A) 1 : 3 (B) 4 : 3 (C) 8 : 24 (D) 1 : 4 (E) 5 : 12

10. In a survey by mail, 30 out of 750 questionnaires were returned. Write the ratio of questionnaires returned to questionnaires mailed (write in simplest form).

(A) 30 : 750 (B) 24 : 25 (C) 3 : 75 (D) 1 : 4 (E) 1 : 25

MASTERING RATIONAL EQUATIONS

A **rational expression** is a ratio that includes at least one monomial or polynomial. For example, the expressions $\frac{3}{4+x}$, $\frac{2x^2}{x^2-4}$, and $\frac{x+3}{(x-1)(x+2)}$ are all rational expressions.

A **rational equation** is an equation that includes at least one rational expression. To solve an equation with a single rational expression, cross-multiply. Then isolate the variable.

PROBLEM

Solve the equation $\frac{x+1}{x+5}=\frac{6}{7}$.

SOLUTION

This equation has a single rational expression. Solve it by cross-multiplying.

$$\frac{x+1}{x+5}=\frac{6}{7}$$

$7(x + 1) = 6(x + 5)$ Cross-multiply.

$7x + 7 = 6x + 30$ Use the Distributive Property.

$x + 7 = 30$ Subtract $6x$ from each side.

$x = 23$ Subtract 7 from each side.

Check that $x = 23$ is the solution to the original equation.

$\frac{23+1}{23+5}=\frac{6}{7}$ Substitute 23 for x.

$\frac{24}{28}=\frac{6}{7}$ Add.

$\frac{6}{7}=\frac{6}{7}$ Simplify.

The solution to the equation is $x = 23$.

When working with rational equations, it is important that you substitute the solution(s) back into the original equation to check that they do not result in a division by zero. For example, if a rational equation contained the expression $\frac{3}{4+x}$, the value of x could not equal –4 because this value would make the denominator of the expression equal to zero. If a solution results in division by zero, that solution is said to be extraneous and should be eliminated from the solution set.

Some rational equations may include a sum or difference of rational expressions. To solve these types of problems, perform the addition or subtraction of the rational expressions first. Then cross-multiply to solve for the variable.

PROBLEM

Solve the equation $\frac{1}{x}-\frac{1}{x+2}=\frac{1}{4}$.

SOLUTION

Rewrite the rational expressions on the left side of the equation with a common denominator. To do so, multiply the numerator and denominator of each rational expression by the denominator of the other rational expression.

Rewrite $\frac{1}{x}$.

$$\frac{1}{x}=\left(\frac{1}{x}\right)\left(\frac{x+2}{x+2}\right)=\frac{x+2}{x(x+2)}$$

Rewrite $\frac{1}{x+2}$.

$$\frac{1}{x+2}=\left(\frac{1}{x+2}\right)\left(\frac{x}{x}\right)=\frac{x}{x(x+2)}$$

Now subtract the rational expressions. Subtract the numerators and write this difference over the common denominator.

$$\frac{x+2}{x(x+2)}-\frac{x}{x(x+2)}=\frac{x+2-x}{x(x+2)}$$ Subtract.

$$=\frac{2}{x(x+2)}$$ Simplify.

Substitute this difference into the original rational equation. Then cross-multiply to solve for *x*.

$$\frac{1}{x}-\frac{1}{x+2}=\frac{1}{4}$$

$$\frac{2}{x(x+2)}=\frac{1}{4}$$ Substitute.

$(2)(4) = x(x + 2)(1)$ Cross-multiply.

$8 = x^2 + 2x$ Simplify.

$0 = x^2 + 2x - 8$ Subtract 8 from each side.

$0 = (x + 4)(x - 2)$ Factor the right side of the equation.

$0 = (x + 4)$ or $0 = (x - 2)$

$x = -4$ or $x = 2$

Check that $x = -4$ or $x = 2$ is the solution to the original equation.

$$\frac{1}{-4}-\frac{1}{-4+2}=\frac{1}{4} \qquad \frac{1}{2}-\frac{1}{2+2}=\frac{1}{4} \qquad \text{Substitute.}$$

$$-\frac{1}{4}-\left(-\frac{1}{2}\right)=\frac{1}{4} \qquad \frac{1}{2}-\frac{1}{4}=\frac{1}{4} \qquad \text{Simplify.}$$

$$\frac{1}{4}=\frac{1}{4} \qquad \frac{1}{4}=\frac{1}{4} \qquad \text{Subtract.}$$

The solution to the equation is $x = -4$ or $x = 2$.

DRILL: RATIONAL EQUATIONS

DIRECTIONS: Solve each equation for x.

1. $\frac{10}{x-2}=\frac{5}{3}$

   (A) $x = 4$ (B) $x = 5$ (C) $x = 6$ (D) $x = 7$ (E) $x = 8$

2. $\frac{6}{x+3}=\frac{8}{x-5}$

   (A) $x = -27$ (B) $x = -4$ (C) $x = -3$ (D) $x = 4$ (E) $x = 27$

3. $\frac{4}{x}=\frac{x-1}{3}$

   (A) $x = -12$ or $x = 6$ (B) $x = -6$ or $x = 2$ (C) $x = -3$ or $x = 4$
   (D) $x = 0$ or $x = 1$ (E) No solution

4. $\frac{8}{x^2}-\frac{7}{x}=1$

   (A) $x = -8$ or $x = 1$ (B) $x = -1$ or $x = 0$ (C) $x = -1$ or $x = 1$
   (D) $x = 7$ or $x = 8$ (E) No solution

5. $\frac{1}{x-3}-\frac{1}{x}=\frac{1}{6}$

   (A) $x = -9$ or $x = 3$ (B) $x = -6$ or $x = 9$ (C) $x = -3$ or $x = 6$
   (D) $x = 2$ or $x = 18$ (E) No solution

MASTERING DIRECT AND INVERSE VARIATION

If two quantities x and y are related to each other so that their ratio is always constant, the two quantities are said to be directly proportional. This type of relationship is called **direct variation** and can be described by the equation $\frac{y}{x} = k$ or $y = kx$, where k is a constant. In this type of relationship, when the value of one quantity increases, the value of the other quantity also increases.

PROBLEM

If y is directly proportional to x, and if y = 15 when x = 3, then what is the value of y when x = 10?

SOLUTION

Because y and x are directly proportional, they are related by the equation $y = kx$. We know that $y = 15$ when $x = 3$. Use this information to determine the constant, k.

$y = kx$

$15 = k(3)$ Substitute the known values.

$5 = k$ Divide both sides by 3.

The constant of variation is 5, so the equation that relates y and x is $y = 5x$. Use this equation to find the value of y when $x = 10$.

$$y = 5x = 5(10) = 50$$

When $x = 10$, $y = 50$.

If two quantities x and y are related to each other so that their product is always constant, the two quantities are said to be inversely proportional. This type of relationship is called **inverse variation** and can be described by the equation $yx = k$ or $y = \frac{k}{x}$, where k is a constant. In this type of relationship, when the value of one quantity increases, the value of the other quantity decreases.

PROBLEM

If y is inversely proportional to x, and if y = 15 when x = 3, then what is the value of y when x = 10?

SOLUTION

Because y and x are inversely proportional, they are related by the equation $y = \frac{k}{x}$. We know that $y = 15$ when $x = 3$. Use this information to determine the constant, k.

$y = \frac{k}{x}$

$15 = \frac{k}{3}$ Substitute the known values.

$45 = k$ Multiply both sides by 3.

The constant of variation is 45, so the equation that relates y and x is $y = \frac{45}{x}$. Use this equation to find the value of y when x = 10.

$$y = \frac{45}{x} = \frac{45}{10} = 4.5$$

When $x = 10$, $y = 4.5$.

DRILL: DIRECT AND INVERSE VARIATION

DIRECTIONS: Solve each problem.

1. If y is directly proportional to x, and if $y = 4$ when $x = 2$, what equation relates y and x?

   (A) $y = \frac{1}{8}x$ (B) $y = \frac{1}{2}x$ (C) $y = 2x$ (D) $y = 8x$ (E) $y = 6x$

2. If a is directly proportional to b, and if $a = 8$ when $b = 12$, then what is the value of a when $b = 20$?

   (A) $6\frac{2}{3}$ (B) $13\frac{1}{3}$ (C) 16 (D) 30 (E) 45

3. If q is inversely proportional to p, and if $q = 14$ when $p = 4$, then what is the value of q when $p = 2$?

   (A) 7 (B) 9 (C) 24 (D) 28 (E) 112

4. The distance a train travels is directly proportional to the number of hours it travels. If the train travels 315 miles in 7 hours, how many miles will the train travel in 12 hours?

   (A) 500 miles (B) 540 miles (C) 840 miles
   (D) 1,575 miles (E) 2,205 miles

5. The number of hours needed to paint a house is inversely proportional to the number of painters. If it takes 4 painters 15 hours to paint a house, how many hours would it take 6 painters to paint the house?

   (A) 5 hours (B) 6 hours (C) 9 hours
   (D) 10 hours (E) 13 hours

MASTERING FUNCTIONS

A **function** is a relation that assigns exactly one output value to every input value. The notation $f(x)$ is often used to name a function f with an input of x. For example, if a function f produces output values that are 3 more than the input values, the equation that describes this function can be written as $f(x) = x + 3$. To determine the value of this function when the input is 6, simply substitute 6 for the input variable x in the function rule.

$$f(6) = 6 + 3 = 9$$

Therefore, $f(6) = 9$, which means that the function f has a value of 9 when the input value is 6.

PROBLEM

Determine whether the following relations are functions.

(1) {(0, 2), (3, 4), (–2, 6), (4, 2)}

(2) {(1, 4), (–1, 0), (6, 0), (1, 7)}

SOLUTION

When a relation is written as a set of ordered pairs, the first numbers in the ordered pairs are the input values, and the second numbers are the output values.

(1) In this relation, the input values are 0, 3, –2, and 4. Each of these input values is different, which means that each input value has exactly one output value. Therefore, this relation is a function.

(2) In this relation, the input values are 1, –1, 6, and 1. The input value 1 appears more than once. There are two output values assigned to the input value of 1: 4 and 7. Because one of the input values has more than one output value, this relation is not a function.

PROBLEM

The function f is defined by $f(x) = x^2 - x$. Determine each of the following.

(1) $f(5)$	(2) $f(-3)$
(3) $f(p)$	(4) $f(x + 1)$

SOLUTION

(1) $f(5) = 5^2 - 5 = 25 - 5 = 20$

(2) $f(-3) = (-3)^2 - (-3) = 9 - (-3) = 12$

(3) The input of a function can be a variable as well as a number. In this case, substitute the variable p for x.

$$f(p) = p^2 - p$$

(4) The input of a function can also be an algebraic expression. In this case, substitute $x + 1$ for x. Then simplify.

$f(x + 1) = (x + 1)^2 - (x + 1)$	Substitute.
$= x^2 + 2x + 1 - (x + 1)$	Determine $(x + 1)^2$.
$= x^2 + 2x + 1 - x - 1$	Use the Distributive Property to remove the parentheses.
$= x^2 + x$	Combine like terms.

Domain and Range

The **domain** of a function is the set of input values for which a function is defined. Sometimes the domain of a function may include all real numbers. At other times, the domain may be restricted to certain values. For example, if the function f is defined as $f(x) = 2x$ for $2 \le x \le 5$, then the domain of the function is restricted. The input value, x, must be greater than or equal to 2 and less than or equal to 5. Therefore, the domain of this function is $2 \le x \le 5$.

If a function is undefined for certain input values, then the domain of the function does not include those values. For instance, if $f(x) = \sqrt{x}$, then the value of x must be greater than or equal to zero for the function to be defined. The domain of this function is therefore $x \ge 0$.

The **range** of a function is the set of output values that result from the input values in the function's domain. Like the domain, the range of a function may sometimes include all real numbers. At other times, the range may be limited to certain values. For example, if $f(x) = x^2$, then the range of f is restricted to numbers greater than or equal to zero because the square of a number is never negative. In this case, the range of f is $f(x) \ge 0$.

PROBLEM

Determine the domain and range of the following function.

{(–1, 3), (0, –1), (2, 4), (6, –1)}

SOLUTION

The input values of the function are –1, 0, 2, and 6. Therefore, the domain is the set {–1, 0, 2, 6}. The output values of the function are 3, –1, 4, and –1. The range of the function is {–1, 3, 4}.

PROBLEM

Determine the domain and range of the function graphed as follows.

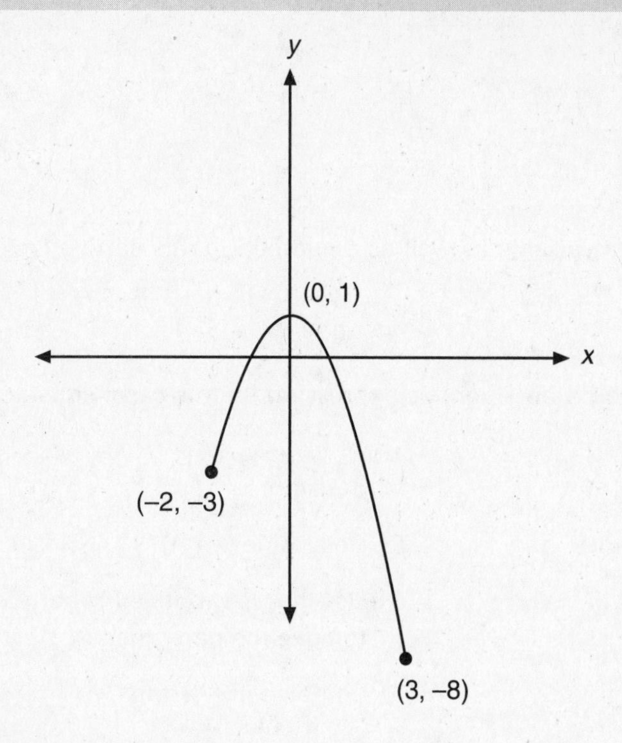

SOLUTION

The domain of the function is the set of *x*-values for which the function is defined. The least *x*-value for which the function is defined is –2, and the greatest *x*-value for which the function is defined is 3. Therefore, the domain of the function is $-2 \leq x \leq 3$.

The range of the function is the set of *y*-values that result from the values in the domain. The *y*-values of the function range from –8 to 1. Therefore, the range of the function is $-8 \leq y \leq 1$.

Linear functions

A **linear function** can be written in the form $y = mx + b$, where *m* and *b* are constants. The graph of any linear function is a line with a slope of *m* and a *y*-intercept of *b*.

The ***y*-intercept** of a line is the *y*-coordinate of the point where the line crosses the *y*-axis. The **slope** of a line is a measure of its steepness. Slope can be determined by using the coordinates of two points on the line and the formula $m = \frac{(y_2 - y_1)}{(x_2 - x_1)}$, where (x_1, y_1) are the coordinates of the first point and (x_2, y_2) are the coordinates of the second point.

When given a graph of a linear function, use the coordinates of two points on the line to determine its slope, *m*. Then identify the coordinates of the point where the line crosses the *y*-axis to determine the *y*-intercept, *b*. Finally, use the values of *m* and *b* to write the equation of the linear function.

PROBLEM

Write the equation of the linear function graphed as follows.

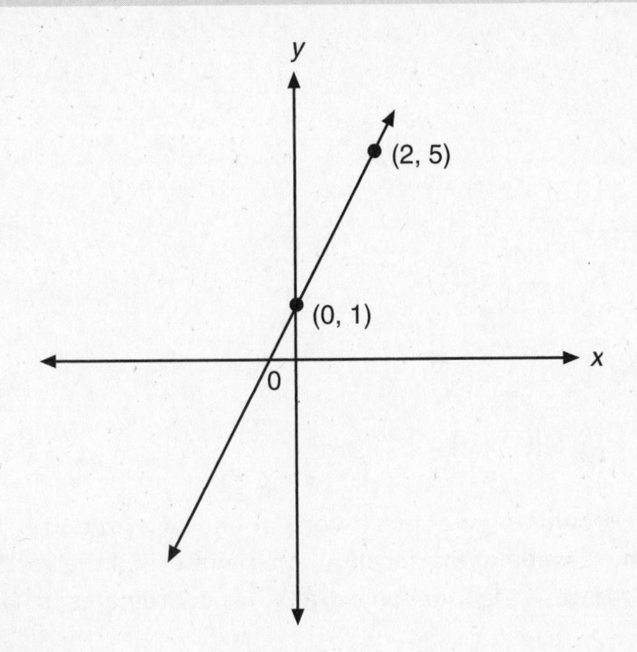

SOLUTION

Use the points (0, 1) and (2, 5) to determine the line's slope.

$$m = \frac{y_2 - y_1}{x_2 - x_1} = \frac{5-1}{2-0} = \frac{4}{2} = 2$$

The line crosses the *y*-axis at point (0, 1), so the *y*-intercept, *b*, is 1. Substitute the values of *m* and *b* into the equation $y = mx + b$. The equation that describes the linear function is $y = 2x + 1$.

PROBLEM

The graph of a line has a slope of –3 and passes through the point (5, –2). What is the y-intercept of the line?

SOLUTION

Because the slope of the line is –3, the equation of the line has the form $y = -3x + b$. Substitute the *x*- and *y*-values of the point (5, –2) into the equation to solve for *b*, the *y*-intercept.

$y = -3x + b$	
$-2 = -3(5) + b$	Substitute.
$-2 = -15 + b$	Simplify.
$13 = b$	Add 15 to each side.

The *y*-intercept of the line is 13.

Quadratic Functions

A **quadratic function** can be written in the form $y = a(x - h)^2 + k$, where a, h, and k are constants. The graph of any quadratic function is a parabola, such as the one shown.

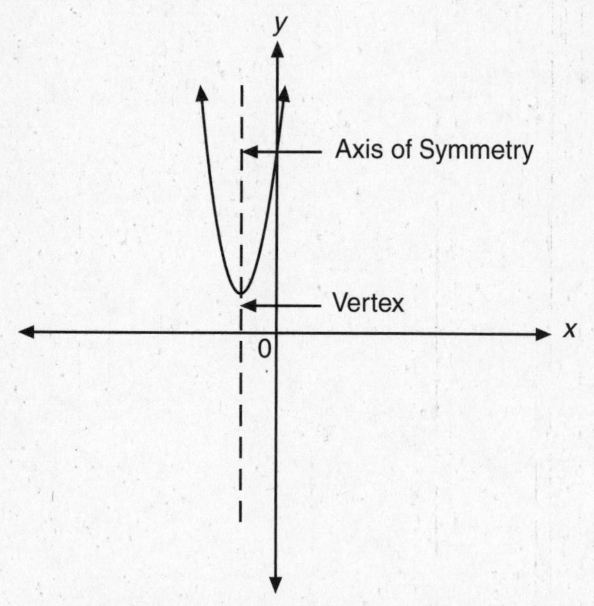

The graph of a quadratic function is symmetric about a vertical **line of symmetry.** This line passes through the maximum or minimum value of the function at a point called the **vertex**. For a quadratic function $y = a(x - h)^2 + k$, the vertex of the function's graph has coordinates (h, k) and the line of symmetry is the line $x = h$.

The sign of the constant a determines whether the parabola opens upward or downward. If $a < 0$, the parabola opens downward. If $a > 0$, the parabola opens upward.

The equation of a quadratic function can be determined if the coordinates of the vertex and one other point on the graph are known.

PROBLEM

Determine the equation of the quadratic function graphed as follows.

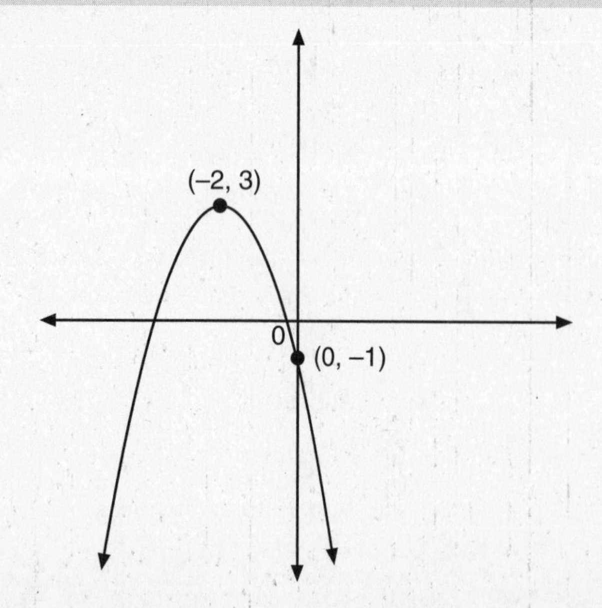

SOLUTION

To determine the equation of the function, we need to determine the values of the constants h, k, and a. The vertex of the parabola has coordinates (–2, 3). Therefore, $h = -2$ and $k = 3$. Substitute these values into the equation of a quadratic function: $y = a(x - (-2))^2 + 3$, which simplifies to $y = a(x + 2)^2 + 3$.

To find the value of a, use the coordinates of another point on the parabola. In this case, use the point (0, –1). Substitute these values for x and y into the equation and solve for a.

$y = a(x + 2)^2 + 3$	
$-1 = a(0 + 2)^2 + 3$	Substitute.
$-1 = 4a + 3$	Simplify.
$-4 = 4a$	Subtract 3 from each side.
$-1 = a$	Divide each side by 4.

Substituting this value for a into the quadratic equation yields $y = -1(x + 2)^2 + 3$, which simplifies to $y = -(x + 2)^2 + 3$.

PROBLEM

How does the graph of the function $y = 3x^2 - 4$ differ from the graph of $y = 3x^2$?

SOLUTION

For both functions, the value of a is 3. However, the vertex of the graph of the first function has coordinates (0, –4), and the vertex of the graph of the second function has coordinates (0, 0). Therefore, the graph of the first function is shifted 4 units down compared to the graph of the second function.

DRILL: FUNCTIONS

Functions

DIRECTIONS: Evaluate each function for the given input.

1. Find $f(-2)$ if $f(x) = 2^x$.

   (A) -4 (B) -1 (C) $-\frac{1}{4}$ (D) $\frac{1}{4}$ (E) 4

2. Find $f(6)$ if $f(x) = \frac{x^2 + 6}{x - 4}$.

   (A) 3 (B) 6 (C) 7 (D) 9 (E) 21

3. Find $f(2a)$ if $f(x) = -5x - 1$.

   (A) $-11a$ (B) $-10a - 1$ (C) $-5a - 1$ (D) $2a - 1$ (E) $2a - 6$

4. Find $f(x+3)$ if $f(x) = 7x + 8$.

(A) $7x + 11$ (B) $7x + 21$ (C) $7x + 29$
(D) $8x + 11$ (E) $8x + 24$

5. Find $f(2x - 4)$ if $f(x) = x^2 + 3x - 1$.

(A) $x^2 + 5x - 5$ (B) $4x^2 - 10x + 3$ (C) $4x^2 - 6x + 3$
(D) $4x^2 + 6x - 21$ (E) $2x^2 + 2x^2 - 12x - 1$

Domain and Range

DIRECTIONS: Choose the best answer for each question.

6. Which **best** describes the range of the function $f(x) = x^2 + 3$?

(A) $f(x) \geq -3$ (B) $f(x) \geq 0$ (C) $f(x) \geq 3$
(D) $-3 \leq f(x) \leq 3$ (E) $0 \leq f(x) \leq 3$

7. Which **best** describes the domain of the function $f(x) = \sqrt{x+5}$?

(A) $x \geq -5$ (B) $x \geq 0$ (C) $x \geq 5$
(D) $x \leq -5$ (E) $x \leq 5$

8. Which value is **not** included in the domain of the function $\frac{5+x}{x^2-4x+4}$?

(A) -5 (B) -4 (C) -2 (D) 2 (E) 4

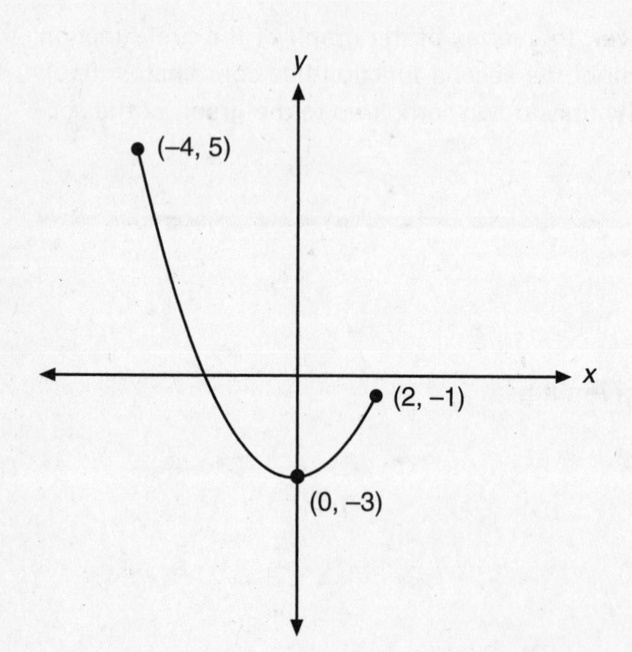

9. What is the domain of the function shown in the graph?

(A) $-4 \leq x \leq 2$ (B) $-4 \leq x \leq 5$ (C) $0 \leq x \leq 4$
(D) $0 \leq x \leq 5$ (E) $2 \leq x \leq 4$

10. What is the range of the function shown in the graph?

(A) $-4 \le y \le -1$ (B) $-3 \le y \le -1$ (C) $-3 \le y \le 5$

(D) $-1 \le y \le 5$ (E) $0 \le y \le 5$

Linear Functions

DIRECTIONS: Write the equation of each line or linear function.

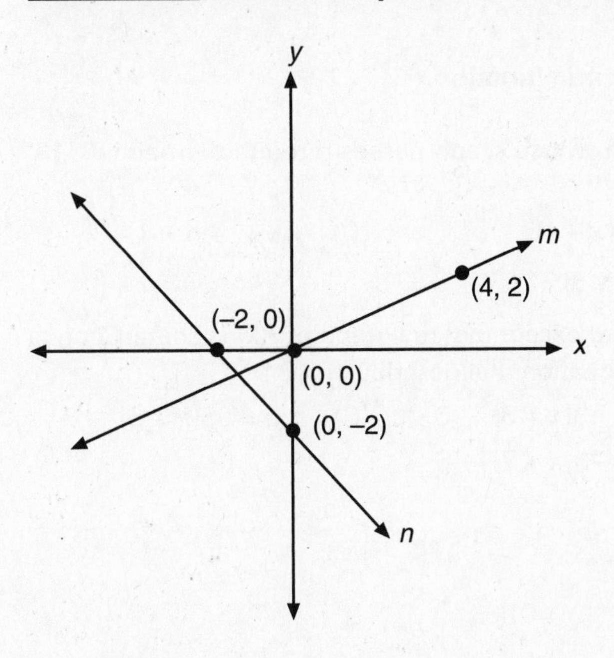

11. What is the equation of line *m*?

(A) $y = \frac{1}{2}x$ (B) $y = \frac{1}{2}x + 2$ (C) $y = 2x$

(D) $y = 2x + 1$ (E) $y = 4x + 2$

12. What is the equation of line *n*?

(A) $y = -x - 1$ (B) $y = -x - 2$ (C) $y = -2x - 1$

(D) $y = -2x - 2$ (E) $y = x - 2$

13. What is the equation of a linear function whose graph has a slope of –2 and passes through the point (1, 4)?

(A) $y = -2x - 2$ (B) $y = -2x + 6$ (C) $y = -2x + 9$

(D) $y = 4x - 2$ (E) $y = 6x - 2$

14. What is the equation of a linear function whose graph passes through the points (–5, 5) and (10, –4)?

(A) $y = -\frac{7}{5}x + 10$ (B) $y = -\frac{3}{5}x + 2$ (C) $y = -\frac{1}{5}x - 2$

(D) $y = \frac{1}{5}x + 6$ (E) $y = \frac{3}{5}x - 10$

15. What is the equation of a linear function whose graph passes through the point (2, 9) and has a y-intercept of -1?

(A) $y = -x + 11$ (B) $y = \frac{1}{3}x - 1$ (C) $y = 2x - 9$

(D) $y = 5x - 1$ (E) $y = 11x - 1$

Quadratic Functions

DIRECTIONS: Write the equation of each quadratic function.

16. What is the equation of the quadratic function whose graph passes through the point (3, 13) and has a vertex at (0, 4)?

(A) $y = (x - 4)^2$ (B) $y = 3(x - 4)^2$ (C) $y = x^2 + 4$

(D) $y = 3x^2 + 4$ (E) $y = 9x^2 + 4$

17. The graph of $f(x)$ is identical to the graph of $g(x)$ except that the graph of $f(x)$ is shifted 2 units to the left. If $g(x) = -3(x + 5)^2 - 6$, what is the equation that describes $f(x)$?

(A) $f(x) = 6(x + 5)^2 - 6$ (B) $f(x) = -3(x + 3)^2 - 6$ (C) $f(x) = -3(x + 5)^2 - 8$

(D) $f(x) = -3(x + 5)^2 - 4$ (E) $f(x) = -3(x + 7)^2 - 6$

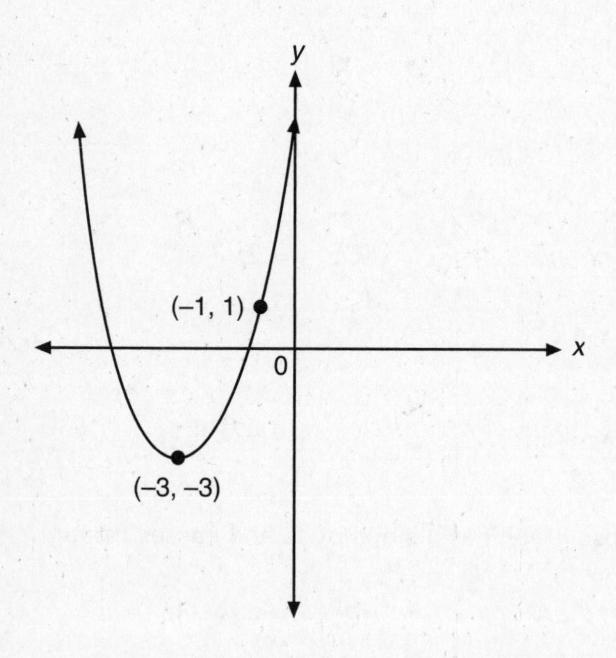

18. What is the equation of the function graphed above?

(A) $y = -(x - 3)^2 - 3$ (B) $y = (x - 3)^2 - 3$ (C) $y = 2(x - 3)^2 - 3$

(D) $y = (x + 3)^2 - 3$ (E) $y = 2(x + 3)^2 - 3$

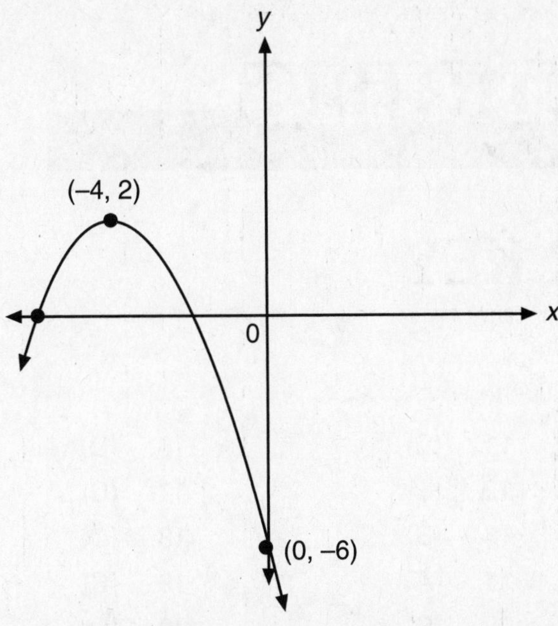

19. What is the equation of the function graphed above?

(A) $y = -3(x+4)^2 + 2$
(B) $y = -2(x-4)^2 + 2$
(C) $y = -(x+4)^2 + 2$
(D) $y = -\frac{1}{2}(x+4)^2 + 2$
(E) $y = \frac{1}{2}(x-4)^2 + 2$

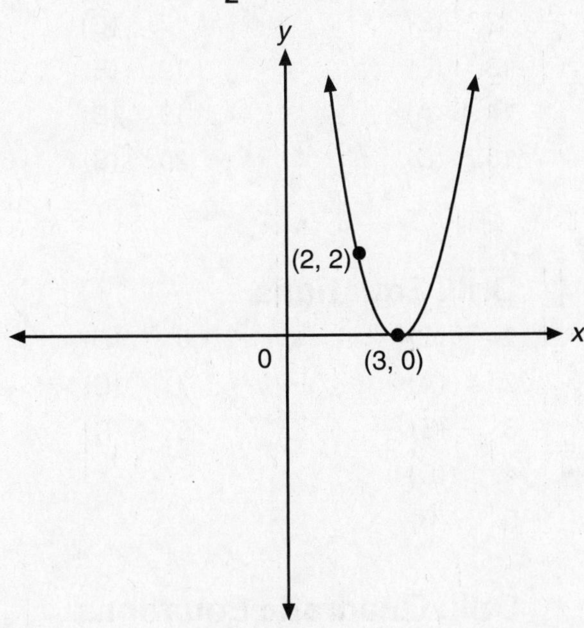

20. What is the equation of the function graphed above?

(A) $y = \frac{3}{4}(x-3)^2$
(B) $y = \frac{3}{4}x^2 + 3$
(C) $y = 2(x-3)^2$
(D) $y = 2(x+3)^2$
(E) $y = 2x^2 + 3$

ALGEBRA DRILLS

ANSWER KEY

Drill: Exponents

1.	(B)	6.	(E)	11.	(C)	16.	(D)
2.	(A)	7.	(B)	12.	(E)	17.	(C)
3.	(C)	8.	(C)	13.	(B)	18.	(D)
4.	(D)	9.	(B)	14.	(A)	19.	(C)
5.	(B)	10.	(D)	15.	(D)	20.	(E)

Drill: Operations with Polynomials

1.	(B)	6.	(B)	11.	(C)	16.	(C)
2.	(C)	7.	(C)	12.	(B)	17.	(D)
3.	(C)	8.	(E)	13.	(E)	18.	(E)
4.	(D)	9.	(A)	14.	(A)	19.	(B)
5.	(A)	10.	(D)	15.	(D)	20.	(B)

Drill: Simplifying Algebraic Expressions

1.	(C)	6.	(B)
2.	(D)	7.	(A)
3.	(B)	8.	(B)
4.	(A)	9.	(A)
5.	(D)	10.	(E)

Drill: Equations

1.	(C)	6.	(D)
2.	(A)	7.	(C)
3.	(E)	8.	(D)
4.	(D)		
5.	(B)		

Drill: Two Linear Equations

1.	(E)	6.	(B)
2.	(B)	7.	(D)
3.	(A)	8.	(E)
4.	(D)	9.	(A)
5.	(C)	10.	(B)

Drill: Quadratic Equations

1.	(A)	6.	(B)
2.	(D)	7.	(B)
3.	(B)	8.	(D)
4.	(C)	9.	(C)
5.	(E)	10.	(E)

Drill: Absolute Value Equations

1.	(B)	6.	(B)
2.	(D)	7.	(E)
3.	(C)	8.	(A)
4.	(E)	9.	(C)
5.	(D)	10.	(D)

Drill: Inequalities

1. (E)
2. (C)
3. (D)
4. (B)
5. (A)

Drill: Ratios and Proportions

1.	(D)	6.	(C)
2.	(B)	7.	(A)
3.	(C)	8.	(B)
4.	(A)	9.	(D)
5.	(B)	10.	(E)

Drill: Rational Equations

1. (E)
2. (A)
3. (C)
4. (A)
5. (C)

Drill: Direct and Inverse Variation

1. (C)
2. (B)
3. (D)
4. (B)
5. (D)

Drill: Functions

1.	(D)	6.	(C)	11.	(A)	16.	(C)
2.	(E)	7.	(A)	12.	(B)	17.	(E)
3.	(B)	8.	(D)	13.	(B)	18.	(D)
4.	(C)	9.	(A)	14.	(B)	19.	(D)
5.	(B)	10.	(C)	15.	(D)	20.	(C)

III. GEOMETRY

MASTERING POINTS, LINES, AND ANGLES

Geometry is built upon a series of undefined terms. These terms are those which we accept as known in order to define other undefined terms.

A) **Point**: Although we represent points on paper with small dots, a point has no size, thickness, or width. A point is named by using a letter, such as *A* or *B*.

B) **Line**: A line is a series of adjacent points that extends indefinitely. A line can be either curved or straight; however, unless otherwise stated, the term "line" refers to a straight line. A line can be named by using two points that lie on the line. For example, a line passing through points *A* and *B* can be named line *AB*, or $\overline{AB}$.

C) **Plane**: A plane is a collection of points lying on a flat surface that extends indefinitely in all directions.

If *A* and *B* are two points on a line, then the **line segment** $\overline{AB}$ is the set of points on that line between *A* and *B* and including *A* and *B*, which are endpoints. The line segment is referred to as $\overline{AB}$. The length of $\overline{AB}$ is represented by *AB*. If $\overline{AB}$ has a length of 2 units, then $AB = 2$.

A **ray** is a series of points that lie to one side of a single endpoint. A ray is named by its endpoint and one other point on the ray. For example, $\overrightarrow{AB}$ is the ray with its endpoint at *A* and passing through point *B*. Likewise, $\overrightarrow{BA}$ is the ray with its endpoint at *B* and passing through point *A*.

PROBLEM

How many lines can be found that contain (a) one given point, (b) two given points, and (c) three given points?

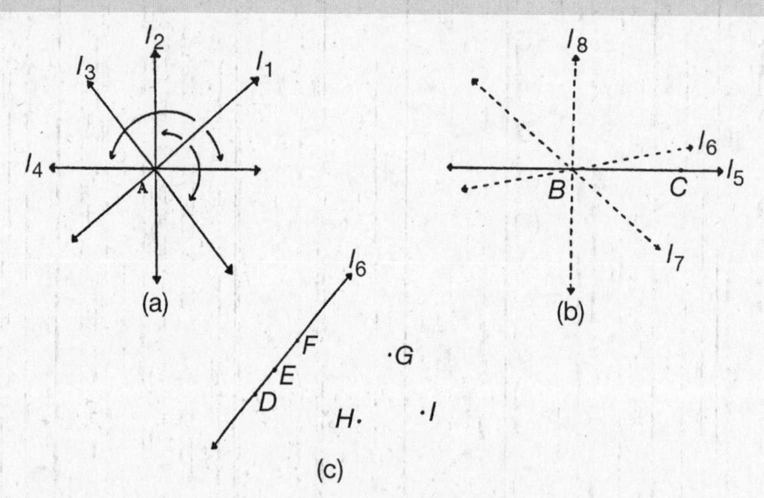

SOLUTION

(a) *Given one point A*, there are an infinite number of distinct lines that contain the given point. To see this, consider line l_1 passing through point *A*. By rotating l_1 around *A* like the hands of a clock, we obtain different lines l_2, l_3, and so on. Since we can rotate l_1 in infinitely many ways, there are infinitely many lines containing *A*.

(b) *Given two distinct points B and C,* there is one and only one straight line passing through both. To see this, consider all the lines containing point *B*: l_5, l_6, l_7, and l_8. Only l_5 contains both points *B* and *C*. Thus, there is only one line containing both points *B* and *C*. Since there is always at least one line containing two distinct points and never more than one, the line passing through the two points is said to be determined by the two points.

(c) *Given three distinct points,* there may be one line or none. If a line exists that contains the three points, such as *D, E,* and *F*, then the points are said to be **collinear**. If no such line exists (as in the case of points *G, H,* and *I*) then the points are said to be **noncollinear**.

Intersection Lines and Angles

An **angle** is a collection of points that is the union of two rays having the same endpoint. An angle such as the one illustrated below can be referred to in any of the following ways:

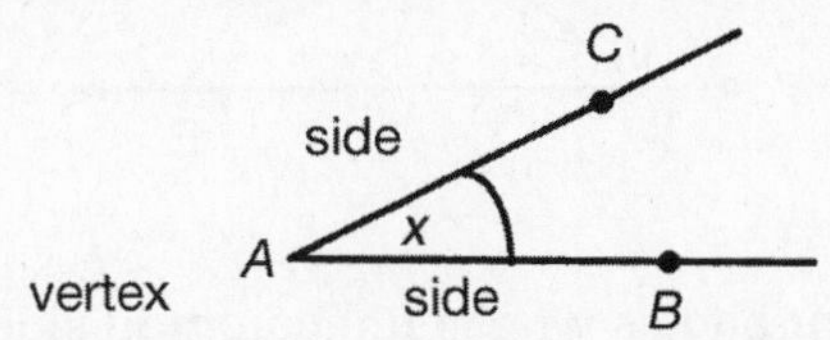

A) by a capital letter that names its vertex, for example, $\angle A$;

B) by a lowercase letter or number placed inside the angle, for example, $\angle x$;

C) by three capital letters, where the middle letter is the vertex and the other two letters are not on the same ray, for example, $\angle CAB$ or $\angle BAC$, both of which represent the angle illustrated in the figure.

Types of Angles

A) **Vertical angles** are formed when two lines intersect. These angles are equal.

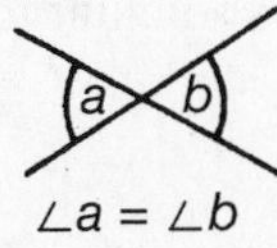

$\angle a = \angle b$

B) **Adjacent angles** are two angles with a common vertex and a common side, but no common interior points. In the following figure, $\angle DAC$ and $\angle BAC$ are adjacent angles. $\angle DAB$ and $\angle BAC$ are not.

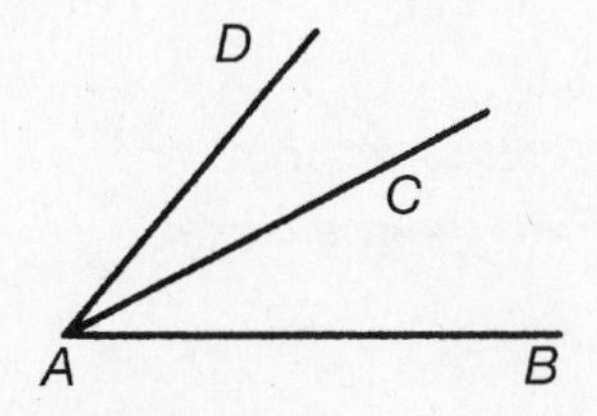

C) A **right angle** is an angle whose measure is 90°.

D) An **acute angle** is an angle whose measure is larger than 0°, but less than 90°.

E) An **obtuse angle** is an angle whose measure is larger than 90° but less than 180°.

F) A **straight angle** is an angle whose measure is 180°. Such an angle is, in fact, a straight line.

G) A **reflex angle** is an angle whose measure is greater than 180° but less than 360°.

H) **Complementary angles** are two angles whose measures total 90°.

I) **Supplementary angles** are two angles whose measures total 180°.

J) **Congruent angles** are angles of equal measure. The symbol ≅ means "congruent to." If $\angle P$ and $\angle Q$ have the same measure, then $\angle P \cong \angle Q$.

PROBLEM

In the figure, we are given $\overline{AB}$ and triangle ABC. We are told that the measure of $\angle 1$ is five times the measure of $\angle 2$. Determine the measures of $\angle 1$ and $\angle 2$.

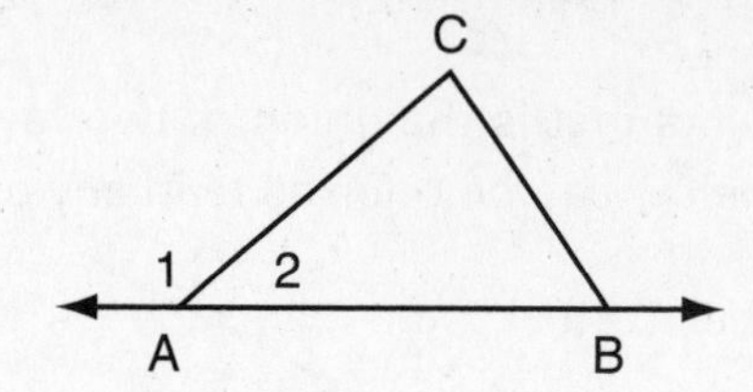

SOLUTION

Since $\angle 1$ and $\angle 2$ are adjacent angles whose noncommon sides lie on a straight line, they are, by definition, supplementary. As supplements, their measures must total 180°.

If we let x = the measure of $\angle 2$, then $5x$ = the measure of $\angle 1$.

To determine the respective angle measures, set $x + 5x = 180$ and solve for x. $6x = 180$. Therefore, $x = 30$ and $5x = 150$.

Therefore, the measure of $\angle 1 = 150$ and the measure of $\angle 2 = 30$.

Perpendicular Lines

Two lines are said to be **perpendicular** if they intersect and form right angles. The symbol for perpendicular (or, is therefore perpendicular to) is ⊥; $\overline{AB}$ is perpendicular to $\overline{CD}$ is written $\overline{AB} \perp \overline{CD}$.

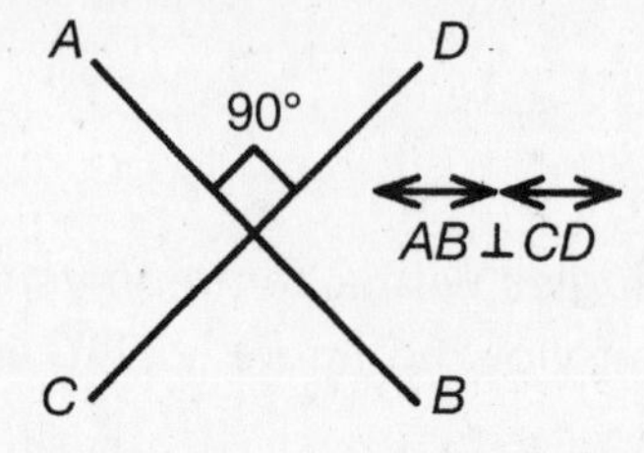

PROBLEM

We are given straight lines $\overline{AB}$ and $\overline{CD}$ intersecting at point P. $\overline{PR} \perp \overline{AB}$ and the measure of $\angle APD$ is 170°. Find the measures of $\angle 1$, $\angle 2$, $\angle 3$, and $\angle 4$.

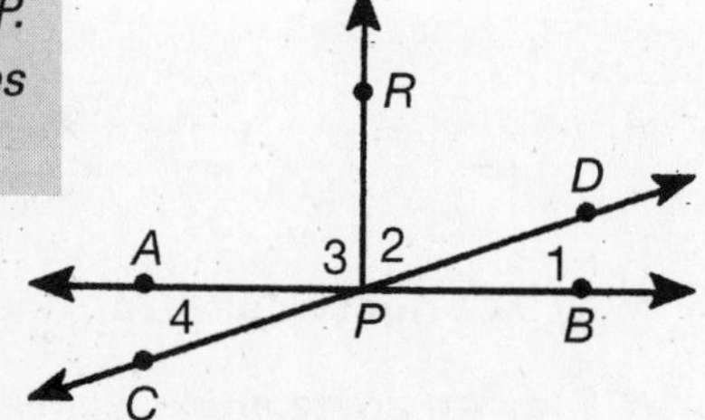

SOLUTION

This problem will involve making use of several of the properties of supplementary and vertical angles, as well as perpendicular lines.

$\angle APD$ and $\angle 1$ are adjacent angles whose noncommon sides lie on a straight line, $\overline{AB}$. Therefore, they are supplements and their measures total 180°.

$$m\angle APD + m\angle 1 = 180°.$$

We know $m\angle APD = 170°$. Therefore, by substitution, $170° + m\angle 1 = 180°$. This implies $m\angle 1 = 10°$.

$\angle 1$ and $\angle 4$ are vertical angles because they are formed by the intersection of two straight lines, $\overline{CD}$ and $\overline{AB}$, and their sides form two pairs of opposite rays. As vertical angles, they are, by theorem, of equal measure. Since $m\angle 1 = 10°$, then $m\angle 4 = 10°$.

Since $\overline{PR} \perp \overline{AB}$, at their intersection the angles formed must be right angles. Therefore, $\angle 3$ is a right angle and its measure is 90°.

The figure shows us that $\angle APD$ is composed of $\angle 3$ and $\angle 2$. Since the measure of the whole must be equal to the sum of the measures of its parts, $m\angle APD = m\angle 3 + m\angle 2$. We know that the $m\angle APD = 170°$ and $m\angle 3 = 90°$, therefore, by substitution, we can solve for $m\angle 2$, our last unknown.

$$170° = 90° + m\angle 2$$

$$80° = m\angle 2$$

Therefore, $m\angle 1 = 10°$, $m\angle 2 = 80°$,

$m\angle 3 = 90°$, $m\angle 4 = 10°$.

PROBLEM

In the accompanying figure $\overline{SM}$ is the perpendicular bisector of $\overline{QR}$, and $\overline{SN}$ is the perpendicular bisector of $\overline{QP}$. Prove that $SR = SP$.

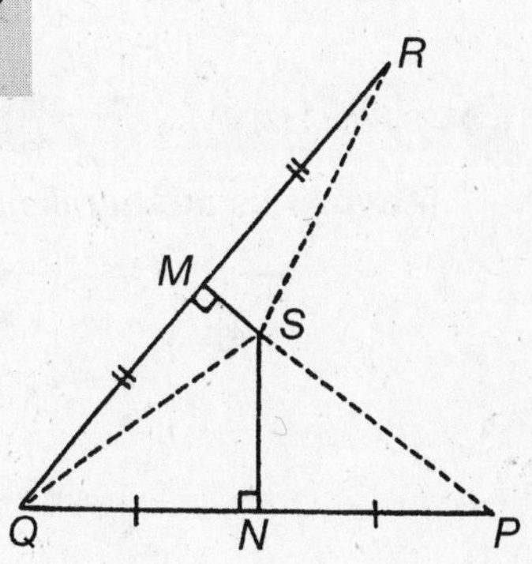

SOLUTION

Every point on the perpendicular bisector of a segment is equidistant from the endpoints of the segment.

Since point S is on the perpendicular bisector of $\overline{QR}$,

$$SR = SQ \tag{1}$$

Also, since point S is on the perpendicular bisector of $\overline{QP}$,

$$SQ = SP \tag{2}$$

By the transitive property (quantities equal to the same quantity are equal), we have

$$SR = SP. \tag{3}$$

Parallel Lines

Two lines are called **parallel lines** if, and only if, they are in the same plane (coplanar) and do not intersect. The symbol for parallel, or is parallel to, is $\parallel$; $\overline{AB}$ is parallel to $\overline{CD}$ is written $\overline{AB} \parallel \overline{CD}$.

The distance between two parallel lines is the length of the perpendicular segment from any point on one line to the other line.

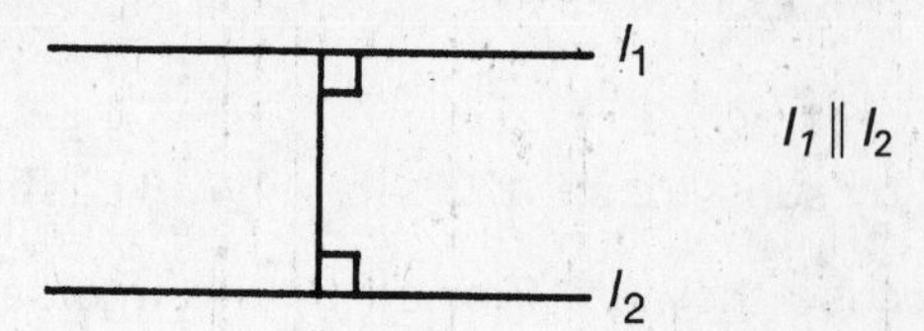

Given a line l and a point P not on line l, there is one and only one line through point P that is parallel to line l.

Two coplanar lines are either intersecting lines or parallel lines.

If two (or more) lines are perpendicular to the same line, then they are parallel to each other.

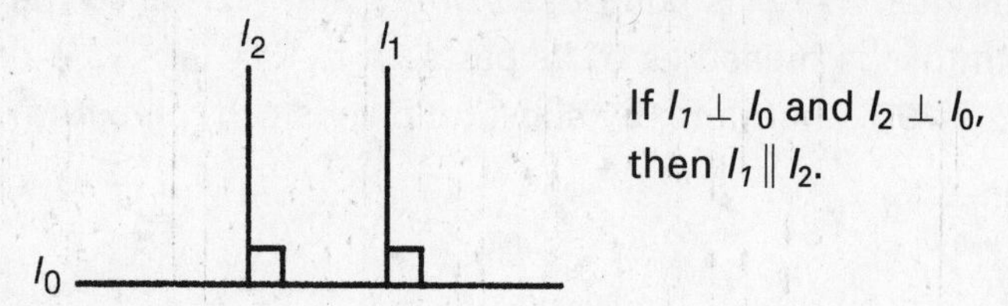

If two lines are cut by a transversal (a line intersecting two or more other lines) so that alternate interior angles are equal, the lines are parallel.

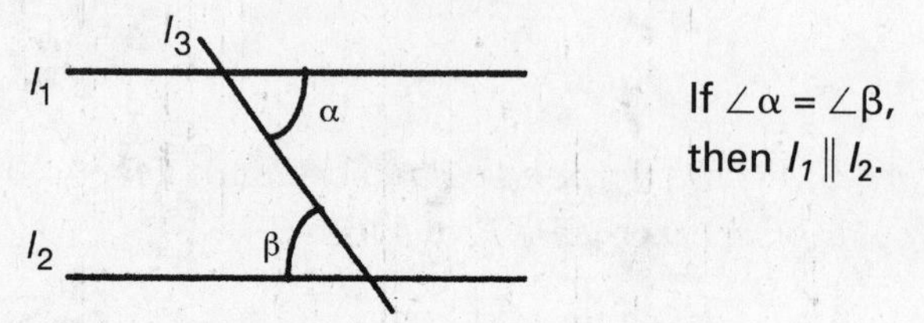

If two lines are parallel to the same line, then they are parallel to each other.

l_1

l_2

l_0

If $l_1 \parallel l_0$ and $l_2 \parallel l_0$,
then $l_1 \parallel l_2$.

If a line is perpendicular to one of two parallel lines, then it is perpendicular to the other line, too.

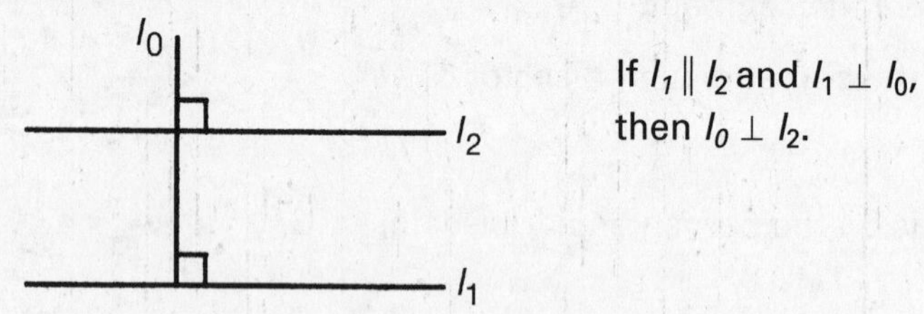

If two lines being cut by a transversal form congruent corresponding angles, then the two lines are parallel.

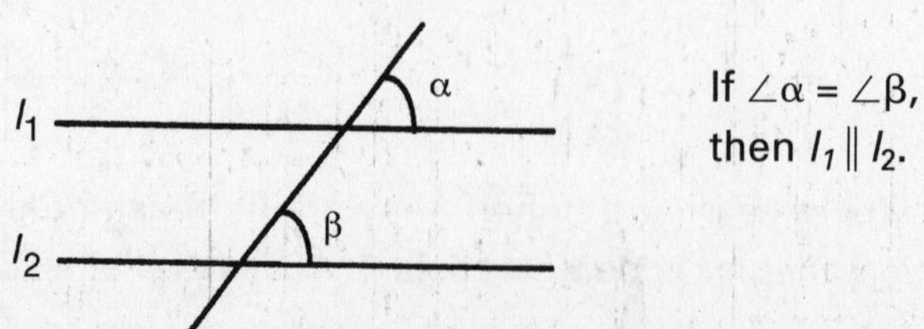

If two lines being cut by a transversal form interior angles on the same side of the transversal that are supplementary, then the two lines are parallel.

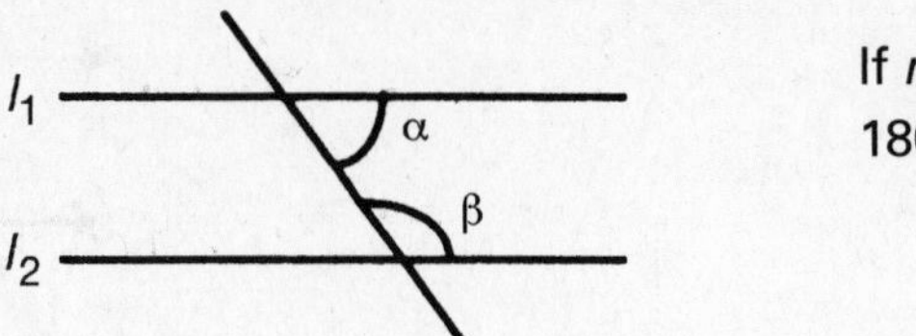

If $m\angle\alpha + m\angle\beta = 180°$ then $l_1 \parallel l_2$.

If a line is parallel to one of two parallel lines, it is also parallel to the other line.

If $l_1 \parallel l_2$ and $l_0 \parallel l_1$, then $l_0 \parallel l_2$.

If two parallel lines are cut by a transversal, then:

A) The alternate interior angles are congruent.

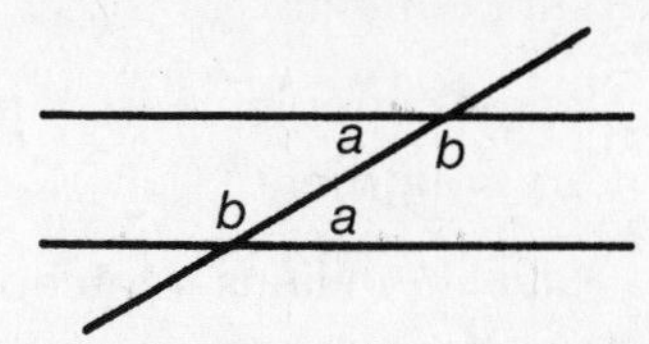

B) The corresponding angles are congruent.

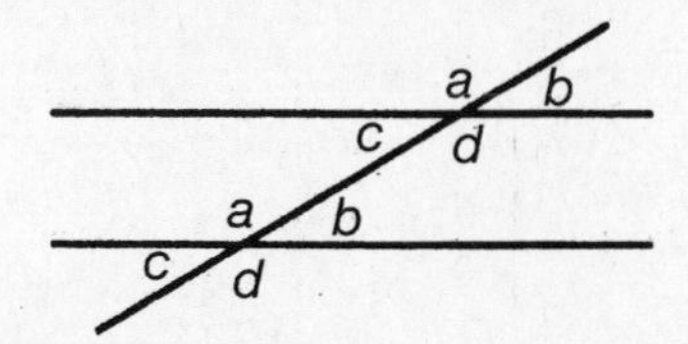

C) The consecutive interior angles are supplementary.

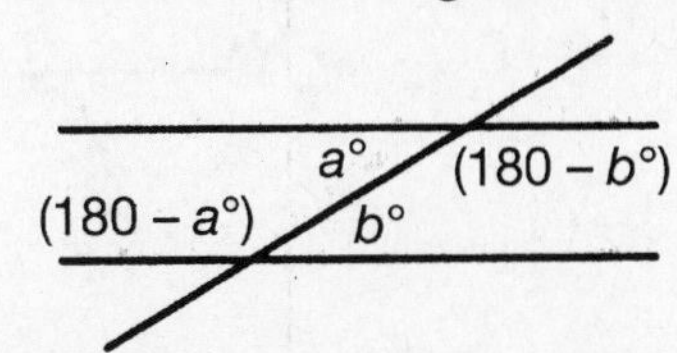

D) The alternate exterior angles are congruent.

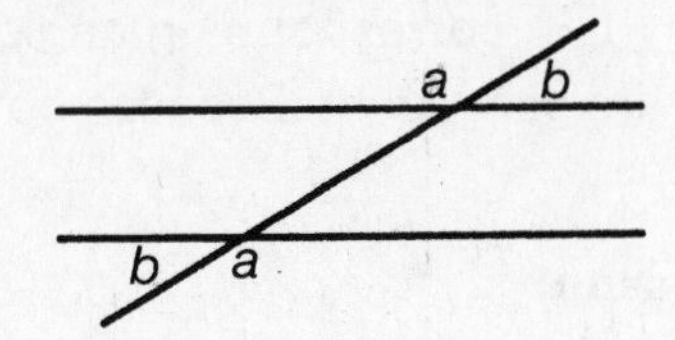

PROBLEM

Given: $\angle 2$ is supplementary to $\angle 3$.

Prove: $l_1 \parallel l_2$.

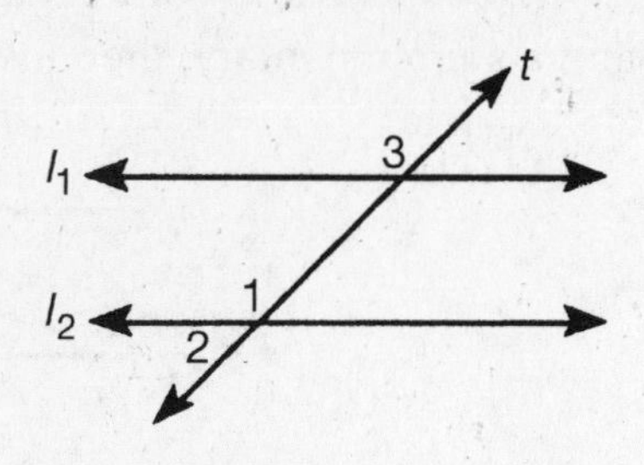

SOLUTION

Given two lines intercepted by a transversal, if a pair of corresponding angles are congruent, then the two lines are parallel. In this problem, we will show that since $\angle 1$ and $\angle 2$ are supplementary and $\angle 2$ and $\angle 3$ are supplementary, $\angle 1$ and $\angle 3$ are congruent. Since corresponding angles $\angle 1$ and $\angle 3$ are congruent, it follows $l_1 \parallel l_2$.

	Statement		Reason
1.	$\angle 2$ is supplementary to $\angle 3$	1.	Given.
2.	$\angle 1$ is supplementary to $\angle 2$	2.	Two angles that form a linear pair are supplementary.
3.	$\angle 1 \cong \angle 3$	3.	Angles supplementary to the same angle are congruent.
4.	$l_1 \parallel l_2$	4.	Given two lines intercepted by a transversal, if a pair of corresponding angles are congruent, then the two lines are parallel.

PROBLEM

If line $\overline{AB}$ is parallel to line $\overline{CD}$ and line $\overline{EF}$ is parallel to line $\overline{GH}$, prove that $m\angle 1 = m\angle 2$.

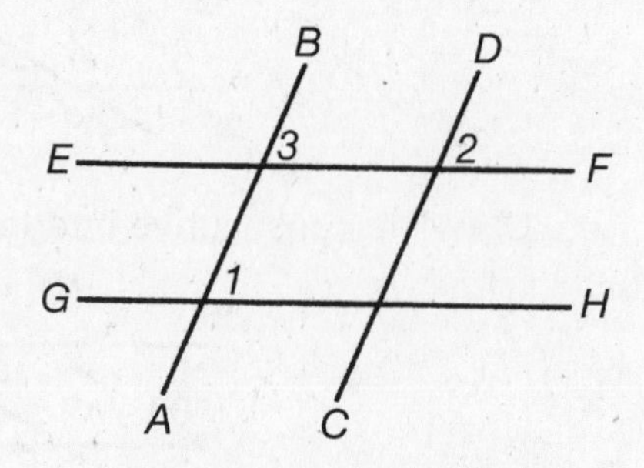

SOLUTION

To show $\angle 1 \cong \angle 2$, we relate both to $\angle 3$. Because $\overline{EF} \parallel \overline{GH}$, corresponding angles 1 and 3 are congruent. Since $\overline{AB} \parallel \overline{CD}$, corresponding angles 3 and 2 are congruent. Because both $\angle 1$ and $\angle 2$ are congruent to the same angle, it follows that $\angle 1 \cong \angle 2$.

	Statement		Reason
1.	$\overline{EF} \parallel \overline{GH}$	1.	Given.
2.	$m\angle 1 = m\angle 3$	2.	If two parallel lines are cut by a transversal, corresponding angles are of equal measure.
3.	$\overline{AB} \parallel \overline{CD}$	3.	Given.

4. $m\angle 2 = m\angle 3$	4. If two parallel lines are cut by a transversal, corresponding angles are equal in measure.
5. $m\angle 1 = m\angle 2$	5. If two quantities are equal to the same quantity, they are equal to each other.

DRILL: LINES AND ANGLES

Intersecting Lines

DIRECTIONS: Refer to the diagram and find the appropriate solution.

1. Find a.

   (A) 38° (B) 68° (C) 78°
   (D) 90° (E) 112°

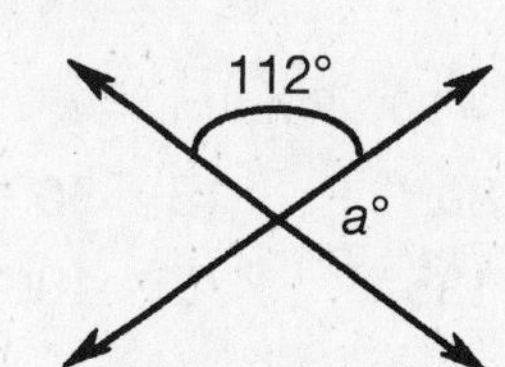

2. Find c.

   (A) 32° (B) 48° (C) 58°
   (D) 82° (E) 148°

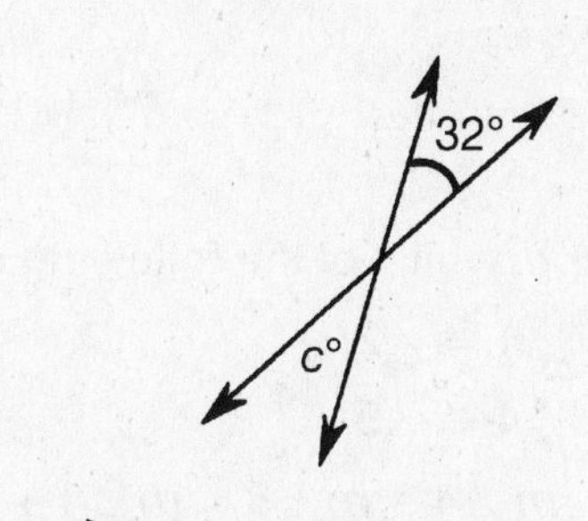

3. Determine x.

   (A) 21° (B) 23° (C) 51°
   (D) 102° (E) 153°

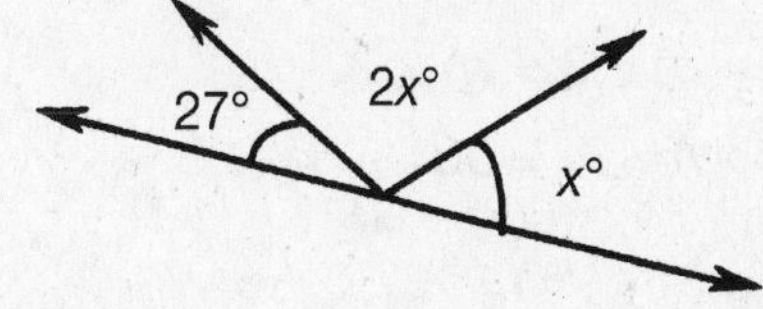

4. Find x.

   (A) 8 (B) 11.75 (C) 21
   (D) 26 (E) 32

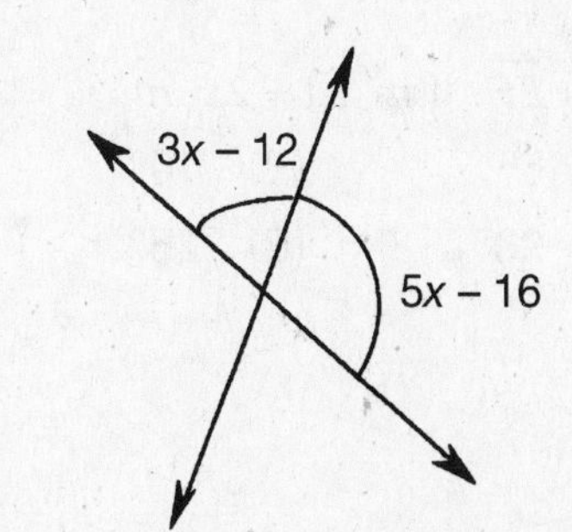

5. Find z.

   (A) 29° (B) 54° (C) 61°
   (D) 88° (E) 92°

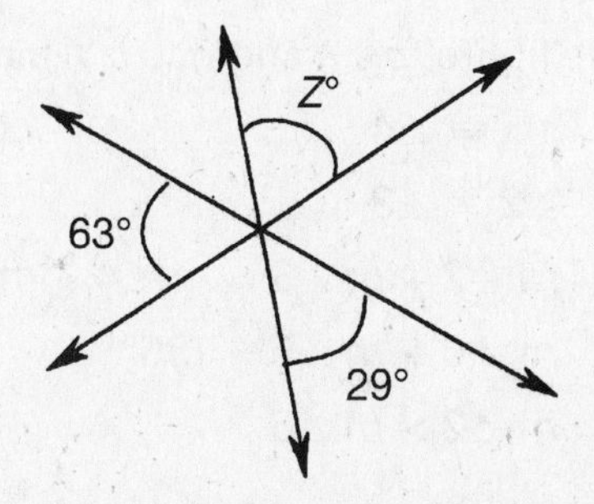

Perpendicular Lines

DIRECTIONS: Refer to the diagram and find the appropriate solution.

6. $\overrightarrow{BA} \perp \overrightarrow{BC}$ and $m\angle DBC = 53°$. Find $m\angle ABD$.

   (A) 27° (B) 33° (C) 37°

   (D) 53° (E) 90°

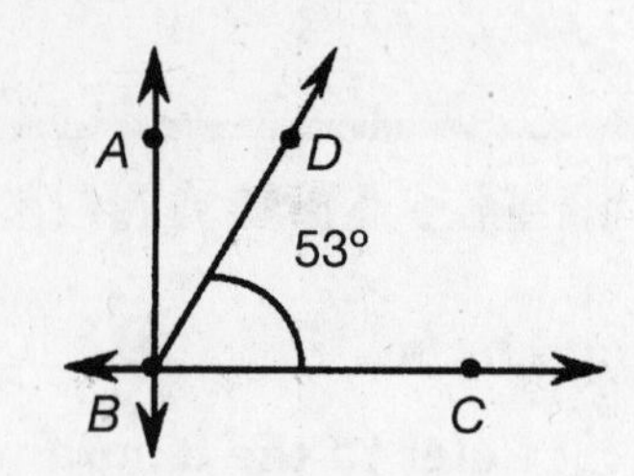

7. $m\angle 1 = 90°$. Find $m\angle 2$.

   (A) 80° (B) 90° (C) 100°

   (D) 135° (E) 180°

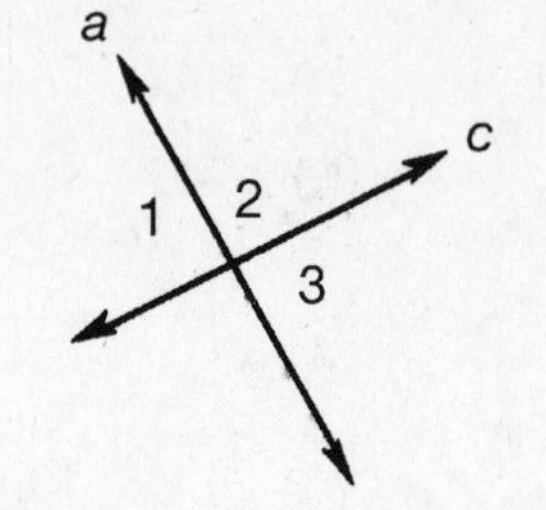

8. If $n \perp p$, which of the following statements is true?

   (A) $\angle 1 \cong \angle 2$

   (B) $\angle 4 \cong \angle 5$

   (C) $m\angle 4 + m\angle 5 > m\angle 1 + m\angle 2$

   (D) $m\angle 3 > m\angle 2$

   (E) $m\angle 4 = 90°$

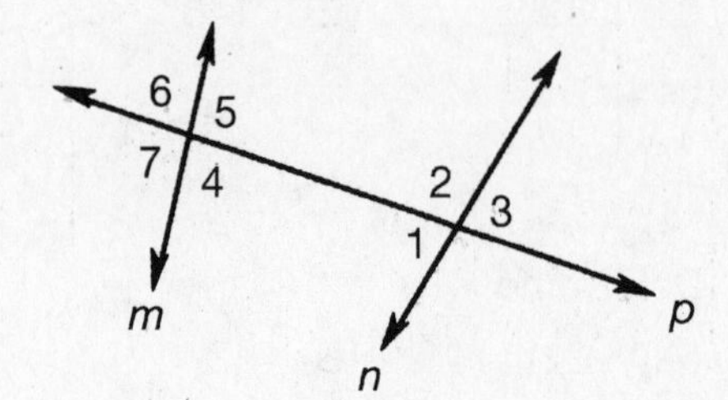

9. $\overline{CD} \perp \overline{EF}$. If $m\angle 1 = 2x$, $m\angle 2 = 30°$, and $m\angle 3 = x$, find x.

   (A) 5° (B) 10° (C) 12°

   (D) 20° (E) 25°

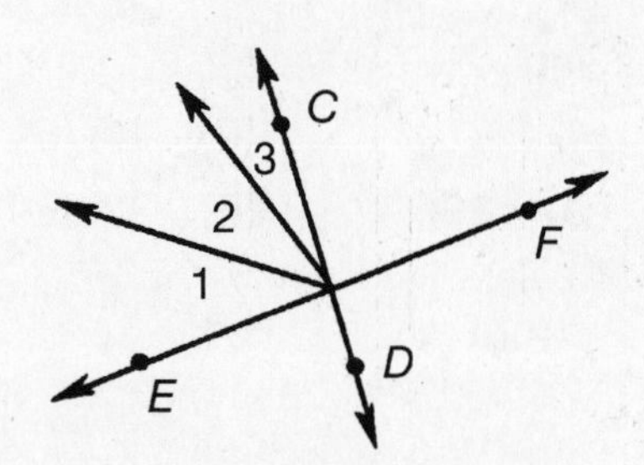

10. In the figure, $p \perp t$ and $q \perp t$. Which of the following statements is false?

   (A) $\angle 1 \cong \angle 4$

   (B) $\angle 2 \cong \angle 3$

   (C) $m\angle 2 + m\angle 3 = m\angle 4 + m\angle 6$

   (D) $m\angle 5 + m\angle 6 = 180°$

   (E) $m\angle 2 > m\angle 5$

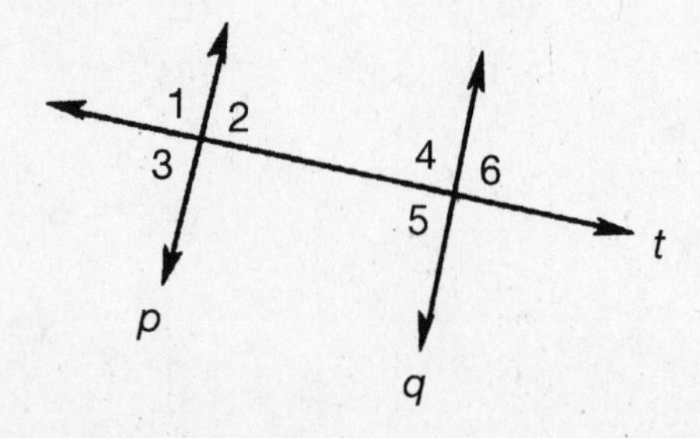

Parallel Lines

DIRECTIONS: Refer to the diagram and find the appropriate solution.

11. If $a \parallel b$, find z.

    (A) 26° (B) 32° (C) 64°
    (D) 86° (E) 116°

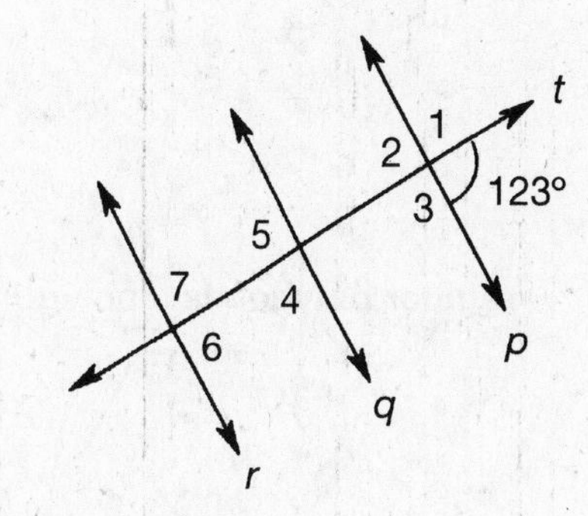

12. In the figure, $p \parallel q \parallel r$. Find $m\angle 7$.

    (A) 27° (B) 33° (C) 47°
    (D) 57° (E) 64°

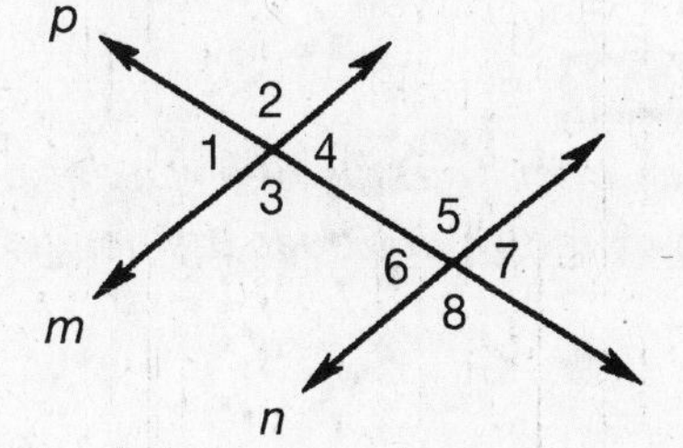

13. If $m \parallel n$, which of the following statements is not necessarily true?

    (A) $\angle 2 \cong \angle 5$
    (B) $\angle 3 \cong \angle 6$
    (C) $m\angle 4 + m\angle 5 = 180°$
    (D) $\angle 1 \cong \angle 6$
    (E) $m\angle 7 + m\angle 3 = 180°$

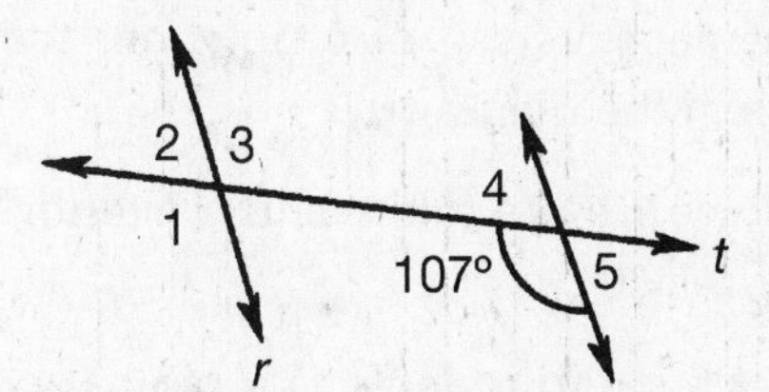

14. If $r \parallel s$, find $m\angle 2$.

    (A) 17° (B) 27° (C) 43°
    (D) 67° (E) 73°

15. If $a \parallel b$ and $c \parallel d$, find $m\angle 5$.

    (A) 55° (B) 65° (C) 75°
    (D) 95° (E) 125°

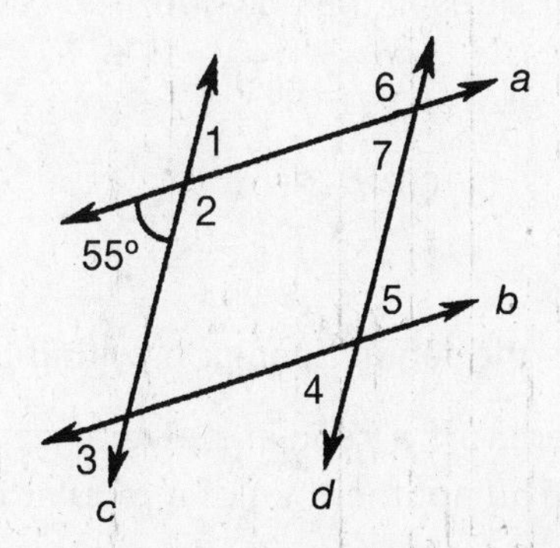

MASTERING POLYGONS(CONVEX)

A **polygon** is a figure with the same number of sides as angles.

An **equilateral polygon** is a polygon all of whose sides are of equal measure.

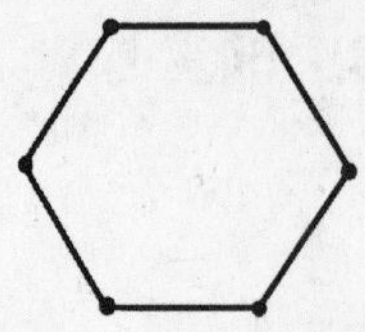

An **equiangular polygon** is a polygon all of whose angles are of equal measure.

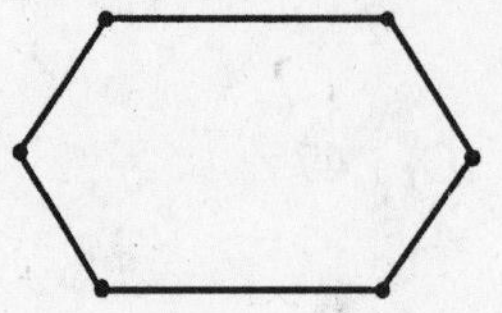

A **regular polygon** is a polygon that is both equilateral and equiangular.

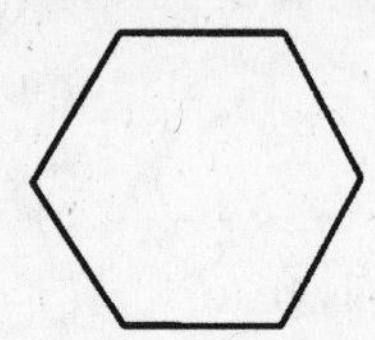
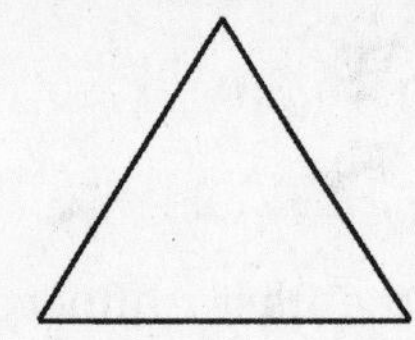
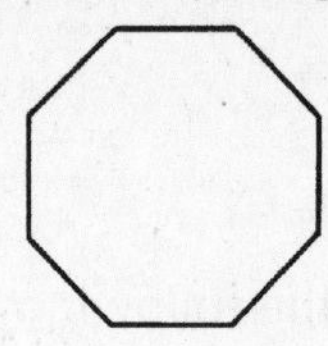

PROBLEM

Each interior angle of a regular polygon contains 120°. How many sides does the polygon have?

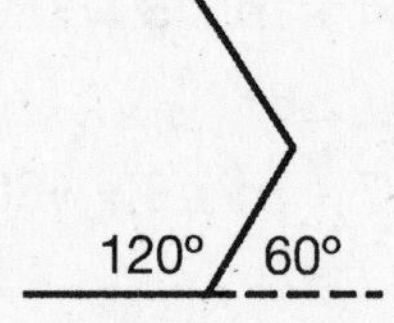

SOLUTION

At each vertex of a polygon, the exterior angle is supplementary to the interior angle, as shown in the diagram.

Since we are told that the interior angles measure 120°, we can deduce that the exterior angle measures 60°.

Each exterior angle of a regular polygon of n sides measure $\frac{360°}{n}$ degrees. We know that each exterior angle measures 60°, and, therefore, by setting $\frac{360°}{n}$ equal to 60°, we can determine the number of sides in the polygon. The calculation is as follows:

$$\frac{360°}{n} = 60°$$

$$60°n = 360°$$

$$n = 6$$

Therefore, the regular polygon, with interior angles of 120°, has six sides and is called a hexagon.

The area of a regular polygon can be determined by using the **apothem** and **radius** of the polygon. The apothem (a) of a regular polygon is the segment from the center of the polygon perpendicular to a side of the polygon. The radius (r) of a regular polygon is the segment joining any vertex of a regular polygon with the center of that polygon.

(1) All radii of a regular polygon are congruent.

(2) All apothems of a regular polygon are congruent.

(3) The radius of a regular hexagon is congruent to a side.

The **area** of a regular polygon equals one-half the product of the length of the apothem and the perimeter.

$$\text{Area} = \frac{1}{2} a \times p$$

PROBLEM

Find the area of the regular pentagon whose radius is 5 and whose apothem is 4.

SOLUTION

If the radius is 5 and the length of the apothem is 4, the length of a side is 6.

$$a^2 + b^2 = c^2$$

$$a^2 + (4)^2 = (5)^2$$

$$a^2 = 25 - 16$$

$$a^2 = 9$$

$$a = 3$$

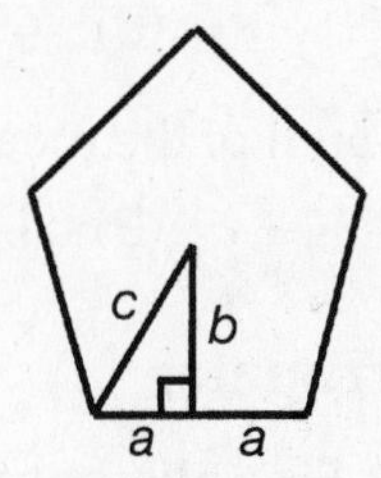

Therefore, the perimeter of the polygon is 30.

$$A = \frac{1}{2} a \times p$$

$$A = \frac{1}{2}(4)(30)$$

$$A = 60$$

PROBLEM

Find the area of a regular hexagon if one side has length 6.

SOLUTION

Since the length of a side equals 6, the radius also equals 6 and the perimeter equals 36. The base of the right triangle, formed by the radius and apothem, is half the length of a side, or 3. You can find the length of the apothem by using what is known as the Pythagorean Theorem (discussed further in the next section).

$$a^2 + b^2 = c^2$$

$$a^2 + (3)^2 = (6)^2$$

$$a^2 = 36 - 9$$

$$a^2 = 27$$

$$a = 3\sqrt{3}$$

The apothem equals $3\sqrt{3}$. Therefore, the area of the hexagon

$$= \frac{1}{2} a \times p$$

$$= \frac{1}{2}(3\sqrt{3})(36)$$

$$= 54\sqrt{3}$$

DRILL: REGULAR POLYGONS

Angle Measures

DIRECTIONS: Find the appropriate solution.

1. Find the measure of an interior angle of a regular pentagon.
   (A) 55° (B) 72° (C) 90° (D) 108° (E) 540°

2. Find the measure of an exterior angle of a regular octagon.
   (A) 40° (B) 45° (C) 135° (D) 540° (E) 1,080°

3. Find the sum of the measures of the exterior angles of a regular triangle.
   (A) 90° (B) 115° (C) 180° (D) 250° (E) 360°

Area and Perimeter

DIRECTIONS: Find the appropriate solution.

4. Find the area of a square with a perimeter of 12 cm.
   (A) 9 cm^2 (B) 12 cm^2 (C) 48 cm^2 (D) 96 cm^2 (E) 144 cm^2

5. A regular triangle has sides of 24 mm. If the apothem is $4\sqrt{3}$ mm, find the area of the triangle.
   (A) 72 mm^2 (B) $96\sqrt{3}$ mm^2 (C) 144 mm^2
   (D) $144\sqrt{3}$ mm^2 (E) 576 mm^2

6. Find the area of a regular hexagon with sides of 4 cm.
   (A) $12\sqrt{3}$ cm^2 (B) 24 cm^2 (C) $24\sqrt{3}$ cm^2
   (D) 48 cm^2 (E) $48\sqrt{3}$ cm^2

7. Find the area of a regular decagon with sides of length 6 cm and an apothem of length 9.2 cm.
   (A) 55.2 cm^2 (B) 60 cm^2 (C) 138 cm^2
   (D) 138.3 cm^2 (E) 276 cm^2

8. The perimeter of a regular heptagon (7-gon) is 36.4 cm. Find the length of each side.
   (A) 4.8 cm (B) 5.2 cm (C) 6.7 cm (D) 7 cm (E) 10.4 cm

9. The apothem of a regular quadrilateral is 4 in. Find the perimeter.
   (A) 12 in. (B) 16 in. (C) 24 in. (D) 32 in. (E) 64 in.

10. A regular triangle has a perimeter of 18 cm; a regular pentagon has a perimeter of 30 cm; a regular hexagon has a perimeter of 33 cm. Which figure (or figures) have sides with the longest measure?

(A) Regular triangle

(B) Regular triangle and regular pentagon

(C) Regular pentagon

(D) Regular pentagon and regular hexagon

(E) Regular hexagon

MASTERING TRIANGLES

A closed three-sided geometric figure is called a **triangle**. The points of the intersection of the sides of a triangle are called the **vertices** of the triangle.

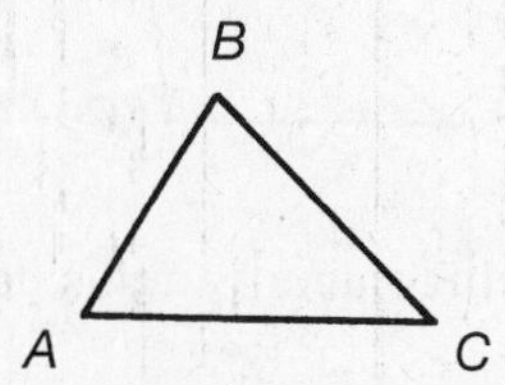

The **perimeter** of a triangle is the sum of the measures of the sides of the triangle.

A triangle with no equal sides is called a **scalene** triangle.

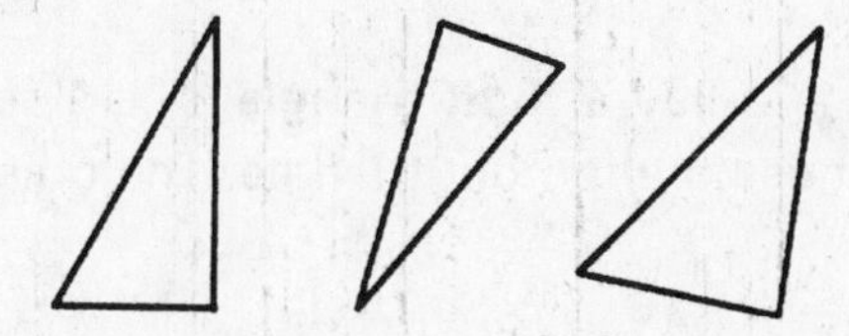

A triangle having at least two equal sides is called an **isosceles** triangle. The third side is called the **base** of the triangle.

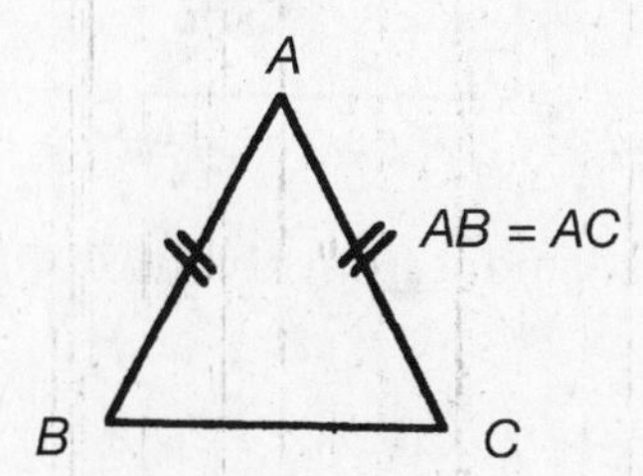

A side of a triangle is a line segment whose endpoints are the vertices of two angles of the triangle.

An **interior angle** of a triangle is an angle formed by two sides and includes the third side within its collection of points.

An **equilateral triangle** is a triangle having three equal sides. $\overline{AB} = \overline{AC} = \overline{BC}$.

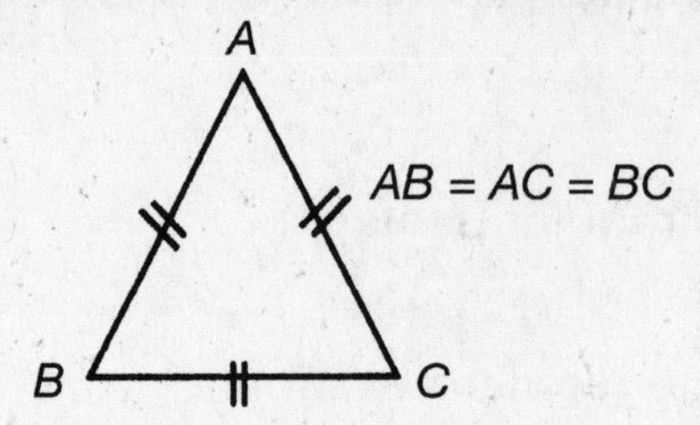

The sum of the measures of the interior angles of a triangle is 180°.

A triangle with one obtuse angle greater than 90° is called an **obtuse triangle.**

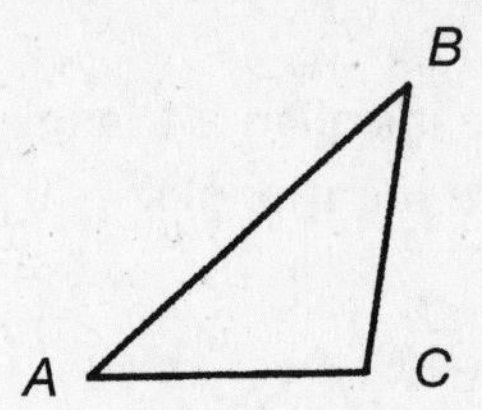

An **acute triangle** is a triangle with three acute angles (less than 90°).

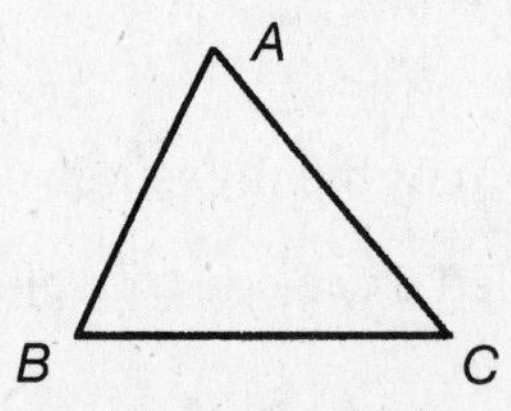

A triangle with a right angle is called a **right triangle**. The side opposite the right angle in a right triangle is called the hypotenuse of the right triangle. The other two sides are called arms or legs of the right triangle.

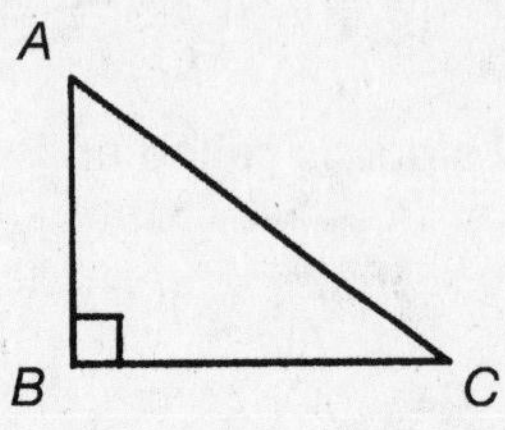

An **altitude** of a triangle is a line segment from a vertex of the triangle perpendicular to the opposite side.

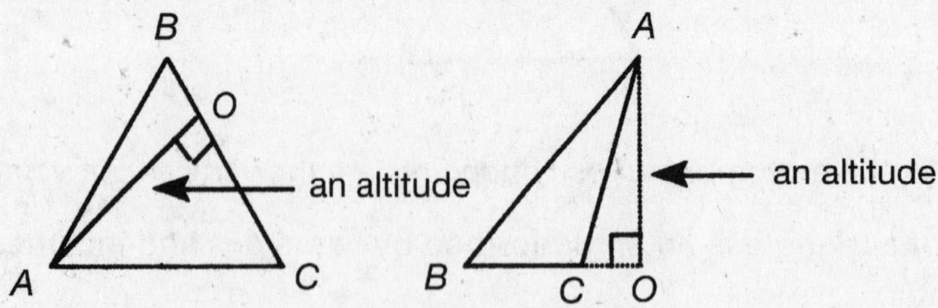

A line segment connecting a vertex of a triangle and the midpoint of the opposite side is called a **median** of the triangle.

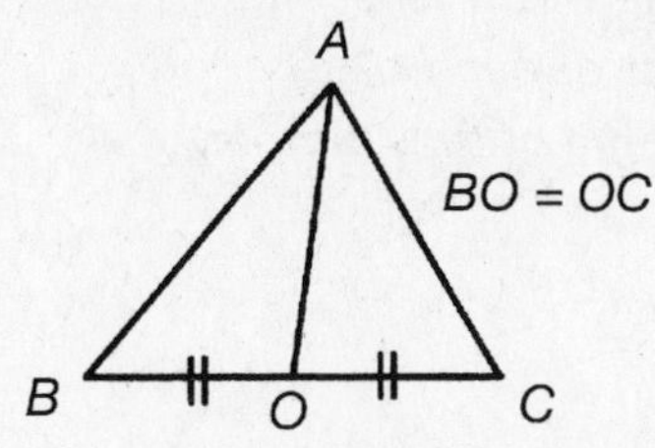

A line that bisects and is perpendicular to a side of a triangle is called a **perpendicular bisector** of that side.

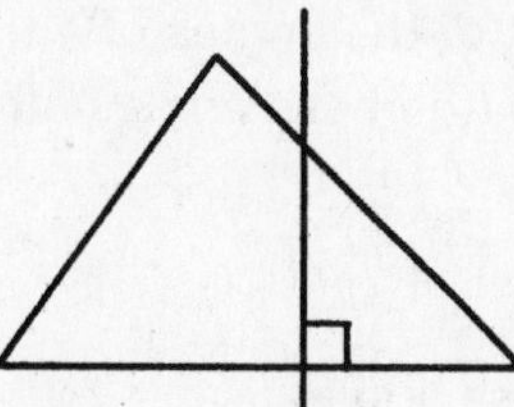

An **angle bisector** of a triangle is a line that bisects an angle and extends to the opposite side of the triangle.

A

α β

B C

$\angle\alpha = \angle\beta$

The line segment that joins the midpoints of two sides of a triangle is called a **midline** of the triangle.

A

$AD = DC$
$BE = EC$

D

midline: DE

B E C

An **exterior angle** of a triangle is an angle formed outside a triangle by one side of the triangle and the extension of an adjacent side.

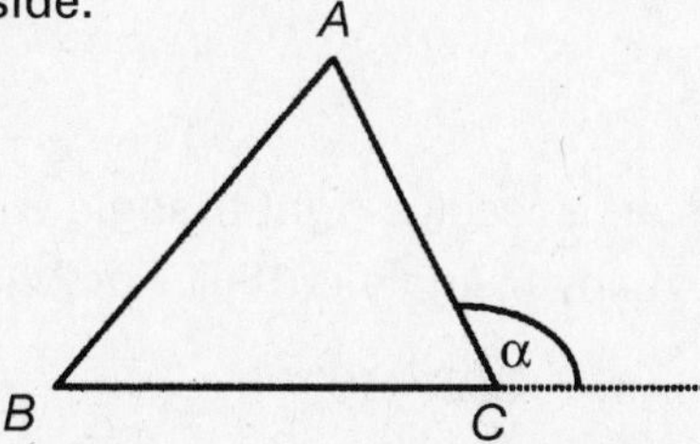

A triangle whose three interior angles have equal measure (60° each) is said to be **equiangular**.

Three or more lines (or rays or segments) are concurrent if there exists one point common to all of them, that is, if they all intersect at the same point.

In a right triangle, the square of the hypotenuse is equal to the sum of the squares of the other two sides. This is commonly known as the theorem of Pythagoras or the Pythagorean Theorem.

PROBLEM

The measure of the vertex angle of an isosceles triangle exceeds the measurement of each base angle by 30°. Find the value of each angle of the triangle.

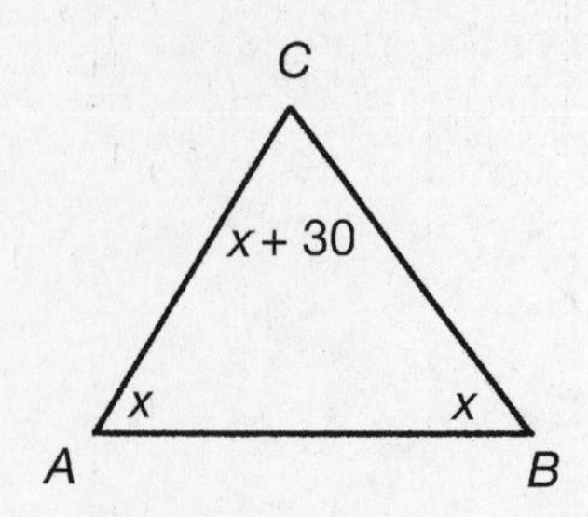

SOLUTION

We know that the sum of the values of the angles of a triangle is 180°. In an isosceles triangle, the angles opposite the congruent sides (the base angles) are, themselves, congruent and of equal value.

Therefore,

(1) Let x = the measure of each base angle.

(2) Then $x + 30$ = the measure of the vertex angle.

We can solve for x algebraically by keeping in mind the sum of all the measures will be 180°.

$$x + x + (x + 30) = 180$$

$$3x + 30 = 180$$

$$3x = 150$$

$$x = 50$$

Therefore, the base angles each measure 50°, and the vertex angle measures 80°.

PROBLEM

Prove that the base angles of an isosceles right triangle measure 45° each.

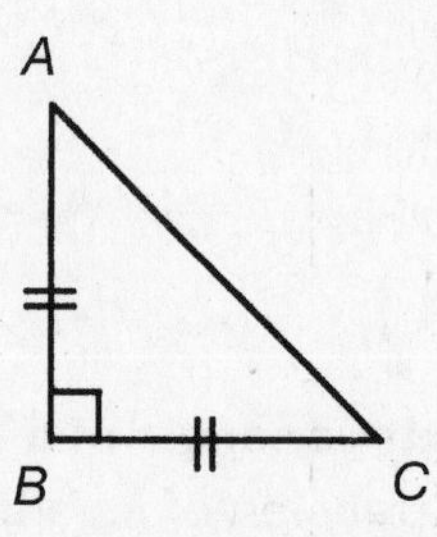

SOLUTION

As drawn in the figure, ΔABC is an isosceles right triangle with base angles *BAC* and *BCA*. The sum of the measures of the angles of any triangle is 180°. For ΔABC, this means

$$m\angle BAC + m\angle BCA + m\angle ABC = 180° \quad (1)$$

But $m\angle ABC = 90°$ because *ABC* is a right triangle. Furthermore, $m\angle BCA = m\angle BAC$, since the base angles of an isosceles triangle are congruent. Using these facts in equation (1)

$$m\angle BAC + m\angle BCA + 90° = 180°$$

or $$2m\angle BAC = 2m\angle BCA = 90°$$

or $$m\angle BAC = m\angle BCA = 45°.$$

Therefore, the base angles of an isosceles right triangle measure 45° each.

The area of a triangle is given by the formula $A = \frac{1}{2}bh$, where b is the length of a base, which can be any side of the triangle, and h is the corresponding height of the triangle, which is the perpendicular line segment that is drawn from the vertex opposite the base to the base itself.

$$A = \frac{1}{2}bh$$

$$A = \frac{1}{2}(10)(3)$$

$$A = 15$$

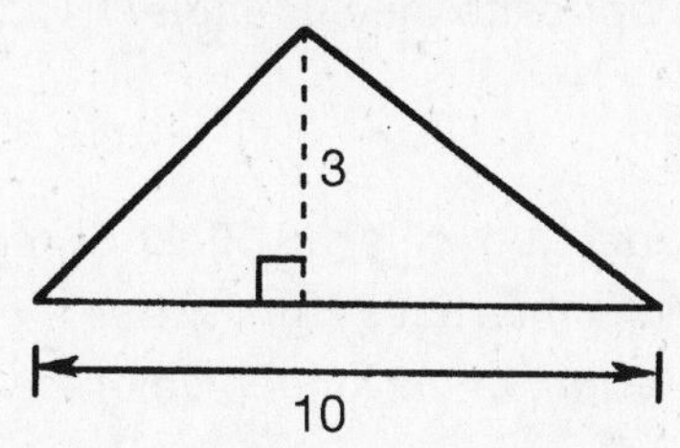

The area of a right triangle is found by taking $\frac{1}{2}$ the product of the lengths of its two arms.

$$A = \frac{1}{2}(5)(12)$$

$$A = 30$$

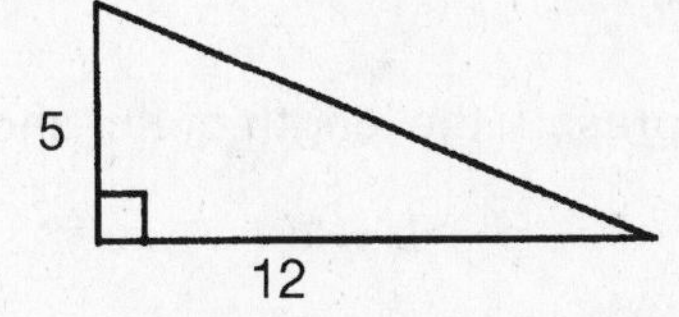

Special Right Triangles

A 45-45-90 triangle has angles that measure 45°, 45°, and 90°. A special relationship exists among the lengths of the sides of a 45-45-90 triangle. If you know the length of one side of a 45-45-90 triangle, you can use this relationship to determine the lengths of the other two sides.

Because a 45-45-90 triangle has two angles with the same measure, the sides opposite these angles have the same length. Therefore, the legs of a 45-45-90 triangle are congruent. In addition, the hypotenuse of a 45-45-90 triangle is equal to the length of one of its legs multiplied by $\sqrt{2}$, as shown in the figure below.

PROBLEM

A leg of a 45-45-90 triangle measures 5 units. Use the Pythagorean Theorem to show that the hypotenuse of the triangle measures $5\sqrt{2}$ units.

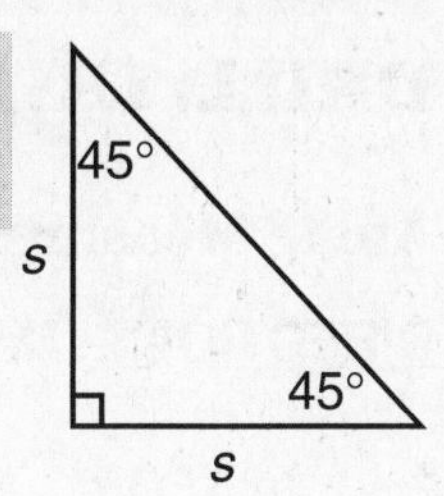

SOLUTION

The Pythagorean Theorem states that for a right triangle, $a^2 + b^2 = c^2$, where a and b are the lengths of the legs and c is the length of the hypotenuse. Because the triangle is a 45-45-90 triangle, its legs are congruent, and the value of both a and b is 5. Use this information to solve for c, the length of the hypotenuse.

$a^2 + b^2 = c^2$

$5^2 + 5^2 = c^2$ Substitute.

$50 = c^2$ Simplify.

$\sqrt{50} = \sqrt{c^2}$ Take the square root of each side.

$5\sqrt{2} = c$ Simplify.

The length of the hypotenuse is equal to the length of a leg multiplied by $\sqrt{2}$, or $5\sqrt{2}$ units.

A special relationship also exists among the side lengths of a 30-60-90 triangle. In this type of triangle, the length of the longer leg is equal to the length of the shorter leg multiplied by $\sqrt{3}$. In addition, the hypotenuse of a 30-60-90 triangle is twice as long as the shorter leg, as shown in the following figure. Remember that the shorter leg is always opposite the 30° angle and the longer leg is always opposite the 60° angle.

PROBLEM

The shorter leg of a 30-60-90 triangle measures 4 units, and the hypotenuse measures 8 units. Use the Pythagorean Theorem to show that the length of the longer leg of the triangle measures $4\sqrt{3}$ units.

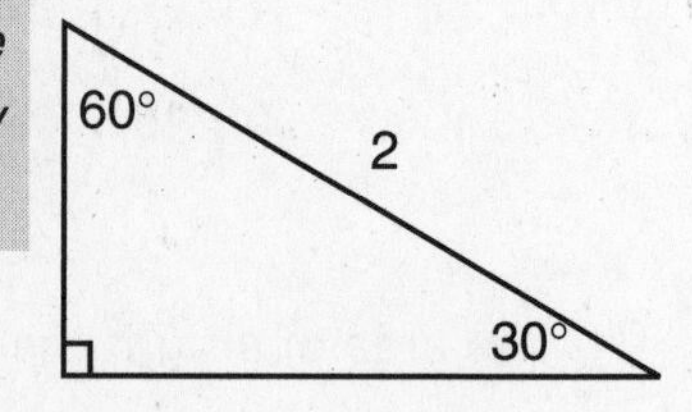

SOLUTION

Let *a* represent the length of the shorter leg and *b* represent the length of the longer leg.

$a^2 + b^2 = c^2$	
$4^2 + b^2 = 8^2$	Substitute.
$16 + b^2 = 64$	Simplify.
$b^2 = 48$	Subtract 16 from each side.
$\sqrt{b^2} = \sqrt{48}$	Take the square root of each side.
$b = \sqrt{(16)(3)} = 4\sqrt{3}$	Simplify.

The length of the longer leg is equal to the length of the shorter leg multiplied by $\sqrt{3}$, or $4\sqrt{3}$ units.

DRILL: TRIANGLES

Angle Measures

DIRECTIONS: Refer to the diagram and find the appropriate solution.

1. In ΔPQR, $\angle Q$ is a right angle. Find $m\angle R$.

   (A) 27° (B) 33° (C) 54°

   (D) 67° (E) 157°

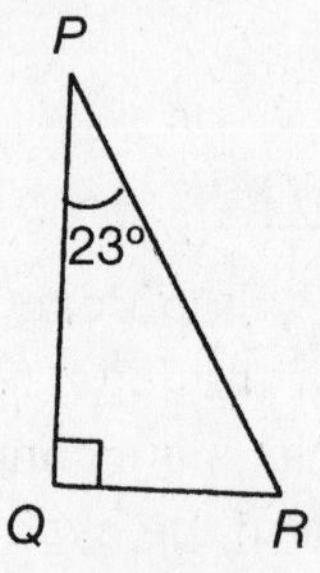

2. ΔMNO is isosceles. If the vertex angle, $\angle N$, has a measure of 96°, find the measure of $\angle M$.

   (A) 21° (B) 42° (C) 64°

   (D) 84° (E) 96°

N

M O

3. Find x.

(A) 15° (B) 25° (C) 30°

(D) 45° (E) 90°

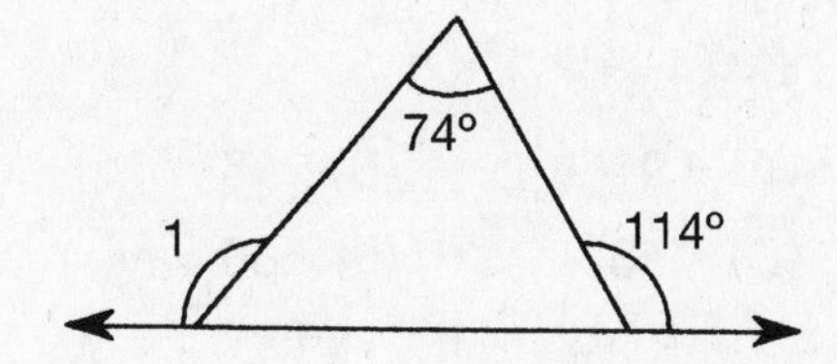

4. Find $m\angle 1$.

(A) 40° (B) 66° (C) 74°

(D) 114° (E) 140°

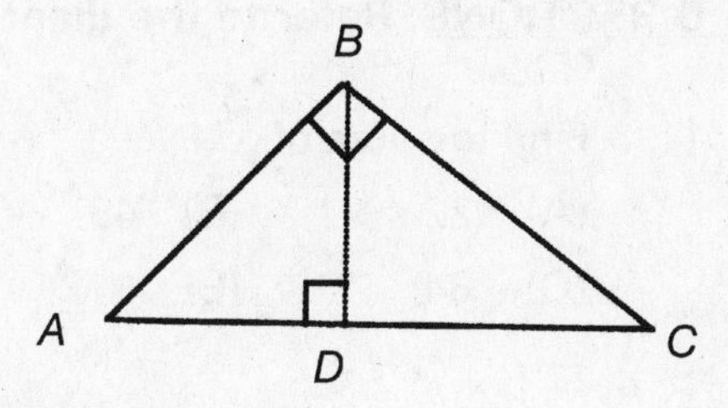

5. ΔABC is a right triangle with a right angle at B. ΔBDC is a right triangle with right angle $\angle BDC$. If $m\angle C = 36°$. Find $m\angle A$.

(A) 18° (B) 36° (C) 54°

(D) 72° (E) 180°

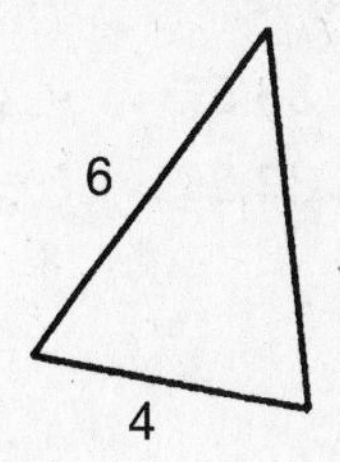
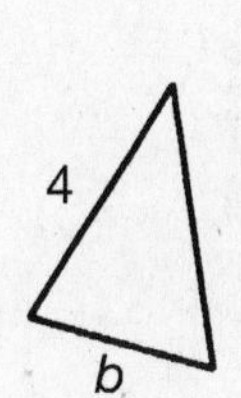

Similar Triangles

<u>DIRECTIONS</u>: Refer to the diagram and find the appropriate solution.

6. The two triangles shown are similar. Find b.

(A) $2\frac{2}{3}$ (B) 3 (C) 4

(D) 16 (E) 24

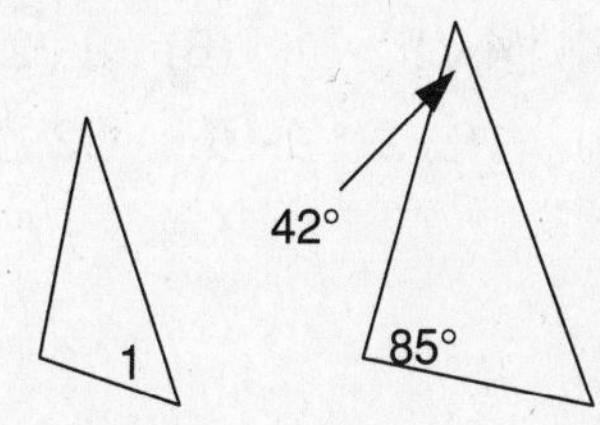

7. The two triangles shown are similar. Find $m\angle 1$.

(A) 48° (B) 53° (C) 74°

(D) 127° (E) 180°

8. The two triangles shown are similar. Find a and b.

(A) 5 and 10 (B) 4 and 8 (C) $4\frac{2}{3}$ and $7\frac{1}{3}$

(D) 5 and 8 (E) $5\frac{1}{3}$ and 8

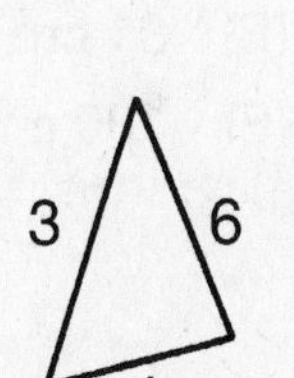

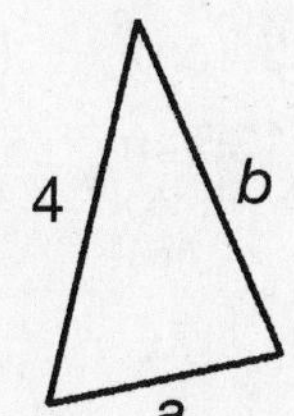

9. The perimeter of ΔLXR is 45 and the perimeter of ΔABC is 27. If $\overline{LX} = 15$, find the length of $\overline{AB}$.

(A) 9 (B) 15 (C) 27
(D) 45 (E) 72

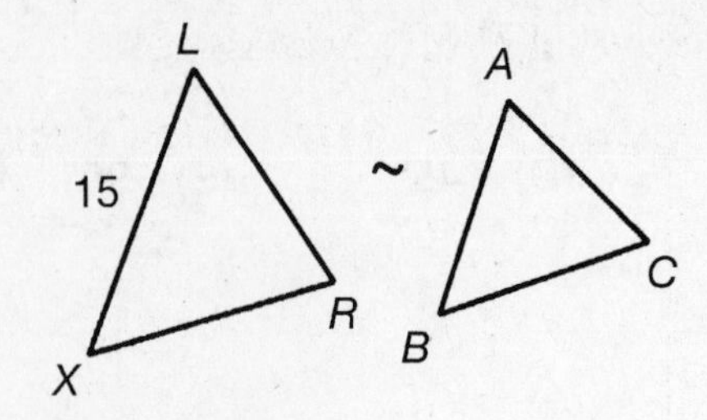

10. Find b.

(A) 9 (B) 15 (C) 20
(D) 45 (E) 60

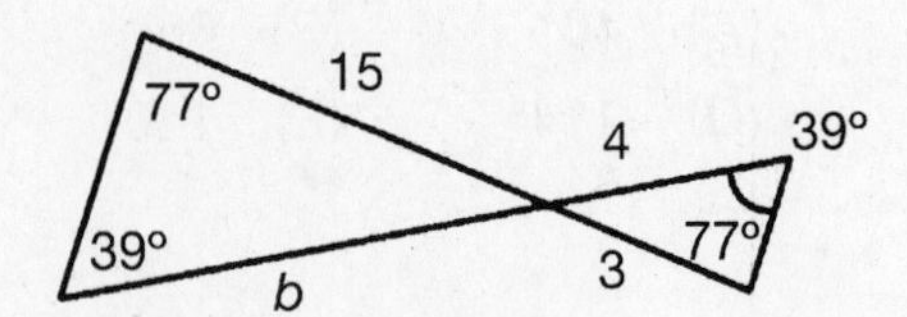

Area

DIRECTIONS: Refer to the diagram and find the appropriate solution.

11. Find the area of ΔMNO.

(A) 22 (B) 49 (C) 56
(D) 84 (E) 112

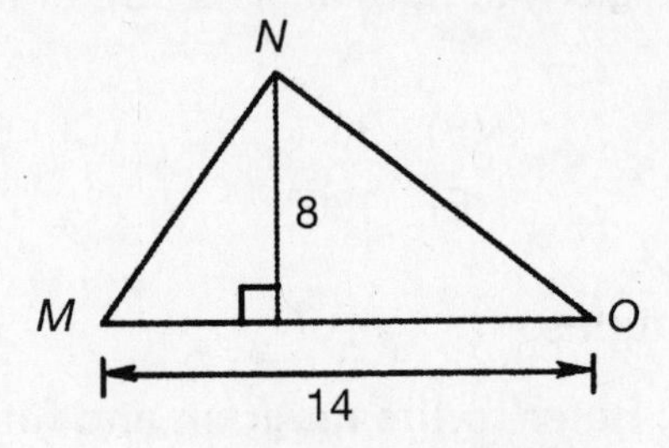

12. Find the area of ΔPQR.

(A) 31.5 (B) 38.5 (C) 53
(D) 77 (E) 82.5

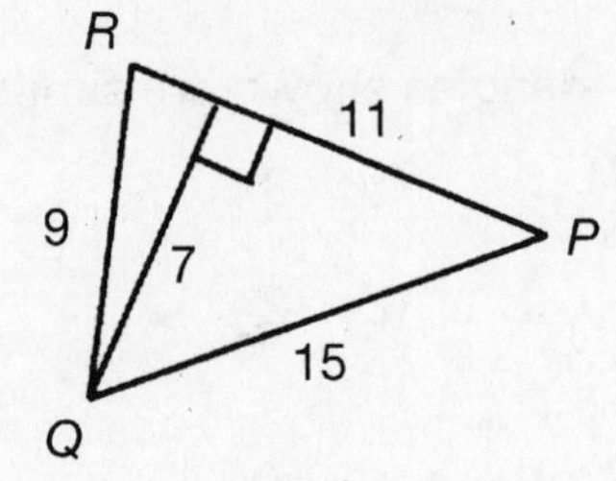

13. Find the area of ΔSTU.

(A) $4\sqrt{2}$ (B) $8\sqrt{2}$ (C) $12\sqrt{2}$
(D) $16\sqrt{2}$ (E) $32\sqrt{2}$

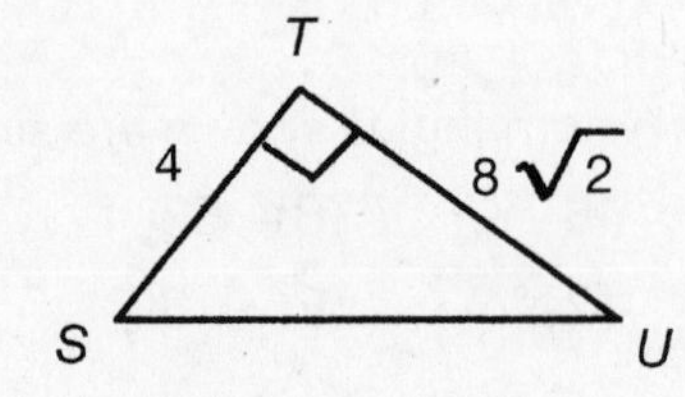

14. Find the area of ΔABC.

(A) 54 cm^2 (B) 81 cm^2 (C) 108 cm^2
(D) 135 cm^2 (E) 180 cm^2

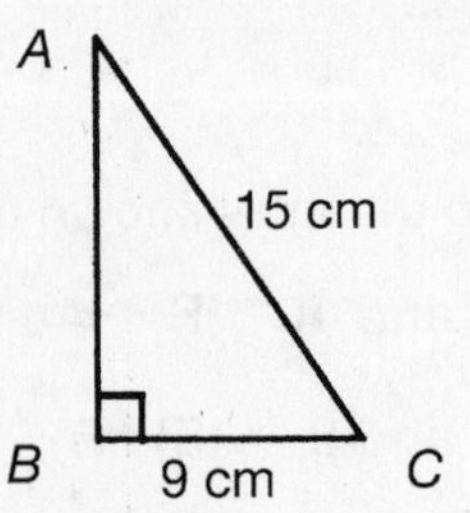

15. Find the area of ΔXYZ.

(A) 20 cm^2 (B) 50 cm^2 (C) $50\sqrt{2}$ cm^2
(D) 100 cm^2 (E) 200 cm^2

45°
$10\sqrt{2}$ cm
45°

Special Right Triangles

DIRECTIONS: Refer to the diagram and find the value of the variable.

16. Find the value of x.

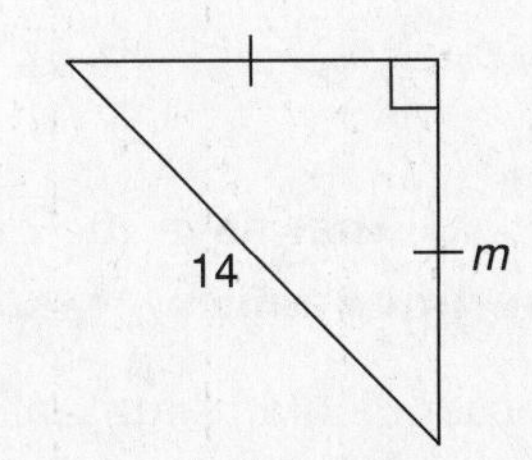

(A) 8 (B) 12 (C) $6\sqrt{2}$
(D) $6\sqrt{3}$ (E) $12\sqrt{2}$

17. Find the value of m.

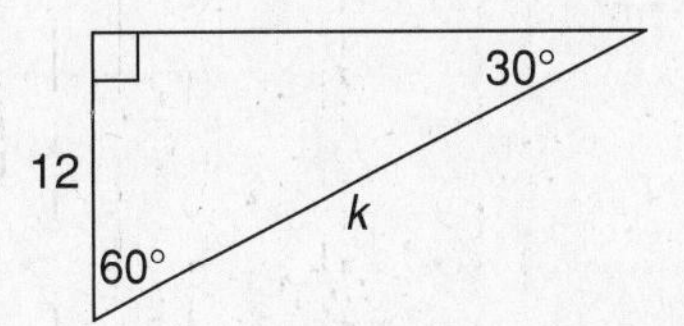

(A) 7 (B) 10 (C) $\sqrt{14}$
(D) $7\sqrt{2}$ (E) $14\sqrt{2}$

18. Find the value of k.

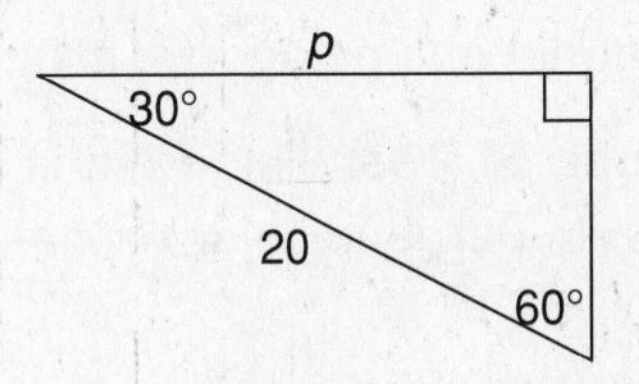

(A) 12 (B) 18 (C) 24
(D) $24\sqrt{3}$ (E) $24\sqrt{2}$

19. Find the value of p.

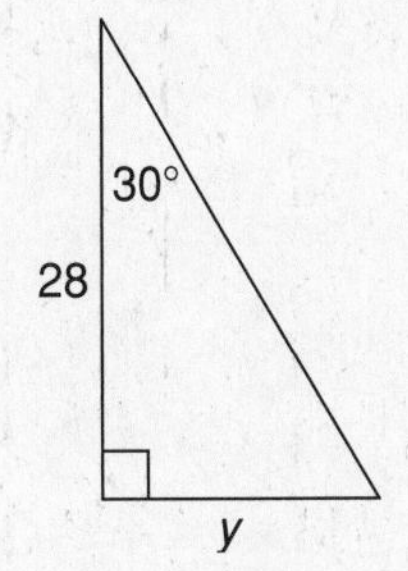

(A) 10 (B) $10\sqrt{3}$ (C) $20\sqrt{3}$
(D) $\frac{10\sqrt{3}}{3}$ (E) $\frac{20\sqrt{3}}{3}$

20. Find the value of y.

(A) 14 (B) $14\sqrt{3}$ (C) $28\sqrt{3}$
(D) $\frac{14\sqrt{3}}{3}$ (E) $\frac{28\sqrt{3}}{3}$

MASTERING QUADRILATERALS

A **quadrilateral** is a polygon with four sides.

Parallelograms

A **parallelogram** is a quadrilateral whose opposite sides are parallel.

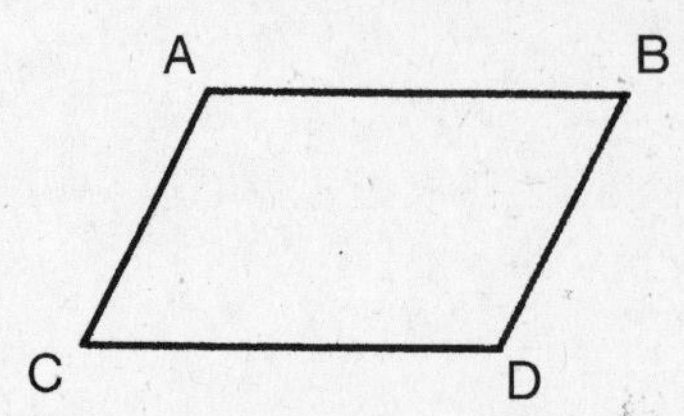

Two angles that have their vertices at the endpoints of the same side of a parallelogram are called **consecutive angles**.

The perpendicular segment connecting any point of a line containing one side of the parallelogram to the line containing the opposite side of the parallelogram is called the **altitude** of the parallelogram.

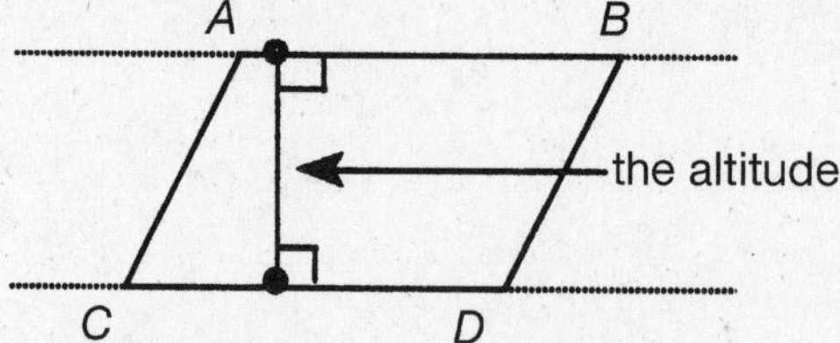

A diagonal of a polygon is a line segment joining any two nonconsecutive vertices.

The area of a parallelogram is given by the formula $A = bh$, where b is the base and h is the height drawn perpendicular to that base. Note that the height equals the altitude of the parallelogram.

$A = bh$

$A = (10)(3)$

$A = 30$

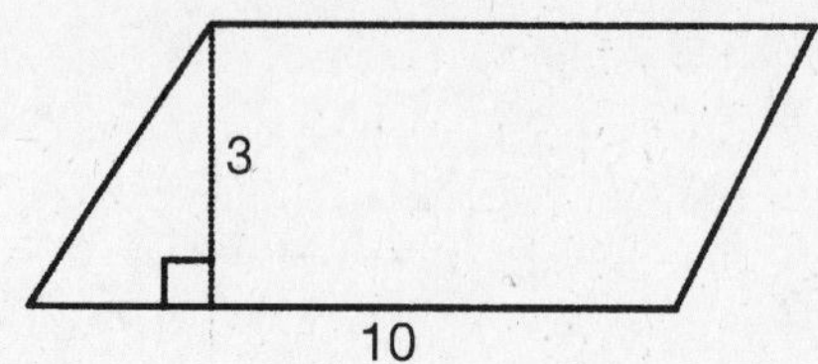

Rectangles

A rectangle is a parallelogram with right angles.

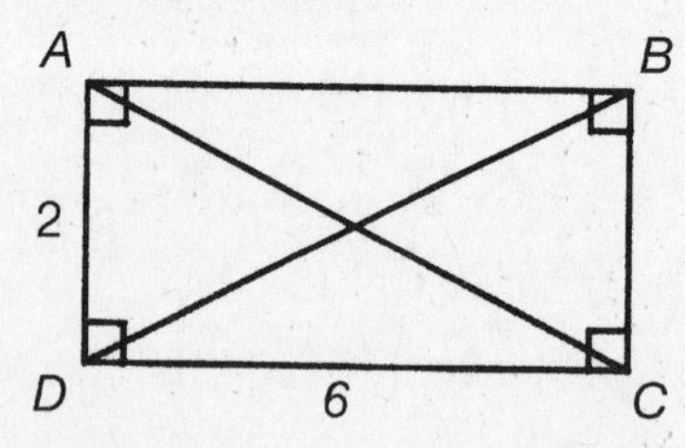

The diagonals of a rectangle are equal.

If the diagonals of a parallelogram are equal, the parallelogram is a rectangle.

If a quadrilateral has four right angles, then it is a rectangle.

The area of a rectangle is given by the formula $A = lw$, where l is the length and w is the width.

$A = lw$

$A = (3)(10)$

$A = 30$

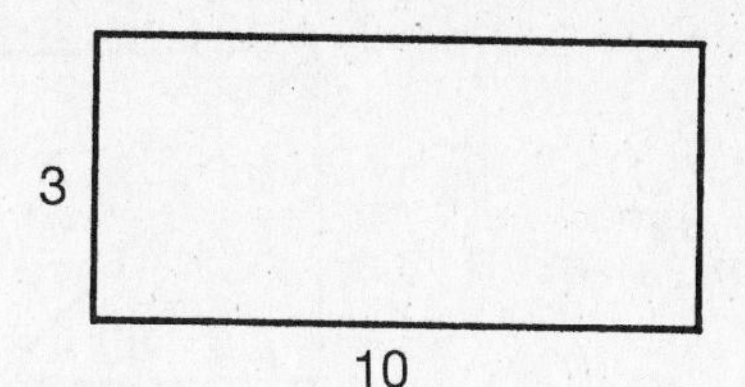

Rhombi

A rhombus is a parallelogram that has two adjacent sides that are equal.

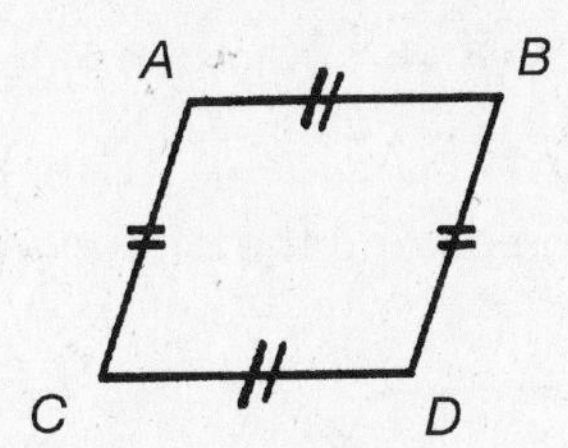

All sides of a rhombus are equal.

The diagonals of a rhombus are perpendicular to each other.

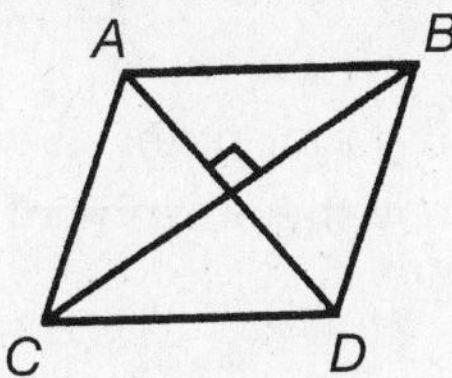

The diagonals of a rhombus bisect the angles of the rhombus.

If the diagonals of a parallelogram are perpendicular, the parallelogram is a rhombus.

If a quadrilateral has four equal sides, then it is a rhombus.

A parallelogram is a rhombus if either diagonal of the parallelogram bisects the angles of the vertices it joins.

Squares

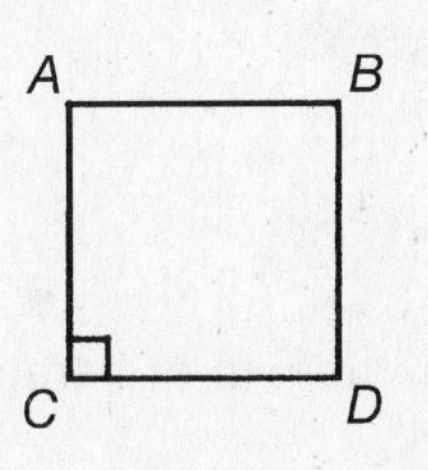

A square is a rhombus with a right angle.

A square is an equilateral quadrilateral.

A square has all the properties of parallelograms and rectangles.

A rhombus is a square if one of its interior angles is a right angle.

In a square, the measure of either diagonal can be calculated by multiplying the length of any side by the square root of 2.

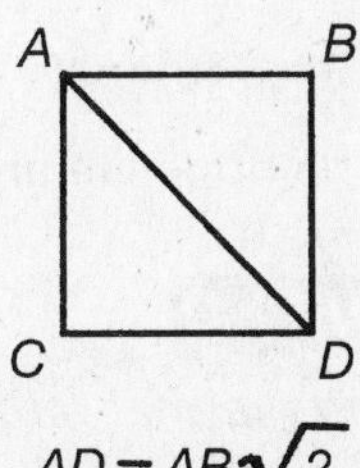

$AD = AB\sqrt{2}$

The area of a square is given by the formula $A = s^2$, where s is the side of the square. Since all sides of a square are equal, it does not matter which side is used.

$A = s^2$

$A = 6^2$

$A = 36$

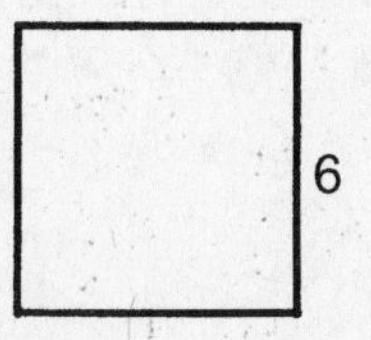

The area of a square can also be found by taking $\frac{1}{2}$ the product of the length of the diagonal squared.

$$A = \frac{1}{2}d^2$$

$$A = \frac{1}{2}(8)^2$$

$$A = 32$$

Trapezoids

A **trapezoid** is a quadrilateral with two and only two sides parallel. The parallel sides of a trapezoid are called **bases**.

The **median** of a trapezoid is the line joining the midpoints of the nonparallel sides.

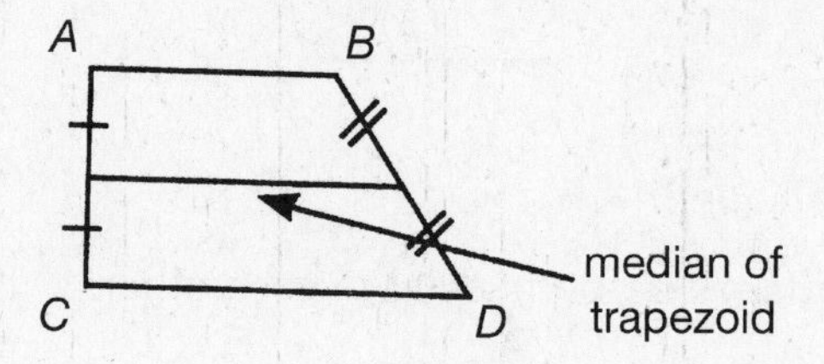

The perpendicular segment connecting any point in the line containing one base of the trapezoid to the line containing the other base is the **altitude** of the trapezoid.

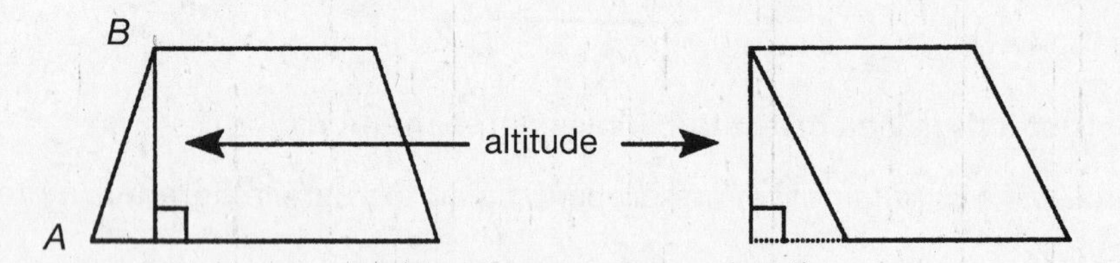

An **isosceles trapezoid** is a trapezoid whose nonparallel sides are equal. A pair of angles including only one of the parallel sides is called **a pair of base angles**.

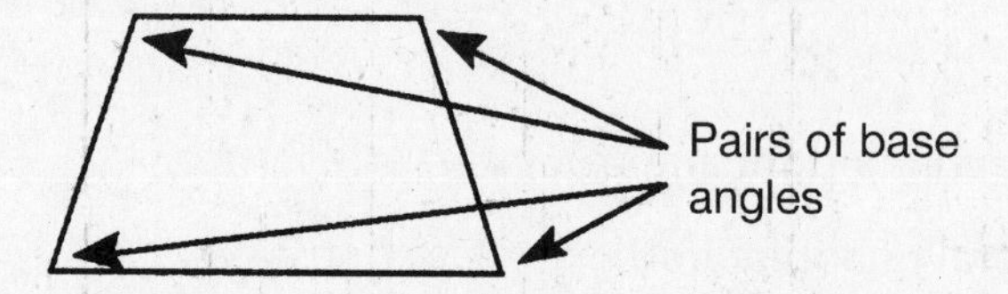

The median of a trapezoid is parallel to the bases and equal to one-half their sum.

The base angles of an isosceles trapezoid are equal.

The diagonals of an isosceles trapezoid are equal.

The opposite angles of an isosceles trapezoid are supplementary.

PROBLEM

Prove that all pairs of consecutive angles of a parallelogram are supplementary.

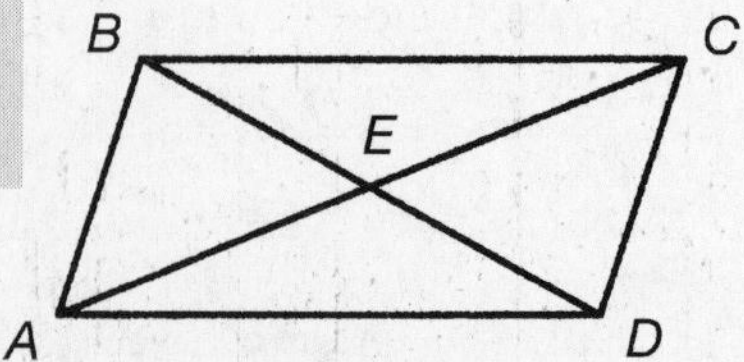

SOLUTION

We must prove that the pairs of angles $\angle BAD$ and $\angle ADC$, $\angle ADC$ and $\angle DCB$, $\angle DCB$ and $\angle CBA$, and $\angle CBA$ and $\angle BAD$ are supplementary. (This means that the sum of their measures is 180°.)

Because $ABCD$ is a parallelogram, $\overline{AB} \parallel \overline{CD}$. Angles BAD and ADC are consecutive interior angles, as are $\angle CBA$ and $\angle DCB$. Since the consecutive interior angles formed by two parallel lines and a transversal are supplementary, $\angle BAD$ and $\angle ADC$ are supplementary, as are $\angle CBA$ and $\angle DCB$.

Similarly, $\overline{AD} \parallel \overline{BC}$. Angles ADC and DCB are consecutive interior angles, as are $\angle CBA$ and $\angle BAD$. Since the consecutive interior angles formed by two parallel lines and a transversal are supplementary, $\angle CBA$ and $\angle BAD$ are supplementary, as are $\angle ADC$ and $\angle DCB$.

PROBLEM

In the accompanying figure, ΔABC is given to be an isosceles right triangle with $\angle ABC$ a right angle and $\overline{AB} \cong \overline{BC}$. Line segment $\overline{BD}$, which bisects $\overline{CA}$, is extended to E, so that $\overline{BD} \cong \overline{DE}$. Prove BAEC is a square.

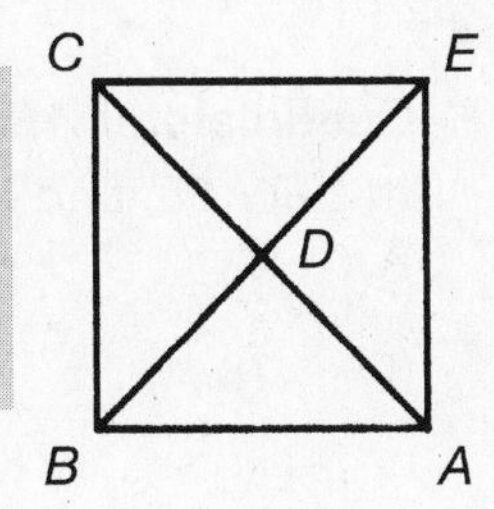

SOLUTION

A square is a rectangle in which two consecutive sides are congruent. This definition will provide the framework for the proof in this problem. We will prove that $BAEC$ is a parallelogram that is specifically a rectangle with consecutive sides congruent, namely a square.

	Statement		**Reason**
1.	$\overline{BD} \cong \overline{DE}$ and $\overline{AD} \cong \overline{DC}$	1.	Given ($\overline{BD}$ bisects $\overline{CA}$).
2.	$BAEC$ is a parallelogram	2.	If diagonals of a quadrilateral bisect each other, then the quadrilateral is a parallelogram.
3.	$\angle ABC$ is a right angle	3.	Given.
4.	$BAEC$ is a rectangle	4.	A parallelogram, one of whose angles is a right angle, is a rectangle.
5.	$\overline{AB} \cong \overline{BC}$	5.	Given.
6.	$BAEC$ is a square	6.	If a rectangle has two congruent consecutive sides, then the rectangle is a square.

DRILL: QUADRILATERALS

Parallelograms, Rectangles, Rhombi, Squares, and Trapezoids

DIRECTIONS: Refer to the diagram and find the appropriate solution.

1. In parallelogram $WXYZ$, $\overline{WX} = 14$, $\overline{WZ} = 6$, $\overline{ZY} = 3x + 5$, and $\overline{XY} = 2y - 4$. Find x and y.

   (A) 3 and 5 (B) 4 and 5 (C) 4 and 6
   (D) 6 and 10 (E) 6 and 14

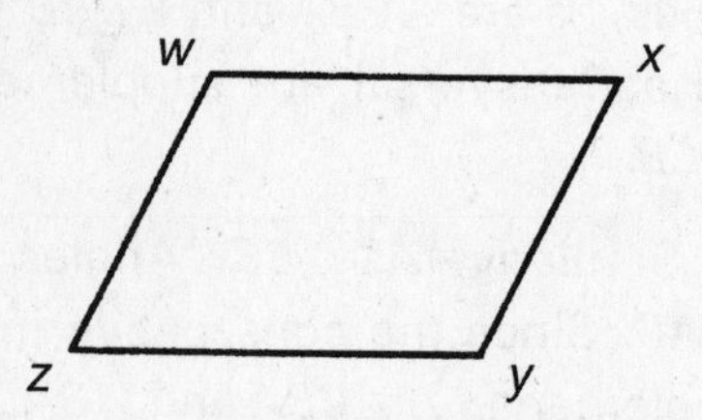

2. Quadrilateral $ABCD$ is a parellelogram. If $m\angle B = 6x + 2$ and $m\angle D = 98$, find x.

   (A) 12 (B) 16 (C) $16\frac{2}{3}$
   (D) 18 (E) 20

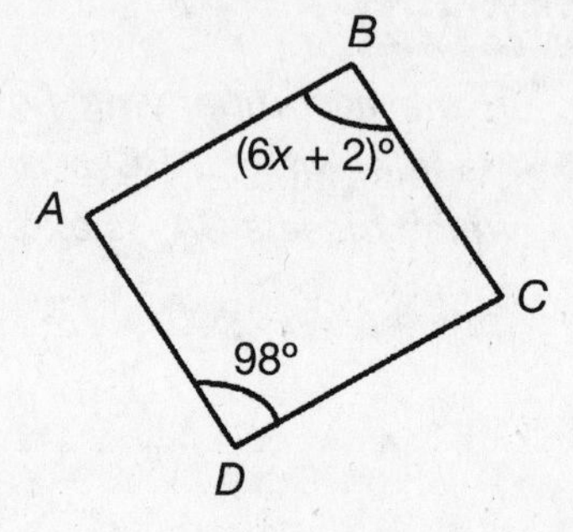

3. Find the area of parallelogram $STUV$.

   (A) 56 (B) 90 (C) 108
   (D) 162 (E) 180

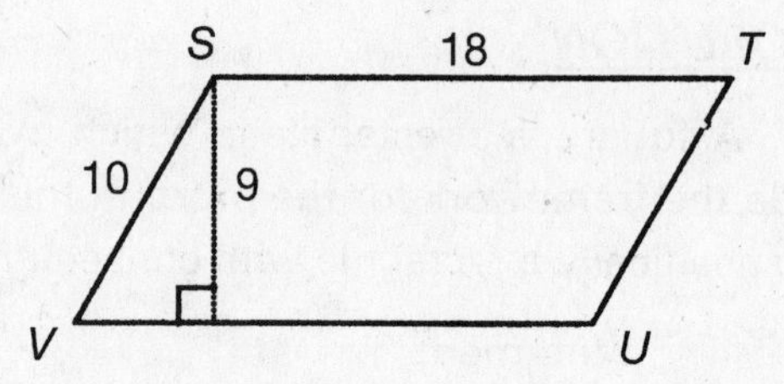

4. Find the area of parallelogram $MNOP$.

   (A) 19 (B) 32 (C) $32\sqrt{3}$
   (D) 44 (E) $44\sqrt{3}$

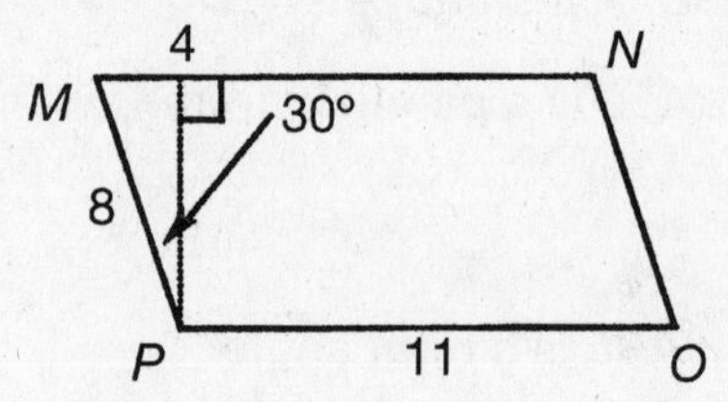

5. If the perimeter of rectangle $PQRS$ is 40, find x.

   (A) 31 in. (B) 38 in. (C) 2 in.
   (D) 44 in. (E) 121 in.

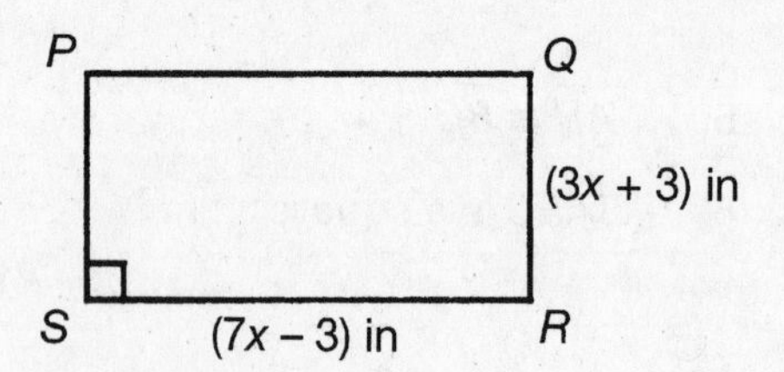

6. In rectangle $ABCD$, $\overline{AD} = 6$ cm and $\overline{DC} = 8$ cm. Find the length of the diagonal $\overline{AC}$.

   (A) 10 cm (B) 12 cm (C) 20 cm
   (D) 28 cm (E) 48 cm

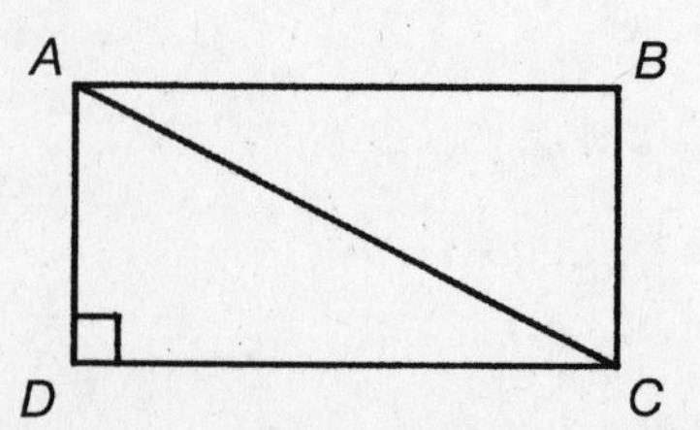

7. Find the area of rectangle *UVXY*.

(A) 17 cm^2 (B) 34 cm^2 (C) 35 cm^2
(D) 70 cm^2 (E) 140 cm^2

8. Find the length of $\overline{BO}$ in rectangle *BCDE* if the diagonal $\overline{EC}$ is 17 mm.

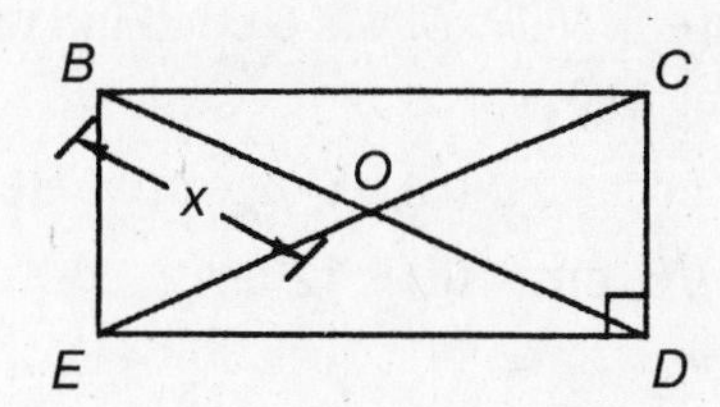

(A) 6.55 mm (B) 8 mm (C) 8.5 mm
(D) 17 mm (E) 34 mm

9. In rhombus *DEFG*, $\overline{DE}$ = 7 cm. Find the perimeter of the rhombus.

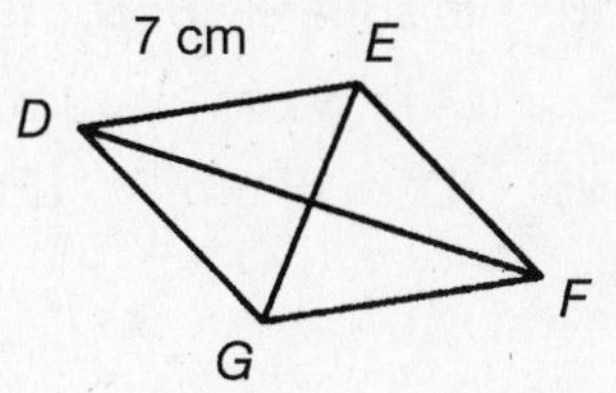

(A) 14 cm (B) 28 cm (C) 42 cm
(D) 49 cm (E) 56 cm

10. In rhombus *RHOM*, the diagonal $\overline{RO}$ is 8 cm and the diagonal $\overline{HM}$ is 12 cm. Find the area of the rhombus.

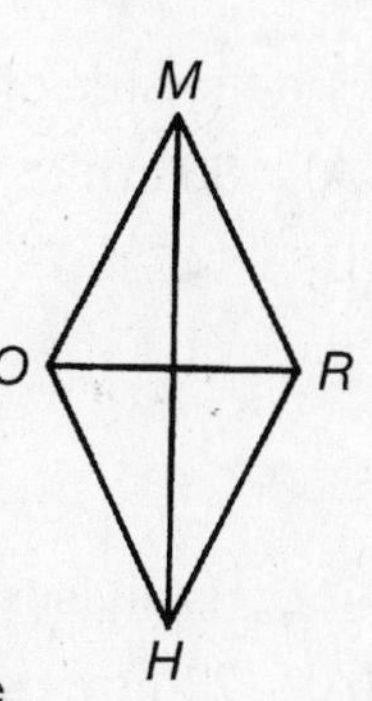

(A) 20 cm^2 (B) 40 cm^2 (C) 48 cm^2
(D) 68 cm^2 (E) 96 cm^2

11. In rhombus *GHIJ*, $\overline{GI}$ = 6 cm and $\overline{HJ}$ = 8 cm. Find the length of $\overline{GH}$.

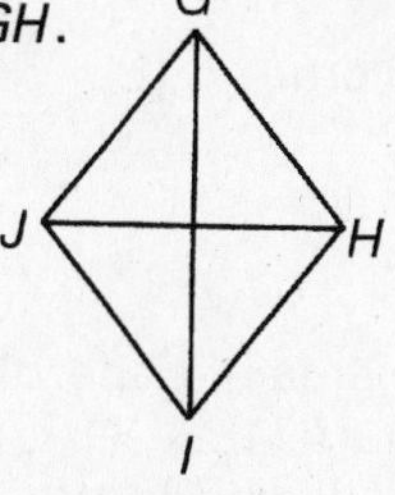

(A) 3 cm (B) 4 cm (C) 5 cm
(D) 4 cm (E) 14 cm

12. In rhombus *CDEF*, $\overline{CD}$ is 13 mm and $\overline{DX}$ is 5 mm. Find the area of the rhombus.

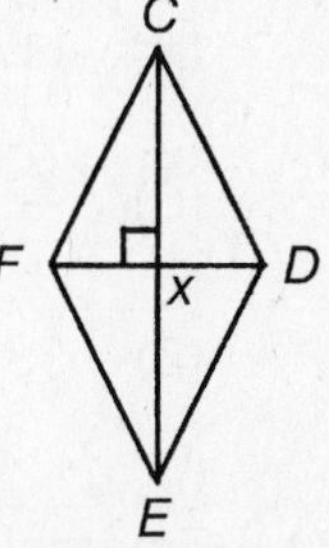

(A) 31 mm^2 (B) 60 mm^2 (C) 78 mm^2
(D) 120 mm^2 (E) 260 mm^2

13. Quadrilateral *ATUV* is a square. If the perimeter of the square is 44 cm, find the length of $\overline{AT}$.

(A) 4 cm (B) 11 cm (C) 22 cm (D) 30 cm (E) 40 cm

14. The area of square $XYZW$ is 196 cm^2. Find the perimeter of the square.

(A) 28 cm (B) 42 cm (C) 56 cm

(D) 98 cm (E) 196 cm

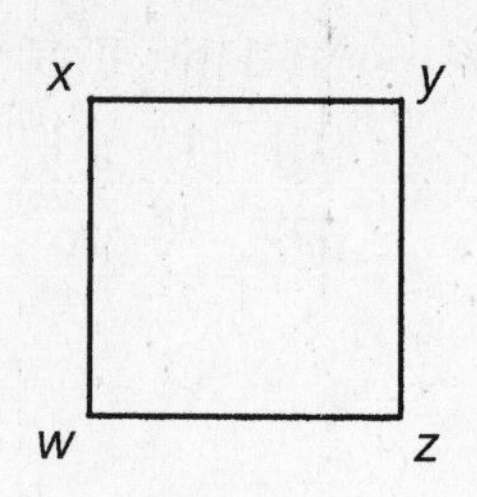

15. In square $MNOP$, $\overline{MN}$ is 6 cm. Find the length of diagonal $\overline{MO}$.

(A) 6 cm (B) $6\sqrt{2}$ cm (C) $6\sqrt{3}$ cm

(D) $6\sqrt{6}$ cm (E) 12 cm

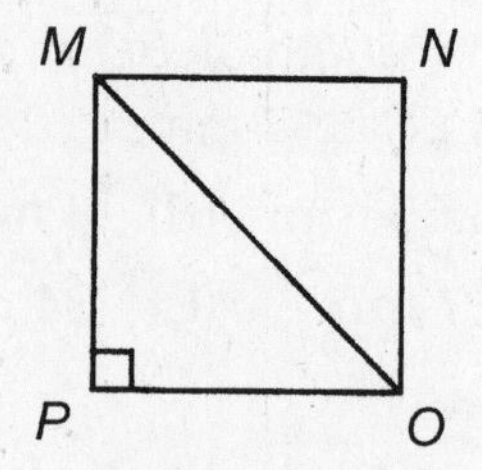

16. In square $ABCD$, $\overline{AB} = 3$ cm. Find the area of the square.

(A) 9 cm^2 (B) 12 cm^2 (C) 15 cm^2

(D) 18 cm^2 (E) 21 cm^2

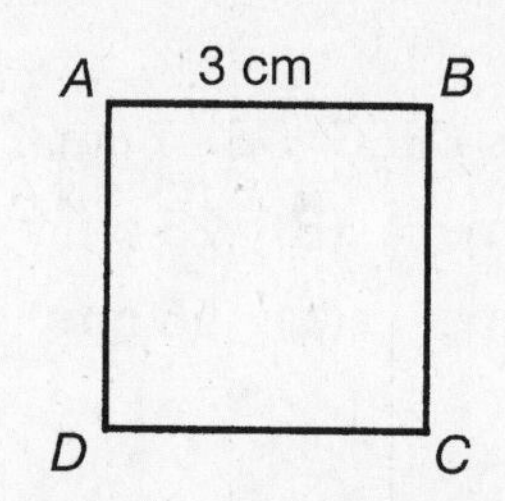

17. Find the area of trapezoid $RSTU$.

(A) 80 cm^2 (B) 87.5 cm^2 (C) 140 cm^2

(D) 147 cm^2 (E) 175 cm^2

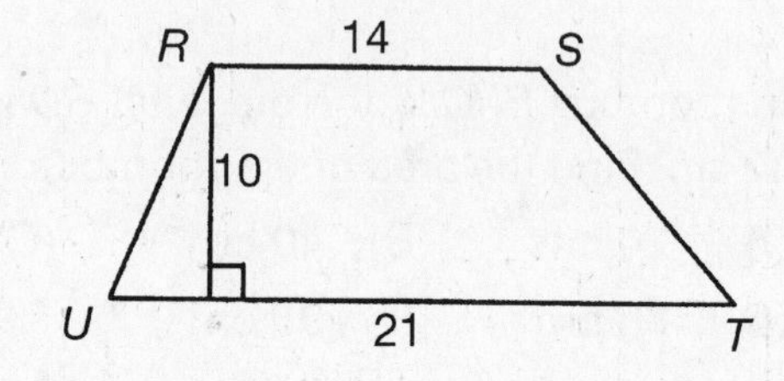

18. $ABCD$ is an isosceles trapezoid. Find the perimeter.

(A) 21 cm (B) 27 cm (C) 30 cm

(D) 50 cm (E) 54 cm

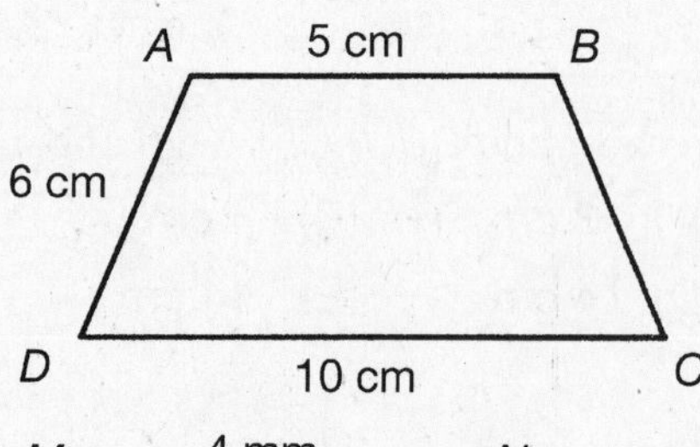

19. Find the area of trapezoid $MNOP$.

(A) $(17 + 3\sqrt{3})$ mm^2 (B) $\frac{33}{2}$ mm^2

(C) $\frac{33\sqrt{3}}{2}$ mm^2 (D) 33 mm^2

(E) $33\sqrt{3}$ mm^2

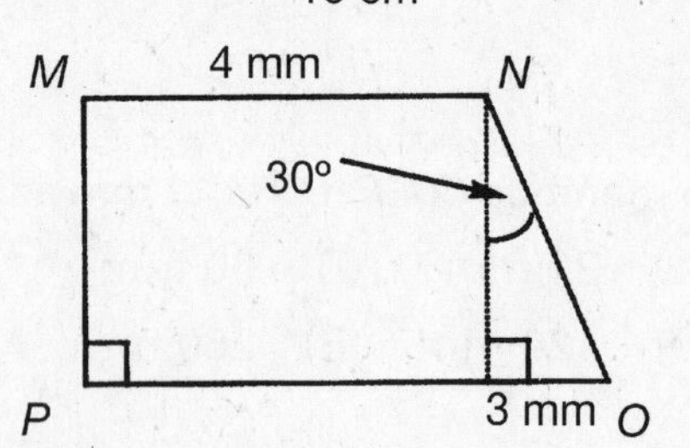

20. Trapezoid $XYZW$ is isosceles. If $m\angle W = 58°$ and $m\angle Z = (4x - 6)$, find $x°$.

(A) 8° (B) 12° (C) 13°

(D) 16° (E) 58°

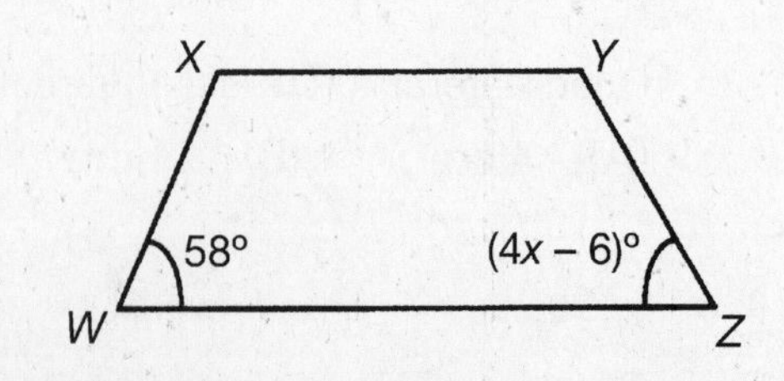

MASTERING CIRCLES

A **circle** is a set of points in the same plane equidistant from a fixed point, called its center.

A **radius** of a circle is a line segment drawn from the center of the circle to any point on the circle.

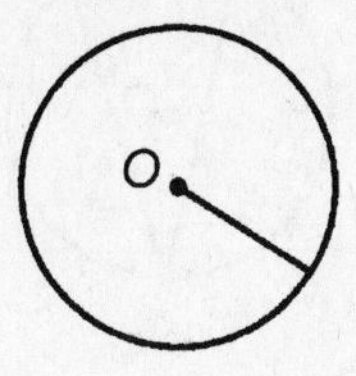

A portion of a circle is called an **arc** of the circle.

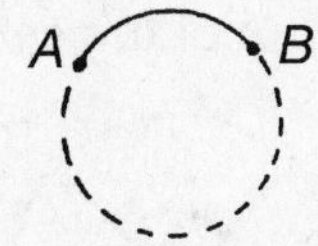

A line that intersects a circle in two points is called a **secant**.

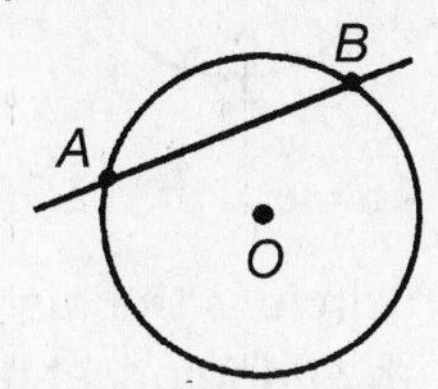

A line segment joining two points on a circle is called a **chord** of the circle.

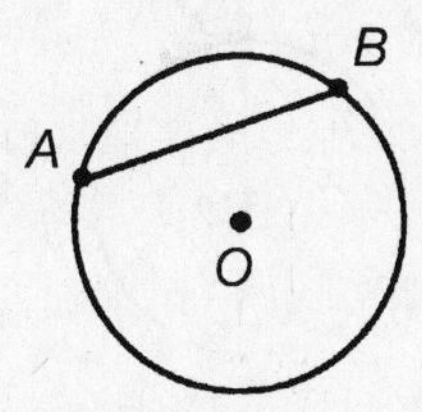

A chord that passes through the center of the circle is called a **diameter** of the circle.

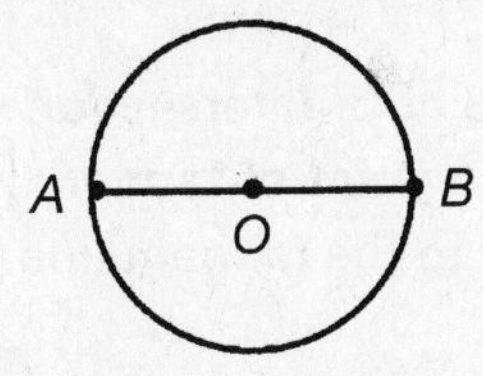

The line passing through the centers of two (or more) circles is called the **line of centers**.

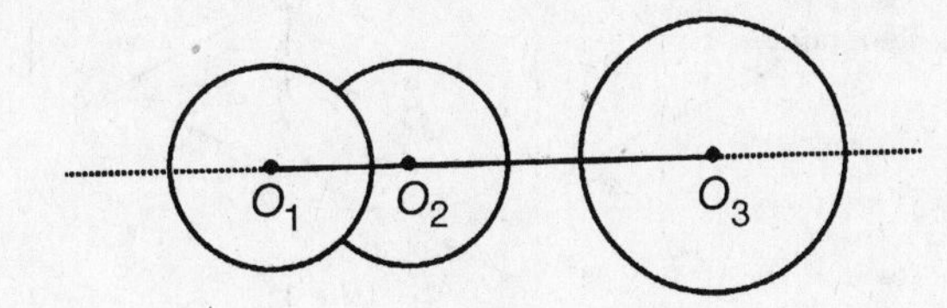

An angle whose vertex is on the circle and whose sides are chords of the circle is called an **inscribed angle**.

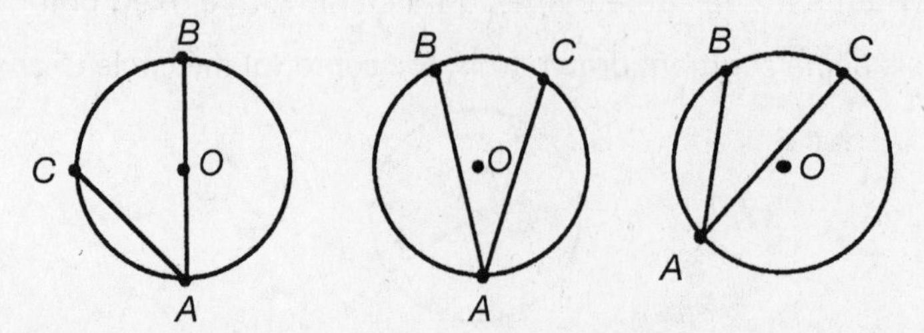

An angle whose vertex is at the center of a circle and whose sides are radii is called a **central angle**.

The measure of a minor arc is the measure of the central angle that intercepts that arc.

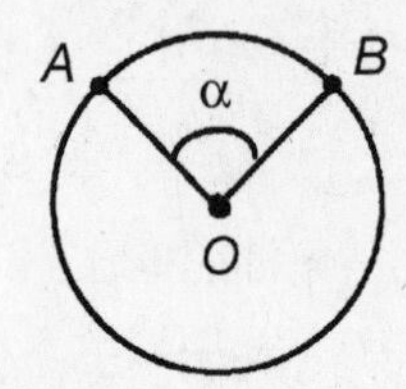

$$m\,\widehat{AB} = \alpha = m\,\angle AOB$$

The distance from a point P to a given circle is the distance from that point to the point where the circle intersects with a line segment with endpoints at the center of the circle and point P.

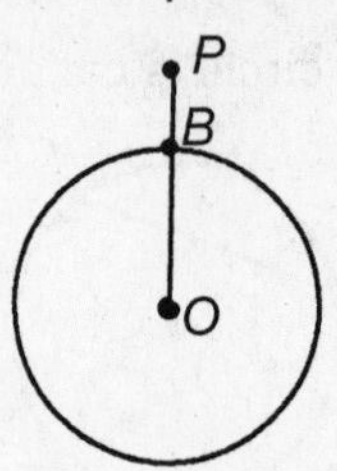

The distance of point P to the diagrammed circle with center O is the line segment $\overline{PB}$ of line segment $\overline{PO}$.

A line that has one and only one point of intersection with a circle is called a **tangent** to that circle, and their common point is called a **point of tangency**. A radius that intersects a tangent line at the point of tangency is perpendicular to the tangent line.

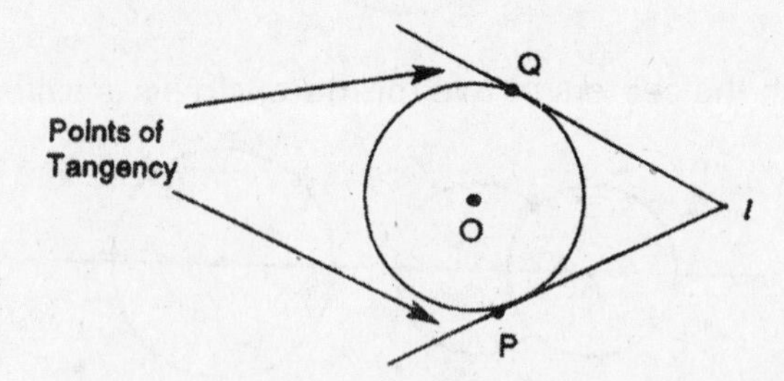

Congruent circles are circles whose radii are congruent.

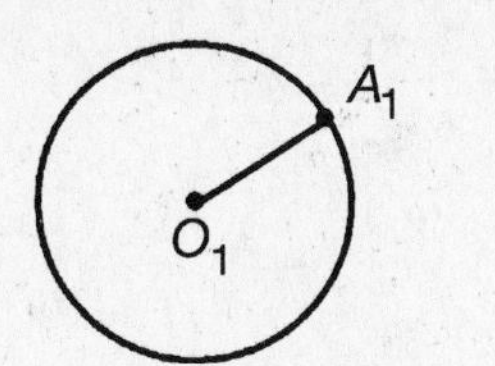

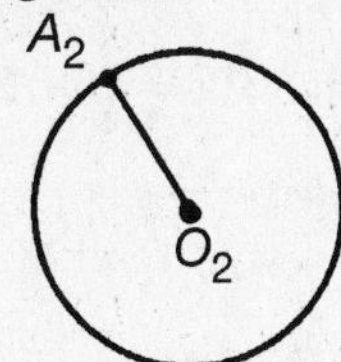

If $O_1A_1 \cong O_2A_2$, then $O_1 \cong O_2$.

The measure of a semicircle is 180°.

A **circumscribed circle** is a circle passing through all the vertices of a polygon.

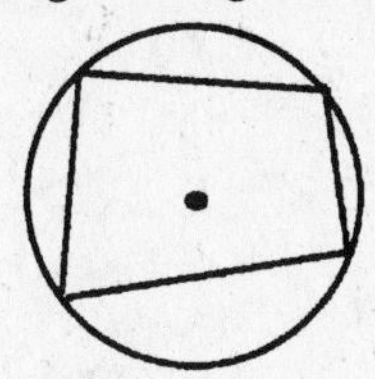

Circles that have the same center and unequal radii are called **concentric circles.**

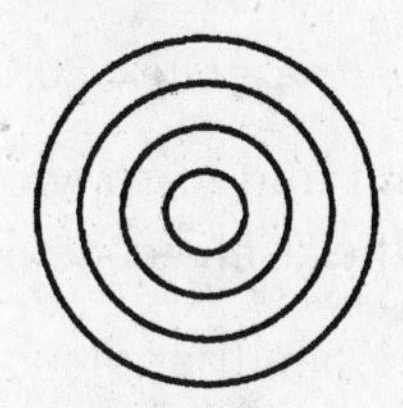

Concentric Circles

PROBLEM

A and B are points on circle Q such that ΔAQB is equilateral. If the length of side $\overline{AB} = 12$, find the length of arc AB.

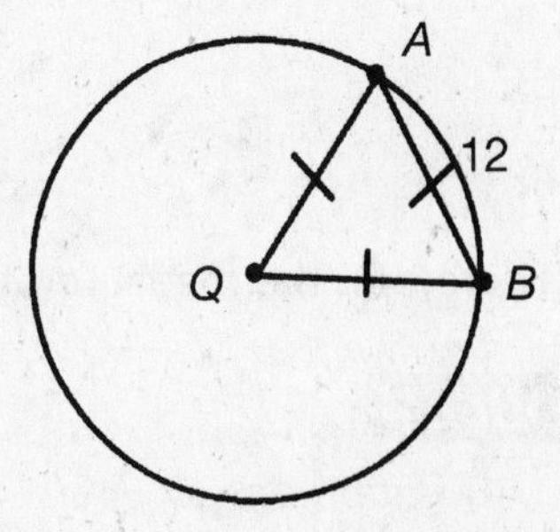

SOLUTION

To find the arc length of arc AB, we must find the measure of the central angle $\angle AQB$ and the measure of the radius $\overline{QA}$. $\angle AQB$ is an interior angle of the equilateral triangle ΔAQB. Therefore,

$$m\angle AQB = 60°.$$

Similarly, in the equilateral ΔAQB,

$$\overline{AQ} = \overline{AB} = \overline{QB} = 12.$$

Given the radius, r, and the central angle, n, the arc length is given by

$$\frac{n}{360} \times 2\pi r$$

Therefore, by substitution,

$$\angle AQB = \frac{60}{360} \times 2\pi \times 12 = \frac{1}{6} \times 2\pi \times 12 = 4\pi.$$

Therefore, the length of arc $AB = 4\pi$.

PROBLEM

In circle O, the measure of arc AB is 80°. Find the measure of $\angle A$.

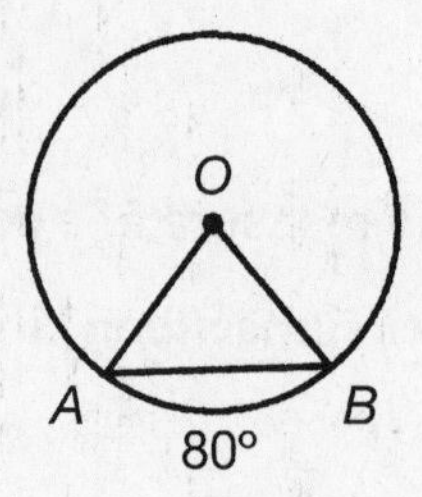

SOLUTION

The accompanying figure shows that arc *AB* is intercepted by central angle *AOB*. By definition, we know that the measure of the central angle is the measure of its intercepted arc. In this case,

$$\text{arc } AB = m\angle AOB = 80°.$$

Radius $\overline{OA}$ and radius $\overline{OB}$ are congruent and form two sides of ΔOAB. By a theorem, the angles opposite these two congruent sides must, themselves, be congruent. Therefore, $m\angle A = m\angle B$.

The sum of the measures of the angles of a triangle is 180°. Therefore,

$$m\angle A + m\angle B + m\angle AOB = 180°.$$

Since $m\angle A = m\angle B$, we can write

$$m\angle A + m\angle A + 80° = 180°$$

or

$$2m\angle A = 100°$$

or

$$m\angle A = 50°.$$

Therefore, the measure of $\angle A$ is 50°.

PROBLEM

$\overline{AB}$ is tangent to circle C at point A. If AB = 15 and BC = 17, what is the length of the radius of circle C?

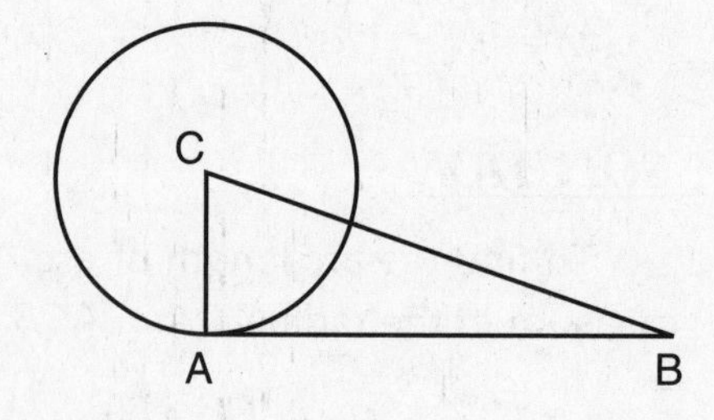

SOLUTION

Because $\overline{AB}$ is tangent to circle C at point A, radius AC is perpendicular to $\overline{AB}$. Therefore, ΔABC is a right triangle with a right angle at A. Use the Pythagorean Theorem to find the length of radius AC.

$AB^2 + AC^2 = BC^2$	
$15^2 + AC^2 = 17^2$	Substitute the known values.
$225 + AC^2 = 289$	Simplify.
$AC^2 = 64$	Subtract 225 from each side.
$AC = 8$	Take the square root of each side.

Because $AC = 8$, circle C has a radius of length 8.

DRILL: CIRCLES

Circumference, Area, Concentric Circles, Tangent Lines

DIRECTIONS: Determine the accurate measure.

1. Find the circumference of circle A if its radius is 3 mm.

   (A) 3π mm (B) 6π mm (C) 9π mm (D) 12π mm (E) 15π mm

2. The circumference of circle H is 20π cm. Find the length of the radius.

   (A) 10 cm (B) 20 cm (C) 10π cm (D) 15π cm (E) 20π cm

3. The circumference of circle A is how many millimeters larger than the circumference of circle B?

   (A) 3 mm (B) 6 mm (C) 3π mm
   (D) 6π mm (E) 7π mm

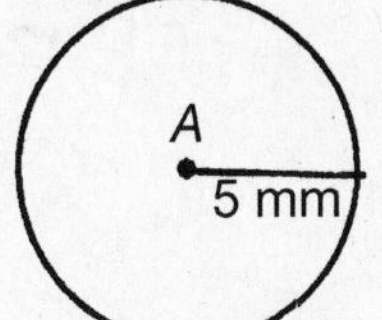

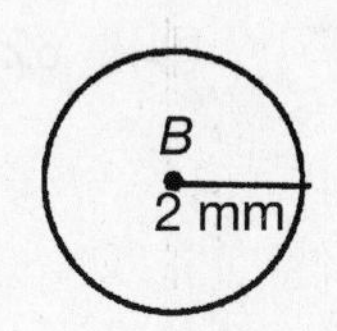

4. If the diameter of circle X is 9 cm and if $\pi = 3.14$, find the circumference of the circle to the nearest tenth.

   (A) 9 cm (B) 14.1 cm (C) 21.1 cm (D) 24.6 cm (E) 28.3 cm

5. Find the area of circle I.

   (A) 22 mm^2 (B) 121 mm^2
   (C) 121π mm^2 (D) 132 mm^2
   (E) 132π mm^2

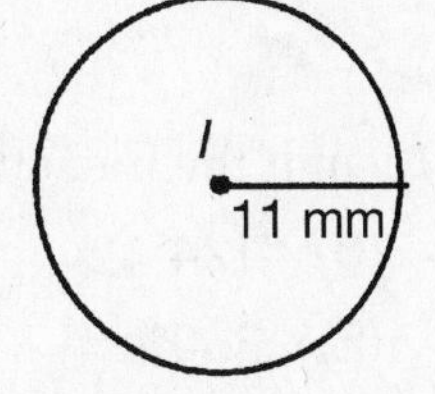

6. The diameter of circle Z is 27 mm. Find the area of the circle.

   (A) 91.125 mm^2 (B) 182.25 mm^2 (C) 191.5π mm^2
   (D) 182.25π mm^2 (E) 729 mm^2

7. The area of circle B is 225π cm^2. Find the length of the diameter of the circle.

(A) 15 cm (B) 20 cm (C) 30 cm (D) 20π cm (E) 25π cm

8. The area of circle X is 144π mm^2 while the area of circle Y is 81π mm^2. Write the ratio of the radius of circle X to that of circle Y.

(A) 3 : 4 (B) 4 : 3 (C) 9 : 12 (D) 27 : 12 (E) 18 : 24

9. The circumference of circle M is 18π cm. Find the area of the circle.

(A) 18π cm^2 (B) 81 cm^2 (C) 36 cm^2 (D) 36π cm^2 (E) 81π cm^2

10. In two concentric circles, the smaller circle has a radius of 3 mm while the larger circle has a radius of 5 mm. Find the area of the shaded region.

(A) 2π mm^2 (B) 8π mm^2
(C) 13π mm^2 (D) 16π mm^2
(E) 26π mm^2

11. The radius of the smaller of two concentric circles is 5 cm while the radius of the larger circle is 7 cm. Determine the area of the shaded region.

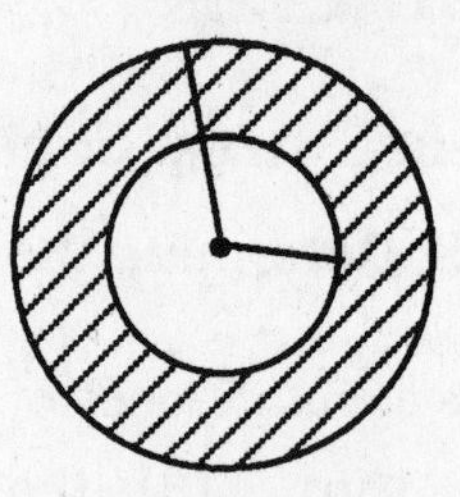

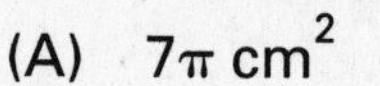

(A) 7π cm^2 (B) 24π cm^2
(C) 25π cm^2 (D) 36π cm^2
(E) 49π cm^2

12. Find the measure of arc MN if $m\angle MON = 62°$.

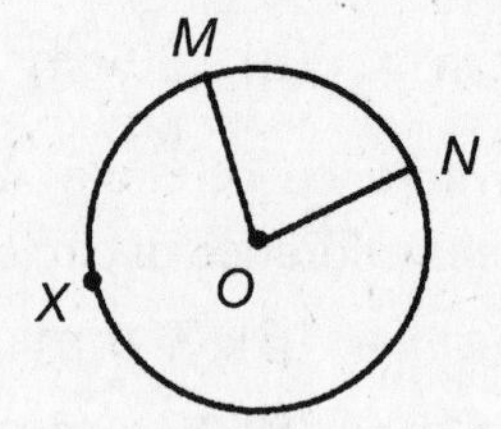

(A) 16° (B) 32° (C) 59°
(D) 62° (E) 124°

13. Find the measure of arc AXC.

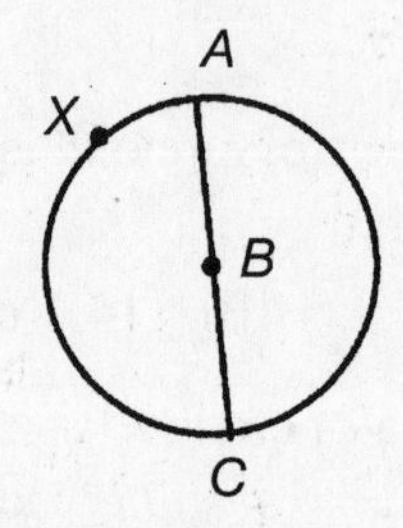

(A) 150° (B) 160° (C) 180°
(D) 270° (E) 360°

14. If arc MXP = 236°, find the measure of arc MP.

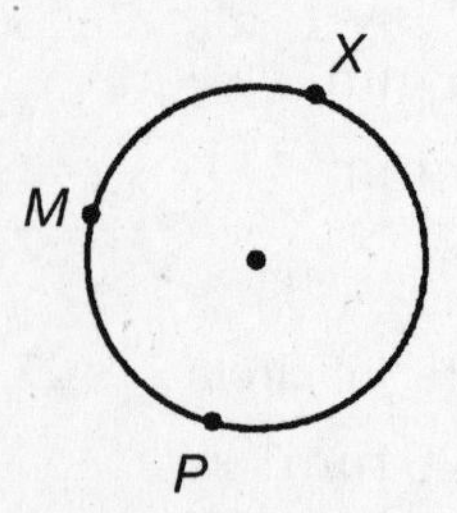

(A) 62° (B) 124° (C) 236°
(D) 270° (E) 360°

15. In circle S, major arc PQR has a measure of 298°. Find the measure of the central angle $\angle PSR$.

(A) 62° (B) 124° (C) 149°
(D) 298° (E) 360°

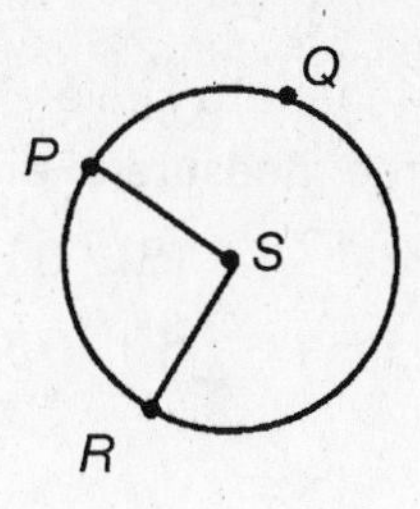

16. Find the measure of arc XY in circle W.

(A) 40° (B) 120° (C) 140°
(D) 180° (E) 220°

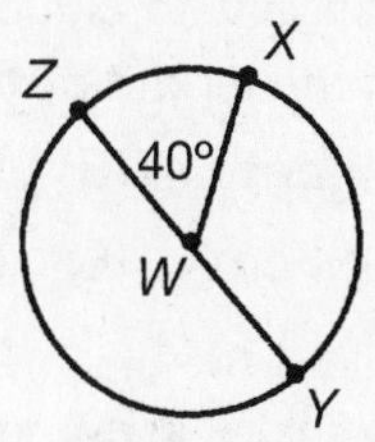

17. Find the area of the sector shown.

(A) 4 cm^2 (B) $2\pi \text{ cm}^2$ (C) 16 cm^2
(D) $8\pi \text{ cm}^2$ (E) $16\pi \text{ cm}^2$

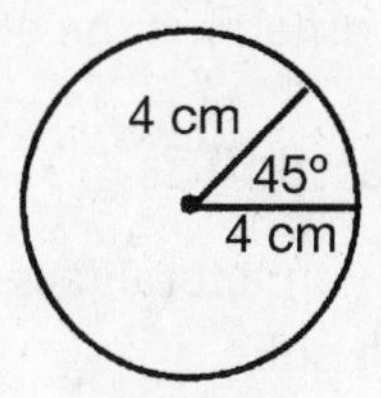

18. Find the area of the shaded region.

(A) 10 (B) 5π (C) 25
(D) 20π (E) 25π

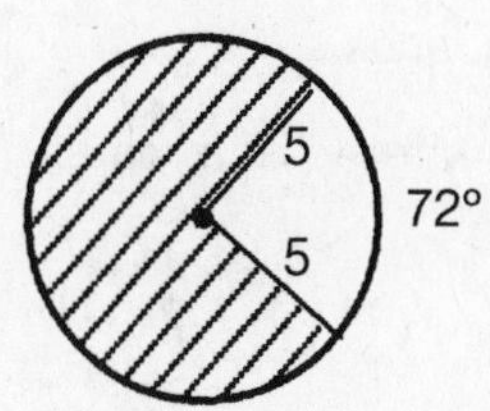

19. Find the area of the shaded sector shown.

(A) $\dfrac{9\pi \text{ mm}^2}{4}$ (B) $\dfrac{9\pi \text{ mm}^2}{2}$ (C) 18 mm^2
(D) $6\pi \text{ mm}^2$ (E) $9\pi \text{ mm}^2$

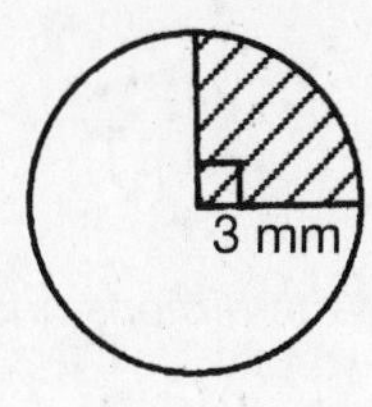

20. If the area of the square is 100 cm^2, find the area of the shaded sector.

(A) $10\pi \text{ cm}^2$ (B) 25 cm^2 (C) $25\pi \text{ cm}^2$
(D) 100 cm^2 (E) $100\pi \text{ cm}^2$

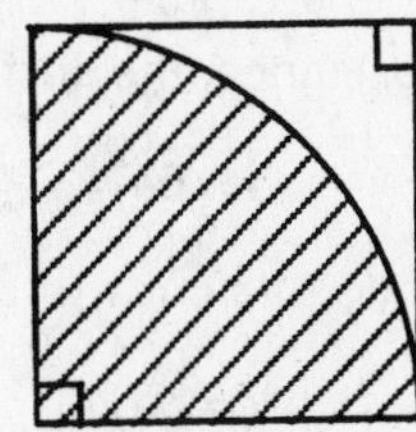

21. $\overline{LM}$ is tangent to circle P at point L. If the measure of $\angle M$ is 28°, what is the measure of arc LN?

(A) 28° (B) 31° (C) 56°

(D) 62° (E) 76°

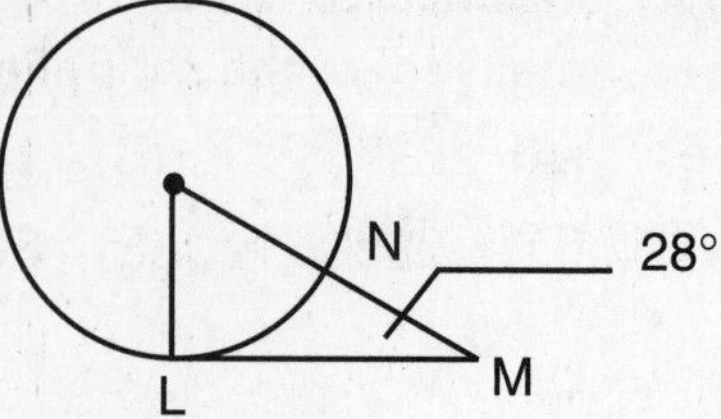

22. $\overline{KL}$ is tangent to circle J at point K. If the measure of $\angle KJL$ is 45° and the length of $\overline{KL}$ is 4 cm, what is the area of circle J?

(A) 4π cm^2 (B) 8π cm^2 (C) 16π cm^2

(D) 32π cm^2 (E) 64π cm^2

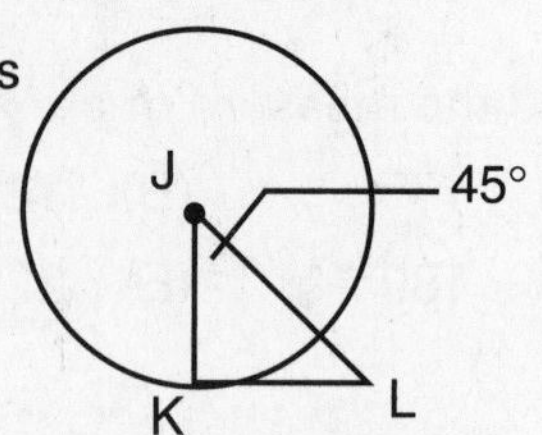

MASTERING SOLIDS

Solid geometry is the study of figures that consist of points not all in the same plane.

Rectangular Solids

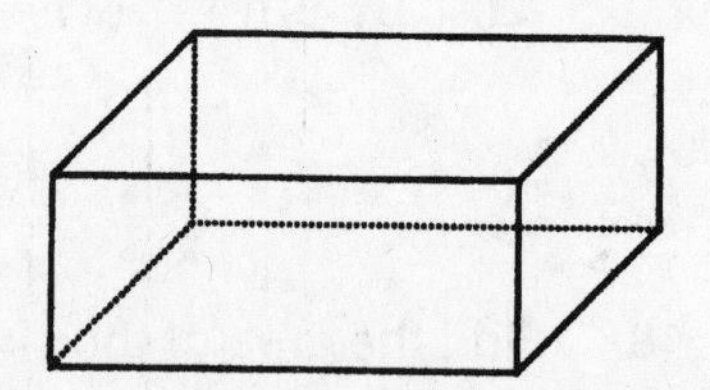

A solid with lateral faces and bases that are rectangles is called a **rectangular solid.**

The surface area of a rectangular solid is the sum of the areas of all the faces.

The volume of a rectangular solid is equal to the product of its length, width, and height.

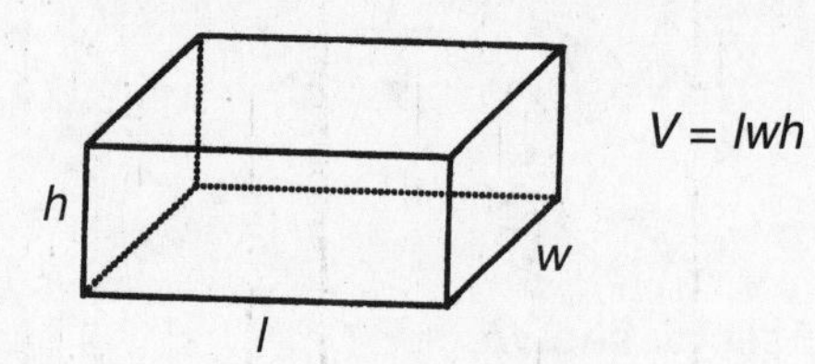

$V = lwh$

PROBLEM

What are the dimensions of a solid cube whose surface area is numerically equal to its volume?

SOLUTION

The surface area of a cube of edge length a is equal to the sum of the areas of its six faces. Since a cube is a regular polygon, all six faces are congruent. Each face of a cube is a square of edge length a. Hence, the surface area of a cube of edge length a is

$$S = 6a^2$$

The volume of a cube of edge length a is

$$V = a^3$$

We require that $A = V$, or that

$$6a^2 = a^3 \quad \text{or} \quad a = 6$$

Hence, if a cube has edge length 6, its surface area will be numerically equal to its volume.

DRILL: SOLIDS

Area and Volume

DIRECTIONS: Refer to the diagram and find the appropriate solution.

1. Find the surface area of the rectangular prism shown.

   (A) 138 cm^2 (B) 336 cm^2 (C) 381 cm^2

   (D) 426 cm^2 (E) 540 cm^2

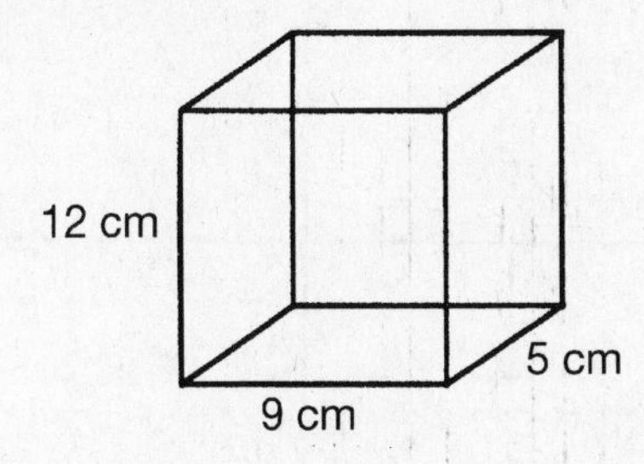

2. Find the volume of the rectangular storage tank shown.

   (A) 24 m^3 (B) 36 m^3 (C) 38 m^3

   (D) 42 m^3 (E) 45 m^3

1.5 m
4 m
6 m

3. The area of a side of a cube is 100 cm^2. Find the length of an edge of the cube.

   (A) 4 cm (B) 5 cm (C) 10 cm (D) 12 cm (E) 15 cm

MASTERING COORDINATE GEOMETRY

Coordinate geometry refers to the study of geometric figures using algebraic principles.

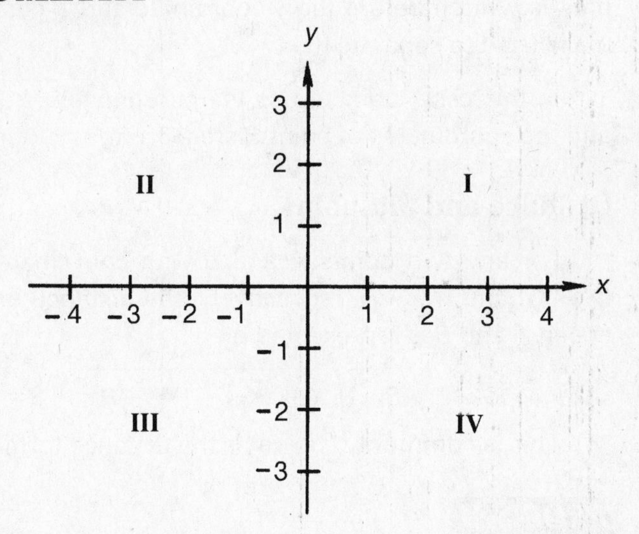

The graph shown is called the Cartesian coordinate plane. The graph consists of a pair of perpendicular lines called **coordinate axes.** The **vertical axis** is the *y*-axis and the **horizontal axis** is the *x*-axis. The point of intersection of these two axes is called the **origin**; it is the zero point of both axes. Furthermore, points to the right of the origin on the *x*-axis and above the origin on the *y*-axis represent positive real numbers. Points to the left of the origin on the *x*-axis or below the origin on the *y*-axis represent negative real numbers.

The four regions cut off by the coordinate axes are, in counterclockwise direction from the top right, called the first, second, third, and fourth quadrants, respectively. The first quadrant contains all points with two positive coordinates.

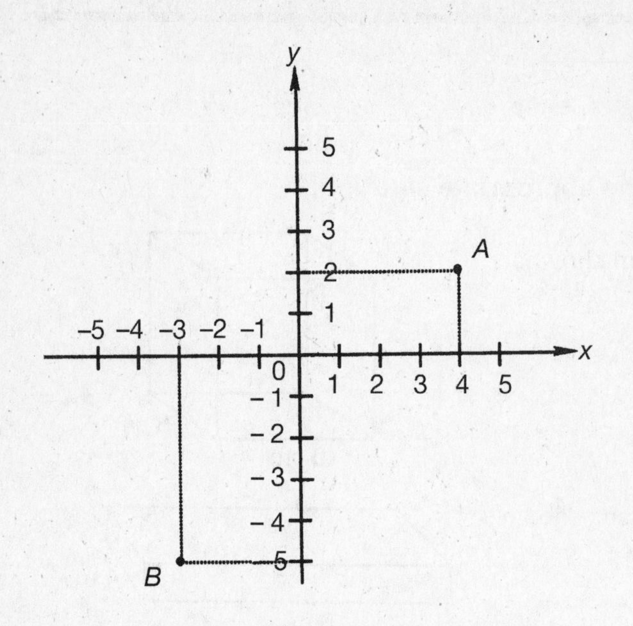

In the graph shown, two points are identified by the ordered pair, (x, y) of numbers. The x-coordinate is the first number and the y-coordinate is the second number.

To plot a point on the graph when given the coordinates, draw perpendicular lines from the number-line coordinates to the point where the two lines intersect.

To find the coordinates of a given point on the graph, draw perpendicular lines from the point to the coordinates on the number line. The x-coordinate is written before the y-coordinate and a comma is used to separate the two.

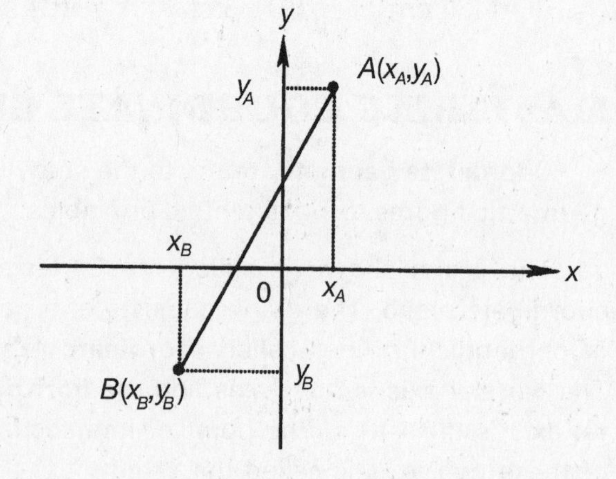

In this case, point A has the coordinates (4, 2) and the coordinates of point B are (−3, −5).

Distance and Midpoint

For any two points A and B with coordinates (X_A, Y_A) and (X_B, Y_B), respectively, the distance between A and B is represented by:

$$AB = \sqrt{(X_A - X_B)^2 + (Y_A - Y_B)^2}$$

This is commonly known as the distance formula.

PROBLEM

Find the distance between points A(1, 3) and B(5, 3).

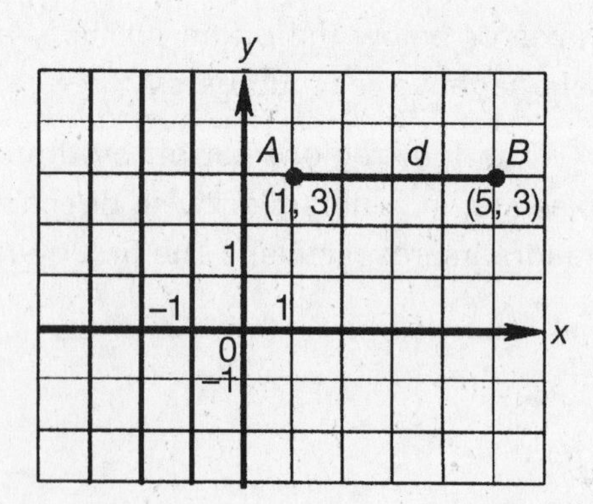

SOLUTION

In this case, where the ordinate of both points is the same, the distance between the two points is given by the absolute value of the difference between the two abscissas. In fact, this case reduces to merely counting boxes as the figure shows.

Let x_1 = abscissa of A y_1 = ordinate of A

x_2 = abscissa of B y_2 = ordinate of B

d = the distance

Therefore, $d = | x_1 - x_2 |$. By substitution, $d = | 1 - 5 | = | - 4 | = 4$. This answer can also be obtained by applying the general formula for distance between any two points.

$$d = \sqrt{(x_1 - x_2)^2 + (y_1 - y_2)^2}$$

By substitution,

$$d = \sqrt{(1-5)^2 + (3-3)^2}$$
$$= \sqrt{(-4)^2 + (0)^2}$$
$$= \sqrt{16}$$
$$= 4$$

The distance is 4.

To find the midpoint of a segment between the two given endpoints, use the formula

$$MP = \left(\frac{x_1 + x_2}{2}, \frac{y_1 + y_2}{2} \right)$$

where x_1 and y_1 are the coordinates of one point; x_2 and y_2 are the coordinates of the other point.

Lines in the Coordinate plane

If two lines in the coordinate plane are parallel, then the lines have the same slope but different y-intercepts. If two lines in the coordinate plane are perpendicular, then the product of their slopes is –1.

PROBLEM

Determine the slope of a line perpendicular to line l.

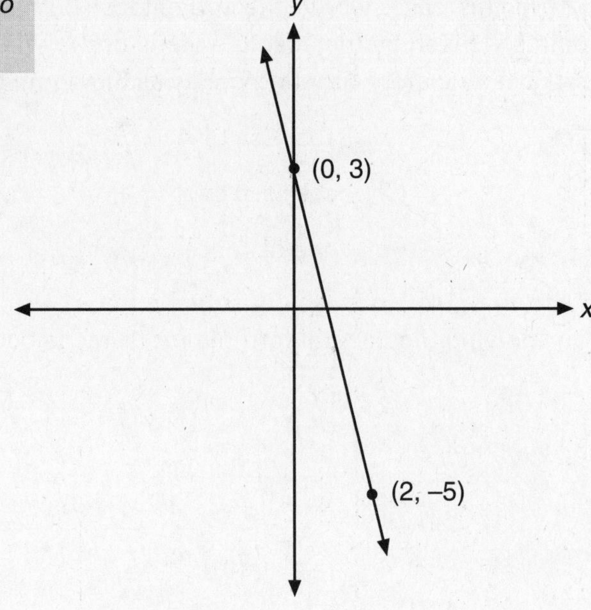

SOLUTION

Use points (0, 3) and (2, –5) to determine the slope of line *l*.

$$m = \frac{y_2 - y_1}{x_2 - x_1} = \frac{-5-3}{2-0} = \frac{-8}{2} = -4$$

The slope of line *l* is –4. Because the product of the slopes of two perpendicular lines is –1, divide –1 by the slope of line *l* to find the slope of a line perpendicular to *l*.

$$-1 \div (-4) = \frac{1}{4}$$

A line perpendicular to line *l* will have a slope of $\frac{1}{4}$.

PROBLEM

What is the equation of the line that is parallel to line l and that passes through the point (2, 4)?

SOLUTION

From the previous problem, we know that line *l* has a slope of –4. Therefore, any line parallel to line *l* will also have a slope of –4. Its equation has the form $y = -4x + b$, where b is the y-intercept. Substitute the x- and y-values of the point (2, 4) into the equation to find the value of b.

$y = -4x + b$	
$4 = -4(2) + b$	Substitute.
$4 = -8 + b$	Simplify.
$12 = b$	Add 8 to each side.

The y-intercept is 8, so the equation of the line parallel to line *l* and passing through (2, 4) is $y = -4x + 8$.

DRILL: COORDINATE GEOMETRY

Coordinates

DIRECTIONS: Refer to the diagram and find the appropriate solution.

1. Which point shown has the coordinates (–3, 2)?

   (A) A (B) B (C) C
   (D) D (E) E

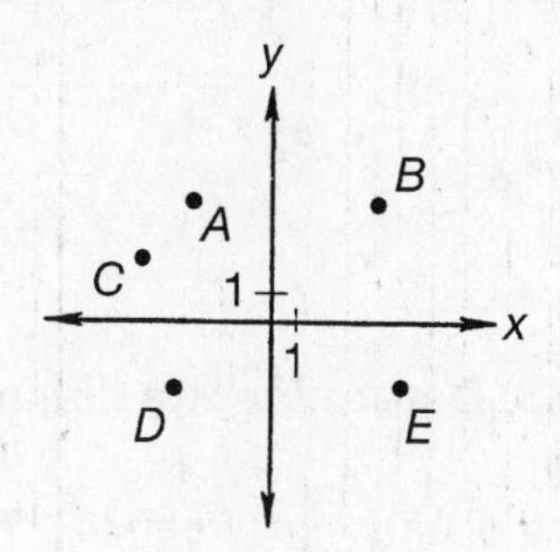

2. Name the coordinates of point *A*.

   (A) (4, 3) (B) (3, –4) (C) (3, 4)
   (D) (–4, 3) (E) (4, –3)

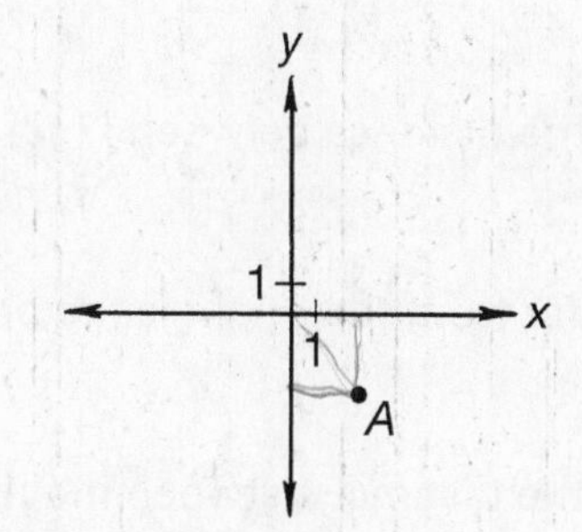

3. Which point shown has the coordinates (2.5, –1)?

   (A) M (B) N (C) P
   (D) Q (E) R

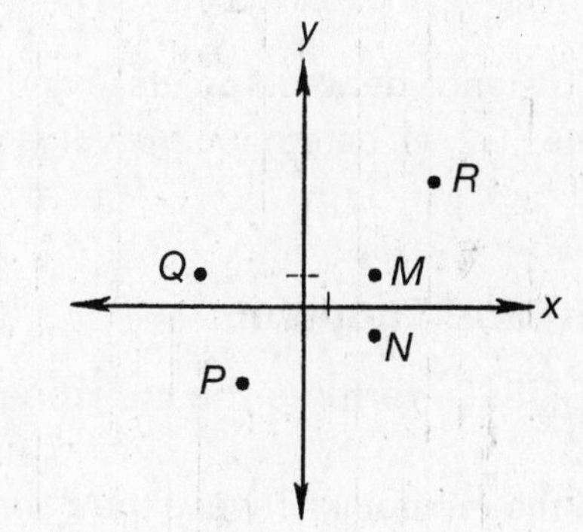

4. The correct *x*-coordinate for point *H* is what number?

   (A) 3 (B) 4 (C) –3
   (D) –4 (E) –5

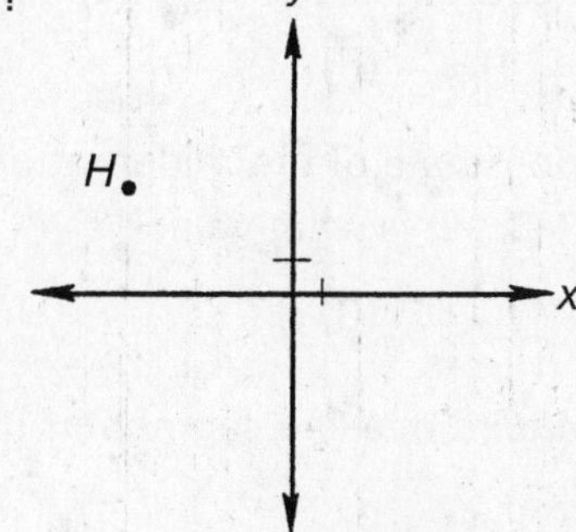

5. The correct y-coordinate for point R is what number?

(A) −7 (B) 2 (C) −2
(D) 7 (E) 8

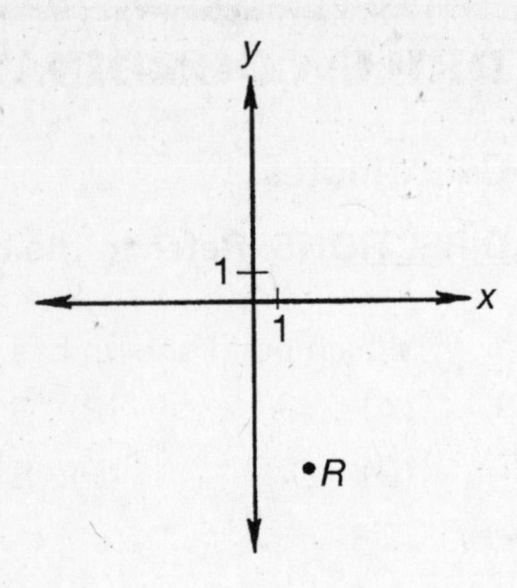

Distance

DIRECTIONS: Determine the distance or value as appropriate.

6. Find the distance between (4, −7) and (−2, −7).

(A) 4 (B) 6 (C) 7 (D) 14 (E) 15

7. Find the distance between (3, 8) and (5, 11).

(A) 2 (B) 3 (C) $\sqrt{13}$ (D) $\sqrt{15}$ (E) $3\sqrt{3}$

8. How far from the origin is the point (3, 4)?

(A) 3 (B) 4 (C) 5 (D) $5\sqrt{3}$ (E) $4\sqrt{5}$

9. Find the distance between the point (−4, 2) and (3, −5).

(A) 3 (B) $3\sqrt{3}$ (C) 7 (D) $7\sqrt{2}$ (E) $7\sqrt{3}$

10. The distance between points A and B is 10 units. If A has coordinates (4, −6) and B has coordinates (−2, y), determine the value of y.

(A) −6 (B) −2 (C) 0 (D) 1 (E) 2

Midpoints and Endpoints

DIRECTIONS: Determine the coordinates or value as appropriate.

11. Find the midpoint between the points (−2, 6) and (4, 8).

(A) (3, 7) (B) (1, 7) (C) (3, 1) (D) (1, 1) (E) (−3, 7)

12. Find the coordinates of the midpoint between the points (−5, 7) and (3, −1).

(A) (−4, 4) (B) (3, −1) (C) (1, −3) (D) (−1, 3) (E) (4, −4)

13. The y-coordinate of the midpoint of segment $\overline{AB}$ if A has coordinates (−3, 7) and B has coordinates (−3, −2) is what value?

(A) $\frac{5}{2}$ (B) 3 (C) $\frac{7}{2}$ (D) 5 (E) $\frac{15}{2}$

14. One endpoint of a line segment is (5, −3). The midpoint is (−1, 6). What is the other endpoint?

(A) (7, 3) (B) (2, 1.5) (C) (−7, 15)
(D) (−2, 1.5) (E) (−7, 12)

15. The point (−2, 6) is the midpoint for which of the following pairs of points?

(A) (1, 4) and (−3, 8) (B) (−1, −3) and (5, 9) (C) (1, 4) and (5, 9)
(D) (−1, 4) and (3, −8) (E) (1, 3) and (−5, 9)

Lines in the Coordinate Plane

DIRECTIONS: Use the diagrams to answer each question.

16. Which of the following lines is parallel to line m?

(A) $y = -2x - 4$ (B) $y = -\frac{1}{2}x - 4$

(C) $y = \frac{1}{2}x + 4$ (D) $y = x - 4$

(E) $y = 2x + 4$

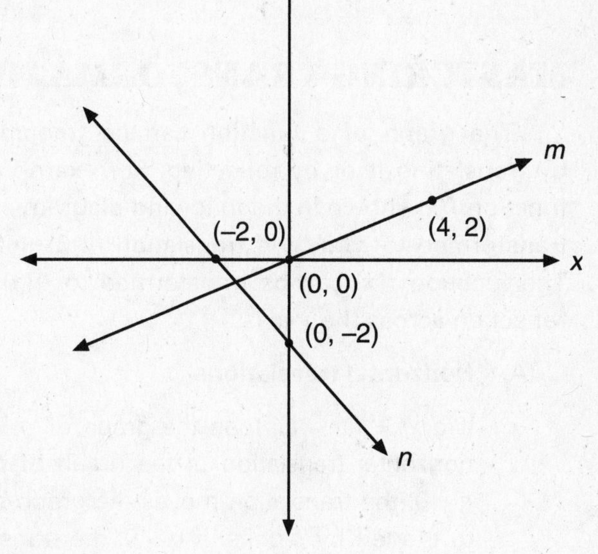

17. Which of the following lines is perpendicular to line m?

(A) $y = -2x - 4$ (B) $y = -\frac{1}{2}x - 4$ (C) $y = \frac{1}{2}x + 4$

(D) $y = x - 4$ (E) $y = 2x + 4$

18. Which of the following lines is perpendicular to line n?

(A) $y = -2x - 6$ (B) $y = -x - 6$ (C) $y = \frac{1}{2}x - 6$

(D) $y = x + 6$ (E) $y = 2x + 6$

19. What is the equation of the line that is parallel to line p and that passes through point (1, 4)?

(A) $y = \frac{1}{2}x - 1$ (B) $y = \frac{1}{2}x + \frac{7}{2}$

(C) $y = 2x - 7$ (D) $y = 2x + 2$

(E) $y = 2x + 4$

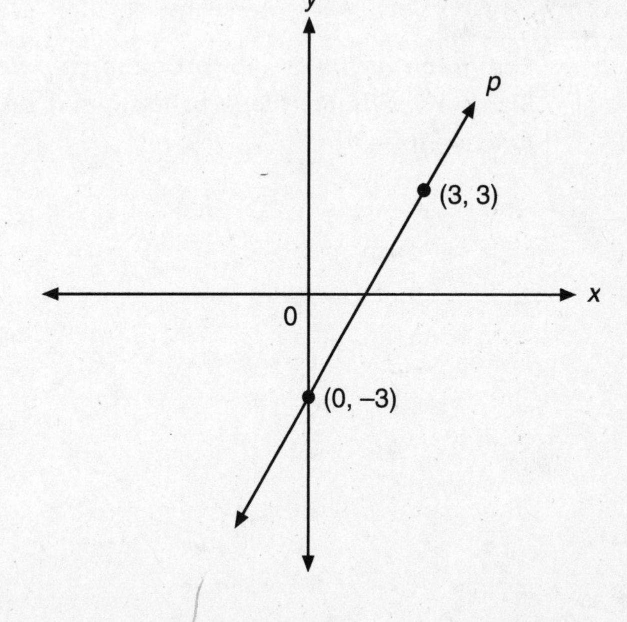

20. What is the equation of the line that is perpendicular to line p and that passes through point $(-2, 0)$?

(A) $y = -2x - 4$ (B) $y = -2x - 2$ (C) $y = -2x + 4$

(D) $y = -\frac{1}{2}x - 2$ (E) $y = -\frac{1}{2}x - 1$

MASTERING TRANSFORMATIONS OF GRAPHS OF FUNCTIONS

The graph of a function can be transformed by translating it or by reflecting. For example, the function $f(x)$ shown in the following diagram can be transformed to $g(x)$ by a translation of 2 units up. The function $f(x)$ can be transformed to $h(x)$ by a reflection across the y-axis.

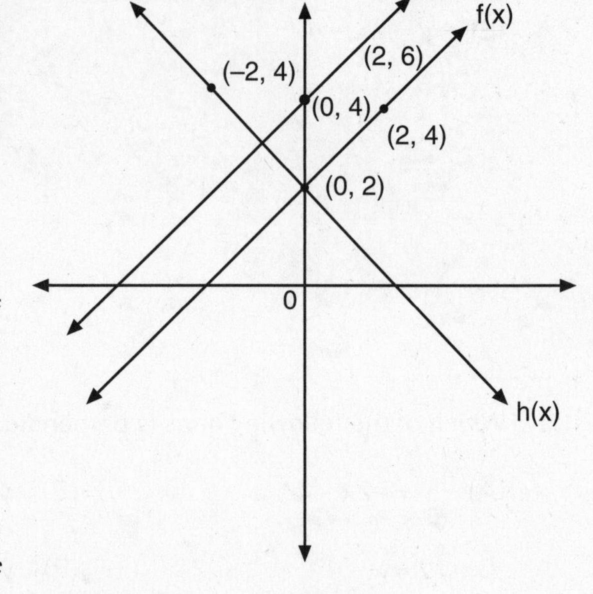

(A) **Horizontal translations**

If $g(x) = f(x - c)$, then the graph of $g(x)$ is a horizontal translation of the graph of $f(x)$. If $c < 0$, the translation moves the graph of $f(x)$ to the left by c units. If $c > 0$, the translation moves the graph of $f(x)$ to the right by c units.

(B) **Vertical translations**

If $g(x) = f(x) + c$, then the graph of $g(x)$ is a vertical translation of the graph of $f(x)$. If $c < 0$, the translation moves the graph of $f(x)$ down by c units. If $c > 0$, the translation moves the graph of $f(x)$ up by c units.

(C) **Reflections**

If $g(x) = f(-x)$, then the graph of $g(x)$ is a reflection of the graph of $f(x)$ across the y-axis. If $g(x) = -f(x)$, then the graph of $g(x)$ is a reflection of the graph of $f(x)$ across the x-axis.

PROBLEM

The graph of $f(x)$ is shown on the coordinate plane. What is the slope of the graph of the function $f(x - 3)$?

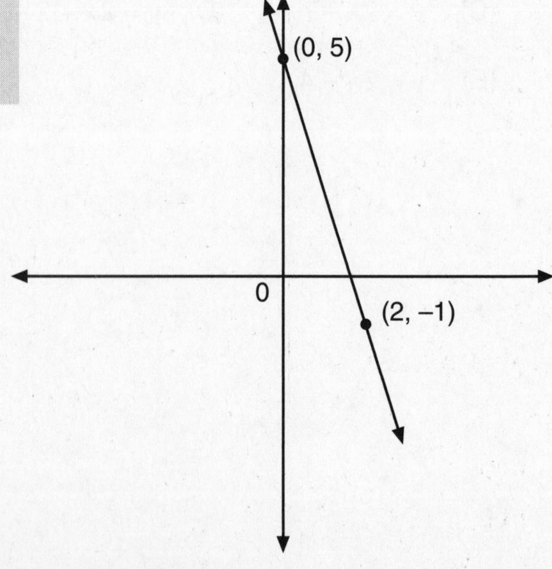

SOLUTION

The function $f(x-3)$ is a horizontal translation of $f(x)$. A horizontal translation does not change the slope of the graph of a linear function. Therefore, the graph of $f(x+3)$ has the same slope as the graph of $f(x)$. Use the points (0, 5) and (2, –3) to determine the slope of the graph of $f(x)$.

$$m=\frac{y_2-y_1}{x_2-x_1}=\frac{-1-5}{2-0}=\frac{-6}{2}=-3$$

Because the graph of $f(x)$ has a slope of –3, the graph of $f(x-3)$ also has a slope of –3.

PROBLEM

The graph of f(x) is shown on the coordinate plane. What are the coordinates of the vertex of the graph of g(x) if g(x) = f(x) + 2?

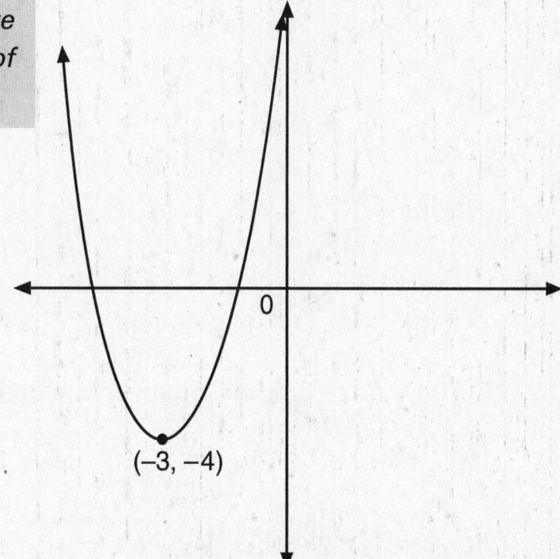

SOLUTION

Because $g(x) = f(x) + 2$, the graph of $g(x)$ is a translation 2 units up of the graph of $f(x)$. To locate the vertex of the graph of $g(x)$, translate the vertex of the graph of $f(x)$ 2 units up. The vertex of the graph of $f(x)$ is located at (–3, –4). Add 2 to the y-coordinate to find the vertex of the graph of $g(x)$. The vertex of the graph of $g(x)$ is located at (–3, –4 + 2), or (–3, –2).

DRILL: TRANSFORMATIONS OF FUNCTIONS

Transformations

DIRECTIONS: Refer to the diagrams to answer each question.

1. Which function translates the graph of $f(x)$ to the right 3 units?

(A) $g(x) = f(x + 3)$ (B) $g(x) = f(x - 3)$
(C) $g(x) = f(x) + 3$ (D) $g(x) = f(x) - 3$
(E) $g(x) = f(3x)$

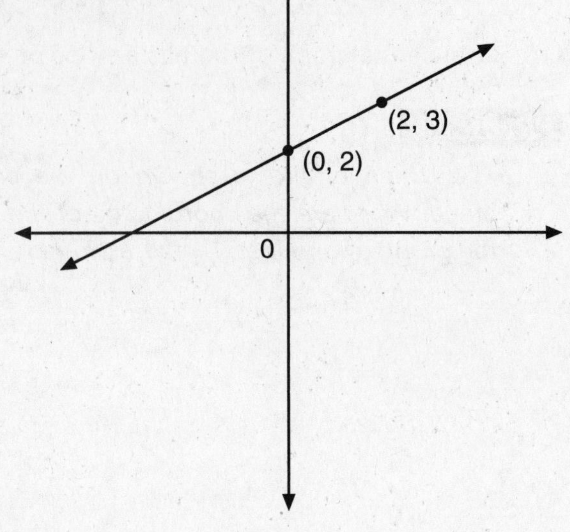

2. If $h(x) = f(x) - 2$, what is the y-intercept of $h(x)$?

(A) -2 (B) $-\frac{1}{2}$ (C) 0 (D) $\frac{1}{2}$ (E) 2

3. If $j(x) = -f(x)$, what is $j(2)$?

(A) -3 (B) -1 (C) 0 (D) 1 (E) 3

4. What are the coordinates of the vertex of the graph of $p(x - 1)$?

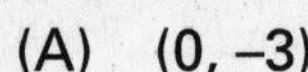

(A) $(0, -3)$ (B) $(1, -4)$
(C) $(1, -3)$ (D) $(1, -2)$
(E) $(2, -3)$

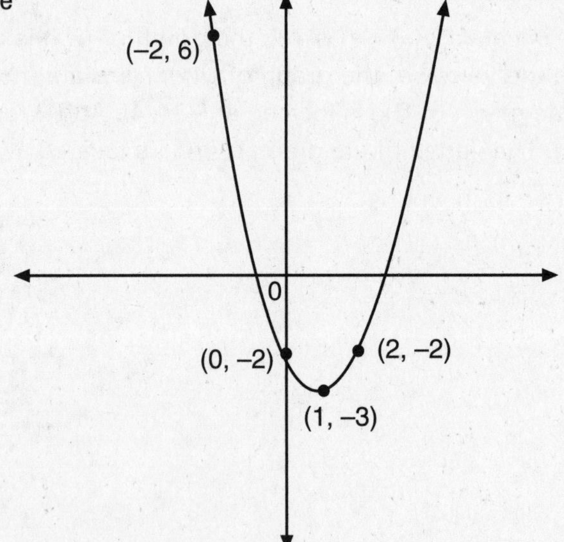

5. If $q(x) = p(-x)$, what is $q(2)$?

(A) -6 (B) -2 (C) 0 (D) 2 (E) 6

GEOMETRY DRILLS

ANSWER KEY

Drill: Lines and Angles

1. (B)
2. (A)
3. (C)
4. (D)
5. (D)
6. (C)
7. (B)
8. (A)
9. (D)
10. (E)
11. (C)
12. (D)
13. (B)
14. (E)
15. (A)

Drill: Regular Polygons

1. (D)
2. (B)
3. (E)
4. (A)
5. (D)
6. (C)
7. (E)
8. (B)
9. (D)
10. (B)

Drill: Triangles

1. (D)
2. (B)
3. (C)
4. (E)
5. (C)
6. (A)
7. (B)
8. (E)
9. (A)
10. (C)
11. (C)
12. (B)
13. (D)
14. (A)
15. (B)
16. (C)
17. (D)
18. (C)
19. (B)
20. (E)

Drill: Quadrilaterals

1. (A)
2. (B)
3. (D)
4. (E)
5. (C)
6. (A)
7. (D)
8. (C)
9. (B)
10. (C)
11. (C)
12. (D)
13. (B)
14. (C)
15. (B)
16. (A)
17. (E)
18. (B)
19. (C)
20. (D)

Drill: Circles

1. (B)
2. (A)
3. (D)
4. (E)
5. (C)
6. (D)
7. (C)
8. (B)
9. (E)
10. (D)
11. (B)
12. (D)
13. (C)
14. (B)
15. (A)
16. (C)
17. (B)
18. (D)
19. (A)
20. (C)
21. (D)
22. (C)

Drill: Solids

1.	(D)	2.	(B)	3.	(C)

Drill: Coordinate Geometry

1.	(C)	6.	(B)	11.	(B)	16.	(C)
2.	(E)	7.	(C)	12.	(D)	17.	(A)
3.	(B)	8.	(C)	13.	(A)	18.	(D)
4.	(D)	9.	(D)	14.	(C)	19.	(D)
5.	(A)	10.	(E)	15.	(E)	20.	(E)

Drill: Transformations of Functions

1. (B)
2. (C)
3. (A)
4. (E)
5. (E)

IV. WORD PROBLEMS

One of the main problems students have in mathematics involves solving word problems. The secret to solving these problems is being able to convert words into numbers and variables in the form of an algebraic equation.

The easiest way to approach a word problem is to read the question and ask yourself what you are trying to find. This unknown quantity can be represented by a variable.

Next, determine how the variable relates to the other quantities in the problem. More than likely, these quantities can be explained in terms of the original variable. If not, a separate variable may have to be used to represent a quantity.

Using these variables and the relationships determined among them, an equation can be written. Solve for a particular variable and then plug this number in for each relationship that involves this variable in order to find any unknown quantities.

Lastly, re-read the problem to be sure that you have answered the questions correctly and fully.

MASTERING ALGEBRAIC WORD PROBLEMS

The following illustrates how to formulate an equation and solve the problem.

EXAMPLE

Find two consecutive odd integers whose sum is 36.

Let x = the first odd integer

Let $x + 2$ = the second odd integer

The sum of the two numbers is 36. Therefore,

$$x + (x + 2) = 36$$

Simplifying,

$$2x + 2 = 36$$

$$2x + 2 + (-2) = 36 + (-2)$$

$$2x = 34$$

$$x = 17$$

Plugging 17 in for x, we find the second odd integer = $(x + 2) = (17 + 2) = 19$. Therefore, we find that the two consecutive odd integers whose sum is 36 are 17 and 19, respectively.

DRILL: ALGEBRAIC

Algebraic Word Problems

DIRECTIONS: Solve the following word problems algebraically.

1. The sum of two numbers is 41. One number is one less than twice the other. Find the larger of the two numbers.

   (A) 13 (B) 14 (C) 21 (D) 27 (E) 41

2. The sum of two consecutive integers is 111. Three times the larger integer less two times the smaller integer is 58. Find the value of the smaller integer.

   (A) 55 (B) 56 (C) 58 (D) 111 (E) 112

3. The difference between two integers is 12. The sum of the two integers is 2. Find both integers.

   (A) 7 and 5 (B) 7 and –5 (C) –7 and 5
   (D) 2 and 12 (E) –2 and 12

RATE

One of the formulas you will use for rate problems will be

Rate × Time = Distance

PROBLEM

If a plane travels five hours from New York to California at a speed of 600 miles per hour, how many miles does the plane travel?

SOLUTION

Using the formula rate × time = distance, multiply 600 mph × 5 hours = 3,000 miles.

The average rate at which an object travels can be solved by dividing the total distance traveled by the total amount of time.

PROBLEM

On a 40-mile bicycle trip, Cathy rode half the distance at 20 mph and the other half at 10 mph. What was Cathy's average speed on the bike trip?

SOLUTION

First you need to break down the problem. On half of the trip, which would be 20 miles, Cathy rode 20 mph. Using the rate formula,

$$\frac{\text{distance}}{\text{rate}} = \text{time},$$

you would compute,

$$\frac{20 \text{ miles}}{20 \text{ miles per hour}} = 1 \text{ hour}$$

to travel the first 20 miles. During the second 20 miles, Cathy traveled at 10 miles per hour, which would be

$$\frac{20 \text{ miles}}{10 \text{ miles per hour}} = 2 \text{ hours}$$

Thus, the average speed Cathy traveled would be $\frac{40}{3}$ = 13.3 miles per hour.

In solving for some rate problems you can use cross multiplication involving ratios to solve for x.

PROBLEM

If 2 pairs of shoes cost $52, then what is the cost of 10 pairs of shoes at this rate?

SOLUTION

$$\frac{2}{52} = \frac{10}{x}, 2x = 52 \times 10, x = \frac{520}{2}, x = \$260$$

DRILL: RATE

Rate Word Problems

DIRECTIONS: Solve to find the rate.

1. Two towns are 420 miles apart. A car leaves the first town traveling toward the second town at 55 mph. At the same time, a second car leaves the other town and heads toward the first town at 65 mph. How long will it take for the two cars to meet?

   (A) 2 hr (B) 3 hr (C) 3.5 hr (D) 4 hr (E) 4.25 hr

2. A camper leaves the campsite walking due east at a rate of 3.5 mph. Another camper leaves the campsite at the same time but travels due west. In two hours the two campers will be 15 miles apart. What is the walking rate of the second camper?

   (A) 2.5 mph (B) 3 mph (C) 3.25 mph
   (D) 3.5 mph (E) 4 mph

3. A bicycle racer covers a 75-mile training route to prepare for an upcoming race. If the racer could increase his speed by 5 mph, he could complete the same course in $\frac{3}{4}$ of the time. Find his average speed.

   (A) 15 mph (B) 15.5 mph (C) 16 mph
   (D) 18 mph (E) 20 mph

WORK

In work problems, one of the basic formulas is

$$\frac{1}{x}+\frac{1}{y}=\frac{1}{z}$$

where x and y represent the number of hours it takes two objects or people to complete the work and z is the total number of hours when both are working together.

PROBLEM

Otis can seal and stamp 400 envelopes in 2 hours whereas Elizabeth seals and stamps 400 envelopes in 1 hour. In how many hours can Otis and Elizabeth, working together, complete a 400-piece mailing at these rates?

SOLUTION

$$\frac{1}{2}+\frac{1}{1}=\frac{1}{z}$$

$$\frac{1}{2}+\frac{2}{2}=\frac{3}{2}$$

$$\frac{3}{2}=\frac{1}{z}$$

$$3z = 2$$

$z = \frac{2}{3}$ of an hour or 40 minutes. Working together, Otis and Elizabeth can seal and stamp 400 envelopes in 40 minutes.

DRILL: WORK

Work Word Problems

DIRECTIONS: Solve to find amount of work.

1. It takes Marty 3 hours to type the address labels for his club's newsletter. It only takes Pat $2\frac{1}{4}$ hours to type the same amount of labels. How long would it take them working together to complete the address labels?

   (A) $\frac{7}{9}$ hr (B) $1\frac{2}{7}$ hr (C) $1\frac{4}{5}$ hr (D) $2\frac{5}{8}$ hr (E) $5\frac{1}{4}$ hr

2. It takes Troy 3 hours to mow his family's large lawn. With his little brother's help, he can finish the job in only 2 hours. How long would it take the little brother to mow the entire lawn alone?

   (A) 4 hr (B) 5 hr (C) 5.5 hr (D) 6 hr (E) 6.75 hr

3. A tank can be filled by one inlet pipe in 15 minutes. It takes an outlet pipe 75 minutes to drain the tank. If the outlet pipe is left open by accident, how long would it take to fill the tank?

   (A) 15.5 min (B) 15.9 min (C) 16.8 min (D) 18.75 min (E) 19.3 min

MASTERING MIXTURES

Mixture problems present the combination of different products and ask you to solve for different parts of the mixture.

PROBLEM

A chemist has an 18% solution and a 45% solution of a disinfectant. How many ounces of each should be used to make 12 ounces of a 36% solution?

SOLUTION

Let x = Number of ounces from the 18% solution, and

y = Number of ounces from the 45% solution.

$$x + y = 12 \quad (1)$$

$$0.18x + 0.45y = 0.36(12) \quad (2)$$

Note that 0.18 of the first solution is pure disinfectant and that 0.45 of the second solution is pure disinfectant. When the proper quantities are drawn from each mixture the result is 12 ounces of mixture which is 0.36 pure disinfectant.

The second equation cannot be solved with two unknowns. Therefore, write one variable in terms of the other and plug it into the second equation.

$$x = 12 - y \quad (1)$$

$$0.18(12 - y) + 0.45y = 0.36(12) \quad (2)$$

Simplifying,

$$2.16 - 0.18y + 0.45y = 4.32$$

$$0.27y = 4.32 - 2.16$$

$$0.27y = 2.16$$

$$y = 8$$

Plugging in for y in the first equation,

$$x + 8 = 12$$

$$x = 4$$

Therefore, 4 ounces of the first and 8 ounces of the second solution should be used.

PROBLEM

Clark pays $2.00 per pound for 3 pounds of peanut butter chocolates and then decides to buy 2 pounds of chocolate covered raisins at $2.50 per pound. If Clark mixes both together, what is the cost per pound of the mixture?

SOLUTION

The total mixture is 5 pounds and the total value of the chocolates is

$$3(\$2.00) + 2(\$2.50) = \$11.00$$

The price per pound of the chocolates is

$$\frac{\$11.00}{5 \text{ pounds}} = \$2.20$$

DRILL: MIXTURES

Mixture Word Problems

DIRECTIONS: Find the appropriate solution.

1. How many liters of a 20% alcohol solution must be added to 80 liters of a 50% alcohol solution to form a 45% solution?

   (A) 4 (B) 8 (C) 16 (D) 20 (E) 32

2. How many kilograms of water must be evaporated from 50 kg of a 10% salt solution to obtain a 15% salt solution?

   (A) 15 (B) 15.75 (C) 16 (D) $16.\overline{66}$ (E) 16.75

3. How many pounds of coffee *A* at \$3.00 a pound should be mixed with 2.5 pounds of coffee *B* at \$4.20 a pound to form a mixture selling for \$3.75 a pound?

   (A) 1 (B) 1.5 (C) 1.75 (D) 2 (E) 2.25

MASTERING INTEREST

If the problem calls for computing simple interest, the interest is computed on the principal alone. If the problem involves compounded interest, then the interest on the principal is taken into account in addition to the interest earned before.

PROBLEM

How much interest will Jerry pay on his loan of \$400 for 60 days at 6% per year?

SOLUTION

Use the formula:

$$\text{Interest} = \text{Principal} \times \text{Rate} \times \text{Time } (I = P \times R \times T)$$

$$= \$400 \times 6\%/\text{year} \times 60 \text{ days} = \$400 \times 0.06$$

$$= \$400 \times 0.00986 = \$3.94$$

Jerry will pay \$4.00.

PROBLEM

Mr. Smith wishes to find out how much interest he will receive on $300 if the rate is 3% compounded annually for three years.

SOLUTION

Compound interest is interest computed on both the principal and the interest it has previously earned. The interest is added to the principal at the end of every year. The interest on the first year is found by multiplying the rate by the principal. Hence, the interest for the first year is

$$3\% \times \$300 = 0.03 \times \$300 = \$9.00$$

The principal for the second year is now $309, the old principal ($300) plus the interest ($9). The interest for the second year is found by multiplying the rate by the new principal. Hence, the interest for the second year is

$$3\% \times \$309 = 0.03 \times \$309 = \$9.27$$

The principal now becomes $309 + $9.27 = $318.27.

The interest for the third year is found using this new principal. It is

$$3\% \times \$318.27 = 0.03 \times \$318.27 = \$9.55$$

At the end of the third year his principal is $318.27 + 9.55 = $327.82. To find how much interest was earned, we subtract his starting principal ($300) from his ending principal ($327.82), to obtain

$$\$327.82 - \$300.00 = \$27.82$$

DRILL: INTEREST

Interest Word Problems

DIRECTIONS: Determine the amount of money invested or possible amount earned as appropriate.

1. A man invests $3,000, part in a 12-month certificate of deposit paying 8% and the rest in municipal bonds that pay 7% a year. If the yearly return from both investments is $220, how much was invested in bonds?

   (A) $80 (B) $140 (C) $220 (D) $1,000 (E) $2,000

2. A sum of money was invested at 11% a year. Four times that amount was invested at 7.5%. How much was invested at 11% if the total annual return was $1,025?

   (A) $112.75 (B) $1,025 (C) $2,500 (D) $3,400 (E) $10,000

3. One bank pays 6.5% a year simple interest on a savings account while a credit union pays 7.2% a year. If you had $1,500 to invest for three years, how much more would you earn by putting the money in the credit union?

   (A) $10.50 (B) $31.50 (C) $97.50 (D) $108 (E) $1,500

MASTERING DISCOUNTS

If the discount problem asks to find the final price after the discount, first multiply the original price by the percent of discount. Then subtract this result from the original price.

If the problem asks to find the original price when only the percent of discount and the discounted price are given, simply subtract the percent of discount from 100% and divide this percent into the sale price. This will give you the original price.

PROBLEM

A popular bookstore gives 10% discount to students. What does a student actually pay for a book that costs $24.00?

SOLUTION

10% of $24 is $2.40, hence the student pays

$$\$24 - \$2.40 = \$21.60$$

PROBLEM

Eugene paid $100 for a business suit. The suit's price included a 25% discount. What was the original price of the suit?

SOLUTION

Let x represent the original price of the suit and take the complement of 0.25 (discount price), which is 0.75.

$$0.75x = \$100 \text{ or } x = 133.34$$

So, the original price of the suit is $133.34.

DRILL: DISCOUNTS

Discount Word Problems

DIRECTIONS: Find cost, price, or discount as appropriate.

1. A man bought a coat marked 20% off for $156. How much had the coat cost originally?
   (A) $136 (B) $156 (C) $175 (D) $195 (E) $205

2. A woman saved $225 on the new sofa which was on sale for 30% off. What was the original price of the sofa?
   (A) $25 (B) $200 (C) $225 (D) $525 (E) $750

3. At an office supply store, customers are given a discount if they pay in cash. If a customer is given a discount of $9.66 on a total order of $276, what is the percent of discount?
   (A) 2% (B) 3.5% (C) 4.5% (D) 9.66% (E) 276%

MASTERING PROFIT

The formula used for the profit problems is

Profit = Revenue – Cost

or Profit = Selling Price – Expenses.

PROBLEM

Four high school and college friends started a business of remodeling and selling old automobiles during the summer. For this purpose they paid $600 to rent an empty barn for the summer. They obtained the cars from a dealer for $250 each, and it takes an average of $410 in materials to remodel each car. How many automobiles must the students sell at $1,440 each to obtain a gross profit of $7,000?

SOLUTION

Total Revenues – Total Cost = Gross Profit

Revenue – [Variable Cost + Fixed Cost] = Gross Profit

Let a = number of cars

$$\text{Revenue} = \$1{,}440a$$

$$\text{Variable Cost} = (\$250 + 410)a$$

$$\text{Fixed Cost} = \$600$$

The desired gross profit is $7,000.

Using the equation for the gross profit,

$$1{,}440a - [660a + 600] = 7{,}000$$

$$1{,}440a - 660a - 600 = 7{,}000$$

$$780a = 7{,}000 + 600$$

$$780a = 7{,}600$$

$$a = 9.74$$

or to the nearest car, $a = 10$.

PROBLEM

A glass vase sells for $25.00. The net profit is 7%, and the operating expenses are 39%. Find the gross profit on the vase.

SOLUTION

The gross profit is equal to the net profit plus the operating expenses. The net profit is 7% of the selling cost; thus, it is equal to

$$7\% \times \$25.00 = 0.07 \times \$25 = \$1.75$$

The operating expenses are 39% of the selling price, thus equal to

$$39\% \times \$25 = 0.39 \times \$25 = \$9.75$$

$1.75	net profit
+ $9.75	operating expenses
$11.50	gross profit

DRILL: PROFIT

Profit Word Problems

DIRECTIONS: Determine profit or stock worth as appropriate.

1. An item cost a store owner $50. She marked it up 40% and advertised it at that price. How much profit did she make if she later sold it at 15% off the advertised price?
   (A) $7.50 (B) $9.50 (C) $10.50 (D) $39.50 (E) $50

2. An antique dealer makes a profit of 115% on the sale of an oak desk. If the desk cost her $200, how much profit did she make on the sale?
   (A) $230 (B) $315 (C) $430 (D) $445 (E) $475

3. As a graduation gift, a young man was given 100 shares of stock worth $27.50 apiece. Within a year the price of the stock had risen by 8%. How much more were the stocks worth at the end of the first year than when they were given to the young man?
   (A) $110 (B) $220 (C) $1,220 (D) $2,750 (E) $2,970

MASTERING GEOMETRY

PROBLEM

A boy knows that his height is 6 ft. and his shadow is 4 ft. long. At the same time of day, a tree's shadow is 24 ft. long. How high is the tree? (See the figure.)

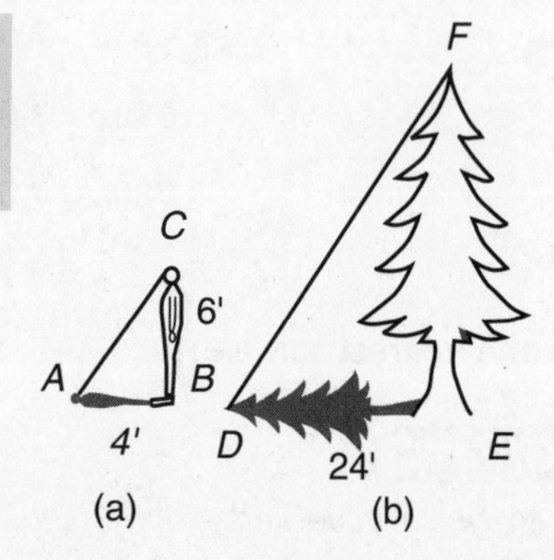

SOLUTION

Show that $\Delta ABC \approx \Delta DEF$, and then set up a proportion between the known sides $\overline{AB}$ and $\overline{DE}$, and the sides $\overline{BC}$ and $\overline{EF}$.

First, assume that both the boy and the tree are $\perp$ to the earth. Then, $\overline{BC} \perp \overline{BA}$ and $\overline{EF} \perp \overline{ED}$. Hence,

$$\angle ABC \cong \angle DEF$$

Since it is the same time of day, the rays of light from the sun are incident on both the tree and the boy at the same angle, relative to the earth's surface. Therefore,

$$\angle BAC \cong \angle EDF$$

We have shown, so far, that two pairs of corresponding angles are congruent. Since the sum of the angles of any triangle is 180°, the third pair of corresponding angles is congruent (i.e., $\angle ACB \cong \angle DFE$). By the Angle-Angle-Angle Theorem

$$\angle ABC \approx \angle DEF.$$

By definition of similarity,

$$\frac{\overline{FE}}{\overline{CB}} = \frac{\overline{ED}}{\overline{BA}}$$

$\overline{CB} = 6'$, $\overline{ED} = 24'$, and $\overline{BA} = 4'$. Therefore,

$$FE = (6')\left(\frac{24'}{4'}\right) = 36'.$$

DRILL: GEOMETRY

Geometry Word Problems

<u>DIRECTIONS</u>: Find the appropriate measurements.

1. ΔPQR is a scalene triangle. The measure of $\angle P$ is 8 more than twice the measure of $\angle R$. The measure of $\angle Q$ is two less than three times the measure of $\angle R$. Determine the measure of $\angle Q$.

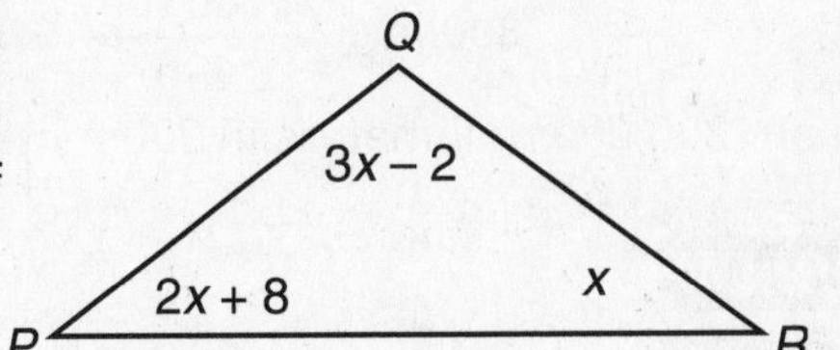

(A) 29° (B) 53° (C) 60°
(D) 85° (E) 174°

2. Angle A and angle B are supplementary. The measure of angle B is 5 more than four times the measure of angle A. Find the measure of angle B.

(A) 35° (B) 125° (C) 140° (D) 145° (E) 155°

3. Triangle RUS is isosceles with base $\overline{SU}$. Each leg is 3 less than 5 times the length of the base. If the perimeter of the triangle is 60 cm, find the length of a leg.

(A) 6 cm (B) 12 cm (C) 27 cm (D) 30 cm (E) 33 cm

MASTERING MEASUREMENT

When measurement problems are presented in either metric or English units that involve conversion of units, the appropriate data will be given in the problem.

PROBLEM

The Eiffel Tower is 984 feet high. Express this height in meters, in kilometers, in centimeters, and in millimeters.

SOLUTION

A meter is equivalent to 39.370 inches. In this problem, the height of the tower in feet must be converted to inches and then the inches can be converted to meters. There are 12 inches in 1 foot. Therefore, feet can be converted to inches by using the factor $\frac{12 \text{ inches}}{1 \text{ foot}}$.

$$984 \text{ feet} \times \frac{12 \text{ in.}}{1 \text{ ft}} = 118 \times 10^2 \text{ in.}$$

Once the height is found in inches, this can be converted to meters by the factor $\frac{1 \text{ m}}{39.370 \text{ in.}}$.

$$11{,}808 \text{ in.} \times \frac{1 \text{ m}}{39.370 \text{ in.}} = 300 \text{ m}$$

Therefore, the height in meters is 300 m.

There are 1,000 meters in 1 kilometer. Meters can be converted to kilometers by using the factor $\frac{1 \text{ km}}{1{,}000 \text{ m}}$.

$$300 \text{ m} \times \frac{1 \text{ km}}{1{,}000 \text{ m}} = 0.300 \text{ km}$$

As such, there are 0.300 kilometers in 300 meters.

There are 100 centimeters in 1 meter, thus meters can be converted to centimeters by multiplying by the factor $\frac{100 \text{ cm}}{1 \text{ m}}$.

$$300 \text{ m} \times \frac{100 \text{ cm}}{1 \text{ m}} = 300 \times 10^2 \text{ cm}$$

There are 30,000 centimeters in 300 meters.

There are 1,000 millimeters in 1 meter; therefore, meters can be converted to millimeters by the factor $\frac{1{,}000 \text{ mm}}{1 \text{ m}}$.

$$300 \text{ m} \times \frac{1{,}000 \text{ mm}}{1 \text{ m}} = 300 \times 10^3 \text{ mm}$$

There are 300,000 millimeters in 300 meters.

PROBLEM

The unaided eye can perceive objects that have a diameter of 0.1 mm. What is the diameter in inches?

SOLUTION

From a standard table of conversion factors, one can find that 1 in. = 2.54 cm. Thus, centimeters can be converted to inches by multiplying by 1 in./2.54 cm. Here, one is given the diameter in millimeters, which equals 0.1 cm. Millimeters are converted to centimeters by multiplying the number of millimeters by $\frac{0.1 \text{ cm}}{1 \text{ mm}}$. Solving for centimeters, you obtain

$$0.1 \text{ mm} \times \frac{0.1 \text{ cm}}{1 \text{ mm}} = 0.01 \text{ cm}$$

Solving for inches:

$$0.01 \text{ cm} \times = 3.94 = 10 - 3 \text{ in.}$$

DRILL: MEASUREMENT

Measurement Word Problems

DIRECTIONS: Determine the appropriate solution from the information provided.

1. A brick walkway measuring 3 feet by 11 feet is to be built. The bricks measure 4 inches by 6 inches. How many bricks will it take to complete the walkway?

   (A) 132 (B) 198 (C) 330 (D) 1,927 (E) 4,752

2. A wall to be papered is three times as long as it is wide. The total area to be covered is 192 ft^2. Wallpaper comes in rolls that are 2 feet wide by 8 feet long. How many rolls will it take to cover the wall?

   (A) 8 (B) 12 (C) 16 (D) 24 (E) 32

3. A bottle of medicine containing 2 kg is to be poured into smaller containers that hold 8 grams each. How many of these smaller containers can be filled from the 2 kg bottle?

   (A) 0.5 (B) 1 (C) 5 (D) 50 (E) 250

MASTERING DATA INTERPRETATION

Some of the problems test ability to apply information given in graphs and tables.

PROBLEM

In which year was the least number of bushels of wheat produced? (See the following figure.)

SOLUTION

By inspection of the graph, we find that the shortest bar representing wheat production is the one representing the wheat production for 1976. Thus, the least number of bushels of wheat was produced in 1976.

Number of bushels (to the nearest 5 bushels) of wheat and corn produced by farm RQS from 1975 – 1985

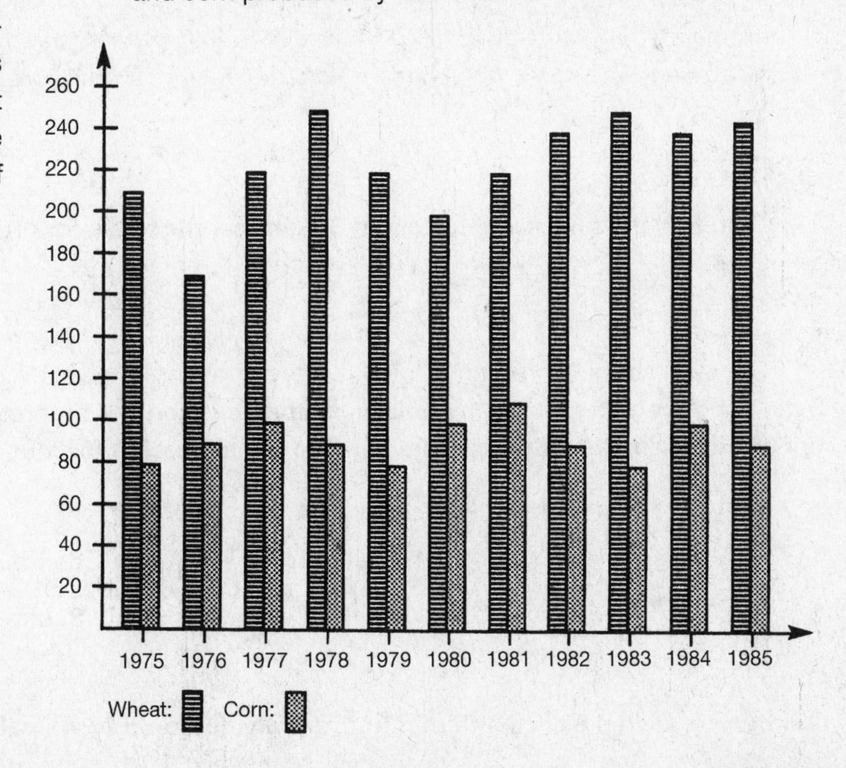

PROBLEM

What was the ratio of wheat production in 1985 to that of 1975?

SOLUTION

From the graph representing wheat production, the number of bushels of wheat produced in 1975 is equal to 210 bushels. This number can be found by locating the bar on the graph representing wheat production in 1975 and then drawing a horizontal line from the top of that bar to the vertical axis. The point where this horizontal line meets the vertical axis represents the number of bushels of wheat produced in 1975. This number on the vertical axis is 210. Similarly, the graph indicates that the number of bushels of wheat produced in 1985 is equal to 245 bushels.

Thus, the ratio of wheat production in 1985 to that of 1975 is 245 to 210, which can be written as $\frac{245}{210}$. Simplifying this ratio to its simplest form yields

$$\frac{245}{210} = \frac{5 \times 7 \times 7}{2 \times 3 \times 5 \times 7}$$

$$= \frac{7}{2 \times 3}$$

$$= \frac{7}{6} \text{ or } 7{:}6$$

PROBLEM

Golf Scores				
Player	**Round 1**	**Round 2**	**Round 3**	**Total**
Benicio	85	83		
Ryan	84	87		
Total	169	170		518

Benicio and Ryan played three rounds of golf. If Ryan's score in the third round was 5 more than Benicio's score, how many points did each of them score in the third round?

SOLUTION

First, find the total score for round 3. Subtract the totals for rounds 1 and 2 from the grand total of 518.

$$518 - 169 - 170 = 179$$

The total score for round 3 is 179. Let x represent Benicio's score in the third round. Because Ryan's score was 5 more than Benicio's, the expression $x + 5$ represents Ryan's score. Set the sum of Benicio's score and Ryan's score equal to the total score for round 3. Then solve for x.

$$x + (x + 5) = 179$$

$$2x + 5 = 179$$ Combine like terms.

$$2x = 174$$ Subtract 5 from each side.

$$x = 87$$ Divide both sides by 2.

Benicio scored 87 in the third round, and Ryan scored 87 + 5 = 92 in the third round.

DRILL: DATA INTERPRETATION

Date Interpretation Word Problems

DIRECTIONS: Determine the correct response from the information provided.

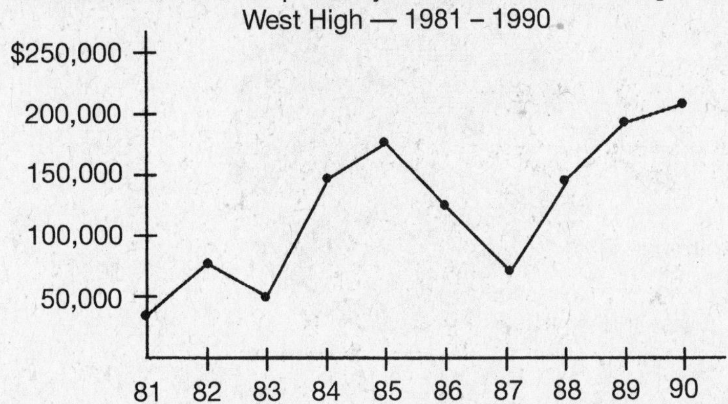

1. What was the approximate amount of scholarship money awarded in 1985?

   (A) $150,000 (B) $155,000 (C) $165,000
   (D) $175,000 (E) $190,000

2. By how much did the scholarship money increase between 1987 and 1988?

   (A) $25,000 (B) $30,000 (C) $50,000
   (D) $55,000 (E) $75,000

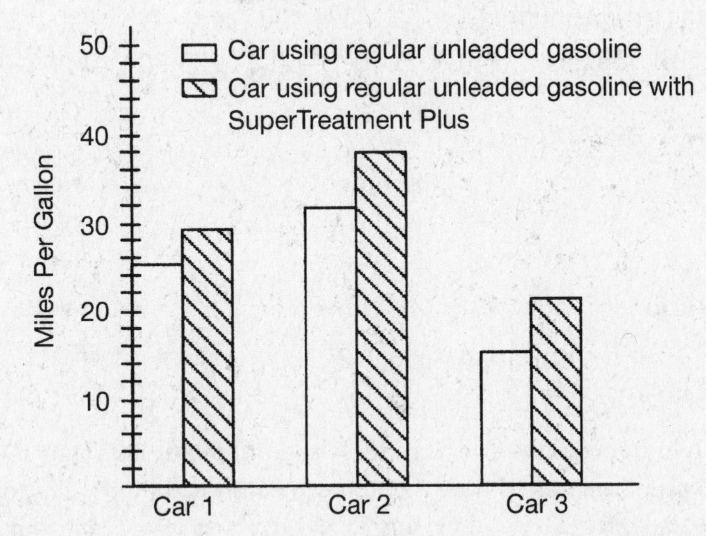

3. By how much did the mileage increase for Car 2 when the new product was used?

   (A) 5 mpg (B) 6 mpg (C) 7 mpg (D) 10 mpg (E) 12 mpg

4. Which car's mileage increased the most in this test?

   (A) Car 1 (B) Car 2 (C) Car 3
   (D) Cars 1 and 2 (E) Cars 2 and 3

5. According to the bar graph, if your car averages 25 mpg, what mileage might you expect with the new product?

(A) 21 mpg (B) 30 mpg (C) 31 mpg (D) 35 mpg (E) 37 mpg

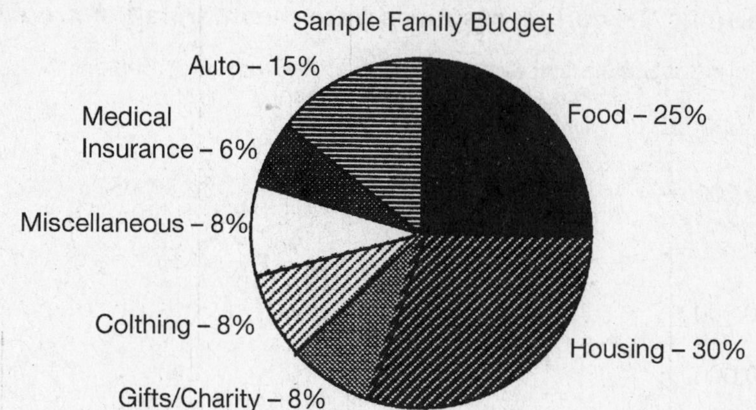

6. Using the budget shown, a family with an income of $1,500 a month would plan to spend what amount on housing?

(A) $300 (B) $375 (C) $450 (D) $490 (E) $520

7. In this sample family budget, how does the amount spent on an automobile compare to the amount spent on housing?

(A) $\frac{1}{3}$ (B) $\frac{1}{2}$ (C) $\frac{2}{3}$ (D) $1\frac{1}{2}$ (E) 2

8. A family with a monthly income of $1,240 spends $125 a month on clothing. By what amount do they exceed the sample budget?

(A) $1.00 (B) $5.20 (C) $10.00 (D) $25.80 (E) $31.75

CALORIE CHART — BREADS

Bread	Amount	Calories
French Bread	2 oz	140
Bran Bread	1 oz	95
Whole Wheat Bread	1 oz	115
Oatmeal Bread	0.5 oz	55
Raisin Bread	1 oz	125

9. One dieter eats two ounces of french bread. A second dieter eats two ounces of bran bread. The second dieter has consumed how many more calories than the first dieter?

(A) 40 (B) 45 (C) 50 (D) 55 (E) 65

10. One ounce of whole wheat bread has how many more calories than an ounce of oatmeal bread?

(A) 5 (B) 15 (C) 60 (D) 75 (E) 125

MASTERING SCATTERPLOTS

A **scatterplot** shows the relationship between two sets of data. Some word problems may ask you to interpret the data in a scatterplot or draw conclusions about its line of best fit.

PROBLEM

The scatterplot shows the relationship between the grams of fiber and the grams of fat in several brands of granola. Describe the slope of the line of best fit for the data in this scatterplot.

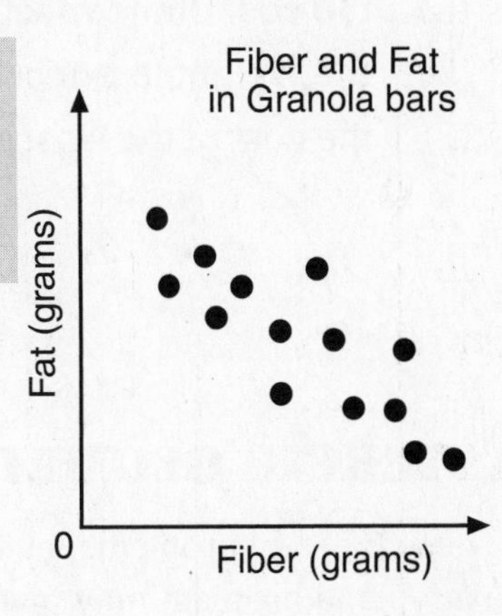

SOLUTION

The data in the scatterplot show a trend. As the number of grams of fiber in the granola increases, the number of grams of fat tends to decrease. The line of best fit for the data moves downward as it moves from left to right. Therefore, the slope of the line of best fit is negative.

DRILL: SCATTERPLOTS

Scatterplot Word Problems

DIRECTIONS: Use the scatterplots to answer each question.

Employee Earnings

Total Earnings

A C E D B

0

Hours Worked

1. The scatterplot shows the relationship between the number of hours worked and the total earnings of the employees at a company during one week. Of the labeled points, which one corresponds to the worker who earned the most per hour?

   (A) A (B) B (C) C (D) D (E) E

2. The scatterplot shows the relationship between the weight and cost of jars of various brands of peanut butter. What does the slope of the line of best fit of this scatterplot represent?

(A) the weight of the largest jar of peanut butter

(B) the cost per ounce of a typical jar of peanut butter

(C) the cost of the most popular brand of peanut butter

(D) the weight in ounces of a typical jar of peanut butter

(E) the cost of the least expensive brand of peanut butter

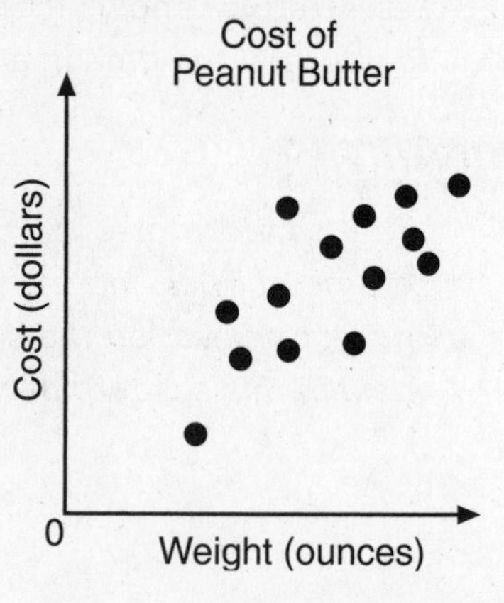

MASTERING GEOMETRIC PROBABILITY

These types of problems involve the probability of events that are related to geometric figures. For example, a problem may involve finding the probability that a randomly selected point will lie in a specific area within a larger geometric figure.

PROBLEM

The circular target below has a radius of 12 in. The inner circle of the target has a radius of 6 in. If an arrow strikes the target at a random point, what is the probability that the arrow will strike the shaded region?

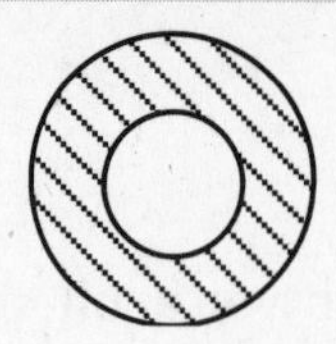

SOLUTION

The probability that the arrow will strike the shaded region is the ratio of the area of the shaded region to the area of the entire target. First, find the area of the entire target, which has a radius of 12 in.

$$\text{Area of entire target: } A = \pi r^2 = \pi(12)^2 = 144\pi \text{ in.}^2$$

The area of the shaded region is equal to the area of the entire target minus the area of the inner circle.

$$\text{Area of inner circle: } A = \pi r^2 = \pi(6)^2 = 36\pi \text{ in.}^2$$

$$\text{Area of shaded region: } 144\pi - 36\pi = 108\pi \text{ in.}^2$$

The probability that the arrow will strike the shaded region is equal to the ratio $\frac{180\pi}{144\pi}$, which simplifies to $\frac{3}{4}$.

DRILL: GEOMETRIC PROBABILITY

Geometric Probability Word Problems

DIRECTIONS: Determine the appropriate solution from the information provided.

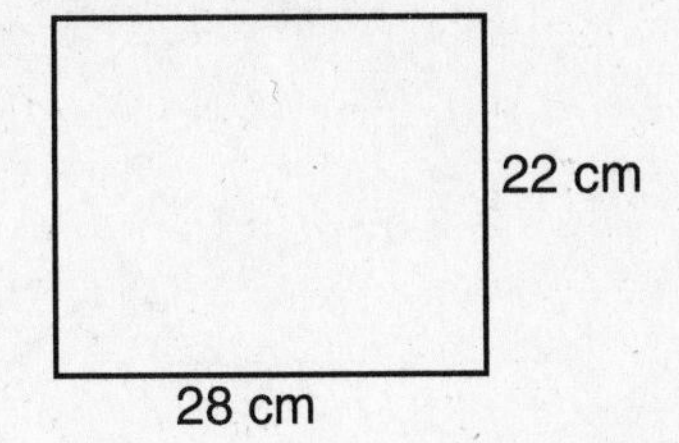

1. A computer is programmed to randomly select a point from the rectangular screen shown in the diagram. What is the probability that the point will be at least 3 cm from each side of the screen?

   (A) $\frac{3}{14}$ (B) $\frac{3}{7}$ (C) $\frac{4}{7}$

   (D) $\frac{8}{11}$ (E) $\frac{11}{14}$

2. If a forest fire starts at a random point within a 40-mile radius of a ranger station, what is the probability that the starting point will occur within 2 miles of the station?

   (A) $\frac{1}{400}$ (B) $\frac{1}{200}$ (C) $\frac{1}{80}$ (D) $\frac{1}{20}$ (E) $\frac{1}{10}$

WORD PROBLEM DRILLS

ANSWER KEY

Drill: Algebraic
1. (D)
2. (A)
3. (B)

Drill: Rate
1. (C)
2. (E)
3. (A)

Drill: Work
1. (B)
2. (D)
3. (D)

Drill: Mixture
1. (C)
2. (D)
3. (B)

Drill: Interest
1. (E)
2. (C)
3. (B)

Drill: Discount
1. (D)
2. (E)
3. (B)

Drill: Profit
1. (B)
2. (A)
3. (B)

Drill: Geometry
5. (D)
6. (D)
7. (C)

Drill: Measurement
9. (B)
10. (B)
11. (E)

Drill: Data Interpretation
1. (D)
2. (E)
3. (B)
4. (E)
5. (B)
6. (C)
7. (B)
8. (D)
9. (C)
10. (A)

Drill: Scatterplots
1. (A)
2. (B)

Drill: Geometric Probability
1. (C)
2. (A)

CHAPTER 8

MASTERING REGULAR MATH QUESTIONS

MASTERING REGULAR MATH QUESTIONS

The Regular Math questions of the SAT are designed to test your ability to solve problems involving arithmetic, algebra, and geometry. A few of the problems may be similar to those found in a math textbook and will require nothing more than the use of basic rules and formulas. Most of the problems, however, will require more than that. Regular Math questions will ask you to think creatively and apply basic skills to solve problems.

All Regular Math questions are in a multiple-choice format with five possible responses. There are a number of advantages and disadvantages associated with multiple-choice math tests. Learning what some of these advantages and disadvantages are can help you improve your test performance.

The greatest disadvantage of a multiple-choice math test is that every question presents you with four wrong answers. These wrong answers are not randomly chosen numbers—they are the answers that students are most likely to get if they make certain mistakes. They also tend to be answers that "look right" to someone who does not know how to solve the problem. Thus, on a particular problem, you may be relieved to find "your" answer among the answer choices, only to discover later that you fell into a common error trap. Wrong answer choices can also distract or confuse you when you are attempting to solve a problem correctly, causing you to question your answer even though it is right.

The greatest advantage of a multiple-choice math test is that the right answer is also presented to you. This means that you may be able to spot the right answer even if you do not understand a problem completely or do not have time to finish it. It means that you may be able to pick the right answer by guessing intelligently. It also means that you may be saved from getting a problem wrong when the answer you obtain is not among the answer choices—and you have to go back and work the problem again.

Keep in mind, also, that the use of a calculator is permitted during the test. Do not be tempted, however, to use this as a crutch. Some problems can actually be solved more quickly without a calculator, and you still have to work through the problem to know what numbers to punch. No calculator in the world can solve a problem for you.

ABOUT THE DIRECTIONS

The directions found at the beginning of each Regular Math section are simple—solve each problem, then mark the best of five answer choices on your answer sheet. Following these instructions, however, is important information that you should understand thoroughly before you attempt to take a test. This information includes definitions of standard symbols and formulas that you may need in order to solve Regular Math problems. The formulas are given so that you don't have to memorize them—however, in order to benefit from this information, you need to know what is and what is not included. Otherwise, you may waste time looking for a formula that is not listed, or you may fail to look for a formula that is listed. The formulas given to you at the beginning of a Regular Math section include the following:

- The number of degrees in a straight line
- Area and circumference of a circle; number of degrees in a circle
- Area of a triangle; Pythagorean Theorem for a right triangle; sum of angle measures of a triangle

Following the formulas and definitions of symbols is a very important statement about the diagrams, or figures, that may accompany Regular Math questions. This statement tells you that, unless stated otherwise in a specific question, figures are drawn to scale.

ABOUT THE QUESTIONS

Most Multiple-Choice questions on the SAT fall into one of three categories: arithmetic, algebra, and geometry. In the following three sections, we will review the kinds of questions you will encounter on the actual test.

MASTERING ARITHMETIC QUESTIONS

Most Regular Math questions on the SAT fall into one of the following four question types. For each question type, an example and solution will be given, highlighting strategies and techniques for completing the problems as quickly as possible.

Question Type 1: Evaluating Expressions

Arithmetic questions on the SAT often ask you to find the value of an arithmetic expression or to find the value of a missing term in an expression. The temptation when you see one of these expressions is to calculate its value—a process that is time-consuming and can easily lead to an error. A better way to approach an arithmetic expression is to use your knowledge of properties of numbers to spot shortcuts.

PROBLEM

$7(8 + 4) - (3 \times 12) =$

(A) 24
(B) 48
(C) 110
(D) 144
(E) 3,024

SOLUTION

Before you jump into multiplication, look at the numbers inside the first parentheses. This is the sum $(8 + 4) = 12$, which makes the entire expression equal to $7(12) - (3 \times 12)$. The distributive property tells you that $a(b + c) = ab + ac$ and $a(b - c) = ab - ac$. The expression $7(12) - (3 \times 12)$ can be made to fit the second formula, with a equal to 12 and 7 and 3 equal to b and c, respectively. Thus, $7(12) - (3 \times 12)$ becomes $12(7 - 3)$, and the answer is simply 12×4, or 48.

Question Type 2: Undefined Symbols

Most SAT math sections include problems that involve undefined symbols. In some problems, these symbols define a value by asking you to perform several arithmetic operations. For example, the symbol [x] may tell you to square some number x then subtract 3: $[x] = x^2 - 3$. In other problems, a symbol may represent a missing numeral, such as $10 - \Delta = 7$. By looking at the arithmetic, you can see that Δ must equal 3.

PROBLEM

Let $[n] = n^2 + 1$ for all numbers n. Which of the following is equal to the product of [2] and [3]?

(A) [6]
(B) [7]
(C) [8]
(D) [9]
(E) [11]

SOLUTION

The newly defined symbol is [ ]. To find the values for [2] and [3], plug them into the formula $[n] = n^2 + 1$:

$$[2] = 2^2 + 1 = 4 + 1 = 5$$

$$[3] = 3^2 + 1 = 9 + 1 = 10$$

Because [2] = 5, and [3] = 10, we can compute the product: $5 \times 10 = 50$.

Now look at the answers. You will see that the answers are given in terms of [ ]. Once again, you must plug them into the formula $[n] = n^2 + 1$. If we plug each answer choice into the equation, we get:

$$[6] = 6^2 + 1 = 36 + 1 = 37$$

$$[7] = 7^2 + 1 = 49 + 1 = 50$$

$$[8] = 8^2 + 1 = 64 + 1 = 65$$

$$[9] = 9^2 + 1 = 81 + 1 = 82$$

$$[11] = 11^2 + 1 = 121 + 1 = 122$$

Question Type 3: Averages

Some SAT math problems ask you to simply compute the average of a given set of values. More challenging problems ask you to apply the definition of average. You will recall that the average of a given set of values is equal to the sum of the values divided by the number of values in the set.

PROBLEM

The average of 10 numbers is 53. What is the sum of the numbers?

(A) 106 (B) 350 (C) 363 (D) 530 (E) 615

SOLUTION

You can solve this problem using the formula:

$$\text{Average} = \frac{\text{sum of values}}{\text{number of values}}$$

In this question,

$$53 = \frac{\text{sum of values}}{10}$$

Therefore, the sum of the numbers is 530:

$$10 \times 53 = \text{sum of values}$$

$$530 = \text{sum of values}$$

PROBLEM

The average of three numbers is 16. If one of the numbers is 5, what is the sum of the other two?

(A) 11 (B) 24 (C) 27 (D) 38 (E) 43

SOLUTION

In this problem, if we plug in the numbers we know we get

$$16 = \frac{a+b+c}{3}$$

This can be converted to

$$3 \times 16 = a + b + c$$

which means that

$$48 = a + b + c$$

If $a = 5$, we can solve:

$$48 = 5 + b + c, \text{ or } 48 - 5 = b + c$$

and finally $43 = b + c$. Therefore, the sum of the other two numbers is 43, which is choice (E). Notice that choice (A) is waiting for the person who fails to work the formula and simply subtracts 5 from 16.

Question Type 4: Data Interpretation

Data interpretation problems usually require two basic steps. First, you have to read a chart or graph in order to obtain certain information. Then you have to apply or manipulate the information in order to obtain an answer.

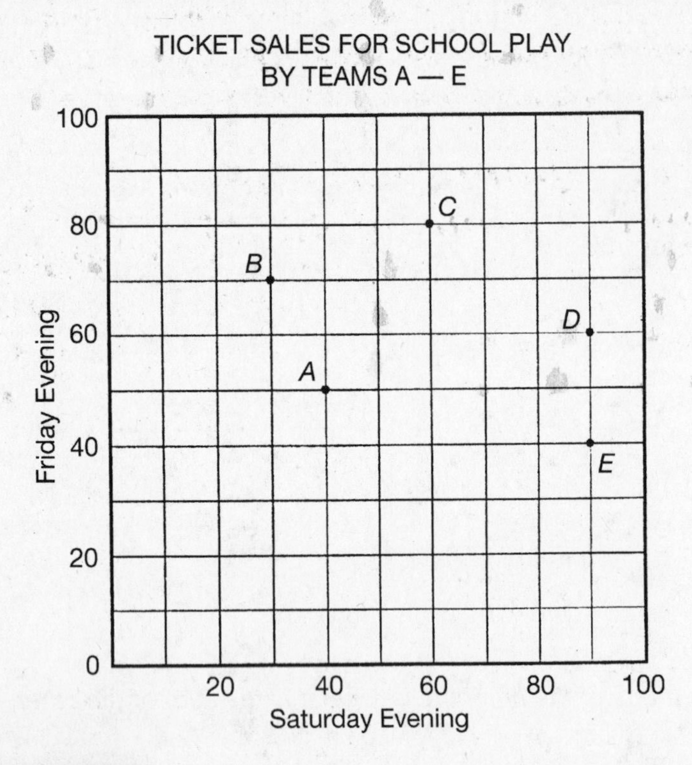

PROBLEM

Which team sold the greatest number of tickets for Friday evening and Saturday evening combined?

(A) Team A (D) Team D
(B) Team B (E) Team E
(C) Team C

SOLUTION

Glancing over the data, you see that the number of tickets sold for Friday evening is represented vertically, whereas the number of tickets sold for Saturday evening is represented horizontally. Points placed on the grid represent each team's ticket sales for the two evenings.

Read the graph to determine the number of tickets sold by each team for Friday evening and Saturday evening.

Add each pair of numbers to find the total number of tickets sold by each team for both evenings.

Compare the totals to see which team sold the most tickets.

The answer is (D) because Team D sold 60 tickets for Friday evening and 90 tickets for Saturday evening for a highest total of 150.

MASTERING ALGEBRA QUESTIONS

Algebra problems use letters or variables to represent numbers. In these types of problems, you will be required to solve existing algebraic expressions or translate word problems into algebraic expressions.

Question Type 1: Algebraic Expressions

Problems involving algebraic expressions often contain hidden shortcuts. You can find these shortcuts by asking yourself, "How can this expression be rearranged?" Often rearrangement will cause an answer to appear almost magically.

There are three basic ways in which you can rearrange an algebraic expression. You can

1. combine like terms
2. factor the expression
3. multiply out the expression

PROBLEM

If $x = \frac{1}{2}$, which of the following equals $x^2 - x + \frac{1}{4}$?

(A) $-\frac{1}{2}$
(B) 0
(C) $\frac{1}{4}$
(D) $\frac{1}{2}$
(E) 1

SOLUTION

You can find the answer by substituting $x = \frac{1}{2}$ into the given expression, but there is an easier way. Look at $x^2 - x + \frac{1}{4}$. Remembering the strategy tip, this expression is equal to the trinomial square $(x - \frac{1}{2})^2$. Now you can see at a glance that since you are told that

$$x = \frac{1}{2}, (x - \frac{1}{2})^2$$

must equal 0. Thus, the answer is (B).

Question Type 2: Word Problems

Among the most common types of word problems found on the SAT are age problems, mixture problems, distance problems, and percent problems. You can find detailed explanations of these and other problem types in the Basic Math Skills Review. There is a strategy, however, that can help you to solve all types of word problems—learning to recognize "keywords."

Keywords are words or phrases that can be translated directly into a mathematical symbol, expression, or operation. As you know, you usually cannot solve a word problem without writing some kind of equation. Learning to spot keywords will enable you to write the equations you need more easily.

Listed below are some of the most common keywords. As you practice solving word problems, you will probably find others.

KEYWORD	MATHEMATICAL EQUIVALENT
is	equals
sum	add
plus	add
more than, older than	add
difference	subtract
less than, younger than	subtract
twice, double	multiply by 2
half as many	divided by 2
increase by 3	add 3
decrease by 3	subtract 3

PROBLEM

Adam has 50 more than twice the number of "frequent flier" miles that Erica has. If Adam has 200 frequent flier miles, how many does Erica have?

(A) 25
(B) 60
(C) 75
(D) 100
(E) 250

SOLUTION

The keywords in this problem are "more" and "twice." If you let a = the number of frequent flier miles that Adam has and e = the number of frequent flier miles that Erica has, you can write: a = 50 + 2*e*. Since a = 200, the solution becomes: 200 = 50 + 2*e*. Therefore, 200 – 50 = 2*e*, or 150 = 2*e*, and $\frac{150}{2} = e$ or $75 = e$, which is choice (C).

Question Type 3: Functions as Models

Some questions on the SAT will involve functions that model real-world situations. In some cases, you may be asked to apply a given function rule to solve a problem. In others, you may be asked to select a graph or an equation that models a given set of data.

Once you have identified a function that seems to model a situation, it is important to check that the function gives the correct output values for all available input values.

PROBLEM

COST OF SLEEPING BAG RENTAL				
Number of Days (x)	2	4	6	8
Cost (y)	$17	$27	$37	$47

The table shows a linear relationship between the cost of renting a sleeping bag and the number of days the sleeping bag is rented. Which of the following models the rental cost y as a function of the number of days x?

(A) $y = x + 15$

(B) $y = 2x + 10$

(C) $y = 5x + 7$

(D) $y = 6x + 5$

(E) $y = 9x - 1$

SOLUTION

The pattern in the table shows that every increase of 2 days results in an increase of $10 in cost. Therefore, the slope of the linear function that models this relationship is $\frac{10}{2}$, or 5. The only answer choice with a slope of 5 is choice (C), $y = 5x + 7$. You can check that this equation is correct by mentally substituting each x-value from the table and verifying that the equation produces the correct set of y-values.

Notice that choices (A), (D), and (E) each give the correct y-value when $x = 2$. These choices are incorrect, however, because they fail to produce the correct y-values when x equals 4, 6, or 8.

MASTERING GEOMETRY QUESTIONS

SAT geometry questions require you to find the area or missing sides of figures given certain information. These problems require you to use "if . . . then" reasoning or to draw figures based on given information.

Question Type 1: "If . . . Then" Reasoning

You will not have to work with geometric proofs on the SAT, but the logic used in proofs can help you enormously when it comes to solving SAT geometry problems. This type of logic is often referred to as "if . . . then" reasoning. In "if . . . then" reasoning, you say to yourself, "If *A* is true, then *B* must be true." By using "if . . . then" reasoning, you can draw conclusions based on the rules and definitions that you know. For example, you might say, "If *ABC* is a triangle, then the sum of its angles must equal 180°."

PROBLEM

If triangle QRS is an equilateral triangle, what is the value of a + b?

(A) 60° (D) 100°
(B) 80° (E) 120°
(C) 85°

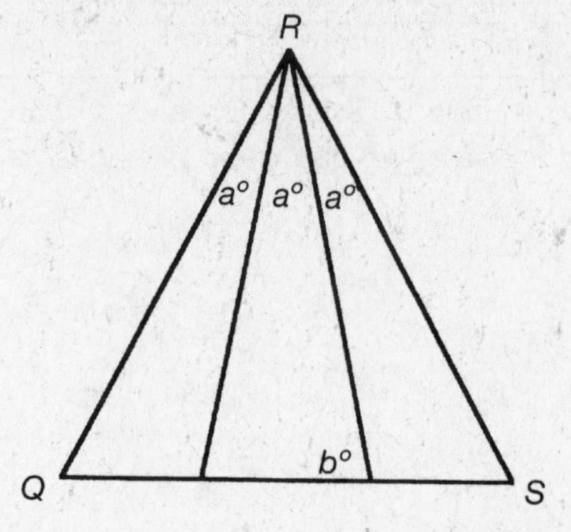

SOLUTION

You can obtain the information that you need to solve this problem by using a series of "if . . . then" statements: "If *QRS* is an equilateral triangle, then each angle must equal 60°." "If angle $R = 60°$, then a must equal 20°." "If $a = 20°$, then b must equal half of (180° – 20°) or 80°." "If $a = 20°$ and $b = 80°$, then $a + b = 100°$." Therefore, answer (D) is correct.

Question Type 2: Drawing Diagrams

Among the most difficult geometry problems on the SAT are those that describe a geometric situation without providing a diagram. For these problems, you must learn to draw your own diagram based on the information that is given. The best way to do this is step-by-step, using each piece of information that the problem provides. As you draw, you should always remember to:

1. label all points, angles, and line segments according to the information provided.
2. indicate parallel or perpendicular lines.
3. write in any measures that you are given.

PROBLEM

If vertical line segment $\overline{AB}$ is perpendicular to line segment $\overline{CD}$ at point O and if ray OE bisects angle BOD, what is the value of angle AOE?

(A) 45° (D) 135°
(B) 90° (E) 180°
(C) 120°

SOLUTION

Draw as follows:

Draw and label vertical line segment $\overline{AB}$.

Draw and label line segment $\overline{CD}$ perpendicular to $\overline{AB}$. Label the right angle that is formed. Label point *O*.

Locate angle *BOD*. Draw and label ray *OE* so that it bisects, or cuts into two equal parts, angle *BOD*. Use equal marks to show that the two parts of the angle are equal. Because you are bisecting a right angle, you can write in the measure 45°.

Your diagram should resemble that shown below. Now you can evaluate your drawing to answer the question. Angle *AOE* is equal to 90° + 45°, or 135°. The answer is (D).

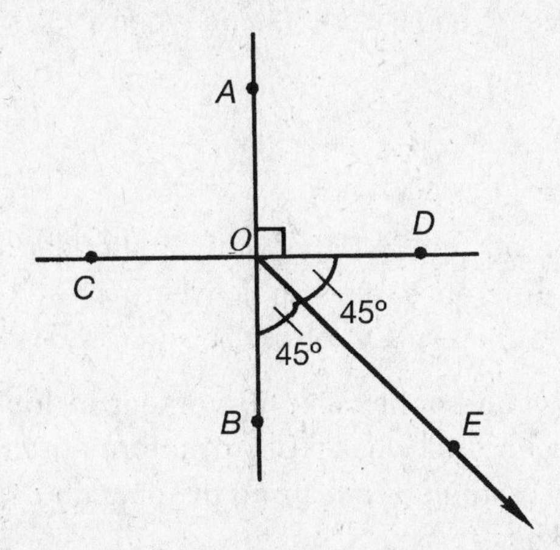

Question Type 3: Behavior of Graphs of Functions

A new type of question on the SAT will involve graphs of functions that are neither linear nor quadratic. You will be presented with a graph of such a function and asked to draw a conclusion about it.

To answer these types of questions, you do not need to know how to write the equation of the graphed function, but you will need to be able to identify the coordinates of points that lie on the function's graph.

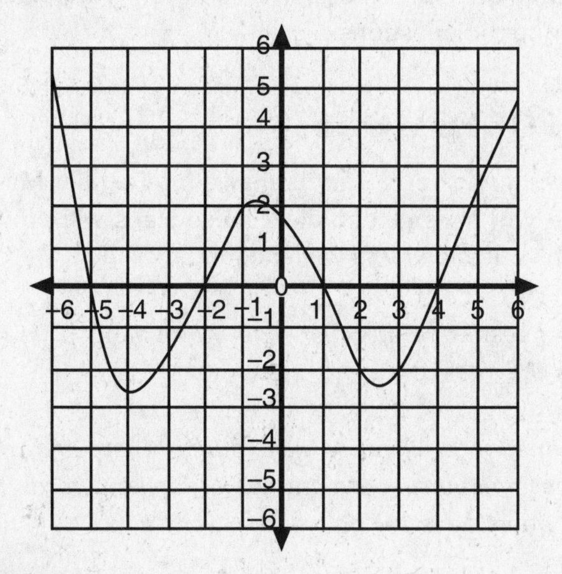

PROBLEM

A function f is graphed as shown. Based on this graph, for what values of x is the value of f(x) negative?

(A) $x < -5$ and $x > 4$

(B) $x < -2$ and $x > 1$

(C) $-2 < x < 1$

(D) $-4 < x < -3$ and $-2 < x < 3$

(E) $-5 < x < -2$ and $1 < x < 4$

SOLUTION

The value of the function $f(x)$ is negative whenever the y-value of the graph of the function is negative. The y-values are negative for sections of the graph of the function that lie below the x-axis. There are two sections of the function's graph that lie below the x-axis. The first section occurs when the x-values are between –5 and –2. The second section occurs when the x-values are between 1 and 4. Therefore, the value of $f(x)$ is negative when $-5 < x < -2$ and when $1 < x < 4$, making choice (E) the correct answer.

POINTS TO REMEMBER

- Certain patterns of factoring appear so often on the SAT that you should learn to spot them in every possible form. These include: difference of squares, where $x^2 - y^2 = (x + y)(x - y)$; and trinomial squares, where $(x + y)^2 = x^2 + 2xy + y^2$ or $(x - y)^2 = x^2 - 2xy + y^2$.
- When solving data interpretation problems, you should take a moment to look at the graph or chart in order to see how it is set up and what types of information are displayed. You should not, however, waste time reading or analyzing the data until you read the question(s) and find out what is being asked.
- If you are dealing with a geometry question that refers to a figure and the question does not provide the figure, draw the figure. This will help you visualize the problem. In addition, if a figure is provided, note whether or not it is drawn to size.
- Do not use your calculator for simple mathematic calculations. Using your calculator for such purposes may waste more time than it saves.
- Be sure to review the reference information. It provides some common formulas that may prove valuable when you are answering the questions.

MASTERING REGULAR MATH QUESTIONS

The following steps should be used to help guide you through answering Regular Math questions. Combined with the review material that you have just studied, these steps will provide you with the tools necessary to correctly answer the questions you will encounter.

Try to determine the type of question with which you are dealing. This will help you focus in on how to attack the question.

Carefully read all of the information presented. Make sure you are answering the question and not reading the question incorrectly. Look for key words that can help you determine what the question is asking.

Perform the operations indicated, but be sure you are taking the easiest approach. Simplify all expressions and equations before performing your calculations. Draw your own figures if a question refers to them but does not provide them.

Try to work backward from the answer choices if you are having difficulty determining an answer.

STEP 5 *If you are still having difficulty determining an answer, use the process of elimination. If you can eliminate at least two choices, you will greatly increase your chances of correctly answering the question. Eliminating three choices means that you have a fifty-fifty chance of correctly answering the question if you guess.*

STEP 6 *Once you have chosen an answer, fill in the oval that corresponds to the question and answer that you have chosen. Beware of stray lines on your answer sheet, as they may cause your answers to be scored incorrectly.*

Now, use the information you have just learned to answer the following drill questions. Doing so will help reinforce the material learned and will better prepare you for the actual test.

DRILL: REGULAR MATH

DIRECTIONS: Choose the best answer choice and fill in the corresponding oval on the answer sheet.

1. If $8 - (7 - 6 - 5) = 8 - 7 - (x - 5)$, what is the value of x?

   (A) −6 (D) 4
   (B) −16 (E) 6
   (C) −18

2. The average (arithmetic mean) volume of 4 containers is 40 liters. If 3 of these containers each have a volume of 35 liters, what is the volume in liters of the fourth container?

   (A) 40 (D) 105
   (B) 50 (E) 160
   (C) 55

3. $$\begin{array}{r} 2\diamond5 \\ \times\quad 7 \\ \hline 1{,}8\oplus5 \end{array}$$

   In the correctly computed multiplication problem above, if $\diamond$ and $\oplus$ are different digits, then $\diamond =$

   (A) 2 (D) 4
   (B) 6 (E) 7
   (C) 5

4. If for any number n, $[n]$ is defined as the least whole number that is greater than or equal to n, then $[-3.7] + 14 =$

   (A) 10.3 (D) 10
   (B) 17.7 (E) 11
   (C) 18

5. How many kilometers will a cyclist travel in 1 minute if she cycles at a rate of 30 km/hour?

(A) 0.5
(B) 1
(C) 2
(D) 5
(E) $\frac{1}{30}$

6. If $\overline{DE}$ is parallel to $\overline{ST}$ and triangle RST is an isoceles triangle, then $\angle E$ is

(A) 50°
(B) 40°
(C) 70°
(D) 90°
(E) 75°

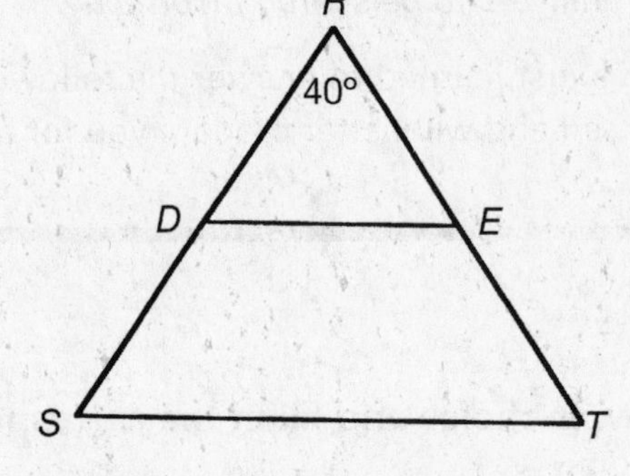

7. If 20 students share the cost equally of a gift for Coach Brown, what percent of the total cost do 5 of the students share?

(A) 5%
(B) 10%
(C) 25%
(D) 40%
(E) 50%

8. If $n \div 5 = 150 \div 25$, then $n =$

(A) 5
(B) 6
(C) 25
(D) 30
(E) 125

9. If the average (arithmetic mean) of –9 and s is –9, then $s =$

(A) –9
(B) 0
(C) 9
(D) –18
(E) –4.5

10. If $x \diamond y = (x - 3)y$, then $8 \diamond 3 + 5 \diamond 2 =$

(A) 34
(B) 50
(C) 0
(D) –5
(E) 19

11. If $22 \times 3 \times R = 6$, the $R =$

(A) $\frac{1}{9}$
(B) $\frac{1}{11}$
(C) $\frac{1}{8}$
(D) 11
(E) 9

12. What is the total area, in square meters, of two adjacent square garden plots with sides 4 meters and 1 meter, respectively?

(A) 1
(B) 4
(C) 16
(D) 17
(E) 20

13. If line m is parallel to line n as shown, then $\angle b =$

(A) 30°
(B) 45°
(C) 55°
(D) 60°
(E) 120°

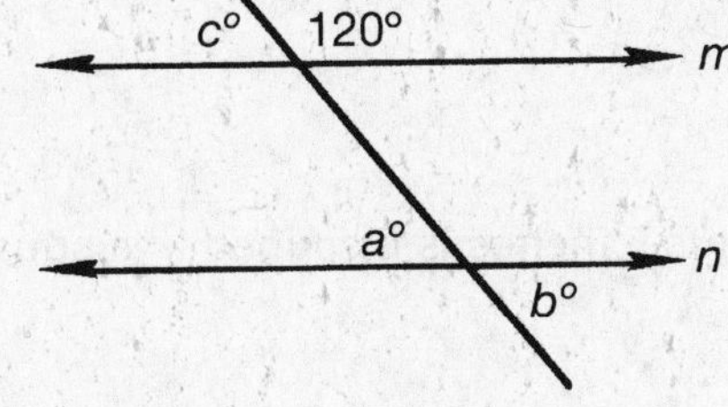

14. If $\frac{6m}{9} = 4$, then $18m =$

(A) 6
(B) 24
(C) 36
(D) 72
(E) 108

15. During a local high school track meet, all three contestants earn points. Olivia earns 12 times more points than Barbara, and Diane earns 8 times more points than Barbara. What is the ratio of Diane's points to Olivia's points?

(A) 96 : 1
(B) 8 : 1
(C) 3 : 2
(D) 2 : 3
(E) 1 : 8

16. If $x^2 - z^2 = 130$, and $x - z = 10$, then $x + z =$

(A) 10
(B) 13
(C) 36
(D) 30
(E) 120

17. In the figure shown here, what does x equal?

(A) 80
(B) 90
(C) 100
(D) 110
(E) 120

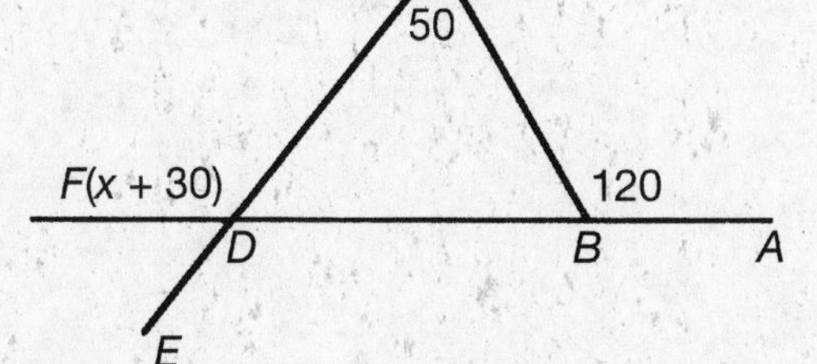

18. In triangle ABC, $\angle ABC$ is 53° and $\angle BCA$ is 37°, $\overline{BC} = 5$ inches and $\overline{AC} = 4$ inches. What is the length of the shortest side?

(A) 2 inches
(B) 3 inches
(C) 4 inches
(D) 5 inches
(E) 6 inches

19. If 22 is the average of n, n, n, 15, and 35, then $n =$

(A) 16
(B) 18
(C) 20
(D) 50
(E) 110

20. If $7x + 18 = 16x$, then $x =$

(A) -2
(B) $\frac{18}{23}$
(C) $\frac{23}{18}$
(D) 2
(E) 36

21. A circular piece of turquoise with radius 5 inches is inscribed in a square piece of tile. What is the area in square inches of the tile?

(A) 25
(B) 78.5
(C) 50
(D) 10
(E) 100

22. If the area of the shaded region is 24π, then the radius of circle O is

(A) 4
(B) 16
(C) 2
(D) 4
(E) 2

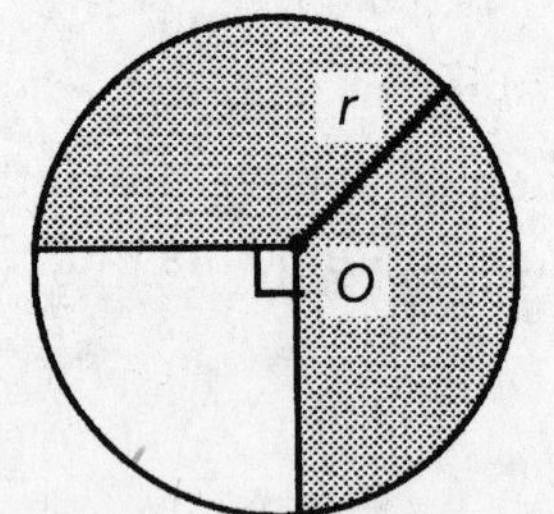

23. If, in the triangle shown here, $\overline{AS} < \overline{ST}$, which of the following cannot be the value of t?

(A) 20
(B) 36
(C) 65
(D) 71
(E) 80

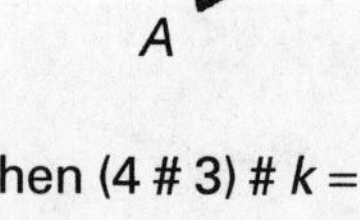

24. If $R \# S = \frac{R+2}{S}$, then $(4 \# 3) \# k =$

(A) $\frac{5}{4}$
(B) 2
(C) $\frac{1}{3}$
(D) $\frac{4}{k}$
(E) $4k$

25. If $48x - 6y = 1$, then $16x - 2y =$

(A) -3
(B) 0
(C) $\frac{1}{3}$
(D) $\frac{1}{2}$
(E) 1

26. Sam is 5 years older than Marcia, and Marcia is 4 years older than Jack will be 2 years from now. How many years older than Jack is Sam?

(A) 11 (D) 6

(B) 9 (E) 2

(C) 7

27. A cat ate $\frac{1}{2}$ of the food in its dish in the morning, and 2 ounces of food in the afternoon. the cat's owner came home to find $\frac{2}{5}$ of the original amount of food in the dish. How many ounces of food were in the dish at the start of the day?

(A) 4 ounces (D) 16 ounces

(B) 5 ounces (E) 20 ounces

(C) 8 ounces

28. The average value of A, B, and C is 8, and $2(A + C) = 24$. What is the value of B?

(A) 0 (D) 32

(B) 3 (E) 12

(C) –16

29. If S is the sum of the positive integers from 1 to n, inclusive, then the average (arithmetic mean), k, of these integers can be represented by the formula

(A) $S \times n$ (D) $S + n$

(B) $\frac{S}{n}$ (E) $n + 1$

(C) $\frac{k}{n}$

30. If $3a + 4a = 14$, and $b + 6b = 20$, then $7(a + b) =$

(A) 21 (D) 40

(B) 28 (E) 42

(C) 34

31. The perimeter of an equilateral triangle is 18. What is the altitude of the triangle?

(A) 3 (D) $\sqrt{3}$

(B) 6 (E) $3\sqrt{3}$

(C) 9

32. A truck driver covered 450 miles during a 9-hour period, stopped for one hour, then drove 90 miles in 2 hours. What was the driver's average rate, in miles per hour, for the total distance traveled?

(A) 45 (D) 54

(B) 49 (E) 60

(C) 50

33. If $x + y = 7$ and $x - y = 3$, then $x^2 - y^2 =$

(A) 4 (D) 21

(B) 10 (E) 40

(C) 16

34. If $\sqrt{x-1} = 2$ then $(x - 1)^2 =$

(A) 4 (D) 10

(B) 6 (E) 16

(C) 8

35. If $m < n < 0$, which expression must be < 0?

(A) The product of m and n

(B) The square of the product of m and n

(C) The sum of m and n

(D) The result of subtracting m from n

(E) The quotient of dividing n by m

36. In the figure shown, if $\overline{BD}$ is the bisector of angle ABC, and angle ABD is one-fourth the size of angle XYZ, what is the size of angle ABC?

(A) 21°

(B) 28°

(C) 42°

(D) 63°

(E) 168°

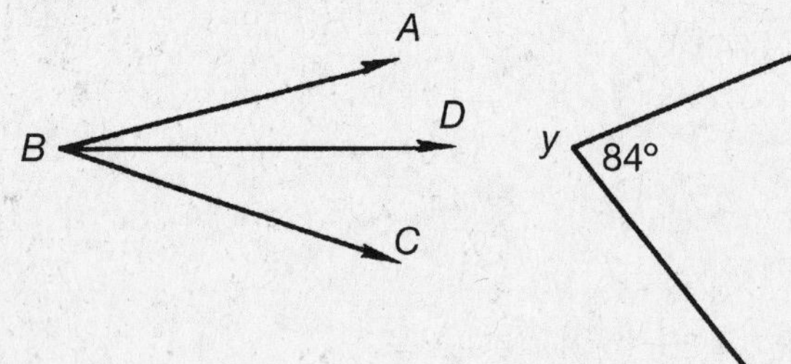

37. A cube of volume 8 cubic centimeters is placed directly next to a cube of volume 125 cubic centimeters. What is the perpendicular distance in centimeters from the top of the larger cube to the top of the smaller cube?

(A) 7 (D) 2

(B) 5 (E) 0

(C) 3

38. If Company A's November income is $1 million less than its average income of July and September, what will its November income be?

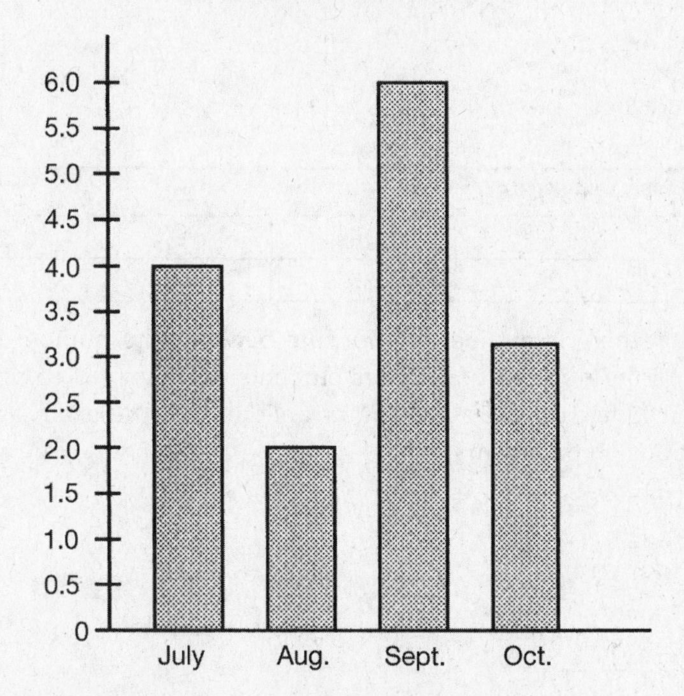

(A) $2 million
(B) $3 million
(C) $4 million
(D) $5 million
(E) $10 million

39. If $21 - (a - b) = 2(b + 9)$, and $a = 8$, what is the value of b?

(A) 31
(B) 13
(C) 4
(D) $-\frac{5}{3}$
(E) -5

40.

EMPLOYEES IN DIVISION *R* OF CORPORATION *S*					
Year	2001	2002	2003	2004	2005
Number of Employees	1,900	2,200	2,500	2,800	3,100

Based on the increase in employees over the past 5 years, how many employees can Division *R* expect to have by 2006?

(A) 3,100
(B) 3,400
(C) 3,550
(D) 3,700
(E) 4,000

41. The approximate number of people n attending an amusement park on a particular day is related to t, the daily the high temperature in degrees Fahrenheit. The function $n(t) = -7t^2 + 1{,}190t + c$ models this situation, where c is a constant. If 6,425 people attend on a day when the high temperature is 70°F, approximately how many people can be expected to attend when the daily high temperature is 90°?

(A) 7,710 (D) 43,975

(B) 7,825 (E) 50,400

(C) 8,260

42.

TIME ALLOTTED FOR CHEMISTRY EXAMS				
Number of Questions (n)	5	10	15	20
Time in Minutes (t)	20	35	50	65

The preceding table shows that there is a linear relationship between the number of questions on a chemistry exam and the number of minutes students are given to complete the exam. Which of the following models the time t in minutes allotted for the exam as a function of n, the number of questions on the exam?

(A) $t(n) = n + 15$ (D) $t(n) = 4n$

(B) $t(n) = 2n + 10$ (E) $t(n) = 5n - 5$

(C) $t(n) = 3n + 5$

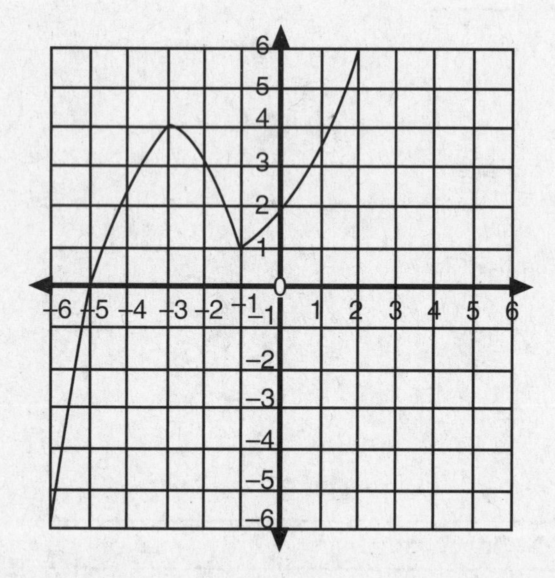

43. The graph shows a portion of the function f. For how many values of x is $f(x) = 2$ for the portion of the graph shown?

(A) 0 (D) 3

(B) 1 (E) an infinite number

(C) 2

44. The graph shows a portion of the function g. What is the greatest value of $g(x)$ for the portion of the graph shown?

(A) 2
(B) 3
(C) 4
(D) 5
(E) 6

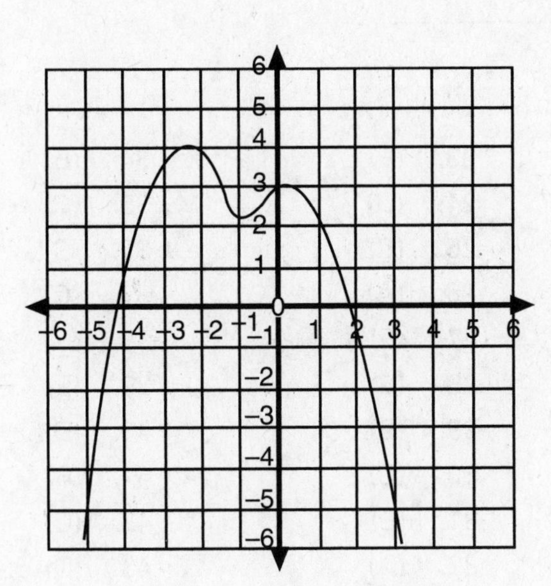

REGULAR MATH DRILL

ANSWER KEY

Drill: Regular Math

1.	(A)	12.	(D)	23.	(E)	34.	(E)
2.	(C)	13.	(D)	24.	(D)	35.	(C)
3.	(B)	14.	(E)	25.	(C)	36.	(C)
4.	(E)	15.	(D)	26.	(A)	37.	(C)
5.	(A)	16.	(B)	27.	(E)	38.	(C)
6.	(C)	17.	(A)	28.	(E)	39.	(E)
7.	(C)	18.	(B)	29.	(B)	40.	(D)
8.	(D)	19.	(C)	30.	(C)	41.	(B)
9.	(A)	20.	(D)	31.	(E)	42.	(C)
10.	(E)	21.	(E)	32.	(A)	43.	(D)
11.	(B)	22.	(D)	33.	(D)	44.	(C)

CHAPTER 9

MASTERING STUDENT-PRODUCED RESPONSE QUESTIONS

MASTERING STUDENT-PRODUCED RESPONSE QUESTIONS

The Student-Produced Response format of the SAT is designed to give the student a certain amount of flexibility in answering questions. In this section the student must calculate the answer to a given question and then enter the solution into a grid. The grid is constructed so that a solution can be given in either decimal or fraction form. Either form is acceptable unless otherwise stated.

The problems in the Student-Produced Response section try to reflect situations arising in the real world. Calculations will involve objects occurring in everyday life. There is also an emphasis on problems involving data interpretation. In keeping with this emphasis, students will be allowed the use of a calculator during the exam.

Through this review, you will learn how to successfully attack Student-Produced Response questions. Familiarity with the test format combined with solid math strategies will prove invaluable in answering the questions quickly and accurately.

ABOUT THE DIRECTIONS

Each Student-Produced Response question will require you to solve the problem and enter your answer in a grid. There are specific rules you will need to know for entering your solution. If you enter your answer in an incorrect form, you will not receive credit, even if you originally solved the problem correctly. Therefore, you should carefully study the following rules now, so you don't have to waste valuable time during the actual test.

DIRECTIONS FOR STUDENT-PRODUCED RESPONSE QUESTIONS

For each of the questions below (15-24), solve the problem and indicate your answer by marking the ovals in the special grid, as shown in the examples below.

Answer: $\frac{9}{5}$ **or 9/5 or 1.8**

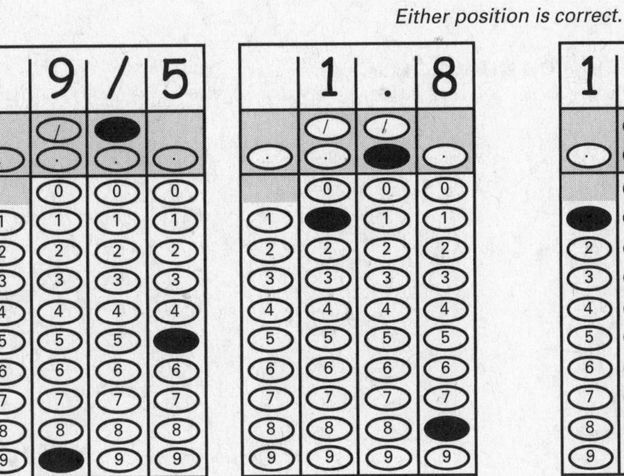

NOTE: You may start your anwers in any column, space permitting. Columns not needed should be left blank.

- Mark no more than one oval in any column.
- Because the answer sheet will be machine scored, you will receive credit only if the ovals are filled in correctly.
- Although not required, it is suggested that you write your answer in the boxes at the top of the columns to help you fill in the ovals accurately.
- Some problems may have more than one correct answer. In such cases, grid only one answer.
- No question has a negative answer.
- Mixed numbers such as $3\frac{1}{2}$ must be gridded as 3.5 or 7/2.

(If 3 1 / 2 is gridded, it will be interpreted as $\frac{31}{2}$, not $3\frac{1}{2}$.)

- **Decimal Accuracy:** If you obtain a decimal answer, enter the most accurate value the grid will accommodate. For example, if you obtain an answer such as 0.6666 ..., you should record the result as .666 or .667. Less accurate values such as .66 or .67 are not acceptable.

Acceptable ways to grid $\frac{2}{3}$ **= .6666...**

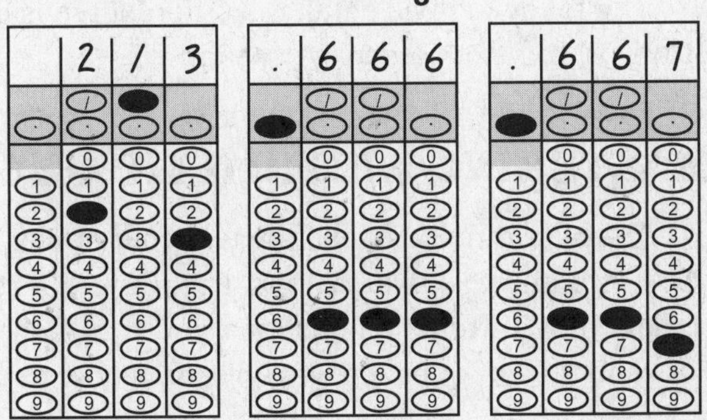

SAMPLE QUESTION

How many pounds of apples can be bought with $5.00 if apples cost $0.40 a pound?

SOLUTION

Converting dollars to cents we obtain the equation

$$x = 500 \div 40$$

$$x = 12.5$$

The solution to this problem would be gridded as

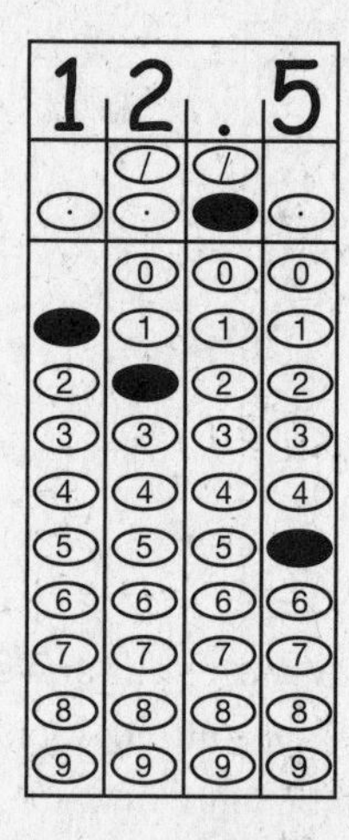

ABOUT THE QUESTIONS

Within the SAT Student-Produced Response section you will be given 5–10 questions. You will have 20 minutes to answer these questions in addition to some multiple-choice questions. This means you will be required to answer about 16 questions. Therefore, you should work quickly.

The Student-Produced Response questions will come from the areas of arithmetic, algebra, and geometry. There is an emphasis on word problems and on data interpretation, which usually involves reading tables to answer questions. Many of the geometry questions will refer to diagrams or will ask you to create a figure from information given in the question.

The following will detail the different types of questions you should expect to encounter on the Student-Produced Response section.

MASTERING ARITHMETIC QUESTIONS

These arithmetic questions test your ability to perform standard manipulations and simplifications of arithmetic expressions. For some questions, there is more than one approach. There are six kinds of arithmetic questions you may encounter in the Student-Produced Response section. For each type of question, we will show how to solve the problem and grid your answer.

Question Type 1: Properties of a Whole Number *N*

This problem tests your ability to find a whole number with a given set of properties. You will be given a list of properties of a whole number and asked to find that number.

PROBLEM

The properties of a whole number N are

(A) N is a perfect square.

(B) N is divisible by 2.

(C) N is divisible by 3.

Grid in the second smallest whole number with the above properties.

SOLUTION

Try to first obtain the smallest number with the above properties. The smallest number with properties (B) and (C) is 6. Because property (A) says the number must be a perfect square, the smallest number with properties (A), (B), and (C) is 36.

$6^2 = 36$ is the smallest whole number with the above properties. The second smallest whole number (the solution) is

$2^2 6^2 = 144.$

The correct answer entered into the grid is

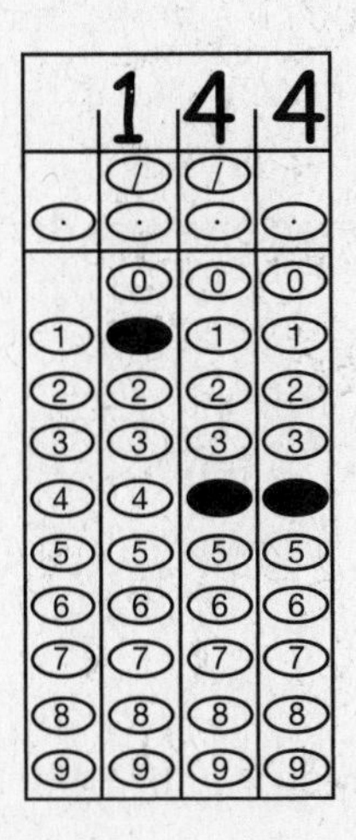

Question Type 2: Simplifying Fractions

This type of question requires you to simplify fractional expressions and grid the answer in the format specified. By canceling out terms common to both the numerator and denominator, we can simplify complex fractional expressions.

PROBLEM

Change $\frac{1}{2} \times \frac{3}{7} \times \frac{2}{8} \times \frac{14}{10} \times \frac{1}{3}$ *to decimal form.*

SOLUTION

The point here is to cancel out terms common to both the numerator and denominator. Once the fraction is brought down to lowest terms, the result is entered into the grid as a decimal.

After cancellation we are left with the fraction $\frac{1}{40}$. Equivalently,

$$\frac{1}{40} = \frac{1}{10} \times \frac{1}{4} = \frac{1}{10}(0.25) = 0.025$$

Hence, in our grid we enter

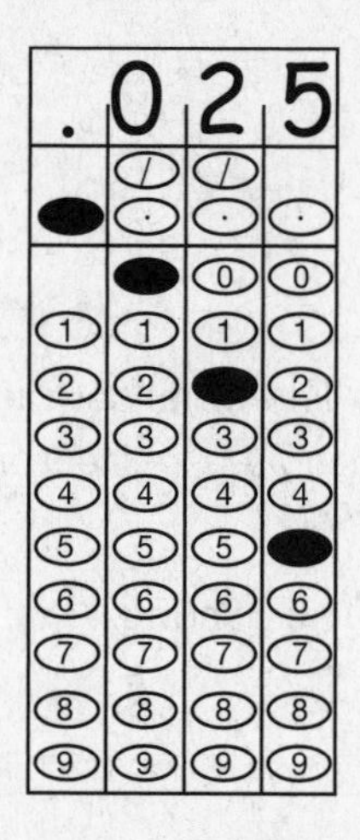

Note: If "decimal" was not specified, any correct version of the answer could be entered into the grid.

Question Type 3: Prime Numbers of a Particular Form

Here, you will be asked to find a prime number with certain characteristics. Remember—a prime number is a number that can only be divided by itself and 1.

PROBLEM

Find a prime number of the form $7k + 1 < 50$.

SOLUTION

This is simply a counting problem. The key is to list all the numbers of the form $7k + 1$ starting with $k = 0$. The first one that is prime is the solution to the problem.

The whole numbers of the form $7k + 1$ which are less than 50 are 1, 8, 15, 22, 29, 36, and 43. Of these, 29 and 43 are prime numbers. The possible solutions are 29 and 43.

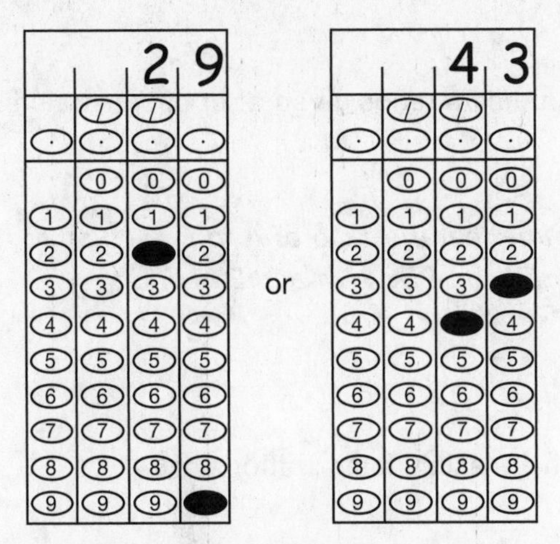

Question Type 4: Order of Operations

The following question type tests your knowledge of the arithmetic order of operations. Always work within the parentheses or with absolute values first, while keeping in mind that multiplication and division are carried out before addition and subtraction.

PROBLEM

Find a solution to the equation $x \div 3 \times 4 \div 2 = 6$.

SOLUTION

The key here is to recall the order of precedence for arithmetic operations. After simplifying the expression one can solve for x.

Because multiplication and division have the same level of precedence, we simplify the equation from left to right to obtain

$$\frac{x}{3} \times 4 \div 2 = 6$$

$$\frac{4x}{3} \div 2 = 6$$

$$\frac{2x}{3} = 6$$

$$x = 9$$

Because 9 solves the above problem, our entry in the grid is

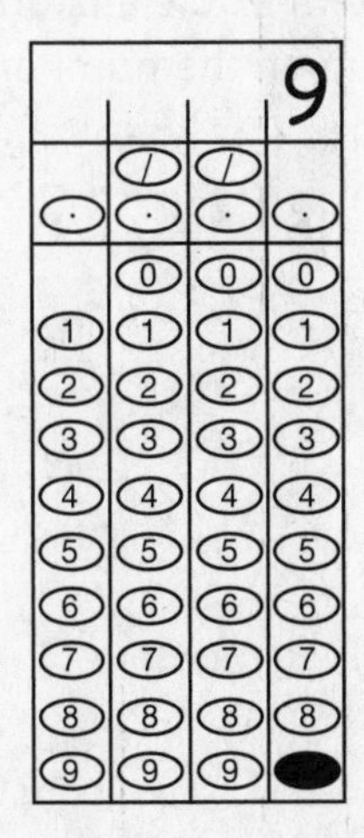

Question Type 5: Solving for Ratios

This type of question tests your ability to manipulate ratios given a set of constraints.

PROBLEM

Let A, B, C, and D be positive integers. Assume that the ratio of A to B is equal to the ratio of C to D. Find a possible value for A if the product of BC = 24 and D is odd.

SOLUTION

The quickest way to find a solution is to list the possible factorizations of 24:

1×24

2×12

3×8

4×6

Since $AD = BC = 24$ and D is odd, the only possible solution is $A = 8$ (corresponding to $D = 3$).

In the following grid we enter

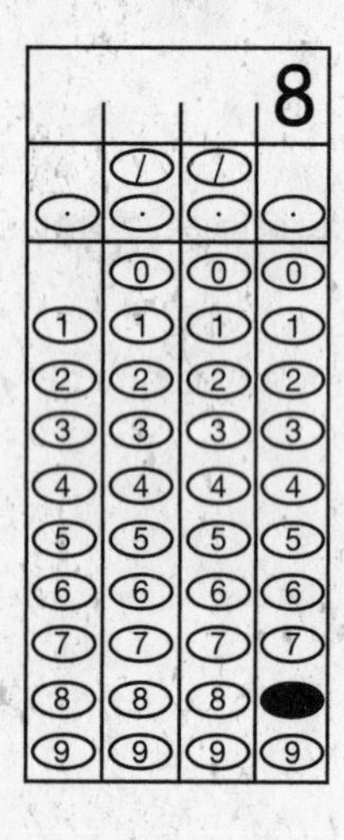

Question Type 6: Simplifying Arithmetic Expressions

Here you will be given an arithmetic problem that is easier to solve if you transform it into a basic algebra problem. This strategy saves valuable time by cutting down on the number and complexity of computations involved.

PROBLEM

Simplify $1-\left(\frac{1}{2}+\frac{1}{4}+\frac{1}{8}+\frac{1}{16}+\frac{1}{32}+\frac{1}{64}\right)$.

SOLUTION

This problem can be done one of two ways. The "brute force" approach would be to get a common denominator and simplify. An approach involving less computation is given below.

$$\text{Set } S = 1 - \left(\frac{1}{2} + \frac{1}{4} + \frac{1}{8} + \frac{1}{16} + \frac{1}{32} + \frac{1}{64}\right)$$

Multiplying this equation by 2 we obtain

$$2S = 2 - \left(1 + \frac{1}{2} + \frac{1}{4} + \frac{1}{8} + \frac{1}{16} + \frac{1}{32}\right)$$

$$2S = 1 - \left(\frac{1}{2} + \frac{1}{4} + \frac{1}{8} + \frac{1}{16} + \frac{1}{32}\right)$$

$$2S = 1 - \left(\frac{1}{2} + \frac{1}{4} + \frac{1}{8} + \frac{1}{16} + \frac{1}{32} + \frac{1}{64}\right) + \frac{1}{64}$$

$$2S = S + \frac{1}{64}$$

$$S = \frac{1}{64}$$

We enter into the grid

1 / 6 4

MASTERING ALGEBRA QUESTIONS

Within the Student-Produced Response section, you will also encounter algebra questions that will test your ability to solve algebraic expressions in the setting of word problems. You may encounter the following six types of algebra questions during the SAT. As in the previous section, we provide methods for approaching each type of problem.

Question Type 1: Solving a System of Linear Equations

This is a standard question that will ask you to find the solution to a system of two linear equations with two unknowns.

PROBLEM

Consider the system of simultaneous equations given by

$y - 2 = x - 4$

$y + 3 = 6 - x$

Solve for the quantity 6y + 3.

SOLUTION

This problem can be solved by taking the first equation given and solving for *x*. This would yield

$$x = y + 2$$

Next, we plug this value for *x* into the second equation, giving us

$$y + 3 = 6 - (y + 2)$$

Solve this equation for *y* and we get

$$y = \frac{1}{2}$$

We are asked to solve for $6y + 3$, so we can plug our value for *y* in and get

$$6\left(\frac{1}{2}\right) + 3 = 6$$

Our answer is 6 and gridded correctly it is

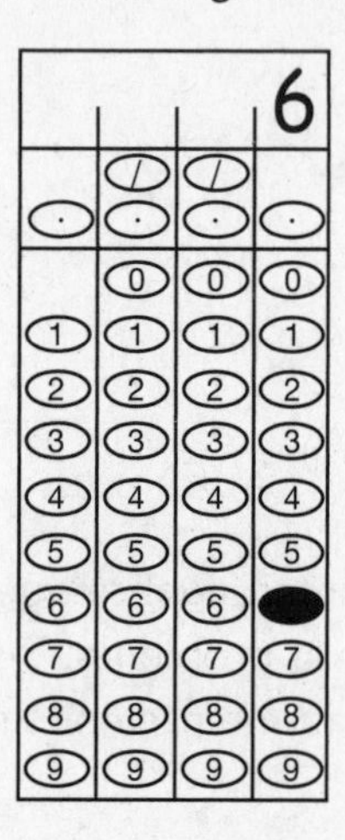

Question Type 2: Word Problems Involving Age

When dealing with this type of question, you will be asked to solve for the age of a particular person. The question may require you to determine how much older one person is, how much younger one person is, or the specific age of the person.

PROBLEM

Tim is 2 years older than Jane and Joe is 4 years younger than Jane. If the sum of the ages of Jane, Joe, and Tim is 28, how old is Joe?

SOLUTION

Define Jane's age to be the variable x and work from there.

Let

$$\begin{aligned} \text{Jane's age} &= x \\ \text{Tim's age} &= x + 2 \\ \text{Joe's age} &= x - 4 \end{aligned}$$

Summing up the ages we get

$$\begin{aligned} x + x + 2 + x - 4 &= 28 \\ 3x - 2 &= 28 \\ 3x &= 30 \\ x &= 10 \end{aligned}$$

Joe's age = 10 – 4 = 6.

Hence, we enter into the grid

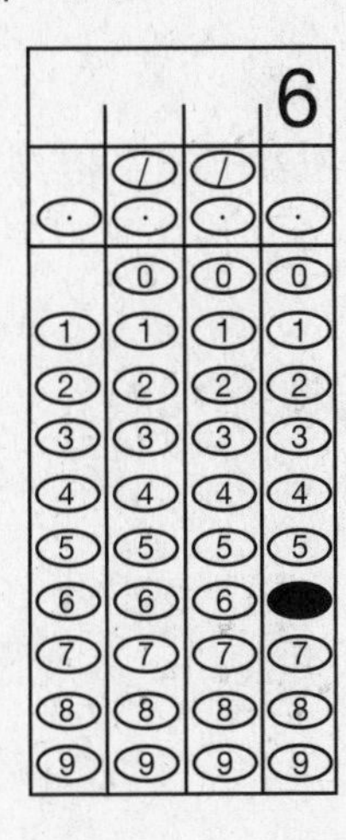

Question Type 3: Word Problems Involving Money

Word problems involving money will test your ability to translate the information given into an algebraic statement. You will also be required to solve your algebraic statement.

PROBLEM

After receiving his weekly paycheck on Friday, a man buys a television for \$100, a suit for \$200, and a radio for \$50. If the total money he spent amounts to 40% of his paycheck, what is his weekly salary?

SOLUTION

Simply set up an equation involving the man's expenditures and the percentage of his paycheck that he used to buy them.

Let the amount of the man's paycheck equal x. We then have the equation

$$40\%x = 100 + 200 + 50$$

$$0.4x = 350$$

$$x = \$875$$

In the grid we enter

	8	7	5

Question Type 4: Systems of Nonlinear Equations

This type of question will test your ability to perform the correct algebraic operations for a given set of equations in order to find the desired quantity.

PROBLEM

Consider the system of equations

$$x^2 + y^2 = 8$$

$$xy = 4$$

Solve for the quantity $3x + 3y$.

SOLUTION

Solve for the quantity $x + y$ and not for x or y individually.

First, multiply the equation $xy = 4$ by 2 to get $2xy = 8$. Adding this to $x^2 + y^2 = 8$ we obtain

$$x^2 + 2xy + y^2 = 16$$

$$(x + y)^2 = 16$$

$$x + y = 4$$

or $$x + y = -4$$

Hence, $3x + 3y = 12$ or $3x + 3y = -12$. We enter 12 for a solution because -12 cannot be entered into the grid.

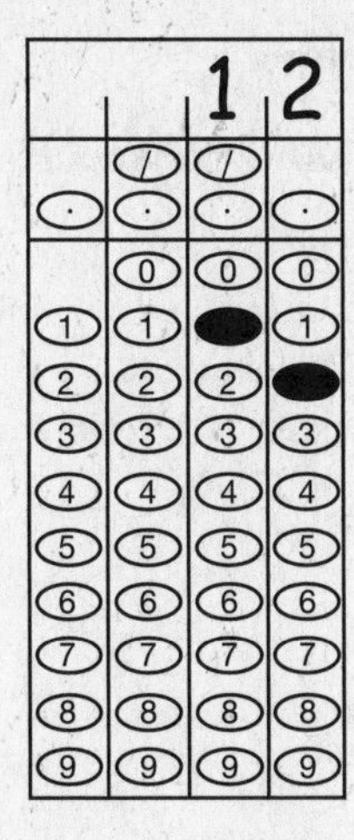

Question Type 5: Word Problems Involving Hourly Wage

When dealing with this type of question, you will be required to form an algebraic expression from the information based on a person's wages. You will then solve the expression to determine the person's wages (i.e., hourly, daily, annually).

PROBLEM

Jim works 25 hours a week earning $10 an hour. Sally works 50 hours a week earning y dollars an hour. If their combined income every two weeks is $2,000, find the amount of money Sally makes an hour.

SOLUTION

Be careful. The combined income is given over a two-week period.

Simply set up an equation involving income. We obtain

$$2[(25)(10) + (50)(y)] = 2{,}000$$

$$[(25)(10) + (50)(y)] = 1{,}000$$

$$250 + 50y = 1{,}000$$

$$50y = 750$$

$$y = \$15 \text{ an hour}$$

We enter in the grid

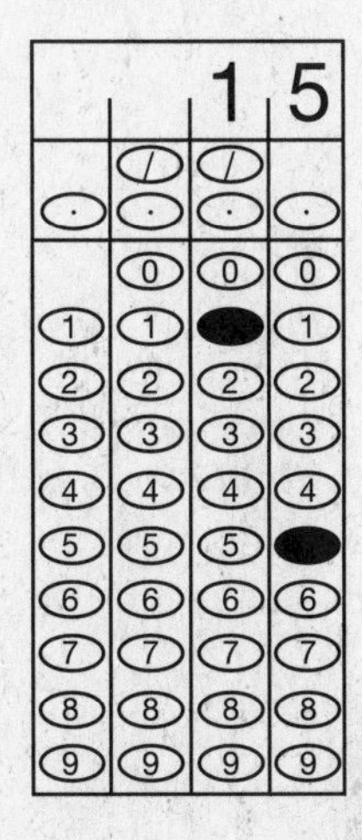

Question Type 6: Word Problems Involving Consecutive Integers

In this type of question, you will need to set up an equation involving consecutive integers based on the product of the integers, which is given.

PROBLEM

Consider two positive consecutive odd integers such that their product is 143. Find their sum.

SOLUTION

Be careful. Notice x and y are consecutive odd integers.

Let

$$\text{1st odd integer} = x$$

$$\text{2nd odd integer} = x + 2$$

We get

$$x(x + 2) = 143$$

$$x^2 + 2x - 143 = 0$$

$$(x - 11)(x + 13) = 0$$

Hence $\quad x = 11$

and $\quad x = -13$

From the preceding we obtain the solution sets {11, 13} and {–13, –11} whose sums are 24 and –24, respectively. Because the problem specifies that the integers are positive, we enter 24.

		2	4
	/	/	
.	.	.	.
	0	0	0
1	1	1	1
2	2	●	2
3	3	3	3
4	4	4	●
5	5	5	5
6	6	6	6
7	7	7	7
8	8	8	8
9	9	9	9

MASTERING GEOMETRY QUESTIONS

In this section, we will explain how to solve questions that test your ability to find the area of various geometric figures. There are six types of questions you may encounter.

Question Type 1: Area of an Inscribed Triangle

This question asks you to find the area of a triangle that is inscribed in a square. By knowing certain properties of both triangles and squares, we can deduce the necessary information.

PROBLEM

Consider the triangle inscribed in the square.

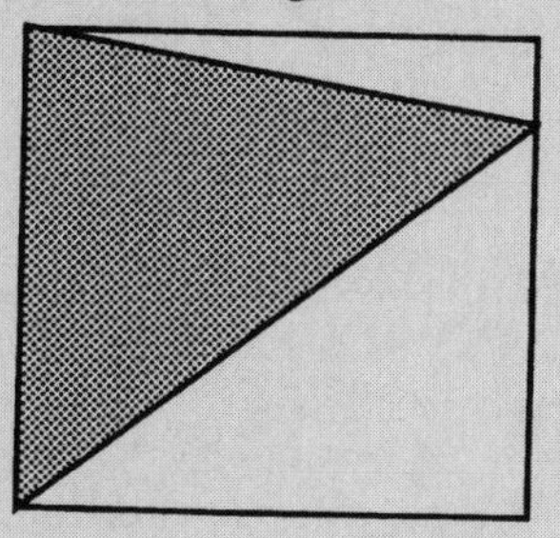

If the area of the square is 36, find the area of the triangle.

SOLUTION

Find the height of the triangle.

Let x be the length of the square. The four sides of a square are equal, and the area of a square is the length of a side squared, so $x^2 = 36$. Therefore, $x = 6$.

The area of a triangle is given by

$$\frac{1}{2}(\text{base})(\text{height})$$

Here x is both the base and height of the triangle. The area of the triangle is

$$\frac{1}{2}(6)(6) = 18$$

This is how the answer would be gridded.

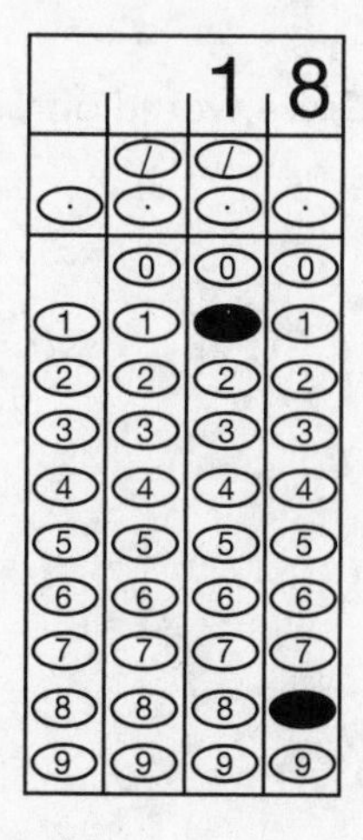

Question Type 2: Length of the Side of a Triangle

For this type of question, one must find the length of a right triangle given information about the other sides. The key here is to apply the Pythagorean Theorem, which states that the square of the hypotenuse of a right triangle is equal to the sum of the squares of the other two sides.

PROBLEM

Consider the line given below

A B C D

where $\overline{AD} = 30$ *and* $\overline{AB} = 5$. *What length is* $\overline{BC}$ *if the sides* $\overline{AB}$, $\overline{BC}$, *and* $\overline{CD}$ *form the sides of a right triangle?*

SOLUTION

Draw a diagram and fill in the known information.

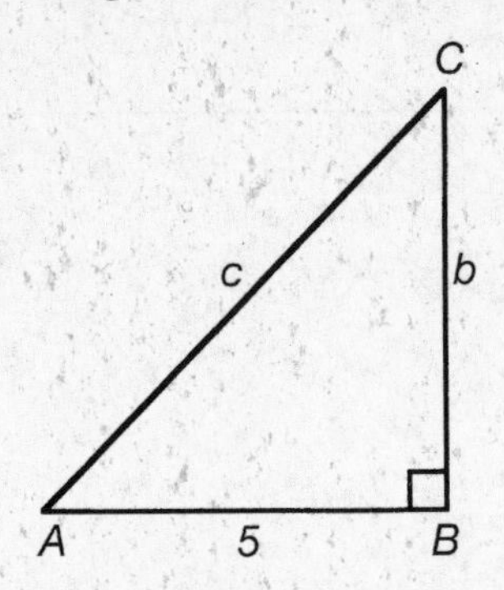

Next, apply the Pythagorean Theorem ($a^2 + b^2 = c^2$), filling in the known variables. Here we are solving for BC (b in our equation). We know that $a = 5$, and because $AD = 30$ and $AB = 5$, $BD = 25$. Filling in these values, we obtain this equation:

$$5^2 + x^2 = (25 - x)^2$$

$$25 + x^2 = 625 - 50x + x^2$$

$$50x = 600$$

$$x = 12$$

This is one possible solution. If we had chosen $x = CD$ and $25 - x = BC$, we would obtain $BC = 13$, which is another possible solution. The possible grid entries are shown here.

		1	2

or

		1	3

Question Type 3: Solving for the Degree of an Angle

Here you will be given a figure with certain information provided. You will need to deduce the measure of an angle based on both this information and other geometric principles. The easiest way to do this is by setting up an algebraic expression.

PROBLEM

Find the angle y in the following diagram.

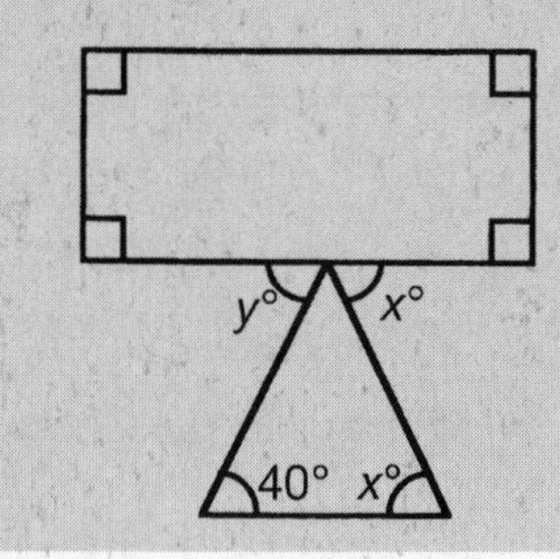

SOLUTION

Use the fact that the sum of the angles on the bottom side of the box is 180°.

Let z be the angle at the top of the triangle. Because we know the sum of the angles of a triangle is 180°,

$$z = 180 - (x + 40)$$

Summing all the angles at the bottom of the square, we get

$$y + [180 - (x + 40)] + x = 180$$

$$y + 140 - x + x = 180$$

$$y + 140 = 180$$

$$y = 40$$

In the grid we enter

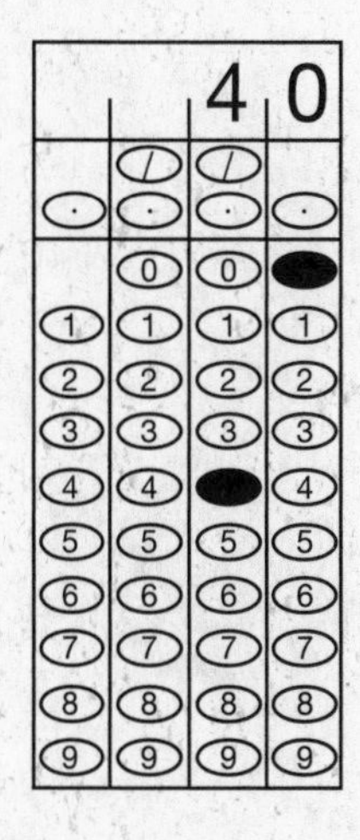

Question Type 4: Solving for the Length of a Side

For this type of question, you will be given a figure with certain measures of sides filled in. You will need to apply geometric principles to find the missing side.

PROBLEM

Consider the following figure.

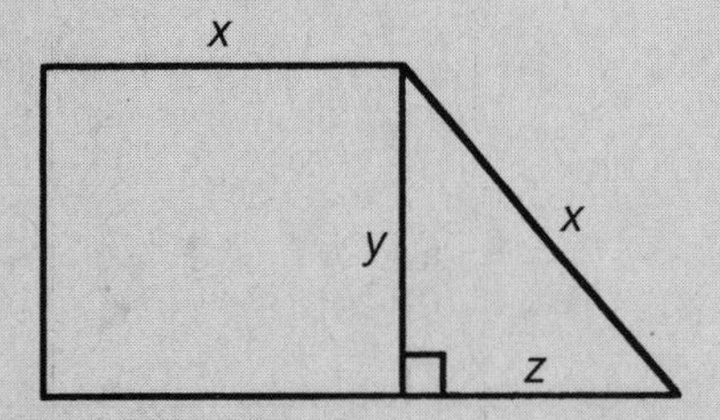

In the figure let x and y be whole numbers, where xy = 65. Also assume the area of the whole figure is 95 square inches. Find y.

SOLUTION

The key point here is that x and y are whole numbers. Using the figure we only have a finite number of possibilities for z.

The equation for the area of the preceding figure is

$$xy + \frac{1}{2}yz = 95.$$

Substituting $xy = 65$ into the preceding equation we get

$$\frac{1}{2}yz = 30;\ yz = 60$$

Using the fact that $xy = 65$, we know y can be either 1, 5, or 13. As $y = 13$ does not yield a factorization for $yz = 60$, y is either 1 or 5. If $y = 1$ this implies $x = 65$ and $z = 60$, which contradicts the Pythagorean Theorem (i.e., $1^2 + 60^2 = 13^2$). If $y = 5$ this implies $x = 13$ and $z = 12$, which satisfies $y^2 + z^2 = x^2$; hence, the solution is $y = 5$.

In our grid we enter

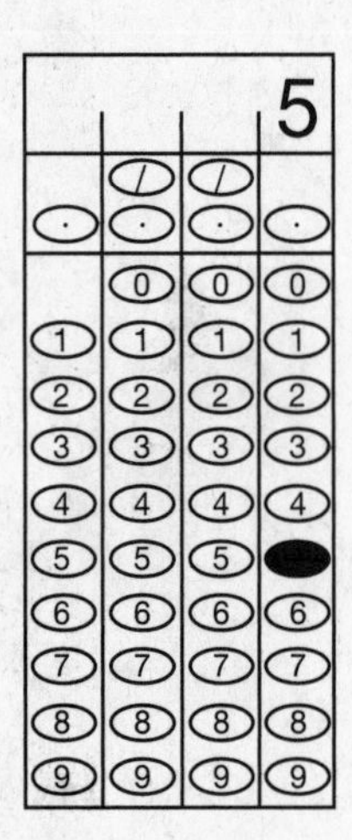

Question Type 5: Solving for the Area of a Region

Here, you will be given a figure with a shaded region. Given certain information, you will need to solve for the area of that region.

PROBLEM

Consider the concentric squares drawn below.

Assume that the side of the larger square is length 1. Also assume that the smaller square's perimeter is equal to the diagonal of the larger square. Find the area of the shaded region.

SOLUTION

The key here is to find the length of the side for the smaller square.

By the Pythagorean Theorem the diameter of the square is

$$d^2 = 1^2 + 1^2$$

which yields $d = \sqrt{2}$. Similarly, the smaller square's perimeter is $\sqrt{2}$; hence, the smaller square's side

$$= \frac{\sqrt{2}}{4}$$

Calculating the area for the shaded region we get

$$A = A_{large} - A_{small}$$

$$A = 1 - \left(\frac{\sqrt{2}}{4}\right)^2$$

$$A = 1 - \frac{2}{16}$$

$$A = \frac{7}{8}$$

In the grid we enter

7	/	8	

Question Type 6: Solve for a Sum of Lengths

The question here involves solving for a sum of lengths in the figure given knowledge pertaining to its area.

PROBLEM

Consider the following figure.

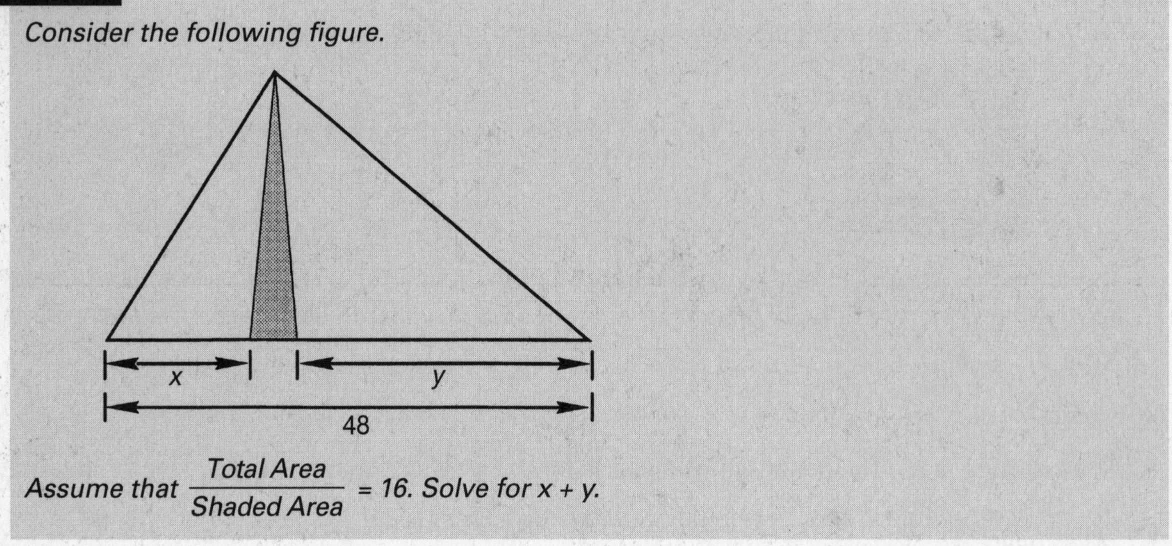

Assume that $\frac{\text{Total Area}}{\text{Shaded Area}} = 16$. *Solve for* $x + y$.

SOLUTION

Solve for $x + y$ and not for x or y individually. Use s to denote the base of the shaded triangle. Then

$$48 - s = x + y$$

From the information given

$$\frac{\frac{48h}{2}}{\frac{sh}{2}} = 16$$

$$\frac{48h}{2} = 16\left(\frac{sh}{2}\right)$$

Simplify

$$24h = 8sh$$

$$24 = 8s$$

$$3 = s$$

Inserting the value for s into our original equation for $x + y$:

$$48 - s = x + y$$

$$48 - 3 = x + y$$

$$45 = x + y$$

We enter in the grid

4	5		
	/	/	
.	.	.	.
	0	0	0
1	1	1	1
2	2	2	2
3	3	3	3
●	4	4	4
5	●	5	5
6	6	6	6
7	7	7	7
8	8	8	8
9	9	9	9

POINTS TO REMEMBER

- Be careful not to overuse your calculator. While it may be helpful in computing large sums and decimals, many problems, such as those involving common denominators, would be more efficiently solved without one.
- Immediately plug the values of all specific information into your equation.
- To visualize exactly what the question is asking and what information you need to find, draw a sketch of the problem.
- Even though you may think you have the equation worked out, make sure your result is actually answering the question. For example, although you may solve an equation for a specific variable, this may not be the final answer. You may have to use this answer to find another quantity.
- Make sure to work in only one unit, and convert if more than one is presented in the problem. For example, if a problem gives numbers in decimals and fractions, convert one in terms of the other and then work out the problem.
- Eliminate all dollar and percent signs when gridding your answers.

MASTERING STUDENT-PRODUCED RESPONSE QUESTIONS

When answering Student-Produced Response questions, you should follow these steps.

Identify the type of question with which you are presented (i.e., arithmetic, algebra, or geometry).

Once you have determined if the question deals with arithmetic, algebra, or geometry, further classify the question. Then, try to determine what type of arithmetic (or algebra or geometry) question is being presented.

Solve the question using the techniques explained in this review. Make sure your answer can be gridded.

STEP 4 *Grid your answer in the question's corresponding answer grid. Make sure you are filling in the correct grid. Keep in mind that it is not mandatory to begin gridding your answer on any particular side of the grid. Fill in the ovals as completely as possible, and beware of stray lines—stray lines may cause your answer to be marked incorrect.*

The drill questions that follow should be completed to help reinforce the material that you have just studied. Be sure to refer back to the review if you need help answering the questions.

DRILL: STUDENT-PRODUCED RESPONSE

(An answer sheet is provided at the end of this section.)

1. At the end of the month, a woman pays $714 in rent. If the rent constitutes 21% of her monthly income, what is her hourly wage given the fact that she works 34 hours per week?

2. Find the largest integer that is less than 100 and divisible by 3 and 7.

3. The radius of the smaller of two concentric circles is 5 cm, and the radius of the larger circle is 7 cm. Determine the area of the shaded region.

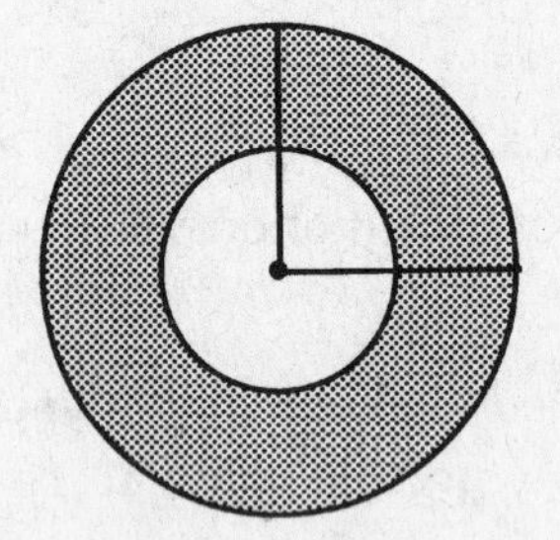

4. $\frac{1}{6}+\frac{2}{3}+\frac{1}{6}-\frac{1}{3}+1-\frac{3}{4}-\frac{1}{4}=$

5. The sum of the squares of two consecutive integers is 41. What is the sum of their cubes?

6. Find x.

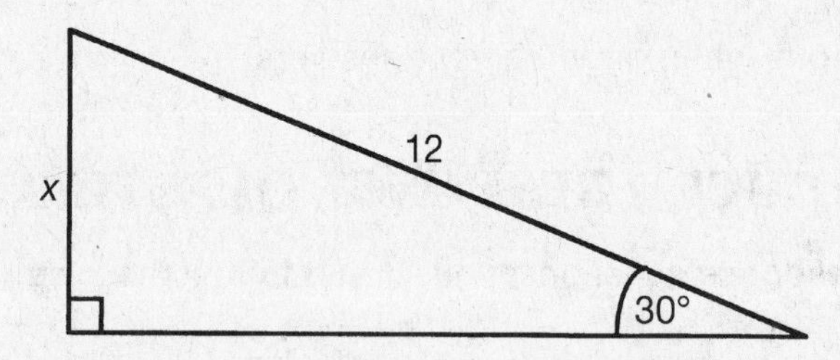

7. $|-8-4|\div 3\times 6+(-4)=$

8. A class of 24 students contains 16 males. What is the ratio of females to males?

9. At an office supply store, customers are given a discount if they pay in cash. If a customer is given a discount of $9.66 on a total order of $276, what is the percent of discount?

10. Let $\overline{RO} = 16$, $\overline{HM} = 30$. Find the perimeter of rhombus *HOMR.*

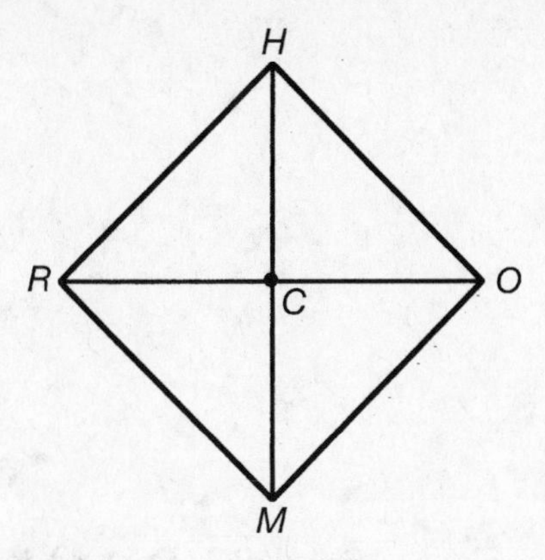

11. Solve for *x.*

$x + 2y = 8$

$3x + 4y = 20$

12. Six years ago, Henry's mother was thirteen times as old as Henry. Now she is only four times as old as Henry. How old is Henry now?

13. Find a prime number less than 40 that is of the form $5k + 1$.

14. Find the area of the shaded triangles.

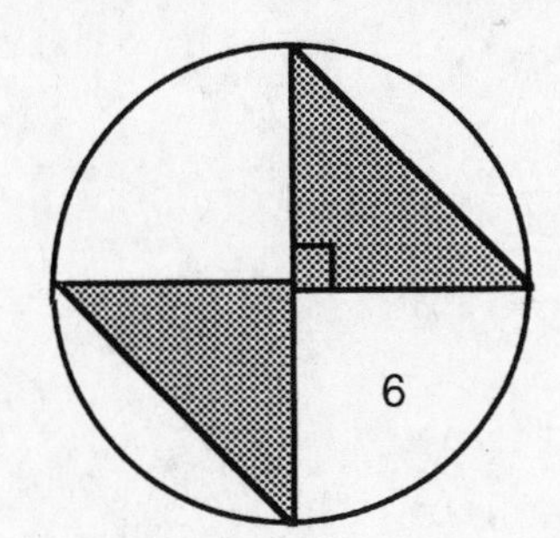

15. $\frac{7}{10} \times \frac{4}{21} \times \frac{25}{36} =$

16. Find the solution for x in the pair of equations.

$x + y = 7$

$x = y - 3$

17. Given the rhombus *RHOM,* find the length of the diagonal $\overline{RO}$ with side $\overline{OM} = 17$.

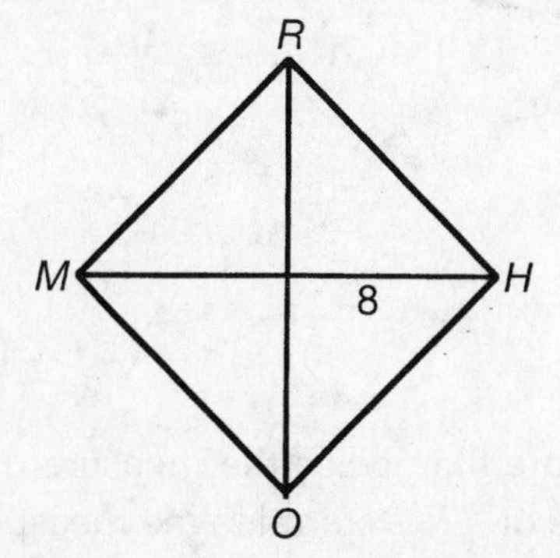

18. ΔMNO is isosceles. If the vertex angle, $\angle N$, has a measure of 96°, find the measure of $\angle M$.

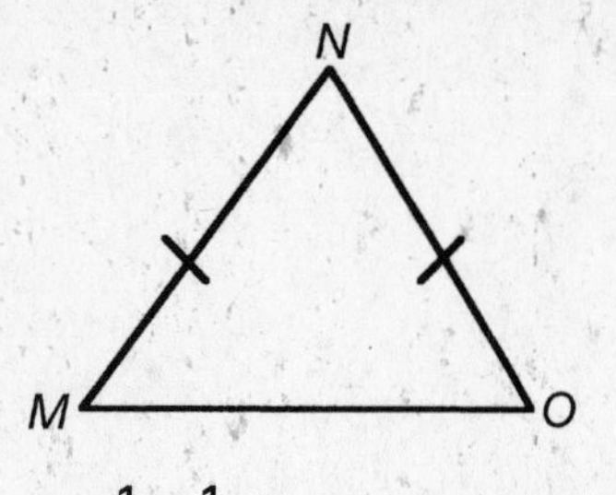

19. Simplify $\dfrac{\frac{1}{2}+\frac{1}{3}}{\frac{1}{6}}$.

20. In the diagram shown, ABC is an isosceles triangle. Sides $\overline{AC}$ and $\overline{BC}$ are extended through C to E and D to form triangle CDE. What is the sum of the measures of angles D and E?

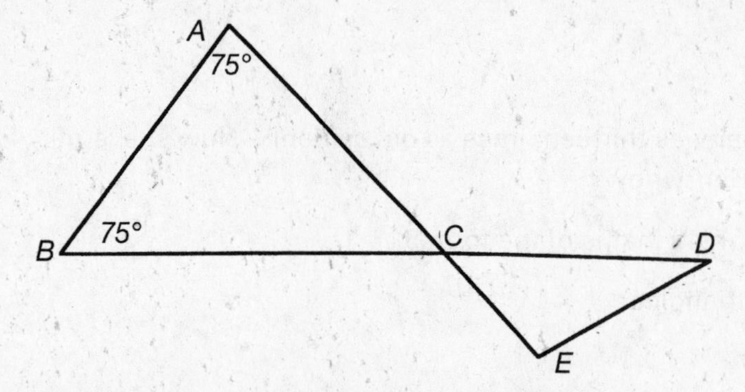

21.

Number of Muffins	Total Price
1	\$0.55
Box of 4	\$2.10
Box of 8	\$4.00

According to the information in the preceding table, what would be the *least* amount of money needed to purchase exactly 19 muffins? (Disregard the dollar sign when gridding your answer.)

22. Several people rented a van for \$30, sharing the cost equally. If there had been one more person in the group, it would have cost each \$1 less. How many people were there in the group originally?

23. For the triangle pictured next, the degree measures of the three angles are x, $3x$, and $3x + 5$. Find x.

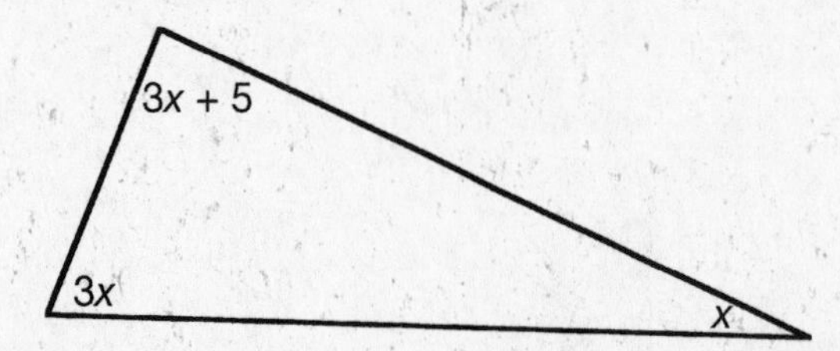

24. ΔPQR is a scalene triangle. The measure of $\angle P$ is 8 more than twice the measure of $\angle R$. The measure of $\angle Q$ is two less than three times the measure of $\angle R$. Determine the measure of $\angle Q$.

25. A mother is now 24 years older than her daughter. In 4 years, the mother will be 3 times as old as the daughter. What is the present age of the daughter?

26. John is 4 times as old as Harry. In six years John will be twice as old as Harry. What is Harry's age now?

27. What is the area of the shaded region?

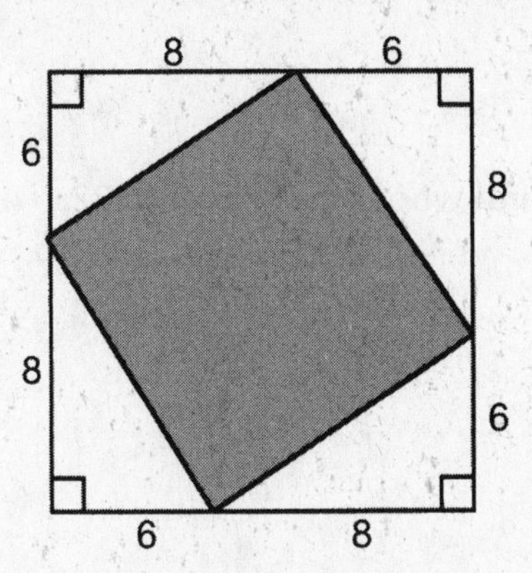

28. In an apartment building there are 9 apartments having terraces for every 16 apartments. If the apartment building has a total of 144 apartments, how many apartments have terraces?

29. Solve the equation $2x^3 - 5x + 3 = 0$.

30. Find the length of a side of an equilateral triangle whose area is $4\sqrt{3}$.

31. Solve $\frac{3}{x-1} + \frac{1}{x-2} = \frac{5}{(x-1)(x-2)}$.

32. Solve the proportion $\frac{x+1}{4} = \frac{15}{12}$.

33. In an isosceles triangle, the length of each of the congruent sides is 10 and the length of the base is 12. Find the length of the altitude drawn to the base.

34. The following are three students' scores on Mr. Page's Music Fundamentals midterm. The given score is the number of correct answers out of 55 total questions.

Liz	48
Jay	45
Carl	25

What is the *average* percentage of questions correct for the three students?

35. Reserved seat tickets to a football game are \$6 more than general admission tickets. Mr. Jones finds that he can buy general admission tickets for his whole family of five for only \$3 more than the cost of reserved seat tickets for himself and Mrs. Jones. How much do the general admission tickets cost?

36. The sum of three numbers is 96. The ratio of the first to the second is 1:2, and the ratio of the second to the third is 2:3. What is the third number?

37. What is the smallest even integer n for which $(0.5)^n$ is less than 0.01?

38. The mean (average) of the numbers 50, 60, 65, 75, x, and y is 65. What is the mean of x and y?

39. The ages of the students enrolled at XYZ University are given in the following table:

Age	Number of students
18	750
19	1,600
20	1,200
21	450

What percent of students are 19 and 20 years old?

40. Find the larger side of a rectangle whose area is 24 and whose perimeter is 22.

STUDENT-PRODUCED RESPONSE DRILL

ANSWER KEY

1.

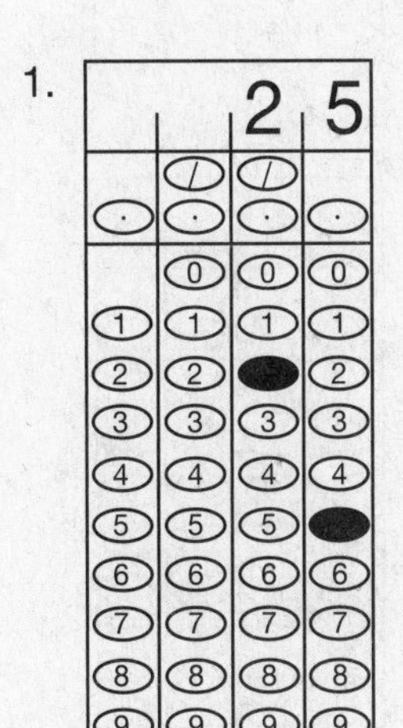

2.

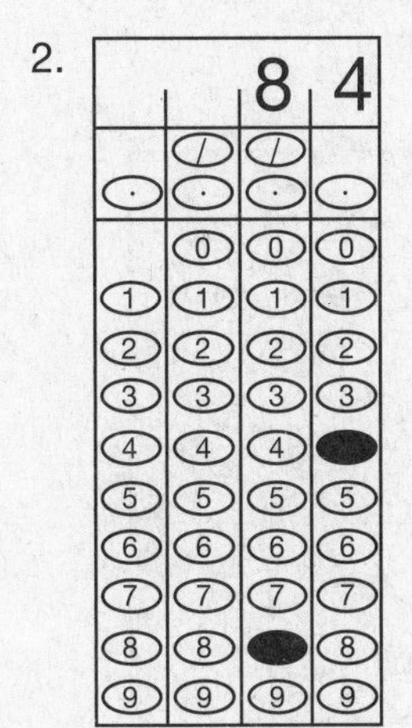

3.

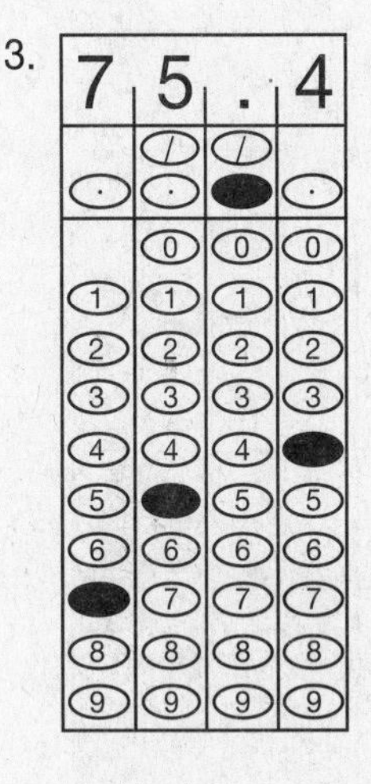

4. 2 / 3

5.

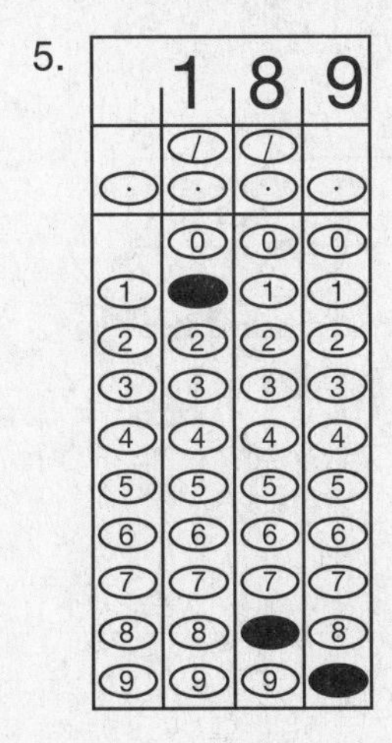

6. 6

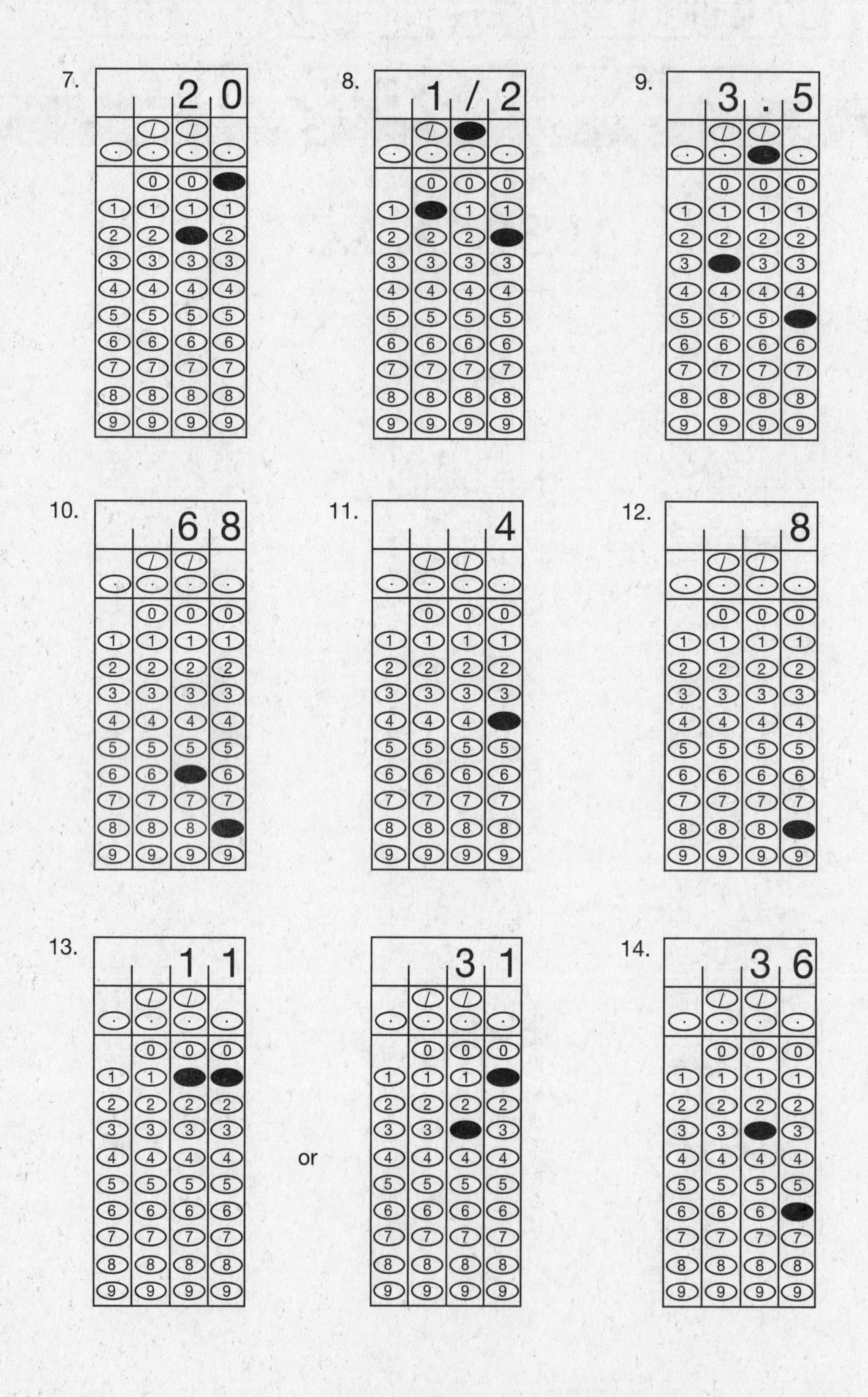
7.
2 0
8.
1 / 2
9.
3 . 5
10.
6 8
11.
4
12.
8
13.
1 1
or
3 1
14.
3 6

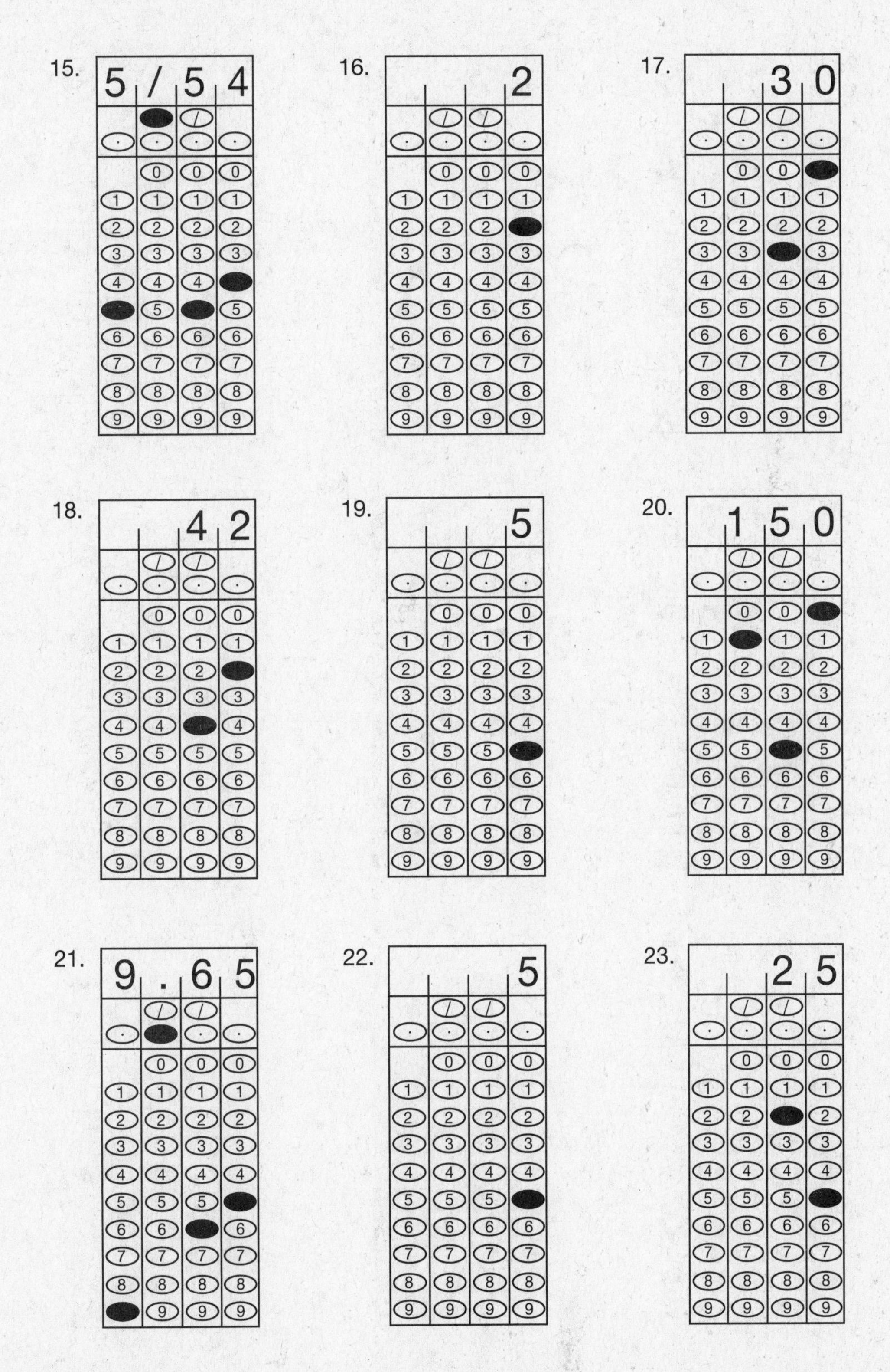
15.
5 / 5 4
16.
2
17.
3 0
18.
4 2
19.
5
20.
1 5 0
21.
9 . 6 5
22.
5
23.
2 5

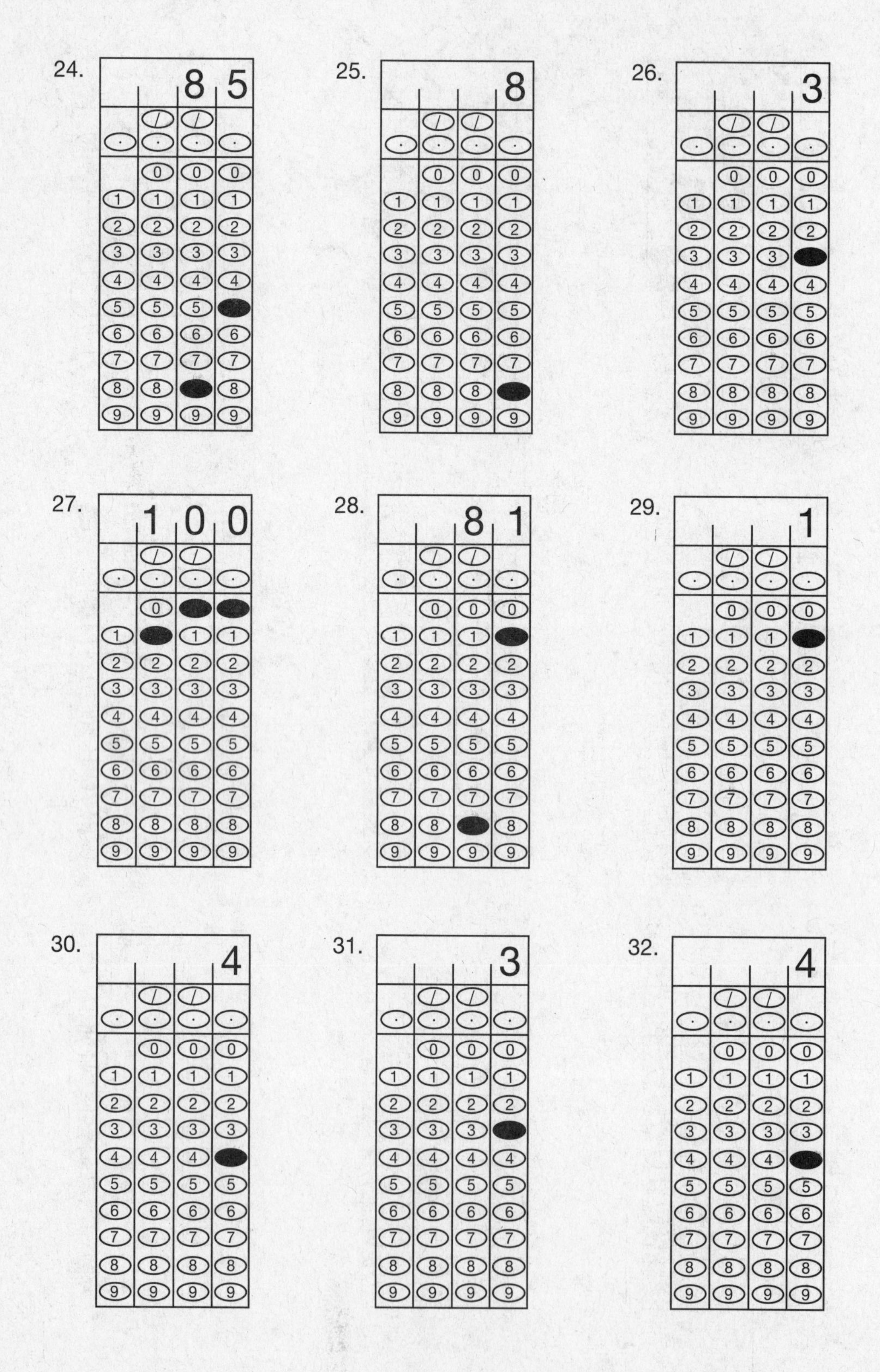
24.
8
5
25.
8
26.
3
27.
1
0
0
28.
8
1
29.
1
30.
4
31.
3
32.
4

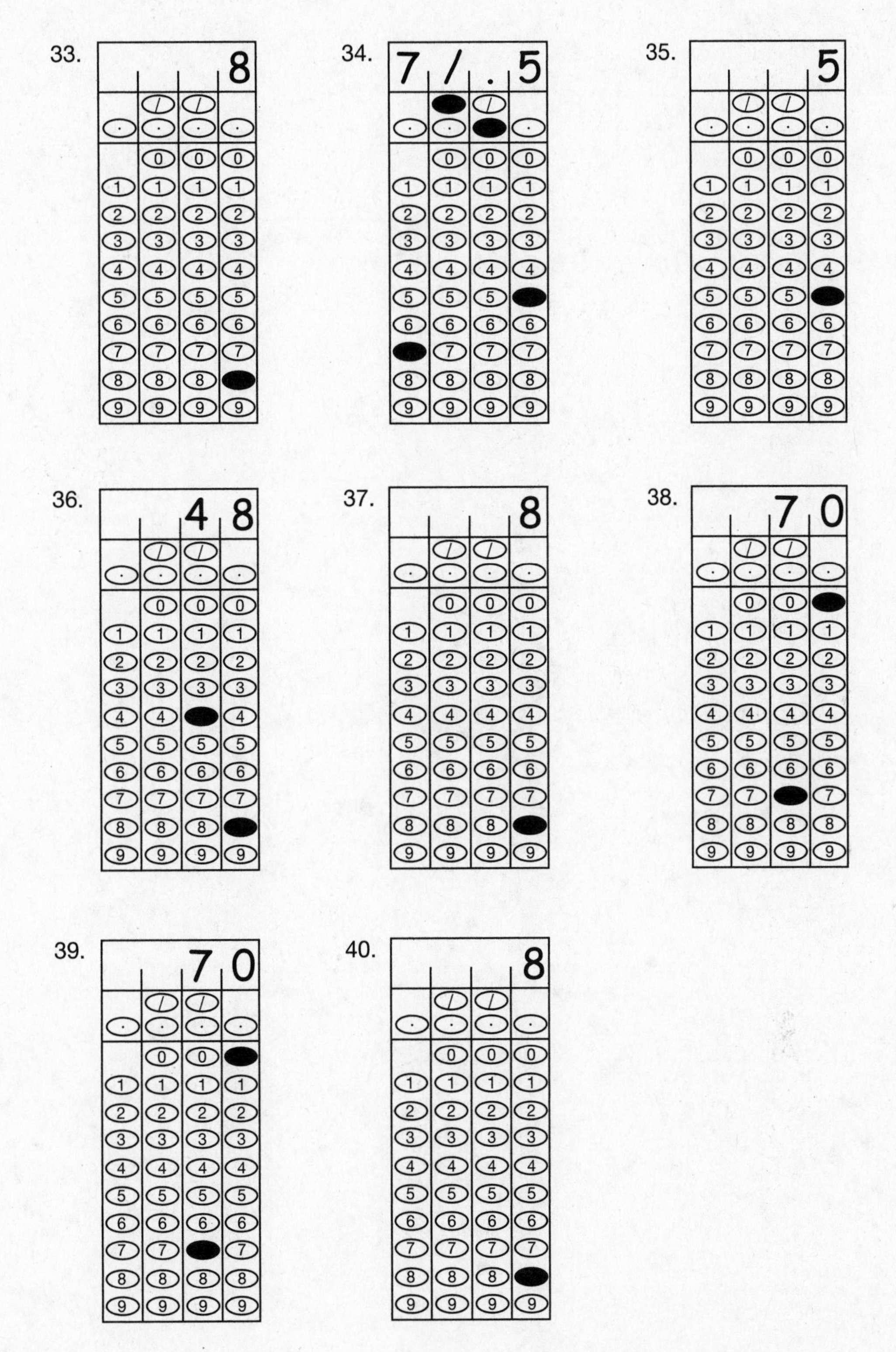
33.
8
34.
7 / . 5
35.
5
36.
4 8
37.
8
38.
7 0
39.
7 0
40.
8

PRACTICE TEST 1

Answer sheets for this test start on page 845.

TIME: 25 Minutes

ESSAY

DIRECTIONS: You have 25 minutes to plan and write an essay on the following topic. You may write on only the assigned topic.

Make sure to give examples to support your thesis. Proofread your essay carefully and take care to express your ideas clearly and effectively.

ESSAY TOPIC

He who sets high goals for himself only sets himself up for failure.

ASSIGNMENT: Consider the preceding statement. Do people who set high goals for themselves only set themselves up for failure? Plan and write an essay in which you develop your point of view on this issue. Support your position with reasoning and examples taken from your reading, studies, experience, or observation.

TIME: 25 Minutes
20 Questions

MATH

DIRECTIONS: In this section solve each problem, using any available space on the page for scratchwork. Then decide which is the best of the choices given and fill in the corresponding oval on the answer sheet.

NOTES

(1) The use of a calculator is permitted.

(2) All numbers used are real numbers.

(3) Figures that accompany problems in this test are intended to provide information useful in solving the problems. They are drawn as accurately as possible EXCEPT when it is stated in a specific problem that the figure is not drawn to scale. All figures lie in a plane unless otherwise indicated.

(4) Unless otherwise specified, the domain of any function f is assumed to be the set of all real numbers x for which $f(x)$ is a real number.

REFERENCE INFORMATION

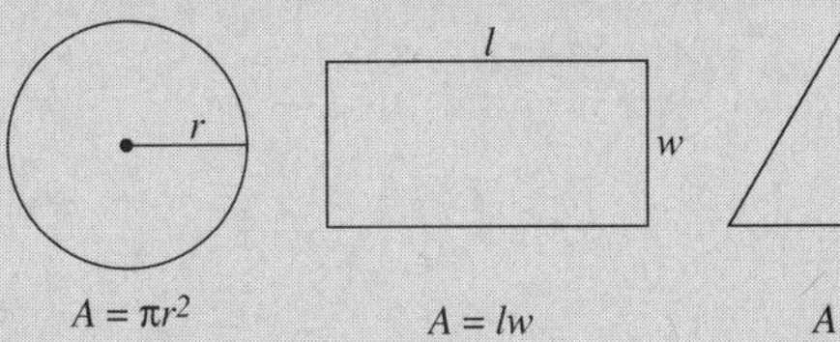

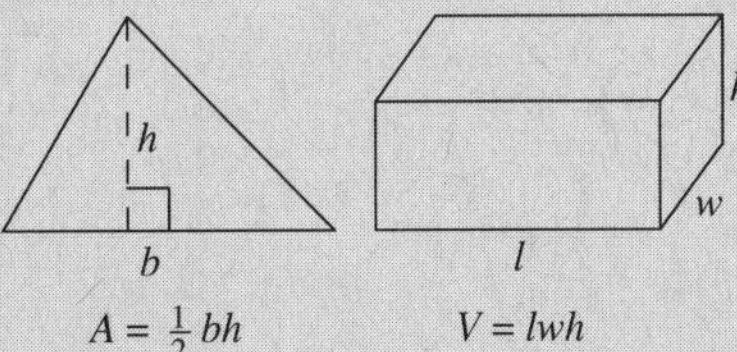

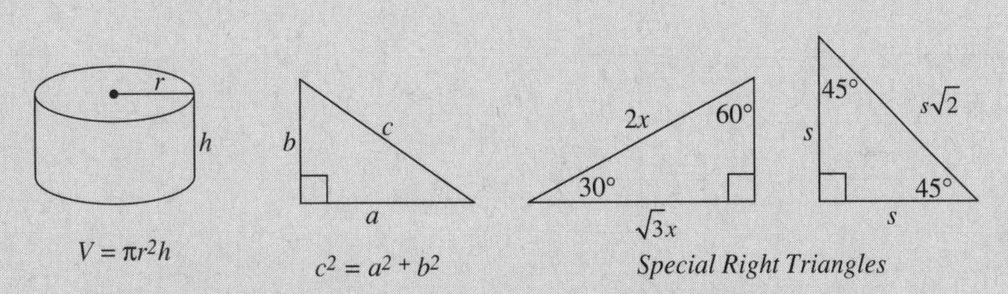

The number of degrees of arc in a circle is 360.

The sum of the measures in degrees of the angles of a triangle is 180.

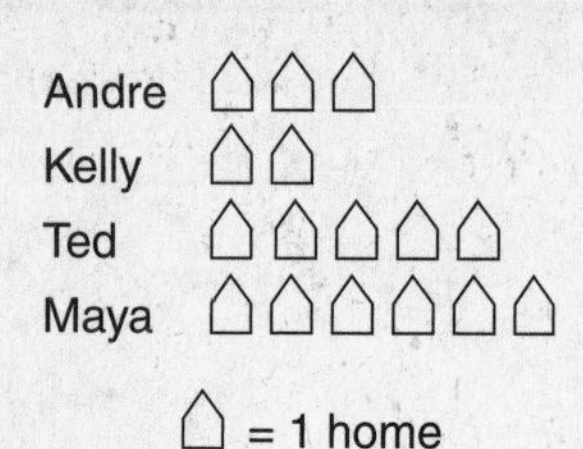

1. Using the chart above, suppose that for this month, Kelly can improve her sales by 50% and Andre can improve his sales by 100%. Assume that Ted's and Maya's sales remain the same. For this month, what fraction of the total number of homes sold by A-Plus realty will Kelly sell?

(A) $\frac{3}{14}$

(B) $\frac{1}{5}$

(C) $\frac{3}{17}$

(D) $\frac{3}{20}$

(E) $\frac{1}{8}$

2. If $0 < w < y\frac{1}{2}$, which of the following <u>must</u> be greater than 1?

(A) $w + y$

(B) $y - w$

(C) $\frac{y}{w}$

(D) wy

(E) $\frac{w}{y}$

3. If $f(x) = (x - 2)^2$ and $g(x) = x^2 - 2$, what is the value of $g(f[-2])$?

(A) -2

(B) 0

(C) 30

(D) 254

(E) 256

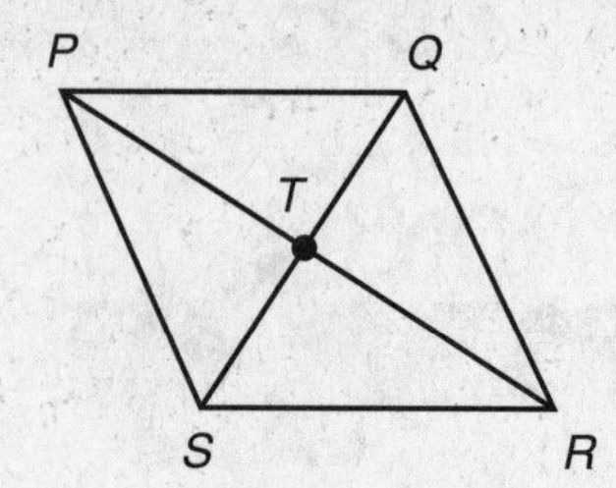

4. In the figure above, which of the following describes sufficient conditions in order for *PQRS* to be classified as a rhombus?

(A) $\overline{PQ} = \overline{QR} = \overline{RS} = \overline{SP}$

(B) $\overline{SP}$ is parallel to $\overline{QR}$ and $\overline{PQ}$ is parallel to $\overline{RS}$

(C) $\overline{PR}$ and $\overline{QS}$ are perpendicular bisectors of each other

(D) $\angle QPS \cong \angle QRS$ and $\angle PSR \cong \angle PQR$

(E) All four angles at point *T* are equal

5. If a certain number *n* is quadrupled, the result is one-third as much as nine less than twice the original number. Which equation could be used to find the value of *n*?

(A) $4n = \frac{2n}{3} - 9$

(B) $4n = \frac{2n-9}{3}$

(C) $4n = \frac{9-2n}{3}$

(D) $4n = (9-2n) - \frac{1}{3}$

(E) $4n = (2n+9) - \frac{1}{3}$

GO ON TO THE NEXT PAGE

6. If $\frac{(6-3x)}{4}$ and x must be positive, which of the following describes *all* possible values of x?

(A) $0 < x < 14$

(B) $x > 14$

(C) $10 < x < 14$

(D) $x < 10$

(E) $x > 10$

7. Given that X = {all prime numbers between 10 and 20} and $X \cup Y$ = {all positive odd numbers less than 20}, what is the minimum number of elements in Y?

(A) 5

(B) 6

(C) 7

(D) 8

(E) 9

8. Which of the following is equivalent to $\sqrt{\frac{x^{-7} \cdot x^{2}}{x^{-3}}}$?

(A) x

(B) x^4

(C) x^6

(D) $\frac{1}{x}$

(E) $\frac{1}{x^4}$

9. If $f(x) = |9 - 2x|$, for which values of x will $f(x) < 3$?

(A) $-3 < x < 3$

(B) $-3 < x < 6$

(C) $-6 < x < 3$

(D) $-6 < x < -3$

(E) $3 < x < 6$

10. Suppose that $f(x)$ is a linear function of x, with a slope of –4. If f(x) is undefined at x = 2, which of the following statements is completely true about the graph of $f(x + 4)$?

(A) It has a slope of 0 and is undefined at $x = 2$.

(B) It has a slope of –4 and is undefined at $x = 6$.

(C) It has a slope of –4 and is undefined at $x = -2$.

(D) It has a slope of 4 and is undefined at $x = 6$.

(E) It has a slope of 4 and is undefined at $x = -2$.

GO ON TO THE NEXT PAGE

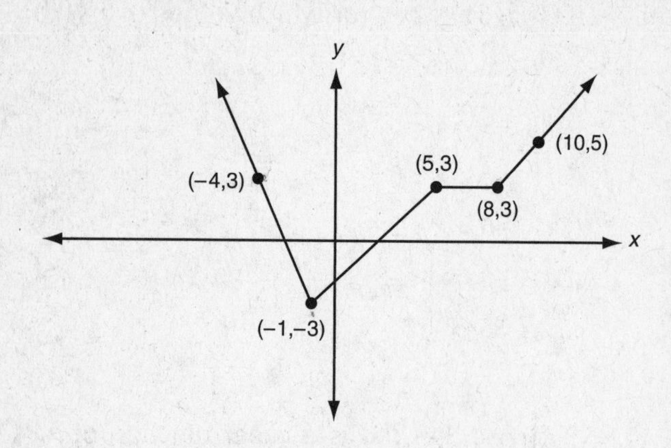

11. In the graph of $f(x)$ shown above, which values of x represent the complete solution to the inequality $|f(x)| \leq 3$?

 (A) $-4 \leq x \leq -1$
 (B) $-4 \leq x \leq 5$
 (C) $-1 \leq x \leq 5$
 (D) $-1 \leq x \leq 8$
 (E) $-4 \leq x \leq 8$

12. What is the range of the function $g(x) = \sqrt{16 - x^2}$?

 (A) All numbers between 0 and 4, inclusive
 (B) All numbers between –4 and 4, inclusive
 (C) All numbers less than or equal to 4
 (D) All numbers greater than or equal to 0
 (E) All numbers that are either at least 4 or at most –4

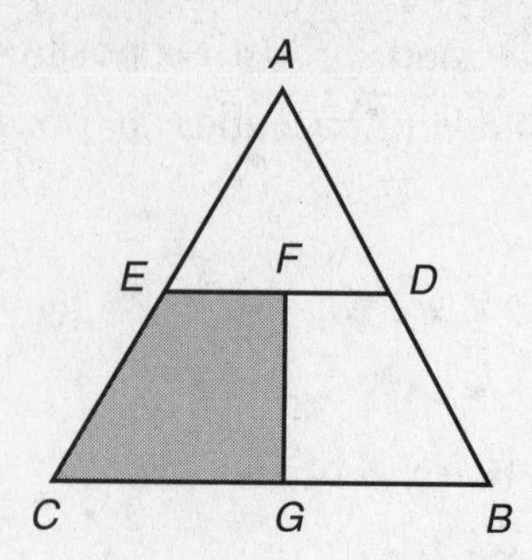

13. The figure above represents a dartboard in which ΔABC is equilateral. E is the midpoint of $\overline{AC}$, D is the midpoint of $\overline{AB}$, F is the midpoint of $\overline{ED}$, and G is the midpoint of $\overline{BC}$. If a dart is thrown and lands on the dartboard, what is the probability that it lands in the shaded area?

 (A) $\frac{1}{4}$
 (B) $\frac{3}{10}$
 (C) $\frac{1}{3}$
 (D) $\frac{3}{8}$
 (E) $\frac{2}{5}$

14. Line L_1 is parallel to line L_2, and each is parallel to the line represented by the equation $y = x + 3$. L_1 contains the point (11, 0). If L_2 lies midway between L_1 and the line represented by $y = x + 3$, what is the y-intercept of L_2?

 (A) (0, 11)
 (B) (0, 8)
 (C) (0, –3)
 (D) (0, –4)
 (E) (0, –11)

GO ON TO THE NEXT PAGE

15. If the length of a rectangle is increased by 30% and the width is decreased by 20%, what is the increase for the area?

(A) 4%

(B) 5%

(C) 10%

(D) 20%

(E) 25%

16. If $\frac{20}{(x+2)} < \frac{x}{4}$ and $x < -2$, which of the following describes all possible values of x?

(A) $-10 < x < 8$

(B) $x < 10$

(C) $-10 < x < -2$

(D) $x < -10$ or $x > 8$

(E) $-8 < x < 10$

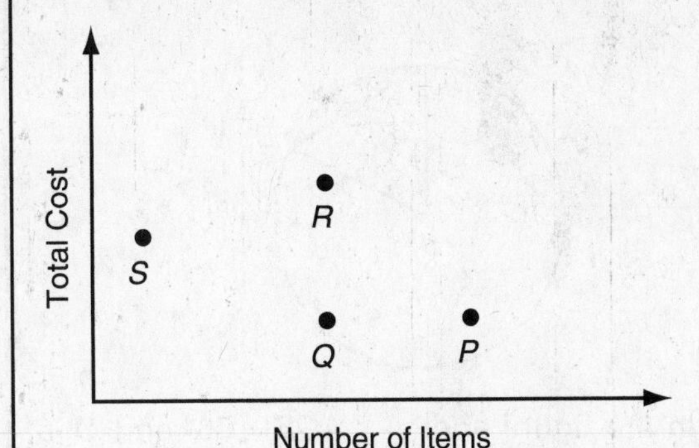

17. Susan went shopping and bought a number of several different items, identified as P, Q, R, and S in the chart above. She also bought some of another item, T, which is not shown. If the correct location of item T is directly below S, which of the following statements is <u>completely</u> correct about the unit cost for T?

(A) It must be less than the unit cost for each of the other items.

(B) It is less than the unit cost for S and can be the same as the unit cost for R.

(C) It can equal the unit cost for both P and Q.

(D) It can equal the unit cost for both Q and R.

(E) It can equal the unit cost for R and be less than the unit cost for P.

18. If x is divisible by 3 and y is divisible by 6, which of the following <u>must</u> be divisible by 4?

(A) $\frac{x}{2y}$

(B) $\frac{y}{x}$

(C) $3xy$

(D) $8x + 2y$

(E) $3x + 2y$

GO ON TO THE NEXT PAGE

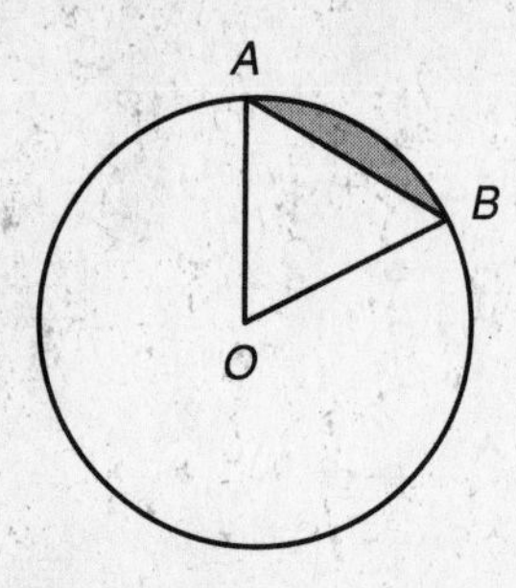

19. In the figure above, $\angle AOB = 60°$ and O is the center of the circle. If $AO = 6$, what is the area of the shaded region?

(A) 6π

(B) $6\pi - 2\sqrt{3}$

(C) $6\pi - 4\sqrt{3}$

(D) $6\pi - 6\sqrt{3}$

(E) $6\pi - 9\sqrt{3}$

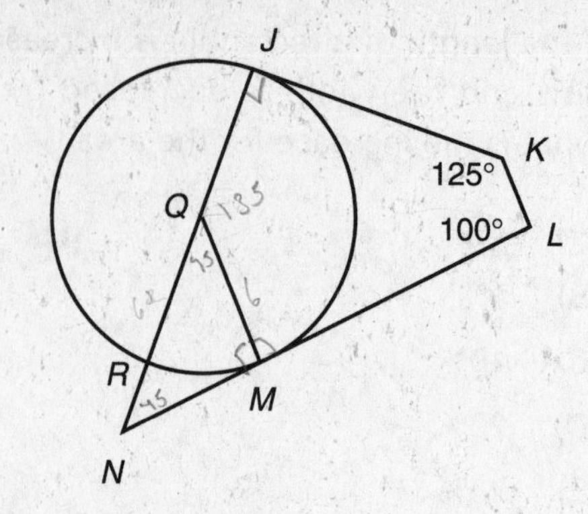

Note: Figure not drawn to scale.

20. In the figure above, Q is the center of the circle. $\overline{JK}$ and $\overline{LM}$ are tangents to the circle. J, Q, N, and R are collinear. Also, L, M, and N are collinear. If $\overline{JQ} = 6$, what is the length of $\overline{RN}$?

(A) $6\sqrt{2} - 6$

(B) $6\sqrt{3} - 6$

(C) $6 - \sqrt{2}$

(D) $6\sqrt{2} - 3$

(E) $6\sqrt{3} - 3$

TIME: 25 Minutes
24 Questions

CRITICAL READING

DIRECTIONS: Each sentence that follows has one or two blanks, each blank indicating that something has been omitted. Beneath the sentence are five lettered words or sets of words. Choose the word or set of words that BEST fits the meaning of the sentence as a whole.

EXAMPLE

Although critics found the book _______ , many readers found it rather _______ .

(A) obnoxious . . . perfect
(B) spectacular . . . interesting
(C) boring . . . intriguing
(D) comical . . . persuasive
(E) popular . . . rare

EXAMPLE ANSWER

1. The new teacher's explanation of the answer to the physics question only served to _______ the students who had learned a completely different method of solving the problem, to the point that they were in danger of failing the class.

(A) discombobulate
(B) mollify
(C) predicate
(D) invigorate
(E) juxtapose

2. The minister believed that _______ from drinking and other immoral practices was the only acceptable lifestyle.

(A) atrophy
(B) defamation
(C) fervor
(D) volatility
(E) abstinence

3. She was overwhelmed at the _______ of gifts that came from _______ community members whose only motive was to help her in time of crisis.

(A) audacity . . . chagrined
(B) paucity . . . entrepreneurial
(C) plethora . . . empathic
(D) glut . . . squalid
(E) dearth . . . tenacious

4. Behind closed doors Robert worked on the painting for over a year, promising to reveal his _______ to us only when he was satisfied it was good enough to be displayed at the most prestigious galleries.

(A) microcosm
(B) conundrum
(C) facsimile
(D) magnum opus
(E) soliloquy

5. The _______ odor of the _______ garbage was so strong that we could hardly force ourselves to pick up the trash bags.

(A) potent . . . redolent
(B) acrid . . . phlegmatic
(C) suppressed . . . toxic
(D) pungent . . . rancid
(E) noxious . . . pristine

6. Only with _______ study and _______ resolve was she able to earn her diploma despite the many obstacles she faced.

(A) quiescent . . . dilatory
(B) pedantic . . . indeterminate
(C) sedulous . . . steadfast
(D) sporadic . . . efficacious
(E) indefatigable . . . irresolute

GO ON TO THE NEXT PAGE

7. Although I sometimes _______ what he says, I _______ uphold his right to right to say it.

   (A) accept. . . . amicably

   (B) abhor . . . unequivocably

   (C) underscore . . . sardonically

   (D) recant . . . inadvertently

   (E) comprehend . . . sanctimoniously

8. On stage he is such a(n) _______ dancer; it is difficult to believe that he is a(n) _______ in his everyday life.

   (A) ungainly . . . oaf

   (B) perspicacious . . . sage

   (C) supple . . . racketeer

   (D) lithe . . . lummox

   (E) wimpish . . . milksop

DIRECTIONS: Each passage in the next section is followed by questions based on its content. Answer the questions on the basis of what is stated or implied in each passage and in any introductory material that may be provided.

Questions 9–10 are based on the following passage.

With contemporary society's urgency to fight aging and maintain a youthful appearance, we seem to have lost sight of reality. One of the most beautiful people in my life has lived more years than anyone else I know. The wrinkles on her face attest to the ready smile that has been so much a part of her for nearly ten decades. Her sharp memory and her willingness to share her experiences have afforded me a far more meaningful education than any textbook could offer. Her unqualified appreciation for life and for almost anyone she encounters imparts a beauty far greater than any plastic surgery, wrinkle cream, or other alleged cosmetic miracles could ever provide.

9. The main idea of the passage is

   (A) today's availability of surgery, creams, and cosmetics helps us remain young and beautiful.

   (B) contemporary society has lost touch with reality.

   (C) people who do not want to have wrinkles should not smile.

   (D) elderly people are the most beautiful people.

   (E) beauty and youth are not synonymous.

GO ON TO THE NEXT PAGE

10. The author's attitude toward the woman described in this passage can best be described as

    (A) disregard.
    (B) admiration.
    (C) confusion.
    (D) indifferent.
    (E) sarcastic.

Questions 11–12 are based on the following passage.

1 We often tend to regard play and education as two separate entities. Not so! Although not yet old enough to be enrolled in school, a young child's learning is ceaseless. Even
5 before his first birthday, he uses every means possible to study whatever ends up in his hand. Eyes completely focused on it, he squeezes and rubs it as if to consume every detail. If it makes any noise at all when he shakes it, the shaking
10 continues until the noise becomes familiar. We know it smells bad if his nose wrinkles disgustedly as the object and his face come into contact. And—at last—it goes to his mouth as a final attempt to learn everything there is to know
15 about the object of study. If only we could keep that level of curiosity—that desire to study the minute details of everything we encounter—for a lifetime! How much more knowledgeable we would be—formally educated or not!

11. Which of the following most accurately reflects the author's message about learning and/or education?

    (A) Education is not confined to the walls of a school building.
    (B) Learning is not complete without formal education.
    (C) Children are the only people who thoroughly learn things.
    (D) We're never too old to learn.
    (E) Formal education stifles learning.

12. In line 8, "consume" most nearly means

    (A) waste.
    (B) exhaust.
    (C) eat.
    (D) absorb.
    (E) use up.

GO ON TO THE NEXT PAGE

Questions 13–24 are based on the following passage.

Almost everyone appreciates the beauty of the monarch butterfly, but few people realize the complexity of this orange and black phenomenon. Every autumn, these butterflies migrate from northern Canada to central Mexico. The small but amazing insects travel 50 to 80 miles per day, at 12 to 20 miles per hour. Some travel almost 3,000 miles, overcoming weather, predators, and fatigue. The primary destination of 300 million fall-migrating butterflies is the Oyamel forests of Michoacan in central Mexico, where they roost and feed on the oyamel fir trees. When early spring arrives, those that survived the winter begin traveling north, lay their eggs, and die. Few if any butterflies live long enough to make a complete migratory trip. So each year the monarchs travel thousands of miles to a place where they have never been.

The female monarchs are in reproductive dormancy during the fall migration. But during their spring northerly migration, they lay eggs on milkweed en route. This succeeding generation will follow their parents' migratory path. The butterflies that are born in the spring live only a month—just long enough to form, mate, and lay their eggs. The reproductive cycle continues along the northward migratory path all the way to Canada, so that some of the butterflies involved in the migration the following autumn will be three or four generations removed from those that spent the winter in Mexico and California the previous year.

One of the mysteries of the monarch migration is why they migrate. For the past several years, researchers have been seeking the answer to this question. The most logical explanation is that they are attempting to escape the cold weather and food shortage of the winter months. However, since they live only one year they have no experience of previous winter hardships. However, they do instinctively sense the decreasing amount of daylight and the lowering temperatures as summer ends. Some say the monarchs follow the direction of the sun; as it moves southward in the sky, the monarchs travel southward.

Also intriguing is the monarchs' migratory path, which follows a route that runs from northeast to southwest in the fall and retraces the same path, in reverse, in the spring. Since no butterfly lives long enough to make the round trip, how and why the butterflies travel the same path every year is a mystery, with several acceptable theories proposing possible answers. Research points to the fact that during the spring migration the monarchs leave their eggs on only milkweed; the larvae feed exclusively on these plants. Thus, the migration pattern follows the milkweed distribution through Texas, Louisiana, and Alabama. Other theories claim that monarchs instinctively use the sun, moon, stars, sounds, and smells as their guide. It has been hypothesized that they use the sun's position as a compass; however, they are able to maintain their course even when the sun is not visible, so another hypothesis is that their bodies contain a small amount of magnetite and therefore they can sense changes in the magnetic field of the earth.

13. The primary purpose of this passage is to

(A) persuade the reader that the monarch butterfly is not a simple insect.

(B) describe the monarch butterflies' migration.

(C) prove the need for additional research on the monarchs' migration.

(D) show how the monarch butterfly is different from other butterflies.

(E) discuss how the monarch butterfly is different from other animals that migrate.

GO ON TO THE NEXT PAGE

14. The author of this passage seems to believe that most people

(A) pay little if any attention to the monarch butterfly.

(B) do not appreciate the monarch butterfly.

(C) see the monarch as more simple than it actually is.

(D) see the monarch as more complex than it actually is.

(E) have a realistic perception of the monarch butterfly.

15. The monarch butterfly can best be described as

(A) small and weak.

(B) small and strong.

(C) rare and beautiful.

(D) near extinction.

(E) slow but steady.

16. In lines 19–20, "reproductive dormancy" most nearly means

(A) being asleep.

(B) not laying eggs.

(C) hibernating.

(D) not eating or drinking.

(E) not moving.

17. The life span of the monarch butterflies

(A) is approximately a month for all of them.

(B) varies from a month to almost a year.

(C) is approximately a year for all of them.

(D) varies according to the length of time the migration requires.

(E) lasts for one to five reproductive cycles.

18. The female butterflies lay their eggs

(A) during both fall and spring migration.

(B) only during fall migration.

(C) only during spring migration.

(D) between fall and spring migration.

(E) while they are roosting in the Oyamel forests of central Mexico.

19. In line 30, "removed" most nearly means

(A) far away.

(B) obscure.

(C) distant in relationship.

(D) alone.

(E) taken away.

20. Monarch butterflies probably migrate because

(A) their instincts tell them to escape the upcoming cold winter.

(B) their "memory" of the past winter tells them to avoid the hardships of the upcoming winter.

(C) the milkweed crop begins to diminish as the fall begins.

(D) they need more daylight than the Canadian falls and winters provide.

(E) the females' reproductive systems become dormant in the winter.

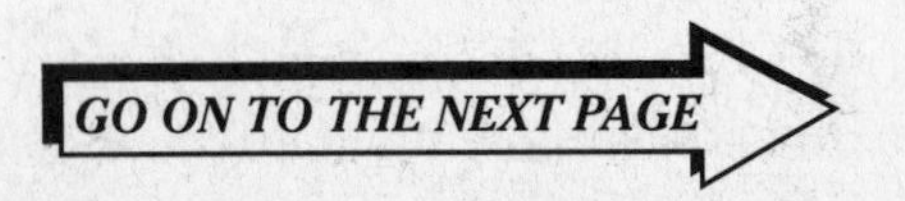

21. Which of the following is the most appropriate description of the direction of the monarchs' migration?

    (A) East to west only
    (B) North to south only
    (C) East to west, and west to east
    (D) Northeast to southwest, and southwest to northeast
    (E) Northwest to south, and southeast to northwest

22. In line 57, "exclusively" most nearly means

    (A) necessarily.
    (B) entirely, only.
    (C) preferably.
    (D) ravenously.
    (E) healthily.

23. Which of the following is <u>not</u> an acceptable theory to explain why monarchs follow the same migratory path year after year?

    (A) They remember the path from the previous year.
    (B) They must go where milkweed is available.
    (C) They use the sun to guide their direction.
    (D) Their instincts and senses guide them.
    (E) They use the earth's magnetic field as their guide.

24. The author seems to find monarch butterflies

    (A) boring.
    (B) frustrating.
    (C) inferior.
    (D) interesting.
    (E) predatory.

STOP

If time remains, you may go back and check your work. When the time allotted is up, you may go on to the next section.

TIME: 25 Minutes
35 Questions

WRITING

DIRECTIONS: In each of the following sentences, some portion of the sentence is underlined. Under each sentence are five choices. The first choice has the same wording as the original. The other four choices are reworded. Sometimes the first choice containing the original wording is the best; sometimes one of the other choices is the best. Choose the letter of the best choice. Your choice should produce a sentence that is not ambiguous or awkward and that is correct, clear, and precise.

This is a test of correct and effective English expression. Keep in mind the standards of English usage, punctuation, grammar, word choice, and construction.

EXAMPLE

When you listen to opera, a person may not appreciate it.

(A) a person may not appreciate it.

(B) it may not be appreciated by a person.

(C) you may not appreciate it.

(D) which may not be appreciated by you.

(E) appreciating it may be a problem for you.

EXAMPLE ANSWER

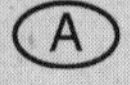

1. Women live longer and have fewer illnesses than men, which proves that women are the strongest sex.

   (A) which proves that women are the strongest sex.
   (B) which proves that women are the stronger sex.
   (C) facts which prove that women are the stronger sex.
   (D) proving that women are the strongest sex.
   (E) a proof that women are the stronger sex.

2. Wealthy citizens often protest about the building of low-cost housing in the affluent communities where they reside.

   (A) about the building of
   (B) whether they should build
   (C) if builders should build
   (D) the building of
   (E) whether or not they should build

3. Siblings growing up in a family do not necessarily have equal opportunities to achieve, the difference being their placement in the family, their innate abilities, and their personalities.

   (A) the difference being their placement in the family, their innate abilities, and their personalities.
   (B) because of their placement in the family, their innate abilities, and their personalities.
   (C) and the difference is their placement in the family, their innate abilities, and their personalities.
   (D) they have different placements in the family, different innate abilities, and different personalities.
   (E) their placement in the family, their innate abilities, and their personalities being different.

4. Two major provisions of the United States Bill of Rights is freedom of speech and that citizens are guaranteed a trial by jury.

   (A) is freedom of speech and that citizens are guaranteed a trial by jury.
   (B) is that citizens have freedom of speech and a guaranteed trial by jury.
   (C) is freedom of speech and the guarantee of a trial by jury.
   (D) are freedom of speech and that citizens are guaranteed a trial by jury.
   (E) are freedom of speech and the guarantee of a trial by jury.

GO ON TO THE NEXT PAGE

5. <u>Whether Leif Erickson was the first to discover America or not</u> is still a debatable issue, but there is general agreement that there probably were a number of "discoveries" through the years.

(A) Whether Leif Erickson was the first to discover America or not

(B) That Leif Erickson was the first to discover America

(C) The Leif Erickson may have been the first to have discovered America

(D) Whether Leif Erickson is the first to discover America or he is not

(E) Whether or not Leif Erickson was or was not the first to discover America

6. <u>People who charge too much are likely to develop</u> a bad credit rating.

(A) People who charge too much are likely to develop

(B) People's charging too much are likely to develop

(C) When people charge too much, likely to develop

(D) That people charge too much is likely to develop

(E) Charging too much is likely to develop for people

7. The museum of natural science has a special exhibit of gems and minerals, <u>and the fifth graders went to see it on a field trip</u>.

(A) and the fifth graders went to see it on a field trip.

(B) and seeing it were the fifth graders on a field trip.

(C) when the fifth graders took a field trip to see it.

(D) which the fifth graders took a field trip to see.

(E) where the fifth graders took their field trip to see.

8. <u>When the case is decided, he plans appealing</u> if the verdict is unfavorable.

(A) When the case is decided, he plans appealing

(B) When deciding the case, he plans appealing

(C) After the case is decided, he is appealing

(D) After deciding the case, he is planning to appeal

(E) When the case is decided, he plans to appeal

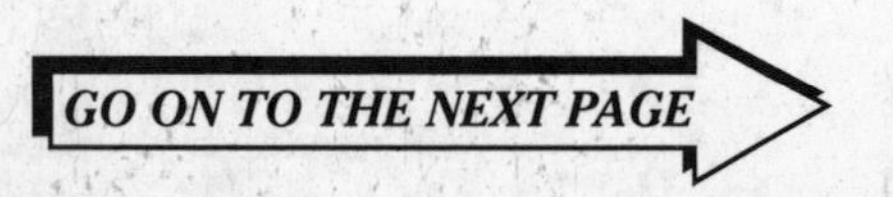

9. <u>We decided there was hardly any reason for his allowing us</u> to stay up later on weeknights.

   (A) We decided there was hardly any reason for his allowing us
   (B) We, deciding there was hardly any reason for his allowing us,
   (C) Deciding there was hardly any reason, we allowed
   (D) We decided there were none of the reasons for him to allow us
   (E) For him to allow us there was hardly any reason we decided

10. At this time <u>it is difficult for me agreeing with your plan of having everyone</u> in the club working on the same project.

   (A) it is difficult for me agreeing with your plan of having everyone
   (B) I find it difficult to agree to your plan of having everyone
   (C) for my agreement with your plan is difficult for everyone
   (D) an agreement to your plan seems difficult for everyone
   (E) finding it difficult for me to agree to your plan of having everyone

11. When the Whites hired a contractor to do remodeling on their home, he <u>promised to completely finish the work inside of three months</u>.

   (A) promised to completely finish the work inside of three months.
   (B) promised to complete the work within three months.
   (C) completely promised to finish the work inside of three months' line span.
   (D) promising to completely finish the work in three months.
   (E) completely finished the work within three months.

GO ON TO THE NEXT PAGE

DIRECTIONS: Each of the following sentences may contain an error in diction, usage, idiom, or grammar. Some sentences are correct. Some sentences contain one error. No sentence contains more than one error.

If there is an error, it will appear in one of the underlined portions labeled A, B, C, or D. If there is no error, choose the portion labeled E. If there is an error, select the letter of the portion that must be changed in order to correct the sentence.

EXAMPLE

He drove slowly (A) and cautiously (B) in order to hopefully (C) avoid having an accident (D). No error (E).

EXAMPLE ANSWER

(A) (B) ● (D) (E)

12. Which (A) suspension bridge is (B) the longest, (C) the Verrazano-Narrows Bridge in New York City or (D) the Golden Gate Bridge in San Francisco? No error. (E)

13. A main function of proteins, (A) whether they (B) come from plant or animal (C) sources, is (D) the building of body tissue. No error. (E)

14. Recognizing (A) that we had worked (B) very hard to complete our project, the teacher told Janice and I (C) that we could give it to her tomorrow (D). No error. (E)

15. They (A) are very grateful the city (B) has set up (C) a special fund that helps pay (D) for electric bills of the elderly and the handicapped. No error. (E)

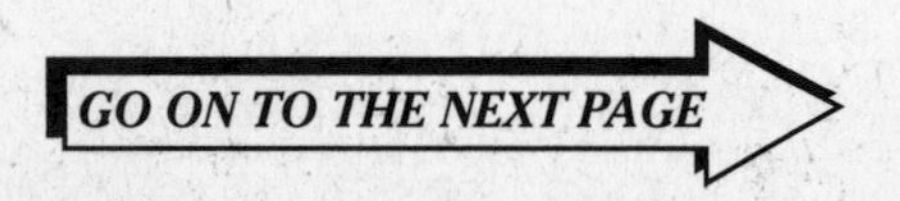
GO ON TO THE NEXT PAGE

16. In order to (A) stay cool during the summer months, Americans not only (B) are using ceiling fans, but they are also using devices to add humidity (C) to the air in particularly (D) arid climates such as Arizona. No error (E).

17. After seeing (A) the technique demonstrated on television, Janie baked homemade bread for the first time yesterday, and her (B) brother thought it tasted good (C), an opinion everyone agreed with (D). No error (E).

18. Although not so (A) prevalent as they once were, hood (B) ornaments still exist, some of which (C) are quite distinctive, such as (D) the symbol for Mercedes-Benz and Jaguar. No error (E).

19. If I were (A) that tourist, I would not argue with (B) those two members of the Guardia Civil because, although they are speaking politely (C), it is obvious they are becoming angry (D). No error (E).

20. Mr. Burns is fully aware of (A) statistics proving the harmful (B) consequences of smoking; irregardless (C), he persists (D) in his habit. No error (E).

21. David was not capable to win (A) the singles tennis match because he had been (B) injured in (C) a game last week and the doctor prohibited him from (D) playing for two weeks. No error (E).

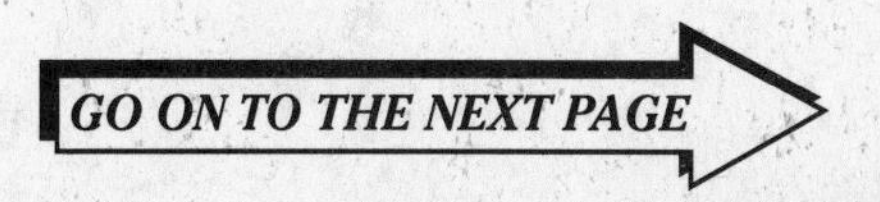

22. In spite (A) of the doctor's orders, David is playing tennis today because he is one of those (B) athletes who are (C) determined to play, no matter what the coach and she (D) say. No error. (E)

23. Although unequal in (A) ability compared to other team members, Wilt practices (B) his free throws every afternoon (C) so he would become a member of (D) the varsity squad. No error. (E)

24. If you would have (A) listened to me carefully, (B) you would have heard me advise against (C) your subscribing (D) to the magazine of that ultraconservative political group. No error. (E)

25. A graduating high school senior who wants to attend (A) a university must attend to (B) many details, such as taking the SAT or ACT, sending an official transcript to the university, arranging for (C) a dormitory room, and etc. (D) etc. No error. (E)

26. Running (A) errands for upperclassmen, pushing pennies, and being thrown in (B) mud holes are all factors of (C) being hazed as (D) a freshman. No error. (E)

27. The majority of parents are concerned (A) about the effects of (B) watching (C) too much television upon the development of their (D) children. No error. (E)

GO ON TO THE NEXT PAGE

28. Whenever anyone who is (A) a toy collector thinks (B) about Theodore Roosevelt, they usually remember (C) that one of the most popular (D) toys, the teddy bear, was named for him. No error. (E)

29. Horses named (A) Gato del Sol, Sunny's Halo, and Swale have been (B) a winner (C) in the Kentucky Derby in recent years (D). No error (E).

Questions 30–35 are based on the following passage.

Dear Senator Simon,

(1) I am writing in support of your bill that, if passed, will be instrumental in getting legislation which will put a warning label on violent television programs. (2) Violence needs to be de-glamorized. (3) One must detest and deplore violence of excess and other such excesses in one's viewing choices.

(4) Unfortunately, violence sells. (5) One of the main reasons is because violent shows are easily translated and marketed to other countries. (6) Network executives have actually requested their script writers to include more violence in certain shows with a steady audience, as well as to create new violent shows for the late evening slot just before the news.

(7) The National Institute of Health did a study. (8) In this study children viewed violent scenes. (9) After this, children are more prone to violent acts. (10) Maybe parents will pay more attention to their children's viewing if this labeling system is enacted. (11) Maybe commercial sponsors will hesitate to sponsor programs that are labeled violent, so these programs will diminish in number and children will have fewer such programs to view.

(12) Yes, I think people need to be aware of violent events happening around the world and within our own country. (13) We need to know what is happening in the Balkans, Somalia, South Africa, as well in Los Angeles riots and the bombings in New York, are just to name two examples. (14) However, these are real events, not glamorizations.

(15) Please keep up your campaign to get rid of excessive violence!

Sincerely,

Sue Chan

GO ON TO THE NEXT PAGE

30. Which of the following is the best revision of the underlined portion of sentence 1 as follows?

    I am writing in support of your bill <u>that, if passed, will be instrumental in getting legislation which will put</u> a warning label on violent television programs.

    (A) in order to pass out legislation in order to put
    (B) passing legislation which will require putting
    (C) to pass a law that will legislate putting
    (D) requiring passage of legislation requiring
    (E) for a law requiring

31. In the context of the sentences preceding and following sentence 3, which of the following is the best revision of sentence 3?

    (A) One should agree with me that excessive violence should be detested and deplored.
    (B) I detest and deplore excessive violence.
    (C) You can see that I think wanton violence should be detested and deplored.
    (D) Detesting and deploring wanton violence and other such excesses is how I feel.
    (E) Excessive violence should be detested and deplored.

32. In relation to the passage as a whole, which of the following best describes the writer's intention in paragraph 2?

    (A) To present background information
    (B) To contradict popular opinion
    (C) To provide supporting evidence
    (D) To outline a specific category
    (E) To rouse the emotions of the reader

33. Which of the following is the best revision of the underlined portion of sentence 5 as follows?

    One of the main reasons <u>is because violent shows are</u> easily translated and marketed to other countries.

    (A) is being that violent shows
    (B) being that violent shows are
    (C) is that violent shows are
    (D) is due to the fact that violent shows seem to be
    (E) comes from violent shows containing violence because

34. Which of the following is the best way to combine sentences 7, 8, and 9?

    (A) After doing a study, the children viewing violent scenes at the national Institute of Health were more prone to violent acts.
    (B) Children viewing violent scenes at the National Institute of Health were more prone to violent acts.
    (C) After viewing violent scenes, children at the National Institute of Health were more prone to doing violent acts themselves.
    (D) The National Institute of Health did a study proving that after viewing violent scenes, children are more prone to violent acts.
    (E) The National Institute of Health did a study proving that children viewing violent scenes are more prone to violent acts.

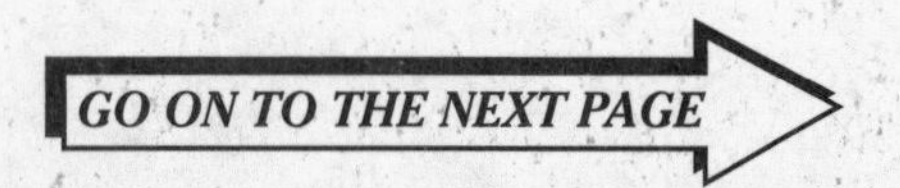

35. Which of the following would be the best revision of the underlined portion of sentence 13?

We need to know what is happening in the Balkans, Somalia, South Africa, as well in Los Angeles riots and the bombings in New York, are just to name two examples.

(A) as well as in Los Angeles riots, and the bombings in new York just to name two examples.

(B) as well as, to name two examples, the Los Angeles riots and the bombings in New York.

(C) as well as riots in Los Angeles and the bombings in New York.

(D) as well as events such as the riots in Los Angeles and the bombings in New York.

(E) Best as it is.

STOP

If time remains, you may go back and check your work. When the time allotted is up, you may go on to the next section.

TIME: 25 Minutes
18 Questions

MATH

DIRECTIONS: In this section solve each problem, using any available space on the page for scratchwork. Then decide which is the best of the choices given and fill in the corresponding oval on the answer sheet.

NOTES

(1) The use of a calculator is permitted.

(2) All numbers used are real numbers.

(3) Figures that accompany problems in this test are intended to provide information useful in solving the problems. They are drawn as accurately as possible EXCEPT when it is stated in a specific problem that the figure is not drawn to scale. All figures lie in a plane unless otherwise indicated.

(4) Unless otherwise specified, the domain of any function f is assumed to be the set of all real numbers x for which $f(x)$ is a real number.

REFERENCE INFORMATION

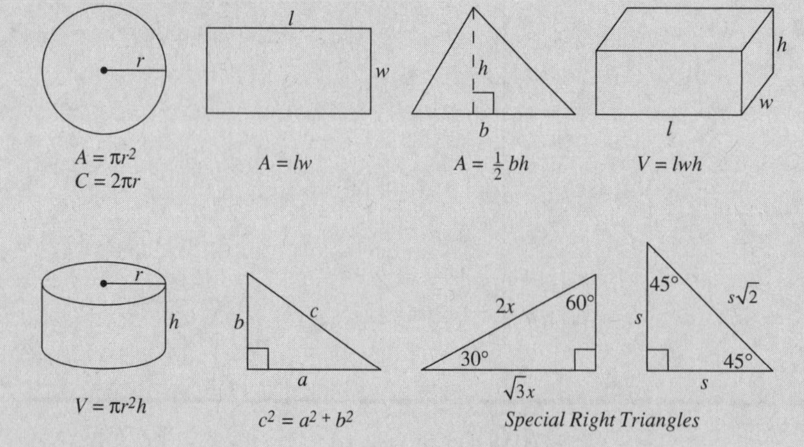

The number of degrees of arc in a circle is 360.

The sum of the measures in degrees of the angles of a triangle is 180.

1. In a certain class, there are twice as many girls as boys. One-half of the boys and two-thirds of the girls are studying physics. What fraction of the entire class is studying physics?

(A) $\frac{1}{3}$

(B) $\frac{2}{5}$

(C) $\frac{9}{20}$

(D) $\frac{7}{12}$

(E) $\frac{11}{18}$

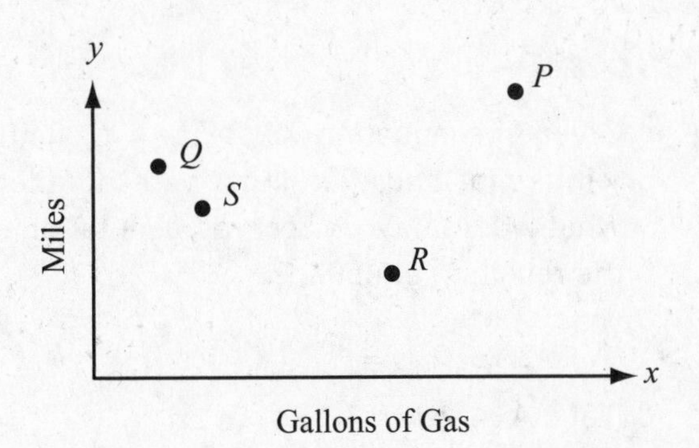

2. David took 4 different cars for a test drive. These cars are identified as *P*, *Q*, *R*, and *S*, as shown in the scatter plot above. If these cars were arranged in order from best mileage per gallon to worst mileage per gallon, how would the list appear?

(A) *R*, *P*, *S*, *Q*

(B) *Q*, *S*, *P*, *R*

(C) *P*, *Q*, *S*, *R*

(D) *R*, *S*, *Q*, *P*

(E) *Q*, *R*, *P*, *S*

3. Erik has a total of 25 t-shirts. Seven are white, four are red, five are brown, and the rest are blue. If he selects one t-shirt at random, what is the probability that it is either blue or red?

(A) 0.08

(B) 0.13

(C) 0.36

(D) 0.44

(E) 0.52

4. What are the values of x in the equation $|-3x + 9| = |5x - 7|$?

(A) -4 and $\frac{1}{4}$

(B) -1 and $\frac{1}{4}$

(C) 1 and $-\frac{1}{4}$

(D) -4 and 2

(E) -1 and 2

5. What is the domain of the function $\frac{(2x^2 + 13x - 15)}{(2x^2 - x - 15)}$?

(A) All numbers except $-\frac{13}{2}$ and 1

(B) All numbers except $\frac{5}{2}$ and -3

(C) All numbers except $-\frac{5}{2}$ and 3

(D) All numbers except $\frac{15}{2}$ and -1

(E) All numbers except $-\frac{15}{2}$ and 1

GO ON TO THE NEXT PAGE

6. Suppose M = {all prime numbers less than 20}, N = {positive even numbers less than 19}, and P = {positive integers less than 20 and greater than 10}. Which of the following has the largest number of elements?

(A) $(M \cap P) \cup N$

(B) $(M \cap P) \cap N$

(C) $(M \cap N) \cup P$

(D) $(M \cap N) \cap P$

(E) $(N \cap P) \cup M$

7. The Alpha Cab Company charges an initial fee of \$3.50 plus \$0.25 for each $\frac{1}{3}$ mile traveled. The Beta Cab Company charges a fee of \$4.50 for the first mile plus \$0.20 for each additional $\frac{1}{2}$ mile traveled. Marie took a ride in an Alpha cab for a distance of 12 miles. If she had taken a ride in a Beta cab and spent the same amount of money, how many miles could she have traveled?

(A) 24

(B) 23

(C) 22

(D) 21

(E) 20

8. If the denominator is not zero, what is the reduced form of the fraction $\frac{(9x^3 - 3x^2 - 30x)}{(12x^3 + 44x^2 + 40x)}$?

(A) $\frac{3(3x-5)}{4(3x+5)}$

(B) $-\frac{3}{4}$

(C) $\frac{3(x-2)}{4(x+2)}$

(D) $\frac{3}{4}$

(E) $\frac{3(x-5)}{4(x+2)}$

9. Given the equation $Ax + By = C$, if the y-intercept and the slope were doubled, what would be a correct representation of the resulting equation?

(A) $2Ax + 2By = C$

(B) $Ax + 2By = 2C$

(C) $Ax + By = C + 2$

(D) $2Ax + By = 2C$

(E) $2Ax + 2By = 2C$

GO ON TO THE NEXT PAGE

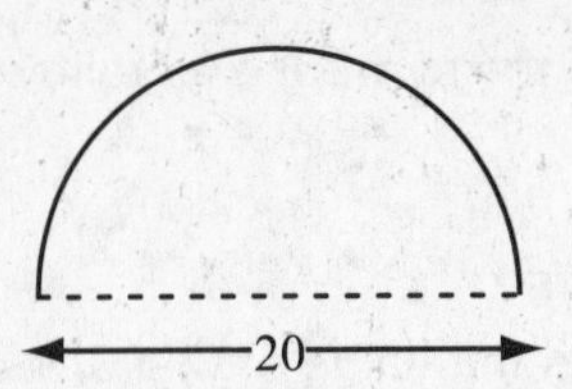

Note: The dotted line represents the diameter.

10. The figure above shows a piece of string in the shape of a semicircle. If this string were bent to form one full circle (using all the string with no overlap) what would be the area of this circle?

(A) 10π
(B) 20π
(C) 25π
(D) 30π
(E) 40π

11. Which of the following can be represented by a graph that has the same vertex as the graph of $f(x) = x^2 + 14x + 51$?

(A) $f(x) = 3(x + 7)^2 + 2$
(B) $f(x) = 4(x - 7)^2 + 51$
(C) $f(x) = 5(x + 7)^2 + 14$
(D) $f(x) = 6(x + 14)^2 + 51$
(E) $f(x) = 7(x - 14)^2 + 2$

12. Suppose that P varies directly as the square root of Q and inversely as the cube of R. How does Q vary with P and R?

(A) Directly as the square root of P and the cube root of R
(B) Directly as the square of P and the sixth power of R
(C) Inversely as the square of P and the cube root of R
(D) Inversely as the square root of P and directly as R
(E) Directly as the square of P and inversely as the fifth power of R

13. Which of the following equations has no solution in real numbers?

(A) $\sqrt{2x-5} - 9 = -4$
(B) $\sqrt{5x-2} + 8 = 5$
(C) $\sqrt{2-3x} + 6 = 12$
(D) $\sqrt{7x+2} - 10 = -10$
(E) $\sqrt{6-2x} - 7 = 3$

14. The population of bacteria that begins with a count of X and triples every n years is given by the formula $Y = (X)(3^{\frac{t}{n}})$, where t represents the number of years of growth and Y represents the growth after t years. Suppose the bacteria population in the year 1895 was 150, and it grew to 36,450 by the year 1935. In what year was the population equal to 1,350?

(A) 1932
(B) 1919
(C) 1911
(D) 1903
(E) 1900

GO ON TO THE NEXT PAGE

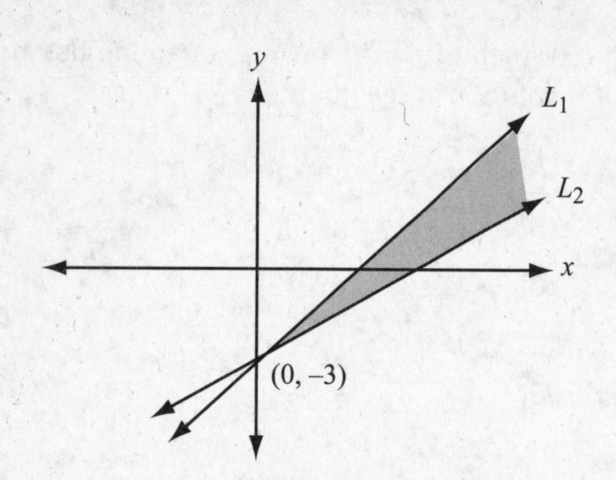

Note: Figure not drawn to scale.

15. In the graph shown above, the slope of L_1 is less than $\frac{2}{3}$, and the slope of L_2 is greater than $\frac{3}{5}$. Both lines contain the point (0,–3). Which of the following points lies in the shaded region? (Note: The shaded region does not include L_1 or L_2.)

(A) (10, 4)

(B) (9, 3)

(C) (25, 11)

(D) (20, 10)

(E) (30, 15)

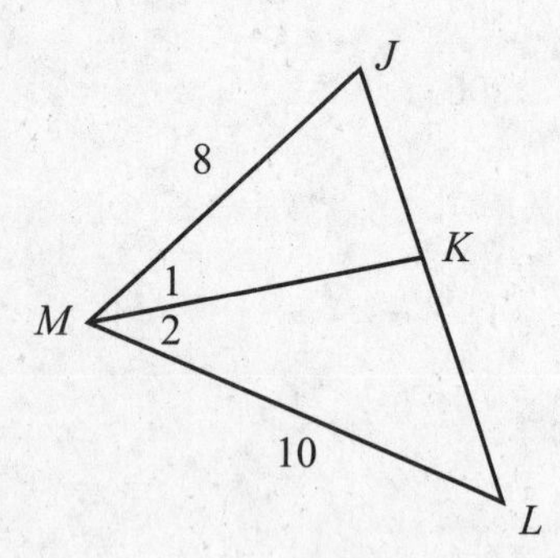

Note: Figure not drawn to scale.

16. In the figure above, $\angle 1 \cong \angle 2$. If $JK = 6$, what is the length of KL?

(A) 1.5

(B) 4.8

(C) 7.5

(D) 12.8

(E) 13.3

17. Define the symbol ♠ as follows:

if $x < y$, $x ♠ y = y^3 - 3x$

if $x \geq y$, $x ♠ y = x^2 + y^2$

What is the value of (2 ♠ 3) ♠ 4?

(A) 15

(B) 425

(C) 457

(D) 9,249

(E) 9,273

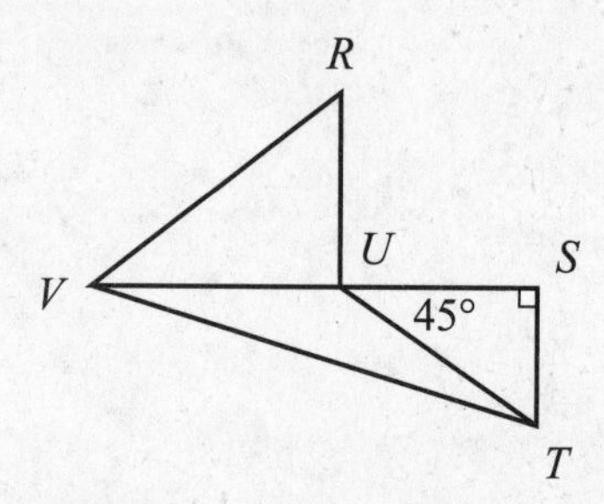

Note: Figure not drawn to scale.

18. In the figure above, $RU = UV = TU$. If $RV = 8$, what is the length of TV?

(A) 4.76

(B) 5.66

(C) 9.66

(D) 10.5

(E) 12.0

TIME: 25 Minutes
24 Questions

CRITICAL READING

DIRECTIONS: Each sentence below has one or two blanks, each blank indicating that something has been omitted. Beneath the sentence are five lettered words or sets of words. Choose the word or set of words that BEST fits the meaning of the sentence as a whole.

EXAMPLE

Although critics found the book _______ , many readers found it rather _______ .

(A) obnoxious . . . perfect
(B) spectacular . . . interesting
(C) boring . . . intriguing
(D) comical . . . persuasive
(E) popular . . . rare

EXAMPLE ANSWER

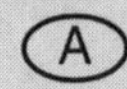   

1. The phenomenon called the "self-fulfilling prophecy" occurs when one holds and acts on a belief that is not true; for example, parents who believe a child is destined to turn out "no good" and treat the child as if he were no good, often have their worst fears realized: False _______ becomes _______ .

    (A) belief . . . reality
    (B) fear . . . truth
    (C) anxiety . . . fact
    (D) presumption . . . certain
    (E) dread . . . proven

2. The computer is a(n) _______ tool, for if one neglects to save a file, it cannot be recalled.

    (A) ominous
    (B) complicated
    (C) essential
    (D) difficult
    (E) unforgiving

3. The gold-studded costume appeared _______ when compared to the _______ of the flannel suit.

    (A) chaste . . . gaudiness
    (B) laconic . . . opulence
    (C) reserved . . . savoir-faire
    (D) ornate . . . simplicity
    (E) feudal . . . raucousness

4. Although her bedroom at home was always in disarray, her office workspace was _______ .

    (A) aloof
    (B) meticulous
    (C) viable
    (D) diligent
    (E) insipid

5. The pitcher tried to _______ the wounded catcher through increased _______ .

    (A) tarry . . . spending
    (B) offset . . . concentration
    (C) mediate . . . dedication
    (D) divulge . . . repugnance
    (E) alleviate . . . preparation

GO ON TO THE NEXT PAGE

DIRECTIONS: Each passage that follows is followed by questions based on its content. Answer the questions on the basis of what is stated or implied in each passage and in any introductory material that may be provided.

Questions 6–9 are based on the following passages.

Passage 1

The traditional approach to written and oral communication requires that we follow an irrefutable set of rules. Every subject and verb must agree, every adverb must modify a verb, and every adjective must describe a noun. Every word must be spelled correctly. Although learning all these rules and practicing them may seem arduous at first, the ability to express oneself correctly is the key to effective communication. Even the most intelligent, most highly educated individual who does not use proper grammar comes across as obtuse.

Passage 2

An effective and popular trend in education condones the holistic approach to writing and speaking. The focus is looking at the "whole"—the main ideas—rather than the "parts"—whether words are spelled correctly, whether commas are appropriately placed, whether every word is used as the correct part of speech. This concept allows students to communicate their thoughts and ideas effectively with little concern for correct grammar, punctuation, and spelling that potentially stifle individuality, creativity, and free thought. "Let it flow" is the philosophy of the holistic approach.

6. Both passages support the idea that

   (A) what we say is less important than how we say it.
   (B) how we say something is less important than what we say.
   (C) what we say and how we say it are equally important.
   (D) effective communication is important.
   (E) communication has many verbal and nonverbal forms.

7. The author of Passage 1 believes that

   (A) proper grammar is stifling but necessary.
   (B) free-flowing expression can effectively accompany proper grammar.
   (C) effective communication is possible only with the use of proper grammar.
   (D) the person who uses proper grammar is well-educated.
   (E) one can be creative with the use of proper grammar.

8. According to the author of Passage 2,

   (A) grammar, spelling, and punctuation rules stifle individuality.
   (B) the communication of all ideas is appropriate in schools.
   (C) schools should disregard grammar rules.
   (D) some highly educated people do not use proper grammar.
   (E) schools should be careful not to be too strict.

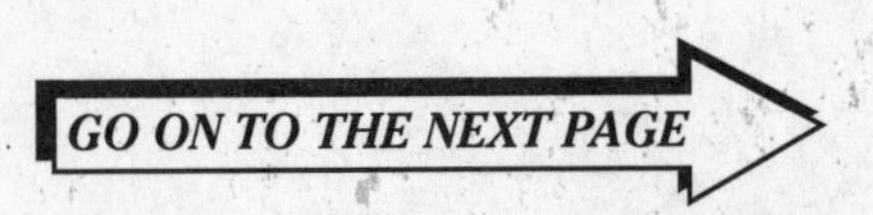

9. Both authors would agree that schools

   (A) have the right to teach what they believe is important.

   (B) should require students to follow rules.

   (C) must teach life skills, including public speaking.

   (D) need to take a closer look at how they teach communication.

   (E) should include communication in their curriculum.

GO ON TO THE NEXT PAGE

Questions 10–15 are based on the following passage.

The following passage is an excerpt from Henry David Thoreau's essay entitled "On the Duty of Civil Disobedience."

1 I heartily accept the motto, "That government is best which governs least"; and I should like to see it acted up to more rapidly and systematically. Carried out, it finally amounts to 5 this, which also I believe—"That government is best which governs not at all"; and when men are prepared for it, that will be the kind of government which they will have. Government is at best but an expedient; but most govern- 10 ments are usually, and all governments are sometimes, inexpedient. . . .

But, to speak practically and as a citizen, unlike those who call themselves no-government men, I ask for, not at once no government, but 15 at once a better government. Let every man make known what kind of government would command his respect, and that will be one step toward obtaining it.

After all, the practical reason why, when 20 the power is once in the hands of the people, a majority are permitted, and for a long period continue, to rule is not because they are most likely to be in the right, nor because this seems fairest to the minority, but because they are physically 25 the strongest. But a government in which the majority ruled in all cases can not be based on justice, even as far as men understand it. Can there not be a government in which the majorities do not virtually decide right and wrong, but 30 conscience?—in which majorities decide only those questions to which the rule of expediency is applicable? Must the citizen ever for a moment, or in the least degree, resign his conscience to the legislator? Why has every man a conscience 35 then? I think that we should be men first, and subjects afterward. It is not desirable to cultivate a respect for the law, so much as for the right. The only obligation which I have a right to assume is to do at any time what I think right. It is truly enough said that a corporation has no con- 40 science; but a corporation of conscientious men is a corporation with a conscience. Law never made men a whit more just; and, by means of their respect for it, even the well-disposed are daily made the agents of injustice. . . . 45

All voting is a sort of gaming, like checkers or backgammon, with a slight moral tinge to it, a playing with right and wrong, with moral questions; and betting naturally accompanies it. The character of the voters is not staked. I 50 cast my vote, perchance, as I think right; but I am not vitally concerned that that right should prevail. I am willing to leave it to the majority. Its obligation, therefore, never exceeds that of expediency. Even voting for the right 55 is doing nothing for it. It is only expressing to men feebly your desire that it should prevail. A wise man will not leave the right to the mercy of chance, nor wish it to prevail through the power of the majority. There is but little virtue 60 in the action of masses of men. When the majority shall at length vote for the abolition of slavery, it will be because they are indifferent to slavery, or because there is but little slavery left to be abolished by their vote. 65

10. Thoreau's desire was for government

(A) to be nonexistent.
(B) to improve.
(C) to remain as it was.
(D) to change its power structure.
(E) to change the election system.

GO ON TO THE NEXT PAGE

11. In line 9, "expedient" most nearly means

(A) something that is successful in achieving a particular purpose.

(B) a person or thing that has a selfish motive.

(C) a popular idea or concept.

(D) democracy.

(E) socialism.

12. Thoreau believed that majority rule

(A) was the most expedient way to govern.

(B) left the minority without power.

(C) allowed the stronger to act for the good of the weaker.

(D) was the most effective way for the individual to express his opinion.

(E) was a just way of governing.

13. The most obvious object of Thoreau's respect was

(A) the law.

(B) elected officials.

(C) what was right.

(D) slaves.

(E) corporations.

14. The basis of Thoreau's vote in an election was what he believed

(A) the law intended.

(B) the majority would decide.

(C) other individual citizens desired.

(D) was right.

(E) the U.S. Constitution intended.

15. The last sentence primarily indicates that Thoreau believed that

(A) slavery was wrong.

(B) eventually American voters would abolish slavery.

(C) most Americans were indifferent to slavery.

(D) the vote of the majority was basically ineffective.

(E) the vote of the majority was basically effective.

Questions 16–24 are based on the following passage.

The following is Susan B. Anthony's address, which she delivered in Philadelphia after illegally voting during the Presidential election of 1872, when women did not yet have voting rights.

1 Friends and fellow citizens: I stand before you tonight under indictment for the alleged crime of having voted at the last presidential election, without having a lawful right to vote. 5 It shall be my work this evening to prove to you that in thus voting, I not only committed no crime, but, instead, simply exercised my citizen's rights, guaranteed to me and all United States citizens by the National Constitution, 10 beyond the power of any state to deny.

The preamble of the Federal Constitution says:

> We the people of the United States, in order to form a more perfect union, establish 15 justice, insure domestic tranquility, provide for the common defense, promote the general welfare, and secure the blessings of liberty to ourselves and our posterity, do ordain and establish this Constitution for 20 the United States of America.

It was we, the people; not we, the white male citizens; nor yet we, the male citizens; but we, the whole people, who formed the Union. And we formed it, not to give the blessings of 25 liberty, but to secure them; not to the half of ourselves and the half of our posterity, but to the whole people—women as well as men. And it is a downright mockery to talk to women of their enjoyment of the blessings of liberty while 30 they are denied the use of the only means of securing them provided by this democratic-republican government—the ballot.

For any state to make sex a qualification that must ever result in the disfranchisement of one entire half of the people, is to pass a bill of 35 attainder, or, an ex post facto law, and is therefore a violation of the supreme law of the land. By it the blessings of liberty are forever withheld from women and their female posterity.

To them this government has no just pow- 40 ers derived from the consent of the governed. To them this government is not a democracy. It is not a republic. It is an odious aristocracy; a hateful oligarchy of sex; the most hateful aristocracy ever established on the face of 45 the globe. An oligarchy of wealth, where the rich govern the poor, an oligarchy of learning, where the educated govern the ignorant, or even an oligarchy of race, where the Saxon rules the African, might be endured; but this 50 oligarchy of sex, which makes father, brothers, husband, sons, the oligarchs over the mother and sisters, the wife and daughters, of every household—which ordains all men sovereigns, all women subjects, carries dissension, discord, 55 and rebellion into every home of the nation. Webster, Worcester, and Bouvier all define a citizen to be a person in the United States, entitled to vote and hold office.

The only question left to be settled now is: 60 Are women persons? And I hardly believe any of our opponents will have the hardihood to say they are not. Being persons, then, women are citizens; and no state has a right to make any law, or to enforce any old law, that shall 65 abridge their privileges or immunities. Hence, every discrimination against women in the constitutions and laws of the several states is today null and void, precisely as is every one against Negroes. 70

GO ON TO THE NEXT PAGE

16. Susan B. Anthony's primary purpose in this presentation was to prove that women

(A) were intelligent enough to vote.
(B) had a legal right to vote.
(C) are as human as men are.
(D) were being treated no better than slaves.
(E) had been treated unfairly throughout history.

17. After having voted illegally in the presidential election, Susan B. Anthony

(A) regretted having broken the law.
(B) denied that she had done anything wrong.
(C) sought to rescind her vote.
(D) planned to appeal to a higher court.
(E) was willing to pay the fine or serve the sentence for her actions.

18. In quoting the Preamble to the U.S. Constitution, Anthony was

(A) exhibiting her knowledge in an effort to prove that she was able to vote intelligently.
(B) indicating which parts she believed should be changed.
(C) emphasizing that it was a privilege to live in the United States.
(D) offering evidence that women had the Constitutional right to vote.
(E) reminding her audience that this was a free country for all citizens.

19. According to Anthony, for women the government of this country was a(n)

(A) democracy.
(B) republic.
(C) aristocracy.
(D) autocracy.
(E) monarchy.

20. With which of the following would Susan B. Anthony be most likely to agree?

(A) It is more acceptable for the rich to rule over the poor than it is for white people to rule over blacks.
(B) It is more acceptable for men to rule over women than it is for the educated to rule over the uneducated.
(C) It is more acceptable for elected officials to rule over common citizens than it is for the rich to rule over the poor.
(D) It is more acceptable for white people to rule over blacks than it is for men to rule over women.
(E) It is more acceptable for the educated to rule over the uneducated than it is for the rich to rule over the poor.

21. In line 57, Anthony's main purpose in referring to Webster, Worcester, and Bouvier was

(A) to use these famous lexicographers' definition of "citizen" to prove that women should be able to exercise their rights as citizens.

(B) to use these famous lexicographers' definition of "citizen" to prove that women were superior human beings.

(C) to use these famous lexicographers' definition to prove that women were just as much "people" as men were.

(D) to use these famous lexicographers as examples of men who supported the right of women to vote.

(E) to prove to her audience that she was educated and intelligent.

22. In line 66, "abridge" most nearly means

(A) expand; broaden.

(B) cross over; traverse.

(C) shorten.

(D) make less; restrict.

(E) condense; shorten but keep the same meaning.

23. Which of the following had the most authority, according to Anthony's speech?

(A) God's law

(B) federal law

(C) state law

(D) city law

(E) personal beliefs

24. How did Anthony probably regard the situation of black people in nineteenth-century America?

(A) She did not take a stand regarding their rights.

(B) She did not believe they should have the right to vote.

(C) She saw parallels between their situation and that of women during her time.

(D) She regarded them as inferior to white people.

(E) She was apparently not concerned about their situation.

STOP

If time remains, you may go back and check your work. When the time allotted is up, you may go on to the next section.

TIME: 20 Minutes
16 Questions

MATH

DIRECTIONS: In this section solve each problem, using any available space on the page for scratchwork. Then decide which is the best of the choices given and fill in the corresponding oval on the answer sheet.

NOTES

(1) The use of a calculator is permitted.

(2) All numbers used are real numbers.

(3) Figures that accompany problems in this test are intended to provide information useful in solving the problems. They are drawn as accurately as possible EXCEPT when it is stated in a specific problem that the figure is not drawn to scale. All figures lie in a plane unless otherwise indicated.

(4) Unless otherwise specified, the domain of any function f is assumed to be the set of all real numbers x for which $f(x)$ is a real number.

REFERENCE INFORMATION

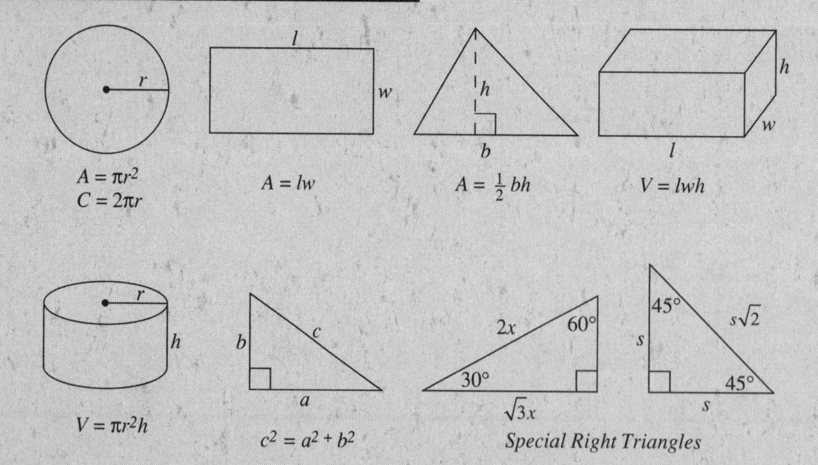

The number of degrees of arc in a circle is 360.

The sum of the measures in degrees of the angles of a triangle is 180.

1. If 60% of x is 50% larger than 20% of y, which of the following describes the correct relationship between x and y?

   (A) $y = 1.3x$

   (B) $y = 0.06x$

   (C) $y = 6x$

   (D) $y = 2x$

   (E) $y = 0.5x$

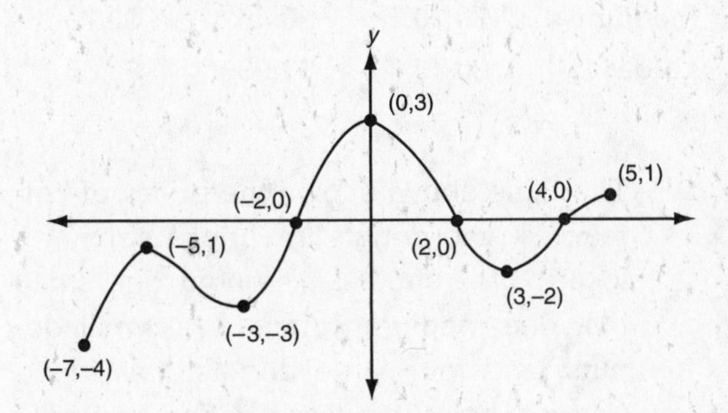

2. The graph of $g(x)$ is shown above. How many zeros would exist for the graph created by shifting $g(x)$ vertically downward by two units?

   (A) 0

   (B) 1

   (C) 2

   (D) 3

   (E) 4

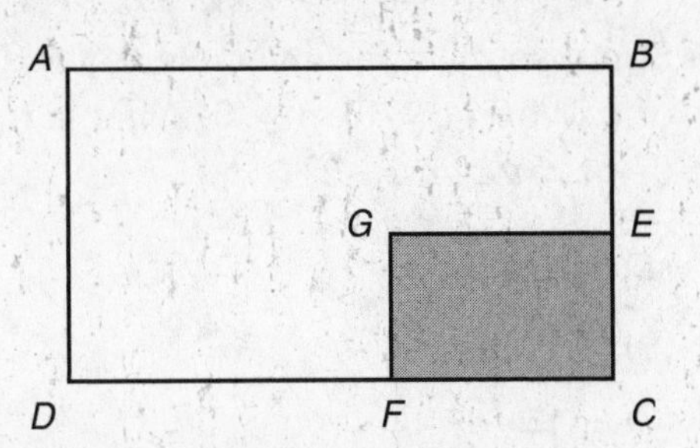

3. The figure above represents a dartboard in which $ABCD$ and $ECFG$ are rectangles. $EC = \frac{1}{2}BC$ and $CF = \frac{1}{3}CD$. If a dart is thrown and lands on the dartboard, what is the probability that it lands in the <u>unshaded</u> area?

   (A) $\frac{1}{6}$

   (B) $\frac{1}{5}$

   (C) $\frac{3}{5}$

   (D) $\frac{4}{5}$

   (E) $\frac{5}{6}$

4. A quadratic function $f(x) = Ax^2 + Bx + C$ contains the points (–9, 0) and (5, 0). If $A < 0$, which of the following could represent the vertex?

   (A) (–2, 1)

   (B) (–4, –1)

   (C) (–2, –3)

   (D) (–9, 5)

   (E) (–4, 2)

GO ON TO THE NEXT PAGE

5. Suppose m varies inversely as the square of c. When $c = 10$, $m = 9$. What is the value of m when $c = 2$?

(A) 0.36

(B) 1

(C) 15

(D) 225

(E) 900

DIRECTIONS: Student-Produced Response questions 6–16 require that you solve the problem and then enter your answer on the special grid on page 849. (For instructions on how to use the grid, see page 69 in the Diagnostic Test.)

6. Peanuts sell for $1.50 per pound and cashews sell for $3.50 per pound. How many pounds of peanuts should be mixed with 12 pounds of cashews in order for the mixture to be worth $2.00 per pound?

7. If $c^3 = 216$, what is the value of $\frac{2}{3}c^2$?

8. The rates at a laundromat are $6.25 for the first piece and $0.35 for each additional piece. For a total charge of $8.35, how many pieces can be laundered?

9. A postal truck leaves its station, averaging 40 miles per hour. An error in the mailing schedule is spotted, and 24 minutes after the truck leaves, a car is sent to overtake the truck. If the car averages 50 miles per hour, how many hours will the car require to catch the postal truck?

10. The mean weight for a group of 5 men is 188 pounds. The mean weight for a group of 10 women is 137 pounds. What is the combined mean weight, in pounds, for these two groups?

11. A car's gas gauge showed that it had $\frac{3}{4}$ of a tank. After using 10 gallons of regular gasoline, the gas gauge showed that $\frac{1}{8}$ of a tank was left. After the tank was emptied, it cost $28.00 to fill it up with regular gasoline. What is the cost per gallon, in dollars, of regular gasoline?

	Coffee	Tea	Donut
Small	$0.40	$0.30	$0.50
Medium	$0.60	$0.45	$0.70
Large	$0.75	$0.60	$0.85

12. The table above shows the prices of coffee, tea, and donuts in various sizes at a local coffee shop. Lin bought one small tea, one medium coffee, five small donuts, and four large donuts. Her friend Carmen bought a large coffee and only wanted to buy medium-sized donuts. If Carmen did not want to spend more than Lin, what is the maximum number of medium-sized donuts she could buy?

GO ON TO THE NEXT PAGE

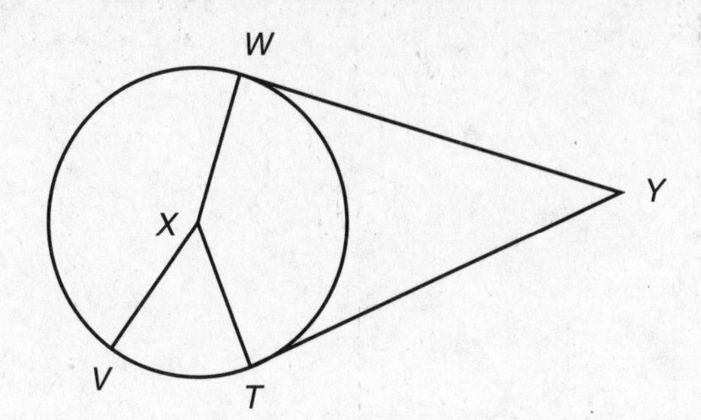

Note: Figure not drawn to scale.

13. In the figure above, X is the center of the circle. $\overline{WY}$ and $\overline{YT}$ are tangents to the circle. If $\angle Y = 40°$ and arc WV is 3 times the length of arc VT, what is the measure, in degrees, of $\angle VXW$?

14. If p is divided by 4 or by 6, the remainder is 2. If q is divided by 5 or by 6, the remainder is 3. If $p < 30$ and $q < 70$, what is one possible value of pq?

15. If x, 27, y, z, 8, . . . represents a geometric sequence for positive numbers, what is the value of $x + y - z$?

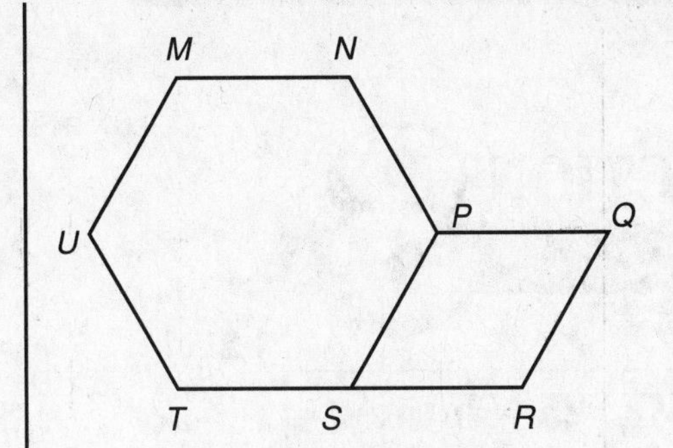

Note: R, S, T are collinear.

16. In the above figure, $MNPSTU$ is a regular hexagon and $PQRS$ is a rhombus. If the perimeter of the hexagon is 24, what is the area of the rhombus?

GO ON TO THE NEXT PAGE

TIME: 25 Minutes
20 Questions

CRITICAL READING

DIRECTIONS: Each sentence below has one or two blanks, each blank indicating that something has been omitted. Beneath the sentence are five lettered words or sets of words. Choose the word or set of words that BEST fits the meaning of the sentence as a whole.

EXAMPLE

Although critics found the book _______, many readers found it rather _______.

(A) obnoxious . . . perfect
(B) spectacular . . . interesting
(C) boring . . . intriguing
(D) comical . . . persuasive
(E) popular . . . rare

EXAMPLE ANSWER

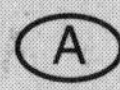   

1. It is difficult to be happy working in such _______ conditions as unfriendly co-workers, outdated equipment, and low pay.

   (A) obvious
   (B) obtuse
   (C) propitious
   (D) adverse
   (E) conducive

2. Evan is five feet five inches tall and inclines toward stoutness, but his erect bearing and quick movements tend to _______ this.

   (A) emphasize
   (B) conceal
   (C) denigrate
   (D) camouflage
   (E) disavow

3. The poet's _______ style caused the publisher to _______ its contract with her.

   (A) sycophantical . . . coil
   (B) slipshod . . . renew
   (C) uninspired . . . cancel
   (D) enduring . . . acclaim
   (E) prosaic . . . praise

4. The spacecraft Voyager, which traveled to Jupiter, was _______ because it was the only space vehicle to _______ a recorded message from our planet to distant star systems.

   (A) blasphemous . . . meander
   (B) egocentric . . . trek
   (C) profound . . . provoke
   (D) unique . . . transport
   (E) vital . . . copy

5. American words and phrases have been added to the lexicon of French and Japanese cultures despite the displeasure of politicians and the _______ of purists.

   (A) concession
   (B) neutrality
   (C) endorsement
   (D) resolution
   (E) denunciation

6. The spelling and pronunciation of some English words are _______ because they don't follow _______ rules.

   (A) infamous . . . prosaic
   (B) erratic . . . inevitable
   (C) fallacious. . . . hypothetical
   (D) hackneyed . . . verbose
   (E) incoherent . . . prudent

GO ON TO THE NEXT PAGE

DIRECTIONS: Each passage that follows is followed by questions based on its content. Answer the questions on the basis of what is stated or implied in each passage and in any introductory material that may be provided.

Questions 7–19 are based on the following passages.

Passage 1

1 Many people deliberately assert their individuality through their appearance. And they certainly have the right to do so. However, too often the people who choose to alter their 5 looks or to dress differently complain that people unfairly or inaccurately judge them by their appearance. And they're probably right. But whose fault is it that people do so?

How are we supposed to get past the 10 purple and orange spiked hair, or the blinding white face paint, or the skimpy shirt that covers almost nothing—especially if we don't already know the person who is flaunting them? It's human nature to assume that the person who 15 exhibits poor grooming and is perhaps somewhat odiferous has little pride and doesn't care if others like him or not. And those who make a show of wearing the most expensive attire, replete with dazzling jewelry, are hardly justified 20 in objecting when we label them pretentious. Granted, following the "norm" is not necessarily the "right" thing to do, but straying to the extreme is a sure way to gain attention—welcome or not. One wonders why people deliber- 25 ately set out to look different if they don't want attention.

People who choose to appear different should not expect to remain unobtrusive, and if they claim a desire for anonymity, they're prob- 30 ably not being completely honest.

Passage 2

People often judge others by their outward 1 appearance, without bothering to see who they really are. It is unfair and prejudicial to assume that the young girl in the low-rise mini-skirt and short, tight-fitting blouse is immoral. In reality, 5 she may simply be wearing the latest fashion, or perhaps that's all she has to wear. Face paint on a teenager may be an intentional attention-getter, but perhaps we need to consider the message he is trying to convey—rather than 10 automatically judging him as "freaky." And the old man who isn't quite as clean or as well groomed as we think he should be may be doing the best he can. Rather than snubbing him as someone who is "beneath" us, we might 15 try getting past the shoddy clothes and the unkempt beard—to the actual individual. Likewise, the lady or gentleman robed in a name-brand suit and wearing a Rolex watch may be the most down-to-earthy, compassionate per- 20 son one could ever meet.

People are first of all human beings, each with a story. Although the façade is what we see first, we must keep in mind that it's only a cover. What's inside is much more important. 25

7. The authors of both passages agree that

(A) appearances do not reflect people's personalities.

(B) appearances are unimportant to most people.

(C) people who choose to look different are simply being honest.

(D) people judge others by their appearances.

(E) our clothes reflect who we are.

GO ON TO THE NEXT PAGE

8. In lines 8 and 9 of Passage 1, the author uses a question

(A) to reveal his/her own openness to both sides of the issue.

(B) as a persuasive tool to lead the reader to the same answer that he/she obviously has.

(C) to show that no one is to blame.

(D) to show that this issue is confusing to almost everyone.

(E) as a way to lead the reader to accept both sides of the issue.

9. In line 20 of Passage 1, "pretentious" most nearly means

(A) exaggerating worth or value, showy.

(B) wealthy, affluent.

(C) sociable, friendly.

(D) sanctimonious, hypocritical.

(E) provocative, alluring.

10. The author of Passage 1 apparently believes that the real reason some people choose to look "different" is that they

(A) cannot afford to look any other way.

(B) want to attract attention.

(C) are exercising their right to express themselves.

(D) don't like the accepted styles.

(E) don't know any other way to look.

11. According to Passage 1, people who purposefully look extremely different should

(A) accept others—even those whose appearance is not unusual.

(B) change their appearance so that it is more acceptable to society.

(C) make sure others understand why they look the way they do.

(D) not be offended when others notice them.

(E) not worry about what others think.

12. What advice would the author of Passage 1 probably give to someone who doesn't want to be noticed?

(A) Follow the latest fad.

(B) Dress to reveal the "real you."

(C) Try to be fashionable but subtle.

(D) Don't worry about what others think.

(E) Wear what you want to wear, but don't talk loudly or put on too much make-up.

13. Which of the following most accurately describes the author of Passage 1?

(A) sympathetic

(B) ambivalent

(C) empathic

(D) angry

(E) judgmental

GO ON TO THE NEXT PAGE

14. In line 23 of Passage 2, "façade" most nearly means

    (A) style.
    (B) behavior.
    (C) attitude.
    (D) facial expression.
    (E) front.

15. How would the author of Passage 2 probably react to a man who was dressed inappropriately by the standards of the others in a group?

    (A) pretend not to notice his appearance
    (B) generously attempt to get to know him
    (C) join the rest of the group in staring at him
    (D) join the others in criticizing him
    (E) ignore him

16. With which of the following would the author of Passage 2 be most likely to agree?

    (A) "Still waters run deep."
    (B) "What you see is what you get."
    (C) "Strike while the iron is hot."
    (D) "Do your own thing."
    (E) "You can't judge a book by its cover."

17. Which of the following describes the attitude of the author of Passage 2?

    (A) friendly
    (B) tolerant
    (C) judgmental
    (D) indifferent
    (E) irritated

18. Which of the following is true about the passages?

    (A) Both use similar examples of attention-getting appearances.
    (B) Both present trends as unusual but attractive.
    (C) Neither offers an opinion about people who look "different."
    (D) Neither is sympathetic toward people whose appearances are "different."
    (E) The two agree on almost every point, although their presentations are different.

GO ON TO THE NEXT PAGE

19. How does the focus of the author of Passage 1 differ from that of the author of Passage 2?

    (A) The author of Passage 1 focuses on the intent behind one's unusual appearance, whereas the author of Passage 2 focuses on each person's "humanness."
    (B) The author of Passage 1 focuses on his/her own anger with people who appear "different," whereas the author of Passage 2 admits irritation but tries to focus on the uniqueness of the individual.
    (C) The author of Passage 1 focuses on the motivation behind the styles we choose, whereas the author of Passage 2 focuses on what we gain from them.
    (D) The author of Passage 1 focuses on others' reactions to people who appear different, whereas the author of Passage 2 focuses on the ways everyone is different.
    (E) The author of Passage 1 focuses on ways to get everyone to look similar, whereas the author of Passage 2 focuses on ways to get each person to look unique.

STOP

If time remains, you may go back and check your work. When the time allotted is up, you may go on to the next section.

TIME: 10 Minutes
14 Questions

WRITING

DIRECTIONS: In each of the following sentences, some portion of the sentence is underlined. Under each sentence are five choices. The first choice has the same wording as the original. The other four choices are reworded. Sometimes the first choice containing the original wording is the best; sometimes one of the other choices is the best. Choose the letter of the best choice. Your choice should produce a sentence that is not ambiguous or awkward and that is correct, clear, and precise.

This is a test of correct and effective English expression. Keep in mind the standards of English usage, punctuation, grammar, word choice, and construction.

EXAMPLE

When you listen to opera, a person may not appreciate it.

(A) a person may not appreciate it.

(B) it may not be appreciated by a person.

(C) you may not appreciate it.

(D) which may not be appreciated by you.

(E) appreciating it may be a problem for you.

EXAMPLE ANSWER

1. She has to walk home after dark, so every evening <u>she would get out her pepper spray in case she encounters trouble</u>.

   (A) she would get out her pepper spray in case she encounters trouble.

   (B) she gets out her pepper spray in case she would encounter trouble.

   (C) she gets out her pepper spray in case she encounters trouble.

   (D) she got out her pepper spray in case she encountered trouble.

   (E) she gets out her pepper spray in case she encountered trouble.

2. John Steinbeck's <u>novel, like F. Scott Fitzgerald, are</u> set in the first half of the twentieth century and depict an important part of American history.

   (A) novel, like F. Scott Fitzgerald, are

   (B) novels, like those of F. Scott Fitzgerald, are

   (C) novel, like those of F. Scott Fitzgerald, is

   (D) novels, like F. Scott Fitzgerald, are

   (E) novel, like that of F. Scott Fitzgerald, are

3. She is going to lose her <u>job because she does not stay focused she never completes the assigned tasks</u>.

   (A) job because she does not stay focused she never completes the assigned tasks.

   (B) job since she does not stay focused she never completes the assigned tasks.

   (C) job, not staying focused she never completes the assigned tasks.

   (D) job; because she does not stay focused, she never completes the assigned tasks.

   (E) job because, not staying focused, never completes the assigned tasks.

4. The store sold matching dinner plates, drinking glasses, and <u>they had serving dishes to match</u>.

   (A) they had serving dishes to match.

   (B) serving dishes.

   (C) matching serving dishes.

   (D) they sold matching serving dishes.

   (E) their serving dishes matched them.

GO ON TO THE NEXT PAGE

5. Of the dozen seashells she picked, only three of them was unusual enough to merit them being included in her collection.

(A) was unusual enough to merit them being included

(B) were unusual enough to merit them being included

(C) was unusual enough to merit their being included

(D) were unusual enough to merit that they be included

(E) were unusual enough to merit their being included

6. I like my brother better than anyone in my family.

(A) I like my brother better than anyone in my family.

(B) My brother is the one I like better than anyone in my family.

(C) I like my brother better than I like anyone else in my family.

(D) In my family I like my brother better than anyone.

(E) In my family my brother is the one I like better than anyone else.

7. When the proper labeling of foods does not occur, they often confuse the cooks in the cafeteria.

(A) When the proper labeling of foods does not occur, they often confuse the cooks

(B) Improper labeling of foods often confuses the cooks

(C) Improper labeling of foods often confuse the cooks

(D) When the foods are not properly labeled, it often confuses the cooks

(E) When the labels of foods not properly labeled, the cooks often become confused

8. In the advanced stages of Alzheimer's, the patient may not recognize even his family members.

(A) Alzheimer's, the patient may not

(B) Alzheimer's may not

(C) Alzheimer's, it may be that the patient may not

(D) Alzheimer's, it may cause the patient not to

(E) Alzheimer's, they may cause the patient not to

9. He does not accept many speaking engagements, it is because of that he has a busy schedule and is not really comfortable speaking to large crowds.

(A) engagements, it is because of that he has a busy schedule and is not really comfortable speaking to large crowds.

(B) engagements because he has a busy schedule and he is not really comfortable speaking to large crowds.

(C) engagements because of his busy schedule and also he is not really comfortable speaking to large crowds.

(D) engagements, it is because he has a busy schedule and his not really being comfortable speaking to large crowds.

(E) engagements, it is because he has a busy schedule and discomfort speaking to large crowds.

10. She accepted his gifts too quickly, without considering the reason behind it.

(A) too quickly, without considering the reason behind it.

(B) too quick, without considering the reason behind them.

(C) too quickly, and there was not enough consideration about the reason behind them.

(D) too quick, and there was not enough consideration about the reason behind it.

(E) too quickly, without considering the reason behind them.

11. The entire class agreed that she would be a good leader, since she has been the only one which understood the project.

(A) has been the only one which understood the project.

(B) was the only one who understood the project.

(C) was the only one which understood the project.

(D) having been the only one to understand the project.

(E) had been the only one who understands the project.

12. You are likely to lose all your friends if one cannot be trusted.

(A) if one cannot be trusted.

(B) when one cannot be trusted.

(C) since they cannot be trusted.

(D) if you cannot be trusted.

(E) if unable to be trusted.

13. He prefers pears over apples because they taste sweeter.

(A) over apples because they taste sweeter.

(B) to apples because of their sweeter taste.

(C) because they taste sweeter than apples.

(D) over apples because pears taste sweeter than apples and so they are better.

(E) over apples because it tastes sweeter.

14. <u>In his essay he wrote that American voters are unthinking robots who base their votes on erroneous campaign ads</u>.

(A) In his essay he wrote that American voters are unthinking robots who base their votes on erroneous campaign ads.

(B) He wrote that American voters are unthinking robots who bases their votes on erroneous campaign ads in his essay.

(C) He wrote that American voters are unthinking robots who base their votes on erroneous campaign ads in his essay.

(D) He wrote that American voters in his essay are unthinking robots who base their votes on erroneous campaign ads.

(E) He wrote in his essay that American voters are unthinking robots of which they base their votes on erroneous campaign ads.

STOP

If time remains, you may go back and check your work. When the time allotted is up, you may go on to the next section.

TEST 1

ANSWER KEY

Section 1—Essay

Refer to the Detailed Explanation for essay analysis

Section 2—Math

1.	(D)	6.	(A)	11.	(E)	16.	(C)
2.	(C)	7.	(B)	12.	(A)	17.	(B)
3.	(D)	8.	(D)	13.	(D)	18.	(D)
4.	(C)	9.	(E)	14.	(D)	19.	(E)
5.	(B)	10.	(C)	15.	(A)	20.	(A)

Section 3—Critical Reading

1.	(A)	7.	(B)	13.	(B)	19.	(C)
2.	(E)	8.	(D)	14.	(C)	20.	(A)
3.	(C)	9.	(E)	15.	(B)	21.	(D)
4.	(D)	10.	(B)	16.	(B)	22.	(B)
5.	(D)	11.	(A)	17.	(B)	23.	(A)
6.	(C)	12.	(D)	18.	(C)	24.	(D)

Section 4—Writing

1.	(C)	10.	(B)	19.	(E)	28.	(C)
2.	(D)	11.	(B)	20.	(C)	29.	(C)
3.	(B)	12.	(C)	21.	(A)	30.	(E)
4.	(E)	13.	(E)	22.	(E)	31.	(B)
5.	(B)	14.	(C)	23.	(B)	32.	(A)
6.	(A)	15.	(A)	24.	(A)	33.	(C)
7.	(D)	16.	(B)	25.	(D)	34.	(D)
8.	(E)	17.	(D)	26.	(C)	35.	(D)
9.	(A)	18.	(A)	27.	(E)		

Section 5—Math

1.	(E)	6.	(A)	11.	(A)	15.	(D)
2.	(D)	7.	(D)	12.	(B)	16.	(C)
3.	(E)	8.	(C)	13.	(B)	17.	(C)
4.	(E)	9.	(D)	14.	(C)	18.	(D)
5.	(C)	10.	(C)				

Section 6—Critical Reading

1.	(A)	7.	(C)	13.	(C)	19.	(C)
2.	(E)	8.	(A)	14.	(D)	20.	(D)
3.	(D)	9.	(E)	15.	(D)	21.	(A)
4.	(B)	10.	(B)	16.	(B)	22.	(D)
5.	(B)	11.	(A)	17.	(B)	23.	(B)
6.	(D)	12.	(B)	18.	(D)	24.	(C)

Section 7—Math

1.	(D)	6.	36	11.	1.75	16.	13.9
2.	(C)	7.	24	12.	8		
3.	(E)	8.	21	13.	165		
4.	(A)	9.	1.6	14.	452, 858, 882, or 1,638		
5.	(D)	10.	154	15.	46.5		

Section 8—Critical Reading

1.	(D)	6.	(B)	11.	(D)	16.	(E)
2.	(D)	7.	(D)	12.	(C)	17.	(B)
3.	(C)	8.	(B)	13.	(E)	18.	(A)
4.	(D)	9.	(A)	14.	(E)	19.	(A)
5.	(E)	10.	(B)	15.	(B)		

Section 9—Writing

1.	(C)	5.	(E)	9.	(B)	13.	(C)
2.	(B)	6.	(C)	10.	(E)	14.	(A)
3.	(D)	7.	(B)	11.	(B)		
4.	(B)	8.	(A)	12.	(D)		

DETAILED EXPLANATIONS

SECTION 1—ESSAY

Sample Essay With Commentary

ESSAY I (Score: 5–6)

Some people aspire to be president. Some strive to earn a college degree. Some set out to become millionaires. Others hope to serve others as teachers, doctors, missionaries, or social workers. And still others seek a life of adventure—perhaps in the military or as a test pilot or a NASCAR driver. All of these goals are admirable, but some would question whether or not they are realistic. However, careful consideration reveals that people who set high goals for themselves are the ones who accomplish the most.

Granted, we may set our sights on goals that may be unreachable. Clearly, not everyone who hopes to become a doctor has the financial or intellectual ability to do so. And although even high-level politicians try their hardest to become the President of the United States, only every four years and only one person has a chance to see this goal to fruition. Likewise, not many of us become millionaires, but that dream is commonly shared by many Americans.

The fact that we may set high goals that we will never achieve does not mean we are wrong to set them. With high goals come several smaller, more attainable goals. The politician who dreams of becoming president may begin that climb by achieving the goal of being elected state representative, then perhaps governor, or U.S. Senator. And with each step the competition becomes stiffer. That person may never actually become U.S. President, but he has achieved many goals anyway. Even if he never attained his "ultimate goal," he definitely did not set himself up for failure by having that ambition.

Without goals, we would accomplish little, if anything. The young man who loves working with and driving cars may never become a NASCAR driver, but learning auto-mechanics can give him job security, and driving locally can be enjoyable and even supplement his income. But if he gives up because he knows NASCAR is too high a goal, he will miss out on both a career and a hobby that would bring him pleasure. The student who knows she can't afford medical school or realizes it is too academically challenging for her may become a nurse, a nurse practitioner, or some other healthcare professional—a rewarding career that requires less money and education but is closely related. People who work hard to move up the corporate ladder and are wise enough to make good investments and set aside some of their earnings may never become millionaires; but a comfortable lifestyle, with some luxuries and pleasure, may be theirs; they are successful. If they had given up because a million-dollar bank account was inaccessible, settling for a low-paying job with no promise of promotion, then they would have been a failure.

We all have heard that things that come easily aren't worth much. This is true when it comes to setting goals. High goals are worth the effort. Even though we may set goals that are unrealistically high—goals that we may never actually reach—those goals require smaller, attainable steps that bring us success along the way. That success is something we would not have if we didn't set a "pie in the sky" goal for ourselves. The only people who set themselves up for failure are those who do not set high goals for themselves.

ANALYSIS

Essay I has a score range of 5–6. It is well-organized, presenting a thesis in the introduction, developing that idea throughout the essay, and repeating it in the conclusion. The author offers examples to prove the main idea. The writing demonstrates a variety of sentence structures and length, and the grammar and punctuation are correct. The vocabulary is fairly sophisticated.

DETAILED EXPLANATIONS

SECTION 2—MATH

1. (D) If Kelly improves her sales by 50%, she will sell 2 + (0.50)(2) = 3 homes. If Andre improves his sales by 100%, he will sell 3 + (1.00)(3) = 6 homes. Ted will still sell 5 homes and Maya will still sell 6 homes. A-Plus will sell a total of 6 + 3 + 5 + 6 = 20 homes, and Kelly will sell 3 of them.

2. (C) The most efficient way to solve this question is to substitute values for w and y. Let $w = \frac{1}{4}$ and $y = \frac{1}{3}$. The answers for (A), (B), (C), (D), and (E) are $\frac{7}{12}$, $\frac{1}{12}$, $\frac{4}{3}$, $\frac{1}{12}$, and $\frac{3}{4}$. Only choice (C) yields a number greater than 1. Another way to solve this is to remember that any positive number divided by another positive number less than itself is always greater than 1.

3. (D) $f(-2) = (-2 - 2)^2 = (-4)^2 = 16$. Then $g(16) = 16^2 - 2 = 256 - 2 = 254$.

4. (C) If $\overline{PR}$ and $\overline{QS}$ are perpendicular bisectors of each other, then all four angles at T are right angles. By using the side-angle-side (SAS) theorem, it can be shown that all four nonoverlapping triangles are congruent to each other. Since corresponding parts of congruent triangles are congruent, $\overline{PQ} = \overline{QR} = \overline{RS} = \overline{PS}$. Thus, $PQRS$ must be a rhombus.

5. (B) Quadrupling a number n means 4 times n or $4n$. Twice the original number is $2n$. Nine less than $2n$ is represented by $2n - 9$. Finally, one-third as much as $2n - 9$ means $(\frac{1}{3})(2n - 9) = \frac{(2n-9)}{3}$.

6. (A) Multiply both sides by 4 to get $6 - 3x > -36$, then $-3x > -42$. Now, divide both sides by -3, which reverses the order of inequality. Thus $x < 14$. Since x must be positive, we know that $x > 0$. So, $0 < x < 14$.

7. (B) X contains four elements, namely, 11, 13, 17, and 19. $X \cup Y$ contains ten elements, namely, 1, 3, 5, 7, 9, 11, 13, 15, 17, and 19. Then Y must contain <u>at least</u> the six elements 1, 3, 5, 7, 9, and 15.

8. (D) Using the rules of exponents, begin by simplifying the numerator, then reduce it to 1 and square the entire expression: $\sqrt{\frac{x^{-7} \cdot x^2}{x^{-3}}} = \sqrt{\frac{x^{-5}}{x^{-3}}} = \sqrt{\frac{1}{x^2}} = \frac{1}{x}$

9. (E) If $f(x) < 3$, then $|9 - 2x| < 3$, which means $-3 < 9 - 2x < 3$. Subtracting 9, we get $-12 < -2x < -6$. Now, divide by -2 and reverse the order of inequality to get $6 > x > 3$, which is equivalent to $3 < x < 6$.

10. (C) The graph of $f(x)$ will be shifted 4 units to the left so that $f(x + 4)$ will be undefined at $x = 2 - 4 = -2$. Also, $f(x + 4)$ will have the same slope as $f(x)$, so its slope is -4.

11. (E) $|f(x)| \leq 3$ means $-3 \leq f(x) \leq 3$. From the graph, when $-4 \leq x \leq 8$, $f(x)$ has a value between -3 and 3, inclusive. When $x < -4$ or when $x > 8$, $f(x) > 3$.

12. (A) The graph of $g(x)$ is a semicircle passing through $(-4, 0)$, $(0, 4)$, and $(4, 0)$. The segment connecting $(-4, 0)$ and $(4, 0)$ is the diameter. The lowest y values are found at $(-4, 0)$ and $(4, 0)$, whereas the highest y value is found at $(0, 4)$. Thus, the range is all numbers between 0 and 4, inclusive.

13. (D) Because ΔADE is equilateral, $\angle A = \angle B = \angle C = 60°$. If we assign the following numerical values for the length of the sides, $\overline{AC} = \overline{AB} = \overline{CB} = 2$, then $\overline{AE} = \overline{AD} = \overline{CG} = \overline{GB} = 1$. In addition, we can find the height of ΔABC to be $\sqrt{3}$ when a perpendicular bisector is dropped from A; note that a 30–60–90 triangle is formed. Thus, the area of $\Delta ADE = \left(\frac{1}{2}\right)(2)(\sqrt{3}) = \sqrt{3}$. Using the same process we find the area of $\Delta ADE = \left(\frac{1}{2}\right)(1)\left(\frac{\sqrt{3}}{2}\right) = \left(\frac{\sqrt{3}}{4}\right)$. To find the area of the shaded region we can subtract half of the area of ΔADE from half of the area of ΔABC, $[\left(\frac{1}{2}\right)(\sqrt{3})] - [\left(\frac{1}{2}\right)\left(\frac{\sqrt{3}}{4}\right)] = \left(\frac{\sqrt{3}}{2}\right) - \left(\frac{\sqrt{3}}{8}\right) = \left(\frac{3\sqrt{3}}{8}\right)$. To find the probability, divide the area of the shaded region by the area of $\left(\frac{3\sqrt{3}}{\sqrt{3}}\right) = \left(\frac{3\sqrt{3}}{8}\right) - \frac{1}{\sqrt{3}} = \frac{3}{8}$.

14. (D) Because L_1 is parallel to the graph of $y = x + 3$, the slope of L_1 must also be 1. The y-intercept of L_1 can be calculated as $(0, -11)$ by substituting 0 for y, 11 for x, and 1 for m in the equation $y = mx + b$ and solving for b. The y-intercept of the graph representing $y = x + 3$ is $(0, 3)$. L_2 lies midway between L_1 and $y = x + 3$, so its y-intercept must be midway between $(0, -11)$ and $(0, 3)$, which is $(0, -4)$.

15. (A) Let x = original length and y = original width. The new rectangle will have a length of $x + 0.3x = 1.3x$ and a width of $y - 0.2y = 0.8y$. The area of the original rectangle is xy, whereas the area of the new rectangle is $(1.3x)(0.8y) = 1.04xy$. Therefore, $1.04xy$ represents a 4% increase over xy.

16. (C) If $x < -2$, then when we multiply both sides of the inequality by $4(x + 2)$, the inequality reverses direction. Thus, $[4(x + 2)][\frac{20}{(x+2)}] > [4(x + 2)][\frac{x}{4}]$. This simplifies to $80 > x^2 + 2x$ or $0 > x^2 + 2x - 80$. Upon factoring, we get $0 > (x + 10)(x - 8)$. The solution is found by setting one factor less than zero and the other factor greater than zero. If we let $x + 10 < 0$ and $x - 8 > 0$, then $x < -10$ and $x > 8$, which is impossible. But if $x + 10 > 0$ and $x - 8 < 0$, then $x > -10$ and $x < 8$, which merges to become $-10 < x < 8$.

17. (B) The unit cost of each item is represented by the slope of the segment connecting each of *P, Q, R, S,* (and the hypothetical *T*) in the figure below.

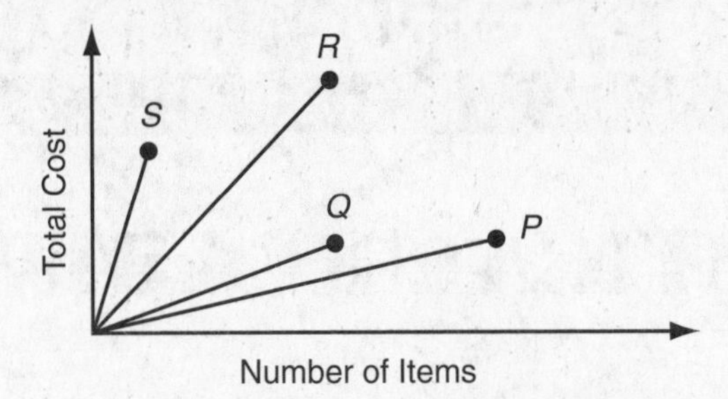

The steeper the slope, the higher the unit cost. If *T* is directly below *S*, then the unit cost of *T* is less than that of *S*; also, it is possible that *T* may lie on the segment connecting the origin and *R*, so that its unit cost would match that of *R*. Choice (A) is not necessarily true because *T* could lie above the segment connecting the origin and *P* or *Q*. Choice (C) is impossible because the unit costs for *P* and *Q* are different. Choice (D) is impossible because the unit costs for Q and R are different. Choice (E) is wrong because if the unit cost of *R* and *T* were equal, *T* would lie on the segment connecting the origin and *R*. Because the unit cost of *R* is already higher than that of *P*, the unit cost of *T* would also be higher than that of *P*.

18. (D) If x is divisible by 3, then $x = 3k_1$, where k_1 is a constant. If y is divisible by 6, then $y = 6k_2$, where k_2 is a constant. Now, $8x + 2y = 2(4x + y) = 2(4 \times 3k_1 + 6k_2) = 4(6k_1 + 3k_2)$. This last expression can be divided by 4 to yield $6k_1 + 3k_2$. To show why the other choices are incorrect, let $x = 3$ and $y = 6$. Then choices (A), (B), (C), and (E) would have values of 21, 2, $\frac{1}{4}$, and 54, respectively; none of these numbers is divisible by 4.

19. (E) The area of any equilateral triangle is $\left(\frac{s^2}{4}\right)(\sqrt{3})$, and since ΔAOB is equilateral, its area is $\left(\frac{36}{4}\right)(\sqrt{3})$. The area of the circle is 36π, so the area of the sector enclosed by AOB is $\left(\frac{60°}{360°}\right)(36\pi) = 6\pi$. The area of the shaded region is the difference of the area of the sector and the area of the triangle, which is $6\pi - 9\sqrt{3}$.

20. (A) $\angle QJK$ and $\angle QML$ are each 90° because tangents to a circle form right angles at points of tangency. In any 5-sided figure, the sum of the interior angles is $(180°)(5 - 2) = 540°$. So $\angle JQM = 540° - 90° - 90° - 125° - 100° = 135°$; this means that $\angle MQN = 180° - 135° = 45°$. This means that $\angle N = 45°$, by noting that the sum of the angles of ΔQMN is 180°. Because ΔQMN is a 45°–45°–90° right triangle, $\overline{QN} = \left(\overline{QM}\right)\left(\sqrt{2}\right) = 6\sqrt{2}$. Now $\overline{RN} = \overline{QN} - \overline{QR}$, and because $\overline{QR}$ is a radius, $\overline{RN} = 6\sqrt{2} - 6$.

DETAILED EXPLANATIONS

SECTION 3—CRITICAL READING

1. (A) You are looking for a word that describes the effect a new teacher's explanation would have on students who had learned a different method of solving the physics problems—to the extent that this explanation would cause them to be in danger of failing the class. "Discombobulate" (upset, confuse) is such a word. Choice (B) is incorrect because it does not make sense that an explanation of a physics problem would mollify (soothe, appease) these students to the point that they might fail the class. Nor does it make sense to say that the teacher's explanation would "predicate" (declare, assert to be a quality of) these students; so choice (C) is wrong. And although a new explanation might "invigorate" physics students, such an explanation would not cause them to be in danger of failing the class; thus, choice (D) is wrong. Choice (E) is wrong because it does not make sense to say that the teacher's explanation served to "juxtapose" the students (place them side by side).

2. (E) A minister is likely to believe that "abstinence" (the practice of voluntarily refraining) from immoral behavior is the only acceptable lifestyle. Choice (A) is wrong, since a minister would not be likely to believe that "atrophy" (wasting away from lack of nutrition) was acceptable, nor would it be likely to result from drinking and other immoral practices. Choice (B) is incorrect because a minister is not likely to find "defamation" (the telling of lies about others) acceptable, and this word does not fit semantically in this sentence. Choices (C) and (D) are wrong because a minister would not be likely to approve of a lifestyle that involved "fervor" (enthusiasm, fervor) or "volatility" (lightheartedness, liveliness) from drinking and immoral practices.

3. (C) You are looking for a description of gifts that overwhelm a person in crisis, and that would come from people who simply wanted to help her. The second word you are seeking should be an adjective that describes such people. A "plethora" (excess, superabundance) of gifts would be overwhelming. Also, it makes sense that gift-giving, helpful community members would be "empathic" (vicariously understanding and sympathetic). Choice (A) is incorrect because "audacity" (boldness, daring) does not make sense in reference to gifts, and "chagrined" (disappointed) community members probably would not give gifts. A "paucity" (shortage, scarcity) of gifts is not likely to be the result of the efforts of "affable" (friendly, pleasant) community members; thus, choice (C) is wrong. Choice (D) is incorrect because "squalid" (poverty-stricken) community members are not likely to give a "glut" (oversupply) of gifts. And the efforts of tenacious" (persistent) community members will probably not result in a "dearth" (inadequate supply) of gifts in a crisis. Therefore, choice (E) is incorrect.

4. (D) You are looking for a term that describes a painting on which the artist might devote years behind closed doors, with the hope that it will be of the quality required by prestigious galleries. A "magnum opus" (great work of art) is an appropriate term. Choice (A) is incorrect because "microcosm" (small universe) does not fit the sentence. Neither does choice (B), "conundrum" (riddle, puzzling question), choice (C), "facsimile" (copy, replication, reproduction), or choice (D), "soliloquy" (a talk one has with oneself, especially in theatre).

5. (D) You are looking for two adjectives with related meanings—one to describe a strong odor and one to describe garbage that would have such an odor. And you know the odor was probably not pleasant since "we could hardly force ourselves to pick up the trash bags." Trash bags with the "pungent" (sharp, stinging) odor of "rancid" (rank, rotten) garbage would be difficult to pick up. If you chose choice (A), you were accurate in selecting words that are related to smell; however, with reference to garbage and trash bags the sentence implies that the odor was not a pleasant one. The smell might very well be "potent" (very strong; powerful); however, the adjective "redolent" (fragrant, aromatic) has a connotation of pleasantness—not appropriate for trash that was difficult to pick up because of its odor. Therefore, choice (A) is wrong. Similarly, in choice (B) the word "acrid" (sharp; bitter; foul) is an appropriate description of the odor, but the word "phlegmatic" (unemotional; disinterested) does not fit in this context. Choice (C) is incorrect because a "suppressed" (held back; subdued) odor would not necessarily keep one from wanting to pick up the trash bags, although the fact that the contents were "toxic" (poisonous) might make a person reluctant to pick them up. And choice (E) is incorrect because "pristine" (pure, unspoiled) hardly describes garbage that has a "noxious" (harmful; unwholesome) odor.

6. (C) Here you are seeking two words that have similar meanings, both of which describe the efforts of a person who faced many obstacles but succeeded anyway. "Sedulous" (diligent, hard-working) and "steadfast" (unwavering, enduring) are two such words. "Quiescent" (inactive) and "dilatory" (slow, inclined to delay) have similar meanings, but they do not describe the study and resolve of a person who earned a diploma despite many obstacles. Thus, choice (A) is incorrect. Choice (B) is wrong because someone whose study is "pedantic" (focusing on trivial details) and whose resolve is "indeterminate" (indefinite) is not likely to succeed, especially if that person is facing many obstacles. With "sporadic" (on-again-off-again) study, even the most "efficacious" (effective) resolve is not likely to lead to success; therefore, choice (D) is wrong. And choice (E) is wrong because "indefatigable" (untiring) and "irresolute" (wavering, indecisive, vacillating) do not have similar meanings.

7. (B) Your clue here is "although," indicating that the two phrases express opposing or contrary reactions. One might "abhor" (hate, despise) what someone says but "unequivocably" (wholeheartedly, unquestionably) uphold his right to say it. Choice (A) is incorrect because "accept" (pardon, overlook) does not contradict the idea of "amicably" (agreeably) upholding something. Choice (C) is incorrect because it makes little sense to say that I "underscore" (emphasize) what he says but "sardonically" (mockingly) uphold his right to say it. Likewise, it doesn't make sense to say that someone "recants" (takes back, withdraws) what someone else says. And one is not likely to "inadvertently" (unintentionally) uphold someone's right to say something. Therefore, choice (D) is wrong. And although I may "comprehend" (understand) what one says, it is questionable that I would "sanctimoniously" (hypocritically, self-righteously) uphold his right to say it.

Thus, choice (E) is wrong.

8. (D) Here you are looking at contrasting ideas. As a dancer, the person is one way; in his everyday life he is the opposite way. The adjective "lithe" (graceful, flexible) and the noun "lummox" (clumsy person) have opposing or contrasting connotations. Choice (A) is wrong because an "oaf" (clumsy, slow-witted person) is often "ungainly" (clumsy, awkward); the two do not contrast. Likewise, choice (E) is incorrect because "wimpish" (weak, ineffectual) and "milksop" (weakling) do not contrast. One does not ordinarily use "perspicacious" (keen, intelligent) to describe a dancer; but even if that adjective were used, "sage" (master, wise person) does not present a contrasting idea; so choice (B) is wrong. And choice (C) is incorrect because although a dancer may be "supple" (limber, not stiff), the idea that he might be a "racketeer" (criminal) has little to do with that description.

9. (E) The paragraph focuses on the beauty of a woman ("one of the most beautiful people in my life") who has lived almost a hundred years ("more years than anyone else I know . . . nearly ten decades."). Therefore, the author is showing that beauty and youth are not the same thing. Although the passage does mention the availability of surgery, creams, and cosmetics—and today's urgency to look young—the focus is on the beauty of an older acquaintance of the author. Therefore, choice (A) is incorrect. If you chose choice (B), you probably did so because the first sentence says "we seem to have lost touch with reality." However, as you read further, you will see that the second sentence is actually the topic sentence of the paragraph because the rest of the passage focuses on the beautiful aspects of "one of the most beautiful people in my life" who "has lived more years than anyone else I know." And although some of the woman's wrinkles are evidently the result of her life-long ready smile, the author mentions this fact to exemplify her beauty—not to advise against smiling. Thus, choice (C) is wrong. Choice (D) is incorrect because although the author does describe the beauty of an elderly woman, the passage does not imply that elderly people are the most beautiful; at the same time, it does not say that a young person cannot be beautiful. Instead, the author is illustrating that beauty and youth are not synonymous by showing that an elderly individual may be beautiful in ways that are not cosmetic.

10. (B) Everything in the author's description of the elderly woman conveys admiration—"one of the most beautiful people in my life," "sharp memory and her willingness to share . . . far more meaningful education than any textbook," "a beauty far greater than. . . ." Choices (B) and (D) are incorrect because the author clearly has much regard (as opposed to disregard or indifference) for the woman—the purpose of writing the passage. Likewise, choice (C) is incorrect because nothing in the passage indicates that the author experiences any sort of confusion toward the woman; his/her admiration for her is clear and direct. Choice (E) is incorrect because a "sarcastic" attitude is a negative one, and the author conveys nothing negative about the woman. Sarcasm is implied in the author's attitude toward the "plastic surgery, wrinkle cream, or other alleged cosmetic miracles" that people seek in order to preserve the so-called beauty of youth—but not in the author's attitude toward the elderly woman whose beauty transcends cosmetic beauty.

11. (A) The passage is devoted to depicting the behavior of a young child ("not yet old enough to be enrolled in school. . . . Even before his first birthday") as he studies and learns about any object he encounters. So, clearly, the author does not believe that education is confined to the walls of a school building. Choice (B) is incorrect because the author indicates that learning occurs far before a person is exposed to formal education, and in the last sentence the author implies that knowledge is possible with or without formal education. Choice (C) is incorrect because the author indicates only that learning begins very early; nothing in the passage indicates that children are the only ones who learn thoroughly. Likewise, choice (D) is wrong because the passage includes nothing to indicate whether or not we become too old to learn. And although the passage indicates that formal education is not necessary for learning, it does not say that formal education stifles learning. Therefore, choice (E) is wrong.

12. (D) The passage focuses on how a small child absorbs every aspect of anything he picks up. One of the definitions of "consume" is "absorb, take in." The dictionary indicates that "consume" can also mean choice (A) "waste," choice (B) "exhaust," choice (C) "eat," and choice (E) "use up"; however, none of these definitions fit the context of the sentence or the passage.

13. (B) The focus of the entire passage is to describe the fall migration of the monarch butterflies from Canada to Mexico and their spring return. Although the passage does convey the idea that the monarch butterfly is complex (as opposed to simple), the passage is devoted primarily to a description of this insect's migration and the tone is not one of persuasiveness; thus, choice (A) is wrong. The passage does mention research on the butterflies' migration; however, it does not indicate the need for additional research. So choice (C) is incorrect. Choices (D) and (E) are wrong because no mention is made of the difference between the monarch butterfly and other butterflies or between the monarch butterfly and other animals that migrate.

14. (C) The author begins the passage with "Almost everyone appreciates the beauty of the monarch butterfly, but few people realize the complexity of this orange and black phenomenon." This statement implies that people tend to see only the physical beauty of the monarch, without being aware of its complexity. Choices (A) and (B) are incorrect because the author acknowledges that people do notice and appreciate the monarch's beauty. Choice (D) is wrong because the opening statement presents the opposite idea. And choice (E) is incorrect because the author indicates that few people have a realistic picture of the monarch's level of complexity.

15. (B) The passage refers to monarch butterflies as "small but amazing insects" that "travel 50 to 80 miles a day, at 12 to 20 miles an hour . . . almost 3,000 miles, overcoming weather, predators, and fatigue." Obviously only strong creatures could do so. Choice (A) is wrong because the monarch is not weak. And although the passage refers to the familiar beauty of the monarch, it does not indicate that the butterfly is rare; in fact, it indicates that 300 million monarch butterflies migrate to Mexico. Because the butterflies are neither rare nor nearing extinction, choices (C) and (D) are wrong. And butterflies that "travel 50 to 80 miles a day, at 12 to 20 miles an hour" may be steady, but they certainly are not slow; therefore, choice (E) is wrong.

16. (B) "Dormancy" means "inactivity, suspension of biological tendencies or functions," so "reproductive dormancy" means "inactivity of reproduction." In reference to the butterflies, this would mean "not laying eggs." Although "dormancy" might involve "being asleep," "hibernating," "not eating or drinking," and "not moving," none of these definitions include anything related to "reproductive" activity; therefore, choices (A), (C), (D), and (E) are wrong.

17. (B) The butterflies that are born in the spring live only a month—just long enough to form, mate, and lay their eggs. However, the fall-migrating butterflies are born sometime in the summer and, assuming they survive the perils of the southerly migration and the winter in Mexico, begin traveling north, leaving their eggs as they travel. Although the passage indicates that few if any survive the entire migratory cycle, these fall-migrating would have to live several months—nearly a year—to go through these phases. Choice (A) is wrong because many live much longer than a month. Choice (C) is wrong because many do not live nearly a year. Choice (D) is incorrect because nothing in the passage indicates that the length of time the migration requires has anything to do with the life span of the monarch. And choice (E) is wrong because nothing in the passage indicates that any monarch lives through five reproductive cycles, although it does indicate that some butterflies die as soon as they lay their eggs.

18. (C) The passage indicates that "during their spring northerly migration, they lay eggs on milkweed en route." Choices (A) and (B) are incorrect because "the female monarchs are in reproductive dormancy during the fall migration." And choices (D) and (E) are incorrect because the passage says that "when early spring arrives, those that survived the winter begin traveling north, lay their eggs," indicating that they do not lay their eggs until they have left the Oyamel forests and spring migration has begun.

19. (C) Because "butterflies that are born in the spring live only a month—just long enough to form, mate, and lay their eggs," four or more generations are born and die just during the spring migration. Therefore, the fall-migrating butterflies are several generations distant in relationship—or "removed"—from those of a year earlier. The dictionary indicates that "removed" can mean choice (A) "far away," choice (B) "obscure," choice (D) "alone," or choice (E) "taken away"; however, none of these definitions fit the context of the sentence or the passage.

20. (A) The passage says that the butterflies are probably "attempting to escape the cold weather and food shortage of the winter months . . . and instinctively sense the decreasing amount of daylight and the lowering temperatures as summer ends." Choice (B) is incorrect because few if any monarchs live long enough to remember a past winter. And choice (C) is incorrect because although the passage does mention "the food shortage of the winter months," it does not specify the diminishing of the milkweed crop. In fact, the passage indicates only that the butterflies lay their eggs on the milkweed and that larvae feed on it; it does not indicate that the butterflies eat this plant. Although the passage does mention that the butterflies may instinctively sense the decreasing daylight hours as a sign of upcoming winter, no mention is made of the butterflies' need for a certain amount of daylight; so choice (D) is incorrect. And although the female monarchs' reproductive systems do become dormant in the winter, the passage does not present this fact as a probable reason for their migration; therefore, choice (E) is wrong.

21. (D) Lines 27 and 28 say "the monarchs' migratory path . . . follows a route that runs from northeast to southwest in the fall and retraces the same path, in reverse, in the spring." Choices (A) and (B) are incorrect because the migration is "round trip," not just in one direction, and the migration is northeast to southwest, and vice versa, rather than east to west or north to south. Choice (C) is wrong because the migration is not simply east to west, or vice versa, but northeast to southwest and southwest to northeast. Choice (E) is wrong because the migratory path is northeast to southwest and southwest to northeast, rather than northwest to south and southeast to northwest.

22. (B) The statement that the larvae feed exclusively on milkweed means that milkweed is the only thing the larvae eat. Choice (A) is incorrect because "exclusively" does not mean "necessarily." Choice (C) is incorrect because "preferably" is not a definition of "exclusively," and nothing in the passage indicates that milkweed is the *preferred* choice of food for the larvae; instead, it indicates that milkweed is the *only* food for the larvae. Choice (D) is wrong because "ravenously" is not a definition of "exclusively" and because nothing in the passage tells us that the larvae feed "ravenously" (greedily) on the milkweed. And although milkweed probably is healthy for the larvae, "healthily" is not a definition of "exclusively," so choice (E) is incorrect.

23. (A) Keep in mind that the question asks you to select the choice that is **NOT** acceptable. The typical monarch's life span is too short for it to live through a migratory round trip; therefore, it would have no memory of a previous year. So choice (A) is not an acceptable theory. Choice (B) is incorrect as the answer because it is an acceptable theory. Milkweed is the only source of nutrition for monarch larvae; "thus, the migration pattern follows the milkweed distribution." Choice (C) is incorrect as the answer because it is an acceptable theory. The passage indicates, "Some say the monarchs follow the direction of the sun; as it moves southward in the sky, the monarchs travel southward" and "Other theories claim that monarchs use the sun . . . as their guide. It has been hypothesized that they use the sun's position as a compass." And choice (D) is incorrect as the answer because it is an acceptable theory "that monarchs instinctively use . . . sounds, and smells as their guide." Likewise, the theory that "their bodies contain a small amount of magnetite and so they can sense changes in the magnetic field of the earth" is also acceptable; so choice (E) is not a correct answer.

24. (D) Such descriptive words as "phenomenon," "intriguing," and "amazing" appear throughout the passage, indicating that the author finds these insects interesting. Nothing in the passage indicates that the author finds them choice (A) "boring," choice (B) "frustrating," choice (C) "inferior," or choice (E) "predatory."

DETAILED EXPLANATIONS

SECTION 4—WRITING

1. (C) General reference should be avoided. The pronoun "which" does not have a clear reference in choices (A) or (B). In choice (C) "which" clearly refers to "facts." The reference "proving" in choice (D) is too general. In choice (E) "a proof" is an incorrect number to refer to the two strengths of women.

2. (D) Because the verb "protest" can be transitive and have a direct object, choice (D) avoids awkward wordiness and use of the unnecessary preposition "about." Choices (B) and (E) include unnecessary words and use the pronoun "they" that has no clear antecedent; choice (C) is also unnecessarily wordy and contains the repetitious words "builders should build."

3. (B) Choice (B) best shows the causal relationship between sibling opportunities and their placement in the family, their abilities, and their personalities and retains the subordination of the original sentence. Choices (A) and (E) provide dangling phrases. Choice (C) with its use of the coordinating conjunction "and" treats the lack of opportunity and its cause as if they are equal ideas and does not show the causal relationship between them, and choice (D) results in a run-on sentence.

4. (E) Only choice (E) corrects the two major problems in the sentence, the lack of subject-verb agreement and the lack of parallelism. In choices (A), (B), and (C), the verb "is" does not agree with its plural subject, "provisions." Choices (A) and (D) have unlike constructions serving as predicate nominatives, the noun "freedom" and the clause "that citizens are guaranteed a trial by jury." Choice (E) correctly uses the plural verb "are" to agree with the plural subject, and the predicate nominative is composed of two parallel nouns, "freedom" and "Guarantee."

5. (B) Choice (B) clearly and precisely states the issue of debate. Choice (C) is eliminated because it is too wordy and not the precise issue under debate. The correlative conjunctions, "whether . . . or," should be followed by parallel structures. Choice (A) follows "Whether" with a subject-verb combination not seen after "not." Choice (D) is parallel but in the wrong tense. Choice (E) has "Whether or not" run together and uses poor wording in the rest of the sentence.

6. (A) Choice (A) has both correct agreement and clear reference. Choice (B) has a subject-verb agreement problem, "charging . . . are." Choice (C) produces a fragment. It is unclear in choice (D) who will have the bad credit rating, and the wording of choice (E) has the obvious subject, "people," in a prepositional phrase.

7. (D) Choice (D) correctly presents the fifth grade field trip in a subordinate clause modifying "exhibit." Choices (A) and (B) have the coordinating conjunction "and," but the first part of the sentence is not equal in meaning or importance to the second part of the sentence. Choice (C) introduces "when" with no antecedent. Choice (E) uses "where" as the subordinating conjunction, but is too far from its antecedent and is not the important idea of the sentence.

8. (E) In choice (E) the present infinitive is correctly used to express an action following another action: "plans to appeal." Choices (A) and (B) use the wrong form, "appealing." Choice (C) uses the wrong tense, "is appealing." Choice (D) sounds as if the same person is deciding the case and appealing the case.

9. (A) Choice (A) has clear wording. Choice (B) is a fragment because it puts the verb in a nonessential phrase. Choices (C), (D), and (E) produce twisted wording. Choice (C) has no object for the word "allowed" and sounds as if the speakers were allowed to stay up later. Choice (E) needs commas and sounds as if the speakers decided to stay up later.

10. (B) Choice (B) plainly states the subject and the verb, "I find." Choices (A) and (E) have the subject in a prepositional phrase, "for me." Choice (E) produces a fragment. Choice (C), a fragment, has no subject because both potential subjects are in prepositional phases: "agreement" and "plan." Choices (C) and (D) imply "everyone" as the main subject.

11. (B) Choice (B) avoids the split infinitive and the incorrect expression, "inside of." Choices (A) and (D) split the infinitive "to finish" with the adverb "completely." Choice (C) uses "inside of," an expression that is incorrect to use because it is redundant ("of" should be deleted) and because it should not be used with measuring time. Choice (E) erroneously changes the idea and would imply two verbs in simple past tense: hired" and "finished."

12. (C) As you read the sentence, you should recognize that choice (C) presents an error in comparison. The comparison of two bridges requires the comparative form "longer." All of the other choices are acceptable in standard written English. Choice (A), the interrogative adjective "Which," introduces the question; choice (B), "is," agrees with its singular subject "bridge"; choice (D), "or," is a coordinating conjunction joining the names of the two bridges.

13. (E) The correct response to this question is choice (E). All labeled elements are choices acceptable in standard written English. Choice (A), "of proteins," is a prepositional phrase that modifies the word "function"; in choice (B), the pronoun "they" is plural to agree with its antecedent, "proteins," and the verb "come" is also plural to agree with its subject, "they"; choice (C), "plant or animal sources," is idiomatic; and Choice (D), "is," is singular to agree with its singular subject.

14. (C) The error is choice (C), "I," which is in the nominative case. Because the words "Janice" and "I" serve as indirect objects in the sentence, the correct pronoun is the first person objective form, "me." Choice (A), "Recognizing," is a participle introducing an introductory participle phrase modifying "teacher;" choice (B), "had worked," is a verb in the past perfect tense because the action in the phrase was completed before the action in the main clause occurred; choice (D), "tomorrow," is an adverb modifying the verb "could give."

15. (A) There is no antecedent for this pronoun, although it is implied that the elderly and the handicapped are the logical ones to be grateful. Choices (B) and (D) are elliptical constructions, "grateful [that] the city" and "helps [to] pay." Choice (C) is a verb.

16. (B) This sentence is not parallel. Two actions are mentioned, connected by "not only" and "but also." The sentence should read, "not only are Americans . . . but also they are." Choice (A) and (C) contain properly used infinitives. Choice (D), "particularly," is the adverb form modifying the adjective "arid."

17. (D) Sentences should not end with a preposition; the sentence should read, "an opinion with which everyone agreed." Choice (A) is a gerund as the object of the preposition. Choice (B) is an appropriate conjunction, and choice (C), "good," is the positive form of the adjective to follow the linking verb "tasted."

18. (A) The expression should read "as prevalent as" for the proper comparison. Choice (B) is a noun used as an adjective. Choice (C), "some of which," has clear reference to "ornaments." Choice (D), "such as," is correct to mean "for example."

19. (E) Choice (A) is subjunctive mood to indicate a condition contrary to fact. Choice (B) is a correct idiom. Choice (C) uses the adverb "politely" to modify "speaking," and choice (D) uses the adjective "angry" to follow the lining verb "becoming."

20. (C) "Irregardless," an incorrect expression, is a combination of "irrespective" and "regardless." (Taking into account the prefix and the suffix of "irregardless," the combination would mean, "not, not regarding," and so would be redundant.) Usually, "regardless" is used, although "irrespective" is also correct. The idioms in choices (A) and (D) are correct. Choice (B) is an adjective.

21. (A) The idiom should be "capable of winning." The verb in choice (B) indicates the prior action of two past actions; the prepositions in choices (C) and (D) are correct.

22. (E) Choice (A) is a correct expression. Choice (B) uses an indefinite pronoun as the predicate nominative and "those" as the object of a preposition. The verb in (C), "are," is plural because a plural verb must be used in a subordinate clause following the phrase "one of those." Choice (D) uses the nominative form of the pronoun as the subject of "say."

23. (B) The verb should be in the past tense, "practiced," as signaled by "would become." Choice (A) is a correct idiom. Choice (C) is one word as used in this sentence. Choice (D) contains a correct preposition.

24. (A) In the prior of two past actions, the verb should be past perfect: "has listened" comes before "would have heard." Choice (B), "carefully," is an adverb modifying "would have heard." Choices (C) and (D) are correct idioms.

25. (D) The expression "and etc." is redundant; it would mean "and, and." In this case, "etc." or "and so on" would be acceptable. Choice (A) is an infinitive; choice (B) and choice (C) are correct idioms.

26. (C) A "factor" is something that contributes toward a result. This phrase could be reworded to read, "are part of being hazed as a freshman" or "are part of freshman hazing." Choice (A) is a gerund parallel to the others: "pushing" and "being." Choices (B) and (D) are correct.

27. (E) All choices in this sentence are correct in standard written English. The passive verb, "are concerned," choice (A), is plural to agree with its subject, the collective noun "majority" that is plural in meaning; choice (B), the noun "effects," represents correct diction; choice (C), "of watching" is a preposition with a gerund serving as its object; and choice (D), "their," is a plural possessive pronoun agreeing in number with its antecedent, "majority.'

28. (C) The error is at choice (C), where the singular "he" or "she" is correct to refer to its singular antecedent, the indefinite pronoun "anyone," and where the verb "remember" must also be changed to the singular form "remembers" to agree with its subject. Choice (A), "is," agrees with its singular subject "who"; choice (B), "thinks," agrees with its singular subject "anyone"; and choice (D), "most popular," is the superlative form of the adjective, correct to compare more than two toys.

29. (C) You should recognize that the problem in the sentence is that the predicate nominative, "winner," does not agree in number with the subject, "Horses." It must be replaced by "winners" and the article "a" must be eliminated. Choice (A), "named," is correct as a participle introducing a phrase that modifies "Horses"; choice (B) is in the present perfect tense, correct for a verb expressing action that occurred in the past and continued over a period of time; and choice (D), "recent years," is idiomatically correct as the object of a preposition and its adjective modifier.

30. (E) Choice (E) is the clearest and most concise rewording of the sentence portion. Choice (A) repeats the phrase "in order to," and choice (D) contains the repetition of "requiring." Choice (B) and choice (C) both contain unnecessary wordiness in the clauses beginning with "which will" and "that will."

31. (B) Choice (B) keeps the first person voice and keeps the main point using parallel structure and concise language. Choices (A), (C), and (D) are still excessively wordy; in addition, choice (A) changes the voice to "one." Choice (E), although somewhat more concise, uses passive voice; because this letter is a call to action, passive voice weakens the argument and intent of the letter.

32. (A) Paragraph 2 provides background information, choice (A), for the reason there is so much violence on television. Choice (B) is incorrect because the argument in paragraph 2 is not a contradiction to anything. Choice (C) would be correct if the paragraph contained support for the main argument. Although the paragraph does give some specifics, choice (D), the evidence cannot be considered as categorizing anything. Choice (E) incorrectly implies that the paragraph is written in an emotionally charged manner.

33. (C) Although choice (C) employs two forms of the verb "to be," it is the best wording. Choices (A) and (B) both use the weak phrase "being that" and create fragments. Choice (D) is far too wordy and uses the weak wording "seem to be"; use of the word "seem" makes the writer appear uncertain and tentative, not precise. Choice (E) is still too wordy, using the phrase "shows containing violence" instead of the more concise phrase "violent shows."

34. (D) Choice (D) is the most concise combination, one which clearly shows time sequence and cause-and-effect. Choice (A), because it has a misplaced modifier, implies that the children conducted the study. Choices (B) and (E) imply that the children were prone to violence only while they were viewing the violent acts, a slight distortion of the correct finding. Choice (C) subtly suggests that only the children at the Institute were affected by this condition, with the implication that other children are not. This failure to indicate an extension of the findings subtly distorts the original meaning.

35. (D) Choice (D) would be the best revision; the phrase "are just to name two examples," besides being grammatically incorrect, is unnecessary, and the structure of (D) sets up the parallel best. Choice (B) is the next best bet, but "to name two examples" is awkward and unnecessary. Choices (A) and (C) are not specific about any particular riots, these would at least need the article "the" as in choices (B) and (D).

DETAILED EXPLANATIONS

SECTION 5—MATH

1. (E) Let x = number of boys, $2x$ = number of girls, and $3x$ = total number of students. The number of boys studying physics is $\left(\frac{1}{2}\right)x$ and the number of girls studying physics is $\left(\frac{2}{3}\right)(2x) = \left(\frac{4}{3}\right)x$. Then the number of students studying physics is $\left(\frac{1}{2}\right)x + \left(\frac{4}{3}\right)x = \left(\frac{11}{6}\right)x$. The required ratio is $\left(\frac{11}{6}\right)x - 3x = \left(\frac{11}{6}\right)\left(\frac{3}{1}\right) = \left(\frac{11}{18}\right)$.

2. (B) Each of P, Q, R, and S would be connected to the origin. The best mileage would be indicated by the segment with the steepest slope and the worst mileage would be indicated by the segment with the smallest slope. The segment connecting the origin to Q would have the steepest slope, whereas the segment connecting the origin to R would have the smallest slope. The correct order is then Q, S, P, R.

3. (E) There are $25 - 7 - 4 - 5 = 9$ blue t-shirts. Because there are 4 red t-shirts, there are a total of $9 + 4 = 13$ t-shirts that are either blue or red. The required probability is $\frac{13}{25} = 0.52$.

4. (E) One equation that will yield a value of x is $-3x + 9 = 5x - 7$, which becomes $-8x = -16$, so $x = 2$. Another equation that will yield a value of x is $-3x + 9 = -(5x - 7)$, which becomes $-3x + 9 = -5x + 7$. Simplifying, we get $2x = -2$, so $x = -1$. When these are checked, we find that they are indeed solutions.

5. (C) The domain is determined solely by the denominator. If the denominator is not zero, then the function has a real value. Thus, $2x^2 - x - 15 = 0$ will yield the excluded values of x. Thus, $2x^2 - x - 15 = (2x + 5)(x - 3)$. Solving $2x + 5 = 0$ yields $x = -5/2$. Solving $x - 3 = 0$ yields $x = 3$. These two numbers represent the excluded values of the domain.

6. (A) $M = \{2, 3, 5, 7, 11, 13, 17, 19\}$, $N = \{2, 4, 6, 8, 10, 12, 14, 16, 18\}$, and $P = \{11, 12, 13, 14, 15, 16, 17, 18, 19\}$. Then, $M \cap P = \{11, 13, 17, 19\}$, $M \cap N = \{2\}$. Choice (A) has the elements 2, 4, 6, 8, 10, 11, 12, 13, 14, 16, 17, 18, 19; choice (B) has no elements; choice (C) has the elements 2, 11, 12, 13, 14, 15, 16, 17, 18, 19; choice (D) has no elements; and choice (E) has the elements 2, 3, 5, 7, 11, 12, 13, 14, 16, 17, 18, 19. Thus choice (A), $(M \cap P) \cup N$, has 13 elements, which is larger than any of the other selections.

7. (D) For the Alpha Cab Company, $C = \$3.50 + \$0.75x$ represents the cost of traveling x miles. A distance of 12 miles costs $3.50 + (0.75)(12) = \$12.50$ (Note that \$0.25 for 1/3 mile is equivalent to \$0.75 for 1 mile.) For the Beta Cab Company, $C = 4.50 + 0.40(x - 1)$ represents the cost of traveling x miles. (Note that \$0.20 for 1/2 mile is equivalent to \$0.40 for 1 mile. Also, the expression $x - 1$ is used because the first mile's cost is already included in the \$4.50 amount.) Now $\$12.50 = \$4.50 + \$0.40(x - 1)$. This simplifies to $8.00 = 0.40x - 0.40$, so $x = \frac{8.40}{0.40} = 21$.

8. (C) The numerator $= 9x^3 - 3x^2 - 30x = 3x(3x^2 - x - 10) = 3x(3x + 5)(x - 2)$. The denominator $= 12x^3 + 44x^2 + 40x = 4x(3x^2 + 11x + 10) = 4x(3x + 5)(x + 2)$. Dividing both numerator and denominator by $x(3x + 5)$ produces the answer $\frac{3(x-2)}{4(x+2)}$.

9. (D) Rewrite the original equation as $By = Ax + C$, which becomes $y = (-\frac{A}{B})x + \frac{C}{B}$. The slope is given by $-\frac{A}{B}$ and the y-intercept is given by $\frac{C}{B}$. If both these quantities are doubled, the new slope and y-intercept become $-\frac{2A}{B}$ and $\frac{2C}{B}$, respectively. Then $y = (-\frac{2A}{B})x + \frac{2C}{B}$ becomes the new equation. Multiply by B to get $By = -2Ax + 2C$, which can then be written as $2Ax + By = 2C$.

10. (C) The length of the string represents one-half the circumference of the circle with radius 10. So, this length must be 10π, which would be the circumference of the smaller circle. Let r represent the radius of the smaller circle. Then $10\pi = (2\pi)(r)$, so $r = 5$. Finally, the area of the smaller circle is $(\pi)(5^2) = 25\pi$.

11. (A) Rewrite the function as $f(x) = (x^2 + 14x + \_\_) + 51$. To complete the (__) so that it is a perfect square, add 49. (Of course, we must add 49 to the left side as well.) Then $f(x) + 49 = (x^2 + 14x + 49) + 51$. This simplifies to $f(x) + 49 = (x + 7)^2 + 51$ and finally to $f(x) = (x + 7)^2 + 2$.

    Alternate method: The vertex is $[-\frac{b}{2a}, f(-\frac{b}{2a})]$ which is $(-7, 2)$. Recall that if a function $f(x) = A(x - h)^2 + k$, its vertex is located at (h, k). Thus the function in this question has its vertex as $(-7, 2)$. Only answer (A) has its vertex at $(-7, 2)$.

12. (B) If k is a nonzero constant, then $P = \frac{k\sqrt{Q}}{R^3}$ would represent the given relationship. To solve for Q, multiply both sides by R^3 to get $PR^3 = k\sqrt{Q}$. Divide by k to get $\frac{PR^3}{k} = \sqrt{Q}$. Now square both sides to get $\frac{(PR^2)^3}{k^2} = (\sqrt{Q})^2$, which can be written as $\frac{1}{k^2}(P^2R^6) = Q$. Note that $\frac{1}{k^2}$ is simply another constant. Thus, Q varies directly as the square of P and the sixth power of R.

13. (B) In choice (B), rewrite as $\sqrt{5x-2} = 5 - 8 = -3$. A square root of any real quantity can never be negative, so this equation has no real solution. Incidentally, the answers for choices (A), (C), (D), and (E) are 15, $\frac{34}{3}$, $\frac{2}{7}$, and -47, respectively.

14. (C) In this example, $X = 150$, $Y = 36{,}450$, and $t = 1935 - 1895$. Then we have $36{,}450 = (150)(3^{\frac{40}{n}})$.
Divide both sides by 150 to get $243 = 3^{\frac{40}{n}}$. By inspection, $3^5 = 243$, so $\frac{40}{n}$. This leads to $n = 8$, which means that the population triples every 8 years. Now use the equation $1{,}350 = (150)(3^{\frac{t}{8}})$. Divide both sides by 150 to get $9 = 3^{\frac{t}{8}}$. This can be written as $3^2 = 3^{\frac{t}{8}}$.
Equating exponents, $\frac{t}{8}$, so $t = 16$. The required year is then $1895 + 16 = 1911$.

15. (D) We seek a point P such that when a segment joins P and $(0, -3)$, the slope must be greater than $\frac{3}{5}$ but less than $\frac{2}{3}$. Using $(20, 10)$ for P, the slope is $\frac{(-3-10)}{(0-20)} = \frac{13}{20}$ and $\frac{3}{5} < \frac{13}{20} < \frac{2}{3}$.
This can be verified by changing all fractions to decimals to get $0.60 < 0.65 < 0.667$. The actual slope values for choices (A), (B), (C), and (E), are $\frac{7}{10}$, $\frac{2}{3}$, $\frac{14}{25}$, and $\frac{3}{5}$, respectively.

16. (C) Let x = the length of KL. An angle bisector always divides the side to which it is drawn into two parts whose ratio is equal to the ratio of the two sides forming the angle bisector. So, $\frac{8}{10} = \frac{6}{x}$. Then $8x = 60$, and $x = 7.5$

17. (C) Because $2 < 3$, use the equation 2 ♠ 3 = 33 – (3)(2) = 27 – 6 = 21. Likewise, because $21 > 4$, 21 ♠ 4= 212 + 42 = 441 + 16 = 457.

18. (D) Triangles RUV and STU are both 45°–45°–90° right triangles. Let $RU = UV = x$. Because $RV = 8$, $x^2 + x^2 = 8^2$. Then $2x^2 = 64$, so $x = \sqrt{32}$. This means that $UT = \sqrt{32}$. Let $ST = SU = y$. Then $y^2 + y^2 = \left(\sqrt{32}\right)^2 = 32$, so $2y^2 = 32$, $y^2 = 16$, thus $y = 4$. Now in ΔSTV, $ST = 4$ and $SV = UV + US = \sqrt{32} + 4$. Finally, in ΔSTV, $VT^2 = SV^2 = 9.657^2 + 4^2 \approx 109.26$. $VT = \sqrt{109.26} \approx 10.5$.

DETAILED EXPLANATIONS

SECTION 6—CRITICAL READING

1. (A) False "belief," when acted upon, becomes "reality." Choice (B) "fear . . . truth" is wrong because it is belief that is acted upon, not fear. Choice (C) "Anxiety . . . fact" is wrong because it is belief that is acted upon, not anxiety. Choice (D) 'presumption . . . certain" is wrong because the presumption itself does not become certain; it remains contrary to fact. The fact that a child turns out to be "no good" does not mean in fact that he is "no good" as an infant. Choice (E) "dread . . . proven" is wrong because, as in choice (D), the first term itself is not what is proven.

2. (E) The second half of the sentence cites an example to support the assertion made in the first part of the sentence. The computer is "unforgiving," choice (E), because if one neglects to save a program, that program is lost. It cannot be recalled. Choice (A) "ominous" (threatening), choice (B) "complicated," choice (C) "essential," and choice (D) "difficult" do not express the idea of no reprieve for a mistake as does choice (E) "unforgiving."

3. (D) The best answer is choice (D). "Ornate" (elaborately decorated) is the most effective way of describing a "gold-studded costume," and "simplicity" (lack of complication) is a good way of describing a "flannel suit," especially when compared to a "gold-studded costume." Choice (A) "chaste" (virtuous, pure) is an incorrect description of something that is "gold-studded," and "gaudiness" (garish, flashy) is not an appropriate choice for describing a "flannel suit." Choice (B) "laconic" (terse) is an inappropriate was to describe a suit; "laconic" is used to describe people, so choice (B) is incorrect. Choice (C) "reserved" (restrained in actions or words) is not a good way to compare a "gold-studded costume" to the "savoir-faire" (knowing how to act) of a "flannel suit." Choice (C) makes no sense when substituted into the sentence. Choice (E) "feudal" (having the characteristics of feudalism) is a meaningless choice in the sentence.

4. (B) "Meticulous" (exacting, precise) is the best answer because the context of the sentence asks for a word that is opposite in meaning to "disarray" (to be out of order, disorganized). The key word here is "although," which implies that whatever follows the first thought will be contradictory in meaning. Choice (A) "aloof" (distant in interest, reserved) is incorrect because this is an adjective that is usually used to describe a person, not a work area. Choice (C) "viable" (capable of maintaining life, possible) is an inappropriate word choice as an antonym for "meticulous." Choice (D) "diligent" (hard working) is a word that describes a person's characteristics, and it is not a good choice for the sentence. Choice (E) "insipid" (uninterested, bland) is not an adjective that meets the conditions of the sentence, where we are looking for a word that is opposite in meaning to "meticulous."

5. (B) The best answer is "offset" (balance, compensate for) and "concentration." The sentence tells us that the catcher is wounded, and therefore the pitcher (who works as a team with the catcher) has to make up for this deficiency. Choice (B) "offset" tells us that the pitcher will compensate for the catcher's injury, and the way he will do that is through "concentration." Choice (A), "tarry" (to delay) and "spending," makes no sense when substituted into the context of the sentence. Choice (C), "mediate" (acting as intermediary to settle a dispute) and "dedication," is not meaningful. The pitcher's increased dedication does not mediate a wounded catcher. Choice (D) "divulge" (reveal, disclose) is not an appropriate choice because it makes no sense when put into the sentence. Choice (E) "alleviate" (to make easier) is not the correct choice. "Alleviate" is used in conjunction with a symptom or a circumstance, not a person.

6. (D) Although the techniques discussed in the two passages differ, both discuss effective communication. Choice (A) is incorrect because it presents an idea that only Passage 2 implies. Choice (B) is wrong because it presents an idea that neither passage actually states or implies, although Passage 1 does emphasize the importance of how we speak and write. Choice (C) is incorrect because neither passage indicates the equality of what we say and how we say it. Choice (E) is incorrect because nothing in either passage refers to nonverbal communication or to the variety of communication forms.

7. (C) The author states, "the ability to express oneself correctly is the key to effective communication." And the author is clear that "to express oneself correctly" entails the use of proper grammar. Choice (A) is wrong because nothing in Passage 1 indicates that proper grammar is "stifling" (suppressing, stultifying). Choice (B) is incorrect because Passage 1 does not address "free-flowing expression." And although the passage does mention that without the use of proper grammar an educated person may appear to "obtuse" (unintelligent), it does not include anything that leads us to assume that the person who uses proper grammar is necessarily well educated. Choice (E) is incorrect because Passage 1 does not address the relationship between creativity and the use of proper grammar.

8. (A) The passage contends that the "effective and popular" holistic concept involves "little concern for correct grammar, punctuation, and spelling that potentially stifle individuality." Choice (B) is wrong because nothing in the passage condones the expression of all ideas in schools; be careful not to infer that the "let it flow" philosophy allows students to communicate inappropriate ideas. And although critical of requiring students to follow strict grammar, spelling, and punctuation rules, the author of Passage 2 does not say that schools should disregard them. Thus, choice (C) is wrong. Choice (D) is incorrect because Passage 2 makes no mention of highly educated people and their grammar. Choice (E) is wrong because the passage says nothing that implies schools should be careful about being too strict; it merely favors allowing students to express themselves without requiring them to follow strict grammar, punctuation, and spelling rules.

9. (E) Both authors address the importance of communication and offer opinions about approaches to communication and how it should be learned. Since such learning usually occurs in schools, both authors would agree that communication should be a part of the curriculum. Choice (A) is incorrect because neither passage includes anything that addresses schools' rights to teach what they regard as important. Choice (B) is wrong because only Passage 1 indicates that following grammar rules is important; it says nothing about other rules. And Passage 2 includes nothing to indicate that schools should require students to follow rules. Choice (C) is wrong because neither passage mentions

life skills or public speaking. And both passages present approaches to communication, but neither implies that schools need to look more closely at how they are teaching it; thus, choice (D) is wrong.

10. (B) Thoreau stated, "I ask for . . . a better government." Choice (A) is proven incorrect by his statement, "I ask for, not at once no government. . . ." Choice (C) is incorrect because Thoreau expressed the desire for a better government, rather than for the government as it was. Thoreau expressed some discontentment with the way power was handled, but he did not propose that the government change its power structure; therefore, choice (D) is incorrect. And choice (E) is incorrect because, although Thoreau was critical of voting, he did not express the desire for a change in the election system.

11. (A) If you are familiar with the word "expedient," you know that choice (A) is its dictionary definition. If you are not familiar with the word, you probably can figure out the correct answer by eliminating the other choices as incorrect. Also, you are looking for a word that describes "government at best," in Thoreau's opinion. After reading the entire passage, you know that Thoreau favored the idea of a goal or a purpose. And in line 18, he used the word "expediency," a related word, to illustrate his desire for a government that had a meaningful purpose. Choice (B) is incorrect because nowhere in the passage did Thoreau support selfishness. Be careful not to confuse individuality with selfishness; they are two completely different concepts. Thoreau did believe individuality was important and necessary, but he was not a supporter of selfishness. And his preference for individuality proves choice (C) to be wrong. Even if you did not know the meaning of the word, Thoreau's comments against going with the "masses" would tell you that he would not favor a government just because it was popular. Choice (D) is incorrect because in a democracy the majority rules—an idea to which Thoreau objected. And choice (E) is wrong because the passage makes no reference to socialism. Also, a look at the ideas Thoreau presented in this essay quickly tells you that he would not favor a government that focused on government rather than on individuals; thus, he would not support a socialistic government.

12. (B) Thoreau believed that the majority rule simply gave power to the strongest and took power away from the individual, to the point that individuals became the "subjects" who were subservient to those in power. Choices (A), (C), (D), and (E) are incorrect because Thoreau criticized majority rule, indicating that it exists "not because they are most likely to be in the right, nor because this seems fairest to the minority, but because they are physically the strongest. But a government in which the majority ruled in all cases can not be based on justice, even as far as men understand it."

13. (C) Thoreau stated, "The only obligation which I have a right to assume is to do at any time what I think right." And he said, "It is not desirable to cultivate a respect for the law, so much as for the right," proving choice (A) wrong. His disdain for elections and elected officials runs throughout the passage, so choice (B) is incorrect. And his brief comments about slaves and corporations did not necessarily imply respect for either; so choices (D) and (E) are wrong.

14. (D) Thoreau stated, "I cast my vote, perchance, as I think right." He was not concerned with what the law intended, since he had little respect for the law ("by means of their respect for it, even the well-disposed are daily made the agents of injustice. . . ."); therefore, choice (A) is incorrect. He expressed no concern for what the majority or other citizens would decide because he believed "there is but little virtue in the action of masses of men"; so choices (B) and (D) are incorrect. And choice (E) is incorrect because this passage makes no reference to the U.S. Constitution.

15. (D) Thoreau was using the slavery issue as an example of his preceding statement "There is but little virtue in the action of masses of men," showing that the majority vote affects only those things about which they are indifferent or issues that are unimportant. Choice (A) is wrong because this passage takes no stand on the slavery issue. Choice (B) is wrong because he did not predict that American voters would abolish slavery; he was merely speculating on the situation if they eventually did vote to abolish it. Likewise, choice (C) is wrong because he did not say Americans were indifferent to slavery; instead, he was saying that would likely be the case if they ever voted to abolish it. And choice (E) is wrong because it reflects an idea that is exactly the opposite of the way he believed.

16. (B) Anthony opened her presentation stating her intention "to prove to you that in thus voting, I not only committed no crime, but, instead, simply exercised my citizen's rights, guaranteed to me and all United States citizens by the National Constitution, beyond the power of any state to deny." And throughout the speech she referred to denying women their voting privileges as violating the U.S. Constitution, as being a "bill of attainder, or, an ex post facto law, and . . . therefore a violation of the supreme law of the land." Choice (A) is incorrect because the passage makes no reference to the intelligence of women or men. And although Anthony did say that women are "persons" and thus were citizens, proving this fact was not her primary purpose or focus; thus, choice (C) is incorrect. In closing, Anthony did contend that discrimination against slaves is just as illegal as discrimination against women; however, she did not imply that women were being treated no better than slaves; so choice (D) is wrong. And although in this presentation she did protest that denying women the right to vote was unfair, she did not trace the history of unfair treatment of women; therefore, choice (E) is wrong.

17. (B) Anthony asserted, "I not only committed no crime, but, instead, simply exercised my citizen's rights, guaranteed to me . . . by the National Constitution. . . ." She expressed no regret, so choice (A) is wrong. Her speech gave no indication that she wanted to rescind (withdraw) her vote, so choice (C) is incorrect. No mention is made of any court or a possible appeal to a higher court; therefore, choice (D) is incorrect. And choice (E) is incorrect because Anthony did not mention a fine or a sentence, and since she admitted to no crime she more than likely would not be willing to comply if she had been fined or sentenced.

18. (D) Anthony was citing the Preamble to point out that the phrase "We, the people of the United States" referred to all citizens, not just male citizens, who formed the Union and therefore had the right to vote—"the only means of securing them provided by this democratic-republican government." Nothing in her presentation indicates that she was trying to impress anyone with her knowledge or intelligence; thus choice (A) is incorrect. Choice (B) is wrong because she was not suggesting that anything in the Preamble should be changed; she was suggesting that it should be upheld. And although she did not specifically say whether or not she believed it was a privilege to live in the U.S., she did say, "it is a downright mockery to talk to women of their enjoyment of the blessings of liberty." Thus, choice (C) is incorrect. And instead of reminding her audience that this was a free country for all citizens, she pointed out that it should be a free country for everyone but was not because women were not free to vote. So choice (E) is incorrect.

19. (C) Anthony said, "It is an odious aristocracy . . . the most hateful aristocracy ever established on the face of the globe. . . ." Choices (A) and (B) are incorrect because she said, "this government is not a democracy. It is not a republic. . . ." And she makes no mention of an autocracy or monarchy; thus choices (D) and (E) are incorrect.

20. (D) Anthony referred to the "hateful oligarchy of sex" as "the most hateful aristocracy ever established on the face of the globe." Among the situations that "might be endured," she listed "an oligarchy of wealth, where the rich govern the poor, an oligarchy of learning, where the educated govern the ignorant, or even an oligarchy of race, where the Saxon rules the African." Of these, she did not say which would be preferable; thus, choices (A), (B), and (E) are incorrect. And choice (D) is wrong because she made no mention of elected officials' ruling over common citizens.

21. (A) Susan B. Anthony pointed out that Webster, Worcester, and Bouvier—all famous lexicographers of the nineteenth century—defined "a citizen to be a person in the United States, entitled to vote and hold office." She went on to pursue the logic that if a woman is a person, then she must be a citizen, with citizens' rights—including the right to vote. She did not set out to prove that women were superior human beings; therefore, choice (B) is wrong. And although she did assert that women were "persons" and that few could disagree with this idea, proving the fact that women were just as much "people" as men were was not her main goal; she wanted to go at least one step further, to prove that because women were "persons" they were, by definition, "citizens." Choice (D) is incorrect because she made no claim that these lexicographers agreed with her regarding women's rights. And although she may have wanted to display her education and intelligence as a means of convincing her audience that she was right, doing so was not her main purpose here; so choice (E) is incorrect.

22. (D) Whether or not you are familiar with this term, it is important to look at its meaning in context. Susan Anthony stated, " . . . no state has a right to make any law, or to enforce any old law, that shall abridge their privileges or immunities." One dictionary definition of "abridge" is "to make less; restrict." This definition fits the word in this context, since Anthony is attempting to prove that no state had a right to lessen or to restrict the privileges or immunities of female citizens of this country. Choice (A) is wrong because it is not a definition of "abridge" and, even if it were, it does not fit with Anthony's concerns. If you selected choice (B), you were probably confusing "abridge" with "bridge." Also, this definition does not fit in this context. Although "abridge" can mean "shorten," this definition does not fit here; so choice (C) is incorrect. Likewise, choice (E) is a dictionary definition of the word, but it does not fit in the context of Anthony's message here.

23. (B) Several times, Anthony referred to the laws established by the U.S. Constitution—federal laws—and contended that the rights established by these laws were "beyond the power of any state to deny" because the federal law is "the supreme law of the land." Anthony made no reference to God's law, so choice (A) is incorrect. She repeatedly emphasized that any state law denying women the rights established by federal law were "null and void"; so choice (C) is incorrect. And she made no reference to city law or to personal beliefs as being superior to the other choices; therefore, choices (D) and (E) are wrong.

24. (C) Anthony asserted, "It was we, the people; not we, the white male citizens; nor yet we, the male citizens; but we, the whole people, who formed the Union," and " every discrimination against women in the constitutions and laws of the several states is today null and void, precisely as is every one against Negroes." Clearly, she saw the plight of black people as parallel to that of women. Choices (A), (B), and (D) are wrong; as these quotes indicate, her stand was that it was not just white male citizens who comprised the Union (indicating that blacks as well as women were citizens as well), and that laws allowing discrimination against Negroes, as well as discrimination against women, were unconstitutional and thus "null and void." Although she did say that "an oligarchy of race, where the Saxon rules the African, might be endured," she made this comment to emphasize how unacceptable male dominance over females was. Nowhere did she say that blacks were inferior to whites. So choice (D) is incorrect.

DETAILED EXPLANATIONS

SECTION 7—MATH

1. (D) 60% of x is $0.60x$ and 20% of y is $0.20y$. If $0.60x$ is 50% larger than $0.20y$, then $0.60x = (1.5)(0.20y)$. Simplifying, we get $0.60x = 0.30y$. Dividing by 0.30, this reduces to $y = 2x$.

2. (C) Real zeros of a function are represented by x-intercepts. Each point of $g(x)$ would have its y value decreased by 2 . The only two places where there would be x-intercepts are: a point on $g(x)$ between (–2, 0) and (0, 3); and a point on $g(x)$ between (0, 3) and (2, 0). Note that any point on $g(x)$ between (4, 0) and (5,1) would lie below the x-axis after the downward shift of 2 units.

3. (E) Let x = length of BC and y = length of CD. The area of $ABCD$ is xy and the area of $ECFG$ = $\left(\frac{x}{2}\right)\left(\frac{y}{3}\right) = \frac{xy}{6}$. Thus, the unshaded area is $xy - \frac{xy}{6} = (\frac{5}{6})(xy)$. The probability that a dart lands in the unshaded area is $\frac{\frac{5}{6}xy}{xy} = \frac{5}{6}$.

4. (A) For any quadratic function, the vertex must lie on the perpendicular bisector of the segment whose endpoints are the x-intercepts. Then, the x value of the vertex must be the average of the x values of the x-intercepts. So, $\frac{(-9+5)}{2} = -2$. Because $A < 0$, we know the parabola is concave down and the vertex must lie above the x-axis; this means that the y coordinate is positive. Only choice (A) satisfies these conditions.

5. (D) $m = \frac{k}{c^2}$, where k is a constant. Simplify $9 = \frac{k}{(10)^2} = \frac{k}{100}$, so $k = 900$. Now, $m = \frac{900}{c^2} = \frac{900}{(2)^2} = \frac{900}{4} = 225$.

6. Answer: 36 — Let x = number of pounds of peanuts. Then 1.50x$ = dollar value of the peanuts and ($3.50)(12) = $42 = dollar value of the cashews. The combined mixture is worth $2 per pound and contains $x + 12$ pounds, which is represented by $2($x$ + 12). Then 1.50x$ + $42 = $2($x$ + 12). Simplifying, 1.50x$ + $42 = 2x$ + $24. This leads to $18 = 0.50x$, so $x = 36$.

7. Answer: 24 — $c = \sqrt{216} = 6$. Now $c^2 = 36$, so $(\frac{2}{3})(c^2) = (\frac{2}{3})(36) = 24$.

8. Answer: 21 — $8.35 – $6.25 = $2.10, which represents the cost for the additional pieces of laundry (above 15). $\frac{\$2.10}{\$0.35} = 6$. Thus the total number of pieces is 15 + 6 = 21.

9. Answer: 1.6 — Let x = number of hours the car travels, so that $x + 0.4$ = number of hours that the truck travels. (Note that 24 minutes = $\frac{24}{60}$ = 0.4 hours.) Their distances will be equal, and because distance equals rate times time, we can write the equation $50x = 40(x + 0.4)$. Simplifying, we get $50x = 40x + 16$. Then $10x = 16$, so $x = 1.6$.

10. Answer: 154 — The mean weight for all 15 people is total weight divided by 15 = $\frac{[(5)(188)+(10)(137)]}{15} = \frac{(940+1,370)}{15} = 154$.

11. Answer: 1.75 — First, calculate the fraction of the gas tank that was used, $\frac{3}{4} - \frac{1}{8} = \frac{5}{8}$. Let x equal the capacity of the tank, so that $(\frac{5}{8})x = 10$. Solving for x reveals that the tank can hold 16 gallons. Then, divide the dollar amount by the number of gallons to get the price per gallon: $\frac{\$28.00}{16} = \$1.75$.

12. Answer: 8 — Lin spent $0.30 on tea, $0.60 on coffee, and (5)($0.50) + (4)($0.85) = $5.90 on donuts, for a total of $6.80. Carmen buys a large coffee for $0.75, so she does not want to exceed $6.80 – $0.75 = $6.05 for her medium-sized donuts. Then, $\frac{\$6.05}{\$0.70} \approx \$8.64$; this means Carmen can only buy 8 medium-sized donuts in order not to spend more than the $6.80 that Lin spent. Note that Carmen's purchase will total to $0.75 + (8)($0.70) = $6.35, but if she were to buy 9 donuts, her total would have been $7.05.

13. Answer: 165 — The angles W and T must be 90° because tangents to a circle form right angles at points of tangency. The sum of all four angles of the quadrilateral $WYTX$ must be 360°, so $\angle WXT = 360° - 40° - 90° - 90° = 140°$. Because $WV = (3)(VT)$, $m\angle VXW = (3)(m\angle VXT)$. Let $n = m\angle VXT$ and $3n = m\angle VXW$. The sum of the three angles at X must be 360°, so $140° + n + 3n = 360°$. Then $4n = 220°$, so $n = 55°$. Finally, $m\angle VXW = (3)(55°) = 165°$.

14. Answer: 462, 858, 882, or 1,638 — To find p, use a common multiple of 4 and 6, then add 2. The common multiples are 12, 24, 36, . . . , so that p could be 14, 26, 38,. . . . However, since $p < 30$, the only available values for p are 14 or 26. To find q, use a common multiple of 5 and 6, and then add 3. The common multiples are 30, 60, 90, . . . , so that q could be 33, 63, 93, . . . However, because $q < 70$, the only available values for q are 33 or 63. Therefore, the only four allowable values of pq are (14)(33) = 462, (26)(33) = 858, (14)(63) = 882, and (26)(63) = 1,638.

15. Answer: 46.5 The formula for a geometric sequence is $L_n = AR^{n-1}$, where L_n is the last term, A is the first term, and R is the common ratio. For simplicity, assume that 27 is the first term, so that 8 is the fourth term. Then $8 = 27 \cdot R^3$, so that $R = \sqrt[3]{\frac{8}{27}} = \frac{2}{3}$. Then $y = (27)\left(\frac{2}{3}\right) = 18$ and $z = (18)\left(\frac{2}{3}\right) = 12$. We now have x, 27, 18, 12, 0, . . . , so $x = \frac{27}{\left(\frac{2}{3}\right)} = 40.5$. Finally, $x + y - z = 40.5 + 18 - 12 = 46.5$.

16. Answer: 13.9 Each side of the hexagon is $\frac{24}{6} = 4$, which is also the length of each side of the rhombus. Each angle of a regular polygon of n sides is given by $\frac{(n-2)(180°)}{n}$. For the hexagon, each angle measures $\frac{(6-2)(180°)}{6} = \frac{720°}{6} = 120°$. Because $\angle PST$ and $\angle PSR$ lie on a straight line, the measure of $\angle PSR = 180° - 120° = 60°$. We already know that $PS = 4$ (from the first sentence). Now consider the following diagram with PL drawn as an altitude to SR.

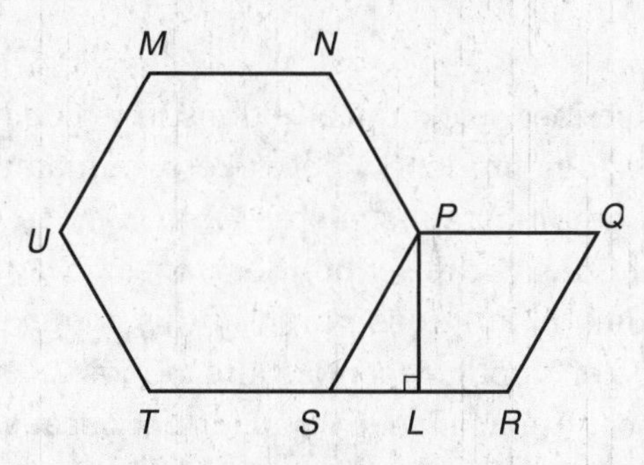

ΔPLS is a 30°–60°–90° right triangle, with $PS = SR = 4$. Then $PL = 2$, because it is opposite $\angle P$ (which is 30°), and $PL = 2 \cdot \sqrt{3}$. The area of a rhombus is its base times its height: $(4)(2 \cdot \sqrt{3}) = 8\sqrt{3} \approx 13.9$.

DETAILED EXPLANATIONS

SECTION 8—CRITICAL READING

1. (D) Here you need a word that describes working conditions in which it would be difficult to be happy. These conditions include unfriendly coworkers, outdated equipment, and low pay. "Adverse" (unfavorable, detrimental) is such a word. Choice (A) is wrong because "obvious" (apparent, evident) does not describe those specific conditions. Choice (B) is incorrect because "obtuse" (dull, blunt) does not fit in this context. Choice (C) is wrong because "propitious" (favorable, good) describes the opposite of the working conditions presented in this sentence. And choice (E) is incorrect because such conditions are not "conducive" (promoting or assisting). Additionally, to be grammatically correct, the word "conducive" should be followed by the preposition "to" and a noun; this sentence does not contain such a prepositional phrase after the blank.

2. (D) "Camouflage" (disguise) is the best answer. Evan is described as relatively short and chubby, but he is "erect" and "quick." Only "camouflage" tends to minimize his lack of height and stoutness through his erect bearing and quick movements. "Emphasize," choice (A), means "to stress," and "conceal," choice (B), means "to remove from view." These words are not appropriate. Although choice (B) might be considered as a possibility, it is not the best choice. His bearing and movements do not "hide" his appearance; they only disguise him. Choice (C), "denigrate," and choice (E), "disavow," are synonyms meaning "to deny." Evan's movements and bearing do not "deny" his appearance, they disguise him.

3. (C) We are looking for an adjective that describes the poet's style and a verb that tells how the publisher reacted to that style. The best answer is choice (C) "uninspired" (common) . . ."cancel." If the poet's style is "uninspired," it is likely that the publisher would "cancel" the contract; there would be no profit in printing a poetry book that would probably inspire no interest. Choice (A), "sycophantical" (servile) . . ."coil" (wind cylindrically) does not fit well in the sentence. Choice (B), "slipshod" (careless) . . ."renew," is not correct. A publisher would not renew a poet's contract if she were careless. Choice (D), "enduring" (lasting) and "acclaim" (praise), does not fit into the sentences as logical choices. Choice (E), "prosaic" (common and uninspired), is a possibility; however, this choice does not fit into the context of the sentence. A poet's uninspired style would not cause a publisher to praise its contract with her.

4. (D) The correct answer contains an adjective that supports the words "only space vehicle" and a verb that does something with a "recorded message" going to "distant star systems." "Unique" means "one of a kind" and is the best choice to describe "only space vehicle"; "transport" (to carry) fits the best choice to describe what happens to "recorded messages" and "distant star systems." Choice (A) is not meaningful because "blasphemous" and "meander" cannot describe a spacecraft. Choice (B) "egocentric" describes a person (self-centered), not an innate object such as a spacecraft. Choice (C) "profound" (knowledgeable) is not a way of describing a spacecraft, for although the message may have been profound, certainly a machine is not. Choice (E) "vital" (important) is a possibility but not within the context of the sentence, where we are looking for a word that means the spacecraft is the "only" one.

5. (E) The correct answer is "denunciation" because we are looking for a word that supports "displeasure." The key word here is "and" between the phrase "displeasure of politicians" and "purists." This means that whatever attitude the politicians take, so would the purists. Choice (A) is incorrect because "concession" (give in) is the opposite of what you would do if you were displeased. Choice (B) is wrong because "neutrality" means that the purists wouldn't take a stand; however, the sentence tells us that the purists are holding the same attitude as the politicians. Choice (C) "endorsement" is wrong because it is the opposite in meaning to "denunciation," and we are looking for a word that has a parallel connotation. Choice (D) is wrong because "resolution" (determination) is not logical to the context of the sentence.

6. (B) The correct answer has two adjectives that are almost opposite in meaning. The best answer is Choice (B), "erratic" (unpredictable, strange) and "inevitable" (sure to happen). The substitution of these words into the sentence means that the spelling and pronunciation of English words are unpredictable because they don't follow rules that are inevitable, or you can't always tell how to spell or pronounce a word based on rules. Choice (A) is incorrect because words cannot be described as "infamous" (having a bad reputation). Choice (C) is not correct because although the words may be "fallacious" (misleading), the rules cannot be "hypothetical" (uncertain). These are not opposite in meaning. Choice (D) is wrong because although "hackneyed" (trite) may fit the first part of the sentence, "verbose" (wordy, talkative) does not fit into the second part. Rules may be "verbose," but that would not explain why the spelling and pronunciation may be "incoherent" (illogical) to some people but not to all; "prudent" (wise) is not an acceptable way to describe spelling rules.

7. (D) The author of Passage 1 writes that "the people who choose to alter their looks or to dress differently complain that people unfairly or inaccurately judge them by their appearance. And they're probably right." And Passage 2 indicates, "People often judge others by their outward appearance." Passage 1 states that some people look different by choice, and Passage 2 asks the reader to consider other reasons—for example, the individual may not have a choice as to what he or she wears—but neither indicates that appearance is not sometimes a reflection of personality. Therefore, choice (A) is incorrect. Choice (B) is incorrect because both passages focus on the fact that people use appearances as a means of judging others; clearly appearances are important in many cases. Choice (C) is incorrect because the author of Passage 1 specifically states, "People who choose to appear different should not expect to remain unobtrusive, and if they claim a desire for anonymity, they're probably not being completely honest." And Passage 2 does not assume the honesty or dishonesty of those who look different. Choice (E) is wrong because Passage 2 indicates that sometimes people do not have a choice as to what they wear, and what people—rich or poor—wear does not reflect who they are.

8. (B) Clearly, the author has a definite opinion, and he/she is using this question to lead the reader to agreeing with that side of the issue. The question is presented as if it has only one (obvious) answer. Choice (A) is wrong because the author is not open to both sides of the issue. Choice (C) is incorrect because the author directs blame toward the people who choose to look different. Choice (D) is wrong because the author implies no confusion about the issue; he or she is quite clear in his or her opinion. And Choice (E) is incorrect because the author clearly sees the issue from only one side and so is not attempting to lead the reader to accept both sides.

9. (A) People who do not try to hide their wealth often come across as "pretentious" (exaggerating their worth or value, showing off). Choice (B) is incorrect because such people probably would not object to being labeled "wealthy" or "affluent" because those terms are not nearly as offensive as "pretentious." Choice (C) is incorrect because others would not label someone as "sociable" or "friendly" just because that person wears expensive clothes and jewelry. And although rich people might be labeled "sanctimonious" or "hypocritical" (self-righteous or deceitful), their clothes and jewelry would not be the reason people saw them that way. And it is generally not offensive to be called "provocative" or "alluring," (captivating or attractive), so choice (E) is wrong.

10. (B) The author says, "If they claim a desire for anonymity, they're probably not being completely honest." In other words, they are seeking attention, not anonymity. Unlike the author of Passage 2, this author does not consider the possibility that the person cannot afford to wear other clothes; so Choice (A) is wrong. No mention is made of their right to express themselves; thus, Choice (C) is wrong. Likewise, nothing is said about whether or not they like accepted styles or whether they know any other way to look. So choices (D) and (E) are wrong.

11. (D) The author says, "People who choose to appear different should not expect to remain unobtrusive," implying that they should not be offended by others' reactions. The passage says nothing about their acceptance of others, so choice (A) is wrong. Nor does it recommend that they change their appearance so that it is more acceptable; so choice (B) is incorrect. The author makes no recommendation that they try to get others to understand them or their appearance; thus, choice (C) is incorrect. And it does not say they should not worry about others' reactions—only that they should not expect to remain unobtrusive; so choice (E) is wrong.

12. (C) Being out of fashion would involve being "different," so a person who doesn't want to be noticed would not want to be anything but fashionable. However, since the author of Passage 1 is obviously offended by anything that is "extreme"—fashionable or not—he/she would recommend being subtle. Choice (A) is incorrect because some fads are attention-getters and extreme; therefore, the author of Passage 1 would find them inadvisable, especially for someone who doesn't want attention. Choice (C) is incorrect because what reveals the "real you" might be offensive to the author and might also be an attention-getter. Choice (D) is incorrect because one of the main ideas of this passage is that people do think about and have opinions about one's appearance, so a person who does not want attention should consider how others might perceive the way he/she looks. And choice (E) is incorrect because the author focuses on clothing perhaps more than anything else, as far as appearance is concerned; no mention is made of how loudly one talks.

13. (E) The author is clearly making a judgment about people who choose to look different and then complain about the attention they receive. This author is not choice (A) sympathetic, choice (B) ambivalent (experiencing two conflicting feelings about a topic), or choice (C) empathic (sensitive to the thoughts and situations of others). And although this author seems impatient, he/she appears to be more judgmental than angry; so choice (D) "angry" is not as appropriate a description as "judgmental."

14. (E) If you are not familiar with the dictionary meaning of this word, you should pick up on context clues. Look again at the sentence in which the word appears: "Although the façade is what we see first, we must keep in mind that it's only a cover." A synonym for "façade" or "cover" is "front." And although choice (A) style, choice (B) behavior, choice (C) attitude, and choice (D) facial expression are all things that we notice about other people, the word "façade" encompasses all these things as well as other aspects of outward appearance. So "front" is a more appropriate choice than any of the others.

15. (B) The author of Passage 2 specifically states, "People are first of all human beings, each with a story. . . . the façade is what we see first . . . it's only a cover. What's inside is much more important." Although this author does not think outward appearance such as clothing is really important, he/she probably would be too genuine to pretend not to notice it—but would see beyond it instead. So choice (A) is wrong. This author would clearly not join the others in choice (C), staring at him, or choice (D), criticizing him, and the author would not choose (E), ignore him.

16. (E) Passage 2 closes with "Although the façade is what we see first, we must keep in mind that it's only a cover. What's inside is much more important." This idea is exactly the same as the one conveyed in the expression "You can't judge a book by its cover." Choice (A) is wrong because it implies that those who are quiet are profound; this idea has nothing to do with what the passage discusses. Choice (B) expresses an idea that is in contrast with the author's message that we can go beyond what we merely see. The main idea of Choice (C) is that we should accept an opportunity while it is available to us—an idea that has nothing to do with Passage 2. And although "do your own thing" is more or less what some people are doing when they choose to be different, Passage 2 contends that not everyone who looks "different" is "doing their own thing," and nothing in the passage implies that they would be wise to do so.

17. (B) The entire passage focuses on trying to understand what's inside people who appear "different" and tolerating outside appearances. Although we may assume the author is friendly to others, there is not as much evidence of friendliness in the passage as there is of the author's tolerance of others. So choice (A) is not as appropriate as choice (B). Choice (C) is incorrect because nothing in the article implies that the author is judgmental of anyone. Also, several of the statements in the passage indicate that the author is concerned rather than indifferent. So choice (D) is incorrect. And nothing in the article implies irritation; thus, choice (E) is wrong.

18. (A) Both passages use face paint, revealing clothes, expensive clothes, and lack of good grooming as examples of appearances that attract attention. Choice (B) is wrong because neither passage presents trends as attractive, although they both point out unusual appearances that may be trendy. Choice (C) is wrong because they both (especially Passage 1) offer opinions about people who look "different." Choice (D) is wrong because Passage 2 is sympathetic toward people who look "different." And choice (E) is incorrect because the authors agree on very little.

19. (A) The author of Passage 1 believes that those who purposely appear different are actually seeking attention, although they may not admit it. However, the author of Passage 2 considers reasons why people may appear different and what is behind their façades, keeping in mind that each person is "human," with his/her own story. Choice (B) is wrong because the author of Passage 1 does not express outright anger as much as irritation or impatience. And the author of Passage 2 expresses no irritation—only an effort to understand. Choice (C) is incorrect because the author of Passage 2 makes no reference to what we may gain by the styles we choose. Choice (D) is wrong because both are alike in that they focus on others' reactions to people who appear different; also, this choice is wrong because Passage 2 does not focus on ways everyone is different as much as emphasizing that people may look the way they do for reasons we do not understand. Choice (E) is incorrect because neither author implies that everyone should look similar, or that each person should look unique.

DETAILED EXPLANATIONS

SECTION 9—WRITING

1. (C) It contains no verb shifts because all the verbs are in present tense ("has . . . gets . . . encounters"). Choice (A) is incorrect because "would get" is inconsistent with the present-tense verbs "has" and "encounters." Likewise, in choice (B) "would encounter" is inconsistent with "has" and "gets." Choice (D) shifts from present tense "has" to past tense "got" and "encountered." And choice (E) shifts from present tense "has" and "gets" to past tense "encountered."

2. (B) It correctly compares "John Steinbeck's novels" to "those of F. Scott Fitzgerald" and uses the plural verb "are" for the plural subject "novels." Choice (A) is incorrect because it illogically compares "John Steinbeck's novels" to "F. Scott Fitzgerald." It also uses a singular subject "novel" with the plural verb "are." Choices (C) and (E) correct the comparison error but still have the subject-verb agreement error. Choice (D) uses correct subject-verb agreement error but still has the comparison error.

3. (D) It correctly uses a semicolon to connect two complete thoughts ("She is going to lose her job" and "because she does not stay focused, she never completes the assigned tasks"). Choices (A) and (B) are run-on sentences, using no punctuation to connect these complete thoughts. Choice (C) contains a comma splice, incorrectly using a comma instead of a semicolon. Choice (E) is illogical with the omission of the subject in the last half of the sentence.

4. (B) It makes the third item in the series a noun phrase, just as the other two are; it contains no unnecessary words. Choice (A) has nonparallel structure: the first two items in the series are noun phrases ("dinner plates" and "drinking glasses"), but the third item is a clause ("they had serving dishes to match). Choice (C) is redundant with the use of the word "matching." Choices (D) and (E) contain the same type of nonparallel structure as Choice (A) does.

5. (E) Its plural verb "were" correctly agrees with the plural subject "three." Also, choice (E) correctly uses the possessive pronoun "their" with the gerund "being." Choice (A) incorrectly uses the singular verb "was" with the plural subject "three" and the pronoun "them" with the gerund "being." Choice (B) uses the correct verb but incorrectly uses "them" instead of "their" before the gerund "being." Choice (C) incorrectly uses the singular verb "was" with the plural subject "three, although it does correctly use "their" with the gerund. Choice (D) is unnecessarily wordy, with the clause "that they be included."

6. (C) It adds the word "else" to show that "my brother" is part of the family. It also clearly shows the comparison. Choice (A) is incorrect because it does not include the word "else," implying that "my brother" is a part of the family. Choice (A) is also confusing because it is difficult to determine whether the sentence means "I like my brother better than anyone else in my family does" or "I like my brother better than I like anyone else in my family." Choice (B) clarifies the comparison issue but is also unnecessarily wordy, with the addition of the phrase "is the one." It also omits the word "else." Likewise, choice (D) omits "else." And choice (E) does use the word "else," improving that part of the sentence; however, it is unnecessarily wordy with the addition of the phrase "is the one."

7. (B) It contains no confusing modifiers, and it correctly uses the singular verb "confuses" with the singular subject "labeling." Choice (A) is incorrect because the introductory clause "when the proper labeling of foods does not occur" incorrectly modifies "they." And the pronoun "they" has no antecedent. Choice (C) incorrectly uses the plural verb "confuse" with the singular subject "labeling." Choice (D) uses the pronoun "it," which has no antecedent. And choice (E) omits the verb in the first clause. Also, the prepositional phrase "in the cafeteria" is improperly placed in this sentence; it is illogical and incorrect to say "the cooks often become confused in the cafeteria."

8. (A) It contains no excess words or confusing modifiers. Choice (B) is incorrect because it contains no subject for the verb "may . . . recognize." Choice (C) is too wordy; the clause "it may be that" is unnecessary. Choice (D) uses the vague pronoun "it," which has no antecedent. And choice (E) uses the vague pronoun "they," which has no antecedent.

9. (B) It does not use unnecessary words, and it shows the cause-and-effect relationship between the main clause and the subordinate clause. Choice (A) is awkward because of the unnecessary words "of that." Also, in this sentence the pronoun "it" has no antecedent. Choice (C) uses nonparallel structure: the phrase "of his busy schedule" is not parallel with the clause "also he is not really comfortable speaking to large crowds." Choice (D) uses the vague pronoun "it." Also, in this sentence the clause "he has a busy schedule" is not parallel with the phrase "his not really being comfortable speaking to large crowds." Choice (E) also contains the vague pronoun "it." This sentence is awkward with its use of "because he has . . . discomfort speaking to large crowds."

10. (E) It uses the correct plural pronoun "them" for the antecedent "gifts." The singular pronoun "it" in choice (A) has no antecedent. Choice (B) incorrectly uses the adjective "quick" instead of the adverb "quickly." Choices (C) and (D) are awkward and wordy, with the clause "there was not enough consideration about. . . ." Additionally, choice (D) incorrectly uses the singular pronoun "it," with no antecedent.

11. (B) It consistently uses past-tense verbs ("agreed," "was," "understood"), and it correctly uses "who" instead of "which" to refer to a person. Choice (A) incorrectly switches from past tense ("agreed") in the first clause to present perfect ("has been") and then back to past tense ("understood") in the rest of the sentence. Choice (C) incorrectly uses "which" instead of "who" to refer to a person. Choice (D) creates a fragment because "having been" has no subject. And choice (E) switches from past tense in the first clause to past perfect ("had been") and then to present tense ("understands") in the last part.

12. (D) It contains no vague pronoun, and its meaning is clear. Choice (A) incorrectly switches from "you" and "your" to "one." Choice (B) contains the same error, and switching from "if" to "when" does not correct any problem that is present in choice (A). In choice (C), the use of the conjunction "since" incorrectly creates a cause-and-effect relationship. It also uses the plural pronoun "they," which has to refer to the plural noun "friends," making the sentence illogical. Choice (E) contains an unclear, awkward adjective phrase that is incorrectly placed.

13. (C) This sentence is not ambiguous in its meaning, and the antecedent of the pronoun "they" is clearly "pears." In choice (A) the pronoun "they" has a vague antecedent; the reader cannot tell if this pronoun refers to pears or apples. Choice (B) has the same problem with the possessive pronoun "their." Choice (D) is far too wordy and awkward. Choice (E) incorrectly uses the singular pronoun "it" to refer to the plural "pears" or "apples," and it is unclear whether this pronoun refers to apples or pears.

14. (A) Because the pronoun "who" refers to "robots," it requires a plural verb. So the plural verb "base" correctly agrees with this subject. Its use of the prepositional phrase at the beginning of the sentence leaves no confusion as to what that phrase modifies; it is easy to understand that he wrote "in his essay." In choices (B) and (C), placing this phrase at the end of the sentence causes it to modify "ads," incorrectly indicating that the "ads" were "in his essay." Also, choice (B) incorrectly uses the singular verb "bases" to go with the plural subject. Choice (D) incorrectly places the prepositional phrase "in his essay" after "voters," illogically indicating that "voters" were "in his essay." In choice (E), the placement of this phrase after "wrote" is correct; however, the use of the phrase "of which" after "robots" is illogical and awkward because "which" has no antecedent.

PRACTICE TEST 2

Answer sheets for this test start on page 851.

TIME: 25 Minutes

ESSAY

The essay gives you an opportunity to show how effectively you can develop and express ideas. You should, therefore, take care to develop your point of view, present your ideas logically and clearly, and use language precisely. Your essay must be written on the lines provided on your answer sheet—you will receive no other paper on which to write. You will have enough space if you write on every line, avoid wide margins, and keep your handwriting to a reasonable size. Remember that people who are not familiar with your handwriting will read what you write. Try to write or print so that what you are writing is legible to those readers.

DIRECTIONS: You have 25 minutes to plan and write an essay on the following topic. You may write on only the assigned topic.

Make sure to give examples to support your thesis. Proofread your essay carefully and take care to express your ideas clearly and effectively.

ESSAY TOPIC

The person who is thankful for the least enjoys the most.

ASSIGNMENT: Do people who are thankful for the least enjoy the most? Plan and write an essay in which you develop your point of view on this issue. Support your position with reasoning and examples taken from your reading, studies, experience, or observations.

TIME: 25 Minutes
20 Questions

MATH

DIRECTIONS: In this section solve each problem, using any available space on the page for scratchwork. Then decide which is the best of the choices given and fill in the corresponding oval on the answer sheet.

NOTES

(1) The use of a calculator is permitted.

(2) All numbers used are real numbers.

(3) Figures that accompany problems in this test are intended to provide information useful in solving the problems. They are drawn as accurately as possible EXCEPT when it is stated in a specific problem that the figure is not drawn to scale. All figures lie in a plane unless otherwise indicated.

(4) Unless otherwise specified, the domain of any function f is assumed to be the set of all real numbers x for which $f(x)$ is a real number.

REFERENCE INFORMATION

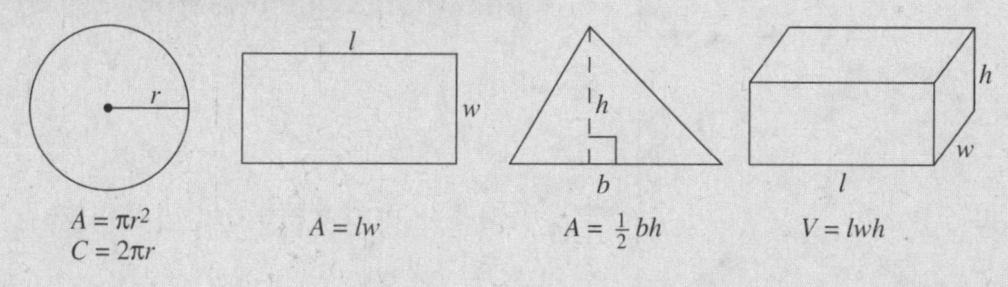

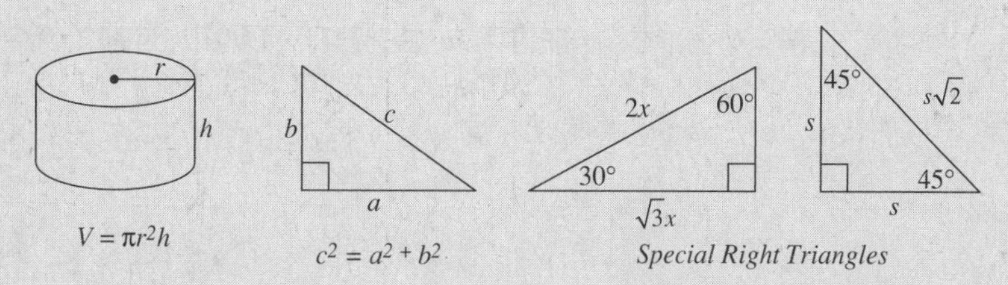

The number of degrees of arc in a circle is 360.

The sum of the measures in degrees of the angles of a triangle is 180.

1. Segment $\overline{AB}$ has one endpoint at (5, –6) and its midpoint is (1, 2). What is the location of the other endpoint?

(A) (–3, 10)

(B) (–2, 4)

(C) (3, –2)

(D) (4, –8)

(E) (6, –4)

2. In a group of ten different numbers, which of the following never affects the value of the median?

(A) Doubling each number

(B) Decreasing each number by 2

(C) Increasing the largest number

(D) Decreasing the largest number

(E) Increasing the smallest number

3. When n is divided by 6, the remainder is 4. Which of the following expressions will yield no remainder when it is divided by 6?

(A) $n + 1$

(B) $n + 2$

(C) $n + 3$

(D) $n + 4$

(E) $n + 5$

4. A jar contains seven white marbles, two red marbles, and three blue marbles. After Joe replaces one white marble with two red marbles, Maria randomly selects one marble from the jar. What is the probability that she will select a red marble?

(A) $\frac{1}{4}$

(B) $\frac{4}{13}$

(C) $\frac{1}{3}$

(D) $\frac{1}{2}$

(E) $\frac{9}{13}$

5. A machine makes 5 bottle caps in 22 seconds. How many bottle caps can it make in 66 minutes?

(A) 15

(B) 290

(C) 900

(D) 1,500

(E) 17,424

6. If $|x - y| > x - y$, which of the following statements must be true?

(A) $x = y$

(B) $x > y$

(C) x and y are both positive

(D) $x < y$

(E) x and y are both negative

GO ON TO THE NEXT PAGE

7. What is the range for the function $f(x) = \frac{6x}{(3x+1)}$?

(A) All numbers

(B) All numbers except 2

(C) All numbers except $-\frac{1}{3}$

(D) All numbers except zero

(E) All numbers except $\frac{1}{6}$

8. If $x \neq 0, -4$, which of the following is the reduced expression for $\frac{(10x^2+30x-40)}{(5x^2+20x)}$?

(A) $\frac{2(x-2)}{(x+2)}$

(B) $\frac{2x}{(x+4)}$

(C) $\frac{(x+2)}{2}$

(D) $\frac{(x-1)(x+4)}{2}$

(E) $\frac{2(x-1)}{x}$

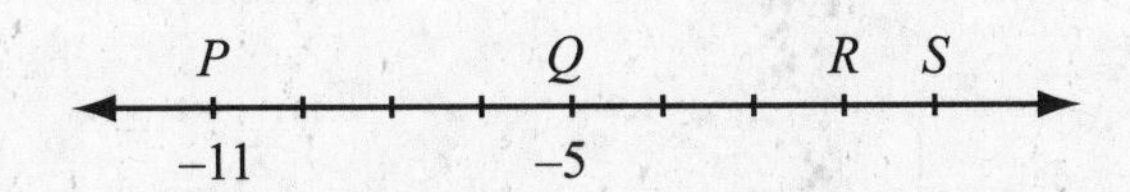

9. In the figure above, all consecutive slash marks are equally spaced on the number line. Point P has a coordinate of –11 and point Q has a coordinate of –5. What is the sum of the coordinates of points R and S?

(A) –2

(B) –1.5

(C) –0.5

(D) 0

(E) 0.5

Age Distribution of People Living in Smallville

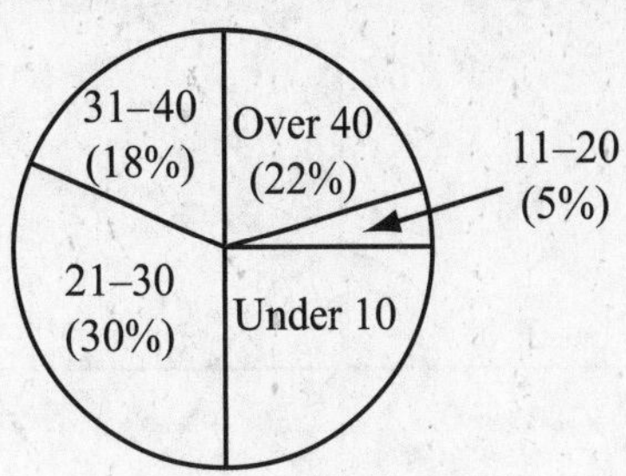

10. Using the above pie chart, if there are a total of 282 people in Smallville who are under 10 years old or over 40 years old, what is the combined total of people who are between the ages of 11 and 30, inclusive?

(A) 180

(B) 210

(C) 330

(D) 460

(E) 600

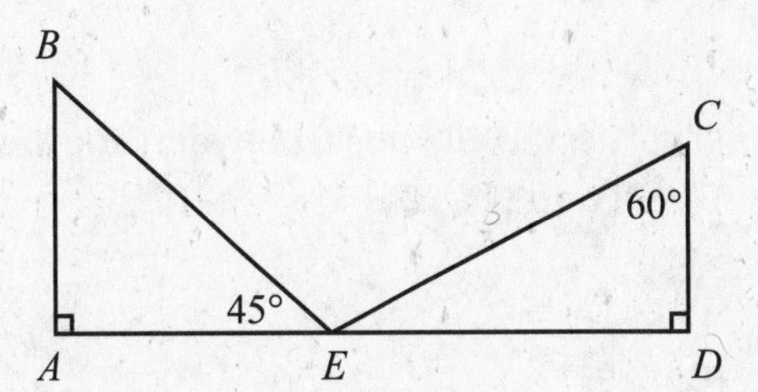

Note: Figure not drawn to scale

11. In the figure above, $\overline{BE} = 10$ and $\overline{CE} = 8$. What is the best approximation to the length of $\overline{AB}$? (A, E, D are collinear points).

(A) 12

(B) 13

(C) 14

(D) 15

(E) 16

GO ON TO THE NEXT PAGE

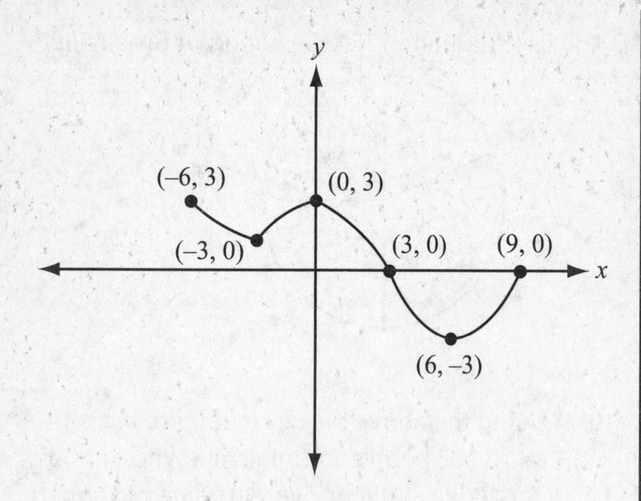

12. Based on the graph of the function $f(x)$ shown above, which values of x satisfy the inequality $0 \le f(x) \le 3$?

(A) $0 < x < 3$

(B) $0 < x < 6$

(C) $0 < x < 9$

(D) $-3 < x < 9$

(E) $-6 < x < 3$

13. What is the distance between the points (7, −3) and (16, 37)?

(A) 7

(B) 9

(C) 40

(D) 40.2

(E) 41

14. The points (2, −8), (6, 14) and (10, −7) lie on the graph of $f(x)$. If $g(x) = f(x + 4)$, which of the following points must lie on the graph of $g(x)$?

(A) (−8, 2)

(B) (−4, −22)

(C) (−2, 8)

(D) (10, 14)

(E) (6, −11)

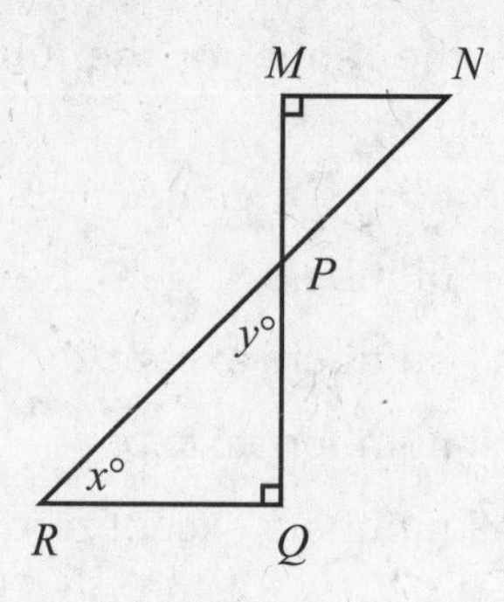

Note: Figure not drawn to scale.

15. In the figure shown above, $x = y$, $\overline{PR} = 6$, and $\overline{NP} = 4$. What is the length of $\overline{MQ}$?

(A) $2\cdot\sqrt{2}$

(B) $3\cdot\sqrt{2}$

(C) $4\cdot\sqrt{2}$

(D) $5\cdot\sqrt{2}$

(E) $6\cdot\sqrt{2}$

16. Let $m \oplus n$ be defined as $m^2 + n$ where $m, n > 0$; $m + n^2$ where m, $n < 0$; and zero otherwise. How many of the following have a positive value?

$2 \oplus 7$
$-4 \oplus -3$
$-7 \oplus 2$
$3 \oplus -2$
$7 \oplus 0$
$4 \oplus 4$

(A) 2

(B) 3

(C) 4

(D) 5

(E) 6

GO ON TO THE NEXT PAGE

17. Suppose $S = \{5, 9, 16, 12, 3\}$ and $T = \{4, x, 12, y\}$. If $S \cap T = \{5, 16, 12\}$, which of the following conditions is a possible set of values for x and y?

(A) $x = 9$ and $y = 5$

(B) $x = 16$ and $y = 3$

(C) $x = 5$ and $y = 16$

(D) $x = 3$ and $y = 9$

(E) $x = 5$ and $y = 3$

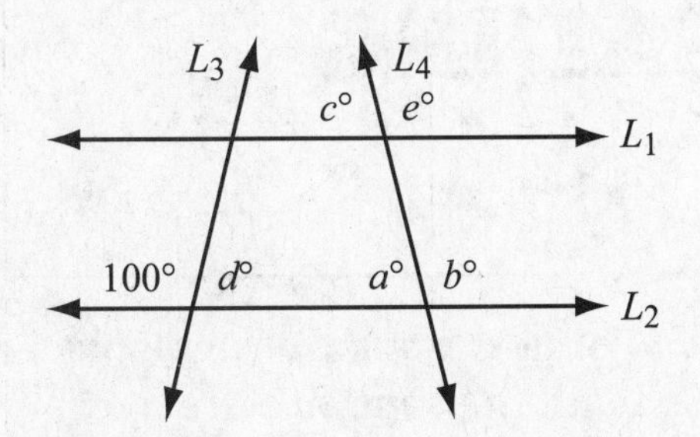

Note: Figure not drawn to scale

18. In the figure shown above, line L_1 is parallel to line L_2, but L_3 is not parallel to L_4. Which group of five angles whose measurements are indicated by $a°$, $b°$, $c°$, $d°$, $e°$ could not equal 100?

(A) d, a, c

(B) d, b, e

(C) d, b, c

(D) d, a, b

(E) d, c, e

19. D varies inversely as the square root of F. When $F = 16$, $D = 9$. What is the value of D when $F = 100$?

(A) 0.003

(B) 3.6

(C) 4.44

(D) 22.5

(E) 351.56

20. An arithmetic sequence is given by 500, 494, 488, . . . What is the 30th term of this sequence?

(A 308

(B) 314

(C) 320

(D) 326

(E) 332

STOP

If time remains, you may go back and check your work. When the time allotted is up, you may go on to the next section.

TIME: 25 Minutes
20 Questions

CRITICAL READING

DIRECTIONS: Each sentence that follows has one or two blanks, each blank indicating that something has been omitted. Beneath the sentence are five lettered words or sets of words. Choose the word or set of words that BEST fits the meaning of the sentence as a whole.

EXAMPLE

Although critics found the book _______ , many readers found it rather _______ .

(A) obnoxious . . . perfect
(B) spectacular . . . interesting
(C) boring . . . intriguing
(D) comical . . . persuasive
(E) popular . . . rare

EXAMPLE ANSWER

1. She did not agree to the speaking engagement because she was not ____________ with the featured topic.

   (A) conversant
   (B) nescient
   (C) discordant
   (D) disconsolate
   (E) indubitable

2. Completing the complex task in such a short time seemed ____________ until my coworkers offered to ____________ me.

   (A) bodacious . . . acclaim
   (B) redundant . . . relegate
   (C) insuperable . . . abet
   (D) facile . . . absolve
   (E) derogatory . . . acknowledge

3. None of his friends understood when he stopped partying and became completely ____________.

   (A) obstinate
   (B) acquiescent
   (C) fractious
   (D) abstemious
   (E) genteel

4. Whereas John's interests lie in racing and mechanics, Susan has a more ____________ taste, preferring classical concerts and art museums.

   (A) acetic
   (B) acerbic
   (C) ascetic
   (D) affable
   (E) aesthetic

5. The political candidate aroused so much ____________ among his constituents that I knew he had no chance of winning the election.

   (A) collusion
   (B) antipathy
   (C) incursion
   (D) impunity
   (E) empathy

6. The teacher's ____________ response to the child's ____________ request for help made me realize how ineffective she really was.

   (A) cursory . . . bona fide
   (B) obliging . . . frustrated
   (C) imperturbable . . . incessant
   (D) astute . . . incoherent
   (E) mindful . . . frivolous

7. I was trying to ____________ my opinion to her; however, I managed only to ____________ her instead.

   (A) delineate . . . enlighten
   (B) mediate . . . implicate
   (C) elucidate . . . confound
   (D) sell . . . actuate
   (E) entangle . . . disconcert

GO ON TO THE NEXT PAGE

8. Those who ____________ to move up the highly competitive corporate ladder must be ____________ all the time, no matter what they are doing.

(A) deign . . . assiduous
(B) aspire . . . sedulous
(C) consort . . . impolitic
(D) condescend . . . immutable
(E) endeavor . . . hebetudinous

<u>DIRECTIONS</u>: Each passage below is followed by questions based on its content. Answer the questions on the basis of what is <u>stated</u> or <u>implied</u> in each passage and in any introductory material that may be provided.

Questions 9–10 are based on the following passage.

1 I long for the days when kids played with pots and pans instead of computer and video games. My fondest memories include the redolence of roast slow-cooking in the oven for 5 hours—an aroma that no pre-cooked meal will ever emanate from the microwave. I miss doors with no locks, clothes hanging outside on the clothesline to dry in the sun, dust flying behind our truck on the unpaved road, mowers with 10 rotary blades. With the disappearance of these simple things came the end of the good life.

9. The author of this passage seems to be

(A) happy.
(B) nostalgic.
(C) anticipatory.
(D) angry.
(E) hungry.

10. Which of the following best describes the author's preferences?

(A) simple and basic
(B) complex and expensive
(C) old and much-used
(D) modern and popular
(E) convenient and quick

Questions 11–12 are based on the following passage.

I sat in the car. I wasn't sure the day could 1 have been any worse, and the ominous rumbling clouds overhead seemed to agree with me. As drops of rain speckled the windshield, the gray of the sky lightened. The sun peeped 5 through the clouds, turning the drops of water into diamonds. I lowered the windows just enough to let some air in. The essence of freshness renewed my spirit. And the chirping birds told me the shower had revitalized them as well. 10

11. The author's tone

(A) remains positive throughout the passage.
(B) changes from positive to negative.
(C) remains negative throughout the passage.
(D) changes from negative to positive.
(E) is neither positive nor negative.

12. In line 4 the use of "diamonds" shows

(A) that the author is dreaming.
(B) that the author values jewels.
(C) the author's appreciation of the sunshine's effect on the raindrops.
(D) the author's appreciation of the world.
(E) that life brings us rewards.

GO ON TO THE NEXT PAGE

Questions 13–24 are based on the following passage.

The weather is important to us. It often determines our plans for weekend activities, and is sometimes the reason we have to cancel or postpone those plans. It affects the workday for those of us who have to travel or who work outside part of the time. And for people whose work is always outdoors, it determines whether or not they must report to work. It sometimes causes schools to close and may result in the cancellation of athletic events.

Because the weather has such an effect on our lives, some of us pay more attention to the weather forecast in the media than we do to the news, music, sports, or television shows. We hang onto every word of the meteorologists on the radio or television, carefully following their explanations of the upcoming front, watching the radar to see how close the precipitation is, and heeding their prognostication of temperatures. And all too frequently, we plan a big outing for the day full of promised sunshine and warm weather, only to realize we should have worn heavy jackets and brought umbrellas. Or we cancel our Christmas shopping plans because of the ninety-percent chance of heavy snow, only to go outside and discover unseasonably warm temperatures and brilliant blue skies. Perhaps we should pay less attention to the scientific speculation of the weather and refer to more time-proven methods of forecasting the weather.

Old-timers will tell you that animals are the best meteorologists in existence. Rain is impending if your cat sneezes, if lightning bugs are flying low, if spiders abandon their webs, if a peacock calls, if a fly bites you, if cows are lying in the pasture, or if rabbits are playing in the middle of a dusty road. You can expect clear skies if swallows are flying low, if your cat is washing its face, if you see an open ant hole, or if the crickets are chirping loudly.

According to generations past, animals even serve as harbingers of the upcoming season. Many swear by the woolly worm, also known as the woolly bear or fuzzy bear caterpillar. Each end of this creature's body is black, and the middle is rust. Claims are that the wider the black on the woolly worm's body in the fall, the more severe the upcoming winter. Likewise, the wider the rust portion, the milder the winter is going to be. And some woolly-worm watchers go so far as to contend that if the head is dark, the beginning of winter will be severe; if the tail end is dark, the end of winter will be very cold. And some say that the thickness of the worm's coat portends the severity of the winter.

Animals are not the only natural weather forecasters. Folklore tells us we can expect rain if our water glasses and pitchers sweat, if our noses itch, if dandelions have closed their blossoms, or if we see rings around the moon. We can expect rain by morning if the grass is dry at night, and rain by evening if the morning sky is red. We can expect clement weather when dew is on the grass in the morning, when we see lightning in the north, or when we see a rainbow or a red sky in the evening.

Odd as it may sound, the weather itself may be telling us what to expect. If it happens to thunder in the fall, you should expect cold weather to arrive soon. Weather-watchers swear that clouds at sunset on Friday mean Sunday will be clear. Snow at Christmas tells us to expect nice weather on Easter. Large snowflakes mean the snow won't last long. And snow before Thanksgiving indicates that we won't have many snows throughout the winter. Lightning in the north is a predictor of dry weather.

GO ON TO THE NEXT PAGE

Folklore methods of predicting the weather
80 are endless. And those who believe in them take much satisfaction in the accuracy of these nature-based forecasts. Obviously, time-honored weather lore did not have the benefit of modern-day science and technology; however,
85 years of observation and experience—two important elements of scientific investigation—serve as the basis for many of these "weather rules." And—as far-fetched as some of them may appear at first glance--many have proven
90 to be scientifically sound. So when the weather is vital to our plans for work or play, it may behoove us to check out nature as well as the local forecast; if they all point in the same direction, we can be pretty certain about what plans
95 we should make. If they are contradictory, it's up to each of us to decide which forecast is most likely to be correct. A few trials will let us know who—or what—we should believe.

13. The main purpose of this passage is to

(A) prove that meteorologists' weather forecasts are usually wrong.

(B) prove that folklore weather predictions are usually right.

(C) share some folklore weather predictions.

(D) give examples of folklore predictions that have a scientific basis.

(E) prove that meteorologists' weather predictions have a folklore basis.

14. The author refers to the closing of school and the cancellation of athletic events to

(A) exemplify the importance of weather.

(B) exemplify the importance of planning.

(C) prove the importance of weather forecasts.

(D) demonstrate how important it is that weather predictions be accurate.

(E) demonstrate that we should believe folklore weather predictions.

15. The author of this passage implies that weather predictions are sometimes more important to us than several things. Which of the following is not among those things?

(A) television shows

(B) music

(C) school

(D) news

(E) ballgames

16. In line 19, "prognostication" most nearly means

(A) guess.

(B) lie.

(C) forecast.

(D) postponement.

(E) broadcast.

17. What would the author of this passage probably say about computers and the contemporary reliance on them?

(A) They are far better than methods of yesterday.

(B) In every area except weather prediction, they are far superior to the methods of yesterday.

(C) They are superior in providing us the convenience of television and radio broadcasts of weather forecasting.

(D) We should not allow them to replace the wisdom acquired in times past.

(E) They provide us with entertainment and education no matter what the weather is like.

45 50 55 60 65

GO ON TO THE NEXT PAGE

18. In line 33, "impending" most nearly means

(A) heavy.
(B) about to occur.
(C) not likely.
(D) questionable.
(E) over.

19. According to folklore, woolly worms are most likely to tell us

(A) if spring will come early.
(B) if we will have much precipitation throughout the winter.
(C) if we will have many summer storms.
(D) how many times it will snow during the winter.
(E) if the winter is going to be harsh.

20. Which of the following is not among the weather predictors discussed in the passage?

(A) people
(B) plants
(C) weather
(D) animals
(E) air

21. According to folklore predictions, what does the sky tell us about the weather?

(A) It tells us nothing.
(B) If the sky is red in the morning, it is going to rain.
(C) If the sky is red in the evening, it is going to rain.
(D) If the sky is blue, the sun will shine.
(E) Lightning in the northern skies mean it will rain soon.

22. Folklore weather predictions apply to

(A) only the weather that is likely to occur within the next few hours.
(B) mainly weather that is likely to occur within the next few days.
(C) only to rain or snow.
(D) not only weather coming soon, but also to upcoming seasons.
(E) only certain areas of the country.

23. What elements do folklore predictions and modern-day science-based predictions have in common?

(A) experience and guesswork
(B) education and observation
(C) observation and documentation
(D) education and documentation
(E) observation and experience

24. The author of this passage implies that if we are making plans for an outdoor activity, we should

(A) expect weather that is the opposite of what the meteorologist predicts.
(B) regard the meteorologist's forecast as accurate.
(C) go outside and look for natural predictors of upcoming weather.
(D) check the meteorologists' forecast as well as natural weather predictors.
(E) pay no attention to either meteorologists' or natural predictions.

STOP

If time remains, you may go back and check your work. When the time allotted is up, you may go on to the next section.

TIME: 25 Minutes
35 Questions

WRITING

<u>DIRECTIONS</u>: In each of the following sentences, some portion of the sentence is underlined. Under each sentence are five choices. The first choice has the same wording as the original. The other four choices are reworded. Sometimes the first choice containing the original wording is the best; sometimes one of the other choices is the best. Choose the letter of the best choice. Your choice should produce a sentence that is not ambiguous or awkward and that is correct, clear, and precise.

This is a test of correct and effective English expression. Keep in mind the standards of English usage, punctuation, grammar, word choice, and construction.

EXAMPLE

When you listen to opera, <u>a person may not appreciate it.</u>

(A) a person may not appreciate it.

(B) it may not be appreciated by a person.

(C) you may not appreciate it.

(D) which may not be appreciated by you.

(E) appreciating it may be a problem for you.

EXAMPLE ANSWER

1. Conducting a survey independent from all company-paid surveys, the students made some surprising discoveries.

   (A) from
   (B) of
   (C) with
   (D) on
   (E) for

2. She checked the refrigerator, made a list, went to the bank, and was going to the grocery store.

   (A) was going
   (B) was going to go
   (C) went
   (D) had gone
   (E) goes

3. There is a list of categories on the chalkboard that are associated with the chapter material.

   (A) There is a list of categories on the chalkboard that are associated with the chapter material.
   (B) A list is on the chalkboard of categories that are associated with the chapter material.
   (C) On the chalkboard is a list of categories that are associated with the chapter material.
   (D) There is a list on the chalkboard of categories that are associated with the chapter material.
   (E) There is a list of categories on the chalkboard that is associated with the chapter material.

4. One of them was hoping to win the sweepstakes.

   (A) was hoping to
   (B) were hoping to
   (C) was hoping for
   (D) hoped for
   (E) were hoping that

5. When we behave thoughtless, we often hurt others' feelings.

   (A) thoughtless, we
   (B) thoughtless; we
   (C) thoughtlessly, we
   (D) thoughtlessly, you
   (E) thoughtless, one

6. Often doctors do not prescribe medicine for flu victims for the reason being that antibiotics have no effect on the virus.

   (A) victims for the reason being that
   (B) victims, it is because
   (C) victims for the reason that
   (D) victims because of
   (E) victims because

GO ON TO THE NEXT PAGE

7. <u>Brandon's mother boasts that he can dance better than any other girl in his class, and it embarrasses him</u>.

(A) Brandon's mother boasts that he can dance better than any other girl in his class, and it embarrasses him.
(B) Brandon's mother boasts that he can dance better than any girl in his class, and it embarrasses him.
(C) Brandon is embarrassed when his mother boasts that he can dance better than any other girl in his class.
(D) Brandon is embarrassed when his mother boasts that he can dance better than any girl in his class.
(E) Brandon's mother boasts he can dance better than any girl in his class embarrasses him.

8. He <u>gave the dogs to John and Ben because he doesn't like them</u>.

(A) He gave the dogs to John and Ben because he doesn't like them.
(B) Because he doesn't like them, he gave the dogs to John and Ben.
(C) Not liking them, he gave the dogs to John and Ben.
(D) He gave to John and Ben the dogs because he doesn't like them.
(E) Because he doesn't like the dogs, he gave them to John and Ben.

9. Apprenticeships provide practical experience for business <u>students who want to become a successful entrepreneur</u>.

(A) students who want to become a successful entrepreneur
(B) students who want to become successful entrepreneurs
(C) students who were wanting to become a successful entrepreneur
(D) students who wants to become successful entrepreneurs.
(E) students in wanting to become a successful entrepreneur.

10. He prefers dark shades of red, black, and navy; he dislikes <u>pastel shades as</u> pink, gray, and pale blue.

(A) pastel shades as
(B) pastel shades that are
(C) shades of such pastel as
(D) such pastel shades as
(E) a shade pastel as

11. <u>Slander is like when someone says something false about someone else, it ruins their reputation</u>.

(A) Slander is like when someone says something false about someone else, it ruins their reputation.
(B) Slander is a false statement that may ruin another person's reputation.
(C) Slander is someone who makes a false statement that may ruin another person's reputation.
(D) Slander is when someone makes a false statement that may ruin another person's reputation.
(E) Slander ruins another person's reputation by making a false statement.

GO ON TO THE NEXT PAGE

DIRECTIONS: Each of the following sentences may contain an error in diction, usage, idiom, or grammar. Some sentences are correct. Some sentences contain one error. No sentence contains more than one error.

If there is an error, it will appear in one of the underlined portions labeled A, B, C, or D. If there is no error, choose the portion labeled E. If there is an error, select the letter of the portion that must be changed in order to correct the sentence.

EXAMPLE

He drove slowly and cautiously in order to hopefully avoid having an accident. No error.
A B C D E

EXAMPLE ANSWER

 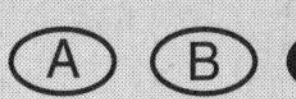  

12. She worked hard to prepare for (A) the final exam, reading (B) every assigned passage as reviewing (C) every worksheet the instructor had given (D) to the class. No error (E)

13. What the policitians say (A) in their campaign speeches are (B) not necessarily what they actually do (C), once they are in office (D). No error (E)

14. Las Vegas, Nevada, (A) has been called (B) the "entertainment capital of the world" because it offering (C) countless drinking and eating establishments, casinos (D), music shows, and other forms of entertainment. No error (E)

15. Over the past year, the little boy has learned (A) to walk, to say several words (B), and is feeding (C) himself with a spoon (D). No error (E)

GO ON TO THE NEXT PAGE

16. The strict rules about students' attire
A
is causing several people to protest that their
B C
civil rights are being violated. No error
D E

17. Being the only mammal that flies, the bat
A B C
intrigues zoologists. No error
D E

18. An effective presenter not only appeals to
A B
his audience's interests and emotions
C
and also uses appropriate humor. No error
D E

19. Some portions of the test are timed;
A
therefore, you must work as quick as possible. No error
B C D E

20. When we listen to political speeches, full of
A
promises for the future and criticism of the opposing parties, you can easily become
B
confused about who is the best candidate.
C D
No error
E

21. Each of the women wants a clearly expressed explanation of what the legislators
A B C
intended when they passed that particular
D
bill. No error
E

22. The discussion between you and I must
A
remain strictly confidential so that no
B C
one's feelings will be hurt. No error
D E

23. Like the other students, Angel's textbooks
(A: the other students)
were outdated and worn, but the school
(B: were; C: worn)
had no money to buy updated materials.
(D: to buy)
No error
(E)

24. Famous American writer Mark Twain is
(A: Famous American writer)
well-known for his profound reflections,
(B: well-known)
sardonic observations, and by his straight-
(C: sardonic observations; D: by his)
forward witticisms. No error
(E)

25. Commonly regarded as simply smelly ani-
(A: Commonly; B: smelly animals)
mals, the skunk is actually an entertaining,
(C: actually; D: entertaining, fun-loving pet)
fun-loving pet. No error
(E)

26. Bigots are often impatient with people
(A: often)
who are open-minded and of those who
(B: who are; C: of those)
have an easy-going demeanor. No error
(D: easy-going demeanor; E)

27. During the Christmas shopping season,
(A: During the Christmas shopping season)
many online merchants hope to attract
(B: merchants hope; C: to attract)
customers by offering complimentary free
(D: complimentary free shipping)
shipping. No error
(E)

28. Research indicating that many Americans
(A: indicating)
are obese, many weighing in excess of 250
(B: are obese; C: weighing)
pounds and having a Body Mass Index
of more than 30. No error
(D: of more than 30; E)

29. Some of we perfectionists become upset
(A: we perfectionists; B: upset with)
with coworkers who are in a hurry and
(C: who are)
settle for inadequate, inferior products.
(D: inadequate, inferior)
No error
(E)

GO ON TO THE NEXT PAGE

DIRECTIONS: The following passage is an early draft of an essay. Some parts of the passage need to be rewritten. Read the passage and select the best answers for the questions that follow. Some questions are about particular sentences or parts of sentences and ask you to improve sentence structure or word choice. Other questions ask you to consider organization and development. In choosing answers, follow the requirements of standard written English.

(1) Stock-car racing is a major spectator sport in this country, but an estimated thirty percent of Americans claiming to be fans of the 53-year-old sport. **(2)** However, racing has not always been so popular. **(3)** As a matter of fact, it had a rather shaky start, with a negative reputation and problems with the law.

(4) During the 1920s, with the Prohibition came moonshining; and with moonshining came bootleggers who were moonshine runners that illegally ran whiskey from the illegal stills to the markets. **(5)** These drivers ran races with the law; the losers were subject to jail time and steep fines. **(6)** Soon racing among the bootleggers became a weekend sport; they used their "whiskey run" cars to prove who was the fastest. **(7)** And of course the area denizens came out to watch. **(8)** These races became popular. **(9)** They continued even after the end of the Prohibition.

(10) In 1938 Bill France organized the first Daytona Beach race. **(11)** Winners received such prizes as rum, cigars, and motor oil. **(12)** After a hiatus during World War II, in the late 1940s France held a meeting of promoters.

30. What is the best way to deal with sentence 1 (reproduced below)?

    Stock-car racing is a major spectator sport in this country, but an estimated thirty percent of Americans claiming to be fans of the 53-year-old sport.

    (A) Leave it as it is.
    (B) Change "but" to "with."
    (C) Change "claiming" to "claim."
    (D) Change "is" to "was."
    (E) Remove the comma and insert a semicolon.

31. Which of the following best describes the relationship between sentences 2 and 3?

    (A) Sentence 3 contradicts the idea presented in Sentence 2.
    (B) Sentence 3 provides examples to illustrate the idea presented in Sentence 2.
    (C) Sentence 3 offers a new idea that is unrelated to Sentence 2.
    (D) Sentence 3 defines a term that Sentence 2 presents.
    (E) Sentence 2 leads to Sentence 3 with an explanation of a situation presented in Sentence 3.

GO ON TO THE NEXT PAGE

32. Which of the following is the best revision of the underlined portion of Sentence 4 (reproduced below)?

During the 1920s, with the Prohibition came moonshining; and with moonshining came bootleggers who were moonshine runners that illegally ran whiskey from the illegal stills to the markets.

(A) bootleggers, who were also called moonshine runners, because they illegally ran whiskey
(B) bootleggers, or moonshine runners, who ran whiskey
(C) the illegal running of whiskey by bootleggers, or moonshine runners,
(D) bootlegging, or moonshine running whiskey
(E) bootleggers who are moonshine runners who ran whiskey

33. What is the best way to revise the underlined wording in order to combine sentences 8 and 9?

These races became popular. They continued even after the end of the Prohibition.

(A) Although these races became popular, they continued even after the end of the Prohibition.
(B) These races became popular, but they continued even after the end of the Prohibition.
(C) These races became popular even after they continued at the end of the Prohibition.
(D) These races became popular, since they continued even after the end of the Prohibition.
(E) These races became so popular that they continued even after the end of the Prohibition.

34. In relation to the passage as a whole, which of the following best describes the writer's intention in paragraph 2?

(A) To provide an example
(B) To provide a summary
(C) To detail a chain of events
(D) To describe a location
(E) To propose a solution

35. Which of the following sentences would be the best sentence to add immediately after Sentence 12?

(A) This three-day meeting resulted in the establishment of the rules, specifications, and official name of the organization—the National Association of Stock Car Auto Racing (NASCAR).
(B) These promoters were interested in racing.
(C) Stock-car racing continues to be popular today, with many fans who attend in person or who watch the races on television.
(D) Although bootlegging is almost non-existent today, much racing still goes on both on and off the track.
(E) Not many people know who Bill France was, but he definitely played an important role in the history of racing.

STOP

If time remains, you may go back and check your work. When the time allotted is up, you may go on to the next section.

TIME: 25 Minutes
18 Questions

MATH

DIRECTIONS: In this section solve each problem, using any available space on the page for scratchwork. Then decide which is the best of the choices given and fill in the corresponding oval on the answer sheet.

NOTES

(1) The use of a calculator is permitted.

(2) All numbers used are real numbers.

(3) Figures that accompany problems in this test are intended to provide information useful in solving the problems. They are drawn as accurately as possible EXCEPT when it is stated in a specific problem that the figure is not drawn to scale. All figures lie in a plane unless otherwise indicated.

(4) Unless otherwise specified, the domain of any function f is assumed to be the set of all real numbers x for which f(x) is a real number.

REFERENCE INFORMATION

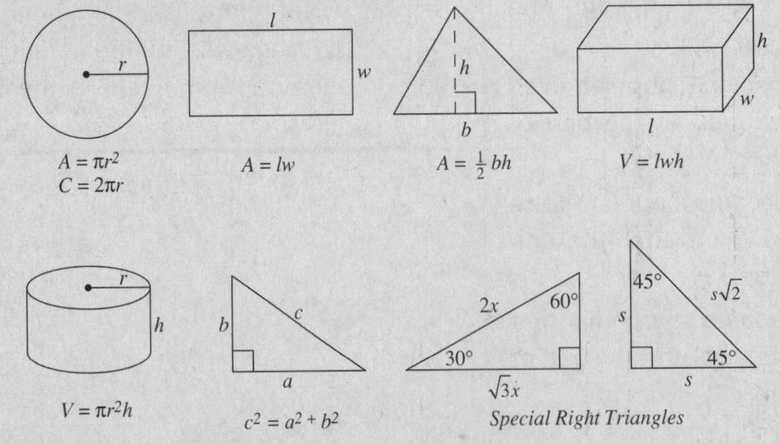

The number of degrees of arc in a circle is 360.

The sum of the measures in degrees of the angles of a triangle is 180.

1. John is older than Mary, Ken is younger than John, and Mary is older than Tracy. Which of the following is not possible?

 (A) Tracy is the youngest.
 (B) Ken is older than Tracy.
 (C) Tracy is older than John.
 (D) Ken is younger than Mary.
 (E) Ken is the youngest.

2. If A = {red, green, purple} and B = {green, white, purple, blue}, which of the following describes the set $A - B$?

 (A) {red, green}
 (B) {red, purple}
 (C) {white, blue}
 (D) {red}
 (E) {green, white, blue}

3. If $a + x = ax + 2$, which of the following expressions is equivalent to x?

 (A) $\frac{(1+a)}{(2+a)}$
 (B) $\frac{(2-a)}{(1-a)}$
 (C) $\frac{1}{(2+a)}$
 (D) $\frac{2}{(1-a)}$
 (E) $2 + a$

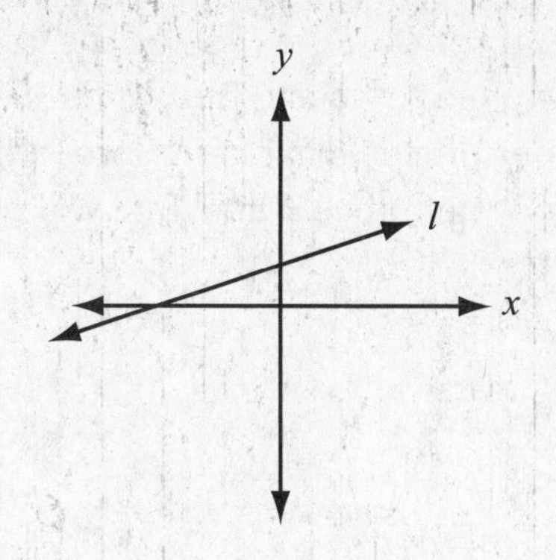

Note: Figure not drawn to scale.

4. In the figure above, line l has a slope of $\frac{2}{7}$ and a y-intercept of 14. What is the x-intercept?

 (A) –49
 (B) –35
 (C) –28
 (D) –14
 (E) –4

5. What is the domain of $f(x)$ if $f(x) = \sqrt{49 - x^2}$?

 (A) $-49 \le x \le 49$
 (B) $0 \le x \le 49$
 (C) $7 \le x \le 49$
 (D) $0 \le x \le 7$
 (E) $-7 \le x \le 7$

GO ON TO THE NEXT PAGE

6. Offices A, B, and C will be randomly assigned to Kasey, Tina, and Lu, one person per office. What is the probability that Lu will get office B and Tina will get office A?

(A) $\frac{1}{2}$

(B) $\frac{1}{3}$

(C) $\frac{1}{4}$

(D) $\frac{1}{6}$

(E) $\frac{1}{9}$

7. $\overline{DE}$ is tangent to a circle at point E, and point A is the center of the circle. If $\angle ADE = 30°$ and $AD = 8$, what is the circumference of the circle?

(A) 4π

(B) 6π

(C) 8π

(D) 16π

(E) 32π

8. The Upbeat Music Shop paid $200 for a clarinet. At what selling price should this item be marked so that the store can offer a 25% discount and still make a 20% profit on the cost?

(A) $240

(B) $250

(C) $290

(D) $320

(E) $360

Hot dogs:	$2.50 each
Soda:	$0.75 per can
Roast Beef:	$6.25 per pound

9. The sign above is displayed at an outdoor hot dog stand. Raul has $40 to spend and he wants to buy 2 pounds of roast beef and 7 hot dogs. What is the maximum number of cans of soda he can buy?

(A) 14

(B) 13

(C) 12

(D) 11

(E) 10

10. If $|m - n| = 5$, and $m < n$, which of the following is not possible?

(A) $m = 17$ and $n = 12$

(B) m and n are both negative numbers.

(C) $m = 12$ and $n = 17$

(D) m and n are both positive numbers.

(E) $n = 0$

11. Joe wants to paint a rectangular ceiling that is 8 feet long and 7 feet wide. One can of paint costs $1.80 and covers 10 square feet. How much will Joe have to spend to buy the minimum number of cans of paint required? (Assume that he cannot buy a fractional portion of a can of paint.)

(A) $16.20

(B) $14.40

(C) $12.60

(D) $10.80

(E) $9.00

GO ON TO THE NEXT PAGE

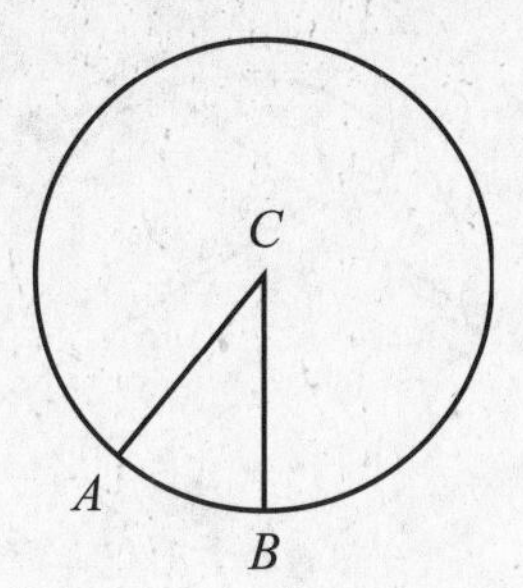

12. The circle shown above has an area of 81π. What is the perimeter of the sector bounded by $\overline{AC}$, $\overline{BC}$, and minor arc AC?

(A) $18 + 18\pi$

(B) $18 + 12\pi$

(C) $18 + 9\pi$

(D) $18 + 6\pi$

(E) $18 + 2\pi$

13. If $g(x) = kx^2 - 5x + 25$ has only one distinct root, what is the value of k?

(A) $\frac{1}{25}$

(B) $\frac{1}{16}$

(C) $\frac{1}{10}$

(D) $\frac{1}{5}$

(E) $\frac{1}{4}$

14. The sum of four consecutive integers is 166. What is the fourth integer?

(A) 40

(B) 41.5

(C) 42

(D) 43

(E) 54

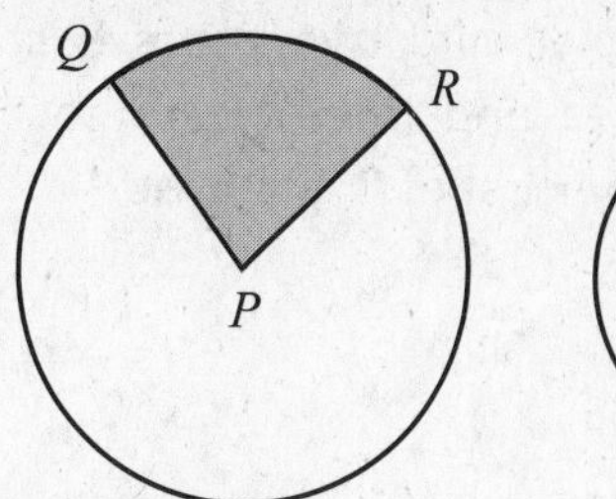

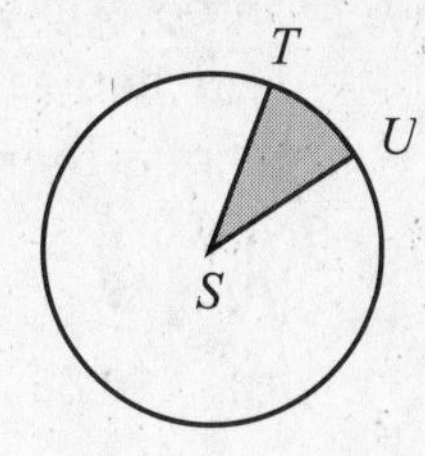

<u>Note</u>: Figures not drawn to scale

15. Points P and S are each the center of a circle, as shown above. $PR = (1.5)(ST)$ and the measure of $\angle P$ is twice the measure of $\angle S$. If the area of the shaded portion of the smaller circle is 60, what is the area of the shaded portion of the larger circle?

(A) 90

(B) 120

(C) 180

(D) 270

(E) 300

16. M varies directly as the square of V. When $V = 4$, $M = 24$. What is the value of M when $V = 6$?

(A) 9

(B) $10\frac{2}{3}$

(C) $12\sqrt{6}$

(D) 36

(E) 54

GO ON TO THE NEXT PAGE

17. Three apples and two pears cost \$1.17. Two apples and three pears cost \$1.33. What is the cost of five apples?

(A) \$0.17

(B) \$0.80

(C) \$0.85

(D) \$1.25

(E) \$2.45

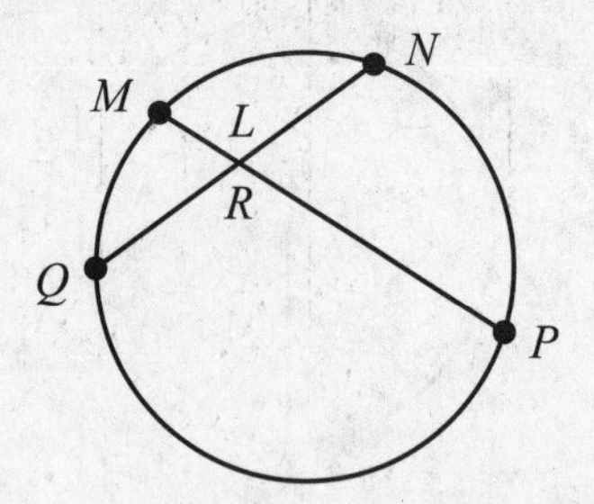

Note: Figure is not drawn to scale.

18. Chords $\overline{MP}$ and $\overline{NQ}$ intersect at point R, which is not the center of the circle. If arc $QM = 50°$ and $\angle L = 110°$, what is the degree measure of arc NP?

(A) 70°

(B) 90°

(C) 105°

(D) 110°

(E) 140°

STOP

If time remains, you may go back and check your work. When the time allotted is up, you may go on to the next section.

TIME: 25 Minutes
24 Questions

CRITICAL READING

DIRECTIONS: Each sentence below has one or two blanks, each blank indicating that something has been omitted. Beneath the sentence are five lettered words or sets of words. Choose the word or set of words that BEST fits the meaning of the sentence as a whole.

EXAMPLE

Although critics found the book _______ , many readers found it rather _______ .

(A) obnoxious . . . perfect

(B) spectacular . . . interesting

(C) boring . . . intriguing

(D) comical . . . persuasive

(E) popular . . . rare

EXAMPLE ANSWER

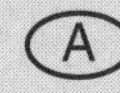

1. The old man was so _______ that even his grandchildren were frightened of him.

   (A) genial
   (B) loquacious
   (C) dyspeptic
   (D) somnolent
   (E) innocuous

2. The man's sense of humor and quick wit were obvious in his ability to come up with a(n) _______ every time a comment or question was directed toward him.

   (A) calumny
   (B) rejoinder
   (C) prevarication
   (D) expletive
   (E) remonstrance

3. It was difficult not to believe even his most blatant lies because of his _______ .

   (A) verisimilitude
   (B) ambience
   (C) ignominy
   (D) petulance
   (E) inveracity

4. The sage counseled his followers that hate _______ ill will; likewise, kindness _______ benevolence.

   (A) provokes . . . impairs
   (B) precedes . . . supersedes
   (C) expunges . . . induces
   (D) adulterates . . . temporizes
   (E) engenders . . . begets

5. Because the perpetrator was _______ , the witnesses were not able to _______ him.

   (A) inconsequential . . . inculcate
   (B) conspicuous . . . incriminate
   (C) pensive . . . vindicate
   (D) nondescript . . .inculpate
   (E) urbane . . . identify

GO ON TO THE NEXT PAGE

DIRECTIONS: Each passage below is followed by questions based on its content. Answer the questions on the basis of what is stated or implied in each passage and in any introductory material that may be provided.

Questions 6–9 are based on the following passages.

Passage 1

1 It's been said that the amount of trash a family sets out weekly for the garbage collector is an indicator of how rich they are. The logic behind this conjecture is that the poor must be
5 as resourceful as possible, whereas the wealthy can afford to throw away what they don't like or need—or perhaps they are just tired of it or want something better or newer. The more trash bags we see, the richer the residents, we assume—
10 and, more often than not, we're correct.

Passage 2

1 We make assumptions about a family's socio-economic status by the condition of their dwelling. A nicely landscaped lawn in front of a three-story brick mansion obviously contrasts
5 with a tiny unpainted shack surrounded by junk cars, unmowed grass and weeds, and debris strewn about. In the judgmental myopia that is so typical of us Americans, we need little to convince us of the veracity of our presupposi-
10 tion as to who has money and who doesn't. It's simple: the more trash we see, the poorer the dwellers are.

6. Both passages support which of the following?

(A) We tend to judge others by the way they look.

(B) "Waste not, want not" is frequently true.

(C) People's assumptions are often based on their own backgrounds.

(D) We tend to make assumptions about the socio-economic status of people based on the amount of trash they have.

(E) Rich people take more pride in their property than poor people do.

7. The author of Passage 1 seems to regard wealthy people as

(A) wasteful.

(B) resourceful.

(C) snobby.

(D) show-offs.

(E) neat.

8. In line 9 of Passage 2, "veracity" most nearly means

(A) short-sighted viewpoint.

(B) narrow-mindedness.

(C) bias.

(D) inaccuracy.

(E) truthfulness.

GO ON TO THE NEXT PAGE

9. Which of the following reflects the differences between the viewpoints of the two authors?

(A) The author of Passage 1 is critical of rich people, while the author of Passage 2 is critical of the poor.

(B) The author of Passage 1 is critical of poor people, while the author of Passage 2 is critical of the wealthy.

(C) The author of Passage 1 believes that trash outside a house indicates wealth, while the author of Passage 2 believes that trash outside a house means poverty.

(D) The author of Passage 1 believes that trash outside a house indicates poverty, while the author of Passage 2 believes that trash outside a house means wealth.

(E) The author of Passage 1 believes Americans are too judgmental, while the author of Passage 2 believes Americans' judgments are inaccurate.

Questions 10–15 are based on the following passage.

The following passage is an excerpt from Booth Tarkington's novel Alice Adams.

. . . he did sleep intermittently, drowsed between times, and even dreamed; but, forgetting his dreams before he opened his eyes, and having some part of him all the while aware of his discomfort, he believed, as usual, that he lay awake the whole night long. He was conscious of the city as of some single great creature resting fitfully in the dark outside his windows. It lay all round about, in the damp cover of its night cloud of smoke, and tried to keep quiet for a few hours after midnight, but was too powerful a growing thing ever to lie altogether still. Even while it strove to sleep it muttered with digestions of the day before, and these already merged with rumblings of the morrow. "Owl" cars, bringing in last passengers over distant trolley-lines, now and then howled on a curve; faraway metallic stirrings could be heard from factories in the sooty suburbs on the plain outside the city; east, west, and south, switch-engines chugged and snorted on sidings; and everywhere in the air there seemed to be a faint, voluminous hum as of innumerable wires trembling overhead to vibration of machinery underground.

In his youth Adams might have been less resentful of sounds such as these when they interfered with his night's sleep: even during an illness he might have taken some pride in them as proof of his citizenship in a "live town"; but at fifty-five he merely hated them because they kept him awake. They "pressed on his nerves," as he put it; and so did almost everything else, for that matter.

GO ON TO THE NEXT PAGE

35 He heard the milk-wagon drive into the cross-street beneath his windows and stop at each house. The milkman carried his jars round to the "back porch," while the horse moved slowly ahead to the gate of the next customer and 40 waited there. "He's gone into Pollocks,'" Adams thought, following this progress. "I hope it'll sour on 'em before breakfast. Delivered the Andersons.' Now he's getting out ours. Listen to the darn brute! What's he care who wants 45 to sleep!" His complaint was of the horse, who casually shifted weight with a clink of steel shoes on the worn brick pavement of the street, and then heartily shook himself in his harness, perhaps to dislodge a fly far ahead of 50 its season. Light had just filmed the windows; and with that the first sparrow woke, chirped instantly, and roused neighbours in the trees of the small yard, including a loud-voiced robin. Vociferations began irregularly, but were soon 55 unanimous.

"Sleep? Dang likely now, ain't it!"

Night sounds were becoming day sounds; the far-away hooting of freight-engines seemed brisker than an hour ago in the dark. A cheerful 60 whistler passed the house, even more careless of sleepers than the milkman's horse had been; then a group of coloured workmen came by, and although it was impossible to be sure whether they were homeward bound from 65 night-work or on their way to day-work, at least it was certain that they were jocose. Loose, aboriginal laughter preceded them afar, and beat on the air long after they had gone by. . . .

In spite of noises without, he drowsed again, 70 not knowing that he did; and when he opened his eyes the nurse was just rising from her cot. He took no pleasure in the sight, it may be said. She exhibited to him a face mismodelled by sleep, and set like a clay face left on its cheek in 75 a hot and dry studio. She was still only in part awake, however, and by the time she had extinguished the night-light and given her patient his tonic, she had recovered enough plasticity. "Well, isn't that grand! We've had another good night," she said as she departed to dress 80 in the bathroom.

"Yes, you had another!" he retorted, though not until after she had closed the door.

10. Throughout this passage, Adams could best be described as

(A) irascible.
(B) sick.
(C) complacent.
(D) tired.
(E) lonely.

11. Which of the following is true about the way Adams spent the night?

(A) He did not sleep at all.
(B) He rested well but dreamed throughout the night.
(C) He slept part of the night and lay awake part of the night.
(D) He did not go to bed.
(E) He remained active throughout the night.

12. Adams clearly perceived the city at night as

(A) dead.
(B) alive.
(C) boring.
(D) exciting.
(E) tantalizing.

GO ON TO THE NEXT PAGE

13. Which of the following describes the difference between Adams when he was younger and the way he was at the time depicted in this passage?

    (A) He used to be in excellent health, but at this point he was very ill.
    (B) He used to work at night, but at this point he is bed-ridden.
    (C) He used to like the night sounds, but at this point he hated them.
    (D) He used to find his nurse attractive, but at this point he thought she was ugly.
    (E) He used to sleep soundly all night long, but at this point he scarcely slept at all.

14. In line 54, "vociferations" most nearly means

    (A) hums.
    (B) chirps.
    (C) whistles.
    (D) cries.
    (E) buzzes.

15. We can assume that the setting of this passage is

    (A) a hotel.
    (B) a nursing home.
    (C) a hospital.
    (D) a house.
    (E) an apartment building.

Questions 16–24 are based on the following passage.

The following is an excerpt from Willa Cather's novel One of Ours.

Claude backed the little Ford car out of its shed, ran it up to the horse-tank, and began to throw water on the mud-crusted wheels and windshield. . . . The two hired men, Dan and Jerry, came shambling down the hill to feed the stock. Jerry was grumbling and swearing about something, but Claude wrung out his wet rags and, beyond a nod, paid no attention to them. . . . Claude had a grievance against Jerry just now, because of his treatment of one of the horses.

Molly was a faithful old mare, the mother of many colts; Claude and his younger brother had learned to ride on her. This man Jerry, taking her out to work one morning, let her step on a board with a nail sticking up in it. He pulled the nail out of her foot, said nothing to anybody, and drove her to the cultivator all day. Now she had been standing in her stall for weeks, patiently suffering, her body wretchedly thin, and her leg swollen until it looked like an elephant's. She would have to stand there, the veterinary said, until her hoof came off and she grew a new one, and she would always be stiff. Jerry had not been discharged, and he exhibited the poor animal as if she were a credit to him.

Mahailey came out on the hilltop and rang the breakfast bell. After the hired men went up to the house, Claude slipped into the barn to see that Molly had got her share of oats. She was eating quietly, her head hanging, and her scaly, dead-looking foot lifted just a little from the ground. When he stroked her neck and talked to her she stopped grinding and gazed at him mournfully. She knew him, and wrinkled her nose and drew her upper lip back from her worn teeth, to show that she liked being petted. She let him touch her foot and examine her leg.

GO ON TO THE NEXT PAGE

When Claude reached the kitchen, his mother was sitting at one end of the breakfast table, pouring weak coffee, his brother and Dan and Jerry were in their chairs, and Mahailey was baking griddle cakes at the stove. A moment later Mr. Wheeler came down the enclosed stairway and walked the length of the table to his own place. . . .

As soon as he was seated, . . . Ralph asked him if he were going to the circus. Mr. Wheeler winked.

"I shouldn't wonder if I happened in town sometime before the elephants get away." . . . his voice was smooth and agreeable. "You boys better start in early, though. You can take the wagon and the mules, and load in the cowhides. The butcher has agreed to take them."

Claude put down his knife. "Can't we have the car? I've washed it on purpose."

"And what about Dan and Jerry? They want to see the circus just as much as you do, and I want the hides should go in; they're bringing a good price now. I don't mind about your washing the car; mud preserves the paint, they say, but it'll be all right this time, Claude."

The hired men haw-hawed and Ralph giggled. Claude's freckled face got very red. The pancake grew stiff and heavy in his mouth and was hard to swallow. His father knew he hated to drive the mules to town, and knew how he hated to go anywhere with Dan and Jerry. As for the hides, they were the skins of four steers that had perished in the blizzard last winter through the wanton carelessness of these same hired men, and the price they would bring would not half pay for the time his father had spent in stripping and curing them. They had lain in a shed loft all summer, and the wagon had been to town a dozen times. But today, when he wanted to go to Frankfort clean and care-free, he must take these stinking hides and two coarse-mouthed men, and drive a pair of mules that always brayed and balked and behaved ridiculously in a crowd. Probably his father had looked out of the window and seen him washing the car, and had put this up on him while he dressed. It was like his father's idea of a joke.

16. In line 5, what does the word "shambling" imply about the men, especially after you have read the rest of the passage?

(A) They were eager to start working.
(B) They were good friends.
(C) They were not well.
(D) They were lazy and unmotivated.
(E) They were disabled.

17. The mare's injury exemplifies

(A) the hard life of farm animals.
(B) the condition of the soil on the Wheelers' farm.
(C) the cruelty of Jerry, the hired hand.
(D) the poor health of the mare.
(E) the relationship between Claude and his father.

18. In line 19, the word "wretchedly" most nearly means

(A) poorly.
(B) angrily.
(C) ordinarily.
(D) fitfully.
(E) miserably.

GO ON TO THE NEXT PAGE

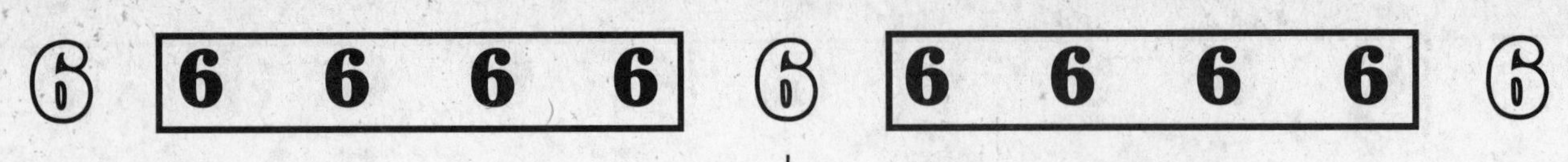

19. Which of the following best describes Mr. Wheeler's reaction to Claude washing the car?

(A) supportive
(B) oblivious
(C) appreciative
(D) mocking
(E) sympathetic

20. In line 63–64, the description of the pancake as "stiff and heavy in his mouth and was hard to swallow" helps reveal

(A) what a bad cook Mahailey was.
(B) how Claude felt.
(C) how poor Claude's family was.
(D) how long it took Claude to eat breakfast.
(E) how late the family had begun eating breakfast.

21. In line 69, the word "wanton" most nearly means

(A) lazy; lackadaisical.
(B) passionate; excited.
(C) merciless; inhumane.
(D) immature; youthful.
(E) restrained; bridled.

22. Which of the following is <u>not</u> something Claude dreaded about the upcoming trip to town?

(A) taking his brother with him
(B) going with the hired hands
(C) driving the mules into town
(D) loading the cowhides
(E) going in the wagon

23. This passage implies which of the following conflicts?

(A) the Wheelers versus their hired hands
(B) the Wheelers versus their farm animals
(C) Claude and his brother versus the rest of the household
(D) Claude versus the hired hands and his family
(E) the Wheelers versus the hired men and Mahailey

24. Which of the following lists the things Claude disliked about the hired men?

(A) their language, clothing, and mistreatment of animals
(B) their language, drinking habits, and attitude
(C) their clothing, drinking habits, and mistreatment of animals
(D) their language, mistreatment of animals, and attitude
(E) their drinking habits, language, and mistreatment of animals

STOP

If time remains, you may go back and check your work. When the time allotted is up, you may go on to the next section.

TIME: 20 Minutes
16 Questions

MATH

DIRECTIONS: In this section solve each problem, using any available space on the page for scratchwork. Then decide which is the best of the choices given and fill in the corresponding oval on the answer sheet.

NOTES

(1) The use of a calculator is permitted.

(2) All numbers used are real numbers.

(3) Figures that accompany problems in this test are intended to provide information useful in solving the problems. They are drawn as accurately as possible EXCEPT when it is stated in a specific problem that the figure is not drawn to scale. All figures lie in a plane unless otherwise indicated.

(4) Unless otherwise specified, the domain of any function f is assumed to be the set of all real numbers x for which $f(x)$ is a real number.

REFERENCE INFORMATION

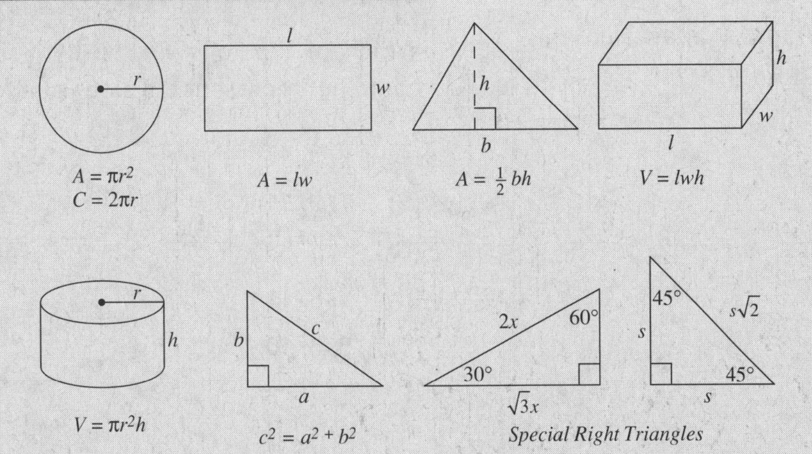

The number of degrees of arc in a circle is 360.

The sum of the measures in degrees of the angles of a triangle is 180.

1. If the chairs in a certain room were arranged in six rows, each row would have an equal number of chairs. However, if the chairs were arranged in five rows, each row could have an equal number of chairs only if two chairs were removed. Which of the following could represent the number of chairs in this room?

(A) 90
(B) 72
(C) 60
(D) 52
(E) 36

If it rains, then Peter will not play golf.

2. Given that the statement above is true, which of the following statements <u>cannot</u> be true?

(A) It is not raining and Peter is not playing golf.
(B) It is not raining or Peter is playing golf.
(C) It is not raining and Peter is bowling.
(D) It is raining and Peter is playing golf.
(E) It is raining and Peter is bowling.

3. $R = \{3, 5, 7\}$ and $S = \{-5, -3, -2, -1\}$. If x is an element of set R and y is an element of set S, which of the following is <u>not</u> a possible value of xy?

(A) –7
(B) –10
(C) –14
(D) –15
(E) –20

4. An article is bought for p dollars and is sold for t dollars. What is the percent profit?

(A) $\frac{100(t-p)}{p}$
(B) $\frac{100(p-t)}{t}$
(C) $100(p-t)$
(D) $\frac{100(t-p)}{t}$
(E) $\frac{(t-p)}{100}$

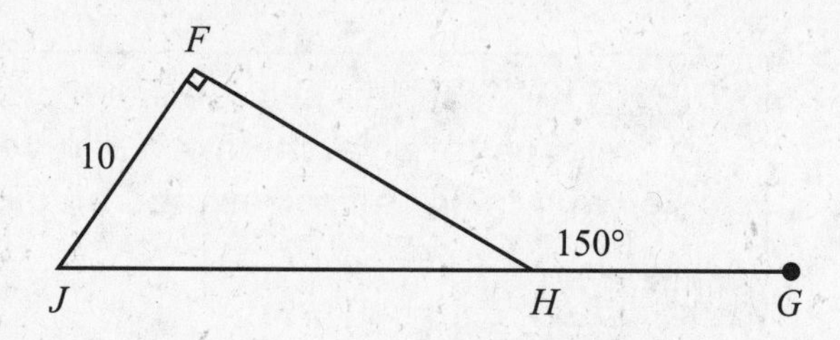

<u>Note</u>: Figure is not drawn to scale.

5. What is the perimeter of ΔFHJ?

(A) $20+\sqrt{2}$
(B) $30+\sqrt{3}$
(C) $20+10\sqrt{2}$
(D) $30+10\sqrt{3}$
(E) $50\sqrt{3}$

6. Which of the following sets of numbers could <u>not</u> represent the sides of a triangle?

(A) 2, 3, 4
(B) 7, 9, 11
(C) 1, 3, 5
(D) 6, 8, 11
(E) 5, 6, 10

GO ON TO THE NEXT PAGE

7. If $f(x) = \frac{2}{(x+1)}$, then which of the following represents $f\left(\frac{1}{[x+2]}\right)$?

(A) $\frac{2}{(x+2)(x+1)}$

(B) $\frac{2(x+2)}{(x+3)}$

(C) $\frac{2}{(x+2)(x+3)}$

(D) $\frac{2(x+1)}{(x+2)}$

(E) $\frac{2}{(x+1)(x+3)}$

8. If $2^n = x$ and $3 = y$, which of the following is the correct expression for 6^{n-1}?

(A) $\frac{xy}{6}$

(B) $xy - 1$

(C) $xy - 6$

(D) $6xy$

(E) $-6xy$

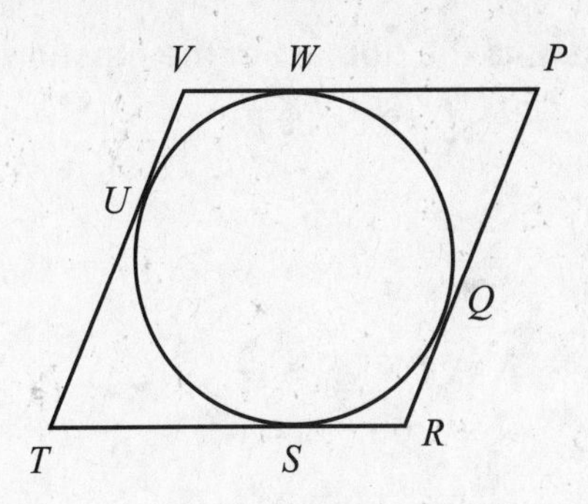

Note: Figure is not drawn to scale.

9. In the figure above, $\overline{PR}$, $\overline{RT}$, $\overline{TV}$, and $\overline{VP}$ are tangents to the circle at points Q, S, U, and W, respectively. The perimeter of $PRTV$ is 28 and $PW = 4$. Which of the following must be true?

(A) $PR = RT$

(B) $VW = 3$

(C) $PR + TV = 14$

(D) $m\angle P = m\angle T$

(E) $m\angle V + m\angle T = 180°$

10. Working alone, Mrs. Jones can paint 5 rooms in 12 hours. When Mrs. Jones and Mr. Smith work together, they can paint these rooms in 9 hours. How long would Mr. Smith require to paint these rooms if he worked alone?

(A) 45 hours

(B) 36 hours

(C) 28.5 hours

(D) 21 hours

(E) 10.5 hours

GO ON TO THE NEXT PAGE

11. What is the domain of the function $f(x) = \sqrt{x^2 - 9x + 20}$?

(A) $-5 \leq x \leq -4$

(B) $4 \leq x \leq 5$

(C) $x \geq -4$ or $x \leq -5$

(D) $x \geq 5$ or $x \leq 4$

(E) all real numbers

Student-Produced Response Questions

DIRECTIONS: Student-Produced Response questions 12–16 require that you solve the problem and then enter your answer on the special grid on page 855. (For instructions on how to use the grid, see page 69 in the Diagnostic Test.)

12. A 10-gallon jar of acid and water is 70% acid. How many gallons of pure water must be added to reduce it to a 20% acid solution?

13. Three-fifths of the people in the town of Anyville are Republicans. Of the remaining individuals, one-third are retired. What fraction of the people in Anyville are both retired and not Republicans?

Age Distribution of University Faculty Members

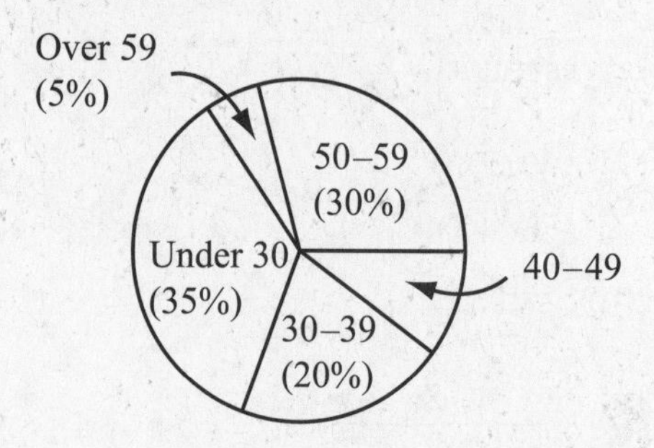

14. If there are 88 faculty members between the ages of 30 and 39, what is the combined total of faculty members who are at least 40 years old?

15. A jar contains blue, red, and green marbles. There are 36 green marbles and the probability of randomly selecting a green marble is 8%. If the probability of randomly selecting a blue marble is 70%, how many red marbles are in the jar?

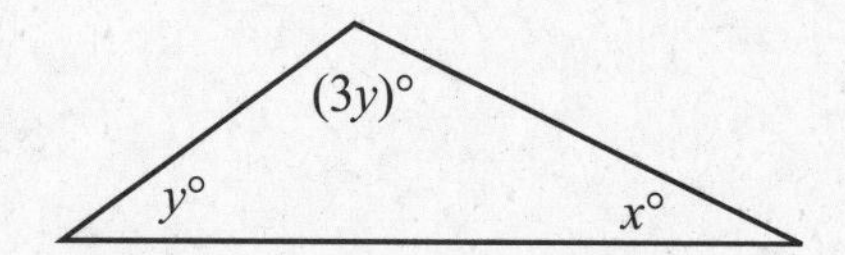

Note: Figure not drawn to scale.

16. In the figure above, x and y are integers and $18 < x < 30$. What is a possible value for y?

STOP

If time remains, you may go back and check your work. When the time allotted is up, you may go on to the next section.

TIME: 20 Minutes
19 Questions

CRITICAL READING

DIRECTIONS: Each sentence below has one or two blanks, each blank indicating that something has been omitted. Beneath the sentence are five lettered words or sets of words. Choose the word or set of words that BEST fits the meaning of the sentence as a whole.

EXAMPLE

Although critics found the book _______ , many readers found it rather _______ .

(A) obnoxious . . . perfect
(B) spectacular . . . interesting
(C) boring . . . intriguing
(D) comical . . . persuasive
(E) popular . . . rare

EXAMPLE ANSWER

1. The __________ expression on his girlfriend's face told him immediately that he would never feel comfortable with his future in-laws.

   (A) jocular
   (B) composed
   (C) hospitable
   (D) compassionate
   (E) supercilious

2. For her, popularity was __________ to success, so she constantly worried about whether or not others liked her and approved of her.

   (A) paramount
   (B) tantamount
   (C) dissimilar
   (D) derogatory
   (E) inconsequential

3. My father has always been __________, so it does little good to try to get him to plan in advance or to see any viewpoint other than his own.

   (A) cosmopolitan
   (B) myopic
   (C) eccentric
   (D) bellicose
   (E) progressive

4. It didn't take long for the panel of experts to regard the newest member's comments as __________; although he certainly had the vocabulary and knew the lingo, he simply didn't have the experience or the education to function at their level.

   (A) sophomoric
   (B) irresolute
   (C) oblivious
   (D) sagacious
   (E) perceptive

5. While shopping in an antique store, she had the good fortune to find a(n) __________ vase that was in unusually __________ condition.

   (A) contemporary . . . meretricious
   (B) exquisite . . . mediocre
   (C) antediluvian . . . .pristine
   (D) illicit . . . ignoble
   (E) Noachian . . . shoddy

6. One of the administrators believed that students could learn better in a(n) __________ group with common abilities and interests; however, the other one contended that more learning occurs when students in a(n) __________ group contribute their unique abilities and share knowledge about their individual interests.

   (A) gregarious . . . asocial
   (B) homogeneous . . . hetrogeneous
   (C) incongruous . . . divisive
   (D) perspicacious . . . moronic
   (E) exclusionary . . . restricted

GO ON TO THE NEXT PAGE

DIRECTIONS: The two passages below are followed by questions based on their content and on the relationship between the two passages. Answer the questions on the basis of what is stated or implied in each passage and in any introductory material that may be provided.

Questions 7–19 are based on the following passages.

Passage 1

This passage is an excerpt of an essay that describes the typical one-room schoolhouse in nineteenth-century America.

1 The rural schoolhouse in nineteenth-century America had limited financial resources. The agrarian lifestyle, with its meager income and demanding time requirements, dictated that 5 the school term last only a few months each year. Little of the farmers' profits could be spared to pay for the teacher's salary or for school supplies. Trying to hold classes during planting or harvest time would have been 10 futile, since the all family members—including the children—were expected to help.

The one-school room was spartan, with homemade desks and chairs. Few decorations adorned the walls. The teacher's tools in the 15 best-equipped classroom included a slate, chalk, and books. The curriculum was limited to the reading, writing, and arithmetic—"the three R's." The teacher evaluated the students by means of daily drills, oral quizzes, and recitation.

20 Establishing and maintaining the school was a community project; thus, the community members strived to provide what they could to support the school. Farmers provided the wood for the stove. Parents crafted the desks and chairs. 25 The teacher often had no permanent residence but rotated from one family to the next throughout the year.

As a public servant, the teacher was expected to meet certain expectations and fulfill specific obligations. For example, it was up to him or 30 her to arrive at the school very early in the morning in order to make sure the wood fire was burning and the schoolhouse was warm by the time the children arrived. The teacher was to haul the water in from the pump. Since the 35 pencil sharpener was unheard-of, the teacher whittled the pencils so that the students could use them. And during the school day, since all the students were in one room, the teacher worked with one age or ability group at a time, 40 making certain that all students were working and not misbehaving.

A teacher's life was not his or her own, even outside of school. The male teacher who attended church regularly was allowed to devote 45 two evenings for "courting"; if he was not a church-goer, he could use only one evening for such purposes. The lady teacher was expected to spend her hours away from school reading the Bible. She was expected to be unmarried 50 and chaste; if she did marry or engaged in other unacceptable behavior, she would more than likely lose her job. Regardless of gender, the teacher who smoked, drank liquor, or visited the pool hall or tavern—and the male teacher 55 who got a shave in a barber shop—was considered to be a poor role model and was in danger of losing his or her position as a teacher.

Students were to treat their teacher with respect. Walking into the schoolroom, the stu- 60 dents were expected to take off their hats, bow to the teacher, and walk straight to their seats. Students were forbidden to communicate with one another during class. They were to focus totally on their school work. Common punish- 65 ments for infractions of these rules included a rap on the knuckles with a ruler, sitting in front of the class while wearing a dunce cap, standing in the corner, and being swatted with a wooden paddle. And, in most cases, the 70 student who was punished at school was punished at home as well.

GO ON TO THE NEXT PAGE

Passage 2

This passage below is an excerpt from a description of and comments about today's American schools.

1 American schools today are becoming more and more impersonal. In many districts, education has become "big business," with multimillion dollar budgets, contracted janitorial and 5 food services, and strict employment policies. Little if any consideration is given to family vacations and other activities when the school calendar—sometimes twelve months long—is planned. For efficiency's sake, even a small 10 community may have several school buildings, separating children by age and grade level. In larger cities, a building or complex, geographically separated from all other such buildings in the district, may serve children of only one 15 grade level.

Schools are required to provide adequate equipment to educate every child. Such equipment includes up-to-date computers and computer programs, textbooks, encyclopedias, calcula- 20 tors, and televisions. Advanced technology has the potential of decreasing—and perhaps someday eliminating—personal teacher-student interactions.

Although community members are required to 25 support the schools by means of paying taxes, many do little else. All too often, even those who have children in attendance never set foot in the school to meet the teachers or participate in school activities. Unfortunately, more 30 common are incidents of parents' supporting their children for breaking the rules or failing to cooperate with their teachers. Corporal punishment is almost nonexistent in schools, and disciplinary action is quite limited. In many dis- 35 tricts, administrators spend hours developing a disciplinary policy, consulting lawyers to make sure it is legal, and hoping they have covered all possibly scenarios. But invariably students and parents will find a loophole and ques- 40 tion—or even file a lawsuit against—the school that tries to enforce that policy strictly. Because educators are public servants, too frequently they must compromise their professional standards to pacify the public.

Today's teachers face few restrictions regard- 45 ing their personal lives. It is illegal not to hire a teacher—or to dismiss a teacher—on the basis of his or her marital status, pregnancy, religious preference or participation, recreational activities, or almost any other reason except 50 breaking the law or having a criminal history.

7. Passage 1 implies that which of the following was most important to the nineteenth-century family?

(A) education
(B) crops
(C) money
(D) church
(E) recreation

8. In line 3, "agrarian" most nearly means

(A) urban.
(B) simple.
(C) agricultural.
(D) complex.
(E) unpredictable.

GO ON TO THE NEXT PAGE

9. In Passage 1, smoking and drinking are examples of

(A) behaviors that were acceptable for community members but not for teachers.
(B) behaviors that were acceptable for male teachers but not for female teachers.
(C) the freedoms that teachers and other community members were able to enjoy in a small town.
(D) activities in which teachers could engage if they were regular churchgoers.
(E) behaviors that were considered unacceptable for both male and female teachers.

10. Passage 2 implies that which of the following is the most important to school districts?

(A) the individual student
(B) the teacher's morals and activities
(C) business-like policies and structure
(D) the students' families
(E) the teachers' interests

11. Which of the following seems to reflect the conflict over disciplinary actions in today's schools?

(A) parents versus teachers
(B) parents versus students
(C) teachers versus administrators
(D) teachers versus teachers
(E) school versus church

12. In line 29 of Passage 2, the use of "unfortunately" suggests that the author

(A) does not approve of the disciplinary policies of today's schools.
(B) does not agree with parents who support their children's misbehavior.
(C) prefers the way schools were run in the past.
(D) supports corporal punishment.
(E) does not agree with administrators' typical approach to discipline.

13. In line 38 of Passage 2, "invariably" most nearly means

(A) annoyingly.
(B) eagerly.
(C) angrily.
(D) litigiously.
(E) consistently.

14. Today, a school can dismiss or refuse to hire a teacher because he or she

(A) does not go to church.
(B) is married.
(C) has a criminal record.
(D) is expecting a baby.
(E) is involved in political activities.

15. Which of the following was more abundant in nineteenth-century schools than in those of today?

(A) janitorial and cooking services
(B) school equipment and supplies
(C) financial backing
(D) community support
(E) number of school days per year

GO ON TO THE NEXT PAGE

16. Compared to today's situation as described in Passage 2, the nineteenth-century family described in Passage 1 was

    (A) less actively involved in school.
    (B) more actively involved in school.
    (C) more critical of the school's policies.
    (D) more legally obligated to support schools.
    (E) more educated.

17. Which of the following reflects the similarity between nineteenth-century teachers and those of today?

    (A) society's view of them as public servants
    (B) society's support of their disciplinary actions
    (C) freedom in their personal lives
    (D) availability of supplies
    (E) living arrangements

18. The modern society described in Passage 2 would most likely regard the nineteenth-century teacher described in Passage 1 as

    (A) too lenient.
    (B) boring.
    (C) uneducated.
    (D) too strict.
    (E) too poor.

19. Which of the following is probably true regarding students' behavior?

    (A) Students today are just as respectful of teachers as students of the nineteenth century were.
    (B) Students today are more respectful of teachers than students of the nineteenth century were.
    (C) Students today are less respectful of teachers than students of the nineteenth century were.
    (D) Students today are more intelligent than students of the nineteenth century were.
    (E) Students today are less intelligent than students of the nineteenth century were.

STOP

If time remains, you may go back and check your work. When the time allotted is up, you may go on to the next section.

TIME: 10 Minutes
14 Questions

WRITING

DIRECTIONS: In each of the following sentences, some portion of the sentence is underlined. Under each sentence are five choices. The first choice has the same wording as the original. The other four choices are reworded. Sometimes the first choice containing the original wording is the best; sometimes one of the other choices is the best. Choose the letter of the best choice. Your choice should produce a sentence that is not ambiguous or awkward and that is correct, clear, and precise.

This is a test of correct and effective English expression. Keep in mind the standards of English usage, punctuation, grammar, word choice, and construction.

EXAMPLE

When you listen to opera, a person may not appreciate it.

(A) a person may not appreciate it.

(B) it may not be appreciated by a person.

(C) you may not appreciate it.

(D) which may not be appreciated by you.

(E) appreciating it may be a problem for you.

EXAMPLE ANSWER

1. William Shakespeare was a prolific playwright of the sixteenth century, and he <u>has also written</u> many poems.

   (A) has also written
   (B) also wrote
   (C) has also been writing
   (D) had also written
   (E) was also writing

2. All of his children are either good <u>hunters or athletic</u>.

   (A) hunters or athletic
   (B) at hunting or athletic
   (C) hunters or athletes
   (D) hunters or good at athletics
   (E) at hunting or athletes

3. His third-grade teacher's classroom was similar <u>to his second-grade teacher</u>.

   (A) to his second-grade teacher.
   (B) to that of his second-grade teacher.
   (C) with his second-grade teacher.
   (D) with that of his second-grade teacher.
   (E) that of his second-grade teacher.

4. As we talked, I was unable to agree <u>to</u> him because his ideas were so irrational.

   (A) to
   (B) upon
   (C) with
   (D) on
   (E) about

5. An effective secretary is able to answer the phone courteously, use the computer proficiently, greet clients <u>cordially, and she also knows how to file accurately, too</u>.

   (A) cordially, and she also knows how to file accurately, too.
   (B) cordially, knowing how to file accurately as well.
   (C) and file accurately.
   (D) and file with accuracy as well.
   (E) and how to file accurately.

6. <u>A person cannot</u> get much work done while you are watching television.

   (A) A person cannot
   (B) One cannot
   (C) You could not
   (D) One could not
   (E) You cannot

7. <u>He wants to travel not only to Paris but also to London</u>.

   (A) He wants to travel not only to Paris but also to London.
   (B) He not only wants to travel to Paris but also to London.
   (C) He wants to travel not only to Paris but also to visit London.
   (D) Not only does he want to travel to Paris but to London also.
   (E) He wants not only to travel to Paris but also to London.

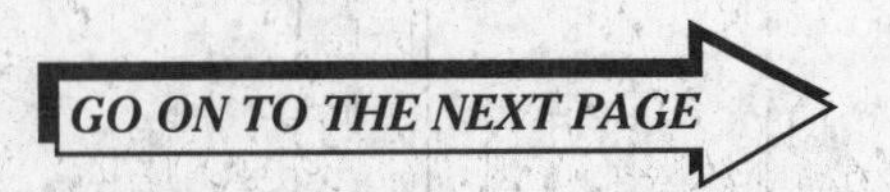

8. Two people in the group wants to discuss the issue, the others prefer to avoid controversy.

(A) wants to discuss the issue, the others
(B) want to discuss the issue; the others
(C) wants to discuss the issue the others
(D) wanting to discuss the issue; the others
(E) want to discus the issue however,

9. The inability to read, it interferes with almost everything we do each day.

(A) read, it interferes with almost everything we do
(B) read, it interferes upon almost everything we do
(C) read interferes upon almost everything we do
(D) read, interfering with almost everything we do
(E) read interferes with almost everything we do

10. Snows in Colorado are often heavy, so they have capitalized on the climate by building elaborate ski resorts.

(A) heavy, so they have
(B) heavy, so the people there have
(C) heavy and have
(D) heavy, and they have
(E) heavy, having

11. Being lazier and less motivated compared with his wife, he depends on her to make enough money to pay the monthly bills.

(A) Being lazier and less motivated compared with his wife, he
(B) Being lazier and less motivated compared to
(C) By being lazier and less motivated than
(D) Being lazier and less motivated than
(E) Lazier as well as less motivated, unlike

12. A fable, illustrating a specific point or moral, often involves supernatural occurrences or animals that act or talk like humans.

(A) A fable, illustrating a specific point or moral, often involves
(B) A fable illustrating a specific point or moral, often involves
(C) A fable, illustrating a specific point or moral, it often involves
(D) Because a fable illustrates a specific point or moral, it often involves
(E) A fable, illustrating a specific point or moral, often involving

GO ON TO THE NEXT PAGE

13. Planaria are unusual animals because they are able to reproduce by dividing in two and regenerating the missing parts their reproduction is considered asexual.

    (A) animals because they are able to reproduce by dividing in two and regenerating the missing parts their reproduction
    (B) animals, because they are able to reproduce by dividing in two and regenerating the missing parts their reproduction
    (C) animals; because they are able to reproduce by dividing in two and regenerating the missing parts, their reproduction
    (D) animals since they are able to reproduce by dividing in two and regenerating the missing parts their reproduction
    (E) animals; because they are able to reproduce by dividing in two and regenerating the missing parts, it

14. Them regarding the young woman as simply a beauty queen caused her to become angry and quitting her job.

    (A) Them regarding the young woman as simply a beauty queen caused her to become angry and quitting her job.
    (B) Their regarding the young woman as simply a beauty queen, causing her to become angry and quit her job.
    (C) They regarded the young woman as simply a beauty queen caused her to become angry, quitting her job.
    (D) Them regarding the young woman, as simply a beauty queen, caused her to become angry, quitting her job.
    (E) Their regarding the young woman as simply a beauty queen caused her to become angry and quit her job.

STOP

If time remains, you may go back and check your work. When the time allotted is up, you may go on to the next section.

TEST 2

ANSWER KEY

Section 1—Essay

Refer to the Detailed Explanation for essay analysis

Section 2—Math

1. (A)
2. (C)
3. (B)
4. (B)
5. (C)
6. (D)
7. (B)
8. (E)
9. (E)
10. (B)
11. (C)
12. (E)
13. (E)
14. (D)
15. (D)
16. (B)
17. (C)
18. (A)
19. (B)
20. (D)

Section 3—Critical Reading

1. (A)
2. (C)
3. (D)
4. (E)
5. (B)
6. (A)
7. (C)
8. (B)
9. (B)
10. (A)
11. (D)
12. (C)
13. (C)
14. (A)
15. (C)
16. (C)
17. (D)
18. (B)
19. (E)
20. (E)
21. (B)
22. (D)
23. (E)
24. (D)

Section 4—Writing

1. (B)
2. (C)
3. (C)
4. (A)
5. (C)
6. (E)
7. (D)
8. (E)
9. (B)
10. (D)
11. (B)
12. (C)
13. (B)
14. (C)
15. (C)
16. (B)
17. (E)
18. (D)
19. (C)
20. (B)
21. (E)
22. (A)
23. (A)
24. (D)
25. (B)
26. (C)
27. (D)
28. (A)
29. (A)
30. (B)
31. (B)
32. (B)
33. (E)
34. (C)
35. (A)

Section 5—Math

1. (C)
2. (D)
3. (B)
4. (A)
5. (E)
6. (D)
7. (C)
8. (D)
9. (S)
10. (A)
11. (D)
12. (E)
13. (E)
14. (D)
15. (D)
16. (E)
17. (C)
18. (B)

Section 6—Critical Reading

1. (C)
2. (B)
3. (A)
4. (E)
5. (D)
6. (D)
7. (A)
8. (E)
9. (C)
10. (A)
11. (C)
12. (B)
13. (C)
14. (D)
15. (D)
16. (D)
17. (C)
18. (E)
19. (D)
20. (B)
21. (C)
22. (A)
23. (D)
24. (D)

Section 7—Math

1. (B)
2. (D)
3. (E)
4. (A)
5. (D)
6. (C)
7. (B)
8. (A)
9. (C)
10. (B)
11. (D)
12. 25
13. 2/15
14. 198
15. 99
16. 38, 39, or 40

Section 8—Critical Reading

1. (E)
2. (B)
3. (B)
4. (A)
5. (C)
6. (B)
7. (B)
8. (C)
9. (E)
10. (C)
11. (A)
12. (B)
13. (E)
14. (C)
15. (D)
16. (B)
17. (A)
18. (D)
19. (C)

Section 9—Writing

1. (B)
2. (C)
3. (B)
4. (C)
5. (C)
6. (E)
7. (A)
8. (B)
9. (E)
10. (B)
11. (D)
12. (A)
13. (C)
14. (E)

DETAILED EXPLANATIONS

SECTION 1—ESSAY

Sample Essay With Commentary

ESSAY I (Score: 5–6)

Everyone is thankful for something. Sometimes our gratitude is elicited by something as simple as a smile or a hug; other times we do not feel grateful until we receive something that is very expensive or rare. Different people are grateful for different things. However, careful analysis reveals that those of us who are thankful for the least are the ones who enjoy the most.

I have a friend who lives in a palatial home, with a swimming pool and a tennis court in the backyard. She has more clothes than any person could ever need, and she has taken trips to places I can only imagine. However, she always seems dissatisfied with life and with what life has brought her. I'm not sure she is truly thankful for anything she has, and she never seems to enjoy any of the luxuries that are so much a part of her life.

My lifestyle is so different from hers! Although my mother works very hard, she struggles to pay the bills each month. When she is able to buy me something—even if it is something as small as my favorite candy bar—I am grateful because I know it requires some kind of sacrifice on her part. We do not have much, but we have each other, and we give thanks each day for that!

About once a month I go to the local shelter for the homeless. I help the people there in any way I can. And I am always amazed that most of them thank me for even the smallest things. When I gave some gloves to a little boy whose coat was dirty and holey, it was as if I had given him the world. The hungry old man whose soup bowl I filled took the time to express his gratitude before putting the first spoonful to his lips. When I handed a young woman a couple of sweaters that had been donated to the shelter, she expressed her joy not only with her heartfelt "thank you," but with shining eyes that said far more than any words could say.

I believe that the pleasure these homeless people felt at receiving such basic things far exceeds any enjoyment that my rich friend—and those like her—will ever experience. I believe that such gratitude as I feel toward my mother for a candy bar is something my friend will never feel. Without question, it is my opinion that people who are thankful for the least enjoy the most.

ANALYSIS

Essay I has a score range of 5–6. It is a well-organized essay, presenting a thesis statement in the introduction, developing that idea throughout the body, and referring to the quote in the conclusion. It demonstrates a clear understanding of the assignment, taking a stand and using personal observations and experiences to support the main idea. It uses a variety of sentence structure and length. It contains no major errors in grammar, usage, or mechanics. The vocabulary is fairly sophisticated.

DETAILED EXPLANATIONS

SECTION 2—MATH

1. (A) The point (1, 2) must represent the average of the endpoints. Since $\frac{(5-3)}{2}$ = 1 and $\frac{(-6+10)}{2}$ = 2, (–3, 10) must be the other endpoint.

2. (C) By increasing the largest number, the median will still be located midway between the fifth and sixth numbers. The median will remain the same. All of the other choices can affect the value of the median.

3. (B) Let $n = 16$, so that n divided by 6 yields a remainder of 4. Then $n + 2 = 18$, and 18 divided by 6 leaves no remainder.

4. (B) After Joe replaces one white marble with two red marbles, there are four red marbles out of a total of 13 marbles. (Originally there were 12 marbles, but removing one marble and replacing it with two marbles brings the new total to 13.)

5. (C) 66 minutes = 3,960 seconds. Let x represent the number of bottle caps. Then $\frac{5}{22} = \frac{x}{3{,}960}$, so $x = \frac{(3{,}960)(5)}{22} = 900$.

6. (D) When $x < y$, $x - y$ is negative, but $|x-y|$ is always positive. Choice (A) is wrong because if $x = y$, $|x-y| = x-y$. Choice (B) is wrong because if $x > y$, $|x-y| = x-y$. Choice (C) can be shown as incorrect by letting $x = 7$ and $y = 9$. Choice (E) can be shown as incorrect by letting $x = -9$ and $y = -7$.

7. (B) This function can be rewritten as $f(x) = 2 - \left(\frac{2}{[3x+1]}\right)$, and since $\frac{2}{[3x+1]}$ can never equal zero, $f(x)$ can never equal 2.

8. (E) The numerator can be factored as $10(x + 4)(x - 1)$, and the denominator can be factored as $5x(x + 4)$. The fraction can be reduced to $\frac{2(x-1)}{x}$.

9. (E) The difference between consecutive slash marks is $-5 - \frac{(-11)}{4} = 1.5$. So point R has a value $-5 + (3)(1.5) = -0.5$ and point S has a value of $-0.5 + 1.5 = 1$. Then $-0.5 + 1 = 0.5$.

10. (B) The percent of people under the age of 10 is 25%. Thus, 25% + 22% = 47% of the people are either under 10 years old or over 40 years old. If x = the total population of Smallville, $0.47x = 282$. Solving, $x = 600$. Because 5% + 30% = 35% of the population is between the ages of 11 and 30 (inclusive), the actual number of people in this age bracket is $(0.35)(600) = 210$.

11. (C) ΔABE is an isosceles right triangle, so AE = $\frac{10}{\sqrt{2}} \approx 7.07$. ΔCDE is a right triangle with acute angles of 30° and 60°, so $ED = 4\sqrt{3} \approx 6.93$. Thus, $AD = 7.07 + 6.93 = 14$.

12. (E) When $-6 \leq x \leq 3$, the value of y, [which is $f(x)$], lies between 0 and 3, inclusive. Note that when $x > 3$, $f(x)$ is negative.

13. (E) The distance is given by $\sqrt{(16-7)^2+(37+3)^2}=\sqrt{81+1{,}600}=\sqrt{1{,}681}=41$.

14. (D) $g(x)$ must contain points derived by adding 4 units to each of the x values of the points of $f(x)$, while leaving alone the y values. Thus $g(x)$ must contain the points (6, –8), (10, 14), and (14, –7).

15. (D) Because $x = y$, $PQ = QR$. This means ΔPQR is an isosceles right triangle so that $PQ = \frac{6}{\sqrt{2}}$. From the diagram, $\angle N = \angle R$, because MN is parallel to QR. Also, the vertical angles at P are equal. This means ΔMNP is an isosceles right triangle so that $MP = \frac{4}{\sqrt{2}}$. Thus, $MQ = \frac{6}{\sqrt{2}} + \frac{4}{\sqrt{2}} = \frac{10}{\sqrt{2}} = 5\sqrt{2}$.

16. (B) 2 + 7 = (2)(2) + 7 = 11. –4 + –3 = –4 + (–3)(–3) = 5. –7 + 2 = 0. 3 + –2 = 0. 7 + 0 = 0. 4 + 4 = (4)(4) + 4 = 20. Thus, three of these computations have a positive value.

17. (C) If T = {4, 5, 12, 16} then the intersection of sets S and T, which means the list of elements common to both sets, is {5, 16, 12}.

18. (A) Angle d must equal 80, since it is supplementary to 100. The angle indicated by $a°$ is a corresponding angle to the 100° angle, but a is not equal to 100 because L_3 is not parallel to L_4. Since L_1 is parallel to L_2, $a = c$, so c is not equal to 100. Note that it is possible for b to equal 100, and because $b = e$, e could also equal 100.

19. (B) $D = \frac{k}{\sqrt{F}}$, where k is a constant. By substitution, $9 = \frac{k}{\sqrt{16}} = \frac{k}{4}$. Solving, $k = 36$. Then $D = \frac{36}{\sqrt{100}} = \frac{36}{10} = 3.6$.

20. (D) A formula for an arithmetic sequence is $L = a + (n - 1)(d)$, where L is the last term, a is the first term, n is the number of terms, and d is the common difference between terms. Thus, $L = 500 + (29)(-6) = 326$.

DETAILED EXPLANATIONS

SECTION 3—CRITICAL READING

1. (A) This is a cause-and-effect sentence due to the word "because." You are looking for an adjective that indicates why she did not agree to the speaking engagement. It makes sense that she would not agree to a speaking engagement because she was not "conversant" (familiar; acquainted) with the featured topic. Choice (B) is wrong, since it doesn't make sense that she would turn down a speaking engagement because she was "nescient" (ignorant; not knowledgeable) with the featured topic. Choice (C) is incorrect, since "discordant" means "in disagreement; not harmonious." Be careful not to misread the sentence and overlook the word *not*. It doesn't make sense that a person would turn down a speaking engagement because she was not discordant (in other words, she was in agreement) with a topic. Since "disconsolate" means "sad; downcast," Choice (D) doesn't fit. And Choice (E) "indubitable" (beyond a doubt; doubtless) doesn't make sense in this context.

2. (C) For the first blank you are seeking an adjective describing a complex task that has to be done in a short time. "Insuperable" (insurmountable; incapable of being completed) is such a word. In the second blank, you are looking for a verb that conveys what co-workers might offer to do to change the situation. If co-workers offered to "abet" (to assist or support in the achievement of a purpose) a person, the situation would change. It doesn't make sense to say that completing a complex task in a short time seemed "bodacious" (noteworthy) until co-workers offered to "acclaim" (praise) one; therefore, Choice (A) is incorrect. Choice (B) is incorrect, since "redundant" (repetitious) does not adequately describe a complex task that is to be completed in a short time, and it hardly makes sense that one's co-workers' offer to "relegate" (banish; expel) him/her would improve the situation. Choice (D) is wrong because completing a complex task in a short time is not "facile" (easy), and it doesn't make sense to say that co-workers offered to "absolve" (pardon; discharge) someone. And Choice (E) is incorrect because "derogatory" (criticizing; disparaging) and "acknowledge" (recognize; take notice of) do not make sense in the context of this sentence.

3. (D) Here, your main clue is "he stopped partying," so you need an adjective that describes such a person. Another clue is that his friends did not understand. "Abstemious" means "temperate; marked by restraint in eating and/or drinking." Although such a person might be "obstinate" (stubborn; pertinacious), this term does not apply specifically to "stopped partying," so Choice (A) is not as good a choice as Choice (D). Choice (B) is incorrect, since "acquiescent" means "inclined to go along; tending to comply without restraint." Choice (C) is incorrect because "fractious" means "unruly; not readily ruled or managed" and would not describe a person who has decided to stop partying. And Choice (E) is wrong because one does not become "genteel" (well-bred; elegant; of the aristocracy or upper class) simply by choosing not to party anymore.

4. (E) Here are looking for a word that is related to the area of classical concerts and museums. The word "whereas" tells you this word will be more or less the opposite of the area of racing and mechanics. "Aesthetic" means "having to do with beauty, art, and taste and with the creation and appreciation of beauty; related to the arts," so it is appropriate. If you chose Choice (A) or Choice (C), you may have simply been confused about the spelling of these words and Choice (E). (A) "acetic" means "vinegary; producing vinegar," and (C) "ascetic" means "practicing strict self-denial as a measure of personal and especially spiritual discipline." Choice (B) is wrong because "acerbic" means "acidic in temper or tone; sarcastic," and does not fit here. And Choice (D) is wrong, since "affable" means "pleasant, at ease" and so does not fit in this context.

5. (B) You need a noun that reflects the emotional reaction that a political candidate might arouse in his constituents—to the point that they would choose not to vote for him. "Antipathy" means "dislike, antagonism, animosity," and thus is an appropriate word here. Choice (A) is incorrect, since "collusion" means "secret agreement or cooperation especially for an illegal or deceitful purpose; conspiracy." "Incursion" (hostile entrance into a territory; raid) does not make sense here; so Choice (C) is wrong. Likewise, "impunity" (freedom from punishment) does not make sense; thus, Choice (D) is wrong. And a candidate who aroused "empathy" (understanding, awareness, sensitivity to the experiences of others) would be more likely to win an election than lose it. So Choice (E) is incorrect.

6. (A) Here, you want a word that conveys the ineffectiveness of a teacher, as conveyed through her reaction to a child's request for help. And you need a word that describes a child's request; this word should be one that contributes to the meaning of the sentence, which is depicting an ineffective teacher who is responding to this request. A teacher's "cursory" (rapid and superficial) response to a child's "bona fide" (sincere; genuine) request for help would reveal her ineffectiveness. Choice (B) is incorrect, since an "obliging" (kind, receptive) response to a "frustrated" (baffled, defeated) request for help would not cause an observer to believe the teacher is ineffective. Choice (C) is wrong: an ineffective teacher's response to a child's "incessant" (endless; ceaseless) request for help would probably not be "imperturbable" (calm; composed; unruffled). Choice (D) is incorrect because an "astute" (shrewd; keen) response to a child's "incoherent" (incomprehensible; lacking logic) would not indicate ineffectiveness on a teacher's part. And Choice (E) is incorrect because an ineffective teacher's response would not be "mindful" (aware; knowledgeable) if a child was making a "frivolous" (silly, mindless) request for help.

7. (C) The words "however" and "instead" tell you that you should look for two verbs with opposite meanings. It makes sense to say "I was trying to 'elucidate' (clarify; to make clear) my opinion to her; however, I managed only to 'confound' (confuse; puzzle) her instead." Choice (A) is incorrect because "delineate" (clarify, define) and "enlighten" (instruct; furnish knowledge to) do not convey contrasting ideas. Choice (B) is incorrect because it doesn't make sense: "mediate" means "to act as an intermediary; to intervene," and "implicate" means "to involve; to bring into a connection." Choice (D) is incorrect because "sell" (to convince of the acceptability of an idea) and "actuate" (motivate to do; encourage) do not have contrasting connotations. Likewise, Choice (E) is wrong, since "entangle" (complicate; muddle) and "disconcert" (confuse; upset) do not have contrasting connotations.

8. (B) Your clues here are "highly competitive" and "cannot afford to." You are looking for a verb that conveys the perspective of those who are moving up a highly competitive corporate ladder, and an adjective that depicts how these people must be at all times. Those who "aspire" (hope; seek) to move up the corporate ladder (or in any highly competitive situation) must be "sedulous" (diligent; industrious) at all times. Although "assiduous" (persistent) fits in the second blank, Choice (A) is incorrect, since "deign" (stoop) does not fit with "to move up the . . . corporate ladder." Choice (C) is incorrect; since it does not make sense to say that they "consort" (unite; associate) to move up the corporate ladder, nor does it make sense to say that they must be "impolitic" (not wise; rash) all the time. They do not "condescend" (descend; give up privileges of rank) to move up, and they should not be "immutable" (unchangeable) all the time; so Choice (D) is wrong. And although the verb "endeavor" (try; attempt) makes sense in this context, the adjective "hebetudinous" (lethargic; dull) does not accurately describe the way these people must be; thus, Choice (E) is incorrect.

9. (B) Everything in the passage confirms the author's desire for times past; thus, the author is nostalgic. Although the author's memories of the past may be happy ones, he/she is clearly not content with the way things are now; thus, Choice (A) is incorrect. Choice (C) is wrong since nothing in the passage indicates that the author is anticipating anything. And although the author does not seem to like the current situation, nothing in the passage indicates anger. Therefore, Choice (D) is wrong. And although the author misses the smells of a slow-cooking roast (as opposed to a pre-prepared, microwaved one), the passage does not imply that the author is hungry.

10. (A) The author clearly indicates a preference for the simple, basic things of the past—pots and pans as toys, slow-cooked food, unlocked doors, clotheslines, unpaved roads, and rotary mowers—as opposed to video games, the computer, and microwaves of modern times. Nothing in the passage implies a preference for complex, expensive, modern, convenient, or popular things; thus, Choices (B),(D), and (E) are incorrect. And although the author prefers things of the past as opposed to things of modern times, nothing in the passage indicates a preference for old and much-used things; so Choice (C) is incorrect.

11. (D) The passage begins with a day that couldn't have been any worse and ends with a sense of renewal and revitalization. Choice (A) is incorrect because of the statement "I wasn't sure the day could have been any worse, and the ominous rumbling clouds overhead seemed to agree with me"—definitely not positive. Choice (B) is incorrect, since it reflects the opposite of what is true. Choice (C) is wrong since the passage ends with "The essence of freshness renewed my spirit. And the chirping birds told me the shower had revitalized them as well"—definitely not negative. And Choice (E) is incorrect since the passage is both negative at the beginning and positive at the end.

12. (C) The diamonds metaphorically reveal the effect of the sun's shining on the drops of rain, and indicate a change in the author's perspective. Choice (A) is incorrect, since nothing in the passage indicates that the author is dreaming. Choice (B) is wrong because nothing in the passage addresses jewels and the author's attitude toward them. And although the author's attitude toward the day changes from negative to positive, nothing in the passage deals with his/her appreciation of the world in general; so Choice (D) is wrong. Likewise, Choice (E) is incorrect because the passage does not address any rewards that life brings us; instead, it focuses on only the author's change in attitude toward that particular day.

13. (C) The passage focuses on various examples of natural weather predictions based on folklore. Although it does mention that meteorologists' weather forecasts are sometimes wrong and suggest that we pay less attention to them, it does not focus on their inaccuracies; so Choice (A) is wrong. Neither does it offer evidence to prove that folklore predictions are usually right, although it does mention that they are sometimes accurate. So Choice (B) is not correct. It does mention that some folklore predictions are scientifically sound, but it does not indicate which examples have scientific basis; thus, Choice (D) is incorrect. And it says nothing about a folklore basis of meteorologists' predictions; therefore, Choice (E) is wrong.

14. (A) The first paragraph of the passage focuses on the importance of weather, and how it affects our lives—especially work, play, and school. Although the passage says that weather affects our plans, it does not indicate that the closing of school and the cancellation of athletic events has anything to do with our planning. Therefore, Choice (B) is incorrect. At the point where the passage mentions school closings and athletic event cancellations, weather forecasts have not yet been discussed. Therefore, Choices (C), (D), and (E) are wrong.

15. (C) Keep in mind that you are looking for the thing that is **not** mentioned among the things over which weather predictions sometimes take priority. The author states, "some of us pay more attention to the weather forecast in the media than we do to the news, music, sports, or television shows." It does not mention weather predictions as being more important than school. Thus, Choice (C) is the only correct answer.

16. (C) The definition of "prognostication" is "forecast." If this term is unfamiliar to you, you should consider carefully the use of the word in context. The sentence presents reasons why "we hang onto every word of the meteorologists on the radio or television" regarding upcoming weather, including "their prognostication of temperatures." Replacing the word "prognostication" with (A) "guess," (B) "lie," (D) "postponement," or (E) "broadcast" makes little sense and sounds awkward.

17. (D) The passage conveys an acceptance of and respect for old timers' folklore weather predictions, and implies that these forecasts have a place alongside the modern-day science-based forecasts that rely on technology. Nowhere is it implied that technology or computers are better than (or superior to) methods of yesterday, regarding the weather or anything else. So Choices (A), (B), and (C) are incorrect. And Choice (E) is incorrect because nothing in the passage refers to the entertainment or education provided by computers.

18. (B) The definition of "impending" is "about to occur; threatening." None of the other choices are possible definitions of this word. If you are not familiar with the term "impending," you might be able to guess its meaning by its use in the context of the passage. You might be familiar enough with folklore weather predictions to recognize at least one of the signs that forecast rain, according to folklore. Another possible clue is that the next sentence gives examples of "clear skies," so it is reasonable to assume that the first sentence is focusing on the opposite type of weather, or rain.

19. (E) The passage says that folklore tells us that the woolly worm's coat predicts the severity of the upcoming winter. The passage makes no connection between the woolly worm and (A) spring, (B) winter precipitation, (C) summer storms, or (D) number of snows.

20. (E) Keep in mind that you are looking for the choice that is **not** mentioned as a weather predictor. Air is the only choice not included in the passage. People (e.g., meteorologists), plants (e.g., dandelions and grass), weather (e.g., lightning, clouds, snow), and animals (e.g., woolly worms, flies, cats, lightning bugs, spiders) all are discussed as predictors of the weather.

21. (B) The passage states, "We can expect . . . rain by evening if the morning sky is red." Therefore, Choice (A) is incorrect. Choice (C) is incorrect because the passage indicates, "We can expect clement (mild; favorable) weather . . . when we see . . . a red sky in the evening." Although a blue sky may mean sunshine, the passage says nothing to indicate that this is the case; therefore, Choice (D) is wrong. And Choice (E) is incorrect because "Lightning in the north is a predictor of dry weather."

22. (D) The passage gives examples of natural signs of weather to occur within the next few hours, all the way to natural predictions of upcoming seasons. Folklore weather predictions are not limited to (A) the next few hours, (B) the next few days, (C) rain or snow, or (E) certain areas of the country.

23. (E) The passage says, "years of observation and experience—two important elements of scientific investigation—serve as the basis for many of these "weather rules." It does not mention guesswork, education, or documentation; thus, Choices (A), (B), (C), and (D) are incorrect.

24. (D) The author states, "when the weather is vital to our plans for work or play, it may behoove us to check out nature as well as the local forecast." Although the passage does indicate that the meteorologists' forecasts are not always accurate, it does not say that we should necessarily expect the opposite of what they predict; so Choice (A) is incorrect. Choice (B) is wrong, since the passage does say that it may be warm and sunny even though meteorologists have predicted cold rainy weather, or vice versa. And Choice (C) is only partially correct, since the author recommends that we pay attention to both nature and meteorologists. Choice (E) is incorrect because nowhere does the author recommend that we pay no attention to either type of prediction.

DETAILED EXPLANATIONS

SECTION 4—WRITING

1. (B) The correct idiom is "independent of." Choices (A) "from," (C) "with," (D) "on," and (E) "for" are all unacceptable.

2. (C) It consistently uses past-tense verbs in the series ("checked," "made," "went," and "went"). Choice (A) shifts from past tense to past progressive ("was going"). Choice (B) also shifts from past tense to past progressive and adds the unnecessary infinitive "to go." Choice (D) shifts from past tense to past perfect, and Choice (E) shifts from past tense to present tense.

3. (C) It includes no unnecessary words, and it correctly places the adjective clause ("that are associated with the chapter material") directly after "categories," the noun it modifies. Choice (A) contains the unnecessary words "there is," and the adjective clause "that are associated with the chapter material" incorrectly and illogically modifies "chalkboard." Choice (B) illogically has the prepositional phrase "of categories . . ." modifying "chalkboard." Choice (D) has the same error; it also contains the unnecessary words "there is." Choice (E) illogically has the adjective clause "that is associated with the chapter material" modifying "chalkboard." This sentence also uses the unnecessary words "there is."

4. (A) The singular verb "was hoping" agrees with the singular subject "one." Choice (B) incorrectly uses the plural verb "were hoping" with the singular subject "one." Choices (C) and (D) contain an inappropriate idiom; the preposition "for" requires that a noun follow it, but these sentences incorrectly have a verb following it. Choice (D) incorrectly uses a plural verb ("were hoping") with a singular subject ("one"). Also, it uses an inappropriate idiom "hoping that." The use of "that" must be followed by a clause; however, in this sentence, it is followed by only a verb phrase ("win the sweepstakes").

5. (C) It consistently uses "we" as the subject of both clauses, and it uses the adverb "thoughtlessly" to modify the verb "behave." It also uses a comma to separate the introductory dependent clause ("When . . . thoughtlessly") from the main clause ("we . . . feelings"). Choices (A) and (B) incorrectly use the adjective "thoughtless" to modify the verb "behave." Choice (B) also creates a fragment by separating the dependent clause "When . . . thoughtless" from the main clause "we . . . feelings." Choice (D) incorrectly switches from the subject "we" in the first clause to the subject "you" in the main clause. Choice (E) switches from the subject "we" in the main clause to the subject "one" in the main clause; it also incorrectly uses the adjective "thoughtless" to modify the verb "behave."

6. (E) It contains no unnecessary words, and it shows proper coordination, introducing the dependent clause with the conjunction "because." Choice (A) is wordy and awkward with the use of "for the reason being that." Choice (B) contains a comma splice, with the use of a comma to connect two independent clauses ("Often . . . victims." And "it . . . virus"). Also, the pronoun "it" has no antecedent. Choice (C) has the same problem as Choice (A), with the wordy and awkward phrase "for the reason that." Choice (D) uses the inappropriate idiom "because of"; the preposition "of" requires a noun or noun phrase as its object; however, this sentence has a clause after it.

7. (D) It correctly omits the unnecessary words "other" and "it," creating a complex sentence with correctly placed dependent clauses ("when . . . boasts" and "that . . . class"). Choices (A) and (C) use faulty comparison with the use of the word "other," implying that Brandon is one of the girls in his class. Choices (A) and (B) use the vague pronoun "it," which has no antecedent. Choice (E) is confusing and illogical, since the verb "embarrasses" has no subject.

8. (E) In this sentence, the antecedent of "them" is obviously "dogs." In Choices (A), (B), (C), and (D), it is difficult to determine if the antecedent of the pronoun "them" is "dogs" or "John and Ben."

9. (B) It consistently uses plural nouns ("students" and "entrepreneurs"). Choices (A) and (C) involve an error in agreement: the singular noun "entrepreneur" cannot correctly refer to the plural noun "students." Choice (C) also incorrectly shifts from present tense ("provide") to past progressive ("were wanting"). Choice (D) incorrectly uses the singular verb "wants" with the plural noun "students." In Choice (E) the use of "in wanting" is illogical and confusing; this sentence also incorrectly uses the singular noun "entrepreneur" to refer to the plural noun "students."

10. (D) It correctly uses "such" with "as." Choice (A) lacks the word "such," which is necessary with "as." Choice (B) involves improper phrasing with the use of "that are" instead of "such as." Choice (C) incorrectly has "such" modifying "pastel" instead of "shades." Choice (E) illogically inverts the adjective and noun to create the awkward phrase "a shade pastel."

11. (B) It specifically defines the term, without any unnecessary or misleading words. Choice (A) is incorrect because slander is a "what," not a "when," and the word "like" is unnecessary. Also, in this sentence, "it" has an unclear antecedent. Choice (C) is incorrect because slander is a "statement," not a person ("someone"). Choice (D) is incorrect because slander is not a "when." Choice (E) is illogical because it has slander "making a false statement."

12. (C) The word "and," instead of "as," should connect two parallel verb forms ("reading" and "reviewing") to form a compound predicate. Choice (A) correctly uses the preposition "for" after "prepare." Choice (B) correctly uses a comma to separate the modifier from the main clause. Choice (D) correctly uses a past perfect verb ("had given") for an action that had occurred prior to the action in the main clause.

13. (B) The subject of the sentence is the singular pronoun "what," which requires the singular verb "is," instead of "are." Choice (A) correctly uses the plural verb "say" to agree with the plural subject "politicians." Choice (C) correctly uses the adverb "actually" to modify the verb "do." Also, the plural verb "do" agrees with the plural subject "they." And Choice (D) correctly uses the prepositional phrase "in office."

14. (C) It correctly uses "offers" as the verb of "it." The participle "offering" cannot stand alone as the verb. Choice (A) correctly uses a comma after the city and another one after the state. Choice (B) correctly uses the present perfect verb "has been called." Choice (D) uses a comma to separate items in a series. The items—both nouns ("establishments" and "casinos")—are parallel.

15. (C) Items in a series must be parallel; the present progressive verb "is feeding" is not parallel with the infinitives "to walk" and "to say." So "is feeding" must be changed to "to feed," an infinitive that is parallel with the other infinitives ("to walk" and "to say"). Choice (A) correctly uses to present-perfect verb "has learned" to show action that began in the past but extends into the present. Choice (B) correctly uses the plural adjective "several" with the plural noun "words." Choice (D) uses an appropriate prepositional phrase "with a spoon" to modify "to feed."

16. (B) This choice contains a subject-verb agreement error; the plural subject "rules" requires the plural verb "are causing," instead of the singular verb "is causing." Choice (A) correctly uses the plural possessive "students,'" with the apostrophe after the "s." Choice (C) uses the subordinating conjunction "that" to introduce a noun clause ("their . . . violated"). Choice (D) appropriately uses the plural present progressive verb "are being violated" to agree with the plural subject "rights" and to show action that is occurring.

17. (E) This sentence has no errors. In Choice (A), the participle "being" correctly serves to introduce an adjective phrase. The singular noun in Choice (B) agrees with the singular noun ("bat") in the main clause. Choice (C) correctly uses a singular verb ("flies") to agree with its singular subject ("mammal"). In Choice (D), a singular verb ("intrigues") agrees with its singular subject ("bat"), and the direct object "zoologists" is correct.

18. (D) The phrase "not only" requires "but also" to create an appropriate correlative construction. Choice (A) correctly uses the adjective "effective" to modify the noun "presenter." Choice (B) uses the singular verb "appeals" to agree with the singular subject "presenter"; it also uses the preposition "to" to create an appropriate idiom. In Choice (C), the plural noun "interests" contains no error.

19. (C) Choice (C) is the correct answer; "quick" must be changed to "quickly," since it serves as an adverb modifying the verb "work." Choice (A) correctly uses a plural verb ("are") to agree with a plural subject ("portions"). With the use of "therefore," Choice (B) appropriately shows the relationship between the two clauses. Choice (D) correctly completes the phrase "as . . . as."

20. (B) "You" must be changed to "we" so that the subjects of the two clauses are both third-person pronouns. Choice (A) shows correct subject-verb agreement, with the plural subject ("we") and its plural verb ("listen"); also the present-tense verb "listen" is consistent with the present-tense verbs ("can become" and "is") in the rest of the sentence. Choice (C) correctly uses the nominative-case pronoun "who" as the subject of "is." Choice (D) correctly uses the adjective "best" to modify "candidate." The phrase "the opposing parties" is a clue that more than two parties are involved; therefore, the superlative adjective "best" should be used, instead of the comparative "better."

21. (E) This sentence has no errors. Choice (A) correctly uses a singular verb ("wants") to agree with the singular subject ("each"). Choice (B) correctly uses the adverb "clearly" to modify the adjective "expressed," which modifies "explanation." In Choice (C), the preposition "of" combines with the relative pronoun "what" to create an appropriate idiom indicating what the women want in the explanation. And in Choice (D), the past-tense verb "passed" is consistent with the verb "intended," indicating that these actions have already occurred.

22. (A) The phrase "you and I" must be changed to "you and me"; since this phrase is the object of the preposition, both pronouns must be in the objective case. In Choice (B), the adverb "strictly" modifies the adjective "confidential." Choice (C) correctly uses the conjunction "so that" to connect the main clause with the dependent clause. And in Choice (D) the verb "will be hurt" contains no errors.

23. (A) This choice illogically compares "other students" with "Angel's textbooks." The correction compares "Angel's textbooks" and "those of the other students," with textbooks being the obvious antecedent of "those." Choice (B) correctly uses the plural verb "were" to go with the plural subject "textbooks." Choice (C) completes the parallel structure, using two adjectives ("outdated" and "worn") connected with "and." In Choice (D), the infinitive "to buy" properly modifies the noun "money."

24. (D) The words "by his" are unnecessary, since all items in a series must be parallel. This phrase must be adjusted so that it—like the other two items—contains an adjective and a noun ("profound reflections," "sardonic observations," and "straightforward witticisms"). Choice (A) correctly contains the adjective "famous" to modify "American author." No commas are required to set the appositive "Mark Twain" apart from the rest of the sentence because the appositive identifies the noun just before it and the noun is not preceded by "a" or "the." In Choice (B), the adjective "well-known" correctly modifies the subject "writer." In Choice (C), the phrase "sardonic observations" contains an adjective and a noun, making it parallel with the other items in the series.

25. (B) For proper noun-noun agreement, the plural noun "animals" must be changed to the singular noun "animal" in order to agree with the singular noun "skunk." Then the article "a" must be added, to create the phrase "a smelly animal." Choice (A) correctly uses the adverb "commonly" to modify the verb "regarded." Choice (C) correctly uses the adverb "actually" to modify "is." And Choice (D) correctly uses a comma to separate two coordinate adjectives ("entertaining" and "fun-loving") modifying the same noun ("pet").

26. (C) The use of the preposition "of" with "impatient" creates an incorrect idiom; thus, the preposition must be changed to "with." In Choice (A), the use of the adverb "often" is correct. Choice (B) correctly uses the nominative-case pronoun "who" as the subject "are." And since the antecedent of "who" is the plural noun "people," the plural verb "are" is correct. Choice (D) correctly uses the adjective "easy-going" to modify the noun "demeanor."

27. (D) "Complimentary" and "free" mean the same thing, so in order to avoid redundancy only one of these adjectives should be used. Choice (A) correctly introduces the sentence with an adverb phrase, using a comma to separate it from the rest of the sentence. Choice (B) uses a plural verb ("hope") to agree with a plural subject ("merchants"). In Choice (C), the infinitive "to attract" indicates what the merchants hope and appropriately introduces its object "customers."

28. (A) A participle ("indicating") cannot serve as the verb of a clause; it must be changed to "indicates." Choice (B) correctly uses the plural verb "are" to agree with the plural subject "Americans." Choice (C) correctly uses the participle "weighing" to modify the pronoun "many." Choice (D) appropriately uses the prepositional phrase "or more than 30" to modify the noun "Body Mass Index."

29. (A) An object of a preposition must be in the objective case; thus, in this sentence "we" must be changed to "us" to create the correct prepositional phrase "of us." Choice (B) correctly uses the adjective "upset" to modify the subject "some," and the idiom "upset with" is appropriate. Choice (C) correctly uses the nominative-case pronoun "who" as the subject of the clause ("who . . . products"). Since the antecedent of "who" is the plural noun "co-workers," the plural verb "are" is correct. Choice (D) uses a comma to separate the two coordinate adjectives "inadequate" and "inferior," which modify the noun "products."

30. (B) The use of "with" reinforces the idea presented in the first clause. In Choice (A), the use of "but" incorrectly implies that the idea in the second half of the sentence contrasts with that of the main clause. Also, the use of "but" makes the second part of the sentence a fragment, since a participle ("claiming") cannot serve as the verb of a clause. Choice (C) incorrectly implies that the idea in the second half of the sentence contrasts with that of the main clause. In Choice (D) the switch to past tense is incorrect because the information is current information. Choice (E) is incorrect, since a comma is the correct punctuation to separate two independent clauses, if the second clause begins with a coordinating conjunction ("but").

31. (B) Sentence 3 gives two examples to show that racing has not always been so popular—its negative reputation and its problems with the law. Choice (A) is incorrect, since the sentences are not contradictory. Choice (C) is wrong, since Sentences 2 and 3 are related, and Sentence 3 does not offer a new idea. Choice (D) is incorrect since Sentence 3 does not define a term. And Choice (E) is incorrect because Sentence 3 does not present a situation that Sentence 3 explains.

32. (B) It uses no unnecessary words and contains no redundancy. It correctly uses an apostrophe to set apart the appositive "or moonshine runners." Choice (A) is redundant with the words "illegally" and "illegal." It contains the unnecessary words "who were also called . . . because." This sentence is confusing because it is difficult to determine what the clause "because . . . markets" modifies. Choice (C) is awkwardly worded and creates a confusing sentence. Choice (D) is illogical with the phrase "moonshine running whiskey." Choice (E) shifts from past tense to present tense, and the repetition of the word "who" is awkward although grammatically correct.

33. (E) The use of "so popular that" correctly reflects the reason for the races' continuing. Choices (A) and (B) are incorrect because the conjunctions "although" and "but" incorrectly implies that the two clauses are contradictory. Choice (C) is illogical, incorrectly implying that the races did not become popular until after the end of Prohibition. In Choice (D), the use of the conjunction "since" creates an illogical relationship between the two clauses, implying that the races became popular *as a result of* their continuation after the end of Prohibition.

34. (C) Paragraph 2 focuses on the progress of racing from illegal whiskey runs to popular entertainment. It does not (A) provide an example, (B) provide a summary, (D) describe a location, or (E) propose a solution.

35. (A) It provides information that is relevant to the specific issue at hand in this paragraph—the meeting that Bill France held with promoters at Daytona Beach in the late 1940s. Although all the other choices are related to the topic of racing, none develop the specific topic that the first part of the paragraph has introduced.

DETAILED EXPLANATIONS

SECTION 5—MATH

1. (C) Because John is older than Mary, who is older than Tracy, Tracy cannot be older than John. Options (A), (B), and (E) are possible because we don't know how Tracy's and Ken's ages compare. Option (D) is possible because we don't know the relationship between Mary's and Ken's ages.

2. (D) Set $A - B$ means all elements in A, with the removal of any elements that are in both A and B. Since green and purple are elements of both A and B, set $A - B = \{\text{red}\}$.

3. (B) Rearranging the terms, we get $x - ax = 2 - a$. The left side of this equation can be factored as $x(1 - a)$. Finally, divide each side by $(1 - a)$ to get $\frac{(2-a)}{(1-a)}$.

4. (A) Let x represent the x-intercept value. This line contains (0, 14) and (x, 0) and has a slope of $\frac{2}{7}$, so $\frac{(0-14)}{(x-0)} = \frac{2}{7}$, which simplifies to $\frac{-14}{x} = \frac{2}{7}$. Then $2x = -98$, so $x = -49$.

5. (E) Since any square root must have a value of at least zero, $49 - x^2 \geq 0$. This means $x^2 \leq 49$, so $-7 \leq x \leq 7$.

6. (D) The probability that Lu gets office B is $\frac{1}{3}$. Then provided that Lu does get office B, the probability that Tina will get office A is $\frac{1}{2}$. Thus, the probability that both events occur is $\left(\frac{1}{3}\right)\left(\frac{1}{2}\right) = \frac{1}{6}$.

7. (C) ΔADE is a 30°–60°–90° right triangle, since a tangent to a circle forms a right angle with the radius of the circle at the point of tangency. In this case, E is the point of tangency and the radius is AE. In addition, $AE = \left(\frac{1}{2}\right)(AD) = 4$. Thus the circumference = $(2\pi)(4) = 8\pi$.

8. (D) Let x = the marked price of the item. After a 25% discount, the item is now $0.75x$. This value must be 20% higher than the cost of \$200, which is \$240. Thus $0.75x = 240$. Solving, x = \$320.

9. (B) Two pounds of roast beef and 7 hot dogs cost (2)(\$6.25) + (7)(\$2.50) = \$30. From the original \$40 that Raul had, he now has \$10 to spend on soda. $\frac{\$10}{\$0.75} = 13.3$, so the maximum number of cans of soda he can buy is 13.

10. (A) Since $m < n$, we cannot have $m = 17$ and $n = 12$. Option (B) is possible if $m = -8$ and $n = -3$ (as an example). Option (C) is automatically possible. Option (D) is possible if $m = 3$ and $n = 8$ (as an example). Option (E) is possible if $m = -5$.

11. (D) The area to be painted is (8)(7) = 56 square feet. 56/10 = 5.6, so he will need to buy 6 cans of paint. The total cost of paint = (6)($1.80) = $10.80.

12. (E) The radius of the circle must be $\sqrt{81} = 9$, which is then the length of each of AC and BC. Note that the circumference of the circle is $(2\pi)(9) = 18\pi$. Since the central angle is 40°, minor arc AB represents $\frac{40}{360} = \frac{1}{9}$ the circumference = 2π. Now, the perimeter of the sector = $9 + 9 + 2\pi = 18 + 2\pi$.

13. (E) For any function $g(x) = Ax^2 + Bx + C$ to have only one distinct root, it is necessary that $B^2 - 4AC = 0$. For this particular function, this means that $(-5)^2 - (4)(k)(25) = 0$. Then $25 - 100k = 0$. Solving, $k = \frac{25}{100} = \frac{1}{4}$.

14. (D) Let x, $x + 1$, $x + 2$, and $x + 3$ represent the numbers. Then $4x + 6 = 166$. Solving, $x = 40$. So the four numbers are 40, 41, 42, and 43.

15. (D) If the central angles of the two circles were equal, the ratio of the areas of the sectors would be $(1.5)^2 : 1 = 2.25 : 1$ Because $\angle P$ is twice as large as $\angle S$, the ratio of these sectors is $(2)(1.5^2) : (2)(1) = 4.5 : 1$. Multiply the area of the sector smaller circle by this ratio to get the area of the larger circle's sector, $(4.5)(60) = 270$.

16. (E) For this direct variation problem, use the formula $M = kV^2$, where k is a constant. By substituting $M = 24$ and $V = 4$, $24 = k(16)$, so $k = 1.5$. Now $M = 1.5V^2$. Substituting $V = 6$, we get $M = (1.5)(36) = 54$.

17. (C) Let x = the cost of an apple and let y = the cost of a pear. Then $3x + 2y = 1.17$ and $2x + 3y = 1.33$. By multiplying the first equation by 3 and the second equation by 2, we get $9x + 6y = 3.51$ and $4x + 6y = 2.66$. Now subtracting the new equations, we get $5x = 0.85$ and this value represents the cost of 5 apples. This problem can also be solved by substitution.

18. (B) Because $L = 110°$, $\angle MRQ = 70°$. A theorem states that $\angle MRQ = \left(\frac{1}{2}\right)(QM + NP)$. Let $NP = x°$. Then $70 = \left(\frac{1}{2}\right)(50 + x)$. Simplifying, $70 = 25 + \left(\frac{1}{2}\right)x$. Solving, $x = (70 - 25)(2) = 90$.

DETAILED EXPLANATIONS

SECTION 6—CRITICAL READING

1. (C) This is a cause-and-effect sentence; you are looking for an adjective that describes a man who frightens even his grandchildren. "Dyspeptic" (disgruntled; grumpy) is such an adjective. Choices (A) and (B) are wrong, since a "genial" (agreeable) or "innocuous" (harmless) grandfather would not be likely to frighten his grandchildren. And the fact that a man was "loquacious" (talkative) or "somnolent" (sleepy) probably wouldn't frighten his grandchildren; thus, Choices (C) and (D) are incorrect.

2. (B) Your clues here are "sense of humor" and "quick wit." You are looking for a noun describing a response that reveals these attributes when someone directs a comment or question toward the man. "Rejoinder" (quick, clever answer; appropriate reply) is such a word. Choice (A) is wrong, since "calumny" (false charge) does not indicate a sense of humor and quick wit. Choice (C) "prevarication" (lie) is also an inappropriate choice. And coming up with an "expletive" (curse word) in response to every comment or question is hardly indicative of a sense of humor or quick wit; therefore, Choice (D) is wrong. And "remonstrance" (presentation of reasons for opposition or grievance) does not fit here; so Choice (E) is incorrect.

3. (A) You need a noun that conveys the quality of a person who is easy to believe. "Verisimilitude" is "the quality of appearing to be true or real; the appearance of being authentic"; it would be easy to believe even the most blatant lies of someone who possesses verisimilitude. Since "ambience" (feeling or mood associated with a place, person, or thing) is not as specific as "verisimilitude," it is not as good a choice. It would be difficult to believe a person because of his "ignominy" (disgrace, bad reputation); thus, Choice (C) is wrong. Choice (D) is incorrect, since one's "petulance" (ill humor; peevishness) would not cause others to believe him. And Choice (E) is incorrect, since it does not make sense to say that it was difficult not to believe a person's lies because of his "inveracity" (lie, falsehood).

4. (E) Because of the obvious connection between "hate" and "ill will," and between "kindness" and "benevolence" (tendency or inclination to do good), you know you are seeking verbs that are synonyms. An additional clue is the conjunction "likewise," which points out the similarity between the two clauses. "Engender" and "beget" both mean "to cause, produce, or develop." It is reasonable to say that hate "provokes" (stirs up; results in) ill will; however, it does not make sense to say that kindness "impairs" (hinders; interferes with) benevolence; so Choice (A) is wrong. Choice (B) is incorrect because it is illogical to say that hate "precedes" (comes before) ill will, or to say that kindness "supersedes" (takes the place of) benevolence. Choice (C) is wrong because hate does not "expunge" (cancel; destroy) ill will, although kindness often does "induce" (cause; result in) benevolence. And Choice (D) is wrong because "adulterate" (corrupt) and "temporize" (to draw out or extend in order to gain time) do not fit in the context of the sentence.

5. (D) When a perpetrator is "nondescript" (not easily described; lacking distinctive characteristics), it is difficult for witnesses to "inculpate" him (incriminate him, show evidence of his involvement in a crime). If you selected Choice (A), you may have been confusing "inculcate" (to teach or impress by repetition) with "inculpate." Also in this choice, the word "inconsequential" (of no significance, unimportant) is not as appropriate as "nondescript." Choice (B) is wrong, since witnesses probably *WOULD* be able to "incriminate" (accuse) a "conspicuous" (noticeable) perpetrator. Choice (C) is incorrect since a perpetrator's being "pensive" (thoughtful; reflective) would have little to do with whether or not witnesses could "vindicate" (acquit, clear) him. And Choice (E) is wrong because a perpetrator's being "urbane" (sophisticated, courteous, polished) would not affect witnesses' ability to "identify" him (point him out; recognize him).

6. (D) Both passages focus on the assumptions we make about people—whether they are rich or poor—based on the amount of trash we see outside their homes. Although the two authors are referring to two different types of trash, both use visible trash to make assumptions about the residents' socio-economic level. Choice (A) is incorrect because neither passage refers to the appearance of the people themselves. Choice (B) is incorrect because neither passage supports this adage; as a matter of fact, Passage 1 tends to support the opposing idea. Choice (C) is wrong, since neither passage addresses people's use of their own backgrounds as the basis of their assumptions. And Choice (E) is incorrect because nothing is said about the amount of pride people have in their property. Be careful not to make assumptions on your own; instead, base your answer selection on what the passages say. For example, here you should not assume that the author is saying the nicely landscaped lawn and the mansion implies pride on the part of the owner, even though you yourself might make this assumption.

7. (A) The whole passage focuses on the fact that wealthy people throw away, and waste, much more than poor people do—simply because they don't want, need, or like something. Choice (B) is incorrect, since the author refers to poor people—not wealthy people—as resourceful. Choices (C) and (D) are wrong because, although one can infer that the author does not approve of the wealthy, the passage does not really refer to the wealthy as snobby or as show-offs. And although the author says that the rich often throw away things, nothing in the passage indicates that they are neat; thus, Choice (E) is wrong.

8. (E) The dictionary definition of "veracity" is "accuracy; truthfulness." However, even if you are not familiar with this word, you should be able to figure out its meaning by its use in the sentence. Try replacing the word "veracity" with each of the choices, and you'll see that "truthfulness" is the only one that makes sense and/or corresponds with the message of the passage. The author is not attempting to prove the (A) short-sightedness, (B) narrow-mindedness, (C) bias, or (D) inaccuracy of the assumption that the more trash we see, the poorer the dwellers are. So all of these choices are wrong.

9. (C) The author of Passage 1 indicates that rich people set out more trash for the garbage collector to pick up than poor people do. The author of Passage 2 indicates that poor people tend to have more trash and junk in their yards than rich people do. Although the author of Passage 1 seems somewhat irritated with the wealthy and the author of Passage 2 seems somewhat impatient with Americans' tendency to judge others, neither Choice (A) or Choice (B) is correct regarding the authors' attitudes toward the rich and the poor. Choice (D) is incorrect because it says the exact opposite of what is true of the passages. And Choice (E) is wrong because Passage 1 includes nothing to imply that Americans are too judgmental, and nothing in Passage 2 indicates that Americans' judgments are inaccurate.

10. (A) Throughout the passage, Adams was easily provoked and irritated by the night and morning sounds and by what he perceived to be his inability to sleep. Everything "pressed on his nerves." "Irascible" (easily provoked, irritated) is the only choice that fits. And his reaction to his nurse exemplifies his irascible attitude as well. Although the fact that he had a nurse implies that he was sick, he did not seem especially sick in this passage; so Choice (B) is incorrect. Choice (C) is incorrect because nothing in the passage indicates that Adams was "complacent" (satisfied; pleased). And you might assume he was tired because it was nighttime and he was not sleeping; however, nothing in the passage indicates that he was tired. Thus, Choice (D) is incorrect. Likewise, Choice (E) is wrong because the passage includes nothing to imply he was lonely.

11. (C) The passage begins with ". . . he did sleep intermittently, drowsed between times, and even dreamed; but, forgetting his dreams before he opened his eyes, . . . he believed, as usual, that he lay awake the whole night long." And later "he drowsed again, not knowing that he did. . . ." Since he did sleep some, Choice (A) is incorrect. Choice (B) is wrong because, although he did dream some, he did not rest well. Choice (D) is wrong because he spent the entire night in bed, even though he slept only intermittently. And Choice (E) is wrong; although the city outside his window remained active throughout the night, he was not a part of this activity, except to hear the noises and visualize the activity that accompanied them.

12. (B) Adams regarded the city at night as "some single great creature resting fitfully in the dark outside his windows. It . . . was too powerful a growing thing ever to lie altogether still. Even while it strove to sleep it muttered. . . ." Such words and phrases as "rumblings," "howled," "metallic stirrings," "chugged and snorted," "a faint, voluminous hum," and "vibration" all depict the activity going on throughout the night, outside Adams' window. He heard clinks, whistles, and chirps. When he was younger, he might have perceived these sounds as "proof of his citizenship in a 'live town' . . ." Obviously, he did not think of the city at night as being dead or boring; so Choices (A) and (C) are incorrect. And although you as a reader might find such sounds (D) exciting or (E) tantalizing, nothing in the passage indicates that Adams saw them that way. They clearly irritated him, but we have no evidence that they excited or tantalized him.

13. (C) "In his youth Adams might have been less resentful of sounds . . . when they interfered with his night's sleep: even during an illness he might have taken some pride in them as proof of his citizenship in a 'live town'; but at fifty-five he merely hated them because they kept him awake. They 'pressed on his nerves.' . . ." Choice (A) is incorrect since this passage is the only place that refers to his earlier health, and it implies that he was at least sometimes ill. Choice (B) is incorrect because the passage does not say whether or not he worked; it does imply that he did not spend his nights working, since the night sounds sometimes "interfered with his night's sleep." As for Choice (D), we know that he found his nurse unattractive early that morning, but the passage gives us no other information on this topic. And Choice (E) is incorrect, since we know that he did not sleep on the night depicted in this passage, and it is implied that he often lay awake during the night; however, we have no way of knowing whether or not he used to sleep soundly all night long.

14. (D) "Cries" is the only selection that reflects a correct definition of "vociferations," although all of the choices are sounds mentioned in the passage.

15. (D) "He heard the milk-wagon drive into the cross-street beneath his windows and stop at each house. The milkman carried his jars round to the 'back porch,' while the horse moved slowly ahead to the gate of the next customer and waited there. 'He's gone into Pollocks' . . . Delivered the Andersons.' Now he's getting out ours.'" Clearly, the milkman was going from house to house, including that of Adams. Later, "A cheerful whistler passed the house. . . . then a group of coloured workmen came by. . . ." Here you see a direct reference to "the house." Therefore, you can assume Adams was not in (A) a hotel or (E) an apartment building. And although he had a nurse, the above passages imply that the setting was not (B) a nursing home or (C) a hospital. Also, it would not be likely that a nurse would sleep on a cot in the same room as the patient or use his bathroom to get dressed if the setting were a nursing home or a hospital.

16. (D) "Shamble" means "shuffle; moving slowly, dragging one's feet." So the fact that the two men were "shambling" implies that they were lazy and not motivated to feed the stock. Choice (A) is incorrect because men who were eager to work would not be shambling. Choice (B) is incorrect because the fact that they were shambling does not indicate whether or not they were friends. If you did not have the rest of the passage to read, you might find Choices (C) and (E) to be possible correct answers; however, reading even a portion of the rest of the passage tells you they were neither ill nor disabled.

17. (C) The description of the way the mare was injured, Jerry's treatment of the mare when she was injured, and the horse's condition as a result of the injury and Jerry's mistreatment are indicative of the type of person Jerry was. Claude, on the other hand, cared about the mare and treated her gently and kindly; so generally, the mare's life had not been a hard one. Therefore, Choice (A) is wrong. The mare was injured because Jerry had let her step on a board with a nail that had been sticking up—not because of the condition of the soil on the farm. Thus, Choice (B) is wrong. The mare was evidently in good health, except for the wounded foot; she was suffering only because of this injury, but was evidently healing and was well enough to respond to Claude's kindness. So Choice (D) is wrong. And since the passage includes nothing that indicates the mare's injury had anything to do with the relationship between Claude and his father, Choice (E) is incorrect.

18. (E) The dictionary definition of "wretchedly" is "miserably; woefully." The description of the mare reveals that she was suffering from her injury, so it makes sense that she would be miserably thin. If you are not familiar with this word, try replacing it with each of the choices, and choose the one that seems most appropriate. For example, describing the injured mare's body as "miserably thin" is appropriate. Choice (A) is incorrect because this word doesn't make sense if you try to put it in the place of "wretchedly" in this sentence. Likewise, saying the mare's body was "angrily thin" makes little sense. So Choice (B) is incorrect. Also, describing her body as "ordinarily thin" is not appropriate here, since the purpose of the descriptive passage is to let the reader know how much the mare was suffering. Therefore, Choice (C) is not appropriate. And since it is illogical to say she was "fitfully thin," Choice (D) is not a good fit.

19. (D) Mr. Wheeler said, "I don't mind about your washing the car; mud preserves the paint, they say, but it'll be all right this time, Claude." As a result, Claude feels embarrassed and humiliated. Mr. Wheeler's comment is "mocking" (ridiculing; treating with contempt). Mr. Wheeler's reaction is not (A) supportive, (B) oblivious (unaware), (C) appreciative, or (E) sympathetic.

20. (B) Claude was upset and embarrassed because his father had decided that Claude would be going into town in the mule wagon with the two hired men whom he disliked, and had ridiculed him for reacting as he had. His emotions affected the way the pancakes tasted to him, as well as his ability to swallow them. Choice (A) is incorrect because the passage indicates nothing about Mahailey's ability to cook, and the focus is on Claude's emotions toward the two hired men. The passage indicates nothing about (C) how poor the family was, (D) how long it took Claude to eat, or (E) when the family had begun eating.

21. (C) The dictionary definition of "wanton" is "merciless; inhumane." Also, by reading the passage, you know that four steers died in a blizzard as a result of Dan and Jerry's "wanton carelessness." This clue tells you that their carelessness was inhumane and without mercy; thus, Choice (C) is the best choice here. Although the men were lazy, their carelessness was more than "lazy" or "lackadaisical"; here, you need a stronger word that more specifically reflects what they had done (or not done). Also, "lazy; lackadaisical" is not a correct definition of "wanton." Choice (B) is wrong because "wanton" does not mean "passionate; excited." But even if you did not know the definition of the word, you can tell by reading the passage nothing the men did was "passionate" or "excited." Choice (D) is wrong because it does not reflect the meaning of "wanton"; additionally, although the men's behavior may be deplorable, nothing in the passage tells us that they were "immature" or "youthful." Choice (E) is wrong because these words are not part of the definition of "wanton," and nothing about these men's carelessness was restrained or bridled (controlled; checked).

22. (A) Be careful to select the choice that is **not** something Claude dreaded. Nothing in the passage indicates that Claude did not want to take his brother with him. And when he asked his father, "Can't we have the car?" he appeared to be including his brother in his plans. Choice (B) is wrong because he did dread going with the hired hands; the passage says, "His father . . . knew how he hated to go anywhere with Dan and Jerry." Choices (C) and (E) are wrong because "His father knew he hated to drive the mules to town. . . . mules that always brayed and balked and behaved ridiculously in a crowd." Choice (D) is incorrect; we know he dreaded loading the cowhides because "he must take these stinking hides" of "four steers that had perished in the blizzard last winter through the wanton carelessness of these same hired men, and the price they would bring would not half pay for the time his father had spent in stripping and curing them."

23. (D) The hired hands and Claude's brother laughed at Claude when Mr. Wheeler said that the boys should go into town with the hired hands and made a sarcastic comment about Claude's washing the car. Clearly, in this scene the conflict has Claude on one side; his brother, his father, and the hired men were on the other side. The passage includes nothing that indicates the Wheeler family—other than Claude—was in conflict with the hired men; in fact, Mr. Wheeler reminded Claude that they probably wanted to go to the circus just as much as he did. So Choice (A) is wrong. Nothing points to any conflict between the Wheelers and the farm animals; only Jerry had been cruel to an animal. Therefore, Choice (B) is wrong. Choice (C) is wrong because nowhere does the passage indicate that Claude and his brother were pitted against the rest of the household; in fact, Claude's brother joined in the laughter at Claude's expense. So Choice (C) is incorrect. And Choice (E) is incorrect because Mahailey was not depicted as being in conflict with anyone.

24. (D) Claude was upset because " . . . today, when he wanted to go to Frankfort clean and care-free, he must take . . . two coarse-mouthed men. . . ." He disliked Dan and Jerry because Jerry had mistreated the mare and because the four steers had died in a blizzard due to the two men's carelessness. And it is clear throughout the passage that Claude disliked the men's attitude in general. Choices (A) and (C) are incorrect because nothing is mentioned about their clothes. Choices (B), (C), and (E) are wrong because the passage does not tell us anything about their drinking habits.

DETAILED EXPLANATIONS

SECTION 7—MATH

1. (B) From the first sentence, the number we seek must be divisible by 6. In the second sentence, when this number is divided by 5, the remainder is 2. Only the number 72 from these answer options would satisfy both these conditions.

2. (D) In an "if . . . then" compound statement, the only way the statement is false is when the antecedent is true but the consequent is false. The antecedent in the example given is "it rains" and the consequent is "Peter will not play golf." So the statement that "it is raining and Peter is playing golf" must be false. Options (A), (B), and (C) could be true, since we are not given any certainties if it is not raining. Option (E) is possible, because all we can say for certain is that Peter will not play golf.

3. (E) There is no multiplication of elements possible, one from R and the other from S, such that the result is –20. Option (A) is possible using 7 and –1. Option (B) is possible using 5 and –2. Option (C) is possible using 7 and –2. Option (D) is possible using 3 and –5 or 5 and –3.

4. (A) The dollar profit is $t - p$. The percent profit is based on the cost, which is the "bought" amount. It is defined as the dollar profit divided by the bought amount, then multiplied by 100. For this example it is $(100)\frac{(t-p)}{p}$.

5. (D) $\angle FHJ = 30°$, so that FHJ is a $30° - 60° - 90°$ right triangle. $JH = (2)(FJ) = 20$, and $FH = 10\sqrt{3}$. The perimeter is then $10 + 10\sqrt{3} + 20 = 30 + 10\sqrt{3}$.

6. (C) The sum of any two sides of a triangle must exceed the length of the third side. Because $1 + 3 < 5$, no triangle can have sides of 1, 3, and 5.

7. (B) $f\left(\frac{1}{[x+2]}\right) = 2$ divided by $\left(\frac{1}{[x+2]} + 1\right) = 2$ divided by $\frac{(x+3)}{(x+2)}$, which simplifies to $\frac{2(x+2)}{(x+3)}$. Note: To simplify $\frac{1}{[x+2]} + 1$, convert 1 to $\frac{(x+2)}{(x+2)}$, so that this expression becomes $\frac{(1+x+2)}{(x+2)}$.

8. (A) $6^n = (2^n)(3^n) = xy$. However, $6^{n-1} = \frac{6^n}{6} = \frac{xy}{6}$.

9. (C) Given an external point from a circle, two tangents drawn to the circle from that point must be equal. So $PQ = PW$, $QR = RS$, $ST = TU$, $UV = VW$. Then $PQ + QR + TU + UV$ must represent one-half the perimeter of $PRTU$, which is 14. By substitution, we get $PR + TV = 14$. Note that $PRTV$ is not any specific kind of quadrilateral.

10. (B) Let x = time required by Mr. Smith if he worked alone. Then, $\frac{9}{x} + \frac{9}{12} = 1$. Multiplying the equation by $12x$, we get $108 + 9x = 12x$. Then $3x = 108$, so $x = 36$.

11. (D) The square root of any quantity must be at least zero, so $x^2 - 9x + 20 \geq 0$. By factoring, $(x-5)(x-4) \geq 0$. This is possible if both factors are greater than or equal to zero, so $x \geq 5$. This is also possible if both factors are less than or equal to zero, so $x \leq 4$.

12. Answer: 25 — Let x represent the number of gallons of pure water needed. The original jar has 7 gallons of acid. When the pure water is added, there will still be 7 gallons of acid; however, the total number of gallons will increase to $10 + x$. Since there will now be a 20% acid concentration, $0.20 = \frac{7}{(10+x)}$. This becomes $0.20(10 + x) = 7$. So, $2 + 0.20x = 7$. Solving, $x = \frac{5}{0.20} = 25$.

13. Answer: 2/15 — The fraction of non-Republicans is $1 - \frac{3}{5} = \frac{2}{5}$. Since we know that one-third of these people are retired, $\left(\frac{2}{5}\right)\left(\frac{1}{3}\right) = \frac{2}{15}$ of the people of Anyville are retired and not Republicans. (Note: we cannot determine the total number of retirees in Anyville since we don't know the percentage of Republicans who are retired.)

14. Answer: 198 — Because 88 faculty members represents the age bracket 30 to 39, there must be a total of $\frac{88}{0.20} = 440$ faculty members. The age bracket 40 to 49 represents $100\% - 30\% - 35\% - 20\% - 5\% = 10\%$ of the total, so that 45% of the faculty is at least 40 years old. (Total of the age brackets 40 to 49, 50 to 59, and over 59). Thus, $(440)(0.45) = 198$ faculty members are at least 40 years old.

15. Answer: 99 — Because 36 green marbles represent 8% of the marbles, there are a total of $\frac{36}{0.08} = 450$ marbles. $100\% - 8\% - 70\% = 22\%$ of the marbles are red, which means there are $(450)(0.22) = 99$ red marbles.

16. Answer: 38, 39, or 40 — The three angles of a triangle must add up to 180°, so $3y + y + x = 180°$. The lowest x possible is 19, for which $4y + 19 = 180$. The solution to this equation is $y = 40.25$, so the highest allowable y value is 40. The highest x possible is 29, for which $4y + 29 = 180$. The solution to this equation is $y = 37.75$, so the lowest allowable y value is 38.

DETAILED EXPLANATIONS

SECTION 8—CRITICAL READING

1. (E) Consider what type of facial expression conveys a message that would cause one to feel comfortable. "Supercilious" means "haughty, scornful, and judgmental, especially as revealed by raised eyebrows and pursed lips." Such an expression—especially on the face of a future mother-in-law—would certainly make one feel uncomfortable. A (A) "jocular" (playful; jolly), (B) "composed" (peaceful), (C) "hospitable" (cordial, welcoming) or (D) "compassionate" (tender; warmhearted) expression would not make him feel that way.

2. (B) Since the sentence tells you that she worried about whether or not others liked her and approved of her, you know that popularity was important to her. So you can assume that she perceived popularity and success as equal. So you are looking for a word that implies equality. "Tantamount" means "equal in value, meaning, or effect; equivalent." So this is an appropriate choice. Choice (A) is incorrect, since "paramount" means "superior to all others" and does not fit in this context. Choice (C) is wrong because "dissimilar" means "unlike; different." And it does not make sense to say that popularity was "derogatory" (belittling; disadvantageous) to success; thus, Choice (D) is wrong. Likewise, it is illogical to say that popularity was "inconsequential" (irrelevant; unrelated) to success, since she regarded them as connected.

3. (B) Here, you need an adjective that describes a person who does not plan in advance and who is not open to viewpoints other than the one he has. A "myopic" person is short-sighted in terms of planning and thoughtfulness, and maintains a narrow view of things; thus, this choice is appropriate. Choice (A) is incorrect, since "cosmopolitan" means "having worldwide rather than limited scope; possessing the polish and experience of wide experience." And although it might be fruitless to try to get an "eccentric" (strange, odd) or "bellicose" (belligerent; warlike) person to plan in advance or see a different viewpoint, the simple fact that he is eccentric or bellicose does not mean that this would be the case. And a "progressive" (moving forward, moving onward) person would be likely to look ahead and to be open to new ideas; so Choice (E) is not a good choice.

4. (A) You are seeking an adjective that conveys the way a panel of experts would see the comments of a newcomer who knew the appropriate vocabulary and lingo (special vocabulary of a particular field of interest) but who didn't have enough experience or education. "Sophomoric" is an adjective that means "overly confident with knowledge but poorly informed; characteristic of immaturity arising from too little learning." Thus, a panel of experts would probably see such a person and what he had to say as "sophomoric." And although the panel might also see him as "irresolute" (wavering; indecisive), nothing in the sentence indicates that they would see him in this way; thus, Choice (B) is incorrect. Likewise, they might also see him and his comments as "oblivious" (forgetful, unaware); however, the sentence offers no clues that they did so; therefore, Choice (C) is not appropriate. And Choices (D) and (E) are incorrect because this panel more than likely would not see him as "sagacious" (intelligent; insightful) or "perceptive" (observant; discerning) since he was unable to function at their level.

5. (C) Your clues are "in an antique store," "good fortune" and "unusually . . . condition." So you want an adjective that describes a vase that one would find in an antique store and feel fortunate to have found it. And you want another adjective that describes the (unusual) condition of such a vase. While shopping in an antique store, one would be fortunate to find an "antediluvian" (very old; primitive) vase that was in "pristine" (like new; unspoiled) condition. One would not be likely to find a "contemporary" (modern, current, modern-day) vase in an antique store, and one would probably not consider finding a "meretricious" (tawdrily and falsely attractive) vase a good fortune. Therefore, Choice (A) is wrong. Choice (B) is incorrect since it is illogical to say that something was in unusually "mediocre" (ordinary) condition, even though one might be fortunate to find an "exquisite" (beautiful) item. And Choice (D) is wrong because "illicit" (illegal) and "ignoble" (dishonorable) do not make sense in this context. As for Choice (E), one might be fortunate to find a "Noachian" (antique, very old) vase, but not if it was in unusually "shoddy" (inferior) condition.

6. (B) Here, you have two contrasting ideas; you are looking for an adjective that describes a group comprised of students with common abilities and interests, and another one that describes a group with unique abilities and interests. "Homogeneous" (composed of elements of the same or a similar kind or nature) and "heterogeneous" (consisting of dissimilar or diverse ingredients or constituents) fit well here. Choice (A) is incorrect because although a "gregarious" (social; sociable; tending to associate with others of one's kind) group might very well have common abilities and interests, an "asocial" (withdrawn; selfish) group probably would not contribute abilities or share knowledge with one another. Choice (C) is incorrect because "incongruous" (not harmonious; out of place) does not describe a group with common abilities and interests, and a "divisive" (causing disunity, disagreement, or discord) group is not likely to contribute their abilities or share with one another. Although students in a "perspicacious" (intelligent) group with common abilities and interests might learn, it does not make sense that more learning would occur in a group that is "dispersed" (scattered, disassembled) or that students in such a group would contribute their abilities and share their knowledge. So Choice (D) is incorrect. And Choice (E) is wrong, since "exclusionary" and "restricted" (limited; selective) are not really contradictory or contrasting.

7. (B) The passage states, "The agrarian lifestyle . . . dictated that the school term last only a few months each year. . . . Trying to hold classes during planting or harvest time would have been futile, since the all family members—including the children—were expected to help." These sentences prove Choice (A) incorrect. And although the passage does indicate that the farmers' income was limited, nowhere does it imply that money was more important than the other choices; so Choice (C) is wrong. Likewise, the passage indicates that the community had certain expectations regarding the teacher's religious practices; however, it says nothing about the importance of church in relation to the other choices listed; so Choice (D) is incorrect. And the passage does not mention recreation; therefore, Choice (E) is wrong.

8. (C) The dictionary definition of "agrarian" is "agricultural, farming." However, even if you did not know the meaning of this word by itself, you should have been able to use the context clues to determine the correct choice. Following the sentence containing this word, the passage mentions "the farmers' profits" and "planting or harvest time." Choice (A) is wrong because it is not the definition of the word; additionally, the passage makes no reference to anything urban. And although an agrarian lifestyle may be a simple one, "simple" is not a definition of "agrarian." Choice (D) is wrong because it is not a definition of the word, and nothing in the passage implies a complex lifestyle. Choice (D) is wrong because the passage does not address anything that is unpredictable about the lifestyle described in the passage; also, this is not a definition of "agrarian."

9. (E) Passage 1 says, "Regardless of gender, the teacher who smoked, drank liquor . . . was considered to be a poor role model and was in danger of losing his or her position as a teacher. " Choice (A) is wrong because nothing in the passage says whether or not such behavior was acceptable for community members. Choice (B) is wrong because the passage says these behaviors were unacceptable for all teachers "regardless of gender." Choice (C) is wrong because these behaviors were *un*acceptable for teachers. And Choice (D) is incorrect because teachers were expected to attend church, but they were expected *not* to drink or smoke.

10. (C) The passage says, "In many districts, education has become 'big business,' with multimillion dollar budgets, contracted janitorial and food services, and strict employment policies." The passage makes no mention of individual students except to say, "Advanced technology has the potential of decreasing—and perhaps someday eliminating—personal teacher-student interactions." Therefore, Choice (A) is wrong. Choice (B) is incorrect because "Today's teacher faces few restrictions regarding their personal lives." Choice (D) is incorrect because "Little if any consideration is given to family vacations and other activities when the school calendar . . . is planned." And nowhere are teachers' interests mentioned; thus, Choice (E) is wrong.

11. (A) The passage says, " . . . more common are incidents of parents' supporting their children for breaking the rules or failing to cooperate with their teachers. . . . . administrators spend hours developing a disciplinary policy, consulting lawyers to make sure it is legal, and hoping they have covered all possible scenarios. But invariably students and parents will find a loophole and question—or even file a lawsuit against—the school that tries to enforce that policy strictly." Nothing in the passage indicates that (B) parents are at odds with students, that teachers are working against (C) administrators or (D) other teachers, or that (E) schools are in disagreement with churches. Thus, all these choices are wrong.

12. (B) The adverb "unfortunately" introduces the idea "more common are incidents of parents' supporting their children for breaking the rules or failing to cooperate with their teachers." This word implies that the author does not agree with such parents' actions. Choice (A) is wrong because nothing in the passage indicates whether or not the author approves of schools' disciplinary policies. And although the passage does say "Corporal punishment is almost non-existent in schools, and disciplinary action is quite limited," nothing indicates that the author (C) prefers the way schools were run in the past or (D) supports corporal punishment. And Choice (E) is incorrect because the author does not address administrators' typical approach to discipline.

13. (E) The dictionary definition of "invariably" is "consistently." However, the context clues should also help you determine the definition. The paragraph contains such phrases as "all too often," "more and more common," and "too frequently"—all of which indicate the consistency of such events. Thus, "consistently" is a logical choice even if you do not know the definition of "invariably." This word does not mean (A) "annoyingly," (B) "eagerly," or (C) "angrily." Thus, all these choices are incorrect. And Choice (D), "litigiously" is redundant, since "file a lawsuit" is used later in the sentence.

14. (C) Passage 2 indicates, "It is illegal not to hire a teacher—or to dismiss a teacher—on the basis of his or her marital status, pregnancy, religious preference or participation, recreational activities, or almost any other reason except breaking the law or having a criminal history." Therefore, Choices (A), (B), and (D) are incorrect. And since the passage does not address a teacher's participation in political activities, Choice (E) is also wrong.

15. (D) Passage 1 indicates that in the nineteenth century, families' contributions included making school furniture, allowing the teacher to live with them, providing fuel for the stove, and reinforcing disciplinary actions. Little if any of this type of community support is evident today, according to Passage 2. Choice (A) is incorrect because the nineteenth-century teacher had to do many of the janitorial chores (e.g., bring in fuel for the stove and start a fire before the children arrived); however, in today's schools such services are frequently contracted, as they are in other big businesses. Choice (B) is wrong because in the nineteenth century "the teacher's tools in the best-equipped classroom included a slate, chalk, and books"; but today's schools have "up-to-date computers and computer programs, textbooks, encyclopedias, calculators, and televisions." Choice (D) is incorrect, since "the rural schoolhouse in nineteenth-century America had limited financial resources." But many districts today have "multi-million dollar budgets." And Choice (E) is incorrect since the nineteenth-century "agrarian lifestyle . . . dictated that the school term last only a few months each year." But today "school calendar" is "sometimes twelve months long."

16. (B) Passage 1 implies that a portion of the farmer's meager income went to support the school. And it says, "Establishing and maintaining the school was a community project; thus, the community members strived to provide what they could to support the school. Farmers provided the wood for the stove. Parents crafted the desks and chairs." And the families even had the teacher live with them. Clearly, the families were involved. By contrast, Passage 1 indicates that today "all too often, those who have children in attendance never set foot in the school to meet the teachers or participate in any school activities." Their involvement tends to be more negative than positive, with criticism and sometimes even lawsuits. Choice (A) is incorrect because it is exactly the opposite of what is indicated in the passages. Choice (C) is incorrect because Passage 1 indicates that the nineteenth-century family supported the school's discipline policy and implies other types of support. And although Passage 1 does say that the community members financially supported the schools, nowhere does it say that they were legally obligated to provide that support; however, Passage 2 says that today's citizens "are required to support the schools by means of paying taxes." So Choice (D) is wrong. And Choice (E) is wrong because neither passage addressed the educational level of the family.

17. (A) In Passage 1, line 18 refers to the teacher as a public servant, as do lines 69 and 70 of Passage 2. Choice (B) is incorrect because Passage 1 implies nineteenth-century public support of the teacher's discipline by stating, " . . . in most cases the student who was punished at school was punished at home as well," but Passage 2 indicates today's lack of public support: "it is not uncommon today for school to face community criticism, especially if teachers attempt to enforce strict disciplinary rules." Choice (C) is wrong because Passage 1 describes the limitations of a teacher's personal life in the nineteenth century, while Passage 2 indicates, "Today's teachers face few restrictions regarding their personal lives." As for school supplies, Passage 1 addresses the nineteenth-century farmer's meager income and the inability to pay for many supplies; by contrast, Passage 2 lists the plentiful supply of equipment required in today's schools. Therefore, Choice (D) is wrong. And Choice (E) is wrong because Passage 2 does not address the modern-day teacher's living arrangements, except to say that a school cannot refuse to hire and cannot dismiss a teacher for specific personal reasons. Yet Passage 1 indicates that a teacher often lived with students' families on a rotating basis.

18. (D) Passage 2 indicates that the nineteenth-century educator taught in a classroom where "students were forbidden to communicate with one another during class. They were to focus totally on their school work. Common punishments for infractions of these rules included a rap on the knuckles with a ruler, sitting in front of the class while wearing a dunce cap, standing in the corner, and being swatted with a wooden paddle." Today, a teacher who maintained such standards would be considered far too strict in a world where "corporal punishment is almost non-existent in schools, and disciplinary action is quite limited." Choice (A) is wrong because it reflects an idea that is exactly the opposite of what Passage 2 indicated. Choices (B), (C), and (E) are incorrect because nothing in the passages indicates whether or not such a teacher would be regarded as "boring," "uneducated," or "poor."

19. (C) In the nineteenth century, "walking into the schoolroom, the students were expected to take off their hats, bow to the teacher, and walk straight to their seats." However, today parents commonly support "their children for breaking the rules or failing to cooperate with their teachers," thus implying lack of respect for teachers on the part of both parents and students. Choice (A) is incorrect because nothing in the passage indicates that today's students show the degree of respect that Passage 1 describes regarding nineteenth-century students. Choice (B) is wrong because the passages show that the opposite is true. Choices (D) and (E) are incorrect because nothing in the passage addresses the intelligence of students in the nineteenth century or today.

DETAILED EXPLANATIONS

SECTION 9—WRITING

1. (B) It consistently uses past tense ("was" and "wrote") in both clauses. Choice (A) incorrectly shifts from past tense "was" to present perfect "has written." Likewise, Choice (C) switches to the progressive form of present perfect ("has been writing"); Choice (D) switches to past perfect ("had written"); and Choice (E) switches to past progressive ("was writing").

2. (C) It contains parallel structure, with the consistent use of plural nouns ("hunters" and "athletes") in the "either . . . or" phrase. Choice (A) involves an error in parallelism, with the use of a noun ("hunters") and an adjective ("athletic"). In Choice (B), the prepositional phrase "at hunting" is not parallel with the adjective "athletic." In Choice (D) the noun "hunters" is not parallel with the adjective phrase "good at athletics." And in Choice (E) the prepositional phrase "at hunting" is not parallel with the noun "athletes."

3. (B) This sentence correctly compares "his third-grade teacher's classroom" and "that of his second-grade teacher." The use of "to" with "similar" creates an appropriate idiom. Choices (A) and (C) illogically compare "his third-grade teacher's classroom" and "his second-grade teacher." Choices (C) and (D) create an inappropriate idiom with the phrase "similar with." And Choice (E) omits the preposition "to" after "similar," creating an illogical and confusing sentence.

4. (C) Although all the choices are acceptable idioms, each has a different meaning, and only Choice (C) is appropriate for this sentence. The objects of Choices (A) "agree to," (B) "agree upon," (D) "agree on," and (E) "agree about" are typically things, not people; thus, none of these choices is correct.

5. (C) All the items in the series are infinitive forms of the verb followed by an adverb ("to answer . . . courteously," "use . . . proficiently," "greet . . . cordially," "file accurately"); thus, they are parallel. Choice (A) involves nonparallel structure, since the last item in the series is a clause ("she . . . too."), instead of a verb and an adverb ("to answer . . . courteously," "use . . . proficiently," "greet . . . cordially") as the other items are. Also, the word "too" is unnecessary; it is redundant, since the word "and" has already been used. Choice (B) has an error in parallelism, since "knowing how to file accurately as well" is not parallel with the other items in the series. In Choice (D), the verb "file" is parallel with the other verbs in the series; however, the prepositional phrase "with accuracy" is not parallel with the adverbs in the other items in the series. Also, the unnecessary phrase "as well" makes the sentence wordy. In Choice (E), the word "how" keeps the phrase from being parallel with the other phrases in the series.

6. (E) It consistently uses "you" as the subject for both clauses, and the present tense verb "cannot" is consistent with the present-progressive verb "are watching." Choice (A) shifts from third person ("a person") to second person ("you"). Choice (B) has the same problem, switching from third person ("one") to second person ("you"). Choice (C) is incorrect because it switches from past tense ("could") to present progressive ("are watching"). Choice (D) has the same problem; it also switches from third person ("one") to second person ("you").

7. (A) The use of "not only . . . but also" requires parallel structure. In Choice (A), "to Paris" and "to London" are parallel; thus, it is correct to say "not only to Paris but also to London." Choice (B) is incorrect because "wants to travel to Paris" is not parallel with "to London." Choice (C) is wrong because "to Paris" and "to visit London" are not parallel. Choice (D) is incorrect because "does he want to travel to Paris" and "to London also" are not parallel. And Choice (E) is incorrect because "to travel to Paris" is not parallel with "to London."

8. (B) The plural verb "want" agrees with the plural subject "people," and a semicolon correctly separates the two complete thoughts ("Two . . . issue" and "the . . . controversy"). Choice (A) incorrectly uses a singular verb ("wants") with a plural subject ("people"); it also contains a comma splice, using only a comma between the two clauses ("Two . . . issue" and "the . . . controversy"). Choice (C) uses a singular verb ("wants") with a plural subject ("people"); also, it is a run-on sentence, using no punctuation between the two clauses ("Two . . . issue" and "the . . . controversy"). Choice (D) is incorrect because "wanting" cannot serve as the main verb of a clause. And Choice (E) is a run-on sentence, with no punctuation between the two clauses; a semicolon should be used before the conjunction "however."

9. (E) It creates a correct sentence by eliminating the comma and the pronoun "it" that are in Choice (A). In Choice (A), "it" is redundant, repeating the subject. Also, in this sentence, a comma is incorrectly placed after the subject. Choice (B) has these same errors, along with an incorrect idiom ("interferes upon"). Choice (C) also uses this incorrect idiom. Choice (D) is a fragment because the subject "inability" has no verb.

10. (B) Choice (B) is the correct answer, with correct subject-verb agreement and no unclear antecedents. In Choice (A) the antecedent of "they" is unclear. Choice (C) illogically has "snows" as the subject of "have capitalized." In Choice (D) the antecedent of the pronoun "they" is unclear. And Choice (E) illogically has the phrase "having . . . resorts" modifying "snows," with the implication that the snows are heavy because they have capitalized on the climate and built elaborate ski resorts.

11. (D) Choice (D) is the correct answer, correctly completing the comparison with the word "than." Choice (A) involves improper comparison and wordiness: "compared with" should be "than." Choice (B) also involves improper comparison and wordiness: "compared to" should be "than." Choice (C) uses the unnecessary word "by" and is awkward. Choice (E) involves improper comparison, lacking "than" and using the unnecessary word "unlike."

12. (A) It uses correct verb forms and punctuation. Choice (B) involves an error in punctuation, omitting the comma at the beginning of the non-essential phrase ("illustrating a specific point or moral"). Choice (C) includes the unnecessary pronoun "it," creating redundancy, since it repeats the subject. If you consider "it" to be the subject of "involves," then the first part of the sentence is a fragment, with no verb for "fable." Choice (D) illogically indicates that a fable involves supernatural occurrences or animals that act or talk like humans *because* they illustrate a specific point or moral. Choice (E) uses an improper verb form with "involving" instead of "involves." The participle "involving" cannot serve as the verb.

13. (C) It uses a semicolon to separate the two complete thoughts ("Planaria . . . animals" and "because . . . asexual") and a comma after the dependent clause ("because . . . parts") in the second part of the sentence. Choice (A) is a run-on sentence, using no punctuation to separate the two complete thoughts ("Planaria . . . animals" and "because . . . asexual"), and it does not use a comma to separate the dependent clause ("because . . . parts") from the main clause ("their . . . asexual") in the second part of the sentence. Choice (B) contains a comma splice, using only a comma to separate the two complete thoughts ("Planaria . . . animals" and "because . . . asexual"). Choice (C) has the same problems as Choice (A); substituting the word "because" with "since" does not correct any of the errors. In Choice (E), the pronoun "it" has no antecedent.

14. (E) It correctly uses "their" to modify the gerund "regarding." Also, the two items connected by "and" are parallel ("become" and "quit"). Choice (A) incorrectly uses the pronoun "them" to modify the gerund "regarding." Also, at the end of the sentence "to become" is not parallel with "quitting," and it does not make sense to say "caused her . . . quitting her job." Choice (B) is a fragment, with no verb for the subject "regarding." In Choice (C), "caused" has no subject. Choice (D) incorrectly uses the pronoun "them" to modify the gerund "regarding"; it also separates the essential phrase "as simply a beauty queen" from the rest of the sentence.

PRACTICE TEST 3

Answer sheets for this test start on page 857.

TIME: 25 Minutes

ESSAY

DIRECTIONS: You have 25 minutes to plan and write an essay on the topic below. You may write on only the assigned topic.

Make sure to give examples to support your thesis. Proofread your essay carefully and take care to express your ideas clearly and effectively.

ESSAY TOPIC

We have all heard the adage "The love of money is the root of all evil." However, Mark Twain said, "The lack of money is the root of all evil."

ASSIGNMENT: Consider the two contrasting statements above. Select the one that most closely reflects your view. Then write an essay that explains your choice. To support your view, use an example or examples from history, literature, the arts, current events, politics, science and technology, or your experience or observation.

TIME: 25 Minutes
20 Questions

MATH

DIRECTIONS: In this section solve each problem, using any available space on the page for scratchwork. Then decide which is the best of the choices given and fill in the corresponding oval on the answer sheet.

NOTES

(1) The use of a calculator is permitted.

(2) All numbers used are real numbers.

(3) Figures that accompany problems in this test are intended to provide information useful in solving the problems. They are drawn as accurately as possible EXCEPT when it is stated in a specific problem that the figure is not drawn to scale. All figures lie in a plane unless otherwise indicated.

(4) Unless otherwise specified, the domain of any function f is assumed to be the set of all real numbers x for which $f(x)$ is a real number.

REFERENCE INFORMATION

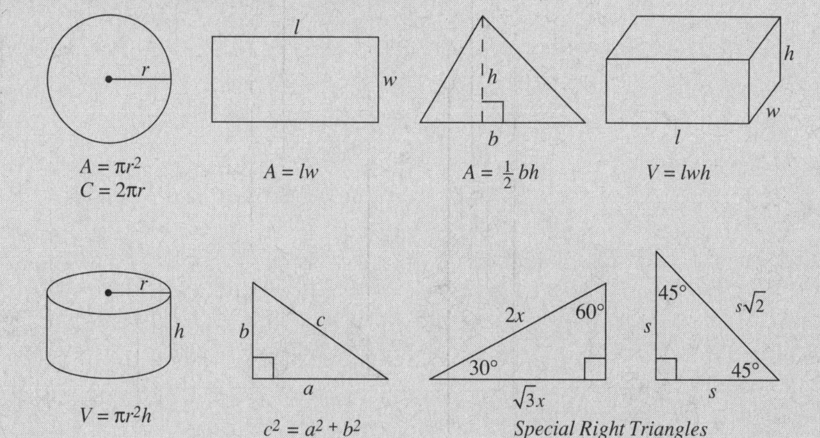

The number of degrees of arc in a circle is 360.

The sum of the measures in degrees of the angles of a triangle is 180.

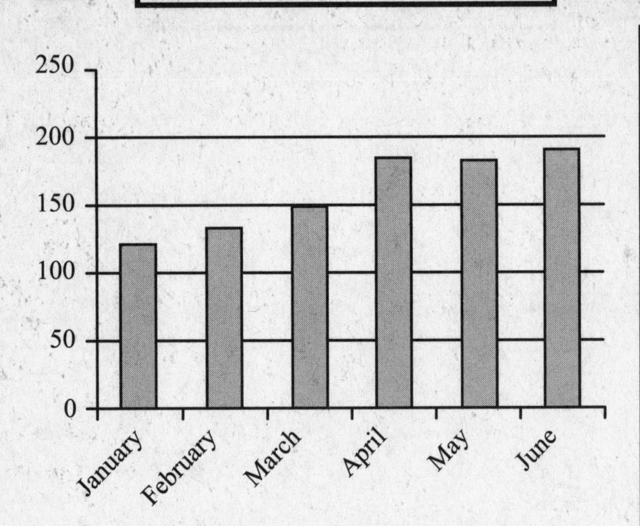

1. The graph above shows the profits of an international company during the first six months of the year in millions of dollars. For which month was there the greatest increase over the previous month?

(A) February
(B) March
(C) April
(D) May
(E) June

2. For a number m, let $\lhd m \rhd = \frac{m^2+1}{m^2-1}$. Which of the following is equal to $\lhd a^2 \rhd - \lhd a \rhd$?

(A) $\frac{2a^2}{a^4-1}$ t
(B) $\frac{-2a^2}{a^4-1}$
(C) $\frac{a^2}{a^4-1}$
(D) $\frac{-a^2}{a^4-1}$
(E) $\frac{-2a^2}{a^2-1}$

3. If $f(x) = x^4 - x^2 + 1$, the graph of which of the following functions is the same as the graph of $f(x)$?

(A) $-f(x)$
(B) $f(x^2)$
(C) $f(-x)$
(D) $-f(-x)$
(E) $f(-x^2)$

4. 456 is 80% of three times what number?

(A) 121.6
(B) 190
(C) 273.6
(D) 570
(E) 1,094.4

$\frac{80}{100} = 456$

5. If $\frac{x^{-2}+y^{-3}}{x^{-3}+y^{-2}} = 1$, which of the following is equal to $\frac{1-x}{1-y}$?

(A) $\left(\frac{y}{x}\right)^2$
(B) $\left(\frac{x}{y}\right)^3$
(C) $\frac{x}{y}$
(D) $\frac{y}{x}$
(E) xy

GO ON TO THE NEXT PAGE

6. Consider the function $f(x) = mnx^2$, where m and n are constants. Which of the following sequences is geometric?

(A) $f(2), f(6), f(8)$

(B) $f(1), f(2), f(3)$

(C) $f(2), f(4), f(8)$

(D) $f(3), f(5), f(7)$

(E) $f(3), f(4), f(6)$

7. Desiree earns $240 per month plus 12% commission on her sales. If x represents her sales per month, which function represents her total income (y) per month?

(A) $y = 240(x + 12)$

(B) $y = 012(x + 240)$

(C) $y = 240 + 0.12x$

(D) $y = 240x + 0.12$

(E) $y = 240x + 12$

8. Maurice rides his bicycle downhill to the store twice as fast as he rides uphill going home. If the total trip is 8 miles and he rides for a total of 30 minutes, how fast does he ride going home?

(A) 5 miles per hour

(B) 8 miles per hour

(C) 10 miles per hour

(D) 12 miles per hour

(E) 24 miles per hour

9. On line segment $\overline{ADEMPQRB}$, M is the midpoint. $\overline{AM}$ is divided into three equal segments by the points D and E. $\overline{MB}$ is divided into four equal segments by the points P, Q, and R. What is the probability that a point chosen on $\overline{AB}$ will be on $\overline{EP}$?

(A) $\frac{5}{12}$

(B) $\frac{5}{24}$

(C) $\frac{7}{12}$

(D) $\frac{7}{24}$

(E) $\frac{12}{14}$

10. The scores of students in Geometry classes are represented by the inequality $|2x - 130| \leq 30$. Which interval shows the range of the scores?

(A) $100 \leq x \leq 80$

(B) $80 \leq x \leq 85$

(C) $80 < x < 85$

(D) $50 < x < 80$

(E) $50 \leq x \leq 80$

11. The set M includes all the integers greater than 12 and less than 24, set N includes all the even integers greater than 4 and less than 20, and set P includes all even integers greater than 8 and less than 28 the sum of whose digits are less than 5. Which set is equal to $M \cap N \cap P$?

(A) {14, 16}

(B) {16, 18}

(C) {18}

(D) {14}

(E) $M \cap N \cap P$ is an empty set

GO ON TO THE NEXT PAGE

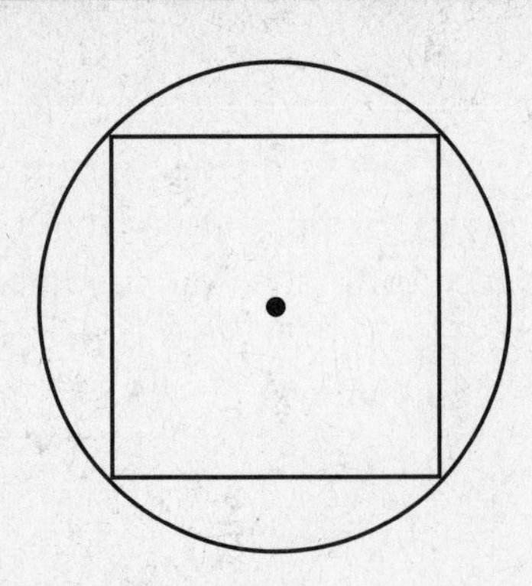

12. In the figure above, a square is inscribed in a circle. Denote the length of the radius of the circle by r, the area of the circle by A, and the area of the square by S. Which of the following statements is true?

(A) S varies directly as r

(B) S varies inversely as r^2

(C) S varies directly as A

(D) S varies inversely as A

(E) S varies inversely as r

13. Line segment $\overline{AB}$ is given by $A(-4, 7)$ and $B(8, -1)$. Which equation passes through the midpoint of $\overline{AB}$?

(A) $3x - y = 12$

(B) $x - 3y = 11$

(C) $-x + y = 7$

(D) $x - 2y = -4$

(E) $-x + 2y = 4$

14. For which of the following functions is the range equal to the set of all positive numbers?

(A) $f(x) = 23x^2 - 12$

(B) $f(x) = |x - 6| - 1$

(C) $f(x) = |x^2 + 3| + 6$

(D) $f(x) = 4 + |x - 11|$

(E) $f(x) = 23|x^2| - 6$

15. If $5^m = a^{\frac{3}{2}}$ and $5^n = b^{\frac{3}{2}}$, then which expression is equivalent to $(ab)^3$?

(A) 5^{m+n}

(B) 5^{2m+n}

(C) 25^{mn}

(D) 25^{m+n}

(E) 2×5^{mn}

16. The Parent Teacher Association sold raffle tickets for $5 each. Three winning tickets will be drawn, one after another, and will be thrown away after they are drawn. If 500 tickets are sold, what is the probability that someone who bought $20 worth of tickets will have all three winning tickets?

(A) $\frac{1}{125}$

(B) $\frac{1}{1{,}500}$

(C) $\frac{9}{41{,}417{,}000}$

(D) $\frac{1}{5{,}177{,}125}$

(E) $\frac{1}{125{,}000{,}000}$

17. For nonzero a, which of the following lists of terms can form a geometric sequence?

(A) $(a - 2), 2a, (a + 2)$

(B) $4a, (2a - 1), (a - 1)$

(C) $4a, 8a, 12a$

(D) $a, 3a, 5a$

(E) $(a + 3), a, (a + 4)$

GO ON TO THE NEXT PAGE

18. Line M passes through the points (0, 0) and (4, 4). Which of the following points is located in the region between the graph of the line M and the positive part of the x-axis?

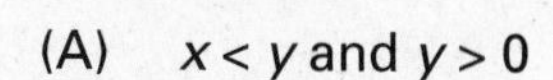

(A) $x < y$ and $y > 0$

(B) $x = y$ and $x > 0$

(C) $x < 2y$ and $y > 0$

(D) $x > y$ and $y > 0$

(E) $x < 2y$ and $y > 0$

19. If y varies directly as $\sqrt{x^3}$ and x varies directly as z^2, then

(A) y varies inversely as z^3.

(B) y varies directly as z^3.

(C) y varies inversely as z^2.

(D) y varies directly as z^2.

(E) y varies directly as z.

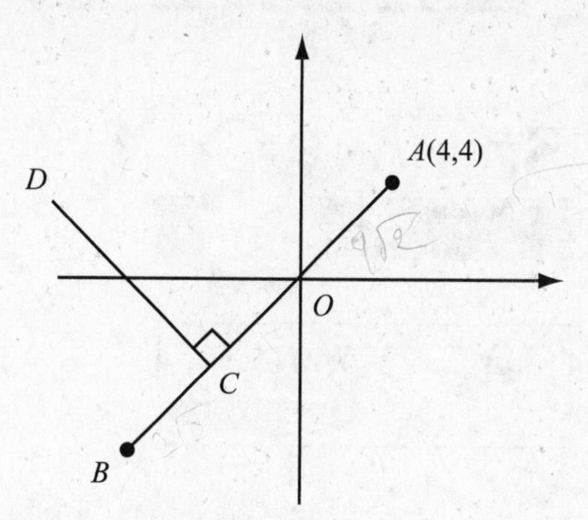

20. In the figure above, $OB = 2OA$ and $\overline{CD}$ is perpendicular to $\overline{AB}$. C is the midpoint of $\overline{OB}$. Which is the equation of $\overline{CD}$?

(A) $y = x + 8$

(B) $y = -x - 8$

(C) $y = x - 8$

(D) $y = -x - 4$

(E) $y = -x + 8$

STOP

If time remains, you may go back and check your work. When the time allotted is up, you may go on to the next section.

TIME: 25 Minutes
20 Questions

CRITICAL READING

DIRECTIONS: Each sentence below has one or two blanks, each blank indicating that something has been omitted. Beneath the sentences are five lettered words or sets of words. Choose the word or set of words that BEST fits the meaning of the sentence as a whole.

EXAMPLE

Although critics found the book _______ , many readers found it rather _______ .

(A) obnoxious . . . perfect
(B) spectacular . . . interesting
(C) boring . . . intriguing
(D) comical . . . persuasive
(E) popular . . . rare

EXAMPLE ANSWER

1. The fish and ships that decorated every one of his rooms conveyed his __________ background.

   (A) aeronautical
   (B) aviary
   (C) nautical
   (D) aviation
   (E) eclectic

2. The __________ of the bright lights, fanciful music, and smell of cotton candy were too much for the little boy to resist.

   (A) motif
   (B) juxtaposition
   (C) alacrity
   (D) allure
   (E) utopia

3. The __________ was __________ in completing such a complex process.

   (A) novice . . . adept
   (B) expert . . . erroneous
   (C) fledgling . . . inept
   (D) mentor . . . livid
   (E) neophyte . . . pertinent

4. The __________ citizens threatened to __________ the senator if he voted against their wishes.

   (A) seething . . . oust
   (B) reclusive . . . imbue
   (C) errant . . . foster
   (D) affable . . . impeach
   (E) rabid . . . captivate

5. Children whose parents treat them as small adults are more likely to be __________ than children whose parents treat them as babies.

   (A) charismatic
   (B) diffident
   (C) petulant
   (D) impecunious
   (E) precocious

6. The __________ reason for his leaving the group was to go home and study, but he really just wanted to be alone.

   (A) veracious
   (B) impetuous
   (C) ostensible
   (D) audacious
   (E) rancorous

7. It is necessary to function within the __________ of socially acceptable standards in order to succeed in the business world.

   (A) perimeters
   (B) parameters
   (C) progeny
   (D) penury
   (E) parlance

GO ON TO THE NEXT PAGE

8. Whether to go along with his peers or follow his parents' advice was a ________ he often faced at school.

(A) quirk

(B) quandary

(C) liaison

(D) hubris

(E) catalyst

9. It does not take a good supervisor long to fire a ________ who ________ instead of working.

(A) laggard . . . malingers

(B) dolt . . . toils

(C) charlatan . . . empathizes

(D) ingrate . . . genuflects

(E) sycophant . . . labors

<u>DIRECTIONS</u>: Each passage is followed by questions based on its content. Answer the questions on the basis of what is <u>stated</u> or <u>implied</u> in each passage and in any introductory material that may be provided.

If you get an "O.K." sign (a circle formed with the thumb and index finger) from a friend, a colleague, or even your boss, you know you've done something well. And you feel good about yourself and your accomplishment—that is, if you're in America. However, if someone gives you that same gesture in Brazil or Germany, you should not be so pleased, since the message is an obscene one in those countries. In Spain and in some areas of Europe and South America, it is rude. In Japan if you use that gesture you're telling a salesperson you want your change in coins instead of bills. And in France, you're conveying that something or someone is a "zero" or "worthless." Clearly, the nonverbal communication that comes so naturally to us and that seems so harmless can actually get us into a great deal of trouble in other countries.

So when you're in the Middle East, don't use your index finger to point. In Europe don't turn your palm toward you while forming a "V" with your fingers. Don't sit with the soles of your shoes showing if you're in Thailand. And while you're visiting Japan, don't use only one hand when giving an item to someone. It may seem that using no gestures at all is the only advisable way to behave when you're traveling internationally; however, the absence of gestures during a conversation is also offensive in some parts of the world.

GO ON TO THE NEXT PAGE

10. The advice implied in this passage is

    (A) don't gesture when you're in a foreign country.
    (B) know the language of the country you're visiting.
    (C) the interpretations of gestures vary from one person to another in any country, so be sure you know the person with whom you are communicating.
    (D) learn in advance what different gestures mean in a country you are going to visit.
    (E) verbal communication is more important than nonverbal communication for the international traveler.

11. An American traveler who uses the thumb and index finger to form a circle is likely to

    (A) receive coins rather than bills as change when in France.
    (B) make a new friend in Germany.
    (C) receive a smile from a fellow American.
    (D) offend someone in Japan.
    (E) get into a fight with someone from the Middle East.

When he entered the auditorium, ten thousand applauding and cheering people jumped to their feet. His perfect posture and self-assured demeanor made his 5'10" frame seem 7' tall. His soft-spoken manner caused his listeners to lean toward him, reluctant to miss even one word of his message that somehow came across loudly and clearly. His ready smile conveyed his unconditional acceptance of everyone he encountered. In his presence all were equal. The man with millions of dollars of assets was no richer than the one with only pocket change. The perfectly shaped supermodel was no more beautiful than the arthritic ninety-year-old woman standing in her shadow. The Nobel Peace Prize winner's opinion was no more important than that of the high school dropout.

12. What word best describes the way the author feels toward the man described in this passage?

    (A) admiration
    (B) anger
    (C) rejection
    (D) ambivalence
    (E) distrust

13. The man portrayed in this passage can best be described as

    (A) patronizing.
    (B) charismatic.
    (C) pompous.
    (D) deceitful.
    (E) pretentious.

GO ON TO THE NEXT PAGE

The following selections discuss the effects of television today.

Passage 1

Television has become the most popular form of entertainment today. Virtually every home has one, and many homes have one for every room—even the bathroom. Clearly, television has an effect on our lives, entertaining us; educating us; and keeping us abreast of world, national, and local events. Television shows apprise us of the latest fads and fashion trends. They offer us a view of lifestyles most of us will never experience firsthand. They reflect—and perhaps even have initiated—morals that have become acceptable in contemporary society. Another less recognizable effect of television is that it affects the career choices of young viewers.

Young people are attracted by the glamour and excitement associated with the careers of the characters they see on TV. During the last decade of the twentieth century, thousands of young boys aspired to become a Texas Ranger, just like the character portrayed on *Walker, Texas Ranger*. Since the inception of *CSI* in 2000, forensic science as a major has become incredibly popular on many college campuses. *ER* made it more acceptable for women to become doctors, just as *The Practice*—perhaps inadvertently—encouraged women to become lawyers. Seldom do plumbers, roofers, or other such artisans have major roles in TV shows, and if they do they are probably on a sitcom, as opposed to the high-drama programs with the glamorous figures that viewers idolize and young people emulate.

Television shows may give young people some ideas about career possibilities; however, the media's depictions of careers are seldom realistic. Viewers of *ER* see EMTs—all handsome—shouting medical terminology at the nurses—beautiful, of course—as they bustle critically ill patients into the emergency room. Surgeons and other specialists are successful in the complicated operation. And it all happens within an hour. Likewise, crime scene investigators gather evidence on the scene, analyze it in the lab, and solve the crime—usually even arresting the suspect—within a sixty-minute span. And lawyers are just as adept at convincing judges and juries to acquit or convict a suspect.

Admittedly, writers and producers of many television shows today do consult with experts in the professions they depict, and the essential elements are based on real aspects of real careers. But the producers are selective in what they choose to show. It's not exciting—and is more than a little disgusting—to watch a doctor lance a boil or clean the wax out of a patient's ear. So although these procedures may be a normal part of a doctor's day, we're not likely to see them on television. Likewise, we're not likely to see the real-world situation of lawyers poring hour after hour over humongous volumes of law and legal cases. Little acknowledgement is given to the fact that—unlike the heroic television detectives and cops—real-life pros work on several cases at once, rather than devoting twenty-four hours a day to one specific case.

It's not likely that young people—or even adults—are ever going to find reality as entertaining and glamorous as the lives depicted on television dramas. But it's also not likely that viewers believe everything they see on TV. So although the media heroes and heroines may give us a biased picture of certain professions, at least they expose us to some aspects of those careers—and, hopefully, inspire young people to explore those possibilities further.

GO ON TO THE NEXT PAGE

Passage 2

1 Television! Americans think they can't live without it! But how do they live with it? In a typical American household the television is on between six and eight hours a day. It's almost 5 normal these days for the television to be playing during meals; in fact, many families have TV sets in the kitchen and dining room. Parents set their children in front of the "boob tube" to keep them quiet and to entertain them. Teen- 10 agers do their homework with the TV blaring. And automobile companies are beginning to offer television sets, along with DVD players, as optional in some vehicles.

Obesity, even among young people, is an in- 15 creasing problem in America. Well, of course it is! The most exercise some of us get is moving our thumb on the remote! Many Americans—young and old—spend seven to eight hours sleeping, seven to eight hours sitting at a desk at school or 20 at work, and the rest of the time watching television. It's little wonder that America is becoming fat and unhealthy. Coinciding with increasing obesity is a rising incidence of diabetes, high blood pressure, high cholesterol, and numerous 25 other health problems associated with a sedentary lifestyle.

An ongoing concern in today's society is the downward turn of morality. Each decade seems to reflect fewer moral standards and 30 greater acceptance of behaviors that, not so long ago, were unacceptable if not forbidden. And although the reasons for today's moral decline are many, television has certainly not helped the problem. Even prime-time program- 35 ming censors few curse words; the use of such words by popular and sometimes idolized TV stars makes them acceptable in everyday life. The plots of many popular shows center on promiscuity and infidelity. Sadly, the message 40 is not that these behaviors are wrong; often, the most promiscuous and unfaithful characters are the ones with whom the viewers empathize and who they sometimes emulate. Granted, not many of us set out purposefully to be like our favorite TV character; however, 45 viewing such lifestyles for hours each week can only cause us to believe they are the norm and thus acceptable. Television is our "window to the world" that most of us would not otherwise experience, so we tend to believe that what we 50 see is how the rest of the world must be living. Does television really reflect how people live, or are people leading lives exemplified by television?

Accompanying contemporary America's 55 moral turpitude is increased violence, especially among young people. Whether or not specific criminal acts are actually attributed to specific television shows viewed by the perpetrators is often debated, even in the courts. But 60 no one can argue that children see so many violent acts on television—even in cartoons—that they become inured to them. And often our heroes use violence as a means of achieving admirable goals. But how admirable are those 65 characters if they must hurt others to achieve what they set out to do? It doesn't take young minds long to embrace the idea that violence and hurtful behaviors are okay, especially if they see those are the standards for the "good 70 guys" on TV.

Proponents say television is educational. But what exactly does it teach? Perhaps the lessons the television audience—especially young viewers—learn are not beneficial to them or to 75 society as a whole. It would be interesting to see what would happen if all televisions were removed from—or at least turned off in—all American homes for six to twelve months. Who knows? The nation's health, morals, and crime 80 rate might actually improve!

GO ON TO THE NEXT PAGE

14. In line 4, "apprise" most nearly means

(A) evaluate.
(B) learn.
(C) inform.
(D) show.
(E) award.

15. In Passage 1, *Walker, Texas Ranger*, *ER*, *The Practice*, and *CSI* are given as examples of television shows that

(A) accurately depict different careers.
(B) unrealistically glamorize different careers.
(C) contribute to the decline of today's morals.
(D) portray life as it really is.
(E) occupy too much of Americans' time.

16. The author of Passage 1 believes that television encourages young people to

(A) explore careers of all skill levels.
(B) seek careers that fit their individual needs and interests.
(C) aspire to careers that offer glamour and excitement.
(D) respect blue-collar jobs as necessary and important careers.
(E) understand that a career is not as important as a lifestyle.

17. In line 18, "emulate" most nearly means

(A) criticize, find fault with.
(B) dislike.
(C) watch.
(D) strive to be like, imitate.
(E) support.

18. The author of Passage 1 seems to believe that television programs depicting certain careers

(A) are a waste of time.
(B) have both positive and negative aspects.
(C) are completely misleading.
(D) are definitely helpful to young people as they consider what to do with their lives.
(E) are boring to most viewers.

19. The attitude of Passage 2's author could be accurately described as

(A) amused.
(B) vindictive.
(C) critical.
(D) benevolent.
(E) sympathetic.

20. In line 75, "turpitude" most nearly means

(A) depravity.
(B) upswing.
(C) stagnation.
(D) behavior.
(E) tenacity.

21. The author of Passage 2 does <u>not</u> believe

(A) television is educational.
(B) television programs promote morality.
(C) television viewing is unhealthy.
(D) television is important to Americans.
(E) television programs promote violence.

22. The author of Passage 2 believes that the immoral and criminal actions of TV characters cause viewers to

    (A) be cautious about engaging in such actions themselves.

    (B) believe that such behavior is normal.

    (C) become frightened and distrustful of others.

    (D) be amused at such behavior both on television and in real life.

    (E) avoid watching shows that portray such behavior.

23. The author of Passage 2 seems to believe that the best solution to the problems that result from our watching television would be to

    (A) censor the language on television shows.

    (B) eliminate violence in television shows.

    (C) limit the amount of time a person can watch television.

    (D) limit the number of televisions in each home.

    (E) eliminate all televisions.

24. Which of the following is not true about the passages?

    (A) Passage 1 is more critical of television shows than Passage 2 is.

    (B) The author of Passage 2 believes that nothing good comes from watching television.

    (C) Both express concern about the effects of some programs on their viewers.

    (D) Both realize how much time Americans spend watching television.

    (E) Both acknowledge that television offers viewers a chance to experience things they might not otherwise experience.

STOP

If time remains, you may go back and check your work. When the time allotted is up, you may go on to the next section.

TIME: 25 Minutes
35 Questions

CRITICAL READING

DIRECTIONS: In each of the following sentences, some portion of the sentence is underlined. Under each sentence are five choices. The first choice has the same wording as the original. The other four choices are reworded. Sometimes the first choice containing the original wording is the best; sometimes one of the other choices is the best. Choose the letter of the best choice. Your choice should produce a sentence that is not ambiguous or awkward and that is correct, clear, and precise.

This is a test of correct and effective English expression. Keep in mind the standards of English usage, punctuation, grammar, word choice, and construction.

EXAMPLE

When you listen to opera, a person may not appreciate it.

(A) a person may not appreciate it.
(B) it may not be appreciated by a person.
(C) you may not appreciate it.
(D) which may not be appreciated by you.
(E) appreciating it may be a problem for you.

EXAMPLE ANSWER

1. Returning to the ancestral home after 12 years, the house itself seemed much smaller to Joe than it had been when he visited it as a child.

    (A) Returning to the ancestral home after 12 years, the house itself seemed much smaller to Joe
    (B) When Joe returned to the ancestral home after 12 years, he thought the house itself much smaller
    (C) Joe returned to the ancestral home after 12 years, and then he thought the house itself much smaller
    (D) After Joe returned to the ancestral home in 12 years, the house itself seemed much smaller
    (E) Having returned to the ancestral home after 12 years, it seemed a much smaller house to Joe

2. Historians say that the New River of North Carolina, Virginia, and West Virginia, which is 2,700 feet above sea level and 2,000 feet above the surrounding foothills, is the oldest river in the United States.

    (A) which is 2,700 feet above sea level and 2,000 feet above
    (B) with a height of 2,700 feet above sea level as well as 2,000 feet above that of
    (C) 2,700 feet higher than sea level and ascending 2,000 feet above
    (D) being 2,700 feet above sea level and 2,000 feet high measure from that of
    (E) located 2,700 feet high above sea level while measuring 2,000 feet above

3. The age of 36 having been reached, the Ukrainian-born Polish sailor Teodor Josef Konrad Korzeniowski changed his name to Joseph Conrad and began a new and successful career as a British novelist and short story writer.

    (A) The age of 36 having been reached
    (B) When having reached the age of 36
    (C) When he reached the age of 36
    (D) The age of 36 being reached
    (E) At 36, when he reached that age

4. During the strike, black South African miners threw a cordon around the gold mine, and they thereby blocked it to all white workers.

    (A) gold mine, and they thereby blocked it to all white workers.
    (B) gold mine, by which all white workers were therefore blocked.
    (C) gold mine, and therefore this had all white workers blocked.
    (D) gold mine and therefore blocking it to all white workers.
    (E) gold mine, thereby blocking it to all white workers.

GO ON TO THE NEXT PAGE

5. Because of the long half-life of low-level nuclear wastes, this means that waste depositories could emit dangerous doses of radiation thousands of years into the future.

(A) wastes, this means that waste depositories could emit dangerous doses of radiation thousands of years into the future.

(B) wastes is the reason why waste depositories could emit dangerous doses of radiation thousand of years into the future.

(C) wastes, this is the reason why waste depositories could emit dangerous doses of radiation thousands of years into the future.

(D) wastes, depositories for these wastes could still emit dangerous doses of radiation thousands of years into the future.

(E) wastes, the future means that waste depositories could emit dangerous doses of radiation for thousand of years.

6. The more you listen to and understand classical music, the more our ears will prefer music for the mind to music for the body.

(A) The more you listen to and understand classical music

(B) The more we listen to and understand classical music

(C) The more classical music is listened to and understood

(D) As understanding and listening to classical music increases

(E) As people listen to and understand classical music

7. As modern archaeologists discover new fossils, biologists are amending Darwin's theory of evolution that once served as the standard.

(A) that once served as the standard.

(B) by which all others were measured.

(C) having served as the standard for over a hundred years.

(D) thereby changing the standard.

(E) and creating a new standard.

8. Poets of the nineteenth century tried to entertain their readers but also with the attempt of teaching them lessons about life.

(A) to entertain their readers but also with the attempt of teaching them

(B) to entertain their readers but also attempt to teach them

(C) to both entertain their readers and to teach them

(D) entertainment of their readers and the attempt to teach them

(E) both to entertain and to teach their readers

9. The city council decided to remove parking meters so as to encourage people to shop in Centerville.

(A) so as to encourage

(B) to encourage

(C) thus encouraging

(D) with the desire

(E) thereby encouraging

GO ON TO THE NEXT PAGE

10. Visiting New York City for the first time, <u>the sites most interesting to Megan were</u> the Statue of Liberty, the Empire State Building, and the Brooklyn Bridge.

    (A) the sites most interesting to Megan were
    (B) the sites that Megan found most interesting were
    (C) Megan found that the site most interesting to her were
    (D) Megan was most interested in
    (E) Megan was most interested in the sites of

11. Although most college professors have expertise in their areas of specialty, <u>some are more interested in continuing their research than in teaching undergraduate students</u>.

    (A) some are more interested in continuing their research than in teaching undergraduate students.
    (B) some are most interested in continuing their research rather than in teaching undergraduate students.
    (C) some prefer continuing their research rather than to teach undergraduate students.
    (D) continuing their research, not teaching undergraduate students, is more interesting to some.
    (E) some are more interested in continuing their research than to teach undergraduate students.

GO ON TO THE NEXT PAGE

DIRECTIONS: Each of the following sentences may contain an error in diction, usage, idiom, or grammar. Some sentences are correct. Some sentences contain one error. No sentence contains more than one error.

If there is an error, it will appear in one of the underlines portions labeled A, B, C, or D. If there is no error, choose the portion labeled E. If there is an error, select the letter of the portion that must be changed in order to correct the sentence.

EXAMPLE

He drive slowly (A) and cautiously (B) in order to hopefully (C) avoid having an accident (D). No error (E)

EXAMPLE ANSWER

12. Less (A) students chose liberal arts and sciences (B) majors in the 1980s than in the 1960s because of (C) the contemporary view that a college education is (D) a ticket to enter the job market. No error (E).

13. Span of control is the term that (A) refers to the limits (B) of a leader's ability for managing (C) those employees under (D) his or her supervision. No error (E).

14. Because some (A) people believe strongly (B) that channeling, the process by which (C) an individual goes into a trancelike state and communicates the thoughts of an ancient warrior or guru or an audience, helps them cope with modern problems, but others condemn the whole idea as (D) mere superstition. No error (E).

GO ON TO THE NEXT PAGE

15. The reed on a woodwind instrument is essential (A) being that (B) it controls the quality (C) of tone and sound (D). No error (E).

16. As far as (A) taking an SAT preparation course, educators discourage it (B) because (C) at best the course may alleviate (D) test anxiety. No error (E).

17. In the South (A), they (B) like to eat cured (C) or smoked pork products (D) such as ham, bacon, and barbecue. No error (E).

18. Both Japan and the United States (A) want (B) to remain a net exporter (C) of goods to avoid unfavorable trade imbalances (D). No error (E).

19. As an avid cyclist (A), Jon rode (B) more miles a day (C) than his friends bicycle (D). No error (E).

20. After the end of the Mesozoic Era (A), dinosaurs, once the dominant (B) species are (C) extinct (D). No error (E).

21. According to (A) the United States Constitution, the legislative branch of the government has (B) powers different than (C) those (D) of the executive branch. No error (E).

GO ON TO THE NEXT PAGE

22. After being studied for the preceding ten years by the National Heart, Lung, and Blood Institute, the relationship of high levels of cholesterol in the blood to the possibility of having heart attacks was reported in 1984. No error.
(A: being studied; B: preceding ten years; C: having; D: was reported; E: No error)

23. The book *Cheaper By the Dozen* demonstrates that each of the children of Frank and Lillian Gilbreth was expected to use his or her time efficiently. No error.
(A: demonstrates; B: was expected; C: to use; D: his or her; E: No error)

24. His aversion with snakes made camping an unpleasant activity for him and one that he tried diligently to avoid. No error.
(A: with; B: for him; C: one; D: that; E: No error)

25. The story of the American pioneers, those who willingly left the safety of their homes to move into unsettled territory, show us great courage in the face of danger. No error.
(A: those; B: their; C: show; D: us; E: No error)

26. Because of the long, cold winters and short summers, farming in high altitudes is more difficult than low latitudes. No errors.
(A: Because of; B: and; C: more difficult; D: than low latitudes; E: No errors)

27. When my sister and I were in Los Angeles, we hoped that both of us could be a contestant on a quiz show. No error.
(A: I; B: were; C: we; D: a contestant; E: No error)

GO ON TO THE NEXT PAGE

28. After <u>he had</u> (A) broke the vase <u>that</u> (B) his mother <u>had purchased</u> (C) in Europe, he tried to buy a new one for his father and <u>her</u> (D). <u>No error</u> (E).

29. Some of the people <u>with whom</u> (A) the witness <u>worked</u> (B) <u>were engaged</u> (C) in covert activities <u>on behalf of</u> (D) the United States government. <u>No error</u> (E).

<u>DIRECTIONS</u>: The following passages are considered early draft efforts of a student. Some sentences need to be rewritten to make the ideas clearer and more precise.

Read each passage carefully and answer the questions that follow. Some of the questions are about particular sentences or parts of sentences and ask you to make decisions about sentence structure, diction, and usage. Some of the questions refer to the entire essay or parts of the essay and ask you to make decisions about organization, development, appropriateness of language, audience, and logic. Choose the answer that most effectively makes the intended meaning clear and follows the requirements of standard written English. After you have chosen your answer, fill in the corresponding oval on your answer sheet.

EXAMPLE

(1) On the one hand, I think television is bad, But it also does some good things for all of us. (2) For instance, my little sister thought she wanted to be a policemen until she saw police shows on television.

Which of the following is the best revision of the underlined portion of sentence (1) below?

One the one hand, I think television <u>is bad, But it also</u> does some good things for all of us.

(A) is bad; But it also
(B) is bad. but it also
(C) is bad, but it also
(D) is bad, and it also
(E) is bad because it also

EXAMPLE ANSWER

Questions 30–35 are based on the following passage.

(1) Dripstone features are called speleotherms, and they can take several beautiful forms. (2) When these structures are highlighted by lanterns or electric lights, they transform the cave into a natural wonderland. (3) Some people feel that electric lights have no place in a cave. (4) The most familiar dripstone features are stalactites and stalagmites. (5) Stalactites hang downward from the ceiling and are formed as drop after drop of water slowly trickles through cracks in the cave roof. (6) Each drop of water hangs from the ceiling. (7) The drop of water loses carbon dioxide. (8) It then deposits a film of calcite. (9) Stalagmites grow upward from the floor of the cave generally as a result of water dripping from overhead stalactites. (10) An impressive column forms when a stalactite and stalagmite grow until they join. (11) A curtain or drapery begins to form an inclined ceiling when the drops of water trickle along a slope. (12) Gradually, a thin sheet of calcite grows downward from the ceiling and hang in graceful decorative folds like a drape.

(13) These impressive and beautiful features appear in caves in almost every state, making for easy access for tourists looking for a thrill. (14) In addition, the size and depth of many caves in the United States also impress even the most experienced tourist accustomed to many very unique sights. (15) Seven caves have more than 15 passage miles, the longest being the Flint-Mammoth Cave system in Kentucky with more than 169 miles. (16) The deepest cave in the United States is Neff Canyon in Utah.

(17) Although many people seem to think that the deepest cave is Carlsbad Caverns, located in New Mexico. (18) However, Carlsbad Caverns boasts the largest room, the Bog Room which covers 14 acres. (19) These are sights not to be missed by those who appreciate the handiwork of Mother Nature.

30. Which of the following sentences breaks the unity of paragraph 1 and should be omitted?

(A) sentence 1
(B) sentence 3
(C) sentence 4
(D) sentence 10
(E) sentence 12

31. Which of the following is the best way to combine sentences 6, 7, and 8?

(A) Each drop of water deposits a film of calcite as it is hanging from the ceiling and losing carbon dioxide.
(B) When hanging, water drops lose carbon dioxide and create a film of calcite.
(C) In the process of losing carbon dioxide, the drops of water hang from the ceiling and deposit calcite.
(D) Hanging from the ceiling, losing carbon dioxide, and depositing calcite are the drops of water.
(E) Hanging from the ceiling, each drop of water loses carbon dioxide and deposits a film of calcite.

32. In relation to the passage as a whole, which of the following best describes the writer's intention in paragraph 2?

(A) To describe some examples
(B) To provide a summary
(C) To convince the reader to change an opinion
(D) To persuade the reader to follow a course of action
(E) To detail a chain of events

33. Which of the following is the best revision of the underlined portion of sentence 14 (reproduced below)?

In addition, the size and depth of many caves in the United States <u>also impress even the most experienced tourist accustomed to very unique sights</u>.

(A) impressed the most experienced tourist also accustomed to very unique sights.
(B) impress even the most experienced tourist also accustomed to very unique sights.
(C) impress even the most experienced tourist accustomed to many unique sights.
(D) also impress very experienced tourists accustomed to very unique sights.
(E) also impress even the most experienced tourist accustomed to many unique sights.

34. In the context of the sentences preceding and following sentence 17, which of the following is the best revision of sentence 17?

(A) Although many people are thinking that the deepest cave is Carlsbad Caverns in New Mexico.
(B) Many people think the deepest cave is Carlsbad Caverns in New Mexico.
(C) As a matter of fact, many people think the deepest cave exists in New Mexico in Carlsbad Caverns.
(D) However, many people think the deepest cave is Carlsbad Caverns in New Mexico.
(E) In addition, many people think the deepest cave is located in New Mexico in Carlsbad Caverns.

35. Which of the following corrects the grammatical error in sentence 12?

(A) Gradually, a thin sheets of calcite grow downward from the ceiling and hang in graceful decorative folds like drapes.
(B) Gradually, a thin sheet of calcite grows downward from the ceiling and hangs in graceful decorative folds like a drape.
(C) Gradually, a thin sheet of calcite grows downward from the ceiling and gracefully hang in decorative folds like a drape.
(D) Gradually, a thin sheet of calcite grows downward from the ceilings and hang in graceful decorative folds like a drape.
(E) Gradually, a thin sheet of calcite grows downward from the ceiling and hang from the ceiling like a drape.

STOP

If time remains, you may go back and check your work. When the time allotted is up, you may go on to the next section.

TIME: 25 Minutes
18 Questions

MATH

<u>DIRECTIONS</u>: In this section solve each problem, using any available space on the page for scratchwork. Then decide which is the best of the choices given and fill in the corresponding oval on the answer sheet.

NOTES

(1) The use of a calculator is permitted.

(2) All numbers used are real numbers.

(3) Figures that accompany problems in this test are intended to provide information useful in solving the problems. They are drawn as accurately as possible EXCEPT when it is stated in a specific problem that the figure is not drawn to scale. All figures lie in a plane unless otherwise indicated.

(4) Unless otherwise specified, the domain of any function f is assumed to be the set of all real numbers x for which $f(x)$ is a real number.

REFERENCE INFORMATION

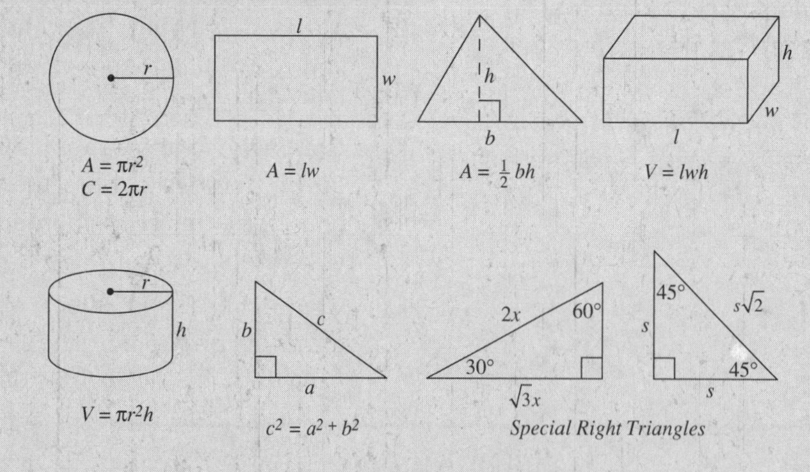

The number of degrees of arc in a circle is 360.

The sum of the measures in degrees of the angles of a triangle is 180.

	Week 1	Week 2	Week 3	Week 4	Week 5
Group A	22	0	38	42	45
Group B	54	0	0	56	0
Group C	48	67	21	0	12
Group D	0	45	0	32	70

1. In a clinical trial, four groups of patients participated, labeled A, B, C, and D in the table above. Each week, a number of patients from each group withdrew from the trial; this number is given in the table. For which week did the most patients withdraw from the trial?

   (A) Week 1
   (B) Week 2
   (C) Week 3
   (D) Week 4
   (E) Week 5

2. If $m = a - 3$ and $n = a + 3$, then what is the value of $mn - a^2$?

   (A) –9
   (B) –6
   (C) 0
   (D) 6
   (E) 9

3. If $f(x) = 3x^2 - 1$ and $g(x) = 3x^2 + 1$, what is the value of $\frac{f(2) - g(-2)}{f(-2) - g(2)}$?

   (A) $-\frac{1}{12}$
   (B) –12
   (C) $\frac{1}{12}$
   (D) 1
   (E) 12

4. An average blue whale weighs 5 tons more than 3 times what an average humpback whale weighs. If B represents the weight of an average blue whale, what is the weight of 5 humpback whales in terms of B?

   (A) $\frac{B-3}{5}$
   (B) $\frac{B+3}{5}$
   (C) $3B - 5$
   (D) $\frac{B-5}{3}$
   (E) $\frac{B+5}{3}$

5. What is the volume of a cylinder with both a height and diameter of 14?

   (A) 14π
   (B) $\frac{49\pi}{2}$
   (C) 98π
   (D) 343π
   (E) 686π

6. If the graphs of the equations $y + mx = 12$ and $y = 2mx + 6$ are perpendicular for $m > 0$, which is the value of m?

   (A) 2
   (B) 1
   (C) $\sqrt{2}$
   (D) $\pm\frac{\sqrt{2}}{2}$
   (E) $\pm 2\sqrt{2}$

GO ON TO THE NEXT PAGE

7. If $25^a = 125^{m-1}$, which of the following is equivalent to m?

(A) $\frac{2a+3}{3}$

(B) $\frac{2a-3}{3}$

(C) $2a - 3$

(D) $2a + 3$

(E) $\frac{a}{3}+1$

8. Set M includes all positive even integers less than 24, set N includes all positive integers less than 30 and divisible by 4, and set P includes all positive integers less than 28 and divisible by 6. Which set is equal to $(M \cap N) \cup (N \cap P)$?

(A) $M \cap P$

(B) $N \cap P$

(C) $M \cap N$

(D) $(M \cup N) \cap (N \cap P)$

(E) $(M \cap P) \cap (N \cup P)$

9. If $f(x) = |x-3|$, what is $f(-x) - f(x)$?

(A) $|2x-6|$

(B) $|x+6|$

(C) $|x-6|$

(D) $|x+3|-|x-3|$

(E) $|x-3|-|x+3|$

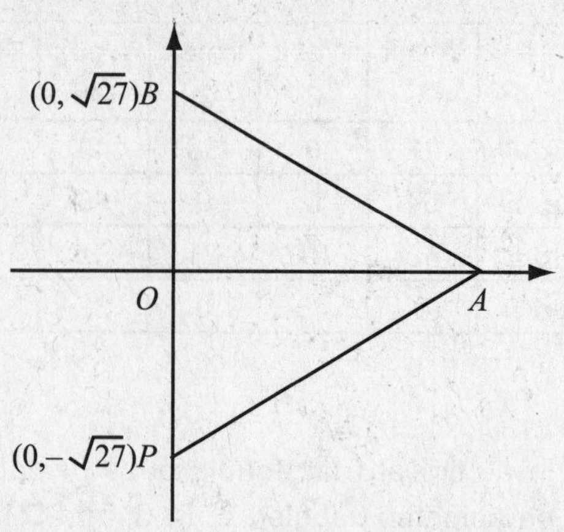

Note: Figure not drawn to scale.

10. In the figure above, ΔABP is an equilateral triangle. What is the length of OA?

(A) 4

(B) $\frac{9}{2}$

(C) $\frac{3\sqrt{3}}{2}$

(D) $3\sqrt{3}$

(E) 9

11. Line M passes through the points (3, 4) and (2, –2). Which of the following lines is perpendicular to M?

(A) $x - 3y = 11$

(B) $4x - 2y = 15$

(C) $x - 6y = 11$

(D) $2x + 12y = 21$

(E) $5x - 15y = 23$

GO ON TO THE NEXT PAGE

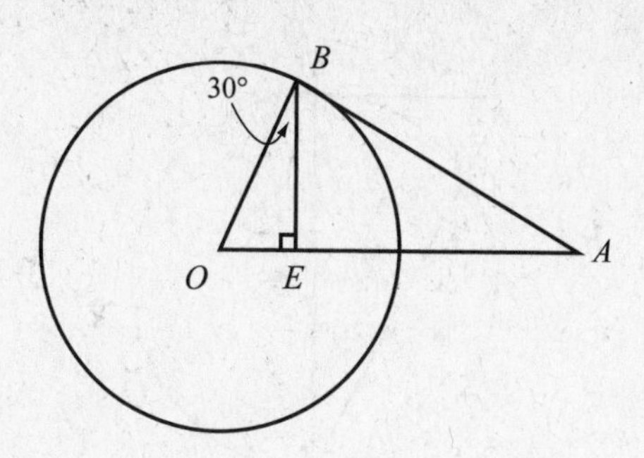

12. In the figure above, $\overline{AB}$ is tangent to the circle O. $\overline{BE}$ is perpendicular to $\overline{OA}$. If $\overline{OE} = 3$, what is OA?

 (A) $\frac{6}{\sqrt{3}}$

 (B) 6

 (C) $6\sqrt{2}$

 (D) 12

 (E) $12\sqrt{3}$

13. Which graph will be obtained if the graph of $f(x) = -3x^2 + 3x - 5$ is shifted 3 units vertically downward?

 (A) $f(x) = -3x^2 - 5$

 (B) $f(x) = -3x^2 + 6x - 5$

 (C) $f(x) = -x^2 + 3x - 5$

 (D) $f(x) = 9x^2 - 9x + 15$

 (E) $f(x) = -3x^2 + 3x - 8$

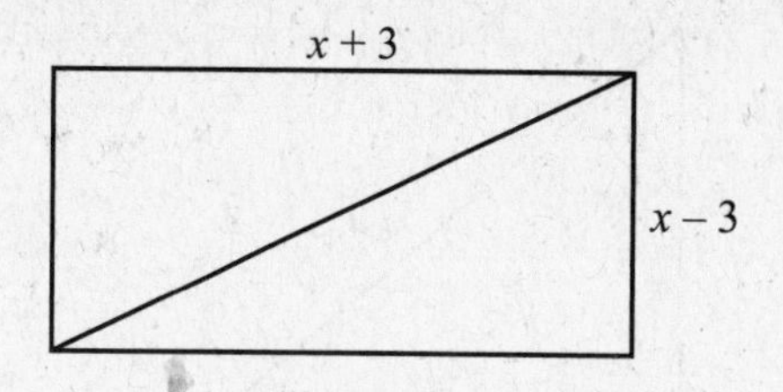

14. In the rectangle above, let L equal length of the diagonal. Which function represents L in terms of x?

 (A) $x - \sqrt{3}$

 (B) $x + \sqrt{3}$

 (C) $\sqrt{x^2 - 3}$

 (D) $\sqrt{x^2 - 6x + 9}$

 (E) $\sqrt{x^2 + 6x + 9}$

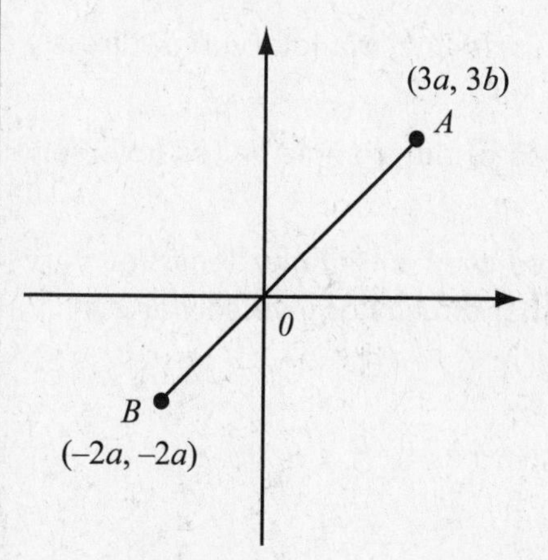

15. In the figure above, what is the probability that a point randomly chosen on $\overline{AB}$ will be on $\overline{OB}$?

 (A) $\frac{2}{5}$

 (B) $\frac{2}{3}$

 (C) $\frac{3}{5}$

 (D) $\frac{5}{6}$

 (E) $\frac{4}{5}$

GO ON TO THE NEXT PAGE

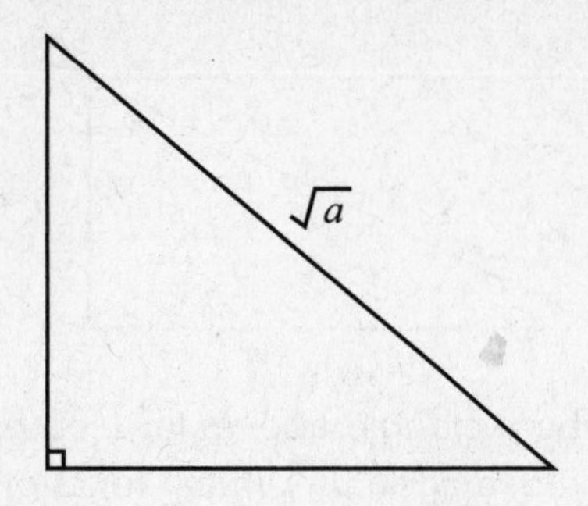

16. In the right isosceles triangle above, the length of the hypotenuse is $\sqrt{a}$. Which statement is true?

(A) Area of the triangle varies directly as a.

(B) Area of the triangle varies inversely as a.

(C) Area of the triangle varies directly as $\sqrt{a}$.

(D) Area of the triangle varies inversely as $\sqrt{a}$.

(E) Area of the triangle does not vary either directly or inversely as a.

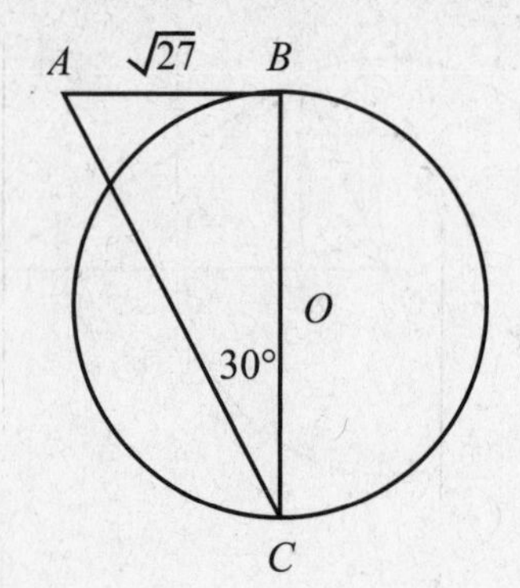

Note: Figure not drawn to scale.

17. In the figure above, $\overline{AB}$ is tangent to the circle with center O. What is the circumference of circle O?

(A) $3\sqrt{3}\pi$

(B) 9π

(C) $6\sqrt{3}\pi$

(D) 18π

(E) $12\sqrt{3}\pi$

18. For which value of k does the equation $|-5x + 3y| = 2k$ have only one solution?

(A) -2

(B) -1

(C) 0

(D) 1

(E) 2

STOP

If time remains, you may go back and check your work. When the time allotted is up, you may go on to the next section.

TIME: 25 Minutes
24 Questions

CRITICAL READING

DIRECTIONS: Each sentence below has one or two blanks, each blank indicating that something has been omitted. Beneath the sentence are five lettered words or sets of words. Choose the word or set of words that BEST fits the meaning of the sentence as a whole.

EXAMPLE

Although critics found the book _______ , many readers found it rather _______ .

(A) obnoxious . . . perfect

(B) spectacular . . . interesting

(C) boring . . . intriguing

(D) comical . . . persuasive

(E) popular . . . rare

EXAMPLE ANSWER

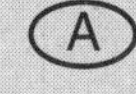

1. Because of the ________ of criticism throughout the state, the politician chose not to run for reelection.

(A) sanctity
(B) paucity
(C) generosity
(D) mediocrity
(E) preponderance

2. The ________ of his behavior caused even his relatives to reject him.

(A) benevolence
(B) turpitude
(C) innocuousness
(D) nescience
(E) obsequiousness

3. It's unfortunate that he cannot earn money at such a(n) ________ as bird-watching, which brings him so much pleasure and occupies so much of his time.

(A) avocation
(B) vocation
(C) equivocation
(D) propensity
(E) nom de plume

4. His ________ disregard for authority offended teachers as well as students.

(A) obsequious
(B) copious
(C) blatant
(D) credulous
(E) tenuous

5. The ________ music never failed to soothe the baby, causing him to fall asleep in a matter of minutes.

(A) raucous
(B) mellifluous
(C) mute
(D) stentorian
(E) loquacious

6. Most people never bothered to get to know him well, mainly because they disliked his ________ .

(A) stature
(B) homage
(C) braggadocio
(D) comeliness
(E) charisma

7. The committee conducted a(n) ________ search of all documents, looking for evidence to ________ his guilt.

(A) tepid . . . sanctify
(B) scant . . . verify
(C) precipitous . . . ameliorate
(D) impecunious . . . undermine
(E) exhaustive . . . corroborate

GO ON TO THE NEXT PAGE

DIRECTIONS: Each passage below is followed by questions based on its content. Answer the questions on the basis of what is stated or implied in each passage and in any introductory material that may be provided.

1 Pandas are currently at risk of becoming extinct. For centuries bamboo—their exclusive food source—was plentiful in pandas' native southern Asia. However, as the Asian
5 population has expanded, heretofore plentiful bamboo forests have steadily declined. Cities have grown up, and people have occupied an increasing amount of space; bamboo forests have been cut down for timber and to make
10 way for agricultural projects. Pandas have been forced to retreat. Not only has the size of their natural habitat drastically decreased; their food supply has become limited to only a few small mountainous regions, which supply a relatively
15 small amount of bamboo. To add to the panda's plight, a few decades ago a severe bamboo die-off occurred, and many starving pandas survived only because they were rescued and taken to zoos. Today, an estimated 1,000
20 pandas still exist in the wild. These numbers are rapidly dwindling, not only because of the decreasing supply of bamboo but also because of poaching, a practice that is still quite common despite the fact that it is illegal.

8. The author implies that the potential extinction of the panda is primarily the result of

(A) man's actions.

(B) the natural decline of the bamboo supply.

(C) man's need to hunt pandas to make up for the decreasing food supply.

(D) pandas' reluctance to stay in their natural habitat.

(E) changing weather patterns.

9. In line 16, "plight" most nearly means

(A) escape.

(B) promise.

(C) starvation.

(D) fear.

(E) predicament.

GO ON TO THE NEXT PAGE

1 Until the early 1900s, Americans were not extremely concerned about their futures as they became older. The major source of economic security was farming, and the ex-
5 tended family cared for the elderly. However, the Industrial Revolution brought an end to this tradition. Farming gave way to more progressive means of earning a living and family ties became looser; as a result, the family was
10 not always available to take care of the older generation. The Great Depression of the 1930s exacerbated these economic security woes. So in 1935 Congress, under the direction of President Franklin D. Roosevelt, signed into law the
15 Social Security Act. This act created a program intended to provide continuing income for retired workers at least 65 years old, partially through the collection of funds from Americans in the work force. Much organization was
20 required to get the program underway, but the first monthly Social Security checks were issued in 1940. Over the years the Social Security Program has metamorphosed into benefits not only for workers but also for the disabled and
25 for survivors of beneficiaries, as well as medical insurance benefits in the form of Medicare.

Today there is some concern that Social Security is not so "secure." Rumors and predictions contend that by 2012, if not sooner, the
30 system will be running in the red, distributing more money in benefits than it is taking in. Life expectancy is lengthening while birth rates are declining, so the number of people receiving benefits is steadily increasing while the num-
35 ber of workers contributing to the Social Security coffers is declining. Fifty years ago, the ratio of workers to Social Security beneficiaries was approximately 16 to 1. In 1998 it was 3 to 1. Theories about how to solve this problem
40 are plentiful, but what approach would be most effective—and what the government decides to do—remains to be seen.

10. The primary purpose of this passage is to

(A) criticize the Social Security Program.

(B) praise the Social Security Program.

(C) offer alternatives to the Social Security Program.

(D) provide background information about the Social Security Program.

(E) warn about the potential failure of the Social Security Program.

11. The passage indicates that the Social Security Program is experiencing problems at least partly because

(A) the people in charge of the program have managed the money poorly.

(B) not enough people have made large enough contributions.

(C) more people are living longer, and fewer people are having children.

(D) too many people are depending on Social Security benefits rather than setting aside their own private retirement funds.

(E) the program has expanded to include more benefits than it can afford to support.

GO ON TO THE NEXT PAGE

Questions 12–17 are based on the following passage.

The following selection discusses the technological advances and negative implications of the Industrial Revolution.

1 It is clear to political historians that the Industrial Revolution did not occur overnight. The formation of the mechanical age was a comparatively slow process, punctuated by fits and 5 starts, and affected only certain manufacturers and specific means of production. For the most part it spread region by region throughout Great Britain and later the whole of Europe and America, until by 1780 its impact could 10 not be ignored. By this time, the changing of European economies from agrarian-based to industrial was, in the words of one noted expert, as significant as the transformation from Paleolithic hunter-gatherer to Neolithic farmer.

15 At the forefront of the "revolution" was the introduction of mill-driven machinery and the way in which running water was converted into mechanical power. In this was was born the era of precise, tireless machines. The ben-20 efits of this type of technology were nothing short of stupendous. Weavers alone by the 1820s increased their output to 20 times that of a handworker, with power-driven spinning machines making clothing a marketable com-25 modity to the general masses for the first time. Similarly, the introduction of the steam-driven locomotive allowed the transportation of goods over long distances to improve dramatically, so much so that political shifts of power within the 30 newly industrialized European communities redefined worldwide alliances.

But the Industrial Revolution, for all its lofty aspirations, far too often engendered neglect and abuse of the individual; the most notable examples are the dehumanization of factory 35 workers and the blatant, heartless abuse of children in the labor force. The introduction of manifold moving parts, gears, logs, coils, and so on necessitated frequent and often costly repairs. Since the mammoth machines required 40 unusually long "start-up" times, manufacturers were reluctant to stop production to fix minor problems in fully functioning machines. As a result, loss of limbs was common and deaths were not infrequent, often with the machines 45 continuing to operate as the brutally mangled "messes" were cleaned up. In the same way, children, paid just a fraction of adult wages, were introduced to the factory labor force. Working as long as 16 hours per day with scant 50 breaks, children often met fates similar to their adult counterparts, some succumbing to the dangers of mechanized production, others suffering from illness brought on by unsanitary working conditions and extreme exhaustion. 55

Governments basking in the heady glow of revitalized economies ignored the gross atrocities being committed against these laborers, placing the onus instead upon the employers, who remained impassive. In the end, the work- 60 ers were forced to fight for more palatable working conditions, proper remunerative compensation, and acceptable safety standards. Though it changed the course of geopolitical relations and vaulted the world into a new age 65 of technology, the dichotomy of the Industrial Revolution prevents it from being a completely benign force.

GO ON TO THE NEXT PAGE

12. The statement "as significant as the transformation from Paleolithic hunter-gatherer to Neolithic farmer" (line 14) conveys a sense of

    (A) how long the Industrial Revolution took place.
    (B) the types of developments the Industrial Revolution produced.
    (C) the profound change the Industrial Revolution had on mankind.
    (D) The dietary predilections of the world's population during the Industrial Revolution.
    (E) the class struggle that took place during the Industrial Revolution.

13. In lines 15–21 and 26–31, the steam engine and mill-driven machinery are presented as primary examples of what?

    (A) Mechanical advances that fueled the Industrial Revolution
    (B) Profitable things to own
    (C) Daring applications of the use of water
    (D) Machines done away with the beginning of the Industrial Revolution
    (E) Failed attempts by companies to develop new products

14. In line 33, "engendered" most nearly means

    (A) prevented.
    (B) brought about.
    (C) forestalled.
    (D) defeated.
    (E) ruled out.

15. In lines 47–55, the author most likely describes a particular experience in order to

    (A) engage the interest of the reader.
    (B) horrify the reader.
    (C) explain how important children were to the Industrial Revolution.
    (D) impress upon the reader the revolutionary nature of the machines.
    (E) make the reader sympathetic to the abuse of children.

16. In line 66, "dichotomy" most nearly means

    (A) hugeness.
    (B) quickness.
    (C) changing nature.
    (D) contradiction.
    (E) success.

17. The author most likely views the development of the Industrial Revolution as

    (A) a comparatively fruitless event in world history.
    (B) a heroic advance that ended the problems of the common man.
    (C) an important advance that created its own distinctive problems.
    (D) a major technological advance that led to the Cold War.
    (E) a predominantly agrarian advance.

GO ON TO THE NEXT PAGE

Questions 18–24 are based on the following passage.

This excerpt from Jack London's South Seas short story "Yah! Yah! Yah!" depicts the character McAllister.

1 . . . the immense thing about him was the power with which he ruled. Oolong Atoll was one hundred and forty miles in circumference. One steered by compass course in its lagoon. 5 It was populated by five thousand Polynesians, all strapping men and women, many of them standing six feet in height and weighing a couple of hundred pounds. Oolong was two hundred and fifty miles from the nearest 10 land. Twice a year a little schooner called to collect copra. The one white man on Oolong was McAllister, petty trader and unintermittent guzzler; and he ruled Oolong and its six thousand savages with an iron hand. He said 15 come, and they came, go, and they went. They never questioned his will nor judgment. He was cantankerous as only an aged Scotchman can be, and interfered continually in their personal affairs. When Nugu, the king's daughter, 20 wanted to marry Haunau from the other end of the atoll, her father said yes; but McAllister said no, and the marriage never came off. When the king wanted to buy a certain islet in the lagoon from the chief priest, McAllister said no. The 25 king was in debt to the Company to the tune of 180,000 cocoanuts, and until that was paid he was not to spend a single cocoanut on anything else.

And yet the king and his people did not 30 love McAllister. In truth, they hated him horribly, and, to my knowledge, the whole population, with the priests at the head, tried vainly for three months to pray him to death. The devil-devils they sent after him were awe- 35 inspiring, but since McAllister did not believe in devil-devils, they were without power over him. With drunken Scotchmen all signs fail. They gathered up scraps of food which had touched his lips, an empty whiskey bottle, a cocoanut 40 from which he had drunk, and even his spittle, and performed all kinds of deviltries over them. But McAllister lived on. His health was superb. He never caught fever; nor coughs nor colds; dysentery passed him by; and the malignant ulcers and vile skin diseases that attack blacks 45 and whites alike in that climate never fastened upon him. He must have been so saturated with alcohol as to defy the lodgment of germs. I used to imagine them falling to the ground in showers of microscopic cinders as fast as they 50 entered his whiskey-sodden aura. No one loved him, not even germs, while he loved only whiskey, and still he lived.

18. The author included a physical description of the Polynesians who lived on Oolong Atoll mainly to show

(A) how powerful they were.

(B) how attractive they were.

(C) how powerful McAllister was.

(D) their lifestyle.

(E) how hard they worked.

19. In line 11, "unintermittent" most nearly means

(A) occasional.

(B) continual.

(C) innocuous.

(D) inconspicuous.

(E) social.

GO ON TO THE NEXT PAGE

20. Why didn't the king's daughter marry the man she wanted to marry?

(A) McAllister didn't like the king.

(B) McAllister didn't like the king's daughter.

(C) McAllister didn't like the man she wanted to marry.

(D) McAllister refused to allow the marriage.

(E) The king had not repaid his debt.

21. The commerce of Oolong Atoll was based upon the exchange of

(A) dollars.

(B) coins.

(C) cocoanuts.

(D) whiskey.

(E) lagoons.

22. Why did the Polynesians collect McAllister's spittle and items that had touched McAllister's lips?

(A) They worshipped him and things associated with him.

(B) They wanted their cocoanut crop to grow.

(C) They wanted souvenirs to remind them of this powerful man and their connection to him.

(D) They wanted to offer sacrifices to the deities.

(E) They wanted to cast spells over him so that he would die.

23. In line 41, the word "deviltries" most nearly means

(A) mischief.

(B) pranks.

(C) worship.

(D) witchcraft.

(E) devotion.

24. Which of the following is <u>not</u> a reason McAllister continued to live, according to the narrator?

(A) He did not believe in the devil-devils.

(B) He drank so much.

(C) The germs hated him.

(D) The gods favored him.

(E) He was in good health.

STOP

If time remains, you may go back and check your work. When the time allotted is up, you may go on to the next section.

TIME: 20 Minutes
16 Questions

MATH

DIRECTIONS: In this section solve each problem, using any available space on the page for scratchwork. Then decide which is the best of the choices given and fill in the corresponding oval on the answer sheet.

NOTES

(1) The use of a calculator is permitted.

(2) All numbers used are real numbers.

(3) Figures that accompany problems in this test are intended to provide information useful in solving the problems. They are drawn as accurately as possible EXCEPT when it is stated in a specific problem that the figure is not drawn to scale. All figures lie in a plane unless otherwise indicated.

(4) Unless otherwise specified, the domain of any function f is assumed to be the set of all real numbers x for which $f(x)$ is a real number.

REFERENCE INFORMATION

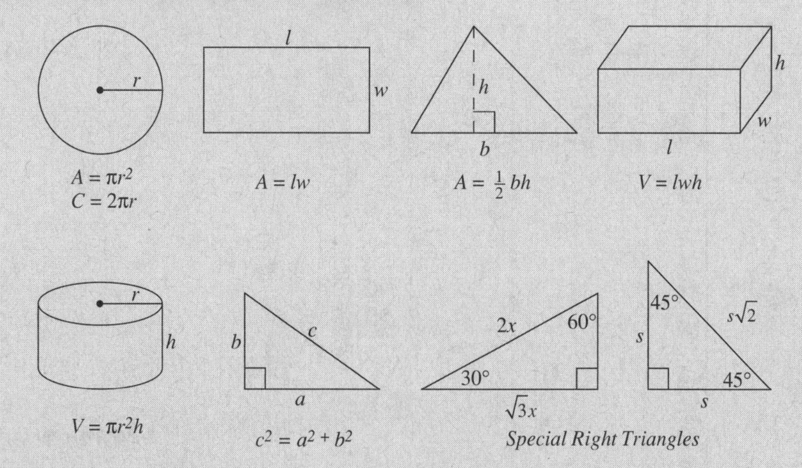

The number of degrees of arc in a circle is 360.

The sum of the measures in degrees of the angles of a triangle is 180.

1. A truck rental agency charges $29.00 per day plus $0.09 per mile driven. If the agency charges a customer for $81.40 for a truck rented for two days, how many miles did the customer drive the truck?

(A) 31.2

(B) 76.8

(C) 260.0

(D) 452.2

(E) 582.2

2. If function ♣ is defined for positive numbers m and n such that $m ♣ n = \frac{m-\sqrt{n}}{\sqrt{m}-n}$, what is 9 ♣ 16?

(A) −13

(B) −5

(C) $\frac{5}{7}$

(D) $\frac{4}{7}$

(E) 5

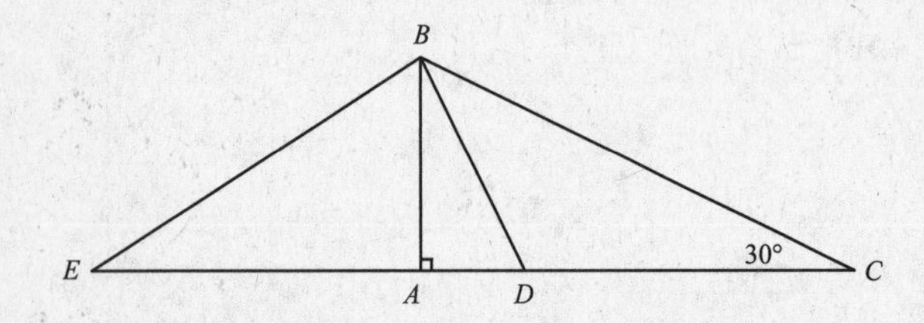

3. In the figure above, $\overline{BC} = 12$ in the right triangle ABC. If the area of ΔBDE = 24, what is the value of $\overline{DE}$?

(A) 4

(B) 6

(C) 8

(D) 12

(E) 24

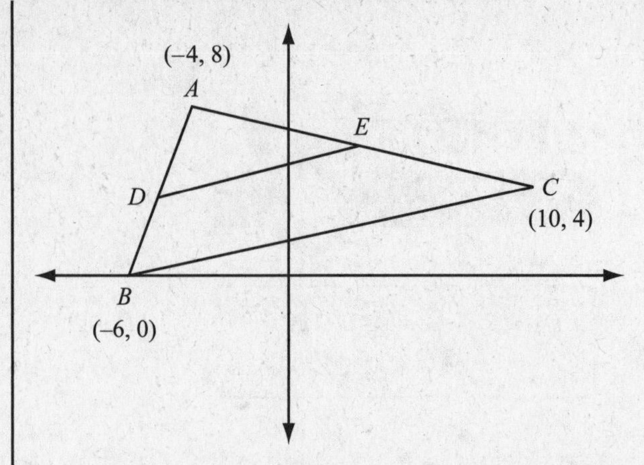

4. In the figure above, D and E are the midpoints of $\overline{AB}$ and $\overline{AC}$, respectively. What is the slope of $\overline{DE}$?

(A) −4

(B) −1

(C) 1/4

(D) 1

(E) 4

5. Which of the following is the solution set of the inequality $|-3x + 5| < 4$?

(A) $3 > x > \frac{1}{3}$

(B) $3 < x < \frac{1}{3}$

(C) $-3x > x > -1$

(D) $9 > x > 1$

(E) $9 > x > 3$

GO ON TO THE NEXT PAGE

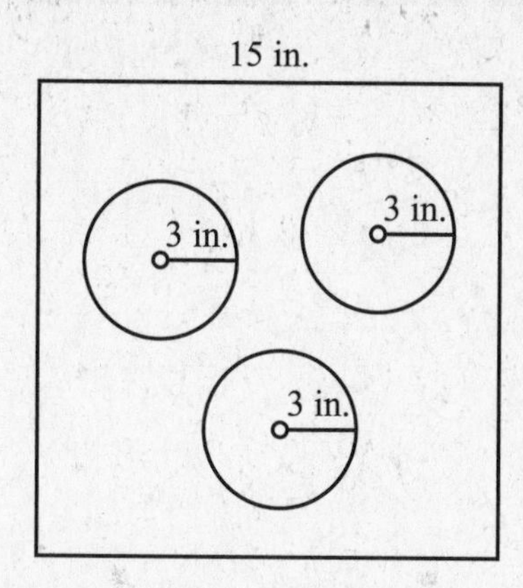

Note: Figure not drawn to scale.

6. In the figure above, a board is shown with three circular targets on it. If an arrow shot toward the board is equally likely to hit any point on the board, what is the probability that the arrow will land in one of the target circles?

(A) $\frac{9\pi}{25}$

(B) $\frac{3\pi}{25}$

(C) $\frac{3\pi}{35}$

(D) $\frac{2\pi}{73}$

(E) $\frac{3\pi}{73}$

DIRECTIONS: Student-Produced Response questions 7–16 require that you solve the problem and then enter your answer on the special grid on page 861. (For instructions on how to use the grid, see page 69 in the Diagnostic Test.)

7. If $f(x) = \frac{3x^2 - 1}{2}$ and $g(x) = \frac{2}{3x^2 - 1}$, what is the value of $f(3) - g(-1)$?

8. In a group of 120 college freshmen, 40 enrolled in Basic Mathematics, 54 enrolled in Writing and Reading, and 25 enrolled in both courses. How many students are not enrolled in either course?

9. If $\left(x^{-\frac{1}{5}}\right) \div \left(y^5\right)^{\frac{1}{5}} = x^{3y^{-3}}$ for positive x and y, what is xy?

$$a, m, an, mn$$

10. For positive numbers, a, m, n, the sequence above is geometric. What is the value of $\frac{a}{m}$?

11. The sum of the three consecutive odd integers is 39 and the sum of the three consecutive even integers is 72. What is the difference between the smallest odd integer and the smallest even integer?

12. If the average (arithmetic mean) of $(x^2 + 3)$, $(12 + 11x - 2x^2)$, and $(x^2 + x)$ is 13, then what is the average of $(x^2 - 1)$ and $(x + 1)$?

13. M varies inversely as the cube of t. When $M = 1$, $t = 2$. What is M when $t = \frac{1}{2}$?

14. The first term of a sequence of integers is 6 and the third term is 16. If the average of the first three terms is 11, what is the fourth term?

15. The dimensions of a rectangle are (x) and $(2x + 3)$. If the area of the rectangle is 27, what is its perimeter?

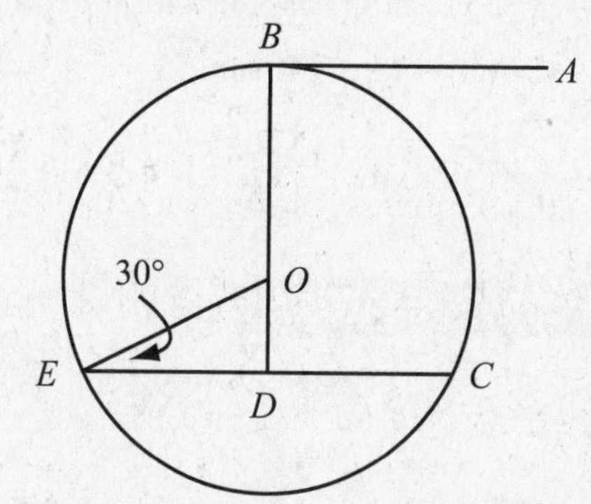

16. In the figure above, $\overline{AB}$ is tangent to the circle O, and $BD = 18$. If $\overline{AB}$ is parallel to $\overline{CE}$, what is the length of a radius of the circle?

STOP

If time remains, you may go back and check your work. When the time allotted is up, you may go on to the next section.

TIME: 19 Minutes
20 Questions

CRITICAL READING

<u>DIRECTIONS</u>: Each sentence below has one or two blanks, each blank indicating that something has been omitted. Beneath the sentence are five lettered words or sets of words. Choose the word or set of words that BEST fits the meaning of the sentence as a whole.

EXAMPLE

Although critics found the book _______ , many readers found it rather _______ .

(A) obnoxious . . . perfect

(B) spectacular . . . interesting

(C) boring . . . intriguing

(D) comical . . . persuasive

(E) popular . . . rare

EXAMPLE ANSWER

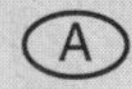   

1. Under __________ conditions, the flowers should bloom all summer long.

   (A) rudimentary
   (B) optimal
   (C) primitive
   (D) unique
   (E) biennial

2. So that the new employee could see what he expected, he offered her a(n) ________ of effective report-writing.

   (A) enigma
   (B) paradox
   (C) paradigm
   (D) panorama
   (E) predilection

3. The entire group was in such _________ over one particular issue that they could not reach a consensus on any other topics that came up during the meeting.

   (A) discord
   (B) alacrity
   (C) optimism
   (D) accord
   (E) cohesiveness

4. The aggressive woman and the shy man seemed so ___________ that everyone was surprised when they announced their plans to get married.

   (A) uniform
   (B) incompatible
   (C) steadfast
   (D) amiable
   (E) fortuitous

5. The bully almost appeared to be ________ in his pursuit of his victims; he reminded me of a cat preying on a bird or a mouse.

   (A) predatory
   (B) esoteric
   (C) pejorative
   (D) timorous
   (E) reticent

6. The art teacher has set up so many rules for the students to follow that she __________ their creativity.

   (A) enhances
   (B) stultifies
   (C) propitiates
   (D) extols
   (E) elucidates

7. Their __________ was enough to ________ her; she cared too much about their opinion to want to displease them ever again.

   (A) chagrin . . . consecrate
   (B) homage . . . devastate
   (C) accolade . . . admonish
   (D) disdain . . . bolster
   (E) derision . . . chastise

GO ON TO THE NEXT PAGE

DIRECTIONS: Each passage below is followed by questions based on its content. Answer the questions on the basis of what is stated or implied in each passage and in any introductory material that may be provided.

The following are taken from the inaugural addresses of Jefferson Davis, President of the Confederacy (February 18, 1861), and Abraham Lincoln, President of the Union (March 4, 1861), at a time when the nation was divided over issues that resulted in the Civil War.

Passage 1

This is an excerpt from the Inaugural Address of Jefferson Davis, delivered at the Confederate capitol on February 18, 1861.

1 Looking forward to the speedy establishment of a permanent government . . . , I enter upon the duties of the office, to which I have been chosen, with the hope that the beginning of our career, as 5 a Confederacy, may not be obstructed by hostile opposition to our enjoyment of the separate existence and independence which we have asserted, and, with the blessing of Providence, intend to maintain. Our present condition . . . illustrates the 10 American idea that governments rest upon the consent of the governed, and that it is the right of the people to alter or abolish governments whenever they become destructive of the ends for which they were established.

15 The declared purpose of the compact of union from which we have withdrawn, was "to establish justice, insure domestic tranquility, provide for the common defense, promote the general welfare;" and when in the judgment of the 20 sovereign States now composing this Confederacy, it had been perverted from the purposes for which it was ordained, and had ceased to answer the ends for which it was established, 25 . . . the government created by that compact should cease to exist. In this they merely asserted a right which the Declaration of Independence of 1776 had defined to be inalienable. . . . He, who knows the hearts of men, will judge of the sincerity with which we labored 30 to preserve the government of our fathers in its spirit. The right solemnly proclaimed at the birth of the States and which has been affirmed and re-affirmed in the bills of rights of States subsequently admitted into the Union of 1789, 35 undeniably recognizes in the people the power to resume the authority delegated for the purposes of government. Thus the sovereign States, here represented, proceeded to form this Confederacy. . . . They formed a new alli- 40 ance, but within each State its government has remained, and the rights of person and property have not been disturbed. The agent, through whom they communicated with foreign nations, is changed; but this does not necessarily 45 interrupt their international relations.

Sustained by the consciousness that the transition from the former Union to the present Confederacy has not proceeded from a disregard on our part of just obligations, or any fail- 50 ure to perform any constitutional duty; moved by no interest or passion to invade the rights of others; anxious to cultivate peace and commerce with all nations, if we may not hope to avoid war, we may at least expect that poster- 55 ity will acquit us of having needlessly engaged in it. Doubly justified by the absence of wrong on our part, and by wanton aggression on the part of others, there can be no cause to doubt that the courage and patriotism of the people 60 of the Confederate States will be found equal to any measures of defense which honor and security may require.

An agricultural people, whose chief interest is the export of a commodity required in every 65 manufacturing country, our true policy is peace and the freest trade which our necessities will permit. It is alike our interest, and that of all those to whom we would sell and from whom we would buy, that there should be fewest 70

GO ON TO THE NEXT PAGE

practicable restrictions upon the interchange of commodities. There can be but little rivalry between ours and any manufacturing or navigating community, such as the northeastern States of the American Union. It must follow, therefore, that a mutual interest would invite good will and kind offices. If, however, passion or the lust of dominion should cloud the judgment or inflame the ambition of those States, we must prepare to meet the emergency, and to maintain, by the final arbitrament of the sword, the position which we have assumed among the nations of the earth. . . . As a necessity, not a choice, we have resorted to the remedy of separation; and henceforth our energies must be directed to the conduct of our own affairs, and the perpetuity of the Confederacy which we have formed. If a just perception of mutual interest shall permit us peaceably to pursue our separate political career, my most earnest desire will have been fulfilled; but if this be denied to us, and the integrity of our territory and jurisdiction be assailed it, it will but remain for us, with firm resolve, to appeal to arms and invoke the blessings of Providence on a just cause. . . .

For purposes of defense, the Confederate States may, under ordinary circumstances, rely mainly upon the militia; but it is deemed advisable, in the present condition of affairs, that there should be a well-instructed and disciplined army, more numerous than would usually be required on a peace establishment. I also suggest that, for the protection of our harbors and commerce on the high seas, a navy adapted to those objects will be required. These necessities have doubtless engaged the attention of Congress.

. . . It is not unreasonable to expect that States from which we have recently parted, may seek to unite their fortunes with ours under the government which we have instituted. For this your constitution makes adequate provision; but beyond this, if I mistake not, the judgment and will of the people, a re-union with the States from which we have separated is neither practicable nor desirable. . . .

Experience in public stations . . . has taught me that care, and toil, and disappointment, are the price of official elevation. You will see many errors to forgive, many deficiencies to tolerate, but you shall not find in me either a want of zeal or fidelity to the cause that is to me highest in hope and of most enduring affection. Your generosity has bestowed upon me an undeserved distinction—one which I never sought nor desired. Upon the continuance of that sentiment, and upon your wisdom and patriotism, I rely to direct and support me in the performance of the duty required at my hands. . . .

Reverently let us invoke the God of our fathers to guide and protect us in our efforts to perpetuate the principles which, by his blessing, they were able to vindicate, establish, and transmit to their posterity, and with a continuance of his favor, ever gratefully acknowledged, we may hopefully look forward to success, to peace, and to prosperity.

Passage 2

This is an excerpt from the First Inaugural Address of Abraham Lincoln on March 4, 1861.

Apprehension seems to exist among the people of the Southern States that by the accession of a Republican Administration their property and their peace and personal security are to be endangered. There has never been any reasonable cause for such apprehension. Indeed, the most ample evidence to the contrary has all the while existed and been open to their inspection. It is found in nearly all the published speeches of him who now addresses you. I do but quote from one of those speeches when I declare that—I have no purpose, directly or indirectly, to interfere with the institution of slavery in the States where it exists. I believe I have no lawful right to do so, and I have no inclination to do so.

GO ON TO THE NEXT PAGE

Those who nominated and elected me did so with full knowledge that I had made this and many similar declarations and had never recanted them; and more than this, they placed in the platform for my acceptance, and as a law to themselves and to me, the clear and emphatic resolution which I now read: *Resolved*, That the maintenance inviolate of the rights of the States, and especially the right of each State to order and control its own domestic institutions according to its own judgment exclusively, is essential to that balance of power on which the perfection and endurance of our political fabric depend; and we denounce the lawless invasion by armed force of the soil of any State or Territory, no matter what pretext, as among the gravest of crimes. . . . I now reiterate these sentiments, and in doing so I only press upon the public attention the most conclusive evidence of which the case is susceptible that the property, peace, and security of no section are to be in any wise endangered by the now incoming Administration. I add, too, that all the protection which, consistently with the Constitution and the laws, can be given will be cheerfully given to all the States when lawfully demanded, for whatever cause—as cheerfully to one section as to another.

. . . Descending from these general principles, we find the proposition that in legal contemplation the Union is perpetual confirmed by the history of the Union itself. The Union is much older than the Constitution. It was formed, in fact, by the Articles of Association in 1774. It was matured and continued by the Declaration of Independence in 1776. It was further matured, and the faith of all the then thirteen States expressly plighted and engaged that it should be perpetual, by the Articles of Confederation in 1778. And finally, in 1787, one of the declared objects for ordaining and establishing the Constitution was "*to form a more perfect Union.*"

But if destruction of the Union by one or by a part only of the States be lawfully possible, the Union is *less* perfect than before the Constitution, having lost the vital element of perpetuity.

It follows from these views that no State upon its own mere motion can lawfully get out of the Union; that *resolves* and *ordinances* to that effect are legally void, and that acts of violence within any State or States against the authority of the United States are insurrectionary or revolutionary, according to circumstances. I therefore consider that in view of the Constitution and the laws the Union is unbroken, and to the extent of my ability, I shall take care, as the Constitution itself expressly enjoins upon me, that the laws of the Union be faithfully executed in all the States. Doing this I deem to be only a simple duty on my part, and I shall perform it so far as practicable unless my rightful masters, the American people, shall withhold the requisite means or in some authoritative manner direct the contrary. I trust this will not be regarded as a menace, but only as the declared purpose of the Union that it *will* constitutionally defend and maintain itself.

In doing this there needs to be no bloodshed or violence, and there shall be none unless it be forced upon the national authority. The power confided to me will be used to hold, occupy, and possess the property and places belonging to the Government and to collect the duties and imposts; but beyond what may be necessary for these objects, there will be no invasion, no using of force against or among the people anywhere. Where hostility to the United States in any interior locality shall be so great and universal as to prevent competent resident citizens from holding the Federal offices, there will be no attempt to force obnoxious strangers among the people for that object. While the strict legal right may exist in the Government to enforce the exercise of these offices, the attempt to do so would be so irritating and so nearly impracticable withal that I deem it better to forego for the time the uses of such offices . . . in every

GO ON TO THE NEXT PAGE

case and exigency my best direction will be exercised . . . with a view and a hope of a peaceful solution of the national troubles and the restoration of fraternal sympathies and affections. . . .

110 All profess to be content in the Union if all constitutional rights can be maintained. Is it true, then, that any right plainly written in the Constitution has been denied? I think not. . . .

One section of our country believes slavery 115 is *right* and ought to be extended, while the other believes it is *wrong* and ought not to be extended. This is the only substantial dispute. . . .

Physically speaking, we can not separate. We can not remove our respective sections from 120 each other nor build an impassable wall between them. A husband and wife may be divorced and go out of the presence and beyond the reach of each other, but the different parts of our country can not do this. They can not but remain face to 125 face, and intercourse, either amicable or hostile, must continue between them. Is it possible, then, to make that intercourse more advantageous or more satisfactory *after* separation than *before*? Can aliens make treaties easier than friends 130 can make laws? Can treaties be more faithfully enforced between aliens than laws can among friends? Suppose you go to war, you can not fight always; and when, after much loss on both sides and no gain on either, you cease fighting, 135 the identical old questions, as to terms of intercourse, are again upon you. . . .

Why should there not be a patient confidence in the ultimate justice of the people? Is there any better or equal hope in the world? In 140 our present differences, is either party without faith of being in the right? If the Almighty Ruler of Nations, with His eternal truth and justice, be on your side of the North, or on yours of the South, that truth and that justice will surely prevail by 145 the judgment of this great tribunal of the American people. . . .

Intelligence, patriotism, Christianity, and a firm reliance on Him who has never yet forsaken this favored land are still competent to 150 adjust in the best way all our present difficulty.

In *your* hands, my dissatisfied fellow-countrymen, and not in *mine*, is the momentous issue of civil war. The Government will not assail *you*. You can have no conflict without being yourselves the aggressors. *You* have no 155 oath registered in heaven to destroy the Government, while I shall have the most solemn one to "preserve, protect, and defend it."

I am loath to close. We are not enemies, but friends. We must not be enemies. Though 160 passion may have strained it must not break our bonds of affection. . . .

8. Jefferson Davis asserted that people had a right to establish their own separate government

(A) for any reason, when they so desired.
(B) when the existing government declared war for reasons with which they did not agree.
(C) when the existing government sought to destroy the principles on which it was established.
(D) when the leader of the existing government was proven weak and misguided.
(E) if their Christian beliefs were threatened.

9. In comparing the Declaration of Independence and the principles of the Confederacy, Jefferson Davis

(A) declared that they were completely different.
(B) declared that they had the same focus.
(C) made no comparison.
(D) believed that they had both similarities and differences.
(E) completely rejected the Declaration of Independence and supported the principles of the Confederacy.

GO ON TO THE NEXT PAGE

10. According to Davis, the North and South shared an interest in

    (A) commerce.
    (B) maintaining slavery.
    (C) the safety of the citizens.
    (D) avoiding war.
    (E) foreign affairs.

11. In line 81, when Davis spoke of the "arbitrament of the sword" he was referring to

    (A) the courts.
    (B) one-on-one dueling.
    (C) war.
    (D) presidential debates.
    (E) slavery.

12. In reference to the Confederacy's reuniting with the Union, Jefferson Davis

    (A) was open to compromise.
    (B) was willing to concede to the Union in order to avoid war.
    (C) expressed a willingness to leave it up to the vote of the Confederate and Union citizens.
    (D) was not agreeable to the idea.
    (E) was hopeful that it would eventually occur.

13. In response to the Southern states' apprehension that their property and their safety were in jeopardy, Lincoln

    (A) confirmed their fears.
    (B) said nothing to confirm their fears or to relieve them.
    (C) offered them a chance to avoid any destruction, by reuniting with the Union.
    (D) reassured them that they had no reason to worry.
    (E) expressed a concern for the North but not for the South.

14. Lincoln believed that by seceding the Southern states were

    (A) exercising their rights as established in the U.S. Constitution.
    (B) violating the principles established in the U.S. Constitution.
    (C) neither exercising nor violating the principles and rights established in the U.S. Constitution.
    (D) declaring war on the North.
    (E) surrendering to the wishes of the North.

15. In line 48 of Passage 2, "perpetual" most nearly means

    (A) moral.
    (B) constitutional.
    (C) religious.
    (D) old.
    (E) everlasting.

GO ON TO THE NEXT PAGE

16. Lincoln saw the basic issue between the North and the South as being an issue of

    (A) religion.
    (B) slavery.
    (C) the right to bear arms.
    (D) freedom of speech.
    (E) foreign trade.

17. Lincoln's speech indicated that he saw war as

    (A) a way to solve problems by forcing the weaker side to accept the beliefs of the stronger side.
    (B) a necessary evil when two side were in strong disagreement.
    (C) serving little purpose, with both sides still in disagreement even after much fighting had occurred.
    (D) a means of ending slavery and negatively affecting the South's economy.
    (E) the only recourse left for the Union.

18. Both Lincoln and Davis

    (A) believed that slavery was acceptable in some situations.
    (B) expressed faith in God.
    (C) believed that the formation of a separate Southern government was necessary.
    (D) were anxious to declare war.
    (E) were reluctant to accept a leadership position.

19. As far as the rights of the states were concerned,

    (A) Davis declared that the rights of the states were secondary to those of the Confederacy, whereas Lincoln supported the rights of the states.
    (B) Lincoln declared that the rights of the states were secondary to those of the Union, whereas Davis supported the rights of the states over those of the Union.
    (C) both Davis and Lincoln supported the rights of the states to act as they saw most beneficial to the states themselves.
    (D) neither Davis nor Lincoln supported the rights of the states.
    (E) neither man's Inaugural Address discussed the rights of the states.

STOP

If time remains, you may go back and check your work. When the time allotted is up, you may go on to the next section.

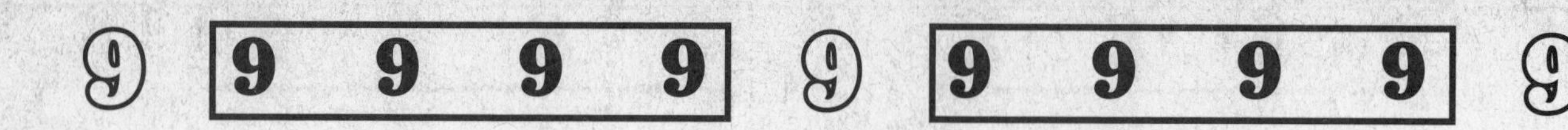

TIME: 10 Minutes
14 Questions

WRITING

DIRECTIONS: In each of the following sentences, some portion of the sentence is underlined. Under each sentence are five choices. The first choice has the same wording as the original. The other four choices are reworded. Sometimes the first choice containing the original wording is the best; sometimes one of the other choices is the best. Choose the letter of the best choice. Your choice should produce a sentence that is not ambiguous or awkward and that is correct, clear, and precise.

This is a test of correct and effective English expression. Keep in mind the standards of English usage, punctuation, grammar, word choice, and construction.

EXAMPLE

When you listen to opera, <u>a person may not appreciate it.</u>

(A) a person may not appreciate it.

(B) it may not be appreciated by a person.

(C) you may not appreciate it.

(D) which may not be appreciated by you.

(E) appreciating it may be a problem for you.

EXAMPLE ANSWER

1. Because only three students in the class wants to write a report, the teacher had decided to give it a test instead.

   (A) wants to write a report, the teacher had decided to give it a test instead.

   (B) want to write a report, the teacher has decided to give them a test instead.

   (C) want to write a report, the teacher had decided to give them a test instead.

   (D) wants to write a report, the teacher has decided to give it a test instead.

   (E) wanting to write a report, the teacher has decided to give them a test instead.

2. The insurance agent gave us a sample policy, he proceeded to explain the fine print.

   (A) The insurance agent gave us a sample policy, he

   (B) The sample policy, which was given to us by the insurance agent, who

   (C) The sample policy was first given to us by the insurance agent, then

   (D) After giving us a sample policy, the insurance agent

   (E) The insurance agent, having given us the sample policy, he

3. After experiencing the catastrophic tornado, she realized that one's material possessions are not as important as your friends and relatives.

   (A) she realized that one's material possessions are not as important as your

   (B) she realized that one's material possessions is not as important as your

   (C) realizing that one's material possessions are not as important as their

   (D) she realized that one's material possessions are not as important as their

   (E) she realized that one's material possessions are not as important as one's

4. Knowing the names of customers is a good idea for a small-town store owner.

   (A) Knowing the names of customers is

   (B) Knowing the names of customers are

   (C) Knowing the names of customer is

   (D) Knowing the name of customers are

   (E) Know the names of customers is

GO ON TO THE NEXT PAGE

5. Bandwagon is where someone uses a type of propaganda and says that the right thing to do is what everyone else is doing.

   (A) Bandwagon is where someone uses a type of propaganda and says that the right thing to do is what everyone else is doing.
   (B) Bandwagon is when they use propaganda that says the right thing to do is what everyone else is doing.
   (C) Bandwagon is a propaganda technique that says the right thing to do is what everyone else is doing.
   (D) Bandwagon is a propaganda technique that says you should do the right thing and that thing is what everyone else is doing.
   (E) Bandwagon, a propaganda technique that says the right thing to do is what everyone else is doing.

6. The woman, wanting her son to graduate at the top of his class, often completed his assignments for him.

   (A) The woman, wanting her son to graduate at the top of his class,
   (B) The woman because wanting her son to graduate at the top of his class,
   (C) The woman wanting her son to graduate at the top of his class and often completing
   (D) With wanting her son to graduate at the top of his class, the woman
   (E) The woman, wanting her son to graduate at the top of his class, she

7. Some doctors argue that a pregnant woman's smoking cigarettes, this causes health problems for the baby to have health problems.

   (A) a pregnant woman's smoking cigarettes, this causes the baby to have
   (B) for a pregnant woman to smoke cigarettes, it causes the baby to have
   (C) the smoking of cigarettes by a pregnant woman would cause the baby to have
   (D) when smoking cigarettes while pregnant, it causes
   (E) a pregnant woman's smoking cigarettes causes the baby to have

8. Their house was an interesting combination of varied décor styles as Victorian, traditional, rustic, retro, and contemporary.

   (A) varied décor styles as
   (B) varied décor styles that are
   (C) décor styles of such variety as
   (D) décor styles as varied as
   (E) a décor style as varied as

9. A healthy lifestyle includes an exercise regimen and to maintain a balanced diet.

   (A) to maintain a balanced diet.
   (B) a balanced diet.
   (C) maintaining a balanced diet.
   (D) balancing your diet.
   (E) to balance your diet.

10. More polite and sociable compared with her sister, Joan received more invitations to social events.

(A) More polite and sociable compared with
(B) Both more polite and more sociable compared to
(C) More polite and social than
(D) By being more polite and social than
(E) More polite as well as more social as

11. Having witnessed the horrible car wreck, getting behind the wheel was something she was fearful of.

(A) getting behind the wheel was something she was fearful of.
(B) she was fearful of getting behind the wheel.
(C) getting behind the wheel frightened her.
(D) getting behind the wheel was something of which she was fearful.
(E) she, being fearful of getting behind the wheel.

12. All the children are in the Christmas program, the boys were dressed as Santas and the girls were dressed as snowmen.

(A) program, the boys were dressed as Santas and the girls were dressed as snowmen.
(B) program, the boys would dress as Santas and the girls would dress as snowmen.
(C) program the boys are dressed as Santas, and the girls are dressed as snowmen.
(D) program; the boys are dressed as Santas, and the girls are dressed as snowmen.
(E) program the boys are dressed as Santas and the girls as snowmen.

13. George Washington's legacy, like several U.S. Presidents, is the desire to maintain freedom and democracy.

(A) like several U. S. Presidents
(B) like several other U. S. Presidents
(C) like that of several U. S. Presidents
(D) like that of several other U. S. Presidents
(E) like that one of several U. S. Presidents

GO ON TO THE NEXT PAGE

14. The weather in Florida can be quite <u>devastating; every year they face destructive hurricanes and tropical storms</u>.

    (A) devastating; every year they face destructive hurricanes and tropical storms.
    (B) devastating, every year facing destructive hurricanes and tropical storms.
    (C) devastating every year the people there face destructive hurricanes and tropical storms.
    (D) devastating, because every year they face destructive hurricanes and tropical storms.
    (E) devastating; every year the people there face destructive hurricanes and tropical storms.

STOP

If time remains, you may go back and check your work.

TEST 3

ANSWER KEY

Section 1—Essay

Refer to the Detailed Explanation for essay analysis

Section 2—Math

1. (C)	6. (C)	11. (E)	16. (D)
2. (B)	7. (C)	12. (C)	17. (E)
3. (C)	8. (D)	13. (D)	18. (D)
4. (B)	9. (D)	14. (C)	19. (B)
5. (C)	10. (E)	15. (D)	20. (B)

Section 3—Critical Reading

1. (C)	7. (B)	13. (C)	19. (D)
2. (D)	8. (B)	14. (C)	20. (A)
3. (C)	9. (A)	15. (B)	21. (B)
4. (A)	10. (D)	16. (C)	22. (B)
5. (E)	11. (C)	17. (D)	23. (E)
6. (C)	12. (A)	18. (B)	24. (A)

Section 4—Writing

1. (B)	10. (D)	19. (D)	28. (A)
2. (A)	11. (A)	20. (C)	29. (E)
3. (C)	12. (A)	21. (C)	30. (B)
4. (E)	13. (C)	22. (A)	31. (E)
5. (D)	14. (A)	23. (E)	32. (D)
6. (B)	15. (B)	24. (A)	33. (C)
7. (A)	16. (A)	25. (C)	34. (B)
8. (E)	17. (B)	26. (D)	35. (B)
9. (B)	18. (C)	27. (D)	

Section 5—Math

1.	(D)	6.	(E)	11.	(D)	15.	(A)
2.	(E)	7.	(A)	12.	(D)	16.	(A)
3.	(D)	8.	(C)	13.	(E)	17.	(B)
4.	(D)	9.	(D)	14.	(C)	18.	(C)
5.	(E)	10.	(E)				

Section 6—Critical Reading

1.	(E)	7.	(A)	13.	(A)	19.	(B)
2.	(B)	8.	(E)	14.	(B)	20.	(D)
3.	(A)	9.	(D)	15.	(E)	21.	(C)
4.	(C)	10.	(D)	16.	(D)	22.	(E)
5.	(B)	11.	(C)	17.	(C)	23.	(D)
6.	(C)	12.	(C)	18.	(C)	24.	(D)

Section 7—Math

1.	(C)	5.	(A)	9.	(1)	13.	(64)
2.	(B)	6.	(B)	10.	(1)	14.	(21)
3.	(C)	7.	(12)	11.	(11)	15.	(27)
4.	(C)	8.	(51)	12.	(3)	16.	(12)

Section 8—Critical Reading

1.	(B)	6.	(B)	11.	(C)	16.	(B)
2.	(C)	7.	(E)	12.	(D)	17.	(C)
3.	(A)	8.	(C)	13.	(D)	18.	(B)
4.	(B)	9.	(B)	14.	(B)	19.	(B)
5.	(A)	10.	(A)	15.	(E)		

Section 9—Writing

1.	(B)	5.	(C)	9.	(B)	12.	(D)
2.	(D)	6.	(A)	10.	(C)	13.	(D)
3.	(E)	7.	(E)	11.	(B)	14.	(E)
4.	(A)	8.	(D)				

DETAILED EXPLANATIONS

SECTION 1—ESSAY

Sample Essay with Commentary

Essay I (Score: 5–6)

Money is obviously necessary for survival today. We need money to meet our basic needs of food, clothing, and shelter. But man's obsession for money has gone far beyond the desire to acquire what is merely essential. It seems that the more we have, the more we want. Thus, material wealth—the possession of so many things that money can buy—has become a common goal. Misguided or not, we regard those who have money as content. Many who want but do not have financial resources will resort to any means to acquire them—thus proving Mark Twain's contention, "The lack of money is the root of all evil."

Even at the high school level, prestige is directly connected to money. Those who can afford to buy the latest fashions, drive the newest cars, and participate in the most activities are the most admired and the most popular. Those who lack the funds for such things are left out and remain unnoticed. Especially at this age, popularity and prestige are extremely important. And for some, the only perceivable way to make up for the lack of money is to follow the "evil" path—steal the Nike shoes and the Old Navy jeans, heist the T-bird convertible, and rob a store to pay the activity fees.

Obviously, not all poor people are evil. But illegal drug sales, prostitution, and robberies—all motivated by the desire to acquire money and escape poverty—are rampant in the slums of America's big cities. A man loses his job and has no means of feeding his family; illicit drugs are a "hot commodity" in the area, and thus are an easy means of obtaining what he lacks—money. So he resorts to drug-dealing. Likewise, a single mother who quit school to have her baby is unable to get a job because of lack of experience and education; she needs—but does not have—money to feed and clothe her baby. The one job that she can get and that will ensure her financial security is prostitution. And for the "have-nots," the cash registers in the countless stores provide easy way to fill the economic void in their lives, and the many products in those stores can be readily sold. "The lack of money is the root of . . . evil (crime)" in each case.

Just as not all poor people are evil, not all rich people are happy. But a careful look at crime statistics reveals that the reason behind many crimes was the perpetrators' perception that they lacked enough money. Ironically, some white-collar crimes—committed by people who should be financially solvent—are efforts to make up for a lack of money. The newspapers are full of stories about once-wealthy CEOs and other "successful people" who squandered all their riches and are desperate at their lack of resources; they then turn to evil—embezzlement, bribery, fraud, tax evasion, and extortion—to rectify their situation.

Would these people become criminals if they did not lack money? Some probably would, but many would not. It is the lack of money that led them to steal, prostitute, and deal drugs. If having money were not so important in our lives today—and lacking money were not so devastating—perhaps the world would be a little less evil.

ANALYSIS

Essay I has a score range of 5–6. It is a well-reasoned essay that takes a stand and supports it. The introductory paragraph sets up the thesis. The three body paragraphs supply information that supports the thesis, and the concluding paragraph restates it. The writing demonstrates a variety of sentence structure and some fairly sophisticated vocabulary; and—importantly—it is grammatically sound.

DETAILED EXPLANATIONS

SECTION 2—MATH

1. (C) The bars in the chart show that the major jump in profits occurred in April.

2. (B) Find $\triangleleft a^2 \triangleright - \triangleleft a \triangleright$ using the given definition as follows:

$$\triangleleft a^2 \triangleright - \triangleleft a \triangleright = \frac{\left(a^2\right)^2+1}{\left(a^2\right)^2-1} - \frac{a^2+1}{a^2-1} = \frac{a^4+1}{a^4-1} - \frac{a^2+1}{a^2-1} =$$

$$\frac{a^4+1}{\left(a^2-1\right)\left(a^2+1\right)} - \frac{a^2+1}{a^2-1} = \frac{a^4+1-\left(a^2+1\right)}{\left(a^2-1\right)\left(a^2+1\right)} = \frac{2a^2}{a^4-1}$$

3. (C) Only the function in (C) is equal to the given function.

Function (A) $-f(x) = -(x^4 - x^2 + 1) = -x^4 + x^2 - 1$

Function (B) $f(x^2) = (x^2)^4 - (x^2)^2 + 1 = x^8 - x^4 + 1$

Function (C) $f(-x) = (-x)^4 - (-x)^2 + 1 = x^4 - x^2 + 1$

Function (D) $-f(-x) = -x^4 + x^2 - 1$

Function (E) $f(-x^2) = (x^2)^4 - (x^2)^2 + 1 = x^8 - x^4 + 1$

4. (B) Let x equal the unknown number. Then, $(0.8)(3x) = 456$ and $2.4x = 456$. Therefore, $x = 190$.

5. (C) Using the definition of rational exponents, convert from negative exponents to rational expressions:

$$\frac{x^{-2}+y^{-3}}{x^{-3}+y^{-2}} = \frac{\frac{1}{x^2}+\frac{1}{y^3}}{\frac{1}{x^3}+\frac{1}{y^2}} = 1$$

Add the fractions in both the numerator and denominator, then simplify.

$$\frac{\frac{x^{-1}+y^{-3}}{x^{-3}y^{-2}}}{\frac{x^2+y^2}{x^3y^2}} = \frac{x^{-2}+y^{-3}}{x^{-3}y^{-2}} \times \frac{x^3y^2}{x^2+y^2} = \frac{x\left(x^2+y^3\right)}{y\left(x^3+y^2\right)} = 1$$

Set the numerator and the denominator equal (because the fraction is equal to 1) and simplify:

$$x^3 + xy^3 = x^3y + y^3$$

Rearrange the terms of the preceding equation as follows and then factor out the common elements in each side.

$$x^3 - x^3y = y^3 - xy^3$$

$$x^3(1 - y) = y^3(1 - x)$$

Cross-multiply again to reach the form given in the question.

$$\frac{1-x}{1-y} = \frac{x^3}{y^3} = \left(\frac{x}{y}\right)^3$$

6. (C) Calculate $f(x)$ for the values of the variable given in each option as follows:

(A) $f(2) = 4mn$, $f(6) = 36mn$, $f(8) = 64mn$

(B) $f(1) = mn$, $f(2) = 4mn$, $f(3) = 9mn$

(C) $f(2) = 4mn$, $f(4) = 16mn$, $f(8) = 64mn$

(D) $f(3) = 9mn$, $f(5) = 25mn$, $f(7) = 49mn$

(E) $f(3) = 9mn$, $f(4) = 16mn$, $f(6) = 36mn$

Among the options given, only terms of (C) form a geometric sequence, because $(16mn)^2 = (4mn)(64mn)$.

7. (C) If x is her sales per month, then $(12\%)x = 0.12x$ in dollars is her earning on commission alone. Adding her fixed salary, \$240, to this earning makes her total monthly income $240 + 0.12x$ in dollars.

8. (D) First, convert the times given in minutes into hours, because the answers are given in miles per *hour*. Therefore, Maurice's total riding time is $\frac{1}{2}$ hour. Next, create a table based on the equation "distance = rate × time" to help set up the problem. Let x = the speed at which Maurice rides home (in miles per hour).

	$D =$	$R \times$	T
Going to store	4	$2x$	t
Going home	4	x	$\frac{1}{2} - t$
Total	8	—	$\frac{1}{2}$

Solve for t by setting the "store" and "home" expressions equal to one another (because the distance is the same:

$$2xt = x\left(\frac{1}{2} - t\right)$$

$$2xt = \frac{1}{2}x - xt$$

$$3xt = \frac{1}{2}x$$

$$t = \frac{1}{6}$$

Next, plug the value for t into the equation for "home":

$$4 = x\left(\frac{1}{2} - \frac{1}{6}\right)$$

$$4 = x\left(\frac{3}{6} - \frac{1}{6}\right)$$

$$4 = \frac{1}{3}x$$

$$x = 12$$

9. (D) Because no figure is given for reference, draw a line and label the points on it to help visualize this problem. Then, let l equal the total length of the line segment $\overline{AB}$.

$$\overline{AM} = \overline{MB} = \frac{1}{2}\overline{AB}$$

One half of the segment, $\overline{AM}$, is divided into three equal sections, each of which is $\frac{1}{2} \div 3 = \frac{1}{6}$ of the total length. The other half of the segment, $\overline{MB}$, is divided into four equal sections, $\frac{1}{2} \div 3 = \frac{1}{8}$ of the total length. Using the line you have sketched, add up the fractional lengths between points E and P:

$$\frac{1}{6} + \frac{1}{8} = \frac{4}{24} + \frac{3}{24} = \frac{7}{24}$$

$\overline{EP}$ is 7/24 of the length of $\overline{AB}$, so that ratio is the probability that a point chosen on $\overline{AB}$ will be on $\overline{EP}$.

10. (E) Write the given inequality as $-30 \leq 2x - 130 \leq 30$. Adding 130 to each side of this inequality, we get $100 \leq 2x \leq 160$. Dividing each side of this inequality by 2 yields $50 \leq x \leq 80$.

11. (E) We are given $M = \{13, 14, 15, 16, 17, 18, 19, 20, 21, 22, 23\}$, $N = \{6, 8, 10, 12, 14, 16, 18\}$, and $P = \{10, 12, 20, 22\}$. We have $M \cap P = \{14, 16, 18\}$. Therefore, $M \cap N \cap P = \varnothing$ (that is, an empty set).

12. (C) Find the areas of both circle and square. The area of the circle is given by the equation $A = \pi r^2$ (which is always given in the Reference Information box at the beginning of each math section on the SAT). Notice that the diameter of the square is $2r$. Let x be the length of a side of the square. Then, using the Pythagorean theorem in a triangle formed by drawing a diameter, we have

$$x^2 + x^2 = (2r)^2$$

$$2x^2 = 4r^2$$

$$x^2 = 2r^2$$

Because $S = x^2$, then $S = 2r^2$. From $A = \pi r^2$, we get $r^2 = \frac{A}{\pi}$. Replacing the equivalent of r^2 in S, we get $S = \frac{2}{\pi}A$, where $\frac{2}{\pi}$ is a constant. So, we can say that S varies directly as A.

13. (D) Find the coordinates of midpoint of $\overline{AB}$. Denote its coordinate by (x, y). Then

$$x = \frac{-4+8}{2} = 2$$

$$y = \frac{7-1}{2} = 3$$

The coordinates of the midpoint only fit in the equation in answer (D), $x - 2y = -4$, because $2 - 2(3) = -4$.

14. (C) Check all the ranges of all answer choices before making your selection.

(A) $f(x) = 23x^2 - 12$

There are some values in the domain, such as $x = \frac{1}{23}$ or $x = -\frac{1}{23}$, for which $f(x)$ is a negative number. So, the range of this function is not just the set of positive numbers.

(B) $f(x) = |x - 6| - 1$

For a value of x such as 5.5, the value of $f(x)$ is a negative number. So, the range of this function is not limited to positive numbers.

(C) $f(x) = |x^2 + 3| + 6$

For all values of x, the total value of the expression inside the absolute value is always a positive number. Adding this number to 1 will result in a positive number. So, x either as a negative or as a positive value gives a positive value for $f(x)$. So, the range of this function is always positive and this is the correct answer.

(D) $f(x) = 4 + |x - 11|$

For this function, $f(x)$ is greater than 4.

(E) $f(x) = 23|x^2| - 6$

This function is a parabola with a y intercept of –6, so $f(x)$ is sometimes less than zero.

15. (D) Multiply corresponding sides of the given equations as follows.

$$5^m \times 5^n = a^{\frac{3}{2}} \times b^{\frac{3}{2}}$$

$$5^{m+n} = (ab)^{\frac{3}{2}}$$

Square each side in the last equation.

$$(5^2)^{m+n} = 25^{m+n} = (ab)^3$$

16. (D) The person bought four tickets, and for the first drawing, the probability that one of these tickets will be drawn is $\frac{4}{500}$. If the person wins this drawing, the probability that he or she will win the next drawing is $\frac{3}{499}$, because the first winning ticket is no longer in play. Likewise, the person's chances in the third drawing would be $\frac{2}{498}$. To get the probability that the person will win all three, multiply the three probabilities:

$$\frac{4}{500} \times \frac{3}{499} \times \frac{2}{498} = \frac{24}{124{,}251{,}000} = \frac{1}{5{,}177{,}125}$$

17. (E) We know that in a geometric sequence, the square of a term is equal to the product of the two terms listed before and after it. Check each sequence against this rule to see whether it is a geometric sequence.

(A) $(2a)^2 = (a - 2)(a + 2)$

$4a^2 = a^2 - 4$

$3a^2 = -4$

The final result is a false statement, so the given list is not a geometric sequence.

(B) $(2a - 1)^2 = (4a)(a - 1)$

$$4a^2 - 4a + 1 = 4a^2 - 4a$$

$$0 = 1$$

The final result is a false statement, so the given list is not a geometric sequence.

(C) $(8a)^2 = (4a)(12a)$

$$64a^2 = 48a^2$$

$$64 = 48$$

The final result is a false statement, so the given list is not a geometric sequence.

(D) $(3a)^2 = (a)(5a)$

$$9a^2 = 5a^2$$

$$9 = 5$$

The final result is a false statement, so the given list is not a geometric sequence.

(E) $a^2 = (a + 3)(a + 4)$

$$a^2 = a^2 + 7a + 12$$

$$7a = -12$$

$$a = -\frac{12}{7}$$

For $a = -\frac{12}{7}$, the given list forms a geometric sequence.

18. (D) Line M passes through (0, 0) and (4, 4). The coordinates of each of these points are the same, so M is the bisector of the first and third quadrants. Therefore, the equation of M is $y = x$ or $y - x = 0$. For any point (x, y) located below this line and above the x-axis, x and y are positive numbers and $x > y$.

19. (B) Using m as the constant of variation for y and $\sqrt{x^3}$, we have

$$y = m\sqrt{x^3}$$

Using n as the constant of variation for x and z^2, we have

$$x = nz^2$$

Set the preceding two equations equal to one another, take each side to the third power, then take the square root of each side:

$$\sqrt{x^3} = \sqrt{(nz^2)^3} = \sqrt{n^3z^6} = z^3\sqrt{n^3}$$

Replace (3) in (1).

$$y = mz^3\sqrt{n^3} = (m\sqrt{n^3})z^3$$

In this equation, $(m\sqrt{n^3})$ is a constant. So, y varies directly as z^3.

20. (B) Because C is the midpoint of $\overline{OB}$ and $OB = 2OA$, then $OA = OC$. The coordinates of A indicate that $\overline{AB}$ is the bisector of the first and third quadrants. Therefore, we have $C(-4, -4)$. The slope of $\overline{AB}$ is 1. Denote the slope of $\overline{CD}$ by m. Then $m(1) = -1$ or $m = -1$. Given the slope and the coordinates of a point of $\overline{CD}$, we can write its equation and simplify the result as follows.

$$y - (-4) = -(x + 4)$$

$$y + 4 = -x - 4$$

$$y = -x - 8$$

DETAILED EXPLANATIONS

SECTION 3—CRITICAL READING

1. (C) You are looking for a word that is related to the clues "fish" and "ships." "Nautical" (marine) is just such a word. Choices (A) and (C) are incorrect because "aeronautical" and "aviation" refer to aircraft, not to fish or ships. Choice (B) "aviary" is a place where birds are kept confined; thus, it is inappropriate here. And since "eclectic" means "varied, from a variety of sources," choice (E) is incorrect.

2. (D) Here you are looking for a word that reflects the irresistibility of bright lights, fanciful music, and cotton candy for a little boy. "Allure" (charm, attraction, appeal) is such a word. At first glance, choice (A) "motif" (dominant idea or central theme) may seem like a good choice, and bright lights, fanciful music, and the smell of cotton candy do have some things in common; however, it is not the motif that attracts him as much as the appeal of these things. Thus, choice (A) is not as good a choice as choice (D). "Juxtaposition" (instance of two or more things being placed side by side) does not fit semantically, so choice (B) is inappropriate. "Alacrity" (promptness in response, readiness) does not work semantically here, since bright lights, fanciful music, and cotton candy smells are things and incapable of responding; thus, choice (C) is incorrect. And although these things may seem to represent an attractive world, they hardly make up a "utopia" (place of ideal social and political perfection); therefore, choice (E) is wrong.

3. (C) Here you are looking for two words that are meaningfully related, a noun and an adjective. It makes sense that a "fledgling" (inexperienced person; beginner) would be "inept" (incompetent; unskilled) in completing a complex process. So choice (C) is appropriate. Choice (A) is incorrect because one would not expect a "novice" (beginner) to be "adept" (skilled; practiced) at a complex process. Nor would an "expert" (one who possesses special skill or knowledge) be "erroneous" (mistaken, in error); thus, choice (B) is wrong. Choice (D) is incorrect because it doesn't make sense to say that a "mentor" (teacher) was "livid" (enraged; extremely angry) in completing a complex process. Likewise, choice (E) is incorrect because a "neophyte" (beginner; newcomer) would not be described as "pertinent" (related to the matter at hand).

4. (A) Here, you are seeking an adjective that describes citizens who are threatening a senator who may vote against their wishes and a verb to convey the action they may take if he does so. Such citizens may be "seething" (agitated, angry), and they may threaten to "oust" him (remove him from office). Such a group is not likely to be "reclusive" (withdrawn), and if they were they would not threaten to "imbue" (inspire; arouse) him if he voted against their wishes. Therefore, choice (B) is incorrect. Describing these citizens as "errant" (wandering) makes little sense, and they would not be likely to threaten to "foster" (encourage; support) a senator if he voted against their wishes. So choice (C) is wrong. Although a group of citizens might threaten to "impeach" (bring an accusation against; charge with a crime or with misconduct) him, such citizens are probably not "affable" (friendly, gracious); therefore, choice (D) is wrong. Choice (E) is wrong because "rabid" (furious, violent) citizens would not threaten to "captivate" (allure; attract) him if he voted against their wishes.

5. (E) Children whose parents treat them as small adults are more likely to be "precocious" (exhibiting mature qualities at an early age) than children whose parents treat them like babies. Although such children may very well be "charismatic" (having a special magnetic appeal or charm), children whose parents have treated them as babies may also have such an appeal; therefore, choice (A) is not a good one. Children who have been treated as small adults are probably not more "diffident" (reserved, unassertive) or "petulant" (insolent, peevish) than children who are used to being treated as babies, so choices (B) and (D) are incorrect. And the word "impecunious" (without money, penniless) is not logical here; thus, choice (D) is incorrect.

6. (C) Here you are seeking a word that describes a reason that may not actually be the real reason for a person's actions. "Ostensible" (plausible, professed, apparent) is an appropriate choice. Choice (B) "veracious" (truthful, honest) does not fit here because his "veracious" reason was that he just wanted to be alone. To go home and study was not an "impetuous" (sudden, rash) reason for leaving, so choice (C) is wrong. Nor was it an "audacious" (adventurous, bold) reason; thus, choice (D) is incorrect. And going home and studying is hardly a "rancorous" (characterized by ill will) reason, even if it was not his real reason for leaving; so choice (E) is wrong.

7. (B) Success in the business world requires that one function within the "parameters" (limits, boundaries) of socially acceptable standards. If you chose choice (A), you may have simply confused the spelling of the two words; however "perimeters" means "measurement around" and usually refers to a physical line or boundary, as with a closed plane figure. Choice (B) is incorrect because "progeny" (descendants, children) does not make sense here. Neither does choice (C) because "penury" means "severe poverty; lack of resources." And choice (E) is incorrect because "parlance" (manner or mode of speaking) does not fit in the context of the sentence.

8. (B) You need a word that reflects the situation of a person facing a conflict between going along with one's peers and following the advice of one's parents. "Quandary" (state of perplexity or doubt) is just such a word. Choice (A) is incorrect because "quirk" (idiosyncrasy; peculiar trait) does not describe the situation. Neither does "liaison" (close bond or connection), "hubris" (arrogance), or "catalyst" (anything that creates a situation in which change can occur), so choices (C), (D), and (E) are incorrect.

9. (A) Here you are looking for a noun that describes a person whom a supervisor would fire and a verb to describe that person's actions—something he would do instead of working. A supervisor would fire a "laggard" (loiterer, slowpoke, lazy person) who "malingers" (pretends incapacity such as illness so as to avoid duty or work). A supervisor would probably not fire even a "dolt" (stupid person) if he "toils" (works hard). So choice (B) is wrong. This choice is also wrong because it does not make sense to say someone would be fired because he "toils" (works hard) instead of working. It does not make sense to say that a supervisor would fire a "charlatan" (one who pretends unscrupulously to have knowledge or skill; imposter) who "empathizes" (understands, is aware of, is sensitive to, and vicariously experiences the feelings, thoughts, and experience of another); therefore, choice (C) is incorrect. Likewise, it makes little sense to say that a supervisor would fire an "ingrate" (ungrateful person) who "genuflects" (kneels, bends at the knee); so choice (D) is wrong. And a "sycophant" (parasite, servile self-serving flatter) may not be someone whom a supervisor likes, but he probably wouldn't fire such a person if he "labors" (exerts one's powers of body or mind especially with painful or strenuous effort, works). And it doesn't make sense to say that someone "labors" instead of working, since "labor" and "work" are synonyms. So choice (E) is wrong.

10. (D) Keep in mind that you are to focus on the advice implied in this passage—not simply on the general wisdom of the choices given. The passage focuses on different countries' interpretations of common gestures and offers advice as to what nonverbal communication to avoid in some countries. Thus, the implied advice is that anyone who is traveling to a foreign country should know beforehand the messages that seemingly harmless gestures may convey. Choice (A) is wrong because the last sentence indicates that the absence of gestures is offensive in some countries. Choice (B) is incorrect because the passage does not address the importance of verbal language for the foreign traveler. Choice (C) is wrong because the passage generalizes the meanings of different gestures in different countries and does not discuss individual interpretations. Choice (E) is wrong because the passage does not discuss the importance of verbal communication versus the importance of nonverbal communication.

11. (C) As the first four lines indicate, Americans regard the formation of a circle with the thumb and the index finger as an "O.K." gesture—a sign of approval that is likely to elicit a smile. Choices (A) and (D) are incorrect because in Japan—not France—this gesture serves as a request for change instead of bills. In France, it is interpreted as meaning "zero" or "worthless." Choice (B) is wrong because Germans regard the gesture as obscene, not as an invitation of friendship. Choice (E) is wrong because the passage does not address the gesture's meaning in the Middle East.

12. (A) The entire passage presents a positive description of the man. Clearly, the author's tone is one of admiration. Choices (B), (C), and (E) are incorrect because they all have negative connotations, and the passage contains nothing negative. Choice (D) is incorrect because none of the author's comments imply ambivalence (simultaneous contradictory feelings).

13. (C) A man who is accepting of everyone—regardless of income, appearance, or education—and who makes everyone feel important is charismatic (appealing, magnetic). Such a person is not choice (A) "patronizing" (condescending; appearing haughty), choice (C) "pompous" (arrogant, self-important), choice (D) "deceitful" (dishonest, deceptive), or choice (E) "pretentious" (flamboyant, showy).

14. (C) If you are familiar with the meaning of the word "apprise" (inform), you had no trouble with this question. However, if you are not familiar with the word, you may be able to determine its meaning by looking at its use in the context of the sentence: *Television shows apprise us of the latest fads and fashion trends.* Try inserting each of the choices in place of "apprise." The only one that makes sense is "inform," creating the sentence: *Television shows inform us of the latest fads and fashion trends.* None of the other choices even sound correct.

15. (B) The passage specifically states, "Young people are attracted by the glamour and excitement associated with the careers of the characters they see on TV . . . however, the media's depictions of careers are seldom realistic." *Walker, Texas Ranger*, *E.R.*, *The Practice*, and *CSI* are examples of such shows. The passage also says, "It's not likely that young people—or even adults—are ever going to find reality as entertaining and glamorous as the lives depicted on television dramas." These sections, as well as several other sections in Passage 1, prove choices (A) and (D) incorrect. Passage 1 does say that "television has an effect on our lives. . . . They reflect—and perhaps even have even initiated—morals that have become acceptable in contemporary society"; however, it does not address any negative effects on today's morals. Therefore, choice (C) is incorrect. And although the first paragraph does indicate that Americans spend a great deal of time watching television, the passage does not say that television shows occupy too much of Americans' time; thus, choice (D) is wrong.

16. (C) The passage says, "Young people are attracted by the glamour and excitement associated with the careers of the characters they see on TV." The author points out that *ER* shows "EMTs—all handsome—shouting medical terminology at the nurses—beautiful, of course—as they bustle critically ill patients into the emergency room. Surgeons and other specialists are successful in the complicated operation." The passage also refers to "the heroic television detectives and cops of the television world" and to "entertaining and glamorous . . . lives depicted on television dramas." Choices (A) and (D) are incorrect because the author points out that "seldom do plumbers, roofers, or other such artisans have major roles in TV shows, and if they do they are probably on a sitcom, as opposed to the high-drama programs with the glamorous figures that viewers idolize and young people emulate." Choice (B) is wrong since the passage says nothing about careers that fit individual needs and interests. And choice (E) is wrong because the passage does not indicate that television distinguishes between the importance of a career and that of lifestyle.

17. (D) Look carefully at the phrase in which this word appears: *the high-drama programs with the glamorous figures that viewers idolize and young people emulate.* Also keep in mind what the rest of the passage—especially this paragraph—says. You know that the word is probably going to mean something similar to "idolize." Try to substitute the word "emulate" with each of the choices. Choice (D) makes sense: *the high-drama programs with the glamorous figures that viewers idolize and young people imitate or strive to be like.* You will see that this choice is the only one that makes sense, in light of the content of the paragraph and the overall passage.

18. (B) The author acknowledges that TV has both positive and negative aspects with the closing remark that, " . . . although the media heroes and heroines may give us a biased picture of certain professions, at least they expose us to some aspects of those careers—and, hopefully, inspire young people to explore those possibilities further." As this sentence proves, choice (A) is incorrect because the author does see some positive aspects of television. And although pointing out that television's portrayal of some careers is misleading, the author also concedes that "writers and producers of many television shows today do consult with experts in the professions they depict, and the essential elements are based on real aspects of real careers." So choice (C) is wrong. If you chose choice (D) you probably overlooked the word "definitely." The author admits that they may be helpful, but the passage is far from definite in this idea. And choice (E) is incorrect because the passage focuses on television's appeal of excitement and glamour—making it far from boring.

19. (D) From the very beginning, the author is critical of television and devotes the entire passage to the negative aspects of television. The author is not choice (A) "amused," choice (D) "benevolent," or choice (E) "sympathetic"; all these words imply a positive attitude, and this author's attitude is anything but positive. At the same time, the author cannot be described as vindictive, since nothing in the passage indicates that the author intends to get revenge.

20. (A) Look carefully at the sentence in which the word appears: *Accompanying contemporary America's moral* <u>*turpitude*</u> *is increased violence, especially among young people.* Also consider the content of the paragraph that precedes this statement. That paragraph expounds on television's contribution to the moral decay or decline of today's society. The only choice that continues to develop this idea is choice (A) "depravity" (corruption). Nothing in the passage supports the idea that morality is on the "upswing"; thus, choice (B) is wrong. The author does not believe that morality has stagnated (remained unchanged); so choice (C) is wrong. Choices (D) and (E) do not make sense here because the word "behavior" is meaningless here, and "tenacity" (persistence) does not develop the author's point.

21. (B) Here, you must consider each choice, and find the one that is <u>*not*</u> consistent with what the author presents in the passage. The author does not believe that television programs promote morality; in fact, he or she strongly supports the idea that television programs promote <u>*immorality*</u>. The third paragraph contends that television encourages cursing, promiscuity, and infidelity; it says, " . . . although the reasons for today's moral decline are many, television has certainly not helped the problem." If you chose choice (A) you may have been thinking of education as exclusively positive; however, as the author points out, TV may teach viewers. But " . . . perhaps the lessons the television audience—especially young viewers—learn are not beneficial to them or to society as a whole." So although the education it offers may not be education in the conventional sense, it does in fact teach lessons. Choice (C) is wrong because the author blames Americans' obsession with television for "increasing obesity, . . . diabetes, high blood pressure, high cholesterol, and numerous other health problems associated with a sedentary lifestyle." Choice (D) is wrong; the author acknowledges that television is important—too important, in fact—to Americans. It is so important to viewers that it is affecting their health, their morality, their lifestyle, and their attitude toward violence. Choice (E) is wrong because the entire fourth paragraph focuses on the point that television models violence and increases our acceptance of such behavior.

22. (B) The author asserts, "viewing such lifestyles for hours each week can only cause us to believe they are the norm and thus acceptable. . . . we tend to believe that what we see is how the rest of the world must be living." Choice (A) is incorrect because the author says nothing to support the idea that televised actions cause viewers to be cautious; instead, the author believes that viewers want to copy those actions: " . . . children see so many violent acts on television . . . that they become inured to them. . . . It doesn't take young minds long to embrace the idea that violence and hurtful behaviors are okay, especially if they see those are the standards for the 'good guys' on TV." Nothing in the passage indicates that viewers become frightened and distrustful of others as a result of watching the immoral and criminal actions of TV characters; so choice (C) is incorrect. Nor does the author indicate that viewers are amused by such behavior or that they avoid watching it. Therefore, choices (D) and (E) are wrong.

23. (E) In the last paragraph, the author contemplates, "It would be interesting to see what would happen if all televisions were removed from—or at least turned off in—all American homes for six to twelve months. Who knows? The nation's health, morals, and crime rate might actually improve!" The author is concerned about the language and the violence aired on television and makes negative comments about the amount of time Americans spend watching television and the number of televisions in the average home; however, nothing in the passage proposes censoring the language, eliminating TV violence, limiting viewing hours, or limiting the number of televisions in each home. Thus, choices (A), (B), (C), and (D) are wrong.

24. (A) The key here is the word "<u>not</u>." You are looking for the statement that is <u>not true</u>. A review of the two passages clearly reveals that Passage 2—not Passage 1—is the more critical of television. Passage 1 says that television does have some positive qualities, whereas Passage 2 admits nothing good about TV. Thus, choice (B) is true, and is not a correct response to this question. Choice (C) is a true statement because Passage 1 is concerned that television unrealistically portrays careers, and Passage 2 is concerned about television's negative effects on viewers. Choice (D) is a true statement because the first paragraph of Passage 1 states, "Television has become the most popular form of entertainment today. Virtually every home has one, and many homes have one for every room—even the bathroom." Passage 2 emphasizes, "In a typical American household the television is on between six and eight hours a day." And choice (E) is a true statement because Passage 1 says, "They offer us a view of lifestyles most of us will never experience firsthand," and Passage 2 points out, "Television is our 'window to the world' that most of us would not otherwise experience. . . ." So, because choices (B), (C), (D), and (E) are true statements and the question asks which one is <u>not true</u>, none is the correct choice for this question.

DETAILED EXPLANATIONS

SECTION 4—WRITING

1. (B) The original sentence, choice (A), has a dangling modifier (participle phrase); it remains that way in choice (E). The house cannot return itself, nor can "it" (pronoun for house). Choice (D) seems to leave something out: "returned to . . . home in 12 years." Choice (C) solves the original problem but is unnecessarily wordy. Choice (B) properly solves the dangling modifier problem by subordinating the return in an adverbial clause.

2. (A) Choice (A) is the only response that makes sense. Each of the others introduces illogical comparisons or structures (nonparallel); choices (B), (D), and (E) are also verbose. Choice (C) is concise but not parallel.

3. (C) This sentence suggests casual relationships between the parts of the sentence that do not belong there. Choices (B) and (D) echo the original choice (A) in that regard. Choice (E) has garbled syntax. Choice (C) shows clearly that the cause-effect relationship is, rather, a time relationship.

4. (E) This is essentially a problem of wordiness. Choice (E) is the shortest and most clear of all the choices. The punctuation is weak in choice (C), and the syntax of choice (B) complicates the idea unnecessarily. Choice (D) does not use the appropriate conjunctive adverb; "thereby" is more precise than "therefore" when referring to an event.

5. (D) All the other responses repeat the cause-effect relationship stated in the phrase "Because of." Choice (D) is the only choice that does not do so.

6. (B) Choice (B) would solve the problem of needless pronoun voice shift from second person to first person ("you" to "our"). It correctly substitutes first person "we" for "you." Choice (C) uses the passive voice awkwardly. Choice (D) introduces nonparallel structure (and incomplete comparison). Choice (E) is a similar voice shift from third to first person.

7. (A) This is the best response from the choices. The sentence implies a change of the standard. For that reason, choice (B) is an incorrect choice. Choice (C) is unnecessarily awkward and wordy. Choice (D) is redundant, and choice (E) states more than the sentence implies.

8. (E) The errors found in the original sentence, choice (A), involve parallelism and redundancy. Choice (E) uses the parallel infinitives "to entertain" and "to teach" as direct objects and eliminates the repetition created in use of both "tried" and "attempt" in the original sentence. Choices (B) and (C) provide parallel construction, but choice (B) retains the redundancy and choice (C) incorrectly splits the infinitive "to entertain;" although choice (D) provides parallelism of the nouns "entertainment" and "attempt," the redundancy still remains, and the word order is not idiomatic.

9. (B) Choice (B) adequately conveys the reason for removal of the parking meters with the least wordiness. Choices (A) and (D) contain unnecessary words; choices (C) and (E) have dangling participial phrases.

10. (D) Choice (D), in which "Megan" correctly follows the phrase, conveys the meaning with the least wordiness. The problem with choice (A) is the introductory participial phrase; it must be eliminated or followed immediately by the word modified. Choice (B) does not solve the problem of the dangling phrase; choices (C) and (E) add words unnecessary to the meaning of the sentence.

11. (A) The given sentence is acceptable in standard written English. Each of the alternate choices introduces a problem. Choice (B) uses the superlative form of the adjective, "most interested," when the comparative form "more interested" is correct for the comparison of two options. Choices (C) and (E) introduce a lack of parallelism, and choice (D) is not idiomatic.

12. (A) This is the classic confusion of "less" for the correct "fewer." "Few(er)" refers to countable things or persons; "little (less)" refers to things that can be measured or estimated but not itemized. The only other choice to examine is choice (D), but it is the appropriate tense referring to hierarchy and responsibility.

13. (C) Choice (A) is a correct use of the relative pronoun. Nothing is unusual about choice (B). Choice (C) is the culprit here: it should be the infinitive form to adhere to the idiom, "ability to (verb)." Choice (D) is an appropriate reference to hierarchy and responsibility.

14. (A) The sentence as it stands is illogical. Removing "Because" will make it sensible. Choice (B) is an appropriate adverb modifying "believe"; choice (C) is a clear and effective subordination of an explanation of a term. Choice (D) uses "as" properly as a preposition.

15. (B) "Being that" is colloquial for "because," which is better for at least the reason that it is shorter, but also that it is more formal. No other choices seem out of bounds.

16. (A) "As far as" is an incomplete comparison that should always include "is" (or variant of "to be") "concerns." "As for" can be substituted for "as far as . . . Is (are concerned)." The others are acceptable.

17. (B) This is the classic vague pronoun reference. Out of context, there is no antecedent for "they." Every pronoun must have an identifiable antecedent. Choice (A) is acceptable, by the way; the geographical region is capitalized. No problems show up in the other choices.

18. (C) This is a problem often called agreement of nominative forms. The compound subject ("Japan and the United States") is plural, and the form to match is "net exporters" in this case. Choice (B) is a verb correctly agreeing with its plural subject. "Imbalances" is correct in choice (D) because all nations, particularly the two mentioned, want an imbalance of trade in their favor.

19. (D) This problem demonstrates a faulty comparison: "Jon" and "his friend's bicycle" are not equivalents and should not be compared. "His friend" would have been correct. Choice (B) is acceptable because it indicates the rides are in the past, which does not conflict with any other part of the sentence.

20. (C) The problem here is appropriate verb tense. Both "After" and "once the dominant species" signal requirement of the past tense. Changing "are" to "became" would solve the problem. Choice (A) is correctly capitalized as a geologic period. Nothing is amiss with either choice (B) or (D).

21. (C) The error occurs at choice (C), where the preposition "from" is idiomatic after the word "different." Although some experts insist on the use of "from" after the adjective "different," others accept the use of "different than" in order to save words. An example would be "different than you thought"; the use of "from" would require the addition of the word "what." Choice (A), "According to," is a preposition correctly introducing a prepositional phrase; choice (B), "has," is third person singular to agree with it's subject "branch"; and choice (D), "those," is a plural pronoun to agree with its antecedent "powers."

22. (A) Choice (A) should be a gerund in the present perfect form (having "been studied") to indicate that the action expressed by the gerund occurred before the relationship was reported. Choice (B), "Preceding ten years," is idiomatic. Choice (C), "having," is a gerund introducing the phrase "having heart attacks," which is the object of the preposition "of"; choice (D), the past tense passive verb "was reported," is singular to agree with its subject.

23. (E) Your answer should be choice (E), indicating that this sentence contains no error in standard written English. Choice (A), "demonstrates," is present tense third person singular to agree with its subject, "book"; choice (B), "was expected," uses the third person singular form of "to be" to agree with its subject "each;" choice (C), the infinitive "to use," is idiomatic after the passive verb "was expected"; and choice (D), "his or her," is singular to agree with its antecedent, the indefinite pronoun "each," and provides gender neutrality.

24. (A) You should recognize in choice (A) that the idiomatically acceptable preposition to follow "aversion" is "to." The other choices in the sentence are acceptable in standard written English. Choice (B), "for him," is a prepositional phrase modifying "activity;" choice (C), "one," is a pronoun appropriate to refer to its antecedent "activity;" and choice (D), "that," is a relative pronoun introducing a restrictive adjective subordinate clause modifying "one."

25. (C) Choice (C) contains the error because the subject of the verb "show" is "story," a singular noun that calls for the third person singular verb, "shows." The rest of the sentence represents correct usage. Choices (A) and (B), "those" and "their," are plural pronouns that agree with the antecedent "pioneers"; choice (D), "us," is in the objective case because it is the indirect object in the sentence.

26. (D) Choice (D) presents an error in comparison, appearing to compare "farming" with "low altitudes" when what is intended is a comparison of "farming in high altitudes." The corrected sentence reads: "Because of the long, cold winters and short summer growing season, farming in high latitudes is more difficult than farming in low latitudes." The other choices all represent appropriate usage in standard written English. Choice (A), "Because of," is idiomatically correct as a preposition; choice (B), "and," is a coordinating conjunction joining the nouns "winter" and "summers." Choice (C), "more difficult," is the comparative form appropriate to compare two items.

27. (D) The error is in choice (D). The word "contestant" is a predicate nominative in the subordinate noun clause, and it must agree in number with the plural subject of the clause, the pronoun "both," to which it refers. The noun clause should, therefore, read: "that both of us could be contestants on a quiz show." Choice (A), "I," is part of the compound subject of the introductory adverb clause and is, correctly, in the nominative case. Choice (B), "were," is plural to agree with its compound subject. Choice (C), "we," is plural to agree with its compound antecedent, "sister and I," and is in the nominative case because it is the subject of the verb "hoped."

28. (A) The error is choice (A). The auxiliary verb "had" calls for the past participle form of the verb "break," which is "broken." All of the other choices are acceptable in standard written English. Choice (B), "that," is the correct relative pronoun to follow "vase" and introduce the subordinate adjective clause; choice (C), "had purchased," is the past perfect form of the verb to indicate action completed in the past before the action of the verb in the main clause; and choice (D), "her," is the object of the preposition "for."

29. (E) This sentence contains no error in standard written English. Choice (A), the prepositional phrase "with whom," introduces an adjective clause modifying the word "people." The relative pronoun "whom" is in the objective case because it served as the object of the preposition "with." The simple past tense "worked" is appropriate for choice (B); choice (C), "were engaged," is plural to agree with its subject "some"; and choice (D) is an isomatic expression replacing the preposition "for."

30. (B) Choice (B) breaks the unity of the paragraph by digressing slightly into an opinion about the presence of the electric lights in a natural setting. As the main thrust of the passage is to discuss the beauty of caves in the United States, this idea is out of place. This sentence also mentions the opinions of "some people," a vague reference. The other choices contain no digressions.

31. (E) Choice (E) concisely combines ideas while effectively showing the sequence of events. Choice (A) puts the cause, "losing carbon dioxide," after the effect, "deposits a film of carbon dioxide." Choice (B) could be better worded by revising, "When hanging, water drops." Choice (C) erroneously presents hanging from the ceiling as part of the process of losing carbon dioxide. Choice (D) completely loses the cause-and-effect and presents the events as an unrelated list.

32. (D) The intent of the second paragraph builds to the last sentence, in which the persuasive choice (D) intent becomes evident. Although the paragraph provides some examples, choice (A), those examples bolster the persuasive intent. Choice (C) could be plausible if the writer had indicated an acknowledgment of the reader's aversion to caves. Choices (B) and (E) are more appropriate for the first paragraph.

33. (C) Choice (C) is the most concise and accurate. Choices (A) and (B) use the incorrect grammatical structure, "very unique" A sight is unique, meaning "one of a kind"; therefore, "very unique" is incorrect. The sentence begins with the phrase, "in addition," so choices (D) and (E) are repetitive because they use the word "also."

34. (B) Choice (B) makes a complete sentence that is correctly punctuated and that fits well as a topic sentence for the paragraph. Choice (A) is a fragment and contains the awkward "people are thinking." Choices (C) and (E) could be moderately acceptable, but they have two prepositional phrases in a row beginning with "in" at the end of the sentence. Choice (D) includes the conjunction "however," which would indicate disagreement with the preceding sentence. Sentence 17 is a continuation of the thought in sentence 16; sentence 18 properly begins with the conjunction "although," indicating a contradiction of the idea presented in sentence 17.

35. (B) Choice (B) is correct because both verbs must agree with the singular noun "sheet." Choice (A) brings the noun in accordance with "grow" and "hang" by changing it to the plural "sheets," but it leaves the singular article "a" instead of removing it. Choice (C) uses "gracefully" as an adverb correctly, but does not correct the verb agreement problem. Choice (D) may appear correct: "hang" would be correct if it modified the plural "ceilings," but it modifies "sheet," which is singular. Similarly, choice (E) does not correct the subject-verb disagreement.

DETAILED EXPLANATIONS

SECTION 5—MATH

1. (D) Calculate number of patients that withdrew each week by adding the numbers in each column:

   Number of patients that withdrew in Week 1 = 22 + 54 + 48 + 0 = 124

   Number of patients that withdrew in Week 2 = 0 + 0 + 67 + 45 = 112

   Number of patients that withdrew in Week 3 = 38 + 0 + 21 + 0 = 59

   Number of patients that withdrew in Week 4 = 42 + 56 + 0 + 32 = 130

   Number of patients that withdrew in Week 5 = 45 + 0 + 12 + 70 = 127

   In Week 4, the highest number of patients withdrew from the trial.

2. (E) Multiply the corresponding sides of the given equations:

   $mn = (a - 3)(a + 3)$

   Simplify the right side of the result, add $-a^2$ to each side:

   $mn = a^2 - 9$

   $mn - a^2 = -9$

3. (D) Using the given functions, we have

   $$\frac{f(2)-g(-2)}{f(-2)-g(2)} = \frac{\left[3(2)^2-1\right]-\left[3(-2)^2+1\right]}{\left[3(-2)^2-1\right]-\left[3(2)^2+1\right]} = \frac{(12-1)-(12+1)}{(12-1)-(12+1)} = 1$$

4. (D) Let H = the weight of an average humpback whale. Then, using the given data,

   $B = 5 + 3H$

   Solve for H by adding –5 to each side of this equation and then dividing each side of this equation by 3:

   $B - 5 = 3H$

   $$H = \frac{B-5}{3}$$

5. (E) The formula for the volume of a cylinder is $\pi r^2 h$, where r is the radius and h is the height. The diameter given in the problem is 14, so the radius is 7. Therefore, the volume (V) of the cylinder is

$V = \pi (7^2)(14)$

$V = 686\pi$

6. (E) To begin, write the first equation in slope-intercept form:

$y = -mx + 12$

The slopes of the lines are $-m$ and $2m$. Because the lines are perpendicular, the product of the slopes equals -1.

$(-m)(2m) = -1$

$m^2 = \frac{1}{2}$

Solve for m by taking the square root of each side, and then rationalize the denominator:

$m = \pm\frac{1}{\sqrt{2}}$

$m = \pm\frac{\sqrt{2}}{2}$

7. (A) Using the rules of exponents, we can write the given equation as $(5^2)^a(5^3)^{m-1}$. Apply the rule of sequential exponents on the resulting equation: $5^{2a} = 5^{3m-3}$. In this equation, the bases on both sides are equal, so their exponents must be equal as well. Thus, $2a = 3m - 3$. Adding 3 to each side of this equation, we get $2a + 3 = 3m$. Divide each side of this equation by 3:

$m = \frac{2a+3}{3}$

8. (C) We are given the following sets.

$M = \{2, 4, 6, 8, 10, 12, 14, 16, 18, 20, 22\}$

$N = \{4, 8, 12, 16, 20, 24, 28\}$

$P = \{6, 12, 18, 24\}$

Therefore, $M \cap N = \{4, 8, 12, 16, 20, 24\}$ and $N \cap P = \{12, 24\}$. So $(M \cap P) \cup (N \cap P) = \{4, 8, 12, 16, 20, 24\} = M \cap N$.

9. (D) We have $f(-x) - f(x) = |(-x) - 3| - |x - 3| = |-(x + 3)| - |x - 3| = |x + 3| - |x - 3|$.

10. (E) Using the coordinates of B and P, we conclude that $BP = 2\sqrt{27} = 6\sqrt{3}$. Because ΔABP is equilateral

$AB = AP = BP = 6\sqrt{3}$

$\overline{OA}$ is both an altitude and a bisector in ΔABP, so ΔOAB is a right triangle. In an equilateral triangle, each interior angle measures 60°. Because $m\angle ABO = 60^\circ$ in ΔOAB, the side opposite the 60° angle measures $\frac{\sqrt{3}}{2}$ times the hypotenuse. Therefore,

$OA = \left(\frac{\sqrt{3}}{2}\right)AB = \left(\frac{\sqrt{3}}{2}\right)(6\sqrt{3}) = 9$.

11. (D) Find the slope of the line M. Denote it by m. Then, $m = \frac{4+2}{3-2}$. The slope of the line perpendicular to this line must be $-\frac{1}{6}$. Among the given equations, only $2x + 12y = 21$ has slope $-\frac{1}{6}$.

12. (D) A tangent to a circle is perpendicular to the radius dropped to the point of tangent. Therefore, ΔOAB is a right triangle. In the right triangle OBE, $\overline{OA}$ is opposite the 30° angle, so $OB = 2OE = 2(3) = 6$. In the right triangle OBA, $m\angle BOE = 60°$. This means that $m\angle BAO = 30°$. In the right triangle OAB, $\overline{OB}$ is opposite the 30° angle. Therefore $OA = 2OB = 2(6) = 12$.

13. (E) To shift the y-intercept of the given parabola, add –3 to the right side of the function:

$$f(x) = -3x^2 + 3x - 5 - 3 = -3x^2 + 3x - 8.$$

14. (C) In fact, in the rectangle, each of the triangles is a right triangle whose legs are $(x + 3)$ and $(x - 3)$ long and whose hypotenuse is L long. Using the Pythagorean Theorem to find the length of the hypotenuse.

$$L^2 = (x - 3)(x + 3)$$

$$L^2 = x^2 - 3$$

Take square root of each side:

$$L = \sqrt{x^2 - 3}$$

15. (A) Using distance formula, we have

$$OB = \sqrt{(-2a)^2 + (-2a)^2} = 2a\sqrt{2} \quad \text{and} \quad OA = \sqrt{(3a)^2 + (3a)^2} = 3a\sqrt{2}$$

Therefore

$$AB = 2a\sqrt{2} + 3a\sqrt{2} = 5a\sqrt{2}.$$

The probability that a point chosen on AB will be on OB is

$$\frac{OB}{AB} = \frac{2a\sqrt{2}}{5a\sqrt{2}} = \frac{2}{5}$$

16. (A) The triangle is isosceles, so both legs are equal. Let m equal the length of each leg. Using the Pythagorean Theorem, solve the equation for m.

$$\left(\sqrt{a}\right)^2 = m^2 + m^2$$

$$a = 2m^2$$

$$m^2 = \frac{1}{2}a$$

Let A equal the area of the triangle. Therefore

$$A = \frac{1}{2}(m)(m) = \frac{1}{2}m^2$$

Substituting the equivalent of m^2 in A, we get

$$A = \frac{1}{2}\left(\frac{1}{2}a\right) = \frac{1}{4}a$$

From this expression, it is evident that the area of the triangle (A) varies directly as a.

17. (B) A radius or a diameter dropped to the point of tangent is perpendicular to the tangent line. So, ΔABC is a right triangle. In this triangle, $\overline{AB}$ is opposite the 30° angle. Using the rules of a 30°–60°–90° special triangle

$$\overline{BC} = \left(\overline{AB}\right) \cdot \sqrt{3}$$

$$\overline{BC} = \sqrt{27} \cdot \sqrt{3} = \sqrt{81} = 9$$

$\overline{BC} = 9$ is the diameter (d) of the circle. Use this information to find the circumference of circle O:

circumference = $2\pi r = \pi d = 9\pi$

18. (C) From the given equation, we have

$-5x + 3 = 2k$ or $-5x + 3 = -2k$

$x = \frac{3-2k}{5}$ or $x = \frac{3+2k}{5}$

In order to have only one solution, these answers must be the same. That is, set the two equations as equal, and then solve for k.

$$\frac{3-2k}{5} = \frac{3+2k}{5}$$

$$0 = 4k$$

$$0 = k$$

For $k = 0$, the given equation has only one solution.

DETAILED EXPLANATIONS

SECTION 6—CRITICAL READING

1. (E) Here you are looking for a noun to help describe criticism that would cause a politician not to seek reelection. It makes sense that a "preponderance" (excess) of criticism would have that effect on a politician. "Sanctity" (holiness, righteousness) does not fit well with "criticism"; thus, choice (A) is wrong. It does not make sense that a "paucity" (scarcity) of criticism would cause him not to seek re-election; thus, choice (B) is wrong. Choice (C) is incorrect because "generosity" (kindness, liberal abundance) of criticism does not make sense. Choice (D) is wrong because "mediocrity" (low quality) of criticism probably would not threaten a politician.

2. (B) Here you need a noun that describes the type of behavior that would repel everyone—even relatives. "Turpitude" (depravity, corruptness) would have that effect. However, choice (A) "benevolence" (kindness), choice (C) "innocuousness" (kindness), choice (D) "nescience" (ignorance, lack of awareness), or choice (E) "obsequiousness" (subservience, submissiveness) would not be likely to cause people—especially relatives—to reject a person.

3. (A) Here you are seeking a noun that describes an activity by which one does not earn money but that one may enjoy and do much of the time. An example is bird watching. "Avocation" (hobby, a calling in life other than one's job) is such a word. Choice (B) is incorrect because one earns money at a "vocation" (trade, profession). Although "equivocation" (deception) might be something at which one cannot earn money but which might bring much pleasure, bird watching is hardly an example of this type of activity. So choice (C) is wrong. The same is true of choice (D) "propensity" (natural inclination). And choice (E) is wrong because it does not make sense; a "nom de plume" is a pen name, or pseudonym.

4. (C) A word that describes such disregard for authority that it is offensive to teachers and students is "blatant" (shameless, brazen). It does not make sense to describe such an attitude as "obsequious" (subservient), so choice (A) is wrong. "Copious" (exuberant, luxurious) is not an appropriate description; therefore, choice (B) is incorrect. And neither "credulous" (gullible, ready to believe anything) nor "tenuous" (flimsy, hesitant) fits in this context, so choices (D) and (E) are not good answers.

5. (B) You need a word that describes music that is so soothing it would put a baby to sleep. "Mellifluous" (sweet and smooth, mellow) is such a word. Such music would not be "raucous" (harsh, squawky) or "stentorian" (extremely loud), so choices (A) and (D) are incorrect. It doesn't make sense to describe music as "mute" (silent); thus, choice (C) is wrong. Nor does it make sense to say music is "loquacious" (wordy, talkative); therefore, choice (E) is wrong.

6. (C) The main clues in this sentence are that people didn't bother to get to know this person well and that they disliked him. People might not bother to get to know a person because they dislike his "braggadocio" (cockiness, arrogance). Choice (A) is wrong because they are not likely to dislike his "stature" (height or status). And it doesn't make sense that people would not bother to get to know a person because they disliked his "homage" (humility, respect), his "comeliness" (pleasant appearance, attractiveness), or his "charisma" (appeal, allure). Thus, choices (B), (D), and (E) are incorrect.

7. (E) A committee might conduct an "exhaustive" (complete and thorough) search of documents, looking for evidence to "corroborate" (confirm) someone's guilt. Choice (A) is incorrect because a "tepid" (half-hearted) search would not serve to "sanctify" (bless, consecrate) someone's guilt. Choice (B) is wrong because they would not conduct a "scant" (short, insufficient) search to "verify" (confirm) his guilt. Choice (C) doesn't make sense because a "precipitous" (hurried) search would not "ameliorate" (improve, make better) someone's guilt. Likewise, choice (D) is wrong because it does not make sense to describe a search of documents as "impecunious" (having no money), nor does it make sense that a committee would search for evidence to "undermine" (damage) someone's guilt.

8. (A) Pandas are becoming extinct primarily as a result of people taking up bamboo forests to build cities, cutting down bamboo forests for timber and to make room for agricultural growth, and poaching. Although the severe bamboo die-off of a few decades ago did add to the panda's extinction, nothing in the passage indicates that the effects of this natural phenomenon were as drastic as these actions of humans; thus, choice (B) is incorrect. Choice (C) is wrong because the passage does not indicate that man needed to hunt pandas to make up for a decreasing food supply. The pandas were forced from their natural habitat, rather than being reluctant to stay there; thus, choice (D) is wrong. And the passage says nothing about changing weather patterns, so choice (E) is incorrect.

9. (D) One definition of "plight" is "predicament" or "dilemma." Although the dictionary indicates that "plight" can also mean "promise," choice (B) is incorrect because this definition does not fit in the context here because the passage is discussing the unfortunate situation of pandas. "Plight" does not mean "escape," "starvation," or "fear"; therefore, choices (A), (C), and (D) are incorrect.

10. (D) The passage traces the development of the Social Security Program from its inception to current speculations about related problems. Although it does acknowledge the existence of concerns that the program currently faces some problems, the passage is not critical; thus, choice (A) is incorrect. At the same time, it does not praise the program, although it does present reasons behind the creation of Social Security. So choice (B) is wrong. Choice (C) is incorrect because the passage does not offer alternatives to the program. And although the passage does acknowledge the predictions and rumors that the system may soon be operating "in the red" and that problems are imminent, it does not include any warnings as to the potential failure of the program. Therefore, choice (E) is not appropriate.

11. (C) As you consider this question, you must be careful to focus on the passage itself rather than thinking about things you may have read elsewhere or about your own opinion. The last paragraph says, "Life expectancy is lengthening while birth rates are declining, so the number of people receiving benefits is steadily increasing while the number of workers contributing to the Social Security coffers is declining." Therefore, choice (C) is correct. Choice (A) is not correct because the passage says nothing to indicate poor management of the money. The passage does indicate, "Rumors and predictions contend that by 2012, if not sooner, the system will be running in the red, distributing more money in benefits than it is taking in." However, nothing in the passage addresses the size of contributions that some people have made; therefore, choice (B) is incorrect. Since no mention is made of private retirement funds, choice (D) is wrong. The passage does mention, "Over the years the Social Security Program has metamorphosed into benefits not only for workers but also for the disabled and for survivors of beneficiaries, as well as medical insurance benefits in the form of Medicare." However, it does not indicate that this expansion is the source of the woes facing the Social Security Program; thus, choice (E) is not the correct choice.

12. (C) The author uses the simile of man's leap from the earliest stage of mankind to the next to show how the Industrial Revolution profoundly changed the fate of the human race. Choice (A) is incorrect because the statement compares significance, not periods of time. Choice (B) is wrong. There is no mention of the types of developments made during the Industrial Revolution. The comparison between two ages is used solely as a way to show the development of mankind, not the development of tools. Choice (D) is incorrect. Though the simile does show a shift from hunting to farming, the meaning of the comparison is rooted in man's social and intellectual advances. Therefore, choice (E) is wrong because it addresses class issues that are never mentioned in the simile.

13. (A) Choice (A) correctly identifies the steam engine and mill-driven machinery as mechanical advances that played a big part in the Industrial Revolution's development. Athough both choices (B) and (C) may be true statements, the article is not primarily concerned with profitability or the daring uses of water. Choices (D) and (E) are wrong. The machines were created, not abandoned, during the Industrial Revolution and were very successful.

14. (B) The Industrial Revolution did in fact bring about the neglect and abuse of individuals. This abuse was a direct result of the advances during the era. Therefore choices (A) and (D) are incorrect. Those in power during the Industrial Revolution made little or no effort to avoid or negate the abuse of the workers. In fact, in many instances gross abuses were ignored. Choice (C) is incorrect. The Industrial Revolution, although it boosted economies and the power of nations, did not stave off or rule out neglect and abuse of workers. Though an impersonal force, the Industrial Revolution in many ways treated men like the precise, tireless machines they operated.

15. (E) Choice (E) is correct because the author's purpose, to show the inhumane treatment of children, is best explained by recounting actual abuses. The graphic nature of the description impresses the seriousness of the situation upon the reader. Choice (A) is incorrect. Though the passage may engage the reader's interest, its purpose is to inform rather than entertain. Choice (C) is wrong. Children were an important part of the Industrial Revolution's labor force, but the incident shows their vulnerability, not their importance. Choice (D) is incorrect. Though the passage itself explains the advantages of the machines, this particular section focuses on the dangers faced by children working them.

16. (D) Choice (D) is correct because dichotomy is defined as "a division into two usually contradictory parts," and the benefits of increased production were contradicted by the damage done to the labor force and children. Choices (A) or (B) are wrong. Neither the Industrial Revolution's size nor the quickness with which it developed affected its contradictory nature. Choice (C) is incorrect because the changing nature of the Industrial Revolution would not prevent it from being a benign force. Choice (E) is also incorrect. The fact that the Industrial Revolution was a success cannot change the nature of the benefits and drawbacks it created.

17. (C) The author shows how the Industrial Revolution both benefited and damaged European life. Choices (A) and (B) both go well beyond the author's scope. Choice (A) underplays the Industrial Revolution's importance, whereas choice (B) completely ignores the many drawbacks of the "machine age." Choice (D) is wrong. The Industrial Revolution occurred almost 100 years before the Cold War, though the birth of the atomic age was a factor in the development of the Cold War. Choice (E) is incorrect because the Industrial Revolution moved European economies away from an agrarian base toward an industrial one.

18. (C) The entire passage focuses on the power that McAllister has over the Polynesians. The fact that they were "all strapping men and women, many of them standing six feet in height and weighing a couple of hundred pounds," did little to weaken the control McAllister had over them. In contrast, the passage shows that they had virtually no power; thus, choice (A) is wrong. And it says little about whether they were attractive, so choice (B) is incorrect. Likewise, choices (D) and (E) are wrong because the passage makes no connections between their lifestyle or their work habits and their physical appearance.

19. (B) If you are not familiar with a word that begins with the prefix "un-" you should consider the root word by dropping the prefix and seeing if you know the definition. The prefix "un-" usually means "not." So if you know the word "intermittent" means "not continual, occasional," you know that "unintermittent" has the opposite meaning and so must mean "continual, not occasional." And at the same time, you know that choice (A) is incorrect, since "continual" is the antonym (opposite) of "unintermittent." If you are not familiar with the word "intermittent," you should use context clues. By reading the passage, you can assume that McAllister is a heavy drinker and thus probably drinks continually. So he is not an "occasional" (not continual) or "innocuous" (harmless, inoffensive) drinker; thus choices (A), (C), and (D) are wrong. And choice (E) is incorrect because the passage makes it clear that he is not a "social drinker" (one who drinks alcoholic beverages in moderation at social gatherings especially as distinguished from one who drinks habitually or to excess).

20. (D) The passage simply says that McAllister said no to the marriage, but it does not say why. It does not say whether McAllister liked or disliked the king, the king's daughter, or the man she wanted to marry; therefore, choices (A), (B), and (C) are incorrect. And although the passage does say that the king owed a debt, it does not say this debt was the reason McAllister refused to allow the king's daughter to marry the man; so choice (E) is incorrect.

21. (C) "Twice a year a little schooner called to collect copra (dried coconut meat yielding coconut oil)." And "the king was in debt to the Company to the tune of 180,000 cocoanuts, and until that was paid he was not to spend a single cocoanut on anything else." The passage does not mention dollars or coins, so choices (A) and (B) are incorrect. Although whiskey and alcohol are mentioned several times, nothing indicates that they are part of the area's commerce; therefore, choice (D) is wrong. And although a lagoon does exist in Oolong Atoll, lagoons are not used as objects of exchange; thus, choice (E) is wrong.

22. (E) The passage states, "the whole population, with the priests at the head, tried vainly for three months to pray him to death," and the Polynesians "gathered up scraps of food which had touched his lips, an empty whiskey bottle, a cocoanut from which he had drunk, and even his spittle, and performed all kinds of deviltries over them" in an effort to kill him. Choice (A) is wrong because the Polynesians hated McAllister rather than worshiping him and things associated with him. And although they probably did want their cocoanut crop to grow, the passage does not address this issue or any connection between McAllister and their desire to have a good crop; thus, choice (B) is wrong. Choice (C) is incorrect because the Polynesians wanted to get rid of McAllister rather than be reminded of him. And choice (D) is incorrect because the passage does not refer to the Polynesians' sacrifices to deities.

23. (D) The passage specifically refers to the "devil-devils" the priests sent after him, giving you a clue the meaning of "deviltries" is "witchcraft" (communication with the devil). They were not using "mischief" or "pranks" because the situation was much more serious than the playfulness implied by these words. So choices (A) and (B) are incorrect. And they were not engaged in worship or devotion, which imply religious respect or reverence; so choices (D) and (E) are incorrect.

24. (D) Keep in mind this question asks which choice is <u>not</u> a reason McAllister continued to live. The idea that the gods favored McAllister is the only choice <u>not</u> mentioned in the passage. The narrator specifically says that the Polynesians "tried vainly for three months to pray him to death . . . , but since McAllister did not believe in devil-devils, they were without power over him." Choice (A) is wrong. The narrator emphasizes the effect McAllister's drinking had on his longevity by saying, "He must have been so saturated with alcohol as to defy the lodgment of germs." Therefore, choice (B) is incorrect. This statement helps prove that choice (C) is wrong, as does the statement, "I used to imagine them falling to the ground in showers of microscopic cinders as fast as they entered his whiskey-sodden aura. No one loved him, not even germs. . . ." And although one would not expect it, the passage indicates " . . . McAllister lived on. His health was superb," thus proving that choice (E) is an inappropriate answer to this question.

DETAILED EXPLANATIONS

SECTION 7—MATH

1. (C) Let t be the number of miles the truck was driven in the first day and s be the number of miles the truck was driven in the second day. Then

   $81.4 = (29 + 0.09t) + (29 + 0.09s)$.

   Simplify this equation and solve for $(t + s)$.

   $$81.4 = 58 + 0.09(t + s)$$

   $$0.09(t + s) = 23.4$$

   $$t + s = \frac{23.4}{0.09} = 260$$

   So, the truck was driven 260 miles in two days.

2. (B) Replace the given values of m and n in the given function.

   $$9 \clubsuit 16 = \frac{9-\sqrt{16}}{\sqrt{9}-16} = \frac{9-4}{3-4} = -5.$$

3. (C) In the right triangle ABC, $\overline{AB}$ is opposite the 30° angle. So, $AB = \frac{BC}{2} = \frac{12}{2} = 6$. Now, the area of $\Delta BED = \frac{AB \times DE}{2}$. Replace the known values in the preceding formula and solve the result for the unknown factor.

   $$24 = \frac{6 \times DE}{2}$$

   $$24 = 3(DE)$$

   $$DE = 8$$

4. (C) First find the slope of $\overline{BC}$. Denote it by m.

   $$m = \frac{4-0}{10+6} = \frac{1}{4}$$

   We know that a segment that joins the midpoints of two sides of a triangle is parallel to the third side. So, $\overline{DE}$ is parallel to $\overline{BC}$, and as a result they have the same slope. Therefore, the slope of $\overline{DE}$ is $\frac{1}{4}$.

5. (A) We can write the given inequality as $-4 < -3x + 5 < 4$. Adding -5 to all sides of this inequality gives $-9 < -3x < -1$. Divide each side of this inequality by -3. Such division changes the directions of the inequality signs: $3 > x > \frac{1}{3}$.

6. (B) Find the sum of the areas of the circular holes. The area of each circular hole is $\pi \times 3^2 = 9\pi$. Therefore, the total area of the three holes is $3 \times 9\pi = 27\pi$. Now, find the area of the square: $15^2 = 225$. The probability that an arrow will pass through a hole is the total area of the holes divided by the area of the board: $\frac{27\pi}{225} = \frac{3\pi}{25}$.

7. Answer: 12 Find $f(3)$ and $g(-1)$ first.

$$f(3) = \frac{3(3^2) - 1}{2} = 13 \quad \text{and} \quad g(-1) = \frac{2}{3(-1)^2 - 1} = 1$$

Therefore, $f(3) - g(-1) = 13 - 1 = 12$.

8. Answer: 51 Denote the set of students enrolled in the mathematics course by A and the set of students enrolled in the Writing and Reading course by B. Denote the number of a set by m. Then

$$m(A \cup B) = m(A) + m(B) - m(A \cap B)$$

$$m(A \cup B) = 40 + 54 - 25 = 69$$

So, the total number of students enrolled in both or either courses is 69. Subtracting this number from the total number of students gives the number of students who aren't enrolled in either class: $120 - 69 = 51$.

9. Answer: 1 Using the rules of rational exponents, we have

$$\left(x^{-\frac{1}{5}}\right)^{-5} \div \left(y^5\right)^{\frac{1}{5}} = \frac{x^3}{y^3}$$

Simplify the right side of this equation.

$$x \div y = \frac{x^3}{y^3}, \quad \frac{x}{y} = \frac{x^3}{y^3}$$

Cross-multiply the last equation, $x^4 = y^4$. Taking the fourth root of each side results in $x = y$. Therefore, $xy = 1$.

10. Answer: 1 Using the definition of a geometric sequence, we have

$a(an) = m(mn)$, $a^2n = m^2n$

Dividing each side of the last equation by n, we get

$a^2 = m^2$ or $a = m$

Divide each side of the last equation by m; we get $\frac{a}{m} = 1$.

11. Answer: 11 Let a be the smallest odd integer and n be the smallest even integer. Then

$(m + 2) + (m + 4) = 39$

$(n + 2) + (n + 4) = 72$

Simplify these equations:

$3m + 6 = 39$

$3n + 6 = 72$

Solving for m and n results in $m = 11$ and $n = 22$. So, $n - m = 11$.

12. Answer: 3 Using the definition of arithmetic mean, we have

$(x^2 + 3) + (12 + 11x - 2x^2) + (x^2 + x) = 3 \times 13$

Simplify this equation and solve for x.

$12x + 15 = 39$

$x = 2$

Then, find the values of the given expressions.

$x^2 - 1 = 2^2 - 1 = 3$

$x + 1 = 2 + 1 = 3$

The average of 3 and 3 is 3.

13. Answer: 64 Because M varies inversely as the cube of t, then $M = \frac{k}{t^3}$, where k is the constant of variation. Replace $M = 1$ and $t = 2$ in this equation to find k.

$$1 = \frac{k}{2^3}$$

$$1 = \frac{k}{8}$$

Cross-multiplying the last equation, we get $k = 8$. So, $m = \frac{8}{t^3}$. Replace $t = \frac{1}{2}$ in this equation and find $M = \frac{8}{\left(\frac{1}{2}\right)^3} = 64$.

14. Answer: 21 Denote the terms by a, b, c, and d. Then, $a = 6$ and $c = 16$. The average of the first three terms is $6 + b + 16 = 3 \times 11$. Solving this equation for b gives $b = 11$. So, the first three terms of the sequence are 6, 11, and 16. This shows that the common ratio is $16 - 11 = 5$. Therefore, the fourth term is $16 + 5 = 21$.

15. Answer: 27 The area of a rectangle is the product of its sides:

$27 = x(2x + 3) = 2x^2 + 3x$

$27 = 2x^2 + 3x$

To solve this quadratic equation, subtract 27 from each side, and then factor the resulting equation:

$0 = 2x^2 + 3x - 27$

$0 = (2x + 9)(x - 3)$

Set each expression equal to zero and solve for x.

$(2x + 9) = 0$ or $(x - 3) = 0$

$x = -\frac{9}{2}$ or $x = 3$

The first value is negative, so it is not acceptable for x as a length variable. So, the only answer is $x = 3$. Replacing the value of x in the expressions of the dimensions, we get the measures of the dimensions as 3 and $(2)(3) + 3 = 9$. The area of the rectangle is therefore $(3)(9) = 27$.

16. Answer: 12

$\overline{AB}$ is tangent to the circle O. Therefore, $\overline{BD}$ is perpendicular to $\overline{AB}$. On the other hand, $\overline{AB}$ and $\overline{CE}$ are parallel. These two conditions imply that $\overline{BD}$ is perpendicular to $\overline{CE}$. That is, OED is a right triangle. The side opposite the 30° angle in this triangle is one-half the hypotenuse. Therefore, $\frac{1}{2}OE = OD$. Denote the length of a radius of the circle by R. Then, $OB = OE = R$ and $OD = \frac{1}{2}R$. We are given $OB + OD = 18$ or $R + \frac{1}{2}R = 18$. Solving this equation for R gives $R = 12$.

DETAILED EXPLANATIONS

SECTION 8—CRITICAL READING

1. (B) It makes sense that under "optimal" (most desirable; satisfactory) conditions, flowers should bloom all summer long. They would not necessarily bloom all summer under "rudimentary" (simplest; basic) conditions; therefore, choice (A) is not appropriate. Nor would they necessarily be expected to bloom all summer because of "primitive" (basic; uncultivated) or "unique" (unusual) conditions; thus, choices (C) and (D) are incorrect. And "biennial" (occurring every two years) does not make sense here, so choice (E) is wrong.

2. (C) It would be helpful to offer a new employee a "paradigm" (model, sample) of what she is expected to do—here, effective report-writing. Choice (A) is incorrect because it would not be helpful to show a new employee an "enigma" (riddle; something difficult to understand or explain) in order to help her understand what she is supposed to do. Choice (B) is incorrect because "paradox" (a statement that is seemingly contradictory or opposed to common sense but apparently true) does not make sense here, nor do choice (D) "panorama" (unobstructed or complete view) and Choice (E) "predilection" (bias; tendency).

3. (A) You want a word that describes the state of a group who cannot reach a consensus (agreement) on any topic of discussion because of their reaction to a particular issue. A group who was in extreme "discord" (disagreement; lack of harmony) over one issue might not be able to reach a consensus on any other topic. A group experiencing "alacrity" (enthusiasm; fervor), "optimism" (anticipation of the best possible outcome), "accord" (agreement), or "cohesiveness" (state of being united) over one topic probably would not have difficulty agreeing on other topics. Therefore, choices (B), (C), (D), and (E) are incorrect.

4. (B) Your main clues are the contrasting adjectives, "aggressive" and "shy," and the fact that everyone was surprised at the match. Everyone would be surprised that a couple who seemed so "incompatible" (not harmonious; disharmonious) would plan to get married; thus, choice (B) is an appropriate choice. Choice (A) is incorrect because "uniform" (consistent, unvaried, unchanging) does not make sense in this context. Choice (C) is wrong because it would not be surprising for a "steadfast" (loyal) couple to announce their engagement. Nor would it be surprising for an "amiable" (friendly) couple to announce their plans to get married; thus, choice (D) is incorrect. And choice (E) "fortuitous" (accidental; happening chance or by luck) does not fit here.

5. (A) Here you are looking for an adjective that describes a bully's pursuit of his victims and that conveys the similarity between his pursuit and that of a cat preying on its victims. A cat is a predator (animal that kills and consumes other animals); "predatory" is the adjective form of "predator" and is therefore descriptive of both the cat and the bully described in this sentence. Choice (B), "esoteric" (incomprehensible; obscure), does not make sense in this context. A bully might have a "pejorative" (disparaging, belittling) attitude, but that kind of approach would not remind one of a cat preying on a bird or a mouse; thus, choice (C) is not appropriate. And a bully—or a cat—pursuing a victim would hardly be "timorous" (shy, shuddering) or "reticent" (reserved; taciturn); so choices (D) and (E) are wrong.

6. (B) You are looking for a verb that describes the effect of too many rules on the creativity of art students. Having to follow too many rules would "stultify" (inhibit; impair) the creativity of art students. It would not "enhance" (strengthen; heighten) their creativity; therefore, choice (A) is incorrect. Choice (C), "propitiates" (appeases; regains the favor or goodwill of), is wrong because it does not make sense in this context. And too many rules would not "extol" (praise; magnify) or "elucidate" (clarify; explain) the art students' creativity; so choices (D) and (E) are incorrect.

7. (E) Your main clue is that she cared about their opinion and did not want to displease them. It makes sense that their "derision" (ridicule; mockery) was enough to "chastise" (punish; discipline) her. It does not make sense to say that their "chagrin" (distress; shame) was enough to "consecrate" (immortalize; sanctify) her; thus, choice (A) is wrong. Nor would their "homage" (respect; honor) "devastate" (overwhelm; upset) her; so choice (B) is incorrect. Their "accolade" (honor, award) would not "admonish" (rebuke; scold); so choice (C) is wrong. And choice (D) is incorrect because their "disdain" (intense dislike) would not "bolster" (encourage; support) her.

8. (C) Davis said, " . . . it is the right of the people to alter or abolish governments whenever they become destructive of the ends for which they were established." He did not say they had a right to establish their own separate government choice (A) for any reason when they so desired; choice (B) when the existing government declared war for reasons with which they did not agree; choice (D) when the government leader was weak and misguided; or choice (E) when their Christian beliefs were threatened.

9. (B) Discussing the Southern states' decision to secede, he said, "In this they merely asserted a right which the Declaration of Independence of 1776 had defined to be inalienable." Choices (A) and (D) are wrong because Davis pointed out the similarity rather than any differences. Choice (C) is wrong because Davis did make the comparison. And choice (E) is wrong because he did not reject the Declaration of Independence but claimed that the Confederacy was upholding it.

10. (A) He said, "It is alike our interest, and that of all those to whom we would sell and from whom we would buy, that there should be fewest practicable restrictions upon the interchange of commodities. There can be but little rivalry between ours and any manufacturing or navigating community, such as the northeastern States of the American Union. It must follow, therefore, that a mutual interest would invite good will and kind offices." Choice (B) is incorrect because Davis did not mention slavery. Choices (C), (D), and (E) are incorrect because he did not mention the North's involvement or interest in citizens' safety, war, or foreign affairs.

11. (C) In this part of his speech, Davis expressed the hope that mutual interests would result in the two sides' being able "peaceable to pursue our separate political career." However, "If . . . passion or the lust of dominion should cloud the judgment or inflame the ambition of those States, we must prepare to meet the emergency, and to maintain, by the final <u>*arbitrament of the sword*</u>, the position which we have assumed among the nations of the earth . . ." and "appeal to arms. . . ." Elsewhere in this speech, he referred to the militia. Clearly, he was expressing his willingness to go to war to defend the Southern cause. Also, the definition of the word "arbitrament" is "the settling of a dispute by an agency of power." Davis was saying that the sword is the agency of power, thus referring to war. He made no reference to the courts, one-on-one dueling, presidential debates, or slavery; thus, choices (A), (B), (D), and (E) are incorrect.

12. (D) Davis said, "a re-union with the States from which we have separated is neither practicable nor desirable. . . ." Clearly, he was not choice (A) open to compromise, choice (B) willing to concede to the Union, choice (C) willing to leave it up to the vote of the citizens, or choice (E) hopeful that a reunion would eventually occur.

13. (D) Lincoln said, "There has never been any reasonable cause for such apprehension. . . . the most ample evidence to the contrary has . . . existed and been open to their inspection. . . . I have no inclination to do so. . . ." and " . . . the property, peace, and security of no section are to be in any wise endangered by the now incoming Administration." He also said, " . . . there needs to be no bloodshed or violence, and there shall be none unless it be forced upon the national authority. . . . . to hold, occupy, and possess the property and places belonging to the Government and to collect the duties and imposts . . . beyond what may be necessary for these objects, there will be no invasion, no using of force against or among the people anywhere." Choices (A) and (B) are wrong because he did not confirm their fears but instead sought to alleviate them. Choice (C) is wrong because he did not mention that reuniting with the Union would result in their avoiding destruction. And choice (E) is wrong because he said, "We are not enemies, but friends. We must not be enemies. Though passion may have strained it must not break our bonds of affection. . . ."

14. (B) Lincoln said, " . . . one of the declared objects for ordaining and establishing the Constitution was '*to form a more perfect Union.*' But if destruction of the Union by one or by a part only of the States be lawfully possible, the Union is *less* perfect than before the Constitution, having lost the vital element of perpetuity. . . . no State upon its own mere motion can lawfully get out of the Union; that *resolves* and *ordinances* to that effect are legally void. . . ." Choices (A) and (C) are incorrect because Lincoln believed the Southern states were violating their Constitutional rights. Although Lincoln did realize the possibility of war, he did not assume that secession was the same as a declaration of war; therefore, choice (D) is incorrect. And choice (E) is wrong because the Southern states seceded because they refused to surrender to the wishes of the North, and Lincoln expressed no belief to the contrary.

15. (E) If you already knew the dictionary definition of "perpetual" (everlasting; continual), you had no problem with this question. If you are not familiar with this word, you should look at the context. "Perpetual" is actually used in the first sentence of the paragraph as well as in the sentence indicated. In this section, Lincoln briefly traced the history of the Union, leading to the Articles of Confederation, the basis of which was the idea that the Union "should be perpetual." And then he went on to say that with the secession of the Southern states, the Union "lost the vital element of perpetuity." "Perpetuity" is the noun form of the adjective "perpetual." In other words, the Union has lost its quality of being everlasting. It has not lost the quality of being choice (A) "moral" (ethical, upright), choice (B) "Constitutional" (relating the U. S. Constitution), choice (C) religious (faithful, believing in a deity), or choice (D) "old."

16. (B) Lincoln said, "One section of our country believes slavery is *right* and ought to be extended, while the other believes it is *wrong* and ought not to be extended. This is the only substantial dispute." Although some of his comments implied religious faith, he did not address this topic as relevant to the split between the North and the South; therefore, choice (A) is wrong. He did express concern that their disagreement would lead to warfare, but he did not state that the right to bear arms was the basic issue between the two sides; so choice (C) is incorrect. Choices (D) and (E) are incorrect because he did not address the issues of freedom of speech or foreign trade.

17. (C) Lincoln said, "Suppose you go to war, you can not fight always; and when, after much loss on both sides and no gain on either, you cease fighting, the identical old questions, as to terms of intercourse, are again upon you." Nothing in his speech support the beliefs stated in choices (A), (B), (D), or (E).

18. (B) Davis inserted the phrases "with the blessing of Providence" and "Reverently let us invoke the God of our fathers to guide and protect us . . . ," "by his blessing," and "with a continuance of his favor." Likewise, Lincoln referred to "the Almighty Ruler of Nations, with His eternal truth and justice . . ." and " . . . Christianity, and a firm reliance on Him who has never yet forsaken this favored land . . ." Choice (A) is wrong because Lincoln stated, "One section of our country believes slavery is *right* . . . while the other believes it is *wrong*. . . ." Choice (C) is wrong because Lincoln said, " . . . no State upon its own mere motion can lawfully get out of the Union," and he did not believe the Southern states' secession was right or justified. As for choice (D), although Davis did seem to believe that war was a definite possibility, he also said that the Confederate states were "anxious to cultivate peace and commerce with all nations." And Lincoln stated, "there needs to be no bloodshed or violence," and he indicated that he would not be the one to declare war: "In *your* hands, my dissatisfied fellow-countrymen, and not in *mine*, is the momentous issue of civil war. The Government will not assail *you*. You can have no conflict without being yourselves the aggressors. *You* have no oath registered in heaven to destroy the Government, while I shall have the most solemn one to 'preserve, protect, and defend it.'" And choice (E) is incorrect because both expressed a willingness to accept their positions of leadership. Davis promised, " . . . you shall not find in me either a want of zeal or fidelity. . . ." And Lincoln conveyed his commitment when he said, "to the extent of my ability, I shall take care, as the Constitution itself expressly enjoins upon me, that the laws of the Union be faithfully executed in all the States."

19. (B) Although both men espoused the rights of the states, Lincoln stated, " . . . no State upon its own mere motion can lawfully get out of the Union; . . . resolves and ordinances to that effect are legally void, and . . . acts . . . against the authority of the United States are insurrectionary or revolutionary. . . ." By contrast, Davis was a proponent of "the right of the people to alter or abolish governments whenever they become destructive of the ends for which they were established." These statements show that choices (A) and (C) are wrong. Both men supported the rights of the states, so choices (D) and (E) are incorrect.

DETAILED EXPLANATIONS

SECTION 9—WRITING

1. (B) This sentence correctly uses the plural verb "want" to agree with the plural subject "students." It correctly uses present perfect "has decided," which is consistent with the present-tense verb "wants" in the first clause. And the pronoun "them" is correct because the antecedent is "class," which in this case is plural because it refers to the individual members or students in the group. Choice (A) is incorrect because it uses the singular verb "wants" with the plural subject "students." It incorrectly uses past perfect "had decided," shifting from the present tense "want" in the first clause. And it incorrectly uses the singular pronoun "it" to refer to the plural noun "class," which refers to the individual members or students in the group. Choice (C) incorrectly shifts to the past perfect verb "had decided." Choice (D) incorrectly uses the singular verb "wants," and it uses the singular pronoun "it," which has a plural antecedent "class." Choice (E) is illogical because "wanting" cannot serve as the verb of a clause.

2. (D) It corrects the error in the original sentence by changing the first complete thought to a phrase and placing that phrase directly before "the insurance agent proceeded" to indicate the correct relationship between the actions in the sentence. Choice (A) contains a comma splice, since only a comma—instead of a semicolon—connects the two complete thoughts ("The . . . policy" and "he . . . print."). Choice (B) is a fragment because it does not contain a main verb. In choices (B) and (C), the use of passive voice ("was . . . given") creates unnecessary wordiness. Also, choice (C) is illogical because it indicates that the "sample policy" "proceeded to explain." Choice (E) is redundant with the unnecessary use of the pronoun "he," which simply repeats the subject.

3. (E) It corrects the error in the original sentence by using the pronoun "one's" consistently. Choices (A) and (B) are incorrect because they switch from the pronoun "one's" to the pronoun "your." And in choice (B), the singular verb "is" does not agree with the plural subject "possessions." Choice (C) has no subject and no main verb; therefore it is a fragment. In choice (D), the use of the plural pronoun "they" is incorrect because "one" is singular; a pronoun must agree in number with its antecedent.

4. (A) The singular verb "is" agrees with the singular subject "knowing." Choices (B) and (D) incorrectly use the plural verb "are." Choice (C) uses the singular noun "customer," which is illogical with the plural noun "names." Likewise, choice (D) uses the singular noun "name" with the plural noun "customers." And choice (E) is incorrect because the verb "know" cannot be used as a subject; the gerund "knowing" must be used instead.

5. (C) It correctly refers to "bandwagon" as "a propaganda technique," and it contains no unnecessary words. Choices (A) and (B) are illogical because "bandwagon" is a "what," not a "where" or a "when." Choice (D) contains unnecessary words ("you should do the right thing and that thing is"). And choice (E) contains no verb for the subject "bandwagon."

6. (A) The phrase "wanting her son to graduate at the top of her class" is correctly located immediately after "woman." Choice (B) is missing the comma that is necessary to separate the nonessential adjective phrase ("wanting her son to graduate at the top of his class") from the rest of the sentence; also, the word "because" is unnecessary and awkward. Choice (C) is a fragment, containing no main verb. In choice (D), the word "because" is unnecessary and awkward. And in choice (E) the word "she" is unnecessary and redundant. The antecedent is "woman," which is the subject of the sentence; it is incorrect to use a pronoun simply to repeat that subject.

7. (E) It contains no unnecessary words separating the subject ("a pregnant woman's smoking cigarettes") from its verb ("causes"). In choice (A), the comma unnecessarily separates the subject from the verb. Also, the word "this" is unnecessarily redundant. In choice (B), the use of an introductory prepositional phrase ("for . . .") is confusing, and the pronoun "it" has an unclear antecedent. Choice (C) is wordy and awkward with the phrase "the smoking of cigarettes by a pregnant woman," and the use of "would cause" instead of "causes" is unnecessary. In choice (D), the verb "smoking" has no subject, and the antecedent of "it" is unclear.

8. (D) Choice (A) involves improper wording; "as" requires the use of "such." Choice (B) also contains improper wording; "that are" should be "such as." Choice (C) contains the improper idiom; "such variety as" should be "variety such as." Choice (E) contains an error in number agreement; the singular phrase "a décor style" cannot refer to the multiple styles listed in the series.

9. (B) It contains the parallel parts "an exercise regimen" and "a balanced diet." Choice (A) involves an error in parallelism, using "and" to connect the noun phrase "an exercise regimen" and the infinitive phrase "to maintain a balanced diet." Choice (C) combines the noun phrase "an exercise regimen" with the gerund phrase "maintaining a balanced diet," creating faulty parallelism. Choice (D) contains the same type of error with the gerund phrase "balancing your diet." And choice (E) has the same type of problem as choice (A): "an exercise regimen" and "to balance your diet" are not parallel. Also, the use of the second-person pronoun "your" in choices (D) and (E) is unnecessary because the rest of the sentence uses third person.

10. (C) It correctly completes the comparison with "than." Choice (A) is incorrect because of the improper comparison "compared with." Choice (B) is wrong because "compared to" is improper wording. In both cases, "than" should replace the incorrect phrases. In choice (D), the use of "by being" is awkward and wordy. And choice (E) lacks "than," which should go with "more," and "as well as . . . as" is awkward and wordy.

11. (B) In this sentence "having witnessed the horrible car wreck" is correctly placed just before "she,'" which it modifies. And the preposition "of" is followed by its object, the gerund "getting." Choice (A) is incorrect because it ends with a preposition. It is also illogical because "having witnessed the horrible car wreck" does not modify "getting behind the wheel." Choices (C) and (D) contain this same error; in each case, "having witnessed the horrible car wreck" is a dangling modifier because it modifies nothing in the sentence. And choice (D) is awkward and wordy, with the phrase "something of which." Choice (E) is wrong because it contains no main verb.

12. (D) It correctly uses a semicolon to separate two complete thoughts ("All . . . program" and "the boys . . . snowmen."), and in the last part of the sentence, it uses a comma and the conjunction "and" to combine two complete thoughts ("the boys are dressed as Santas" and "the girls are dressed as snowmen." Also, it consistently uses present-tense verbs ("are . . . ," "are dressed . . . are dressed"). Choice (A) incorrectly uses a comma instead of a semicolon to separate two complete thoughts ("All . . . program" and "the boys . . . snowmen"), and in the last part it uses the conjunction "and" but no comma. It also shifts from present tense ("are") to past tense ("were dressed" and "were dressed"). Choice (B) contains these same problems; the only difference is that it switches from "are" to "would dress." Choice (C) is a run-on sentence, with no punctuation separating the two complete thoughts ("All . . . program" and "the boys . . . snowmen"). Choice (E) is also a run-on sentence, with no punctuation separating these clauses.

13. (D) It appropriately compares "that of several U.S. Presidents" to George Washington's legacy. Clearly, the pronoun "that" refers to "legacy." It also correctly uses "other," indicating that George Washington was a U.S. President. Choice (A) illogically compares "George Washington's legacy" to "several U.S. Presidents." And the omission of "other" implies that George Washington was not a U.S. President. Choice (B) illogically compares "George Washington's legacy" to "several other U.S. Presidents." Choice (C) omits "other," implying that Washington was not a U.S. President. And choice (E) adds the unnecessary word "one" and omits the necessary word "other."

14. (E) It corrects the pronoun-antecedent error in the original sentence by changing "they" to "the people there." Choice (A) is incorrect because of the vague pronoun "they," which has no antecedent. Choice (B) is illogical because "the weather" cannot be what is "facing destructive hurricanes and tropical storms." Choice (C) is a run-on sentence, with no punctuation between the two independent thoughts ("The weather . . . devastating" and "every year . . . tropical storms"). Choice (D) contains the vague pronoun "they," just as choice (A) does.

PRACTICE TEST 4

Answer sheets for this test start on page 863.

TIME: 25 Minutes

ESSAY

<u>DIRECTIONS</u>: You have 25 minutes to plan and write an essay on the topic below. You may write on only the assigned topic.

Make sure to give examples to support your thesis. Proofread your essay carefully and take care to express your ideas clearly and effectively.

ESSAY TOPIC

The Internet makes doing research much easier.

The Internet makes doing research dangerous.

<u>ASSIGNMENT</u>: Consider the two contrasting statements above. Select the one that more closely reflects your view. Then write an essay that explains your choice. To support your view, use an example or examples from history, literature, the arts, current events, politics, science and technology, or your experience or observation.

TIME: 25 Minutes
20 Questions

MATH

DIRECTIONS: In this section solve each problem, using any available space on the page for scratchwork. Then decide which is the best of the choices given and fill in the corresponding oval on the answer sheet.

NOTES

(1) The use of a calculator is permitted.

(2) All numbers used are real numbers.

(3) Figures that accompany problems in this test are intended to provide information useful in solving the problems. They are drawn as accurately as possible EXCEPT when it is stated in a specific problem that the figure is not drawn to scale. All figures lie in a plane unless otherwise indicated.

(4) Unless otherwise specified, the domain of any function f is assumed to be the set of all real numbers x for which $f(x)$ is a real number.

REFERENCE INFORMATION

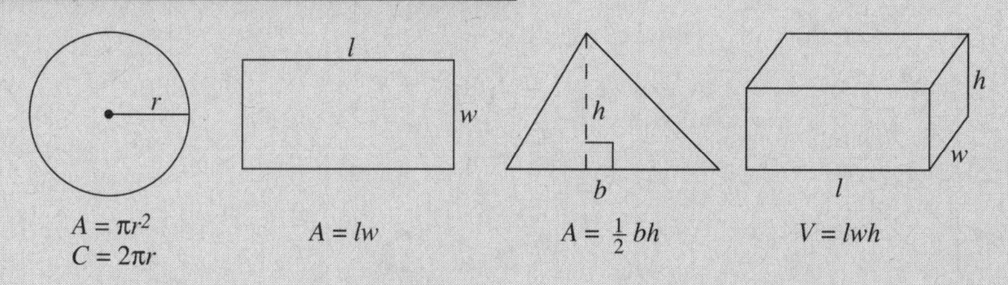

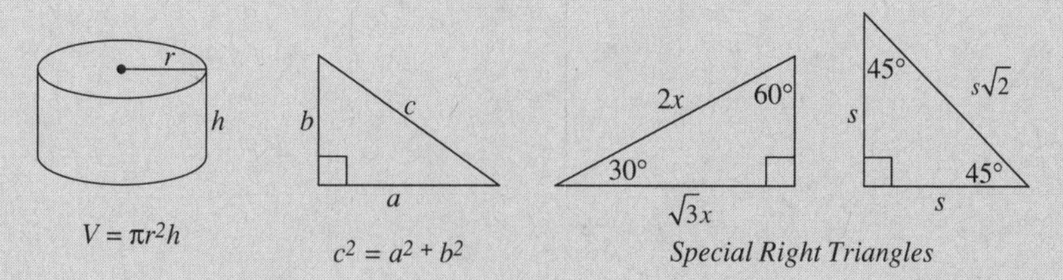

The number of degrees of arc in a circle is 360.

The sum of the measures in degrees of the angles of a triangle is 180.

1. For a positive integer m, $\boxed{m}$ is defined to be the sum of all odd integers less than m. If $\boxed{11} - \boxed{7} = k\boxed{5}$, what is k?

(A) 3
(B) 4
(C) 5
(D) 9
(E) 16

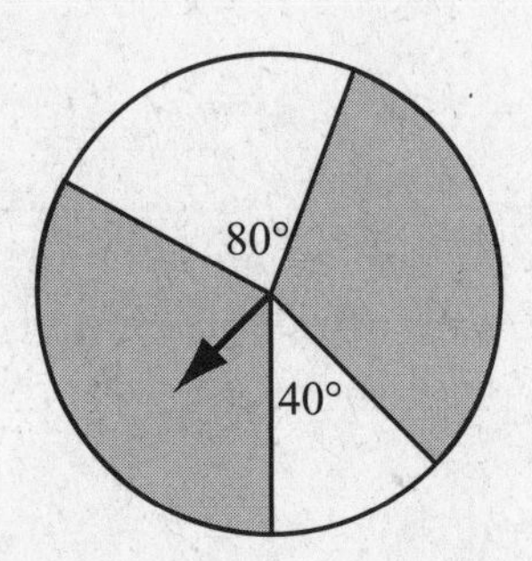

2. The figure above shows a spinner used for a board game. What is the probability that the spinner will stop in the shaded region?

(A) $\frac{2}{3\pi}$
(B) $\frac{1}{3}$
(C) $\frac{2}{\pi}$
(D) $\frac{2}{3}$
(E) $\frac{\pi}{3}$

3. The total population of Kingsport, Jonesborough, and Bristol is n. The population of Kingsport is three times the population of Jonesborough and the population of Bristol is four times the population of Jonesborough. Which is the population of Jonesborough in terms of n?

(A) $\frac{8}{n}$
(B) $\frac{7}{n}$
(C) $\frac{n}{5}$
(D) $\frac{n}{7}$
(E) $\frac{n}{8}$

4. If $x = a^2 - 16$ and $y = a - 4$ for $y \neq 0$ and $m \neq 4$, then what is the value of $\frac{x}{y} - 4$?

(A) a
(B) $-a$
(C) a^2
(D) $a - 4$
(E) $a + 4$

5. Each interior angle of a regular polygon measures 140°. How many sides does the polygon have?

(A) 5
(B) 7
(C) 9
(D) 11
(E) 13

GO ON TO THE NEXT PAGE

6. Rudy wants to build a rectangular pen for his dogs with the maximum possible area. He has 20 yards of fencing and plans to use the side of his garage for one side of the pen. What is the area of the largest pen he can build?

(A) 45 square feet

(B) 100 square feet

(C) 225 square feet

(D) 400 square feet

(E) 900 square feet

Day	Number in audience
Monday	1,500
Tuesday	1,460
Wednesday	1,380
Thursday	1,260
Friday	1,100

7. A play opened on Monday and shows every day of the week. It has been showing for five days, and the size of the audience for each day is recorded in the table above. The theater will stop showing the play after the audience drops below 400 people. If the trend continues, on what day of the week will the play's last performance take place?

(A) Sunday

(B) Monday

(C) Tuesday

(D) Wednesday

(E) Thursday

8. If y varies directly as the square of x and x varies directly as the cube of z, then y is a direct variation of which of the following?

(A) xz^2

(B) xz^3

(C) xz^6

(D) z^5

(E) z^6

9. The elements of set A are evenly divisible either by 3 or by 4. The elements of set B are evenly divisible either by 4 or by 5. The elements of set C are evenly divisible either by 3 or by 5. Which set is empty?

(A) $(A \cup B) \cup C$

(B) $(A \cup B) \cap C$

(C) $(A \cup B) \cap C$

(D) $(A \cap B) \cup C$

(E) $(A \cap B) \cap C$

10. The price of each pen in an office supply store ranges from \$0.82 to \$3.45 and the price of each pencil ranges from \$0.25 to \$1.23. How much money does a person need, in terms of m and n, to be able to buy m pens and n pencils?

(A) $0.82m - 0.25n$

(B) $0.82n - 0.25m$

(C) $0.82m + 0.25n$

(D) $0.82n + 0.25m$

(E) $82m + 25n$

GO ON TO THE NEXT PAGE

11. The kinetic energy of an object (K_E) is determined by $\frac{1}{2}mv^2$, where m represents the mass and v represents the velocity of the object. Which of the following is true about the kinetic energy of an object?

(A) K_E varies directly as m

(B) K_E varies directly as v^2

(C) K_E varies inversely as m

(D) K_E varies inversely as v^2

(E) K_E varies directly as mv^2

12. What is the equation of a line parallel to the line that passes through (1, 3) and (–2, –3)?

(A) $y = \frac{1}{2}x + 1$

(B) $y = -\frac{1}{2}x + 1$

(C) $y = x - 1$

(D) $y = -2x + 12$

(E) $y = 2x - 12$

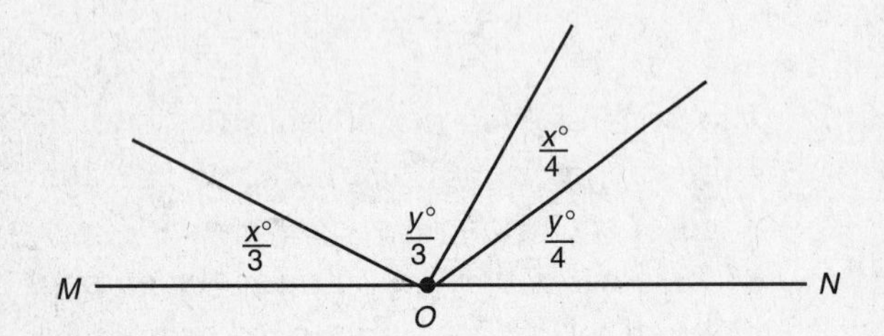

Note: Figure not drawn to scale.

13. In the figure above, O is a point on the line MN. What is the value of $x° + y°$ in degrees?

(A) $\frac{216}{7}$

(B) $\frac{1,260}{12}$

(C) $\frac{1,260}{7}$

(D) $\frac{2,160}{7}$

(E) $\frac{2,160}{12}$

14. If $f(x) = 3x^2 - 4y$ and $g(y) = 3y^2 + 4x$, which of the following is expressions equal to $f(y) - g(x)$?

(A) $3y^2 - 4x - 3x^2 - 4y$

(B) $3y^2 - 4x - 3x^2 + 4y$

(C) $3y^2 - 4y + 3x^2 + 4x$

(D) $3y^2 - 4y - 3x^2 - 4x$

(E) $3y^2 - 4y - 3x^2 + 4x$

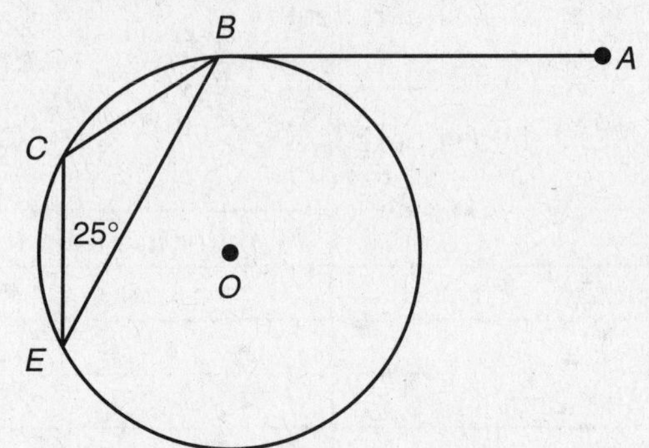

Note: Figure not drawn to scale.

15. In the figure above, $\overline{AB}$ is tangent to circle O and ΔBEC is an isosceles triangle. What is $\angle ABE$?

(A) 100°

(B) 130°

(C) 150°

(D) 165°

(E) 260°

GO ON TO THE NEXT PAGE

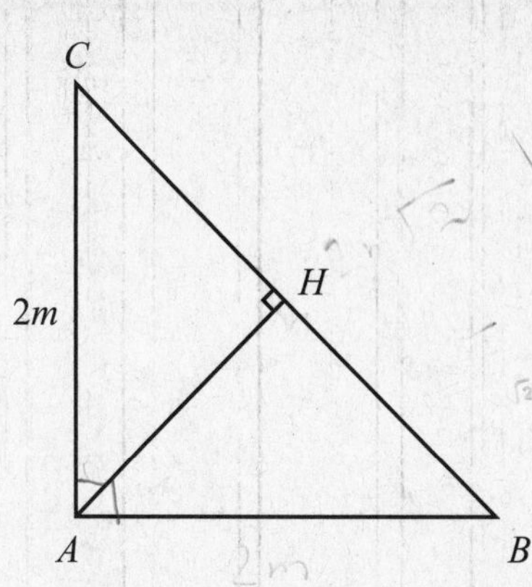

Note: Figure not drawn to scale.

16. In the figure above, ΔABC is an isosceles right triangle and $\overline{AH}$ is an altitude. What is $\overline{AH}$ in terms of m?

    (A) m

    (B) $2m$

    (C) $4m$

    (D) $m\sqrt{2}$

    (E) $2m\sqrt{2}$

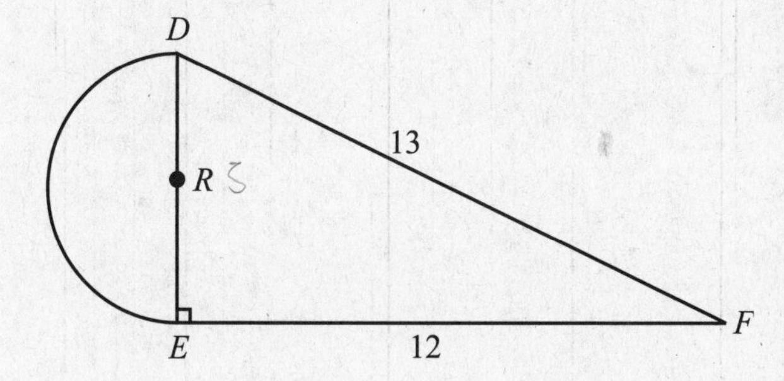

Note: Figure not drawn to scale.

17. In the figure above, ΔDEF is a right triangle. What is the area of the semicircle R?

    (A) 25π

    (B) $25\pi^2$

    (C) $\frac{25\pi}{2}$

    (D) $\frac{25\pi}{4}$

    (E) $\frac{25\pi}{8}$

18. The value of $\frac{1}{x+3}$ is defined and it is a positive number. The value of $\frac{1}{y^2-27}$ is undefined. Which of the following statements is true?

    (A) $x - y > 0$

    (B) $x + y < 0$

    (C) $x + y > 0$

    (D) $x + y > 6$

    (E) $x + y > -6$

19. When x is positive, $\frac{\sqrt{x}-1}{\sqrt{5}+1} = \frac{\sqrt{5}-1}{\sqrt{x}+1}$. What is $\sqrt{x^2-1}$?

    (A) 25

    (B) 24

    (C) 16

    (D) 6

    (E) 4

20. If $2^{p+2q} = 16^q$ and $3^{2p+q} = 9^q$, what is $(p - q)$?

    (A) 4

    (B) 3

    (C) 2

    (D) 1

    (E) 0

STOP

If time remains, you may go back and check your work. When the time allotted is up, you may go on to the next section.

TIME: 25 Minutes
24 Questions

CRITICAL READING

DIRECTIONS: Each sentence that follows has one or two blanks, each blank indicating that something has been omitted. Beneath the sentence are five lettered words or sets of words. Choose the word or set of words that BEST fits the meaning of the sentence as a whole.

EXAMPLE

Although critics found the book _______ , many readers found it rather _______ .

(A) obnoxious . . . perfect

(B) spectacular . . . interesting

(C) boring . . . intriguing

(D) comical . . . persuasive

(E) popular . . . rare

EXAMPLE ANSWER

1. Winston Churchill was such a(n) _______ and _______ speaker that listeners from around the world would postpone whatever they were doing to hear his speeches on the radio.

   (A) gullible . . . deliberate
   (B) impartial . . . obscure
   (C) incoherent . . . pious
   (D) provocative . . . prosaic
   (E) eloquent . . . mesmerizing

2. Although the topic embarrassed him, he spoke _______ about his bankruptcy and financial situation.

   (A) vainly
   (B) rapidly
   (C) cryptically
   (D) bitterly
   (E) candidly

3. Some people believe that flying saucers exist. As far as Jaclyn is concerned, this theory is a(n) _______ and has to be proven.

   (A) obituary
   (B) travesty
   (C) irony
   (D) kaleidoscope
   (E) enigma

4. Eric felt _______ after his _______ remark, but it was too late to retract it.

   (A) respite . . . inadvertent
   (B) irrational . . . incautious
   (C) remorse . . . disparaging
   (D) tentative . . . extravagant
   (E) insensitive . . . premeditated

5. Residents of the town, who normally wouldn't agree on anything, now _______ to _______ the construction of a wood-fired power plant that would cause air pollution.

   (A) proceeded . . . defend
   (B) dispersed . . . hamper
   (C) assembled . . . endorse
   (D) alternated . . . establish
   (E) mobilized . . . resist

6. Zane Gray was able to _______ fictionalize the life of western cowboys even though he lived thousands of miles away in New York and probably never visited his locales.

   (A) benevolently
   (B) authentically
   (C) prosaically
   (D) frivolously
   (E) incoherently

7. The scientist was shocked when all the new data he compiled for his experiment _______ his previous results.

   (A) nullified
   (B) suppressed
   (C) instigated
   (D) validated
   (E) consecrated

GO ON TO THE NEXT PAGE

8. Before an inventor can receive a patent, the U.S. Patent and Trademark Office _______ the application to verify that a _______ invention doesn't exist.

(A) investigates . . . distorted

(B) scrutinizes . . . comparable

(C) expedites . . . compatible

(D) questions . . . probable

(E) analyzes . . . contrasting

<u>DIRECTIONS</u>: Each of the following passages is followed by questions based on its content. Answer the questions on the basis of what is <u>stated</u> or <u>implied</u> in each passage and in any introductory material that may be provided.

1 With the convenience of today's technology, we are readily accessible at all times. Leaving the house or the office no longer means leaving the phone; our cell phones are as mo-
5 bile as we are. And being away from our mailboxes no longer means having no access to mail; electronic correspondence is as close as the nearest connected computer, which may be right at the side of the wireless laptop owner.
10 So is it possible to be available to anyone, anytime, anywhere. However, with this unlimited accessibility comes the inability to escape the complaints of clients or coworkers, orders from a too-demanding boss who wants us to take
15 care of matters immediately, or messages from family members who like to think we are at their disposal. Our free time is no longer free. Our "own time" is no longer our own. Technological progress enhances others' accessibility
20 to us, but—in a world where leisure time is already at a premium—it serves only to stifle our own access to privacy.

9. According to this passage, which of the following is limited by technological progress?

(A) ability to communicate

(B) use of the post office

(C) use of telephones

(D) right to privacy

(E) freedom of speech

10. What does "at a premium" mean in line 21?

(A) expensive

(B) insured

(C) scarce

(D) unappreciated

(E) plentiful

GO ON TO THE NEXT PAGE

1 Today, plastic credit cards and debit cards are almost more common than paper money and coins. Although plastic is a relatively new invention, the concept of credit is over 3,000
5 years old. In those days, Babylonians, Assyrians, and Egyptians could settle one-third of a debt with cash and two-thirds with a "bill of exchange." In 1730 Christopher Thornton first allowed his customers in London to make
10 weekly payments for the furniture they bought from him. And in the eighteenth and nineteenth centuries, tallymen accepted weekly payments for clothing and other articles, using one side of a wooden stick to keep a tally of
15 how much a customer owed and the other side to record how much that customer had paid. In the 1920s, Americans used shopper's plates, in stores that issued them, as a kind of promissory note allowing them a month to pay for
20 their purchases. It was not until the 1950s that Diner's Club and American Express issued the first credit cards. But use of such "plastic money" was quite limited until 1970, when government standards for the magnetic strip were
25 established and credit cards gradually gained widespread acceptance. Today—only a few decades later—people can go shopping and make unlimited purchases without having a penny in their pockets—and often without having the as-
30 sets to pay for them anytime in the near future.

11. People have been able to use credit to acquire the things they want

(A) only in the past century.

(B) for thousands of years.

(C) only since the invention of plastic.

(D) since the government began regulating the use of the magnetic strip.

(E) since Diner's Club and American Express began issuing credit cards.

12. The word "promissory" in line 15–16 means

(A) having to do with money matters; financial.

(B) personally signed.

(C) containing or conveying a promise or assurance.

(D) listing the items purchased on credit.

(E) listing the assets of the consumer.

GO ON TO THE NEXT PAGE

In the following passage, Virginia Woolf discusses the place of Robinson Crusoe *in Daniel Defoe's written work.*

1 The fear which attacks the recorder of centenaries lest he should find himself measuring a diminishing spectre and forced to foretell its approaching dissolution is not only absent in the 5 case of *Robinson Crusoe* but the mere thought of it is ridiculous. It may be true that *Robinson Crusoe* is two hundreds years of age upon the twenty-fifth of April 1919, but far from raising the familiar speculations as to whether people now 10 read it and will continue to read it, the effect of the bi-centenary is to make us marvel that *Robinson Crusoe*, the perennial and immortal, should have been in existence so short a time as that.

The book resembles one of the anonymous 15 productions of the race rather than the effort of a single mind; and as for celebrating its centenary we should as soon think of celebrating the centenaries of Stonehenge itself. Something of this we may attribute to the fact that we have all had 20 *Robinson Crusoe* read aloud to us as children, and were thus much in the same state in mind towards Defoe and his story that the Greeks were in towards Homer. It never occurred to us that there was such a person as Defoe, and to have 25 been told that *Robinson Crusoe* was the work of a man with a pen in his hand would have either disturbed us unpleasantly or meant nothing at all. The impressions of childhood are those that last longest and cut deepest. It still seems that 30 the name of Daniel Defoe has no right to appear upon the title-page of *Robinson Crusoe*, and if we celebrate the bi-centenary of the book we are making a slightly unnecessary allusion to the fact that, like Stonehenge, it is still in existence.

35 The great fame of the book has done its author some injustice; for while it has given him a kind of anonymous glory it has obscured the fact that he was a writer of other works which, it is safe to assert, were not read aloud to us as chil- 40 dren. Thus when the Editor of the *Christian World* in the year 1870 appealed to "the boys and girls of England" to erect a monument upon the grave of Defoe, which a stroke of lightning had mutilated, the marble was inscribed to the memory of the author of *Robinson Crusoe*. No mention was 45 made of *Moll Flanders*. Considering the topics which are dealt with in that book, and in *Roxana*, *Captain Singleton*, *Colonel Jack* and the rest, we need not be surprised, though we may be indignant, at the omission. We may agree with Mr. 50 Wright, the biographer of Defoe, that these "are not works for the drawing-room table." But unless we consent to make that useful piece of furniture the final arbiter of taste, we must deplore that their superficial coarseness, or the universal 55 celebrity of *Robinson Crusoe*, has led them to be far less widely famed that they deserve. On any monument worthy of the name of monument the names of *Moll Flanders* and *Roxana*, at least, should be carved as deeply as the name of 60 Defoe. They stand among the few English novels which we can call indisputably great. The occasion of the bi-centenary of their more famous companion may well lead us to consider in what their greatness, which has so much in common 65 with his, may be found to consist.

Defoe was an elderly man when he turned novelist, many years the predecessor of Richardson and Fielding, and one of the first indeed to shape the novel and launch it on its way. But 70 it is necessary to labour the fact of his precedence, except that he came to his novel-writing with certain conceptions about the art which he derived partly from being himself one of the first to practise it. The novel had to justify its 75 existence by telling a true story and preaching a sound moral. "This supplying a story by invention is certainly a most scandalous crime," he wrote. "It is a sort of lying that makes a great hole in the heart, in which by degrees a 80 habit of lying enters in." Either in the preface or in the text of each of his works, therefore, he takes pains to insist that his purpose has been the highly moral desire to convert the

GO ON TO THE NEXT PAGE

85 vicious or to warn the innocent. Happily these were principles that tallied very well with his natural disposition and endowments. Facts had been drilled into him by sixty years of varying fortunes before he turned his experience 90 to account in fiction. "I have some time ago summed up the Scenes of my life in this distich," he wrote:

No man has tasted differing fortunes more,

And thirteen times I have been rich and 95 poor.

13. By claiming that *Robinson Crusoe* is a monument, the author stresses the novel's

(A) length.
(B) familiarity.
(C) difficulty.
(D) artistic merit.
(E) triviality.

14. In line 12, the word "perennial" most nearly means

(A) recurrent.
(B) nourishing.
(C) popular.
(D) diminishing.
(E) ancient.

15. From Mr. Wright's comment, "these are not works for the drawing-room table," we can infer that Defoe's other books

(A) were not popular.
(B) were not thought suitable for children.
(C) were viewed as inferior writing.
(D) were unjustly ignored.
(E) were superior to *Robinson Crusoe*.

16. The bicentenary should allow us (lines 62–66) to

(A) reevaluate *Robinson Crusoe*.
(B) reconsider *Roxana* and *Captain Singleton*.
(C) establish Defoe as a great novelist.
(D) compare Defoe with Fielding.
(E) erect a new monument to Defoe.

17. Defoe appears to have believed (lines 75–81) that storytelling was potentially

(A) rewarding.
(B) immoral
(C) useful.
(D) painful.
(E) habitual.

18. When the author states "Happily there were principles . . . endowments" (lines 85–87), she refers to Defoe's

(A) careful plotting of stories.
(B) abundant life experience prior to writing.
(C) refusal to use factual material in writing.
(D) tendency to preach a sermon.
(E) ability to create vivid characters.

19. In line 87, the word "endowments" most nearly means

(A) inheritance.
(B) talents.
(C) stature.
(D) donations.
(E) accommodations.

GO ON TO THE NEXT PAGE

20. The author gives examples from the novels to demonstrate Defoe's belief that

(A) good eventually triumphs over evil.
(B) social injustice can be remedied.
(C) existence is a continual struggle.
(D) mankind does not change through the centuries.
(E) one's fortunes decline inevitably.

21. The tone of the passage can be best described as

(A) felicitous.
(B) exonerative.
(C) reverential.
(D) defamatory.
(E) objective.

22. The writer's primary purpose in writing this passage is to

(A) show why *Robinson Crusoe* is Defoe's weakest work.
(B) give a detailed biography of Daniel Defoe.
(C) compare Defoe's style to that of Fielding.
(D) summarize the plot of *Robinson Crusoe.*
(E) show how *Robinson Crusoe* sits apart from Defoe's other novels.

23. In lines 29–31, when the author writes, "It still seems that the name of Daniel Defoe has no right to appear on the title page of *Robinson Crusoe,*" she means

(A) Defoe was not the author's real name, but actually a pseudonym.
(B) Defoe felt that inventing stories, or fiction, was immoral because it was akin to lying, and therefore he had no wish to be cited as the author of the work.
(C) given that the book is so often read aloud to children, it seems odd that it was actually written by someone other than the person who was telling the tale.
(D) Daniel Defoe did not actually write the book, but collaborated with many different authors who have never received proper credit.
(E) the book was actually written by a undisclosed ghostwriter, but Defoe took the credit for it.

24. The author believes that Defoe's other works were not as widely read as *Robinson Crusoe* because

(A) they were poorly written.
(B) they were overshadowed by the popularity of *Robinson Crusoe*, and as a result, overlooked.
(C) the subject matter was questionable and did not reflect the attitudes or values of the times.
(D) they were published posthumously to very little fanfare and did not get much attention.
(E) they were written in a much lighter tone than *Robinson Crusoe.*

STOP

If time still remains, you may go back and check your work. When the time allotted is up, you may go on to the next section.

TIME: 25 Minutes
35 Questions

WRITING

DIRECTIONS: In each of the following sentences, some portion of the sentence is underlined. Under each sentence are five choices. The first choice has the same wording as the original. The other four choices are reworded. Sometimes the first choice containing the original wording is the best; sometimes one of the other choices is the best. Choose the letter of the best choice. Your choice should produce a sentence that is not ambiguous or awkward and that is correct, clear, and precise.

This is a test of correct and effective English expression. Keep in mind the standards of English usage, punctuation, grammar, word choice, and construction.

EXAMPLE

When you listen to opera, a person may not appreciate it.

(A) a person may not appreciate it.

(B) it may not be appreciated by a person.

(C) you may not appreciate it.

(D) which may not be appreciated by you.

(E) appreciating it may be a problem for you.

EXAMPLE ANSWER

1. Eating chips of dried paint, children who live in old houses are at risk for lead poisoning.

   (A) Eating chips of dried paint
   (B) Having eaten chips of dried paint
   (C) Because of eating chips of dried paint
   (D) Chips of dry paint being eaten
   (E) Because they may eat chips of dried paint

2. French architect Pierre Charles L'Enfant was hired to plan the United States capital although he and President Washington had a disagreement, L'Enfant's plan was used in the design of Washington, D.C.

   (A) although he and President Washington had a disagreement,
   (B) and, although he and President Washington had a disagreement,
   (C) who had a disagreement with President Washington, but,
   (D) although having had a disagreement with President Washington,
   (E) he and President Washington had a disagreement.

3. While trying to reduce cholesterol, you should eat lentils in casseroles, salads, and soups, the reason being that lentils provide an excellent source of protein.

   (A) you should eat lentils in casseroles, salads, and soups, the reason being that lentils
   (B) lentils should be eaten in casseroles, salads, and soups, the reason being that lentils
   (C) you should eat lentils in casseroles, salads, and soups because lentils
   (D) eating lentils in casseroles, salads, and soups will
   (E) you should eat lentils in casseroles, salads, and soups, which

4. Being that the first Library of Congress had been destroyed in the War of 1812, Congress purchased the personal library of Thomas Jefferson to replace it.

   (A) Being that the first Library of Congress had been destroyed in the War of 1812,
   (B) Although the first Library of Congress was destroyed in the War of 1812,
   (C) The first Library of Congress was destroyed in 1812, was restored because
   (D) The first Library of Congress was destroyed in the War of 1812, so
   (E) Having destroyed the first Library of Congress in the War of 1812,

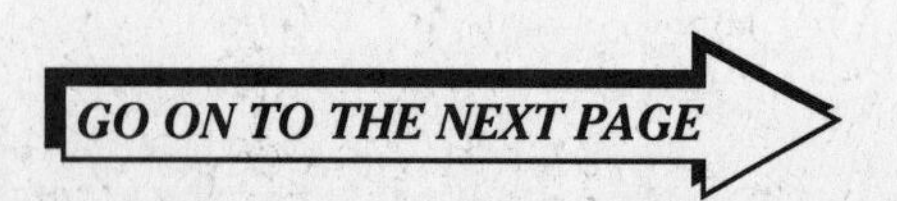

5. The fewer people you tell your secret to, <u>there are fewer people to divulge it to others.</u>

   (A) there are fewer people to divulge it to others.
   (B) there will be fewer people to divulge it to others.
   (C) there being fewer people to divulge it to others.
   (D) the fewer people there are to divulge it to others.
   (E) the fewer people can divulge it to others.

6. Lisbon, Portugal, is a large city <u>whose history goes back</u> to ancient Greek and Roman times.

   (A) whose history goes back
   (B) which history goes back
   (C) the history of which goes back
   (D) its history goes back
   (E) who's history goes back

7. The young entertainer gets unusual opportunities <u>due to the fact that he is a man whose</u> father is famous.

   (A) due to the fact the he is a man whose
   (B) due to the fact that he is a man who's
   (C) due to the fact that his
   (D) because he is a man whose
   (E) because his

8. Because it was cracking, the London Bridge was taken down in <u>1967, and they rebuilt it in Lake Havasu City, Arizona</u>.

   (A) 1967, and they rebuilt it in Lake Havasu City, Arizona.
   (B) 1967 and rebuilt in Lake Havasu City, Arizona.
   (C) 1967, being rebuilt in Lake Havasu City, Arizona.
   (D) 1967, and in Lake Havasu City, Arizona, they rebuilt it.
   (E) 1967, they rebuilt it in Lake Havasu City, Arizona.

9. The instinctive habits of a beaver <u>is that he cuts down trees, stores food, and builds dams and lodges.</u>

   (A) is that he cuts down trees, stores food, and builds dams and lodges.
   (B) is the cutting down trees, storage of food, and dam and lodge building.
   (C) are the cutting down of trees, the storing of food, and the building of dams and lodges.
   (D) are having cut down trees, storing food, and building dams and lodges.
   (E) are trees being cut down, food being stored, and dams and lodges being built.

GO ON TO THE NEXT PAGE

10. Bolivia, a South American country named for Simon Bolivar, who was a leader in freeing the Latin American colonies from Spain, having a climate in which rainfall is seasonal.

(A) having a climate in which rainfall is seasonal.

(B) the climate of which has seasonal rainfall.

(C) has a climate in which rainfall is seasonal.

(D) and it has a climate in which rainfall is seasonal.

(E) The rainfall is seasonal in its climate.

11. Brutus participated in the plot to assassinate Caesar despite the fact that he was pardoned for treason.

(A) despite the fact that he was pardoned for treason.

(B) despite the fact that Brutus had been pardoned by Caesar for treason.

(C) despite the pardoning of Brutus by Caesar for treason.

(D) although he was pardoned for treason.

(E) although Caesar had pardoned Brutus for treason.

GO ON TO THE NEXT PAGE

DIRECTIONS: Each of the following sentences may contain an error in diction, usage, idiom, or grammar. Some sentences are correct. Some sentences contain one error. No sentence contains more than one error.

If there is an error, it will appear in one of the underlined portions labeled A, B, C, or D. If there is no error, choose the portion labeled E. If there is an error, select the letter of the portion that must be changed in order to correct the sentence.

EXAMPLE

He drove slowly and cautiously in order to hopefully avoid having an accident. No error.
A B C D E

EXAMPLE ANSWER

(A) (B)   (E)

12. If anyone plans to visit Yosemite National Park, you should allow several days to see all of the sites. No error.
A: plans B: you C: to see D: of the sites E: No error

13. It is important that medications and cleaning preparations be stored in places that are inaccessible from young children. No error.
A: important B: and C: be stored D: inaccessible from E: No error

14. Although he worked very slow, he was meticulously careful, and his report of the findings of his studies was always very accurate. No error.
A: slow B: meticulously C: was D: accurate E: No error

15. The man giving the radio news is a commentator who radio listeners have heard for many years. No error.
A: giving B: is C: who D: have heard E: No error

GO ON TO THE NEXT PAGE

16. Although she is older than anyone else, she looks very young for her age. No error.
(A: Although; B: older; C: anyone else; D: for her age; E: No error)

17. A large crowd gathered on the president's lawn and shouting their protest about the dismissal of Professor Maxwell. No error.
(A: gathered; B: president's; C: shouting; D: dismissal of; E: No error)

18. The sign in the ice cream parlor read, "Because we make our ice cream fresh daily, the variety of flavors change regularly." No error.
(A: fresh; B: daily; C: of flavors; D: change; E: No error)

19. Although it provided a dangerous way of life, whaling was once an important vocation in New England seaports because whale oil was used for lamps, candles, soap, and as a lubricant. No error.
(A: it provided; B: an important vocation; C: was used; D: as a lubricant; E: No error)

20. He ate greedily because he was literally starved to death after he had been lost in the woods for two days. No error.
(A: greedily; B: literally; C: had been lost; D: for two days; E: No error)

21. Of all the important contributions to history that he might have made, John Montagu, the Earl of Sandwich, is better known for a lunch of meat between two slices of bread that permitted him to continue his gambling while he ate. No error.
(A: might have made; B: better; C: that permitted; D: to continue; E: No error)

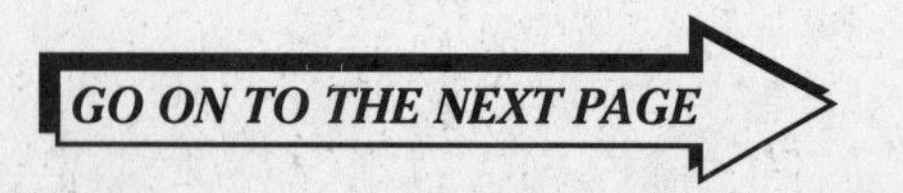

22. When Rutgers had beaten Princeton in the first intercollegiate college football game in America, the 25-man teams made goals by kicking a round ball under the cross bars. No error.
(A) When (B) had beaten (C) made (D) kicking a round ball (E) No error

23. He was one of those kind of people who never feel good about themselves. No error.
(A) kind (B) who (C) good (D) themselves (E) No error

24. After he had been sent to bed, the little boy snuck downstairs and lay down on the kitchen floor. No error.
(A) had been sent (B) snuck (C) downstairs (D) lay (E) No error

25. Although Walt Disney's film *Snow White and the Seven Dwarfs* was originally shown in 1937, most of today's young people hadn't had no chance to see it until its 1987 revival. No error.
(A) was originally shown (B) no chance (C) to see (D) its (E) No error

26. Although Gilbert Stuart, whose portraits of George Washington are recognized by most Americans, and his daughter Jane were both American painters, the father is the most famous of the two. No error.
(A) are recognized (B) by most Americans (C) were (D) most famous (E) No error

GO ON TO THE NEXT PAGE

27. If it <u>were</u> <u>me</u> <u>who</u> had to work that hard, I
A B C
<u>would complain</u> to the manager. <u>No error</u>.
D E

28. Everyone thought it was <u>really tragic</u> when
A
the child <u>fell</u> down the well <u>that</u> <u>should have</u>
B C D
<u>been</u> sealed off years before. <u>No error</u>.
E

29. Only five days <u>after the surrender</u> of Gen-
A
eral Lee to General Grant <u>signalled</u> the
B
end of the Civil War, Abraham Lincoln
<u>has been shot</u> while he <u>was attending</u> a
C D
play at Ford's Theatre. <u>No error</u>.
E

GO ON TO THE NEXT PAGE

DIRECTIONS: The following passages are considered early draft efforts of a student. Some sentences need to be rewritten to make the ideas clearer and more precise.

Read each passage carefully and answer the questions that follow. Some of the questions are about particular sentences or part of a sentence and ask you to make decisions about sentence structure, diction, and usage. Some of the questions refer to the entire essay or parts of the essay and ask you to make decisions about organization, development, appropriateness of language, audience, and logic. Choose the answer that most effectively makes the intended meaning clear and follows the requirements of standard written English. After you have chosen your answer, fill in the corresponding oval on your answer sheet.

EXAMPLE

(1) On the one hand, I think television is bad, But it also does some good things for all of us. (2) For instance, my little sister thought she wanted to be a policemen until she saw police shows on television.

Which of the following is the best revision of the underlined portion of sentence (1) below?

One the one hand, I think television is bad, But it also does some good things for all of us.

(A) is bad; But it also

(B) is bad. but it also

(C) is bad, but it also

(D) is bad, and it also

(E) is bad because it also

EXAMPLE ANSWER

Questions 30–35 are based on the following passage.

(1) In the poem "The Raven" by Edgar Allan Poe, a man has nodded off in his study after reading "many a quaint and curious volume of forgotten lore." (2) His mood is melancholy. (3) He is full of sorrow. (4) He is grieving for "the lost Lenore."

(5) When he hears the tapping at his window, he lets in a raven. (6) The raven perches on the bust of Pallas Athena, a goddess often depicted with a bird on her head by the Greeks who believed that birds were heralds from the dead. (7) At first, the man thinks the bird might be a friend but one who would leave soon, but the raven says, "Nevermore," an affirmation which makes the man smile. (8) However, the bird's repetition of "Nevermore" leads the speaker to the realization that Lenore will never return, the bird becomes an omen of doom and is called "evil" by the mournful speaker. (9) He becomes frantic, imploring the raven to let him know if there is comfort for him or if he will ever again hold close the sainted Lenore. (10) To both questions, the bird replies, "Nevermore." (11) Shrieking with anguish, the bird is ordered to leave, but it replies, "Nevermore." (12) At the end of the poem, the man's soul is trapped in the Raven's shadow "that lies floating on the floor" and "shall be lifted—nevermore."

(13) Thus, the bird evolves into an ominous bird of ill omen. (14) Some argue that the bird deliberately drives the speaker insane. (15) While others feel the bird is innocent of any premeditated wrong doing, and I think the bird doesn't do anything but repeat one word. (16) One thing is certain, however. (17) The poem's haunting refrain is familiar, one that students of American literature will memorize and forget "nevermore."

30. Which of the following is the best way to combine sentences 2, 3, and 4?

(A) He is melancholy and sorrowful and grieving for "the lost Lenore."

(B) He is grieving for "the lost Lenore," full of melancholy and sorrow.

(C) Melancholy and sorrowful, he is grieving for "the lost Lenore."

(D) Melancholy and full of sorrow, he is grieving for "the lost Lenore."

(E) Full of melancholy mood and sorrow, he is grieving for "the lost Lenore."

31. Which of the following is the best revision of the underlined portion of sentence 8 below?

However, the bird's repetition of "Nevermore" leads the speaker to <u>the realization that Lenore will never return, the bird becomes an omen of doom and is called "evil" by the mournful speaker.</u>

(A) realize the following—Lenore will never return, the bird is evil and an omen of doom.

(B) the realization of Lenore's failure to return, the evil and the omen of doom of the bird.

(C) realize that Lenore will never return, so the bird, called "evil" by the mournful speaker, becomes an omen of doom.

(D) the realization that Lenore will never return, the bird becomes an omen of doom and is called "evil" by the mournful speaker.

(E) that Lenore will never return, that the bird is an omen of doom, and that the bird should be called "evil."

32. Which is the best revision of the underlined portion of sentence 11 below?

Shrieking with anguish, <u>the bird is ordered to leave, but it</u> replies, "Nevermore."

(A) the bird is ordered to leave but replies
(B) the man orders the bird to leave, but it
(C) the order is given for the bird to leave, but it
(D) the bird, ordered to leave,
(E) the man orders the bird to leave and

33. In relation to the passage as a whole, which of the following best describes the writer's intention in the second paragraph?

(A) to show specific examples of supernatural effects
(B) to argue that the bird has evil intent
(C) to convince the reader to read the poem
(D) to analyze the progression of ideas in the literature
(E) to describe the appearance of the bird

34. In the context of the sentences preceding and following sentence 15, which of the following is the best revision of sentence 15?

(A) Only repeating one word, others feel the bird is innocent of any premeditated wrongdoing.
(B) Innocent and not doing anything, the bird repeats one word.
(C) Innocent of any premeditated wrongdoing, the bird does not do anything.
(D) One may argue that, innocent of any wrongdoing, the bird does not do anything but repeat one word.
(E) Others feel the bird is innocent of any premeditated wrongdoing because it does not do anything but repeat one word.

GO ON TO THE NEXT PAGE

35. Which of the following would be a better way to end the passage, combining sentences 16 and 17?

(A) One thing is certain, however: the poem's haunting refrain is a familiar one that students of American literature will memorize and forget "nevermore."

(B) One thing is certain, however, that the poem's haunting refrain is familiar, one that students of American literature will memorize and forget "nevermore."

(C) One thing is certain, however; the poem's haunting refrain is familiar, one that students of American literature will memorize and forget "nevermore."

(D) One thing is certain, however: the poem's haunting refrain is familiar, one that students of American literature will memorize and forget "nevermore."

(E) One thing that is certain is that the poem's haunting refrain is familiar, one that students of American literature will memorize and forget "nevermore."

STOP

If time still remains, you may go back and check your work. When the time allotted is up, you may go on to the next section.

TIME: 25 Minutes
18 Questions

MATH

DIRECTIONS: In this section solve each problem, using any available space on the page for scratchwork. Then decide which is the best of the choices given and fill in the corresponding oval on the answer sheet.

NOTES

(1) The use of a calculator is permitted.

(2) All numbers used are real numbers.

(3) Figures that accompany problems in this test are intended to provide information useful in solving the problems. They are drawn as accurately as possible EXCEPT when it is stated in a specific problem that the figure is not drawn to scale. All figures lie in a plane unless otherwise indicated.

(4) Unless otherwise specified, the domain of any function f is assumed to be the set of all real numbers x for which $f(x)$ is a real number.

REFERENCE INFORMATION

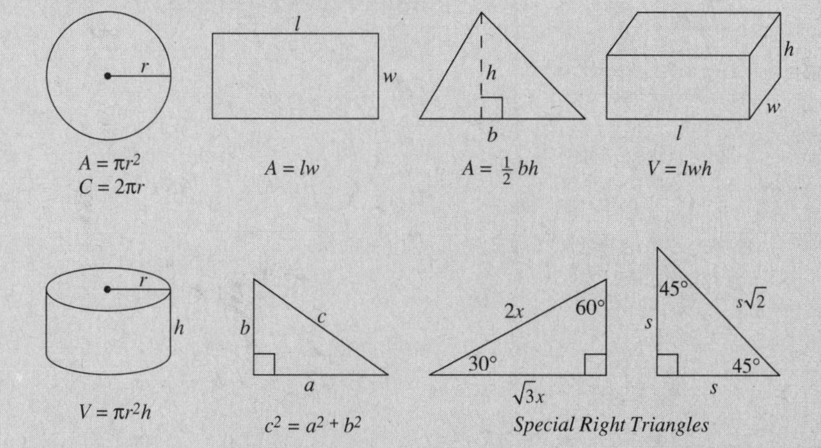

The number of degrees of arc in a circle is 360.

The sum of the measures in degrees of the angles of a triangle is 180.

1. Brenda earns a wage of $18.25 per hour and deposits her paycheck every Friday. The company deducts 12% of her income for taxes and other employment-related fees. If h is the number of hours she works per week and I is her earning after the taxes and fees, which expression represents how much Brenda deposits?

(A) $15.28h = I$

(B) $16.06h = I$

(C) $16.23h = I$

(D) $16.28I = h$

(E) $16.38I = h$

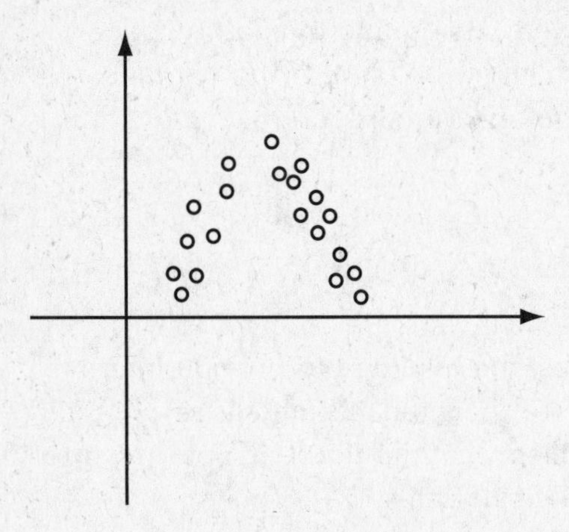

2. Based on the scatterplot above, which of the following is the most logical conclusion?

(A) As x decreases, y decreases.

(B) As x decreases, y increases.

(C) As x increases, y decreases then stays constant.

(D) As x decreases, y decreases and then stays constant.

(E) y varies neither directly nor inversely as x.

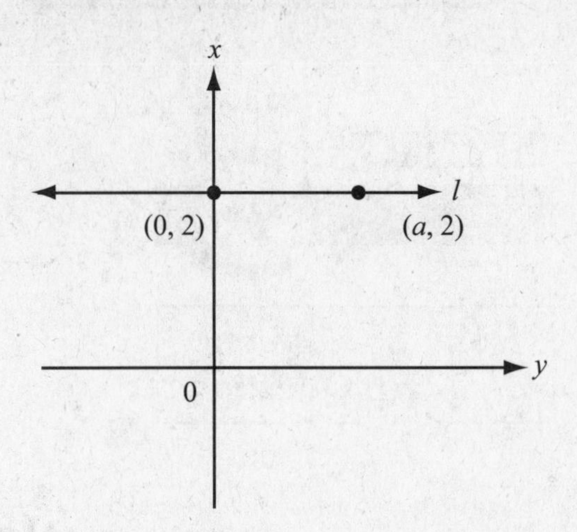

3. Let l represent the equation of the line shown in the figure above. Which of the following is true about line l?

(A) l varies directly as a.

(B) l varies directly as a^2.

(C) l does not vary directly as a.

(D) l varies directly as $-a$.

(E) l varies directly as $2a$

4. For the positive integers m and n, we define [[m]] as the sum of the prime factors of m greater than 1 and define [⌊n⌋] as the sum of the digits of n. Which of the following could be the two-digit integer p if [[35]] − [⌊24⌋] = [⌊p⌋]?

(A) 71

(B) 70

(C) 61

(D) 51

(E) 17

GO ON TO THE NEXT PAGE

Note: Figure not drawn to scale.

5. In the preceding triangle, the values of x, y, and z, respectively, are in the ratio of 9:5:4. What is the length of side AB?

(A) $2\sqrt{3}$

(B) $2\sqrt{6}$

(C) 5

(D) $4\sqrt{3}$

(E) $18\sqrt{2}$

6. What is the value of $\left(\sqrt{150\times2}\right)\left(\sqrt{7^2-1}\right)$?

(A) 24.2

(B) 30.2

(C) 45.8

(D) 120

(E) 210

7. If $7a + 9b = 13$ and $7a + 9 = -5$, what is the value of $(7a - 9b)$?

(A) -2

(B) $-\frac{5}{7}$

(C) 3

(D) 5

(E) 13

8. The number of a certain bacteria (Q) after 4 hours growth in a culture is represented by $Q = 1{,}000e^{0.321(4)}$. Which is the initial number of the bacteria in the culture?

(A) 0.321

(B) 1.284

(C) 321

(D) 1,000

(E) 1,284

9. There are 110 students in Science Hill High School. Of these students, 40 take Biology II class, 25 take Sociology, and 35 take neither Biology II nor Sociology. How many students attend both Biology II and Sociology classes?

(A) 5

(B) 10

(C) 15

(D) 20

(E) 25

10. Which of the following equations is equivalent to the equation $\left|\frac{x-1}{-2}\right| = -6$?

(A) $x - 1 = -12$

(B) $x - 1 = 12$

(C) $|x - 1| = |-6|$

(D) $|x - 1| = 12$

(E) $|x - 1| = -12$

GO ON TO THE NEXT PAGE

<u>Note</u>: Figure not drawn to scale.

11. In parallelogram $ABCD$ above, $\overline{AB} = 14\sqrt{3}$. Which of the following is equal to $\overline{CF}$?

(A) 8

(B) $7\sqrt{3}$

(C) $8\sqrt{3}$

(D) $12\sqrt{3}$

(E) 22

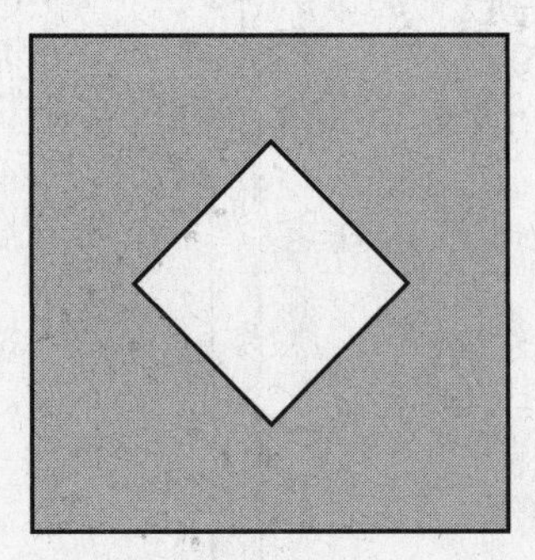

12. The figure above shows a dartboard in which a side of the outer square is twice as long as the length of a side of the inner square. What is the probability that a dart that lands on the board will land in the shaded region?

(A) $\frac{1}{4}$

(B) $\frac{1}{3}$

(C) $\frac{1}{2}$

(D) $\frac{3}{4}$

(E) 1

13. The heights of trees (h) in a national park are represented by the equation $|2h - 10| \leq 6$. Which interval represents the range of the heights of trees?

(A) $2 \leq h \leq 8$

(B) $2 \leq h \leq 16$

(C) $3 \leq h \leq 8$

(D) $4 \leq h \leq 8$

(E) $4 \leq h \leq 16$

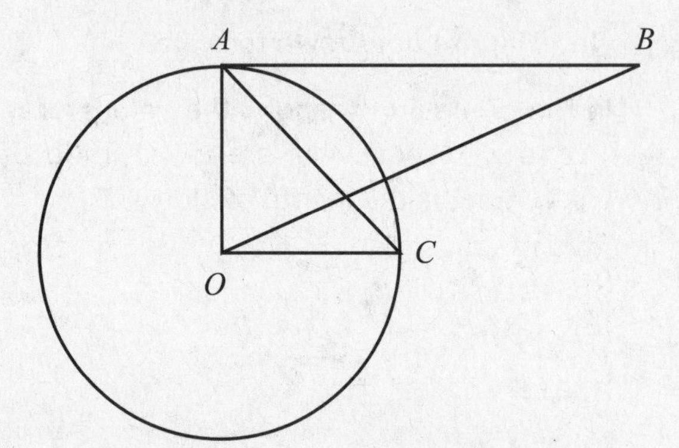

<u>Note</u>: Figure not drawn to scale.

14. In the figure above, $\overline{AB}$ is tangent to the circle O and is parallel to $\overline{OC}$. If $\overline{AC} = 8$, what is the length of a radius of the circle O?

(A) $2\sqrt{2}$

(B) $4\sqrt{2}$

(C) 8

(D) $6\sqrt{2}$

(E) $8\sqrt{2}$

15. The graphs of the equations $mx - 3y = 9$ and $3x - ny = 12$ are perpendicular. Which of the following is the value of $m + n$?

(A) 0

(B) –1

(C) –2

(D) 1

(E) 2

GO ON TO THE NEXT PAGE

16. If $(a - 1)$, m, and $(a + 1)$ form a geometric sequence, then which of the following expressions is equivalent to a^2?

(A) $m^2 - 1$

(B) $m - 1$

(C) $m^2 + 1$

(D) $m + 1$

(E) $ma + 1$

17. Which of the following conditions must be true if $\sqrt{a} + \sqrt{b} = \sqrt{a+b}$?

I. $ab = 0$

II. $\sqrt{a} - \sqrt{b} = 1$

III. $a - b = 0$

(A) I only

(B) II only

(C) III only

(D) I and II

(E) I and III

18. If $(x + \frac{3}{x}) = -3$, what is the value of $(x^2 + \frac{9}{x^2})$?

(A) −15

(B) 3

(C) 6

(D) 10

(E) 15

STOP

If time remains, you may go back and check your work. When the time allotted is up, you may go on to the next section.

TIME: 25 Minutes
20 Questions

CRITICAL READING

DIRECTIONS: Each sentence below has one or two blanks, each blank indicating that something has been omitted. Beneath the sentence are five lettered words or sets of words. Choose the word or set of words that BEST fits the meaning of the sentence as a whole.

EXAMPLE

Although critics found the book _______ , many readers found it rather _______ .

(A) obnoxious . . . perfect
(B) spectacular . . . interesting
(C) boring . . . intriguing
(D) comical . . . persuasive
(E) popular . . . rare

EXAMPLE ANSWER

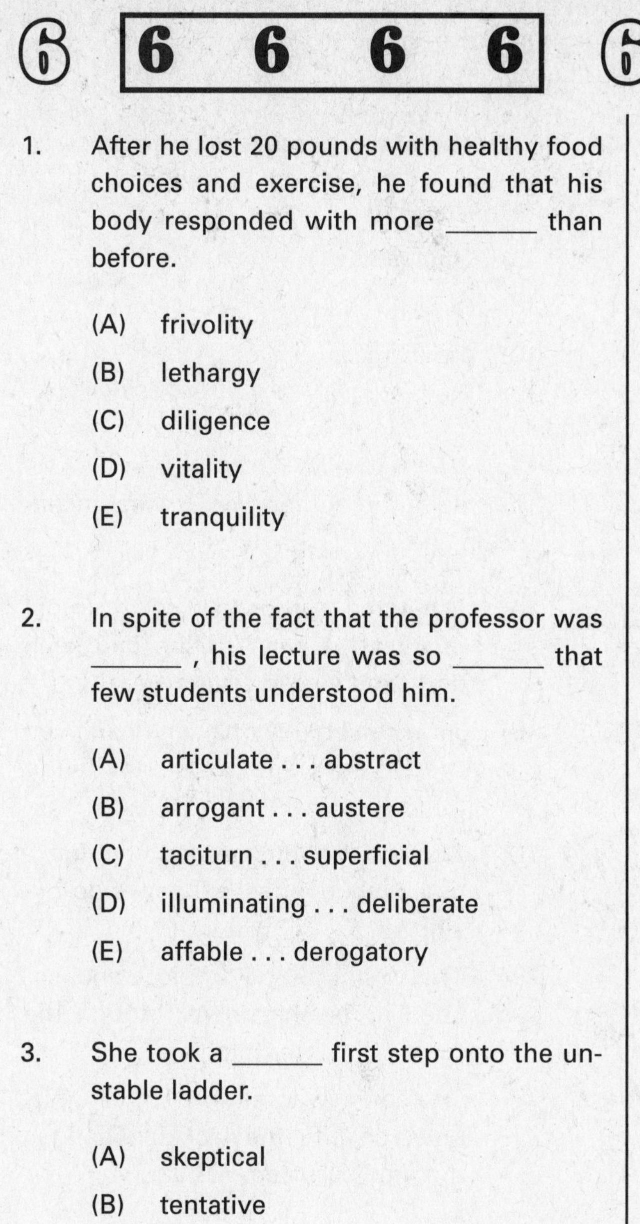

1. After he lost 20 pounds with healthy food choices and exercise, he found that his body responded with more ______ than before.

   (A) frivolity
   (B) lethargy
   (C) diligence
   (D) vitality
   (E) tranquility

2. In spite of the fact that the professor was ______ , his lecture was so ______ that few students understood him.

   (A) articulate . . . abstract
   (B) arrogant . . . austere
   (C) taciturn . . . superficial
   (D) illuminating . . . deliberate
   (E) affable . . . derogatory

3. She took a ______ first step onto the unstable ladder.

   (A) skeptical
   (B) tentative
   (C) limber
   (D) haggard
   (E) wanton

4. Facsimile or fax machines became common in offices because they were ______ and more ______ than express mail and telex.

   (A) faster . . . efficient
   (B) trivial . . . supported
   (C) effusive . . . salutary
   (D) valid . . . impeccable
   (E) sporadic . . . torpid

5. A notary public is sometimes hired by a lawyer to ______ the ______ of certain signatures.

   (A) constrain . . . watermarks
   (B) disdain . . . sanctity
   (C) verify . . . veracity
   (D) efface . . . consonance
   (E) abrade . . . matrix

GO ON TO THE NEXT PAGE

DIRECTIONS: Each of the following passages is followed by questions based on its content. Answer the questions on the basis of what is stated or implied in each passage and in any introductory material that may be provided.

1 Just driving down the quiet shady street was a step back in time. It was uncanny. Many years had passed since ten-year-old Joanie, Jimmy, and I had played tag in that cluster of trees just behind
5 the Cope family's house, but as I drove past the lot three or four squealing children were weaving in and out of the somewhat larger cottonwoods. Taking a second look at the elderly couple gently swinging on the rambling porch of the old Groves'
10 place, I had to remind myself that John and Gertrude had been gone for at least fifteen years. How they had loved to spend their summer evenings sitting side by side on that swing, exchanging greetings and chatting with everyone who walked
15 by! The whitewashed Methodist church seemed somewhat smaller now, but the lawn was just as well maintained as old Jethro used to keep it. And the outhouse was still standing behind the church building, under the huge limbs of the old oak tree.
20 I couldn't help but wonder if the current minister was just a young version of old Reverend Mesler, who refused to agree to the installation of modern facilities. He allegedly believed his church members should listen to every word he had to say,
25 even at the expense of their physical comfort, and anyone who wasn't willing to sit quietly through his two-hour sermons should have to endure the rain, snow, and heat that God was sure to inflict upon such sinners. My short trip through the
30 small business section we always called "town" revealed few empty stores. Kids still stood outside Brown's drugstore, licking the dripping cones that I knew they had bought from the soda fountain that was probably still at the back of the store. Old
35 men still loafed on the benches in front of Vargus's dime store, solving the world's problems, and young mothers still sat under the shade trees in the park, watching their toddlers play in the sand. Did this world really still exist, or was I dreaming?

6. The tone of this passage is primarily

(A) regretful.

(B) sarcastic.

(C) angry.

(D) nostalgic.

(E) reverent.

7. Why does the author say "It was uncanny" in line 2?

(A) What the author saw was almost identical to what he/she had seen and experienced several years ago.

(B) Almost all the people who had lived there years ago were still living there.

(C) Modern-day businesses and long-established ones existed side-by-side.

(D) The author felt out of place and isolated as he/she drove through the town of his/her childhood.

(E) The town was almost unrecognizable, despite the lack of changes that had occurred.

GO ON TO THE NEXT PAGE

1 Obesity in America is on the increase; in fact, statistics indicate that as many as two-thirds of all Americans are overweight, and almost a third of those people are obese. Al-
5 though few disagree that this trend is a health concern, it is also easy to see evidence of what has become known as "size acceptance." Movie theaters are beginning to install wider seats, and airlines are facing pressure to allow
10 heavier people to occupy two seats without having to pay an extra charge. Hospitals are switching to larger and sturdier beds, operating tables, and walkers. Almost every shopping mall has at least one specialty clothing store
15 that sells only large sizes, and almost every department store has a "plus" department. "Size-friendly" resorts and hotels are equipping their swimming pools with plaster steps rather than aluminum ladders and replacing plastic beach
20 chairs with wooden ones. Contending that such accommodations convey the message that being overweight is okay, some health-care professionals find this trend worrisome. However, those who are overweight—especially those
25 who have faced discrimination their entire lives—say these considerations are long overdue. After all, as they are quick to point out, they are just as human, and deserve the same pleasures, as everyone else.

8. The primary purpose of this passage is to

(A) criticize American society for its tendency toward "size acceptance."

(B) criticize American society for ridiculing people who are overweight.

(C) criticize Americans who are overweight or obese.

(D) exemplify America's efforts to encourage overweight people to lead a healthier lifestyle.

(E) discuss America's tendency toward "size acceptance."

9. It can be inferred from this passage that

(A) the number of overweight and obese people is expected to remain stable or increase.

(B) the number of overweight and obese people is expected to decrease.

(C) health-care professionals are going to increase their efforts to educate America about the importance of a healthy lifestyle.

(D) being overweight does not necessarily mean being unhealthy or unhappy.

(E) the definitions of "overweight," "obesity," and "size acceptance" have changed over time.

GO ON TO THE NEXT PAGE

Questions 10–27 are based on the following passage.

Communicable illness and disease were often rampant in the ancient world, incapacitating and killing hundreds on a daily basis. The following passage describes the effects of the Athenian plague on history.

1 The oldest recorded epidemic, often regarded as an outbreak of typhus, is the Athenian plague during the Peloponnesian Wars, which is described in the Second Book of the 5 *History of the Thucydides.*

In trying to make the diagnosis of epidemics from ancient descriptions, when the differentiation of simultaneously occurring diseases was impossible, it is important to remember 10 that in any great outbreak, although the large majority of cases may represent a single type of infection, there is usually a coincident increase of other forms of contagious disease; for the circumstances that favor the spread of 15 one infectious agent often create opportunities for the transmission of others. Very rarely is there a pure transmission of a single malady. It is not unlikely that the description of Thucydides is confused by the fact that a number of 20 diseases were epidemic in Athens at the time of the great plague. The conditions were ripe for it. Early in the summer of 430 B.C., large armies were camped in Attica. The country population swarmed into Athens, which be- 25 came very overcrowded. The disease seems to have started in Ethiopia, traveled through Egypt and Libya, and at length reached the seaport of Piraeus. It spread rapidly. Patients were seized suddenly out of a clear sky. The first 30 symptoms were severe headache and redness of the eyes. These were followed by inflammation of the tongue and pharynx, accompanied by sneezing, hoarseness, and cough. Soon after this, there was acute intestinal involve- 35 ment, with vomiting, diarrhea, and excessive thirst. Delirium was common. The patients who perished usually died between the seventh and ninth days. Many of those who survived the acute stage suffered from extreme weakness and continued diarrhea that yielded to no 40 treatment. At the height of the fever, the body became covered with reddish spots, some of which ulcerated. When one of the severe cases recovered, convalescence was often accompanied by necrosis of the fingers and toes. Some 45 lost their eyesight. In many there was a complete loss of memory. Those who recovered were immune, so that they could nurse the sick without further danger. None of those who, not thoroughly immunized, had it for the second 50 time died of it. Thucydides himself had the disease. After subsiding for a while, when the winter began, the disease reappeared and seriously diminished the strength of the Athenian state. 55

The plague of Athens, whatever it may have been, had a profound effect on historical events. It was one of the main reasons why the Athenian armies, and the advice of Pericles, did not attempt to expel the Lacedaemonians, who 60 were ravaging Attica. Athenian life was completely demoralized, and a spirit of extreme lawlessness resulted. There was no fear of the laws of God or man. Piety and impiety came to the same thing, and no one expected that he 65 would live to be called to account. Finally, the Peloponnesians left Attica in a hurry, not for fear of the Athenians, who were locked up in their cities, but because they were afraid of the disease. 70

10. The point of passage is to demonstrate
    (A) that "pure" outbreaks of disease were common in the ancient world.
    (B) that treatments of epidemic diseases remain relatively ineffective.
    (C) the relation that exists between infectious disease and historical events.
    (D) the relatively poor health conditions of ancient Athens.
    (E) the wisdom of Pericles in dealing with foreign invaders.

11. One of the results of extended sickness in Athens was
    (A) to put their democratic constitution in danger.
    (B) to produce a moral nihilism, where death negated ethical life.
    (C) to invite the Peloponnesians for a prolonged occupation of the city.
    (D) to increase general belief in supernatural powers.
    (E) to clarify the exact cause of the outbreak.

12. The symptom "necrosis" (line 45) probably indicates
    (A) tissue death.
    (B) that victims suffered more at night.
    (C) that patients could not be moved.
    (D) the incompetence of doctors at that time.
    (E) the incredible speed at which victims died.

13. Two of the factors exacerbating the spread of the plague were
    (A) low standards of medical practice and poor army discipline.
    (B) severe overpopulation and outbreaks of several maladies.
    (C) poor political judgment and failure to enforce housing codes.
    (D) an influx of Ethiopians and the departure of the Lacedaemonians.
    (E) the failure to identify the disease and poor nutrition.

14. Which of the following symptoms IS NOT indicative of the plague the author describes?
    (A) amnesia
    (B) ulcers of the skin
    (C) delirium
    (D) death in reinfected patients
    (E) death in one week to nine days

15. The word "convalescence" in line 44 can most nearly be defined as
    (A) languish.
    (B) innocuous.
    (C) contamination.
    (D) recuperation.
    (E) deterioration.

GO ON TO THE NEXT PAGE

Questions 16–24 are based on the following passage.

John Coltrane was one of the world's most influential jazz musicians. The following passage describes the music he made during the last three years of his life.

Coltrane's music of the sixties reflects a programmatic involvement with Africa. He formed his classic quartet with pianist McCoy Tyner, bassist Jimmy Garrison, and the one-man drum-choir, Elvin Jones, and he was playing his own music, free of bop trappings, though still linked to the modal compositions of Miles Davis. Titles of compositions were exact descriptions of African realities. "Liberia" was inspired by meeting Liberians who attended one of his New York club dates. "Dahomey Dance" reflected a serious attempt to translate a Dahomian field recording of musicians making their voices percuss like conga drums into American jazz sounds. "Africa" translated Pygmy vocals into horn chorus elaborations. "Tunji" was a tribute to the influential Nigerian musician Olatunji (who also collaborated with Randy Weston, Yusef Lateef, and Max Roach), and "Ogunde" celebrates another Nigerian musician who contributed to the preservation of the traditional music heritage of his homeland. "Kulu Se Mama" and "Afro Blue" interpreted the African rhythms and textures implicit in Caribbean music.

If Coltrane's African program music attempted to create narrative sequences with moods and events evoking specific African places and persons, his last musical works attempted to spiritualize the African theme, largely through large ensemble units in which each player had considerable freedom to solo. Coltrane's music prior to 1964 is well illuminated by two biographies by Cuthbert Ormond Simpkins and Bill Cole. The music of Coltrane's final three years has seemed to confuse, anger, or simply baffle commentators. I would offer the lens of Africa-as-spiritual-form as a tool for clarifying and understanding this period.

The one album of this period that has proven most accessible to critics and jazz buyers is *A Love Supreme*. It offers inspired Coltrane tenor sax improvisation against a background of a chant, the mantra "a love supreme." Its appearance in 1965 was in a context of a countercultural fascination with Eastern mysticism. George Harrison of the Beatles was learning sitar, and Timothy Leary was offering a reinterpretation of *The Tibetan Book of the Dead* to LSD devotees. Because Coltrane's interest in Indian musical modes and ragas was apparent as far back as 1961 when *India* was recorded at the Village Vanguard concerts, it was easy to hear *A Love Supreme* as a composition reflecting Hindu spirituality.

Coltrane's liner notes to the album suggest the evocation of African and African-American spirituality specifically through the sacred song or psalm. The biblical Book of Psalms could easily have set the tone for Coltrane's recording. Psalm 150 is as follows:

> Praise Him with the sound of the trumpet:
> Praise Him with the psaltery and harp.
>
> Praise Him with the timbrel and dance:
> Praise Him with stringed instruments and organs.
>
> Praise Him upon the loud cymbals:
> Praise Him Upon the high sounding cymbals.
>
> Let everything that hath breath praise the Lord.

This hymn of praise, integral to the Judeo-Christian tradition in the West, also has a counterpart in indigenous African religion. Showering praise upon divinity is paradoxical because the Divine, who has everything and knows everything, would seem to have little need of praise. Yet the implication throughout the biblical songs is that God hears the nature of human spiritual devotion, that faith is made tangible through musical embodiment. Interestingly, the four parts of Coltrane's *A Love Supreme* correspond to

80 this perspective: "Acknowledgement," "Resolution," "Pursuance," and "Psalm." First the acknowledgement of the need for a spiritual relationship with the Divine, then willing it, seeking actively for it, and upon attaining it, raising the
85 voice musically in celebration of the newfound spiritual bond. Coltrane's liner notes to the album described such a progression in his life, from the recognition of the need to find God, through a crisis where he emerged resolved and purpose-
90 ful, and finally into a stage of praise-song recognizing his newly won spiritual communion.

16. The primary emphasis of the entire passage is on the

(A) commitment of John Coltrane to innovation in jazz.

(B) synthesis of jazz and religious music that Coltrane achieved.

(C) voice of Coltrane as a spokesman for the "counterculture."

(D) unbroken musical debt to Africa and the East in Coltrane's music.

(E) popular reception of Coltrane, as opposed to his appreciation by "insiders."

17. We would expect the musical forces employed in *A Love Supreme* to be

(A) spare, in keeping with the chaste sentiments of the psalm.

(B) richly complex and vibrant, because all sounds glorify God.

(C) reserved and dignified, in keeping with Coltrane's personal manner.

(D) "Eastern" only, avoiding any resemblance to "Western" instrumentation.

(E) soft and restrained, in a manner befitting the solemn occasion.

18. The final three years of Coltrane's music-making were

(A) a time of quiet introspection and personal inner-seeking.

(B) a subject of controversy and debate among his listeners.

(C) perceived clearly as an expression of African nationality.

(D) marked by a clear identification of American blues influences.

(E) Dominated by production of the album *India*.

19. The author conceives Coltrane's liner notes (line 53–56) for his album *A Love Supreme* as

(A) an autobiographical narrative and spiritual odyssey.

(B) misleading Coltrane's listeners into thinking his problem has been solved.

(C) essentially unrelated to the real character of the music.

(D) a confession of the frustrated ambitions of the composer.

(E) a journey from Western to Eastern regions of religious thought.

GO ON TO THE NEXT PAGE

20. The social "context" (lines 41–46) of *A Love Supreme* embraces the

    (A) destructive effects of the 1960s drug culture.
    (B) anti-authoritarian environment of the Village Vanguard.
    (C) reaction against non-Western forms of music and instrumentation.
    (D) limited fascination musicians had with early instruments.
    (E) broad interest in Eastern music, philosophy, and culture.

21. The author finds the praise of divinity "paradoxical" (line 73)

    (A) because Coltrane's music is by nature secular and worldly.
    (B) even though the biblical songs try to minimize the element of faith.
    (C) because the Divine tends to ignore the petty efforts of men.
    (D) due to A Love Supreme's celebration of Coltrane's own ego.
    (E) because God ostensibly requires no human acknowledgement of His powers.

22. In lines 1–2, "programmatic" can most nearly be defined as

    (A) predictable.
    (B) thematic.
    (C) unhealthy.
    (D) spiritual.
    (E) uninspired.

23. The tone of the passage can best be described as

    (A) critical.
    (B) awed.
    (C) objective.
    (D) reverential.
    (E) praising.

24. What is the most likely explanation for most audience's negative reaction to Coltrane's later music?

    (A) His involvement with rock musicians angered jazz traditionalists.
    (B) His experimentation with the 1960s drug culture offended his longtime fans.
    (C) Coltrane abandoned the more traditional jazz structure.
    (D) His focus on spiritual themes alienated the secular element of the counterculture.
    (E) His music style became too traditionalized to appeal to more avant-garde audiences.

STOP

If time still remains, you may go back and check your work. When the time allotted is up, you may go on to the next section.

TIME: 20 Minutes
16 Questions

MATH

DIRECTIONS: In this section solve each problem, using any available space on the page for scratchwork. Then decide which is the best of the choices given and fill in the corresponding oval on the answer sheet.

NOTES

(1) The use of a calculator is permitted.

(2) All numbers used are real numbers.

(3) Figures that accompany problems in this test are intended to provide information useful in solving the problems. They are drawn as accurately as possible EXCEPT when it is stated in a specific problem that the figure is not drawn to scale. All figures lie in a plane unless otherwise indicated.

(4) Unless otherwise specified, the domain of any function f is assumed to be the set of all real numbers x for which $f(x)$ is a real number.

REFERENCE INFORMATION

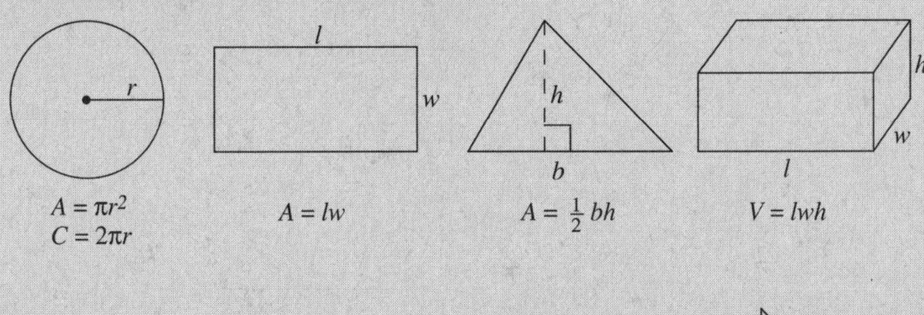

$A = \pi r^2$
$C = 2\pi r$

$A = lw$

$A = \frac{1}{2}bh$

$V = lwh$

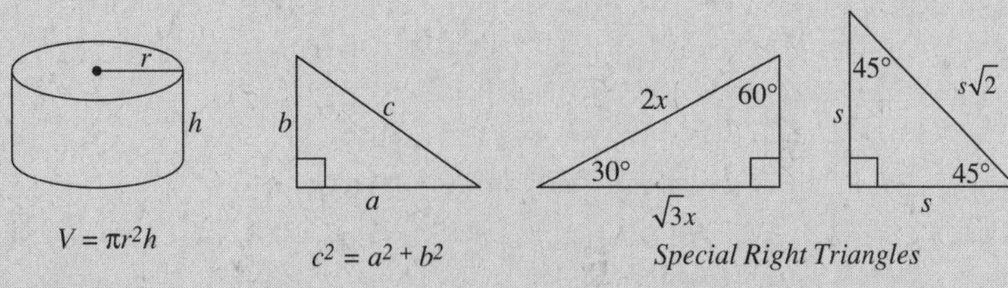

$V = \pi r^2 h$

$c^2 = a^2 + b^2$

Special Right Triangles

The number of degrees of arc in a circle is 360.

The sum of the measures in degrees of the angles of a triangle is 180.

1. In July, a pair of sunglasses cost k dollars. In August, the price of the sunglasses was reduced by 12%. The sunglasses were discounted another \$24 in September. In terms of k, what was the price of the sunglasses in September?

(A) $21.12k$

(B) $24 - 21.12k$

(C) $0.88k - 24$

(D) $24 - 0.88k$

(E) $24 - 1.12k$

2. Let the function ♠ be defined as by m ♠ $n = \frac{m-n}{m+n}$. If m and n are both real numbers, which of the following has the greatest value?

(A) -2 ♠ 3

(B) 2 ♠ -3

(C) -3 ♠ -2

(D) 3 ♠ -4

(E) -3 ♠ 4

3. Studies show that most, but not all, adults sleep x hours per night, with the variation in hours of sleep among adults defined by the equation $|x - 6.5| \leq 1$. What is the range of number of hours that adults sleep?

(A) $4.6 \leq x \leq 5.6$

(B) $5.5 \leq x \leq 7.5$

(C) $5.5 < x < 7.5$

(D) $6 \leq x \leq 8$

(E) $6.4 \leq x \leq 6.6$

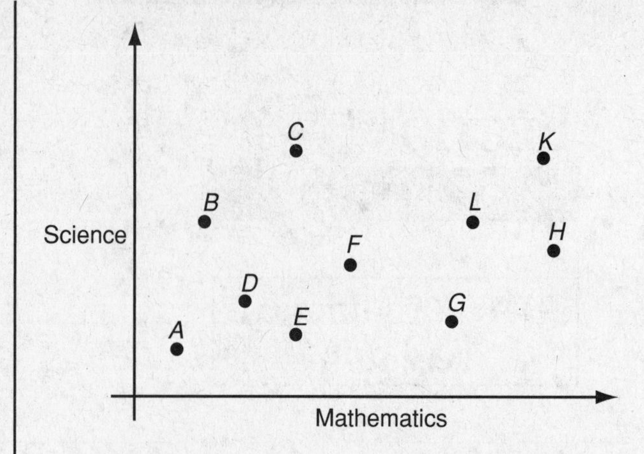

4. The scores of 10 students in mathematics versus their scores in science are shown in the scatterplot above. Which student scored better than half of all students in mathematics but worse than half of all students in science?

(A) Student E

(B) Student F

(C) Student G

(D) Student H

(E) Student L

5. If a fair coin is tossed 6 times, what is the probability of getting heads 1 time and tails 5 times?

(A) $\frac{1}{2}$

(B) $\frac{1}{6}$

(C) $\frac{1}{12}$

(D) $\frac{1}{30}$

(E) $\frac{1}{64}$

GO ON TO THE NEXT PAGE

DIRECTIONS: Student-Produced Response questions 6–16 require that you solve the problem and then enter your answer on the special grid on page 867. (For instructions on how to use the grid, see page 69 in the Diagnostic Test.)

$x, 4, xy, 16y$

6. In the sequence above, if both x and y are positive, what is the value of x?

7. If $x^2 - 1 = 15$ and $y^2 - 1 = 8$, what is the greatest possible value of $(x - y)^3$?

8. The average of five numbers is 12. If another number is added to the set of these numbers, the average of the six numbers increases to 15. What number is added to the original set?

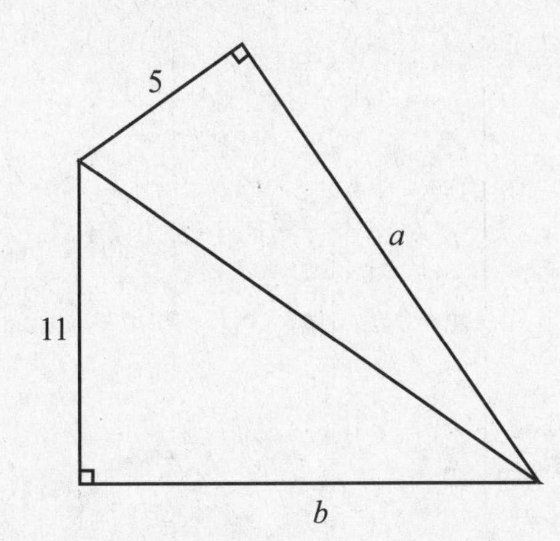

Note: Figure not drawn to scale.

9. In the above figure, both the triangles are right triangles. What is the value of $a^2 - b^2$?

10. Two trains leave a station at the same time, one traveling east and the other traveling west. Train 1 travels at a constant speed of 60 miles per hour, and Train 2 travels at a constant speed of 80 miles per hour. How long will it take them to be 560 miles apart?

11. How many cubes with 2-inch sides will fit in a rectangular solid with a volume of 216 cubic inches?

12. The equation $x^2 - 6x + m + 1 = 0$ has only one solution. What is the value of m?

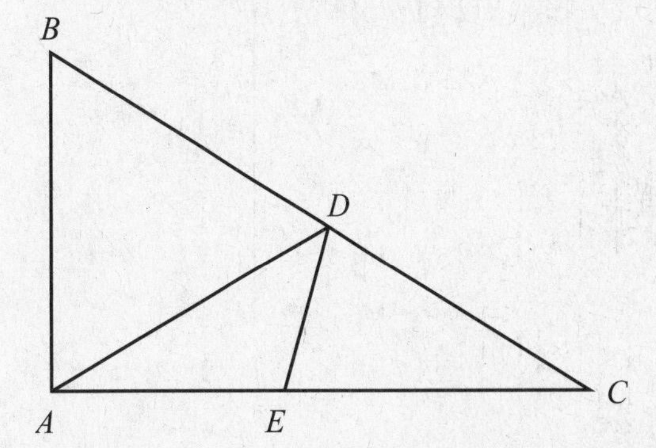

Note: Figure not drawn to scale.

13. In the figure above, $\angle BAC = 90°$, ΔABD is an equilateral triangle, and $\overline{CD} = \overline{CE}$. What is $m\angle AED$?

GO ON TO THE NEXT PAGE

14. If $x - y = \sqrt{11}$ and $xy\sqrt{3} = \sqrt{27}$, what is the value of $x^2 + y^2$?

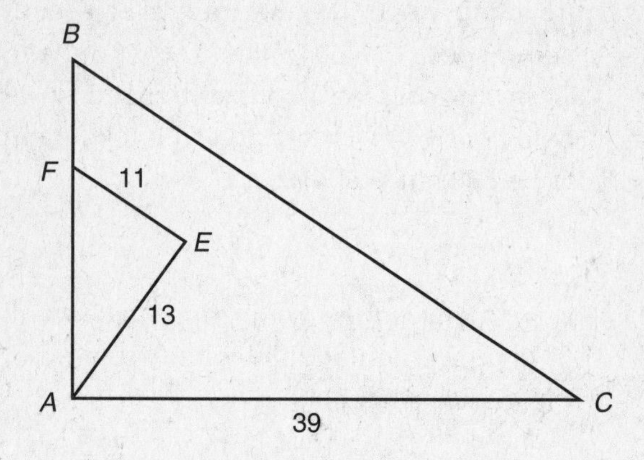

<u>Note</u>: Figure not drawn to scale.

15. In the figure above, $\angle BAC$ and $\angle FEA$ are right angles and $\overline{EF}$ is parallel to $\overline{BC}$. What is the length of $\overline{AB}$?

I. $3x + 2y = 7$

II. $x - ky = 11$

III. $mx + 4y = 14$

16. The graphs of equations I and II above are perpendicular and the graphs of equations II and III are parallel. What is the value of $-km$?

STOP

If time remains, you may go back and check your work. When the time allotted is up, you may go on to the next section.

TIME: 20 Minutes
19 Questions

CRITICAL READING

DIRECTIONS: Each of the following sentences has one or two blanks, each blank indicating that something has been omitted. Beneath the sentences are five lettered words or sets of words. Choose the word or set of words that BEST fits the meaning of the sentence as a whole.

EXAMPLE

Although critics found the book _______ , many readers found it rather _______ .

(A) obnoxious . . . perfect
(B) spectacular . . . interesting
(C) boring . . . intriguing
(D) comical . . . persuasive
(E) popular . . . rare

EXAMPLE ANSWER

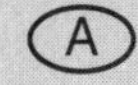    

1. The boy felt _______ when he realized his mother had committed the _______ act.

   (A) inept . . . benevolent
   (B) aghast . . . heinous
   (C) contrite . . . craven
   (D) circumspect . . . expedient
   (E) intangible . . . inadvertent

2. The tornado swept through the town and _______ everything in its path.

   (A) refurbished
   (B) consummated
   (C) depraved
   (D) thwarted
   (E) obliterated

3. The little boy's _______ wheedling often wears his parents down to the point that they finally _______ to his demands.

   (A) altruistic . . . defer
   (B) ambiguous . . . acclaim
   (C) incessant . . . acquiesce
   (D) vacuous . . . genuflect
   (E) errant . . . demur

4. As long as he could remain _______ he could enjoy doing all the things he enjoyed, without fear of being bothered by paparazzi or adoring fans.

   (A) incognito
   (B) incorrigible
   (C) inanimate
   (D) inimical
   (E) inexorable

5. Even those who don't necessarily like her admit that her ability to succeed despite all the obstacles she has faced is _______ .

   (A) infallible
   (B) inequitable
   (C) envious
   (D) enviable
   (E) inimical

6. Out of fear that the sick old man will die if he becomes too _______ , the entire family does all they can to _______ him.

   (A) disconcerted. . . . placate
   (B) taciturn . . . mesmerize
   (C) resilient . . . torment
   (D) innate . . . heed
   (E) ebullient . . . provoke

GO ON TO THE NEXT PAGE

Questions 7–19 are based on the following passages.

The following passages are excerpted from two presidential Inaugural Speeches. Passage 1 is from John F. Kennedy; Passage 2 was delivered by Franklin D. Roosevelt.

Passage 1

1 Let every nation know, whether it wishes us well or ill, that we shall pay any price, bear any burden, meet any hardship, support any friend, oppose any foe to assure the survival 5 and the success of liberty.

To much we pledge—and more.

To those old allies whose cultural and spiritual origins we share, we pledge the loyalty of faithful friends. United, there is little we cannot 10 do in a host of cooperative ventures. Divided, there is little we can do, for we dare not meet a powerful challenge at odds and split asunder.

To those new states whom we welcome to the ranks of the free, we pledge our word 15 that one form of colonial control shall not have passed away merely to be replaced by a far more iron tyranny. We shall not always expect to find them supporting our view, but we shall always hope to find them strongly supporting 20 their own freedom, and to remember that, in the past, those who foolishly sought power by riding the back of the tiger ended up inside.

To those people in the huts and villages of half the globe struggling to break the bonds of 25 mass misery, we pledge our best efforts to help them help themselves, for whatever period is required, not because the Communists may be doing it, not because we seek their votes, but because it is right. If a free society cannot help 30 the many who are poor, it cannot save the few who are rich.

Passage 2

This is pre-eminently the time to speak the 1 truth, the whole truth, frankly and boldly. Nor need we shrink from honestly facing conditions in our country today. This great nation will endure as it has endured, will revive, and will prosper. 5

So first of all let me assert myself, my firm belief that the only thing we have to fear is fear itself—nameless, unreasoning, unjustified terror, which paralyzes needed efforts to convert retreat into advance. 10

In every dark hour of our national life a leadership of frankness and vigor has met with that understanding and support of the people themselves which is essential to victory. I am convinced that you will again give that support 15 to leadership in these critical days.

In such a spirit on my part and yours we face our common difficulties. They concern, thank God, only material things. Values have shrunken to fantastic levels; taxes have risen; 20 our ability to pay has fallen; government of all kinds is faced by serious curtailment of income; the means of exchange are frozen in the currents of trade; the withered leaves of industrial enterprise lie on every side; farmers find no 25 markets for their produce; the savings of many years in thousands of families are gone.

More important, a host of unemployed citizens face the grim problem of existence, and an equally great number toil with no return. 30 Only a foolish optimist can deny the dark realities of the moment.

Yet our distress comes from no failure of substance. We are stricken by no plague of locusts. Compared with the perils which our fore- 35 fathers conquered because they believed and were not afraid, we have still much to be thankful for. Nature still offers her bounty, and human efforts have multiplied it. Plenty is at our

GO ON TO THE NEXT PAGE

40 doorstep, but a generous use of it languishes in the very sight of the supply.

Primarily, this is because the rulers of the exchange of mankind's goods have failed through their own stubbornness and their own incompe-
45 tence, have admitted their failure and abdicated. Practices of the unscrupulous money-changers stand indicted in the court of public opinion, rejected by the hearts and minds of men.

7. In line 5 of Passage 1, "liberty" most nearly means

(A) privilege.
(B) familiarity.
(C) emancipation.
(D) freedom.
(E) cooperation.

8. In line 18 of Passage 1, the use of "supporting" in two ways emphasizes the

(A) idea that liberty itself is more important than the form it takes.
(B) struggle between freedom and tyranny.
(C) power that countries can wield in alliance with one another.
(D) desire of the speaker to influence the politics of weaker countries.
(E) importance the speaker places on individual freedom.

9. In lines 20–21 of Passage 1, the phrase "in the past . . . ended up inside" is a reference to

(A) nations of the past who built great empires through cooperative efforts.
(B) the importance of remembering all those who have lost their lives defending liberty.
(C) the importance of remembering the struggles and hardships that the first free societies fought in order to preserve liberty.
(D) the importance of those who tried to gain by following despotic governments in the past eventually lost their freedom.
(E) the valor of nations that declared their independence against overwhelming odds.

10. The last sentence of Passage 1 (lines 29–31) suggests that the author believes that

(A) supporting the poor is the function of government in a free society.
(B) protecting liberty abroad is a necessary component of a free society.
(C) suppressing communism is the goal of a free society.
(D) democracy is synonymous with a free society.
(E) defending liberty is not a valid reason to go to war.

GO ON TO THE NEXT PAGE

11. In line 4 of Passage 2, the word "endure" most nearly means

    (A) exist.
    (B) linger.
    (C) withstand.
    (D) confirm.
    (E) flourish.

12. The author of Passage 2 uses the phrase "the only thing we have to fear is fear itself" (lines 7–8) to suggest that

    (A) the problems of which he speaks are illusory and inconsequential.
    (B) the dire problems of which he speaks can be overcome if a strong effort is put forth.
    (C) once the unscrupulous people in power are exposed than the economy will improve.
    (D) the nation will eventually prosper despite the grim present outlook.
    (E) speaking the truth will lead to greater prosperity.

13. In line 18 of Passage 2, the author most likely describes "our common difficulties" in order to

    (A) illustrate the near impossibility of conquering these problems.
    (B) back up his earlier statement about the importance of speaking the whole truth.
    (C) portray the powers of observation that a leader must possess.
    (D) show that farmers are the hardest hit by economic difficulties.
    (E) point out that these difficulties are merely transitory in nature.

14. The author of Passage 2 most likely mentions "the perils which our forefathers conquered" (lines 34–35) in order to

    (A) escape from present misery by drawing on the past.
    (B) put current hardships in a historical context to show that no problem is insurmountable.
    (C) illustrate that despite hardship, conditions could be much worse than they are.
    (D) draw strength from the fact that our forefathers were able to overcome adversity.
    (E) remind his audience that perils have existed in every age and will continue to exist.

15. The overall tone of Passage 2 can be said to be one of

    (A) hopeless fatalism.
    (B) desperate fear.
    (C) grim acceptance.
    (D) cautious optimism.
    (E) frank honesty.

16. The authors of both passages are concerned with

    (A) the ability of their nation to thrive despite any adversarial conditions.
    (B) the ability of their nation to overcome economic hardship.
    (C) what it means to be free.
    (D) the role of the president in a democratic society.
    (E) the role of their nation on a global scale.

17. The contrast between the intended audience for the two passages can be best described by which statement?

    (A) Passage 1 addresses the whole free world whereas Passage 2 addresses only the nation.

    (B) Both passages address only one nation.

    (C) Both passages address the entire free world.

    (D) It is impossible to generalize about the intended audiences from the information given.

    (E) Passage 1 addresses a much smaller audience that does Passage 2.

18. Lines 20–21 of Passage 1 echo what theme from Passage 2?

    (A) A strong nation has to have the capacity to help those who are struggling to survive.

    (B) A thriving nation must protect the rich.

    (C) A free society must completely support the poor.

    (D) A clearly defined class structure is necessary if a nation is to survive.

    (E) A nation must be prosperous if it is to influence world events.

19. Both passages mention which of the following as important to the success of a nation as they describe it?

    (A) unity among peoples or countries

    (B) powerful leadership

    (C) a strong foreign policy

    (D) a willingness to protect liberty at all costs

    (E) the ability to overcome material difficulties

STOP

If time still remains, you may go back and check your work. When the time allotted is up, you may go on to the next section.

TIME: 10 Minutes
14 Questions

WRITING

<u>DIRECTIONS</u>: In each of the following sentences, some portion of the sentence is underlined. Under each sentence are five choices. The first choice has the same wording as the original. The other four choices are reworded. Sometimes the first choice containing the original wording is the best; sometimes one of the other choices is the best. Choose the letter of the best choice. Your choice should produce a sentence that is not ambiguous or awkward and that is correct, clear, and precise.

This is a test of correct and effective English expression. Keep in mind the standards of English usage, punctuation, grammar, word choice, and construction.

EXAMPLE

When you listen to opera, <u>a person may not appreciate it.</u>

(A) a person may not appreciate it.

(B) it may not be appreciated by a person.

(C) you may not appreciate it.

(D) which may not be appreciated by you.

(E) appreciating it may be a problem for you.

EXAMPLE ANSWER

1. The presenter spoke so <u>loud that we couldn't understand it.</u>

(A) loud that we couldn't understand it.

(B) loud until we couldn't understand him.

(C) loudly until we couldn't understand his speech.

(D) loud so that we couldn't understand it.

(E) loudly that we couldn't understand his speech.

2. <u>I have three sisters; but he</u> has no siblings.

(A) I have three sisters; but he

(B) I have three sisters, but he

(C) I have three sisters; consequently, he

(D) Because I have three sisters, he

(E) I have three sisters he

3. <u>Thomas Jefferson wrote a draft of the Declaration of Independence he then presented the document to Congress for revision and so that they could approve it.</u>

(A) Thomas Jefferson wrote a draft of the Declaration of Independence he then presented the document to Congress for revision and so that they could approve it.

(B) Thomas Jefferson wrote a draft of the Declaration of Independence which then presented the document to Congress for revision and approval.

(C) Thomas Jefferson, having written a draft of the Declaration of Independence, he then presented the document to Congress for revision and so that they could approve it.

(D) Thomas Jefferson, having written a draft of the Declaration of Independence, he then presented the document to Congress for revision and revision.

(E) After writing a draft of the Declaration of Independence, Thomas Jefferson presented the document to Congress for revision and approval.

GO ON TO THE NEXT PAGE

4. Having grown up with two older brothers, her behavior was often quite aggressive.

(A) Having grown up with two older brothers, her behavior was often quite aggressive.

(B) Her behavior was quite aggressive, having grown up with two older brothers.

(C) As a result of having grown up with two older brothers, her behavior was often quite aggressive.

(D) She, having grown up with two older brothers, her behavior was quite aggressively.

(E) Having grown up with two older brothers, she often behaved quite aggressively.

5. Rappelling is descending a cliff by sliding down a rope passed under one thigh, across the body, and over the opposite shoulder.

(A) descending a cliff by sliding down a rope passed under one thigh, across the body, and over the opposite shoulder.

(B) when you descend a cliff by sliding down a rope passed under one thigh, across the body, and over the opposite shoulder.

(C) a cliff that one descends by sliding down a rope passed under one thigh, across the body, and over the opposite shoulder.

(D) where someone descends a cliff by sliding down a rope passed under one thigh, across the body, and over the opposite shoulder.

(E) a person descending a cliff by sliding down a rope passed under one thigh, across the body, and over the opposite shoulder.

6. Not so many years ago, students were required to memorize facts; however, they were quickly forgotten.

(A) memorize facts; however, they were quickly forgotten.

(B) memorize facts, but they were quickly forgotten.

(C) memorize facts but were quickly forgotten.

(D) memorize facts but quickly forgot them.

(E) although quickly forgotten.

7. Some parents are choosing to home-school their children, it is partly because they want to adapt the lessons to their children's unique abilities and interests.

(A) children, it is partly because they want

(B) children for the reason that they want

(C) children, partly because they want

(D) children, because of wanting

(E) children, they want

8. After she heard the lecture on clinical depression, she realizes she is suffering from depression.

(A) realizes she is suffering

(B) realizes she was suffering

(C) realized she was suffering

(D) realized she is suffering

(E) realizes she had been suffering

9. Abigail Smith Adams was the wife of the second U.S. President and the mother of the sixth U.S. <u>President, she has also been known</u> for her outspokenness on social and political issues.

   (A) President, she has also been known
   (B) President, being known
   (C) President, having been known
   (D) President; she was also known
   (E) President; she is also known

10. On the French holiday <u>*Fête de la Musique*, they come out</u> in the streets and the parks to play their musical instruments.

    (A) <u>*Fête*</u> *de la Musique*, they come out
    (B) <u>*Fête*</u> *de la Musique* is when they come out
    (C) <u>*Fête*</u> *de la Musique*, the people of Paris come out
    (D) <u>*Fête*</u> *de la Musique* is when the people of Paris come out
    (E) <u>*Fête*</u> *de la Musique*, when the people of Paris come out

11. Americans eat beef, poultry, and pork; however, <u>there are not many ants eaten by the people of this country.</u>

    (A) there are not many ants eaten by the people of this country.
    (B) not many of them eat ants.
    (C) not many people eat ants of this country.
    (D) there are not many people eating ants of this country.
    (E) not much eating of ants is done by them.

12. <u>Much like Chicago, the citizens of New York</u> are accustomed to harsh winters, congested traffic, and crowded subways.

    (A) Much like Chicago, the citizens of New York
    (B) Much like Chicago, and the citizens of New York
    (C) The citizens of New York, much like Chicago,
    (D) Much like the citizens of Chicago, New Yorkers
    (E) Much like the citizens of Chicago, New York

13. Sociologists study such variety of subjects as culture, religion, politics, and government, and economics is among one of the topics they study also.

(A) such variety of subjects as culture, religion, politics, and government, and economics is among one of the topics they study also.

(B) varied subjects such as culture, religion, politics, government, and economics.

(C) such a variety of subjects as culture, religion, politics, and government, as well as economics.

(D) culture, religion, politics, and government, and such a variety of subjects as economics, too.

(E) varied subjects such as culture, religion, politics, government, and they also study economics.

14. Only one of the legislators plans to vote against the reform bill that has been proposed.

(A) plans to vote against the reform bill that has

(B) plans to vote against the reform bill that had

(C) plan to vote against the reform bill that has

(D) plan to vote against the reform bill having

(E) he plans to vote against the reform bill that has

STOP

If time still remains, you may go back and check your work.

TEST 4

ANSWER KEY

Section 1—Essay

Refer to the Detailed Explanation for essay analysis

Section 2—Math

1.	(B)	6.	(D)	11.	(E)	16.	(D)
2.	(D)	7.	(B)	12.	(E)	17.	(E)
3.	(E)	8.	(E)	13.	(D)	18.	(C)
4.	(A)	9.	(E)	14.	(D)	19.	(E)
5.	(C)	10.	(C)	15.	(B)	20.	(E)

Section 3—Critical Reading

1.	(E)	7.	(A)	13.	(B)	19.	(B)
2.	(E)	8.	(B)	14.	(A)	20.	(C)
3.	(E)	9.	(D)	15.	(B)	21.	(C)
4.	(C)	10.	(C)	16.	(B)	22.	(E)
5.	(E)	11.	(B)	17.	(B)	23.	(C)
6.	(B)	12.	(C)	18.	(B)	24.	(B)

Section 4—Writing

1.	(E)	10.	(C)	19.	(D)	28.	(D)
2.	(B)	11.	(E)	20.	(B)	29.	(C)
3.	(C)	12.	(B)	21.	(B)	30.	(C)
4.	(D)	13.	(D)	22.	(B)	31.	(C)
5.	(E)	14.	(A)	23.	(A)	32.	(B)
6.	(A)	15.	(C)	24.	(B)	33.	(D)
7.	(E)	16.	(E)	25.	(B)	34.	(E)
8.	(B)	17.	(C)	26.	(D)	35.	(D)
9.	(C)	18.	(D)	27.	(B)		

Section 5—Math

1. (B)
2. (E)
3. (A)
4. (D)
5. (B)
6. (D)
7. (E)
8. (D)
9. (B)
10. (E)
11. (C)
12. (D)
13. (A)
14. (B)
15. (A)
16. (C)
17. (A)
18. (B)

Section 6—Critical Reading

1. (D)
2. (A)
3. (B)
4. (A)
5. (E)
6. (D)
7. (A)
8. (E)
9. (A)
10. (E)
11. (B)
12. (A)
13. (B)
14. (D)
15. (D)
16. (D)
17. (B)
18. (B)
19. (A)
20. (E)
21. (E)
22. (B)
23. (E)
24. (E)

Section 7—Math

1. (C)
2. (C)
3. (B)
4. (D)
5. (E)
6. 8
7. 343
8. 30
9. 96
10. 4
11. 27
12. 8
13. 105
14. 17
15. 33
16. 4

Section 8—Critical Reading

1. (B)
2. (E)
3. (C)
4. (A)
5. (D)
6. (A)
7. (D)
8. (A)
9. (D)
10. (B)
11. (C)
12. (B)
13. (B)
14. (C)
15. (D)
16. (A)
17. (A)
18. (A)
19. (A)

Section 9—Writing

1. (E)
2. (B)
3. (E)
4. (E)
5. (A)
6. (D)
7. (C)
8. (C)
9. (D)
10. (C)
11. (B)
12. (D)
13. (B)
14. (A)

DETAILED EXPLANATIONS

SECTION 1—ESSAY

Sample Essays with Commentary

Essay 1 (Score 5–6)

Although research on the Internet can be very easy and give instant access to many sources, it can also be frustrating, difficult, and dangerous. A search may yield far too many sources, some of which may lead the researcher down some really strange paths, and there is no way to discriminate easily between good and bad.

While a search engine brings up responses to a search, it does not discriminate among the items it finds. A very large topic, say Scottish history, may bring up hundreds of thousands of hits. It can be very difficult to go through the items and decide which are going to be helpful and which are not. When calling up an item that appears to be just right, the website may be found to be "out of service," and there is no way to learn how to get the people behind that website to find out where they are and if they might send information, or if they have published material that can be found in a library.

An individual can put anything he or she likes on the Internet, all it takes is the ability to post and run a website. In contrast, the process of getting a book published by a reputable publisher takes the author through a process that weeds out individuals with half-baked ideas and people who believe they know a lot about a subject even though they haven't done much work in that field.

The Internet is filled with postings by people who have an agenda when it comes to a certain subject, and the unsophisticated researcher might be persuaded by information that is misleading or downright wrong. To take an extreme example, there are individuals who maintain that the Holocaust never happened, and they will give all manner of information that they claim supports their position.

In conclusion, although the Internet is an easy first step in locating material for a research project, the researcher needs to exercise great care in going through what is often an overabundance of information to be able to use those sources that will help.

ANALYSIS

Essay 1 has a score range of 5–6. It is a well-reasoned essay that takes a stand and supports it. The introductory paragraph sets up the thesis. The four body paragraphs supply information that supports the thesis, and the concluding paragraph restates it. The writing demonstrates a variety of sentence structure, some fairly sophisticated vocabulary, and—importantly—it is grammatically correct.

Essay II (Score 4–5)

The Internet has made the process of research a whole lot easier than it has ever been before. Now, if you have a computer and a modem, all you need to do is dial into the web to get stacks of information just begging to be used. Do you want to know about particle physics? Just go to Yahoo.com and type "particle physics" and thousand of websites will come up. All you have to do is click and you're on your way.

A person who lives in a town with a small public library doesn't need to feel inferior to the person in New York City; each goes onto the same Internet and instantly has the same information available. You can download incredible graphics to help illustrate a paper you're writing. For instance, if you are doing a report on exotic birds, you can find pictures of birds from all over the world. If you have a color printer, all you need to do is click the mouse, and you'll have beautiful photos to add interest to your final paper.

The Internet is the wave of the future. Hop on and ride it!

ANALYSIS

Essay II has a score range of 4–5. It contains a thesis and supporting material in its three paragraphs. The conclusion, however, is too brief and the style is too informal. Note the use of "you" and the breezy command that makes up the final sentence. The essay reads more like an advertisement than an essay. The grammar and sentence structure are good, but the informal tone makes the essay seem unsophisticated.

Essay III (Score 3–4)

Back in the bad old days people had to go to the library to read a bunch of stuff out of a bunch of old books. Now you can just download all the information you need right at home if you have a computer and are connected to the net. It's so easy to download anything you need, and you can just cut and paste and put together a paper in no time! Say you want to write a report on folk music. You can find all kinds of stuff and even listen to different songs while your working and deciding what to say. You can read all kinds of information and the use them to put together a good report. You can read stuff from a lot of different places and some of the things you read can really make you think. I think the Internet has made life lots easier for lots of people, including me.

ANALYSIS

Essay III has a score range of 3–4. It doesn't begin by stating the writer's position but dives right in assuming the reader will intuit the thesis. Also, it never mentions the Internet by name. Almost all of the sentences are long, rambling, and similar. There is a conclusion, but it does not repeat the arguments. The tone is too informal—note the use of the pronoun "you" and the word "stuff." There are a few grammatical errors, such as "I think the Internet has made life lots easier for lots of people, including me," and the use of "your" instead of "you're."

Essay IV (Score 1–2)

The Internet has made doing research much easier because now a person can go to the Web and search for almost any subject under sun and come up with a lot of information about that subject. It's a lot easier to do research now than it was back in the days before there was an Internet, and people from all over the world can do research with nothing but a computer and a way to get online.

On the other hand, the Internet makes doing research dangerous because you don't necessarily have a way to trust the information you get from the sources you click on. You might be getting information from someone who has an axe to grind about some subject, and you can be reading material that isn't really true, or is very slanted, and there is no filter on the Internet to help you know what is carefully done and what is propaganda.

All in all, the Internet can make doing research very easy, but it can also be very dangerous. The user has to be careful about what sources he or she uses.

ANALYSIS

Essay IV has a score range of 1–2. The writer did not follow the directions given for this essay. Rather than taking one position and supporting it, both are spoken for. However, neither position is supportd with examples; the essay states both positions but doesn't provide details for either. In the first pragraph particularly, the writer doesn't add any suporting detail but rather just repeats the thesis.

DETAILED EXPLANATIONS

SECTION 2—MATH

1. (B) Using the definition of the given function

$$\boxed{11} = 1 + 3 + 5 + 7 + 9 = 25$$

$$\boxed{7} = 1 + 3 + 5 = 9$$

$$\boxed{5} = 1 + 3 = 4$$

Replace the preceding values in the given equation and solve it for k: $25 - 9 = 4k$, thus $k = 4$.

2. (D) An interior angle from the radius of a circle intercepts an equal number of degrees of arc of the circle's circumference. The nonshaded arc is $40° + 80° = 120°$, so the shaded arc is $360° - 120° = 240°$. The ratio of the shaded arc to the total arc is $\frac{240°}{360°} = \frac{2}{3}$.

3. (E) Let x = population of Kingsport, y = population of Jonesborough, and z = population of Bristol. Then, $n = x + y + z$. We are also given $x = 3y$ and $z = 4y$. Substitute the equivalents of x and y into the first equation, and then solve for y.

$$n = 3y + y + 4y$$

$$n = 8y$$

$$y = \frac{n}{8}$$

4. (A) The first equation is a difference of squares, so rewrite it as $x = (a - 4)(a + 4)$. Divide the sides of this equation by the corresponding sides of the second equation.

$$\frac{x}{y} = \frac{(a-4)(a+4)}{(a-4)}$$

Simplify the right side of the preceding equation.

$$\frac{x}{y} = a + 4$$

Then $\frac{x}{y} - 4 = a$.

5. (C) First, find the measure of the *exterior angles* of the polygon. Exterior angles are supplementary to interior angles, so $180° - 140° = 40°$. The sum of all exterior angles is 360°, so divide 360° by the measure of the exterior angle to find the number of angles: $\frac{360°}{40°} = 9$.

6. (D) The maximum area for a rectangle of a given perimeter will always be in the shape of a square. First, though, convert yards to feet because the answers are given in feet, $20 \cdot 3 = 60$ feet. Because the garage will form one side of the square, Rudy will divide the fencing among 3 equal sides for the pen: $\frac{60}{3} = 20$ feet per side. The area is given by $20^2 = 400$ square feet.

7. (B) We can discover the following pattern between the numbers of audience in each two consecutive days.

Day 2 = 1,500 – 1,460 = 40

Day 3 = 1,460 – 1,380 = 80

Day 4 = 1,380 – 1,260 = 120

Day 5 = 1,260 – 1,100 = 160

Noting that the audience decreases by 40 more people each day, continue the sequence:

Day 6 = 1,100 –200 = 900

Day 7 = 900 – 240 = 660

Day 8 = 660 – 280 = 380

On the eighth day, the number of audience is less than 400, so the play will close. That day is a Monday.

8. (E) Let m be the variation constant in the relationship between y and the square of x and n be the variation constant in the relationship between x and the cube of z. Then, $y = mx^2$ and $x = nz^3$. Replace the equivalent of x from the second equation in the first equation: $y = m(nz^3)^2 = mn^2z^6$. In this equation, mn^2 is a constant number, which can be denoted by a. Then, $y = az^6$.

9. (E) Check each option.

(A) The set $(A \cup B)$ includes all numbers that are divisible by 3, 4, or 5. The union of this set with the set C, $(A \cup B) \cup C$, is the set of all integers that are divisible by 3, 4, or 5.

(B) The intersection of $(A \cup B)$ with the set C is the set of all integers that are divisible by 3 or by 5.

(C) The set $(A \cup C)$ includes all integers that are divisible by 3, 4, or 5. The intersection of this set with the set B is a set of integers that are divisible by 4 or by 5.

(D) The elements of the set $(A \cap B)$ are divisible by 4 only. The union of this set with the set C is the set of all elements that are divisible by 3, 4, or 5.

(E) The set $(A \cap B)$ and the set C, which is divisible either by 3 or 5 have no common elements. Therefore, $(A \cap B) \cap C$ is an empty set.

10. (C) Let x be the total price of pens and y be the total price of pencils. The ranges of x and y are $0.82m \leq x \leq 3.45m$ and $0.25n \leq y \leq 1.23n$. Add the corresponding sides of these inequalities.

$$0.82m + 0.25n \leq x + y \leq 3.45m + 1.23n$$

So, the least amount of money needed for buying m pens and n pencils is $0.82m + 0.25n$ dollars.

11. (E) KE increases as both m and v^2 increase, so KE varies directly as mv^2. In the original expression, $\frac{1}{2}mv^2$, the fraction $\frac{1}{2}$ is a constant.

12. (E) Parallel lines have the same slope, so find the slope of the line that passes through the given points. $\frac{[3-(-3)]}{[1-(-2)]} = \frac{6}{3} = 3$. The only answer choice with a slope of 2 is (E).

13. (D) The sum of the measures of the angles in the figure is 180°.

$$\frac{x}{3}+\frac{y}{3}+\frac{x}{4}+\frac{y}{4}=180°$$

Simplify the preceding equation as follows.

$$\frac{4x+4y+3x+3y}{12}=180°$$

$$\frac{7(x+y)}{12}=180°$$

$$x+y=\frac{2{,}160}{7}$$

14. (D) Find $f(y)$ and $g(x)$ using the given functions, $f(y) = 3y^2 - 4y$ and $g(x) = 3x^2 + 4x$. Then, $f(y) - g(x) = 3y^2 - 4y - (3x^2 + 4x) = 3y^2 - 4y - 3x^2 - 4x$

15. (B) Because ΔBCE is an isosceles triangle, $m\angle CEB = m\angle CBE = 25°$.

Also, we have

$$m\angle CEB=\frac{BC}{2} \text{ and } m\angle CBE=\frac{CE}{2}$$

By substitution, we get $BC = 50°$ and $CE = 50°$. Then,

$$m\angle BDE = 360° - m(BC) - m(CE) = 360° - 50° - 50° = 260°$$

Now

$$\angle ABE=\frac{\angle BDE}{2}=\frac{260}{2}=130°.$$

16. (D) Because ΔABC is an isosceles right triangle, then $AC = AB = 2m$. Using the Pythagorean Theorem, $(BC)^2 = (2m)^2 + (2m)^2$; therefore, $(BC)^2 = 8m^2$, and then $BC=2m\sqrt{2}$. In an isosceles triangle, the altitude dropped to the base is a median as well. So $CH = BH = \frac{2m\sqrt{2}}{2} = m\sqrt{2}$. In the isosceles right triangle AHC, $m\angle ACH = 45°$, so $m\angle CAH = 45°$. This implies that ΔACH is an isosceles triangle. That is, $AH=CH=m\sqrt{2}$.

17. (E) Using the Pythagorean Theorem, find DE:

$$13^2 = (DE)^2 + 12^2,\ (DE)^2 = 169 - 144 = 25$$

$$DE = 5$$

The area of the semicircle is one-half the area of a circle with diameter $DE = 5$. The length of a radius of the semicircle is $\frac{5}{2}$, so the area of the semicircle is

$$\frac{\left(\frac{5}{2}\right)^2\pi}{2}=\frac{25\pi}{8}$$

18. (C) Because the first fraction is defined and it is positive, $x + 3 > 0$ or $x > -3$. The second fraction is undefined. This means that $y^3 - 27 = 0$ or $y = 3$. Adding $x > -3$ and $y = 3$ side by side, we get $x + y > -3 + 3$ or $x + y > 0$.

19. (E) Cross-multiply the given equation.

$$\left(\sqrt{x}-1\right)\left(\sqrt{x}+1\right)=\left(\sqrt{5}-1\right)\left(\sqrt{5}+1\right)$$

$$\left(\sqrt{x}\right)^2-1=\left(\sqrt{5}\right)^2-1$$

$$\sqrt{x^2}-1=5-1=4$$

20. (E) Using the rules of exponents, we can rewrite both equations as follows.

$$2^{p+2q}=(2^4)^q \text{ and } 3^{2p+q}=(3^2)^q$$

Using the rule of sequential exponents, we have

$$2^{p+2q}=2^{4q} \text{ and } 3^{2p+q}=3^{2q}$$

In both equations, the bases are equal, so their exponents are equal as well.

$$p+2q=4q$$

$$2p+q=2q$$

Solving the preceding system of equations, we get $p = 0$ and $q = 0$. So, $p - q = 0$.

DETAILED EXPLANATIONS

SECTION 3—CRITICAL READING

1. (E) The correct choice is "eloquent . . . mesmerizing" because we are looking for two complimentary adjectives that, when inserted into the sentence, support the idea that Churchill was the kind of speaker who made people all over the world stop their tasks and listen to what he had to say. Choice (A) is incorrect because a "gullible" speaker is one who is easily fooled, and this does not fit into the meaning of the sentence. Choice (C) is incorrect because an "incoherent" speaker is "illogical." This is not a complimentary word to use, even if the speaker was "pious." Choice (B) is incorrect because an "impartial" speaker is "unbiased," and although it may be correctly substituted into the sentence, it is unlikely that people would stop and listen to an "obscure" or "unknown" speaker. Choice (D) is also a possibility. "Provocative" means "tempting," and although this is an attractive adjective for a speaker, "prosaic" is not. "Prosaic" means "ordinary" and that is not an appealing word for a speaker, certainly not one who would make people so eagerly attentive.

2. (E) They key word here is "although," which tells us to interpret the sentence as follows: even though the topic was embarrassing to the speaker, he spoke in a way that didn't show his embarrassment. The best choice for this sentence is choice (E) "candidly" (truthful; sincere). Choice (A) is wrong because "vainly" (futile, unsuccessful) implies that he was embarrassed about his situation when he spoke. Choice (B) "rapidly" is a possibility, but doesn't convey the essence of the meaning of the sentence as well as choice (E). Choice (C) "cryptically" (meant to be puzzling) is not meaningful in this sentence; it is semantically wrong. You would expect someone to speak "bitterly" (disagreeably harsh), choice (D), about financial reversals. This choice would ignore the word "although" and is incorrect.

3. (E) To choose a correct answer for this question, we would have to find a noun choice that "has to be proven." An "enigma" (mystery, secret, perplexity) is something that is unknown and has to be proven, just like the existence of flying saucers mentioned in the first sentence. Choice (A) is incorrect because an "obituary" (notice of a person's death) makes no sense at all in the sentence. Choice (B) "travesty" (parody, burlesque) has nothing to do with "theory . . . proven," and makes no sense. Choice (C) "irony" (something that is contradictory, inconsistent, sarcastic) does not have to be proven and does not fit logically into the sentence. Choice (D) "kaleidoscope" (an instrument or toy that forms patterns through the use of mirrors) is incorrect because it is an object that can't "be proven," and it is not related to a "theory." It is a thing.

4. (C) The best answer is "remorse" (regret for one's actions) . . . "disparaging" (degrading, belittling). We have to find a verb and an adjective that would fit the context of the sentence. What was Eric's feeling about his remark and what kind of remark was it that would cause him to want to "retract" (take back) what he said? That Eric felt "remorse" for a "disparaging" remark is the best choice for this sentence. Choice (A), "respite" (temporary delay) and "inadvertent" (heedless, unintentional), is an incorrect choice; although "inadvertent" is a correct adjective, "respite" has no semantic integrity in this sentence. Choice (B), "irrational" (incapable of reasoning) and "incautious" (not careful), is incorrect. Although "incautious" would be a good choice, "irrational" makes no sense in the sentence. Choice (D), "tentative" (hesitant, uncertain) and "extravagant" (excessive), is incorrect. Both words are not logical for inclusion in the sentence. Choice (E), "insensitive" (not caring) and "premeditated" (carefully planned), is wrong. If Eric's remark was "premeditated," he would not want to retract it, even if it was "insensitive."

5. (E) The correct answer would have the residents of the town agreeing on something and reacting to a situation that would cause air pollution. In this case, because air pollution is something negative, the verb would be negative also. Choice (E) "mobilize . . . resist" is the only answer that satisfies both conditions. Choice (A) is not correct because although the first word satisfies the conditions, the residents of the town would not "defend" air pollution. Choice (B) is not correct because the residents would not "disperse" (break up) on this issue. Choice (C) is wrong because although "assembled" satisfies the condition for the residents getting together, they wouldn't "endorse" air pollution in their town. Choice (D) is incorrect because the residents of the town wouldn't "alternate" (take turns) getting together or "establish" a construction that causes air pollution.

6. (B) The correct answer is (B) "authentically" (realistically) because we are searching for an adjective to describe how Zane Gray could "fictionalize" cowboys "even though" he lived in a different place. If Zane Gray lived in a different place (New York), we would expect him to know very little about cowboys; but "even though" tells us that the opposite is true. "Authentically" (in a real or genuine way) is the opposite of not knowing anything about a subject. Choice (A) "benevolently" is incorrect; although Gray could describe cowboys in a "kind" way, that does not fit the context of the sentence, where we are looking for a word that signifies the opposite of "knowing little about a subject." Choice (C) "prosaically" (in an ordinary way) is incorrect because it does not fit into the context of the sentence. Choice (D) "frivolously" put into the sentence would mean that Gray made fun of the cowboys, and that is not implied in the sentence. Choice (E) "incoherently" is wrong because it would mean that Gray did not know what he was writing about, and the sentence implies that the opposite was true.

7. (A) "Nullified" (cancelled) is correct because the sentence implies that whatever happened to the scientist's data caused him to be "horrified." If all the scientist's data were "nullified" by the new data, then the results of his experiment would be incomplete. This is a good reason to be "horrified" (appalled, dismayed). Choice (B) is a possibility because "suppressed" means "to be held back." This is bad for a scientist but is not as terrible as being cancelled, and because the scientist is described as being "horrified," this is not the best answer. Choice (C) "instigated" (provoked) is wrong because if the new data "validated" (made acceptable) the old data, then the scientist would be thrilled, not horrified. Choice (E) is wrong because "consecrated" (make sacred) is not a word associated with a scientific experiment.

8. (B) "Scrutinizes . . . comparable" is the correct answer because we are looking for a verb to tell how the Patent Office looks at an application and an adjective to describe the type of invention for which an inventor wouldn't receive a patent. We know we are looking for this kind of invention because the inventor would receive the patent if a "comparable" (similar) type of invention "doesn't exist." All of the verbs except choice (C) could adequately fit into the sentence; therefore, it is the second word that makes the difference in the other choices (A), (D), and (E). The Patent Office would not issue a patent for one that was "comparable" to one that already exists. Choice (A) is wrong because if an inventor creates a "distorted" (to twist out of normal shape) invention, he does not receive a patent; but this does not make any sense, because no one would create a distorted invention. Choice (C) "expedites" (to make easier) is incorrect because the Patent Office is created to look carefully at an application to "verify" (to confirm) that a similar one doesn't exist. Choice (D) is wrong because "probable" (possible) doesn't fit into the context of the sentence. Choice (E) "contrasting" (the opposite of) is incorrect because it doesn't matter if a contrasting invention exists.

9. (D) The last sentence says, "Technological progress . . . serves only to stifle our own access to privacy." And the entire passage focuses on the fact that because of modern-day technology we are accessible "to anyone, anytime, anywhere." Choice (A) is incorrect because phones and the Internet, by their own nature, encourage communication rather than limiting it. And although the passage does mention that "being away from our mailboxes no longer means having no access to mail," it does not indicate that technology limits our use of the post office. So choice (B) is wrong. Choice (C) is incorrect because the passage includes cell phones as one of the technological devices that make us readily accessible, so phone use certainly isn't limited by technological progress. And no mention is made of limitations on the freedom of speech; therefore, choice (E) is incorrect.

10. (C) Even if you are unfamiliar with this phrase, looking at the context will help you understand the meaning. It may help to make up your own meaning, based on what the rest of the sentence and passage say, and then find the choice that most closely fits your definition. You could say, "Technological progress enhances others' accessibility to us, but—in a world where leisure time is already scarce—it serves only to stifle our own access to privacy." So choice (C) is appropriate. Although in a different context the phrase might mean "expensive," it does not fit here; thus, choice (A) is wrong. If you selected choice (B), you might have been thinking of "premium" in terms of insurance, but that meaning does not fit this context. Choices (D) and (E) are incorrect because their meanings contradict the message of the passage.

11. (B) Lines 2 and 3 say, "the concept of credit is over 3,000 years old." And then the passage explains the use of credit by people who lived in those days, in the 1700s, and on through the centuries until the present time. The passage outlines the historical progress of the use of credit through choice (A) the past century, choice (C) the invention of plastic, choice (D) the governmental regulation of the plastic strip, and choice (E) the issuance of credit cards by Diner's Club and American Express; however, none of these marked the beginning of the use of credit.

12. (C) If you are not familiar with the meaning of the word "promissory" (containing or conveying a promise or assurance) or the phrase "promissory note" (written promise to pay in an agreed-upon manner at a fixed or determinable future time to a specified individual or to bearer), you should look at the use of the word in context. The passage compares shopper's plates to promissory notes, allowing consumers a month to pay for their purchases. In other words, by accepting a shopper's plate (or promissory note), the consumer was promising to pay for the item(s) within a month. Choice (A) is wrong, primarily because "financial" and "promissory" are not synonyms. Although a promissory note often is related to financial matters, by definition it is not limited to finances. Choice (B) is wrong also; although a promissory note usually requires a signature, this is not a part of the definition of "promissory." And although a listing of the purchased items and/or a list of the consumer's assets may be involved in the business agreement between buyer and seller, neither is by definition required in a promissory note. Therefore, choices (D) and (E) are wrong.

13. (B) *Robinson Crusoe* is as well known as Stonehenge (and as unlikely to disappear). The size, choice (A), of the novel is not mentioned. Although Woolf stresses its appropriateness for children, this alone gives no conclusive evidence of its difficulty, choice (C). Artistic merit is implied in subsequent paragraphs, choice (D). Ephemeral works, choice (E), would not be granted monumental status.

14. (A) Returning with each generation is involved in the metaphorical usage. Although *Robinson Crusoe* might be said to "nourish" children, that is not relevant to this context, choice (B). Its popularity is not an issue; its eternal presence in English life is, choice (C). Far from diminishing, choice (D), the novel grows in stature, Woolf stresses that it is "only" 200 years old, hardly "ancient," choice (E).

15. (B) The correct choice depends not on knowledge of drawing rooms, but on recognition of "table" in its context. Placement on a table or a bookshelf has no direct connection to sales of a book, choice (A). Nor is only the best writing placed upon a table in the home, choice (C). The question of appreciation, choice (D), Woolf takes up in the next paragraph. She there argues that Defoe's other novels are comparable to *Robinson Crusoe*, choice (E).

16. (B) Much of this passage is concerned with the reevaluation of all Defoe's work. A reconsideration of *Robinson Crusoe*, unless it were negative, would be pointless, choice (A). Woolf's larger aim is an appreciation, choice (C), of Defoe's entire corpus, but she is chiefly concerned with the lesser read novels. The comparison with Fielding (and Richardson), choice (D), is only an incidental point. There is no physical monument other than the novels themselves, choice (E).

17. (B) The "hole in the heart" leads to lying. This lying, as Defoe notes, does become habitual, choice (E), but that is not its defining characteristic. The creation of fictional plots is, according to Defoe, pointless, choice (A). It is hardly useful; in fact, choice (C), he goes out of his way to use "facts" in writing. Writing would be painful only if the hole, choice (D), were literal.

18. (B) The principles of depending on facts accord with the experience described in the lines. Defoe adjures the authentically fictional story, choice (A). Instead, he insists upon using factual material, choice (C). The tendency to impart a moral is part of the justification discussed in lines 50–51, choice (D). The vivid characters, according to Woolf, grow out of Defoe's experience, choice (E).

19. (B) The best definition within the context of the passage is "talents." The focus of the passage is on the artistic work of Defoe. Both choices (A) and (D) have to do with the financial aspect of the word "endowments," which is not applicable in this context. "Stature," choice (C), is also an acceptable definition of the given word, but it deals with physical size. Choice (E) is not a definition of the "endowments."

20. (C) Woolf shows each character caught in grinding poverty or misery. Good may triumph over evil, choice (A), but only in conclusions that she does not discuss here. Defoe concerns himself with the problem of injustice, choice (B). Man's unchanging nature may be inferred, choice (D), by the modern reader. The suggestion is that Fortune has already exerted her toll before the novels open, without much further dramatic decline, choice (E).

21. (C) The correct answer is choice (C) because the author is clearly treating Defoe and his most famous work as something almost holy, something sacred. "Reverential" can be defined as "feeling profound, adoring or awed respect for," which best describes the tone of the passage. Choice (A) "felicitous" means "pleasant and delightful," and, although this describes the author's fond memories of having *Robinson Crusoe* read to her as a child, it is not the best description of the tone of the passage. Choice (B) "exonerative" means "to relieve from blame," and choice (D) "defamatory" means "to harm the reputation of." Because the author is doing neither of these things, they are both incorrect. Choice (E) is incorrect because the author obviously states her favoritism of the books and therefore is not remaining impartial or objective.

22. (E) The primary purpose in this passage is to show how *Robinson Crusoe* is viewed differently from Defoe's other books because it is often read aloud to children. The author clearly does not consider *Robinson Crusoe* inferior, choice (A); in fact, she feels the opposite. Although she gives some biographical information, choice (B), toward the end of the passage, it is not the primary focus. Nowhere does the author compare literary styles, choice (C), or summarize the plot of *Robinson Crusoe*.

23. (C) The correct answer is choice (C). The author was not implying that Defoe did not write the book but merely that one does not usually consider the author, only the adventure tale itself. This is especially the case with children. Choice (A) is incorrect because Defoe did author the book. Choice (B) reflects Defoe's true feelings, but it did not stop him from claiming authorship. Choices (D) and (E) are incorrect, because, as stated before, Defoe did actually write *Robinson Crusoe*.

24. (B) The focus of the passage is on the popularity of *Robinson Crusoe* and how it affected the reception of Defoe's other novels. The author does not feel that Defoe's other books were poorly written, choice (A). There is no mention in the passage of the subject matter, choice (C), or tone, choice (E). By all indications, the other books were published during his lifetime, choice (D).

DETAILED EXPLANATIONS

SECTION 4—WRITING

1. (E) This sentence starts with a dangling participial phrase that must be followed by the noun or pronoun that it modifies or be eliminated from the sentence all together. Choice (E) correctly replaces the phrase with a subordinate clause that shows the casual relationship between the eating of paint chips and lead poisoning. Choices (B), (C), and (D) are also dangling phrases and create the same basic problem as the phrase in the original sentence.

2. (B) Only choice (B) corrects the run-on sentence in this exercise in an acceptable fashion by adding the conjunction "and" to join the two sentences. Choice (C) corrects the run-on but poses the problem of the relative pronoun "who" that does not follow its antecedent; choice (D) retains the run-on and adds an awkward participial phrase; and choice (E) results in two run-on errors.

3. (C) Choice (C) retains the pronoun "you" after the participial phrase that modifies it but eliminates the wordy dangling phrase, "the reason being that"; the revision expresses clearly the casual relationship. Choices (B) and (D) leave the introductory phrase dangling; and choice (E) results in an unclear antecedent for the pronoun "which" and fails to show the casual relationship.

4. (D) Only choice (D) corrects the opening independent clause while retaining the original meaning of the sentence. Choice (A) is incorrect because "Being" is both an unattached participle and an improper conjugation of "to be," which must always be used as the main verb. Choices (B) and (E) are idiomatically incorrect and change the meaning. Choice (E) also contains an unattached participle. Choice (C) is excessively wordy.

5. (E) The problem here is one of parallel construction; choice (E) completes the comparison in parallel fashion and avoids the wordiness of choice (D). Choice (B) unnecessarily changes the tense but does not eliminate the problem; choice (C) replaces the second clause with a dangling phrase.

6. (A) This sentence is correct in standard written English because "whose," a possessive relative pronoun, is appropriate to refer to a city. Choice (B) eliminates the required possessive pronoun and is not idiomatically correct; choice (C) is wordy and awkward; choice (D) results in a run-on sentence; and choice (E) replaces the possessive pronoun with the contraction that means "who is."

7. (E) The problem is one of wordiness, and choice (E) correctly shows the relationship and eliminates words that add nothing to the meaning of the sentence. Each of the other choices retains unnecessary words.

8. (B) The original sentence is wordy and includes the use of the pronoun "they," for which there is no antecedent. Choice (B) replaces the independent clause with the second part of a compound predicate that is parallel to the predicate in the preceding clause. Choice (C) adds a dangling participial phrase; choice (D) retains the original "they" without an antecedent as well as the unnecessary use of the pronoun "it"; and choice (E) results in a run-on sentence.

9. (C) Choice (C) correctly replaces the singular verb "is" with the plural form "are" to agree with the plural subject, "habits"; in addition, it replaces with parallel gerund phrases the awkward noun clause that serves as predicate nominative. Choices (A) and (B) retain the incorrect singular verb "is"; choice (D) inappropriately changes the gerund to the present perfect form; and choice (E) equates the habits with trees, food, and dams and lodges rather than with the beaver's activities.

10. (C) The words in this exercise result in a sentence fragment not a complete sentence. Only choice (C) provides a predicate for the subject, "Bolivia." Choice (B) replaces the dangling phrase with a subordinate clause; choices (D) and (E) add main clauses but leave the subject, "Bolivia," without a predicate.

11. (E) The problem with the original sentence is ambiguity of "he." Choice (D) does not clarify whether Brutus or Caesar was pardoned. Choices (B) and (C) solve the problem, but each sentence is excessively wordy. Only choice (E) clarifies the meaning economically.

12. (B) As you read the sentence, you should recognize that choice (B) presents a shift of pronoun. The correct choice of pronoun to refer to the antecedent, the indefinite pronoun "anyone," is "he or she." Choice (A), "plans," is third person singular to agree with its subject "anyone"; choice (C), "to see," is an infinitive introducing an adverb phrase; and choice (D), "of the sites," is a prepositional phrase modifying "all."

13. (D) Choice (D), "inaccessible from," is idiomatically incorrect. The preposition that should follow "inaccessible" is "to." Choice (A) "important," is a predicate adjective modifying the subject "it"; choice (B) is a coordinating conjunction correctly joining the subject "it"; choice (B) "and preparations"; and choice (C), "be stored," is the subjunctive form, correctly used is a "that" clause expressing a requirement or recommendation.

14. (A) You should recognize that choice (A) calls for the adverb "slowly" to modify "worked" rather than the adjective "slow." Choice (B), "meticulously," is an adverb modifying the adjective "careful"; choice (C), "was," is third-person singular to agree with the singular subject "report"; and choice (D) is the predicate adjective "accurate" that modifies the subject of the clause, "report."

15. (C) Choice (C) calls for the objective form "whom" because the relative pronoun is the object of the verb "have heard." Choice (A), "giving," is a participle introducing a participial phrase modifying the subject "man"; and choice (D), "have heard," is a verb in the present perfect tense that correctly expresses action that occurred in the past and continues into the present.

16. (E) All labeled elements of this sentence are choices acceptable in standard written English. Choice (A), "although," is a subordinating conjunction introducing a subordinate clause; choice (B), "older," is the comparative form of the adjective, appropriate for the comparison of the singular "she" to the singular indefinite pronoun "anyone else"; choice (C), "anyone else," is idiomatically correct; and choice (D) is an adverb prepositional phrase modifying the adjective "young."

17. (C) The problem with this sentence is parallel structure. Choice (C), "shouting," is a present participle, whereas choice (A) is a simple past tense. Choice (C) should be changed to "shouted" so the tenses will be the same. Choice (B), "president's," is a correct possessive; and choice (D) is idiomatically correct.

18. (D) The verb at choice (D) must agree with its subject, "variety," and, therefore, needs to be the singular form, "changes." Choice (A), "fresh," is an adjective describing the noun "ice cream"; choice (B), "daily," is an adverb modifying the verb "make"; and choice (C), "of flavors," is an adjective prepositional phrase modifying the noun "variety."

19. (D) You should recognize that the word arrangement and punctuation place the prepositional phrase "as a lubricant" as the fourth object of the preposition "for"; the phrase is not parallel to the preceding noun objects. By inserting an "and" between "candles" and "soap" and eliminating the comma after "soap," you revise this sentence by providing parallel phrases after the verb. Choice (A), "it provided," is correct as a pronoun subject, referring to "whaling," and verb of the introductory adverb clause; choice (B) is a predicate nominative, "vocation," and its modifier, the passive verb "was used," choice (C), is past tense and agrees in number with its subject, "oil."

20. (B) The error is one of diction. The word "literally," choice (B), is an adverb that means that the action of the verb really occurred; clearly the subject did not "really" die of starvation if he was able to eat. Choice (A), "greedily," is an adverb appropriately describing how "he" ate; choice (C), the past perfect verb, "had been lost," is correct to refer to an action that occurred before the action of the main verb; and choice (D), "for two days," is an adverb prepositional phrase modifying the verb "had been lost."

21. (B) You should recognize an error in comparison in choice (B). Because the writer is comparing more than two contributions that Montagu might have made, the adverb should be the superlative form, "best." Choice (A), "might have made," is the correct form for the conditional verb in a "that" clause; choice (C), "that permitted," is the subject "that," referring to the antecedent "lunch," and the verb of the adjective subordinate clause; and choice (D), the infinitive "to continue," is idiomatically correct to introduce a noun phrase that serves as direct object of the verb "permitted."

22. (B) You should recognize in choice (B) an error in verb tense. The past perfect, "had beaten," would be appropriate only if the action occurred before the action of the verb in the main clause. Because the two actions occurred simultaneously, the simple past tense form, "beat," is correct. Choice (A), "when," is a subordinating conjunction correctly introducing a subordinate clause; choice (C), "made," is in the simple past tense and agrees in number with its subject, "teams"; and choice (D), "kicking a round ball," is an idiomatically correct phrase serving as object of the preposition "by."

23. (A) You should know that the noun "kind" must agree in number with the adjective preceding it; therefore, in choice (A), the word "kinds" should replace the singular "kind" to agree with the plural adjective "those." Choice (B), "who," introduces the subordinate adjective clause and is in the nominative case because it serves as subject in the clause; choice (C), "good," is the correct predicate adjective to modify "who" when the meaning is "satisfactory"; and choice (D), "themselves," is a reflexive pronoun referring to "who."

24. (B) The error is in choice (B), where the correct past tense form of the verb "sneak" is sneaked," not "snuck." Choice (A), "had been sent," is correct in the past perfect form because the action was completed before the action of the main clause; choice (C), "downstairs," is an adverb modifying the verb; and choice (d), "lay," is the correct past tense form of the verb "lie."

25. (B) The given sentence contains a double negative. The "no" in choice (B) should be replaced by "any" or "a" following the negative verb and adverb "hadn't." Choice (A) is a past tense verb, "was shown," and adverb, "originally," in idiomatic word order; choice (C), "to see," is an infinite introducing an adjective phrase modifying "chance"; choice (D), "its," is a possessive pronoun referring to "cartoon."

26. (D) You should recognize that the error occurs in choice (D), where the correct adjective to compare two people is the comparative form, "more famous." Choice (A), "are recognized," is correct in the present passive form and is plural to agree with its subject, "portraits"; choice (B), "by most Americans," is an adverb prepositional phrase in idiomatic word order; and choice (C), "were," agrees in number with its compound subject, "Gilbert Stuart" and "daughter" and is correct in the past tense.

27. (B) You should recognize that the nominative case pronoun, "I," is needed as predicate nominative following the linking verb, "were," in choice (B). Choice (A), the subjunctive form "were," is correct to show condition contrary to fact; choice (C), "who," is a relative pronoun serving as subject of the subordinate adjective clause it introduces and, therefore, is correct in nominative case; and choice (C) "would complain," uses the auxiliary "would" to express condition.

28. (D) The error in choice (D) is one of diction. The auxiliary verb for the present perfect tense is "have"; "of" is a preposition and does not belong in a verb phrase. Choice (A), the adverb "really" and the predicate adjective "tragic," is idiomatically correct; choice (B), the verb "fell," is correct in the simple past tense; and choice (C), "that," is a relative pronoun that introduces the subordinate adjective clause modifying its antecedent, "well."

29. (C) The correct tense to indicate action that occurred once in the past is the simple past passive form "was shot"; the present perfect form with the auxiliary "has" would be appropriate for action from the past that continues into the present. Choice (A), "after the surrender," is idiomatically correct as the subordinating conjunction, article modifier, and subject of the adverb subordinate clause; choice (B), "signaled," is a verb in the past tense that agrees in number with "surrender" and is appropriate for an action that occurred once in the past; and choice (D), "were attending," is the past progressive verb showing an action that continued for a period of time in the past.

30. (C) Choice (C) correctly and smoothly combines the two adjectives while providing sentence variety. Choice (A) is not parallel; "grieving" is not parallel with the other two adjectives. In choice (B), the adjectives appear to modify the dead Lenore. In choices (D) and (E), the ideas of melancholy and grief are not stated in concise parallel structure.

31. (C) Choice (C) is the most concise expression of the major ideas. Choice (A) is too abrupt and states that the bird "is evil" instead of being "called evil." Choice (B) contains awkward wording; "the realization of" and "the evil omen of doom of the bird." The wording of choice (D) creates a run-on sentence. Choice (E) incorrectly twists the idea of evil.

32. (B) Choice (B) correctly identifies who is doing what. Choices (A) and (C) do not indicate the man as the subject. Choice (D) implies the bird is shrieking with anguish, and choice (E) implies the man is shrieking with anguish as well as replying.

33. (D) This essay is an analysis of Poe's narrative poem "The Raven," and as such shows the progression of ideas in that poem. Choices (A) and (B) might be plausible, but there is too much ambiguity for the events to be classified as supernatural or the bird's intent to be clearly evil. Choice (E) is not fully developed in the paper. Choice (C) might be a logical choice, but it is not the outstanding intent.

34. (E) Choice (E) correctly uses the cause-and-effect construction; in addition, this choice uses a transition word at the beginning of the sentence in order to indicate contrast of ideas. Choice (A) employs a misplaced modifier that implies the "others" are repeating the "one word." Choices (B) and (C) leave out a transition word and present as fact what is considered opinion. Choice (D) is too formal and also omits the transition word.

35. (D) The colon is the most appropriate way to unify the closing thoughts into a single sentence. Choice (A) also utilizes the colon, but the elimination of the comma to create the phrase "is a familiar one" is more clumsy that the original. Choice (B) adds a comma between the two sentences, which creates a run-on. Choice (C) uses a semicolon that is inappropriate. Choice (E), "that is certain is that," both loses the necessary sense of "however" and creates an unnecessarily long and convoluted sentence.

DETAILED EXPLANATIONS

SECTION 5—MATH

1. (B) Before deducting taxes and other fees, her total income is $18.25*h*. The company deducts 12% of this amount. This deduction is [(18.25*h*) × 12%]= $2.19*h*. Deducting this amount from her hourly salary gives her take-home pay:

$$I = (18.25h) - 2.19h = (18.25 - 2.19)h = \$16.06h$$

2. (E) For the first half of the values of x, y increases as x increases; but for the second half, y decreases as x increases. So the variation of y is related to the variation of x neither directly nor inversely.

3. (A) As the coordinates of the points A and B indicate, l is a horizontal line whose length is dependent on a only, so $l = a$. This means that l varies directly as a with a constant variation of 1.

4. (D) The prime factors of 35 are 5 and 7, so

$$\left[\boxed{35}\right] = 5 + 7 = 12$$

Add the two digits of 24 together to get

$$\left[\lfloor 24 \rfloor\right] = 2 + 4 = 6$$

Plug the results of both operations into the given expression to find $\left[\lfloor p \rfloor\right]$:

$$12 - 6 = 6$$

Therefore, the sum of the digits of $\left[\lfloor p \rfloor\right]$ equals 6. Only choice (D) had digits that add up to 6.

5. (B) The sum of the ratio 9:5:4 equals 180°. Solve the equation $9x + 5x + 4x = 180$; $x = 10$. Therefore, ΔABC is a right triangle. Use the Pythagorean Theorem to find AB.

$$7^2 = 5^2 + (AB)^2$$

$$49 = 25 + (AB)^2$$

$$(AB)^2 = 24$$

$$AB = 2\sqrt{6}$$

6. (D) Make the equation easier to solve by using equivalent factors:

$$\sqrt{150 \times 2} \times \sqrt{7^2 - 1} = \sqrt{3 \times 100} \times \sqrt{49 - 1} = \sqrt{3 \times 100} \times \sqrt{16 \times 3}$$

$$= 10\left(\sqrt{3}\right)(4)\left(\sqrt{3}\right) = (10 \times 4)\left(\sqrt{3}\right)^2 = (40)(3) = 120$$

7. (E) Derive a from the second equation:

$7a + 9 - 9 = -5 - 9$

$7a = -14$

$a = -2$

Replace the value of a in the first equation and solve it for b:

$7(-2) + 9b = 13$

$-14 + 9b = 13$

$-14 + 9b + 14 = 13 + 14$

$9b = 27$

$b = 3$

Now, replace the values of a and b in the given expression.

$(7a - 9b) = 7(-2) + 9(3)$

$-14 + 27 = 13$

8. (D) The general form of the exponential function for the growth is $Q(t) = Q_0e^{kt}$. In this equation, Q_0 is the initial amount of the quantity Q. Comparing this function with the given function, we get $Q_0 = 1{,}000$.

9. (B) Denote the set of students taking Biology II by A and the set of students taking Sociology by B. Subtract the number of students taking neither class from the total number of students. This gives the number of students taking either Biology II or Sociology or both. The number of these students is $110 - 35 = 75$. So, $n(A \cup B) = 75$. Then, the number of students attending both classes can be shown by $n(A \cap B)$. So $n(A) = 25$, $n(B) = 40$, and $n(A \cup B) = 75$. Then, $n(A \cap B) = n(A \cup B) - n(A) - n(B) = 75 - 40 - 25 = 10$.

10. (E) Generally, $\left|\frac{A}{B}\right| = \frac{|A|}{|B|}$, where A and B are two algebraic expressions. Using this rule, $\left|\frac{x-1}{-2}\right| = -6$ can be written as $\frac{|x-1|}{|-2|} = -6$. This can be simplified as $\frac{|x-1|}{2} = -6$ or $|x-1| = -12$.

11. (C) In the right triangle ADF, the length of the side opposite the 30° angle is one-half the length of the hypotenuse. So, $\overline{AF} = \frac{AD}{2} = \frac{12}{2} = 6$. Using the Pythagorean Theorem in the right triangle ADF, $12^2 = \left(\overline{DF}\right)^2 + 6^2$. Solve this equation for $\overline{DF}$ as follows: $144 = \left(\overline{DF}\right)^2 + 36$, $\left(\overline{DF}\right)^2 = 108$, $\overline{DF} = 6\sqrt{3}$. Then, $\overline{CF} = \overline{CD} - \overline{DF} = \overline{AB} - \overline{DF} = 14\sqrt{3} - 6\sqrt{3} = 8\sqrt{3}$.

12. (D) Denote the length of a side of the outer square by $2s$ and set up equations for the areas of both squares.

area of outer square = $(2s)^2 = 4s^2$

area of inner square $\left(\frac{2s}{2}\right)^2 = s^2$

Subtracting these areas gives us the area of the shaded region:

area of the shaded region = $4s^2 - s^2 = 3s^2$

Probability of landing dart on the shaded region is

$$\frac{3s^2}{4s^2} = \frac{3}{4}$$

13. (A) Solve the given inequality. We can write $|2h - 10| \leq 6$ as $-6 \leq 2h - 10 \leq 6$. Add 10 to all sides of this inequality: $-6 + 10 \leq 2h - 10 + 10 \leq 6 + 10$. Simplify the result as follows: $4 \leq 2h \leq 16$, $2 \leq h \leq 8$. So, the interval of the heights of the trees is $2 \leq h \leq 8$.

14. (B) Because $\overline{AB}$ is tangent to the circle, the radius OA is perpendicular to it at A. $\overline{AB}$ and $\overline{OC}$ are parallel. These two conditions imply that $\overline{OA}$ is perpendicular to $\overline{OC}$. That is, ΔOAC is a right triangle with two legs that are radii of the circle O. Denote the length of a radius by r and apply the Pythagorean Theorem to find r:

$$8^2 = r^2 + r^2$$

$$64 = 2r^2$$

$$r^2 = 32$$

$$r = 4\sqrt{2}$$

15. (A) Find the slope of each graph first.

$mx - 3y = 9$, $3y = mx - 3$, $y = \frac{m}{3}x - 9$; slope $= \frac{m}{3}$

$3x - ny = 12$, $ny = 3x - 12$, $y = \frac{3}{n}x - 12$; slope $= \frac{3}{n}$

In perpendicular lines, the product of the slopes must be equal to –1. Therefore,

$$\left(\frac{m}{3}\right)\left(\frac{3}{n}\right) = -1$$

$$\frac{m}{n} = -1$$

$$m = -n$$

$$m + n = 0$$

16. (C) Using the definition of a geometric sequence, $m^2 = (a - 1)(a + 1)$, so $m^2 = a^2 - 1$. Adding 1 to each side of this equation results in $m^2 + 1 = a^2$.

17. (A) Square both sides of the given equation:

$$\left(\sqrt{a+b}\right)^2 = \left(\sqrt{a} + \sqrt{b}\right)^2$$

$$a + b = a + b + 2\sqrt{ab}$$

By simplifying, we get $2\sqrt{ab} = 0$ or $\sqrt{ab} = 0$. Squaring both sides of this equation, we get $ab = 0$.

18. (B) Square each side of the given equation: $(x + \frac{3}{x})^2 = (-3)^2$, $x^2 + 2x\left(\frac{3}{x}\right) + \left(\frac{3}{x}\right)^2 = 9$. Simplify the left side of the last equation: $x^2 + 6 + \frac{9}{x^2} = 9$. Add –6 to each side of this equation: $(x^2 + \frac{9}{x^2}) = 9 - 6 = 3$.

DETAILED EXPLANATIONS

SECTION 6—CRITICAL READING

1. (D) This sentence tells us that a person lost weight by eating healthy food and by exercising. We know that this person is healthy and that his body will "respond" in a positive way. So we are looking for a noun that is related to a "healthy" person. Choice (D), "vitality," (having energy) is the best choice. Choice (A) is incorrect because although "frivolity" (fun) is a positive word, it does not relate semantically to a healthy body. Choice (B) is wrong because "lethargy" (lazy, passive) is not related to a healthful way of life. Choice (C) "diligence" (hard work) is related to having a healthy body because it takes hard work to do so, but it does not tell us how the body responds. Choice (E), "tranquility," is incorrect because a healthy body is not usually described as responding in a "peaceful, still, or harmonious" manner.

2. (A) The clue here is "in spite of the fact," which means that no matter how positive or negative the professor was, the lecture would be the opposite in meaning, and relate to what "few students understood." We can conclude that few students would understand a lecture that is "abstract" (vague, having no clear definition), even though the professor is "articulate" (well-spoken). Therefore, choice (A) is the best. Choice (B), "arrogant" (haughty), is possible in describing the professor but does not fit the second part of the sentence. "Austere" (harsh, strict, severe) is not commonly used as an adjective to describe a lecture. Choice (C) is incorrect because it ignores the clue words. A "taciturn" (reserved, quiet) professor who gave a lecture in a "superficial" (cursory, shallow) way would not be understood. Choice (D) is incorrect because although the professor might be "illuminating" (make understandable), students would understand a "deliberate" (consider carefully) lecture. This does not go with "in spite of the fact." Choice (E), "affable" (friendly, amiable, good natured), is a positive way of describing the professor, but "derogatory" (belittling, uncomplimentary) is not an appropriate way of describing a lecture.

3. (B) We are looking for an adjective to describe what kind of step she would take to climb an "unstable ladder." The best answer is choice (B) "tentative," which means "not confirmed, indefinite." Certainly, one would only climb something one was unsure of in a hesitant or uncertain way, which is another way of defining "tentative." Choice (A) is wrong because "skeptical" means "doubtful" and relates more to an attitude than an action. Choice (C), "limber" (flexible, pliant), is obviously wrong in the context of the sentence. No one in her right mind would climb in a limber way up an unstable ladder. Choice (D) is wrong because "haggard" (tired looking, fatigued) describes how a person feels, not the way she would climb a ladder. Choice (E), "wanton" (unruly, excessive), is not contextually accurate because it does not describe how a person would climb a ladder.

4. (A) The best answer is choice (A), "faster . . . efficient," because we are looking for two adjectives that describe why fax machines replaced express mail and telex. Choice (B), "trivial" (minor importance) and "supported," does not make any sense in the sentence. "Effusive" (gushing) and "salutary" (beneficial), choice (C), are not appropriate as reasons why fax machines replaced mail and telex. "Valid" (true) and "impeccable" (faultless), choice (D), do not logically fit into the sentence. "Sporadic" (happening occasionally) and "torpid" (lacking vigor, dull), choice (E), are negative adjectives and if placed in the sentence would mean that fax machines run on an occasional basis and were dull. This makes little sense.

5. (C) A notary public is hired to "verify" (authenticate, confirm) the "veracity" (truthfulness) of signatures. The implication is that the person who is signing the document is he whom he actually says he is. Choice (A), "constrain" (compel) and "watermarks" (a marking in paper visible when one holds it up to the light), does not fit into the sentence. "Disdain" (contempt, scorn) and "sanctity" (holiness), choice (B), have no meaning when substituted into the sentence. "Efface" (wipe out) and "consonance" (agreement), choice (D), are not logically related to the sentence. A lawyer would not hire anyone to "wipe out" a signature. "Abrade" (rub off) and "matrix" (mold), choice (E), are incorrect.

6. (D) Driving through the town of his or her childhood, the author reminisces. The passage is full of the author's pleasant memories and wistful yearning for the past; thus, its tone is nostalgia. The passage contains no hint of choice (A), "regret," choice (B), "sarcasm," or choice (C), "anger." And although mention is made of the preacher and the church, the tone is not really one of "reverence" (honor, respect, worship) as much as fondness for the town and yesteryear. Thus, choice (D) is more accurate than choice (E).

7. (A) The word "uncanny" means "weird" or "unnatural." The entire passage focuses on the similarity between what the author saw while driving through town and what it had been like several years ago. Although the people were different, they looked the same and did the same things. Such similarities are not what one would expect upon returning to one's hometown after several years have passed; thus, one could explain the experience as "uncanny." Choice (B) is incorrect because the author does not indicate that the people who were living in the town were still living there; in fact, the passage states that "John and Gertrude had been gone for at least fifteen years." And the people the author observed were not the people who had lived there years ago because those people would have aged and would not still be children eating ice cream cones and young mothers watching their toddlers. At best, they would be the old men loafing on the benches. Choice (C) is incorrect because the passage makes no mention of modern-day businesses. Choice (D) is wrong because the passage gives no indication that the author felt out of place or isolated. And choice (E) is not correct because the town was definitely recognizable because it had scarcely changed over the years.

8. (E) Much of the passage presents evidence of America's "size acceptance"—providing wider seats in theaters, allowing large people to take up two airplane seats, switching to larger hospital equipment, and so on. Although the passage does mention concerns of health-care professionals, the paragraph does not criticize America's size-acceptance tendency or overweight people themselves; thus, choices (A) and (C) are wrong. Near the end of the paragraph, mention is made of discrimination that overweight people sometimes face, but nothing in the paragraph actually criticizes American society for ridiculing overweight people; therefore, choice (B) is incorrect. Since the passage does not contain information that encourages overweight people to lead a healthier lifestyle, choice (D) is wrong.

9. (A) In this question, be careful to stay focused on the passage itself, not on the things you may have read elsewhere or on your opinion. Choice (A), the many evident accommodations that are being made for people who are overweight or obese, implies a general indication that the need for these accommodations is not going to decline; therefore, the assumption is that the number of obese and overweight people will remain stable or increase. If this situation were believed to be only short-lived, theaters, hospitals, and resorts would not be investing in "size-friendly" equipment. Choice (B) contradicts choice (A) and is therefore incorrect. And although the passage indicates that health-care professionals are concerned, it does not imply that they are increasing their efforts to educate America about the importance of a healthy lifestyle; thus, choice (C) is wrong. Choice (D) is incorrect because the passage does not address the correlation between weight and happiness, and it does nothing to imply that being overweight does not necessarily mean being unhealthy; instead, it implies the opposite idea. And choice (E) is wrong because the passage does not discuss the definitions of "overweight," "obesity," and "size acceptance"—or the changes in the definitions of those terms.

10. (C) The point of this passage is to demonstrate the relationship between infectious disease and historical events. Choice (A) is incorrect because it is made clear that epidemics were complex affairs in lines 16–17. Choice (B) is not correct because the current level of treatment is not discussed in this passage. The level of medical prevention in ancient Athens, choice (D), is also incorrect because it is not the main point. Choice (E) is not correct because the specific wisdom if Pericles is not discussed in great detail.

11. (B) Human morals are neutralized by the continuous presence of death; therefore, choices (A) and (D) are both incorrect because they are not justified by the text. Choice (C) is incorrect; Peloponnesians fled for fear of infection. Choice (E) is not correct because the author discusses the lack of certainty in the cause of the epdidemic.

12. (A) From the context of the sentence, we can infer that "necrosis" means the death of tissue. No other choices fit into the context as well as choice (A).

13. (B) Lines 19–25 identify the two causes of outbreak as overcrowding and coincident epidemics. Choice (A) is not correct because poor army discipline is not given as a factor in spreading the plague. Choice (C) is also incorrect because housing codes are not mentioned in this passage. An influx of Ethiopians, choice (D), would be a misreading of the author's statement that it seems the disease started in Ethiopia. Poor nutrition, choice (E), is not mentioned as a cause for the outbreak.

14. (D) Reinfected patients did not always die, so this is not indicative of the plague. All the other choices are mentioned in the passage and are symptoms indicative of the plague.

15. (D) The context clue is "recover." "Recuperation" and "convalescence" both mean the process of becoming well after an illness, growing strong again. Choice (A) means the opposite; choice (B) "innocuous" means "innocent or harmless." Choice (C) is incorrect because it means "a process of contaminating, soiling, corrupting, or infecting." Choice (E) is wrong because it means "to weaken" and is the opposite of convalescence."

16. (D) Although choice (A) is generally true, the real emphasis of the writer is on Coltrane's African-Asian synthesis, which absorbed music, literature, and philosophy. Again, although choice (B) is broadly true, the writer specifies the African-Eastern as the source of Coltrane's ecstasy. Choice (C) is mentioned incidentally (line 41) in regard to one album. "Popular reception" as opposed to "insiders," choice (E), is a distracter.

17. (B) God rejoices (in the hymn) to all festive sounds; there is nothing "spare," choice (A), or "restrained," choice (E), or "reserved," choice (C), about "loud cymbals" (line 64). Coltrane is a Western jazz artist, so he would not employ merely "Eastern," choice (D), musical means.

18. (B) Coltrane's audience was confounded by his late development (line 33) because his message was not clear, choices (C) and (D), nor particularly quiet, choice (A). *India*, released in 1961, choice (E), falls outside the last-three-year parameter.

19. (A) Lines 86–91 confirm the sense of personal odyssey in Coltrane. The author suggests triumph, not frustration, choice (D). Like any real communication, the music is not detached from the man, choice (C), nor morally deceptive, choice (B). Choice (E) is incorrect because Coltrane did not disown his Western jazz roots.

20. (E) Coltrane's music is a culturally collective experience, according to this author; it is not "limited" to musically archaic instruments, choice (D), nor is it "destructive," choice (A), to anything. The album *India* alone negates choice (C); and Coltrane's odyssey was personal, not motivated by a false claim about the Village Vanguard (B).

21. (E) According to the author, God is self-sufficient, yet He does enjoy human adoration in the form of music and joyful noise. Coltrane's music is described consistently as spiritual, not secular, choice (A), and the chosen hymn for *A Love Supreme* does not minimize faith, choice (B). God does not ignore men's music, choice (C), especially because Coltrane's music transcends his own ego, choice (D), and embraces Africa and Asia.

22. (B) The word "programmatic," in the context of the entire passage, most nearly means "thematic." Coltrane's music in the sixties dealt with African spiritual themes. Nowhere in the passage does the author refer to the music as being predictable, choice (A), or uninspired, choice (E); in fact, the author feels the complete opposite. Choice (C), "unhealthy," makes a judgment about Coltrane's musical focus that is not supported by the tone of the passage. Although the passage does speak of the spiritual, choice (D), aspect of Coltrane's music, the context of the sentence does not support that definition.

23. (E) The author makes a favorable judgment of Coltrane's later music. This is supported by the author's reference to Coltrane's later music, Coltrane's "classic quartet" (line 3), and his "inspired . . . sax improvisation" (lines 39–40). Choice (A) is incorrect because the author is not condemning or criticizing Coltrane's later music. Choices (B) and (D) are incorrect because they imply that the author is placing Coltrane on a pedestal and endowing him with God-like qualities. Choice (C) is incorrect because the author favors the music of John Coltrane, and this fondness is reflected in the passage. If the author were "objective" he would not give opinions either in favor of, or against, the subject of the passage.

24. (C) The correct answer is (C) because the author states, in lines 27–36, that there was a change in Coltrane's later musical style, and in the statement "The music of Coltrane's final three years has seemed to confuse, anger" implies that the more traditional jazz structure he had once employed was abandoned. Nowhere does the passage state or imply that, choice (A), his involvement with rock musicians; choice (B), his involvement with drug culture; or choice (D), his spiritual focus influenced listeners and fans to question or become angered with his style of later music. Choice (E) is incorrect because the opposite is true.

DETAILED EXPLANATIONS

SECTION 7—MATH

1. (C) The price of the sunglasses in August = $k - (12\%)k = 1k - 0.12k = 0.88k$. For the September price, subtract \$24 from the August price. The expression for the price of the sunglasses in September is $0.88k - 24$.

2. (C) Calculate each option.

   (A) $-2 \spadesuit 3 = \frac{-2-3}{-2+3} = -5$

   (B) $2 \spadesuit -3 = \frac{2+3}{2-3} = -5$

   (C) $-3 \spadesuit -2 = \frac{-3+2}{-3-2} = \frac{1}{5}$

   (D) $3 \spadesuit -4 = \frac{3+4}{3-4} = -7$

   (E) $-3 \spadesuit 4 = \frac{-3-4}{-3+4} = -7$

   So the expression in answer (C) has the greatest value.

3. (B) Rewrite the given inequality as the compound inequality $-1 \leq x - 6.5 \leq 1$. Add 6.5 to all sides of this inequality: $-1 + 6.5 \leq x - 6.5 + 6.5 \leq 1 + 6.5$. Simplify all sides of the inequality: $5.5 \leq x \leq 7.5$.

4. (D) There are 10 students in the scatterplot. The plot indicates that students *F, G, H, K,* and *L* scored better than 5 students in mathematics. Among these students, both students *F* and *G* scored worse than 5 students in science.

5. (E) With each toss of the coin, the probability of getting either heads or tails is $\frac{1}{2}$. If this is done six times, the probability is $\frac{1}{2} \cdot \frac{1}{2} \cdot \frac{1}{2} \cdot \frac{1}{2} \cdot \frac{1}{2} \cdot \frac{1}{2} = \frac{1}{2^6} = \frac{1}{64}$.

6. Answer: 8 Using the definition of a geometric sequence, we have

   $x(xy) = 4(16y)$

   $x^2y = 64y$

   Divide each side of the preceding equation by y.

   $x^2 = 64$

   Take the square root of each side of this equation: $x = 8$.

7. Answer: 343 Solving each equation, we get $x = -4$ or 4 and $y = -3$ or 3. These answers can be paired in the following groups:

$x = -4$ and $y = -3$ $x = -4$ and $y = 3$

$x = 4$ and $y = 3$ $x = 4$ and $y = -3$

Of these pairs, only the pair ($x = 4$ and $y = -3$) maximizes $(x - y)^3$. Therefore

$$(x - y)^3 = [4 - (-3)]^3 = 343$$

8. Answer: 30 Denote the sum of the original five numbers by S. Then, $S = 5(12) = 60$. Denote the sixth number added to the set by m, giving the equation $\frac{(60+m)}{6} = 15$. So $60 + m = 90$, and $m = 30$.

9. Answer: 96 Denote the length of the common hypotenuse of the triangles by m. Applying the Pythagorean Theorem to both the right triangles

$$m^2 = b^2 + 112 \text{ and } m^2 = 5^2 + a^2$$

The left sides of these equations are equal. So, their right sides are equal too.

$$b^2 + 11^2 = 5^2 + a^2$$

Simplify and solve this equation for $a^2 - b^2$.

$$b^2 + 121 = 25 + a^2$$

$$a^2 - b^2 = 121 - 25 = 96$$

10. Answer: 4 Let t denote the time it takes the trains to be 560 miles apart. The distance the first train travels after t hours = 60t, and the distance the second train travels after t hours = $80t$. Therefore

$$60t + 80t = 560$$

$$140t = 560$$

$$t = 4 \text{ hours}$$

11. Answer: 27 Maximize the volume of a rectangular solid in the shape of a cube. The dimensions of the cube are given by $\sqrt[3]{216} = 6$. Divide the length of each side by 2 inches to get the number of smaller cubes that fit along each side, $\frac{6}{2} = 3$. Now, cube this number: $3^3 = 27$.

12. Answer: 8 In order to have only one solution, the discriminant of the equation must be zero: $(-6)^2 - 4(m + 1) = 0$. Solving this equation, we get $m = 8$.

13. Answer: 105 Because ΔABD is an equilateral triangle, each of its angles equals 60°. This means that $\angle BAC$ is 30°. ΔCDE is an isosceles triangle, so $\angle CDE$ and $\angle DEC$ are congruent and equal to half of $180° - 30° = \frac{150}{2} = 75°$ each. Therefore, because $\angle DEC$ and $\angle AED$ are supplementary, $\angle AED = 105°$.

14. Answer: 17 Divide each side of the second equation by $\sqrt{3}$:

$$\frac{xy\sqrt{3}}{\sqrt{3}} = \sqrt{\frac{27}{3}}$$

$$xy = 3$$

Square each side of the first equation: $x^2 + y^2 - 2xy = 11$. Replace the value of xy in this equation and solve it for $(x^2 + y^2)$: $x^2 + y^2 - 2(3) = 11$, so $x^2 + y^2 = 17$.

15. Answer: 33

Because $\overline{EF}$ is parallel to $\overline{BC}$, then $m\angle AFE = m\angle B$. Also, the right angles of ΔABC and ΔAEF are equal. Therefore, ΔABC and ΔAEF are similar triangles. Write the similarity proportion among the corresponding sides of the two triangles as follows:

$$\frac{39}{13}=\frac{AB}{11}$$

Solving this equation, we get AB = 33.

16. Answer: 4

Write each equation in slope-intercept form.

I: $3x + 2y = 7,\quad 2y = -3x + 7,\quad y=-\frac{3}{2}x+\frac{7}{2}$

II: $x - ky = 11,\quad ky = x - 11,\quad y=\frac{1}{k}x-\frac{11}{k}$

II: $mx + 4y = 14,\quad 4y = -mx + 14,\quad y=-\frac{m}{4}x+\frac{14}{4}$

Using the given conditions, we have

$$\left(-\frac{3}{2}\right)\left(\frac{1}{k}\right)=-1$$

$$\frac{1}{k}=-\frac{m}{4}$$

Solving the first equation of the preceding system, we get $k = \frac{3}{2}$. Replacing this value in the second equation, we get $m = -\frac{8}{3}$. Then, $(-mn) = -\left(-\frac{8}{3}\right)\left(\frac{3}{2}\right) = 4$.

DETAILED EXPLANATIONS

SECTION 8—CRITICAL READING

1. (B) This item requires that you think in terms of cause and effect. The boy's emotion is caused by the realization that his mother committed a certain kind of act. A person might feel "aghast"(appalled; horrified) upon realizing that his mother had committed a "heinous" (shockingly evil; abominable) act; therefore, choice (B) is correct. He wouldn't feel "inept" (lacking in fitness or aptitude) upon realizing a "benevolent" (generous, magnanimous) act; so choice (A) is incorrect. Choice (C) is wrong because the boy would not feel "contrite" (remorseful, penitent) when realizing that his mother had committed a "craven" (cowardly) act. It doesn't make sense that he would feel "circumspect" (cautious) when he realized his mother had committed an "expedient" (helpful, worthwhile) act; so choice (D) is wrong. Likewise, choice (E) is incorrect because it doesn't make sense to say that he felt "intangible" (immaterial, incapable of being touched) upon realizing that his mother committed an "inadvertent" (unintentional) act.

2. (E) Here you are looking for a word that describes what a tornado would do as it sweeps through a town. A tornado is likely to "obliterate" (destroy completely) everything in its path; thus, choice (E) is appropriate. A tornado does not choice (A), "refurbish" (make new), choice (B) "consummate" (complete; finish), choice (C) "deprave" (make corrupt), or choice (D), "thwart" (prevent from accomplishing) the things that are in its path.

3. (C) You are looking for an adjective to describe the "wheedling" (coaxing, attempts to persuade) of a child who is able to wear his parents down and a verb that describes their reaction to this wheedling. A little boy's "incessant" (uninterrupted) wheedling would wear down his parents to the point that they would finally "acquiesce" (agree, consent) to his demands. Although the choice of "defer" (submit or yield) possibly reflects the parents' reaction, choice (A) is wrong because "altruistic" (unselfish) is not an appropriate adjective to describe the wheedling of a child who wears his parents down. Choice (B) doesn't make sense; "ambiguous" (unclear, vague) does not accurately describe "wheedling," and parents who are worn down by their child would not be likely to "acclaim" (applaud) as a reaction to his demands. Choice (D) is incorrect because "vacuous" (empty) wheedling would not result in the parents' "genuflecting" (kneeling; bending the knee, especially in worship). And choice (E) doesn't make sense; although parents might "demur" (object; react by hesitating) to the demands of a wheedling child, "errant" (wandering) is not an appropriate adjective for wheedling.

4. (A) The key clues here are "he could enjoy doing all the things he enjoyed," "paparazzi," (freelance photographers who aggressively pursue celebrities for the purpose of taking candid photographs), and "adoring fans." You are looking for an adjective that describes a person—probably a person of fame or infamy—who prefers not to be recognized by the media or by fans because such recognition might keep him from doing the things he enjoys. Such a person would want to remain "incognito" (with identify concealed). Remaining choice (B), "incorrigible" (incapable of being reformed or corrected), choice (C), "inanimate" (lifeless; not moving), choice (D), "inimical" (hostile; unfriendly), or choice (E), "inexorable" (relentless; not to be persuaded) has little to do with his being able to do what he wants without the bother of paparazzi or fans.

5. (D) Here you are looking for a word that describes the ability of a person who has faced obstacles but has unquestionably succeeded. Such a person's ability could reasonably be described as "enviable" (desirable; exemplary). Such ability would not appropriately be described as choice (A), "infallible" (not capable of being wrong), choice (B), "inequitable" (unfair), choice (C), "envious" (jealous), or choice (E), "inimical" (hostile, unfriendly).

6. (A) You need a word that conveys a state that might possibly result in the sick old man's death and his family's efforts not to cause his death. Out of fear that he will die if he becomes too "disconcerted" (perturbed, agitated), the family might do all they can to "placate" (pacify; soothe) him. Choice (B) is wrong because becoming too "taciturn" (reserved, quiet) would not cause the sick old man to die, and it would not make sense for the family to do all they can to "mesmerize" (hypnotize) him. Choice (C) is incorrect because the man probably will not die from becoming too "resilient" (flexible; able to handle stress); and if the family is concerned that he will do so, they are not likely to try to "torment" (distress; disturb; cause anguish) him. Choice (D) is wrong; although the family might do all they can to "heed" (obey; attend to) the old man, "innate" (natural; inborn; inherent) does not fit in the context of the first half of the sentence. Choice (E) is wrong because if the family does fear that the man will die if he becomes too "ebullient" (showing excitement, exuberant), they probably aren't going to do all they can to "provoke" (stir the action of feeling of; arouse) him.

7. (D) In the broad sense in which it is employed here, "liberty" most nearly means freedom, (the absence of necessity or constraint in thought or action). Choices (A) and (B) are also synonyms of liberty but make little sense in this context. Choice (C) implies the removal of bondage, which is not necessary for the author's intention, whereas choice (E) has no basis whatsoever.

8. (A) The double use of the word shows that the author feels that specific agreement on how to enact liberty is not as important as enacting it. Choices (B) and (C) are both discussed by the author but do not relate to this portion of the passage. Choice (D) could be inferred by a biased reader, but nothing in the passage backs it up, and choice (E) is wrong because the author does not speak of liberty as it relates to individuals.

9. (D) The metaphor clearly states that following something inherently bloodthirsty, whether a despot or a tiger, will lead to being consumed by the stronger power. With this understanding of the metaphor, choices (A), (B), and (C) are seen to be incorrect. Although choice (E) may express an idea similar to the one implied by the metaphor, the word "valor" makes it incorrect as the author describes such nations as "foolish."

10. (B) This summation restates the author's main idea that it is the duty of a strong nation to support liberty the world over. Choice (A) could be taken to be literally correct, but the author makes no other mention of economics. Choices (C) and (E) are both in direct opposition to the other sections of the passage; the author states that the Communists are not the reason to help other nations and he states that no burden is too great to ensure liberty. Choice (D) is incorrect because the author never mentions democracy.

11. (C) "Withstand" means to oppose with firm determination and to resist successfully. This would lead to the nation reviving and prospering. Both choices (A) and (B) are synonyms, but they mean to continue and do not imply opposition or resistance. Choices (D) and (E) have no basis.

12. (B) The author is eloquently saying that no circumstances warrant fear if they are directly confronted. Choice (A) makes no sense because the author describes the problems in detail. Choices (C), (D), and (E) are all points made by the author elsewhere in the speech but have little to do with this particular phrase.

13. (B) The author is directly stating the grim reality. Choices (A) and (E) stand in direct opposition to the author's ideas. Choice (C) may be true, but it is of only slight, if any, relevance here. Choice (D) is incorrect because the author states that many besides farmers are experiencing hard times.

14. (C) The author directly states that his audience should be thankful that their perils are not as great as their forefathers' were. Choice (A) is too simplistic to be correct. Choice (B) is an attractive answer but not really the purpose of the author's citation, and choices (D) and (E) are merely tangential inferences that have little to do with what is directly stated.

15. (D) The author believes that hardship will be overcome by facing the grim conditionst. Choices (A) and (B) ignore the hope displayed by the author. Choice (C) is close, but the author describes the situation as grim, not his attitude. The author speaks of the importance of honesty, choice (E), but as a means of combating the dire circumstances, not as the overall tone of the passage.

16. (A) Both authors state that their nation can endure any hardship. Choice (B) is a concern only in the second passage. Choices (C) and (D) are not addressed in either passage, and choice (E) is a concern only in the first passage.

17. (A) Passage 1 addresses four separate groups, whereas in Passage 2 the author is speaking directly to the citizens of his "great nation," therefore making choices (B), (C), and (E) incorrect. There is ample evidence to infer the intended audience of both passages, so choice (D) is incorrect.

18. (A) Both passages speak of helping those less well off: in the first, developing countries, in the second, those hard hit by a faltering economy. Choices (B) and (C) are blindly literal interpretations that pay no attention to context, whereas choice (D) has no basis in either passage. Choice (E) could be inferred to be true, but it is not a theme of Passage 2.

19. (A) Both authors speak of the importance of understanding and support that is gained through unity. Passage 2 says that leadership must be frank and vigorous, but no mention is made of being powerful, so choice (B) is incorrect. Choices (C) and (D) are preoccupations of only Passage 1, whereas choice (E) is only a concern of Passage 2.

DETAILED EXPLANATIONS

SECTION 9—WRITING

1. (E) Choice (E) correctly uses the adverb "loudly" to modify the verb "spoke," and it does not use a vague pronoun. Choice (A) incorrectly uses the adjective "loud" to modify the verb "spoke." An adverb modifies a verb, in this case answering the question "how?" Also, in this sentence the pronoun "it" has no antecedent. Choice (B) is incorrect because it uses the adjective "loud" instead of the adverb "loudly," and the use of "until" is illogical. Choice (C) also illogically uses "until." Choice (D) uses the adjective "loud" instead of the adverb "loudly, and it contains the extra word "so." Also, in this sentence, the pronoun "it" has no antecedent.

2. (B) A comma and the conjunction "but" link the two independent clauses ("I have three sisters" and "he has no siblings"). Choice (A) uses a semicolon where it should use a comma. Choices (C) and (D) are illogical sentences, suggesting that "he has no siblings" as a result of the fact that "I have three sisters." Choice (E) is a run-on sentence, using no punctuation or conjunction between the two complete thoughts.

3. (E) Choice (E) clearly indicates the sequence of events with the use of an adverbial phrase ("after . . . Independence") preceding the main clause. It also contains two parallel items "revision" and "approval." Choice (A) is a run-on sentence, using no punctuation between two complete thoughts ("Thomas . . . Independence" and "he . . . it."). Also, the two items joined by "and" are not parallel: "revision" is a noun and "so that they could approve it" is a clause. Choice (B) is illogical because it has the Declaration of Independence presenting the document to Congress. In choices (C) and (D), the pronoun "he" is unnecessary and redundant because it repeats the subject "Thomas Jefferson." Choice (C) also contains the nonparallel structure "revision" and "so that they could approve it."

4. (E) Choice (E) correctly places the adjective phrase "having grown up with two older brothers" just before "she," the word it modifies. Choice (A) illogically has the phrase "having grown up with two older brothers" modifying "her behavior." Choice (B) is confusing because the phrase "having grown up with two older brothers" doesn't modify any part of the sentence. Choice (C) illogically has the introductory phrase ("as a result of having grown . . . brothers") modifying "her behavior." In choice (D), the pronoun "she" has no verb and so the sentence is illogical. Also, this sentence incorrectly uses the adverb "aggressively"—instead of the adjective "aggressive"—to modify "behavior."

5. (A) Choice (A) consistently uses a gerund ("descending") to define a gerund ("rappelling"). Choice (B) is incorrect because "rappelling" is a what not a "when." Choice (C) is wrong because "rappelling" is not "a cliff." Choice (D) is incorrect because "rappelling" is a what, not a "where." And choice (E) is wrong because "rappelling" is not "a person."

6. (D) In this sentence, "students" is clearly the subject of both "were required" and "forgot." No punctuation is required between the two parts of the predicate. Choice (A) is incorrect because the antecedent of "they" is vague; we cannot tell if this pronoun refers to "students" or "facts." Choice (B) has the same problem. Choice (C) illogically has "students" as the subject of "were forgotten." Choice (E) is awkward and confusing because we cannot tell what "although quickly forgotten" modifies.

7. (C) Choice (C) correctly uses a comma and "partly because" to introduce a dependent clause. Choice (A) contains a comma splice, using a comma to connect two independent thoughts ("Some . . . children" and "it . . . interests"). Also, the pronoun "it" has no antecedent. Choice (B) is awkward and wordy with the phrase "for the reason that." Choice (D) is awkward with the use of "because of knowing"; also, this sentence is unclear because we cannot determine what this phrase modifies. Choice (E) is a run-on sentence, using only a comma to connect the two complete thoughts.

8. (C) Choice (C) correctly uses past tense ("heard" and "realized") and past progressive ("was suffering"). Choice (A) shifts from past tense ("heard") to present tense ("realizes) and present progressive ("is suffering"). Choice (B) shifts from past tense ("heard" and "realizes") to past progressive ("was suffering"). Choice (D) shifts from past tense ("heard" and "realized") to present progressive ("is suffering"). Choice (E) shifts from past tense ("heard") to present ("realizes") to past perfect progressive ("had been suffering").

9. (D) In this sentence, a semicolon separates the two complete thoughts ("Abigail . . . President" and "she . . . issues."). Also, it uses past tense consistently ("was" and "was known"). Choice (A) switches from past tense ("was") to present perfect ("has been known"), and it contains a comma splice, using only a comma to connect two complete thoughts. Choice (B) contains a misplace modifier; the placement of "being known for her outspokenness" illogically has this phrase modifying "President." Choice (C) has the same problem with the modifier "having been known." Choice (E) shifts from past tense, "was," to present tense, "is."

10. (C) Choice (C) uses "the people of Paris" to replace the vague pronoun "they" in choice (A). Choice (B) has the same problem as choice (A). Also, this sentence is illogical: the verb "is" has no subject because a prepositional phrase ("On . . . Musique") cannot be a subject. Choice (D) has this same problem. And choice (E) is a fragment because it contains no main clause; the use of "when" creates a dependent clause.

11. (B) Choice (B) contains no unnecessary words; the antecedent of the pronoun "them" is obviously "Americans." Also, it uses the active-voice verb "eat," making the meaning much more distinct than the passive-voice verb "are eaten" in choice (A). Choice (A) is wordy, not only as a result of the passive-voice verb but also because of the use of "there are" and "the people of this country." Choice (C) illogically has the prepositional phrase "of this country" modifying "ants." Choice (D) has the same problem; it also unnecessarily uses the phrase "there are." And choice (E) is awkward, with the use of "eating of ants" and the passive-voice verb "is done."

12. (D) Choice (D) correctly makes the comparison between "citizens of Chicago" and "New Yorkers." Choice (A) illogically compares "Chicago" with "the citizens of New York." Choice (B) is confusing with its use of a comma and the conjunction "and." These errors create a fragment because "much like . . . New York" is a prepositional phrase, and a prepositional phrase cannot be the subject of the verb "are." Choice (C) incorrectly compares "the citizens of New York" with "Chicago." And choice (E) illogically compares "the citizens of Chicago" with "New York."

13. (B) Choice (B) contains no unnecessary words, and the items in the series are all nouns. Choice (A) contains the improper idiom "such variety . . . as," and the last item in the series is a clause instead of a noun. Items in a series must be parallel; here, all the items should be nouns. Choice (C) is wordy, with the unnecessary words "a variety of" and "as well as." Instead of separating "economics" from the rest of the items in the series, this word should be the last item in that series. Choice (D) is also wordy, with the use of "such a variety of subjects as" and "too." Choice (E) contains nonparallel structure because the last item in the series is a clause instead of a noun.

14. (A) Choice (A) correctly uses a singular verb "plans" to agree with the singular subject "one," and it consistently uses present tense ("plans") and present perfect ("has been"). Choice (B) shifts from the present-tense verb "plans" to the past perfect "had been." Choice (C) incorrectly uses the plural verb "plan" to agree with the singular subject "one." Choice (D) has the same problem. It also incorrectly uses "having," which cannot serve as a main verb; thus, this clause has no main verb. Choice (E) unnecessarily uses the pronoun "he," which creates redundancy because it repeats the subject.

LIST OF 125 COMMON AND IMPORTANT PREFIXES, SUFFIXES, AND ROOTS

	Prefix	Meaning	Word
1.	**a-, an-**	not, without	asexual, anarchy
2.	**ad-**	toward	adhesive
3.	**ambi-**	both	ambidextrous
4.	**ante-**	before	antecedent
5.	**anti-**	against	antidote
6.	**apo-**	away from	apostle
7.	**archae-**	ancient	archaeological, archaic
8.	**audio-**	hear	auditory
9.	**auto-**	self	automation
10.	**bene-**	well, good	benign
11.	**bi-**	two	bilingual
12.	**biblio-**	book	bibliophile
13.	**bio-**	life	biodegradable
14.	**capt-**	take, seize	captivated
15.	**cata-**	down	catacombs
16.	**cent-**	one hundred	centennial
17.	**circum-**	around	circumstances
18.	**co-, com-**	together	cohesive
19.	**con-, contra-**	against	contradictory
20.	**de-**	down, away from	devalue
21.	**deci-**	one-tenth	decimate

	Prefix	Meaning	Word
22.	**dia-**	cross, through	diagonal
23.	**dis-**	apart, opposite	disenfranchise, disaffected
24.	**en-**	in, cause to be	entymology, enslave
25.	**epi-**	on, upon, over	epitaph
26.	**eu-**	well, good	eugenics
27.	**ex-**	out	exotic
28.	**hemi-**	half	hemisphere
29.	**hetero-**	other	heterogeneous
30.	**hyper-**	over, excessive	hyperbole
31.	**hypo-**	under, less than normal	hypodermic, hypothermia
32.	**il-, in-, im-**	not	illogical, inconceivable, impractical
33.	**in-**	in, into, on	inspire, incarcerate
34.	**inter-**	between	interfere
35.	**intro-**	inward	introversion
36.	**iso-**	same	isoceles
37.	**mal-**	bad	malevolent
38.	**mis-**	wrong	misconstrue
39.	**mono-**	one	monopoly
40.	**multi-**	much, many	multitude
41.	**non-**	not	nonconformist
42.	**ob-**	in front of	obstruct
43.	**over-**	over, above	overwrought
44.	**pan-**	all, every	pandemic
45.	**para-**	beside	paraprofessional
46.	**photo-**	light	photogenic
47.	**poly-**	many	polytheism
48.	**post-**	after	posterior
49.	**pre-**	before	prefix
50.	**pro-**	forward, in favor of	proceed, proponent
51.	**re-**	again	relapse
52.	**retro-**	backward	retrospect
53.	**semi-**	half, partly	semiskilled
54.	**sub-**	under, below	submerge, subset
55.	**super-**	over, beyond	superimpose, superhuman
56.	**syn-**	together	synthesis
57.	**tele-**	far off	telecommunication
58.	**trans-**	across	transcend
59.	**ultra-**	beyond	ultramodern
60.	**un-**	not	unceasing
61.	**uni-**	one	unitary

	Suffix	Meaning	Word
1.	**-able**	fit for	amenable
2.	**-ac**	affected with	insomniac
3.	**-age**	rate of	dosage
4.	**-al**	relating to	thermal
5.	**-an**	characteristic of	partisan
6.	**-ance**	quality, state of	severance
7.	**-ant**	personal, impersonal agent	vagrant, reactant
8.	**-ary**	place of, for	aviary
9.	**-ate**	action in a specified way	alleviate
10.	**-dom**	quality, condition, domain	freedom, wisdom, fiefdom
11.	**-ence**	action, process	reference, agency
12.	**-en**	made of, cause to be	earthen, lengthen
13.	**-er, -or**	one who performs an action	necromancer, professor
14.	**-escent**	beginning to be, slightly	adolescent, nascent
15.	**-esque**	in the style of, manner of	grotesque
16.	**-ism**	doctrine, quality of	agrarianism
17.	**-ive**	tending toward, indicated action	furtive
18.	**-less**	not having	guileless
19.	**-ment**	object of a specified action	detriment
20.	**-ness**	state, quality, degree	garishness
21.	**-nomy**	law	autonomy
22.	**-ous**	full of, resembling	ludicrous
23.	**-y**	like that of	homey

	Root	Meaning	Word
1.	**anim**	life, mind	animation
2.	**anthrop**	man, human	anthropomorphic
3.	**auto**	self	automobile
4.	**bio**	life	biography
5.	**cap**	take, hold	caprice
6.	**chrom**	color	chromatic
7.	**chron**	time	chronometer
8.	**cycle**	circle	cyclone
9.	**dem**	people	democracy
10.	**eu**	happy	euphoric
11.	**fac**	make	facsimile
12.	**fer**	carry, bring, bear	transfer, ferry
13.	**fin**	end	finite
14.	**flex**	bend	flexible
15.	**gam**	marriage	monogamous

	Root	Meaning	Word
16.	**gen**	produce	generate
17.	**gnos**	knowledge	agnostic
18.	**grade**	go, take steps	retrograde
19.	**graph**	write	graphology
20.	**leg**	law, legal	legalize
21.	**log**	word, speech, science	logistics
22.	**loq, loqy**	speech	loquacious
23.	**meter**	measure	odometer
24.	**mit**	send, throw	mitigate
25.	**mov, mot**	move	motile
26.	**ped**	foot	pediform
27.	**ped**	child	pediatrics
28.	**phil**	love	philanthropy
29.	**phon**	sound	phonetic
30.	**reg**	rule	regimen
31.	**rupt**	break	rupture
32.	**soph**	wisdom	sophisticate
33.	**scrib, scrip**	write	script
34.	**spec**	look	introspective
35.	**stat**	stable	static
36.	**tang, tact**	touch	tangible
37.	**ten**	hold	tenure
38.	**tract**	draw, drag	detract
39.	**vert, vers**	to turn	avert
40.	**vid**	see	provident
41.	**voc**	to call	advocate

HOW TO CHOOSE A COLLEGE

By Lina Miceli, M.A.

As a guidance counselor for a large suburban high school in central New Jersey, Lina Miceli has helped countless students with one of life's biggest choices—selecting the right college.

As junior year comes to an end, and graduation is approaching, many students ask themselves, "Where do I go from here?" All students, especially those interested in furthering their education, will be faced with many questions, options, and decisions.

Selecting a college may not be the most important decision in your life, but it will have a big impact on your future. This process will be exciting, frustrating, confusing, and overwhelming, but once it is complete, you will have a great sense of satisfaction and relief.

To ensure that you end up with the "right" school, you must devote both time and energy to this process. It is best to begin your search in your junior year because this will provide the maximum amount of time to complete all of the steps.

Your biggest challenge in making this decision is to understand what is important to you and what you hope to accomplish in the future. The first step might be to think about your values, goals, dreams, expectations, and career plans. As you start your search, ask yourself these questions:

- "Why do I want to go to college?"
- "What are my career plans?"
- "Am I ready to handle a college program?"
- "Am I willing to work hard?"
- "Am I applying to college because my friends are?"

In answering these questions, did you determine that going to college is truly one of your goals? If so, you are ready to identify what you desire in a university or college. Let's explore some factors that will help you in your search.

HOW DO I FIND COLLEGE INFORMATION?

A valuable source of information is your high school guidance counselor and the counseling office. A guidance counselor can assist you in identifying your goals and strengths and can direct you toward obtaining further information. Guidance counselors are equipped to help you assess your chances of gaining admission to a particular school. A guidance counselor can also pass along information to you about schools from previous graduates.

College catalogues, videos, and handbooks are excellent sources of information. These resources are available in the high school guidance office as well as in school and public libraries. Once you know which Web sites are worth your time, the Internet is an invaluable tool for research. A brief list of informative sites is listed at the end of the chapter, with more at www.REA.com.

Also available are software programs that allow students to take inventories of their values, interests, abilities, and personality and match the results with a list of possible careers and colleges. These computer programs offer information about majors, cost, location, and admission requirements that fit the description of the type of school you wish to attend.

Most colleges have their own Web sites that provide the most current information about their school. Many colleges offer either an electronic application, which you complete online, or the opportunity to download the application for completion.

There are other ways to obtain information about colleges. High schools often sponsor college fairs where admission counselors from colleges can meet with students. In addition, most colleges hold an "open house," providing an excellent opportunity for students to speak with alumni and those that presently attend the college. Also, virtual tours of colleges are available at www.campustours.com.

TYPES OF INSTITUTIONS

There are over 3,500 institutions of higher education throughout the United States. Each has its own unique and distinctive features. Now, let's look at information that will help you determine which type of institution you want to explore.

A **college** offers instruction beyond high school; its programs satisfy the requirements for associate, baccalaureate, or graduate degrees. A **university** offers instruction beyond high school; its programs satisfy requirements for baccalaureate and graduate degrees. Universities generally have several colleges and professional schools.

Private vs. Public

Private colleges and universities are either run by a board of trustees or are church affiliated. Some church-related schools have a strong affiliation that affects curriculum and regulations, whereas others are less stringent. Privately controlled institutions are usually more expensive but have larger endowments and offer larger financial packages to overcome this difference.

Public colleges are usually controlled by a state, county, or municipality. These schools are publicly supported and, consequently, less expensive. Also, tuition is lower for in-state students than for out-of-state students.

Types of Colleges and Universities

Two-Year Colleges—These institutions offer associate degrees. The aim of a two-year college is to provide preparation for students who want to transfer to a four-year college or university, or to provide specialized training for a career in a specific field. Community colleges have open admission.

Liberal Arts Colleges—provide students with a broad foundation in arts and sciences. Students at liberal arts colleges usually major in humanities, social sciences, natural sciences, or fine or performing arts. The degree offered is usually a Bachelor of Arts or a Bachelor of Science.

Technical Institutes—are colleges that offer intensive training in engineering and other scientific fields. Some schools of technology coordinate their programs with liberal arts colleges.

Nursing Schools—There are several avenues that lead to a career in nursing. Some hospitals offer three years of intensive training, leading to state certification as a Registered Nurse. Many junior or community colleges—in conjunction with local hospitals—offer a two-year nursing program, leading to an Associate of Science degree with RN state certification. Many colleges and universities offer a four-year program of liberal arts and nursing training, leading to a Bachelor of Science degree with RN state certification.

Service Academies—are four-year colleges that offer a baccalaureate degree. Their primary purpose is to develop officers for the military. Admission is highly competitive. If you're interested in applying to the service academies, start planning with your guidance counselor in the spring of your junior year.

Career Schools—These are generally private, non-college professional schools that offer specialized training for specific careers such as photography, culinary arts, court reporting, and business. Other career schools offer technical programs such as computer technology, diesel, chemical, and electrical careers.

WHAT AM I LOOKING FOR IN A COLLEGE?

College Majors and Academic Strengths

What do you want to study? This question should guide you in your initial search for a college. Are you looking to enter a liberal arts program, or are you looking for a more specific program such as engineering or pharmacy? Obtain information from colleges that offer a major in this area. If you are not sure what college major you would like to pursue, collect information from any college you think you may want to attend. Look for colleges that offer programs that mesh with your interests.

It is also a good idea to keep in mind whether or not it will be a big problem to change your major. Statistics show that over two-thirds of students change their major; if you decide to do so, you'll want this process to be as easy as possible. This is especially important if you are not sure what major you would like to pursue.

If you are undecided about a major, an institution with a sound liberal arts program is a good choice. Most colleges do not require students to select a major until their junior year. If you are not sure now, you will have plenty of time to experiment and decide what you want to major in.

Do not assume that the harder a college is to get into, the better its quality. Colleges with higher admission standards usually have high academic standards, but many schools that are academically strong may appear more lenient on admissions in order to attract a diverse student body. You should select a college that will be academically challenging, but will also provide an environment in which you can achieve academic success.

You may also want to find out the number of students who go on to graduate school from a college or what the school's academic focus is. This information may be important should you choose to further your education.

Location

If you plan to live at school, you may want to look at schools within a five hundred mile radius from your home. This will ensure that travel cost and time allowance will not inhibit you from coming home, but will eliminate the temptation to return home often.

You should also take the location of the college into consideration. Is the college in a city, suburban area, or rural area? You should choose a location where you feel at ease because the school's location may play an important role in your college experience. There are excellent colleges in all parts of the country, but you need to find one where you will feel most comfortable and that will provide you with the best setting to get an education.

Size

The size of the prospective institution should be considered carefully. Smaller colleges tend to center more on the student as an individual. Because of their size, these types of colleges can often provide the opportunity for closer relations with faculty and professors and can offer a sense of community and belonging.

Larger schools tend to have a wider focus and diversity of programs, activities, faculty, and social opportunities. Classes at these schools tend to be larger and often held in lecture halls.

You need to decide where you think you can get the best experience. Would you feel more comfortable in a small, medium, or large school? Think about it and keep this factor in mind when deciding where you would like to attend college.

Faculty

Quite often the college catalogue proves to be the most informative listing of faculty members and their backgrounds. You should examine this list to see what types of degrees most of the professors have obtained and the amount of experience they have. Also, you might want to find out the ratio of faculty to undergraduate students, the percentage of time devoted to undergraduate teaching and advising, and if the professors make themselves readily available to assist students through seminars, after-class meetings, and office hours.

Cost

In estimating the cost of attending college you should take into account tuition, room and board, additional fees, books, traveling expenses, and personal needs. Just because a school is expensive does not mean it is good; strong colleges exist at all levels of price.

You should place more emphasis on a college's curriculum than its cost. Loans, scholarships, and financial aid are available, and colleges often offer a variety of alternative financing resources. One example of this is cooperative education, which allows students to alternate semesters of work and school.

A high-quality education does not have to be expensive. Attending a community college for a year or two may help reduce tuition expenses and still provide a good education. If you are planning on taking this route, however, it is a good idea to make sure that your credits will transfer to your next school.

Do not let the cost of a college discourage you from applying!

Social Life

Surprisingly, as many students transfer out of a college because of its social atmosphere as do students who transfer for academic reasons. It is important to locate a school that seems to have a personality comparable to your own. Select a school where you think you will feel comfortable for four years.

Look into whether the majority of students go home on the weekend, and if so, whether there are activities planned on the weekend that might interest you.

Other things to consider are the rules and regulations governing campus and dormitory life and the influence of fraternities and sororities on campus. Also, find out if the school provides for the fulfillment of religious obligations.

WHAT DO COLLEGES LOOK FOR?

Knowing the criteria used by colleges for selection of students may help you in your planning. Colleges will base their decision, in part, on the following information:

- **Your high school record**: This includes your grades as well as the courses you have taken in grades nine through twelve. The recommended guideline is as follows:
    - 4 years of English
    - 3 years of social studies
    - 3 or 4 years of mathematics
    - 2 or 3 years of science
    - 2 or 3 years of a foreign language
- Your high school program helps admissions officers determine your level of motivation and shows if you have taken a challenging curriculum. Since entrance requirements vary among colleges, it is best to take the strongest academic program possible, especially if you plan to seek admission to a highly selective college.
- **Class rank**: This is a comparison of your grade point average to that of every other student in your class.
- **Entrance examinations:** An example of an entrance exam is the SAT. Acceptable scoring ranges will vary among institutions. The scores on these tests add another dimension to your high school record.

- **Extracurricular activities:** These include hobbies, community group involvement, sports, and clubs. Colleges are interested more in the quality and depth of your involvement than in the quantity.
- **Recommendations:** These can come from counselors, teachers, employers, or people who know you very well and can write about your special traits and abilities.
- **Essay or personal statement:** This portion of the admission application allows you to express yourself and explain your goals. The essay provides information that does not appear in grades or test scores. It can reveal many things about you that can help an admissions officer determine if you are right for their college.

  The college essay is an opportunity to express your viewpoint, to be creative, and to demonstrate your writing ability. Your essay can provide the admissions officer with insight into how well you think and write and who you are.

VISITING COLLEGES

Should you visit a college campus before applying for admission?

Yes!

Although catalogues and brochures introduce you to a college, there is nothing like the experience of seeing it for yourself. A visit to your college choices can be enjoyable and worthwhile. Write to or call the college for an appointment and try to visit on a day when regular classes are in session because this will give you a realistic picture of the campus. Allow plenty of time for your visit and try to arrange to visit a class and eat at the dining hall if possible.

Before you visit, study the school's literature and write down questions you may have about it. During your visit, don't ask questions that can be answered by reading the catalogue.

Try to talk to some students, not just your assigned tour guide. Find out if you can speak with a student or faculty member in a field in which you are interested. Gather as much information as possible and don't be afraid to ask questions. This is an important decision and the more informed you are, the better equipped you will be to make your decision.

Here's a list of questions you should consider regarding the facilities and services the campus may have:

- Are academic, financial, personal, and career counseling services available?
- What health services are available to students?
- Are tutorial and writing labs available for students?
- Is there Internet access throughout the academic and residential centers?
- Are the science labs equipped with the latest science and technology equipment?
- How extensive are the library's resources, including professionally trained staff to help students with their research?
- What athletic facilities are offered for men and women?
- What intercollegiate and intramural sports are offered?
- What are the quality, nature, and availability of residential housing?

- How are roommates selected?
- Is housing guaranteed for all four years?
- What services are available for career development and job placement?

Look at the campus facilities, walk through as many buildings as you can, and take a tour of the library, dorms, and student centers. You should consider the location of the school. Can it be reached by public transportation? Will traveling to and from school be a problem?

When talking to the students at the college, you should ask how demanding the academic environment is and inquire about the social life and contact with faculty and staff. Find out how many first-year students usually return for the next year.

You should pick up an application, scholarship material, and financial aid information. If you do a thorough job, your campus visit can be vital in telling you if this college seems right for you.

After your visit, you should jot down some notes to yourself about the school; use them when evaluating other schools.

The Interview

Although most colleges still grant interviews, very few require them and many do not consider them in their admission decision. All of the applicants cannot be interviewed, so most colleges have done away with using them to avoid giving an unfair advantage. Even so, you may want to request an interview if you feel that your application does not convey your real strengths and personality, or if you need to explain some personal information that might have affected your scholastic record.

If you wish to be interviewed, you should contact the school and request an appointment. On the day of your interview, be sure to arrive on time and dress neatly and appropriately. Make a list of specific questions you would like to ask, and be prepared to answer some questions about yourself. You should answer all questions completely and honestly. Interviewers often know when someone is exaggerating the truth in an attempt to impress them. You will probably be nervous, but try to relax.

ATHLETICS

No matter how or where you are going to compete in college, you have to start planning your path while you are in high school.

When you hear about NCAA athletics, the references are to Division I athletics and the schools that generally captivate the media. The NCAA is the major governing body of collegiate athletics. This means that you need to learn its rules and live by them.

Most colleges belong to the NCAA in Division I, II, or III. If you are not being actively recruited or contacted by a certain institution, there are ways you can market and promote yourself to let coaches know that you are interested in competing at their schools.

You need to identify your needs and desires and match them first with the academic program and then with the athletic program. You need to base your decision on a combination of both factors.

Learning to market yourself as an athlete is essential. For each school you are considering, ask yourself whether you would be happy at that school if, say, you were injured and could no longer play sports.

The NCAA's requirements change every two years or so; it's important to find out how the current eligibility requirements apply to you. Be sure to check the NCAA's Web site, www.ncaa.org, for the latest eligibility requirements.

Your high school coach can serve as a great resource for college coaches to find out about your training habits, goals, and ability to fit in their program successfully.

Because the NCAA is the governing body of college sports, it is responsible for all sorts of issues, such as recruiting, financial aid, scholarships, and academic standards.

Division I and II Requirements

- Graduation from high school
- Successful completion of a required course curriculum consisting of a minimum number of courses in specified subjects
- Specific minimum SAT or ACT scores

Other Athletic Governing Associations

- National Association of Intercollegiate Athletes (NAIA), www.naia.org
- National Junior College Athletic Association (NJCAA), www.njcaa.org
- NCAA Initial-Eligibility Clearinghouse

If you are planning to enroll in college as a freshman and want to play on a Division I or Division II team, the NCAA Initial-Eligibility Clearinghouse must certify you before you can play. It is your responsibility to make sure the Clearinghouse has everything it needs to certify you, such as the student release form, an application fee, an official high school transcript, and SAT or ACT score reports.

Marketing Yourself

Be honest with yourself about your abilities both academically and athletically. Take a close look at your strengths and weaknesses and how they pertain to playing sports in college. Talk to your high school coach and find out what he or she thinks of your potential.

Over the summer between your junior and senior year, send letters to the coaches of teams for which you are interested in playing. Include a packet and make sure you list your upcoming competitions so they know where and when to look for you in action. Send updates during the year to every coach who has indicated interest in you. Ask your high school coach to call the college coaches on your behalf. Send a videotape of your performances.

Your athletic packet should include information about yourself and your athletic achievements, GPA, SAT/ACT and other standardized test scores, high school transcript, school profile, and a list of relevant references (e.g., coaches, teachers, members of the community). Include their phone numbers and addresses, but remember to ask their permission to use them as references.

Also, include your schedule and your athletic resume.

HOW TO APPLY

After you have made a list of colleges that you feel are right for you, narrow down the list and carefully include one or two colleges for which your scholastic record indicates that you have an excellent chance for admission. Your final choices, probably no more than five colleges, should be schools that you believe you will be happy attending.

Many colleges give students the option of applying on-line or completing a hard copy application. The Common Application is a single application for undergraduate college admission used by a consortium of selective colleges and universities. There are currently 240 member colleges and universities who accept it. You can find the member list on the Common Application Web site, www.commonapp.org.

Whatever you choose, do so early in your senior year. Allow yourself plenty of time to complete the application. Be sure to tell your counselor which schools you are applying to and request your transcript to be sent to those colleges.

Here are some hints for applying to a college:

- Read the application thoroughly before completing it.
- Make a copy of your application and complete a rough draft. When you fill out the actual application, be neat—type it if you can.
- Answer all of the questions accurately and get help from your counselor if you are not sure how to answer a question.
- If an essay is required, make a rough copy before putting it on the application.
- Follow the guidelines given on the application for the essay. Try to be brief but fully answer the question.
- Be sure to check your essay for correct spelling, grammar, punctuation, and clarity.
- Review your completed application with your counselor. Your high school transcript will also be sent to colleges you apply to, so if you have any questions about it, you should ask your counselor.
- Be sure to give your counselor the application well before the deadline so that all of the necessary forms can be completed and mailed to the college.
- Your school record can be sent with your application or mailed separately.

After You Apply

Each college you apply to will send you a letter indicating that you have been accepted, denied admission, or put on the waiting list.

Early decision candidates will usually be notified in December. If you are not accepted during the early decision period, your application will be reconsidered for admission later in the school year. Once you have been accepted to a school and have made a decision to attend, you should withdraw any applications filed at other schools.

If you have been accepted, most colleges require a tuition deposit for enrollment. This fee must be submitted by the candidate's reply date, usually May 1, to secure your place in the entering class. Most acceptances are contingent upon satisfactory completion of senior year course work. Acceptances have been revoked due to failure to maintain academic standards during the remainder of the senior year, so don't slack off.

You may find that you have been put on a waiting list. If so, find out exactly what that means at that particular school. Do not count on being accepted at a school that has wait-listed you. You should try and hold your place at a school where you have been accepted to guarantee a place for you come September.

If you have been denied admission to all the schools to which you have applied, see your guidance counselor. You may still be able to apply at other schools that can meet your educational needs. Do not give up!

When it comes to making your final decision, your best choice is to go with the school that feels most comfortable to you. Be sure to let your guidance counselor know which school you are planning to attend so that your final school records can be sent to your new school.

Good luck!

Visit www.REA.com for a handy pre-college calendar and checklist, and a bevy of Web sites to help you with your college plans. Below you'll find some top picks from this section's author, Lina Miceli, M.A.

FastWeb College Search	www.fastweb.com
Yahoo College Directory	www.dir.yahoo.com/Education/Higher
Campus Tours	www.campustours.com
College Source	www.collegesource.org
Common Application	www.commonapp.org
ECollegeApps	www.ecollegeaps.com
Wired Scholar	www.wiredscholar.com
GoCollege	www.gocollege.com

INDEX

A

B

C

D

E

F

G

H

I

K

L

M

N

O

P

Q

R

S

T

U

V

W

X

Y

ANSWER SHEETS - DIAGNOSTIC TEST

Begin your essay on this page. If necessary, continue on the next page.

Continue on the next page if necessary.

Continuation of your essay from previous page, if necessary.

Begin with number 1 for each section. If a section has fewer questions than answer spaces, leave the extra answer spaces blank. Be sure to erase any errors or stray marks completely.

SECTION 2

1	Ⓐ Ⓑ Ⓒ Ⓓ Ⓔ	11	Ⓐ Ⓑ Ⓒ Ⓓ Ⓔ	21	Ⓐ Ⓑ Ⓒ Ⓓ Ⓔ	31	Ⓐ Ⓑ Ⓒ Ⓓ Ⓔ
2	Ⓐ Ⓑ Ⓒ Ⓓ Ⓔ	12	Ⓐ Ⓑ Ⓒ Ⓓ Ⓔ	22	Ⓐ Ⓑ Ⓒ Ⓓ Ⓔ	32	Ⓐ Ⓑ Ⓒ Ⓓ Ⓔ
3	Ⓐ Ⓑ Ⓒ Ⓓ Ⓔ	13	Ⓐ Ⓑ Ⓒ Ⓓ Ⓔ	23	Ⓐ Ⓑ Ⓒ Ⓓ Ⓔ	33	Ⓐ Ⓑ Ⓒ Ⓓ Ⓔ
4	Ⓐ Ⓑ Ⓒ Ⓓ Ⓔ	14	Ⓐ Ⓑ Ⓒ Ⓓ Ⓔ	24	Ⓐ Ⓑ Ⓒ Ⓓ Ⓔ	34	Ⓐ Ⓑ Ⓒ Ⓓ Ⓔ
5	Ⓐ Ⓑ Ⓒ Ⓓ Ⓔ	15	Ⓐ Ⓑ Ⓒ Ⓓ Ⓔ	25	Ⓐ Ⓑ Ⓒ Ⓓ Ⓔ	35	Ⓐ Ⓑ Ⓒ Ⓓ Ⓔ
6	Ⓐ Ⓑ Ⓒ Ⓓ Ⓔ	16	Ⓐ Ⓑ Ⓒ Ⓓ Ⓔ	26	Ⓐ Ⓑ Ⓒ Ⓓ Ⓔ	36	Ⓐ Ⓑ Ⓒ Ⓓ Ⓔ
7	Ⓐ Ⓑ Ⓒ Ⓓ Ⓔ	17	Ⓐ Ⓑ Ⓒ Ⓓ Ⓔ	27	Ⓐ Ⓑ Ⓒ Ⓓ Ⓔ	37	Ⓐ Ⓑ Ⓒ Ⓓ Ⓔ
8	Ⓐ Ⓑ Ⓒ Ⓓ Ⓔ	18	Ⓐ Ⓑ Ⓒ Ⓓ Ⓔ	28	Ⓐ Ⓑ Ⓒ Ⓓ Ⓔ	38	Ⓐ Ⓑ Ⓒ Ⓓ Ⓔ
9	Ⓐ Ⓑ Ⓒ Ⓓ Ⓔ	19	Ⓐ Ⓑ Ⓒ Ⓓ Ⓔ	29	Ⓐ Ⓑ Ⓒ Ⓓ Ⓔ	39	Ⓐ Ⓑ Ⓒ Ⓓ Ⓔ
10	Ⓐ Ⓑ Ⓒ Ⓓ Ⓔ	20	Ⓐ Ⓑ Ⓒ Ⓓ Ⓔ	30	Ⓐ Ⓑ Ⓒ Ⓓ Ⓔ	40	Ⓐ Ⓑ Ⓒ Ⓓ Ⓔ

SECTION 3

1	Ⓐ Ⓑ Ⓒ Ⓓ Ⓔ	11	Ⓐ Ⓑ Ⓒ Ⓓ Ⓔ	21	Ⓐ Ⓑ Ⓒ Ⓓ Ⓔ	31	Ⓐ Ⓑ Ⓒ Ⓓ Ⓔ
2	Ⓐ Ⓑ Ⓒ Ⓓ Ⓔ	12	Ⓐ Ⓑ Ⓒ Ⓓ Ⓔ	22	Ⓐ Ⓑ Ⓒ Ⓓ Ⓔ	32	Ⓐ Ⓑ Ⓒ Ⓓ Ⓔ
3	Ⓐ Ⓑ Ⓒ Ⓓ Ⓔ	13	Ⓐ Ⓑ Ⓒ Ⓓ Ⓔ	23	Ⓐ Ⓑ Ⓒ Ⓓ Ⓔ	33	Ⓐ Ⓑ Ⓒ Ⓓ Ⓔ
4	Ⓐ Ⓑ Ⓒ Ⓓ Ⓔ	14	Ⓐ Ⓑ Ⓒ Ⓓ Ⓔ	24	Ⓐ Ⓑ Ⓒ Ⓓ Ⓔ	34	Ⓐ Ⓑ Ⓒ Ⓓ Ⓔ
5	Ⓐ Ⓑ Ⓒ Ⓓ Ⓔ	15	Ⓐ Ⓑ Ⓒ Ⓓ Ⓔ	25	Ⓐ Ⓑ Ⓒ Ⓓ Ⓔ	35	Ⓐ Ⓑ Ⓒ Ⓓ Ⓔ
6	Ⓐ Ⓑ Ⓒ Ⓓ Ⓔ	16	Ⓐ Ⓑ Ⓒ Ⓓ Ⓔ	26	Ⓐ Ⓑ Ⓒ Ⓓ Ⓔ	36	Ⓐ Ⓑ Ⓒ Ⓓ Ⓔ
7	Ⓐ Ⓑ Ⓒ Ⓓ Ⓔ	17	Ⓐ Ⓑ Ⓒ Ⓓ Ⓔ	27	Ⓐ Ⓑ Ⓒ Ⓓ Ⓔ	37	Ⓐ Ⓑ Ⓒ Ⓓ Ⓔ
8	Ⓐ Ⓑ Ⓒ Ⓓ Ⓔ	18	Ⓐ Ⓑ Ⓒ Ⓓ Ⓔ	28	Ⓐ Ⓑ Ⓒ Ⓓ Ⓔ	38	Ⓐ Ⓑ Ⓒ Ⓓ Ⓔ
9	Ⓐ Ⓑ Ⓒ Ⓓ Ⓔ	19	Ⓐ Ⓑ Ⓒ Ⓓ Ⓔ	29	Ⓐ Ⓑ Ⓒ Ⓓ Ⓔ	39	Ⓐ Ⓑ Ⓒ Ⓓ Ⓔ
10	Ⓐ Ⓑ Ⓒ Ⓓ Ⓔ	20	Ⓐ Ⓑ Ⓒ Ⓓ Ⓔ	30	Ⓐ Ⓑ Ⓒ Ⓓ Ⓔ	40	Ⓐ Ⓑ Ⓒ Ⓓ Ⓔ

SECTION 4

1	Ⓐ Ⓑ Ⓒ Ⓓ Ⓔ	11	Ⓐ Ⓑ Ⓒ Ⓓ Ⓔ	21	Ⓐ Ⓑ Ⓒ Ⓓ Ⓔ	31	Ⓐ Ⓑ Ⓒ Ⓓ Ⓔ
2	Ⓐ Ⓑ Ⓒ Ⓓ Ⓔ	12	Ⓐ Ⓑ Ⓒ Ⓓ Ⓔ	22	Ⓐ Ⓑ Ⓒ Ⓓ Ⓔ	32	Ⓐ Ⓑ Ⓒ Ⓓ Ⓔ
3	Ⓐ Ⓑ Ⓒ Ⓓ Ⓔ	13	Ⓐ Ⓑ Ⓒ Ⓓ Ⓔ	23	Ⓐ Ⓑ Ⓒ Ⓓ Ⓔ	33	Ⓐ Ⓑ Ⓒ Ⓓ Ⓔ
4	Ⓐ Ⓑ Ⓒ Ⓓ Ⓔ	14	Ⓐ Ⓑ Ⓒ Ⓓ Ⓔ	24	Ⓐ Ⓑ Ⓒ Ⓓ Ⓔ	34	Ⓐ Ⓑ Ⓒ Ⓓ Ⓔ
5	Ⓐ Ⓑ Ⓒ Ⓓ Ⓔ	15	Ⓐ Ⓑ Ⓒ Ⓓ Ⓔ	25	Ⓐ Ⓑ Ⓒ Ⓓ Ⓔ	35	Ⓐ Ⓑ Ⓒ Ⓓ Ⓔ
6	Ⓐ Ⓑ Ⓒ Ⓓ Ⓔ	16	Ⓐ Ⓑ Ⓒ Ⓓ Ⓔ	26	Ⓐ Ⓑ Ⓒ Ⓓ Ⓔ	36	Ⓐ Ⓑ Ⓒ Ⓓ Ⓔ
7	Ⓐ Ⓑ Ⓒ Ⓓ Ⓔ	17	Ⓐ Ⓑ Ⓒ Ⓓ Ⓔ	27	Ⓐ Ⓑ Ⓒ Ⓓ Ⓔ	37	Ⓐ Ⓑ Ⓒ Ⓓ Ⓔ
8	Ⓐ Ⓑ Ⓒ Ⓓ Ⓔ	18	Ⓐ Ⓑ Ⓒ Ⓓ Ⓔ	28	Ⓐ Ⓑ Ⓒ Ⓓ Ⓔ	38	Ⓐ Ⓑ Ⓒ Ⓓ Ⓔ
9	Ⓐ Ⓑ Ⓒ Ⓓ Ⓔ	19	Ⓐ Ⓑ Ⓒ Ⓓ Ⓔ	29	Ⓐ Ⓑ Ⓒ Ⓓ Ⓔ	39	Ⓐ Ⓑ Ⓒ Ⓓ Ⓔ
10	Ⓐ Ⓑ Ⓒ Ⓓ Ⓔ	20	Ⓐ Ⓑ Ⓒ Ⓓ Ⓔ	30	Ⓐ Ⓑ Ⓒ Ⓓ Ⓔ	40	Ⓐ Ⓑ Ⓒ Ⓓ Ⓔ

Begin with number 1 for each section. If a section has fewer questions than answer spaces, leave the extra answer spaces blank. Be sure to erase any errors or stray marks completely.

SECTION 5

1 (A) (B) (C) (D) (E)	11 (A) (B) (C) (D) (E)	21 (A) (B) (C) (D) (E)	31 (A) (B) (C) (D) (E)
2 (A) (B) (C) (D) (E)	12 (A) (B) (C) (D) (E)	22 (A) (B) (C) (D) (E)	32 (A) (B) (C) (D) (E)
3 (A) (B) (C) (D) (E)	13 (A) (B) (C) (D) (E)	23 (A) (B) (C) (D) (E)	33 (A) (B) (C) (D) (E)
4 (A) (B) (C) (D) (E)	14 (A) (B) (C) (D) (E)	24 (A) (B) (C) (D) (E)	34 (A) (B) (C) (D) (E)
5 (A) (B) (C) (D) (E)	15 (A) (B) (C) (D) (E)	25 (A) (B) (C) (D) (E)	35 (A) (B) (C) (D) (E)
6 (A) (B) (C) (D) (E)	16 (A) (B) (C) (D) (E)	26 (A) (B) (C) (D) (E)	36 (A) (B) (C) (D) (E)
7 (A) (B) (C) (D) (E)	17 (A) (B) (C) (D) (E)	27 (A) (B) (C) (D) (E)	37 (A) (B) (C) (D) (E)
8 (A) (B) (C) (D) (E)	18 (A) (B) (C) (D) (E)	28 (A) (B) (C) (D) (E)	38 (A) (B) (C) (D) (E)
9 (A) (B) (C) (D) (E)	19 (A) (B) (C) (D) (E)	29 (A) (B) (C) (D) (E)	39 (A) (B) (C) (D) (E)
10 (A) (B) (C) (D) (E)	20 (A) (B) (C) (D) (E)	30 (A) (B) (C) (D) (E)	40 (A) (B) (C) (D) (E)

SECTION 6

1 (A) (B) (C) (D) (E)	11 (A) (B) (C) (D) (E)	21 (A) (B) (C) (D) (E)	31 (A) (B) (C) (D) (E)
2 (A) (B) (C) (D) (E)	12 (A) (B) (C) (D) (E)	22 (A) (B) (C) (D) (E)	32 (A) (B) (C) (D) (E)
3 (A) (B) (C) (D) (E)	13 (A) (B) (C) (D) (E)	23 (A) (B) (C) (D) (E)	33 (A) (B) (C) (D) (E)
4 (A) (B) (C) (D) (E)	14 (A) (B) (C) (D) (E)	24 (A) (B) (C) (D) (E)	34 (A) (B) (C) (D) (E)
5 (A) (B) (C) (D) (E)	15 (A) (B) (C) (D) (E)	25 (A) (B) (C) (D) (E)	35 (A) (B) (C) (D) (E)
6 (A) (B) (C) (D) (E)	16 (A) (B) (C) (D) (E)	26 (A) (B) (C) (D) (E)	36 (A) (B) (C) (D) (E)
7 (A) (B) (C) (D) (E)	17 (A) (B) (C) (D) (E)	27 (A) (B) (C) (D) (E)	37 (A) (B) (C) (D) (E)
8 (A) (B) (C) (D) (E)	18 (A) (B) (C) (D) (E)	28 (A) (B) (C) (D) (E)	38 (A) (B) (C) (D) (E)
9 (A) (B) (C) (D) (E)	19 (A) (B) (C) (D) (E)	29 (A) (B) (C) (D) (E)	39 (A) (B) (C) (D) (E)
10 (A) (B) (C) (D) (E)	20 (A) (B) (C) (D) (E)	30 (A) (B) (C) (D) (E)	40 (A) (B) (C) (D) (E)

SECTION 7

1 (A) (B) (C) (D) (E)	11 (A) (B) (C) (D) (E)	21 (A) (B) (C) (D) (E)	31 (A) (B) (C) (D) (E)
2 (A) (B) (C) (D) (E)	12 (A) (B) (C) (D) (E)	22 (A) (B) (C) (D) (E)	32 (A) (B) (C) (D) (E)
3 (A) (B) (C) (D) (E)	13 (A) (B) (C) (D) (E)	23 (A) (B) (C) (D) (E)	33 (A) (B) (C) (D) (E)
4 (A) (B) (C) (D) (E)	14 (A) (B) (C) (D) (E)	24 (A) (B) (C) (D) (E)	34 (A) (B) (C) (D) (E)
5 (A) (B) (C) (D) (E)	15 (A) (B) (C) (D) (E)	25 (A) (B) (C) (D) (E)	35 (A) (B) (C) (D) (E)
6 (A) (B) (C) (D) (E)	16 (A) (B) (C) (D) (E)	26 (A) (B) (C) (D) (E)	36 (A) (B) (C) (D) (E)
7 (A) (B) (C) (D) (E)	17 (A) (B) (C) (D) (E)	27 (A) (B) (C) (D) (E)	37 (A) (B) (C) (D) (E)
8 (A) (B) (C) (D) (E)	18 (A) (B) (C) (D) (E)	28 (A) (B) (C) (D) (E)	38 (A) (B) (C) (D) (E)
9 (A) (B) (C) (D) (E)	19 (A) (B) (C) (D) (E)	29 (A) (B) (C) (D) (E)	39 (A) (B) (C) (D) (E)
10 (A) (B) (C) (D) (E)	20 (A) (B) (C) (D) (E)	30 (A) (B) (C) (D) (E)	40 (A) (B) (C) (D) (E)

CONTINUE TO THE GRIDS FOR SECTION 7 ON THE NEXT PAGE WHEN INDICATED IN THE TEST BOOK.

Only answers entered in the ovals in each grid area will be scored.
No credit will be given for anything written in the boxes above the ovals.

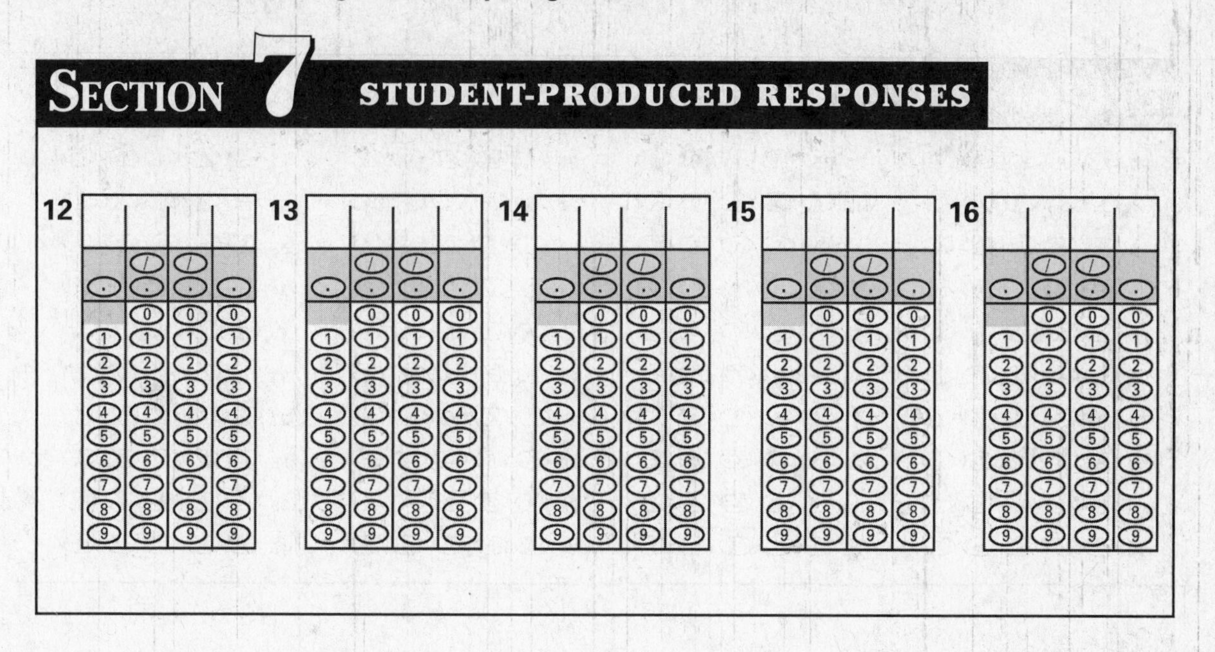

Begin with number 1 for each section. If a section has fewer questions than answer spaces, leave the extra answer spaces blank. Be sure to erase any errors or stray marks completely.

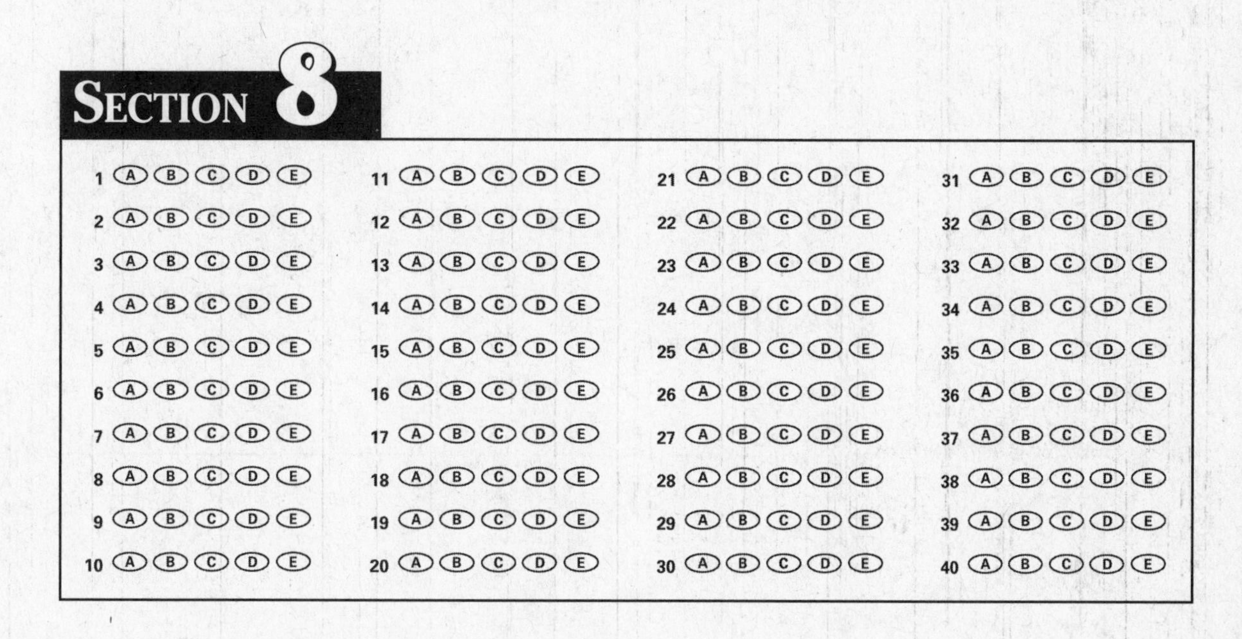

Begin with number 1 for each section. If a section has fewer questions than answer spaces, leave the extra answer spaces blank. Be sure to erase any errors or stray marks completely.

SECTION 9

1 Ⓐ Ⓑ Ⓒ Ⓓ Ⓔ	11 Ⓐ Ⓑ Ⓒ Ⓓ Ⓔ	21 Ⓐ Ⓑ Ⓒ Ⓓ Ⓔ	31 Ⓐ Ⓑ Ⓒ Ⓓ Ⓔ
2 Ⓐ Ⓑ Ⓒ Ⓓ Ⓔ	12 Ⓐ Ⓑ Ⓒ Ⓓ Ⓔ	22 Ⓐ Ⓑ Ⓒ Ⓓ Ⓔ	32 Ⓐ Ⓑ Ⓒ Ⓓ Ⓔ
3 Ⓐ Ⓑ Ⓒ Ⓓ Ⓔ	13 Ⓐ Ⓑ Ⓒ Ⓓ Ⓔ	23 Ⓐ Ⓑ Ⓒ Ⓓ Ⓔ	33 Ⓐ Ⓑ Ⓒ Ⓓ Ⓔ
4 Ⓐ Ⓑ Ⓒ Ⓓ Ⓔ	14 Ⓐ Ⓑ Ⓒ Ⓓ Ⓔ	24 Ⓐ Ⓑ Ⓒ Ⓓ Ⓔ	34 Ⓐ Ⓑ Ⓒ Ⓓ Ⓔ
5 Ⓐ Ⓑ Ⓒ Ⓓ Ⓔ	15 Ⓐ Ⓑ Ⓒ Ⓓ Ⓔ	25 Ⓐ Ⓑ Ⓒ Ⓓ Ⓔ	35 Ⓐ Ⓑ Ⓒ Ⓓ Ⓔ
6 Ⓐ Ⓑ Ⓒ Ⓓ Ⓔ	16 Ⓐ Ⓑ Ⓒ Ⓓ Ⓔ	26 Ⓐ Ⓑ Ⓒ Ⓓ Ⓔ	36 Ⓐ Ⓑ Ⓒ Ⓓ Ⓔ
7 Ⓐ Ⓑ Ⓒ Ⓓ Ⓔ	17 Ⓐ Ⓑ Ⓒ Ⓓ Ⓔ	27 Ⓐ Ⓑ Ⓒ Ⓓ Ⓔ	37 Ⓐ Ⓑ Ⓒ Ⓓ Ⓔ
8 Ⓐ Ⓑ Ⓒ Ⓓ Ⓔ	18 Ⓐ Ⓑ Ⓒ Ⓓ Ⓔ	28 Ⓐ Ⓑ Ⓒ Ⓓ Ⓔ	38 Ⓐ Ⓑ Ⓒ Ⓓ Ⓔ
9 Ⓐ Ⓑ Ⓒ Ⓓ Ⓔ	19 Ⓐ Ⓑ Ⓒ Ⓓ Ⓔ	29 Ⓐ Ⓑ Ⓒ Ⓓ Ⓔ	39 Ⓐ Ⓑ Ⓒ Ⓓ Ⓔ
10 Ⓐ Ⓑ Ⓒ Ⓓ Ⓔ	20 Ⓐ Ⓑ Ⓒ Ⓓ Ⓔ	30 Ⓐ Ⓑ Ⓒ Ⓓ Ⓔ	40 Ⓐ Ⓑ Ⓒ Ⓓ Ⓔ

ANSWER SHEETS - PRACTICE TEST 1

Begin your essay on this page. If necessary, continue on the next page.

Continue on the next page if necessary.

Continuation of your essay from previous page, if necessary.

Begin with number 1 for each section. If a section has fewer questions than answer spaces, leave the extra answer spaces blank. Be sure to erase any errors or stray marks completely.

SECTION 2

1 A B C D E	11 A B C D E	21 A B C D E	31 A B C D E
2 A B C D E	12 A B C D E	22 A B C D E	32 A B C D E
3 A B C D E	13 A B C D E	23 A B C D E	33 A B C D E
4 A B C D E	14 A B C D E	24 A B C D E	34 A B C D E
5 A B C D E	15 A B C D E	25 A B C D E	35 A B C D E
6 A B C D E	16 A B C D E	26 A B C D E	36 A B C D E
7 A B C D E	17 A B C D E	27 A B C D E	37 A B C D E
8 A B C D E	18 A B C D E	28 A B C D E	38 A B C D E
9 A B C D E	19 A B C D E	29 A B C D E	39 A B C D E
10 A B C D E	20 A B C D E	30 A B C D E	40 A B C D E

SECTION 3

1 A B C D E	11 A B C D E	21 A B C D E	31 A B C D E
2 A B C D E	12 A B C D E	22 A B C D E	32 A B C D E
3 A B C D E	13 A B C D E	23 A B C D E	33 A B C D E
4 A B C D E	14 A B C D E	24 A B C D E	34 A B C D E
5 A B C D E	15 A B C D E	25 A B C D E	35 A B C D E
6 A B C D E	16 A B C D E	26 A B C D E	36 A B C D E
7 A B C D E	17 A B C D E	27 A B C D E	37 A B C D E
8 A B C D E	18 A B C D E	28 A B C D E	38 A B C D E
9 A B C D E	19 A B C D E	29 A B C D E	39 A B C D E
10 A B C D E	20 A B C D E	30 A B C D E	40 A B C D E

SECTION 4

1 A B C D E	11 A B C D E	21 A B C D E	31 A B C D E
2 A B C D E	12 A B C D E	22 A B C D E	32 A B C D E
3 A B C D E	13 A B C D E	23 A B C D E	33 A B C D E
4 A B C D E	14 A B C D E	24 A B C D E	34 A B C D E
5 A B C D E	15 A B C D E	25 A B C D E	35 A B C D E
6 A B C D E	16 A B C D E	26 A B C D E	36 A B C D E
7 A B C D E	17 A B C D E	27 A B C D E	37 A B C D E
8 A B C D E	18 A B C D E	28 A B C D E	38 A B C D E
9 A B C D E	19 A B C D E	29 A B C D E	39 A B C D E
10 A B C D E	20 A B C D E	30 A B C D E	40 A B C D E

Begin with number 1 for each section. If a section has fewer questions than answer spaces, leave the extra answer spaces blank. Be sure to erase any errors or stray marks completely.

SECTION 5

1 A B C D E	11 A B C D E	21 A B C D E	31 A B C D E
2 A B C D E	12 A B C D E	22 A B C D E	32 A B C D E
3 A B C D E	13 A B C D E	23 A B C D E	33 A B C D E
4 A B C D E	14 A B C D E	24 A B C D E	34 A B C D E
5 A B C D E	15 A B C D E	25 A B C D E	35 A B C D E
6 A B C D E	16 A B C D E	26 A B C D E	36 A B C D E
7 A B C D E	17 A B C D E	27 A B C D E	37 A B C D E
8 A B C D E	18 A B C D E	28 A B C D E	38 A B C D E
9 A B C D E	19 A B C D E	29 A B C D E	39 A B C D E
10 A B C D E	20 A B C D E	30 A B C D E	40 A B C D E

SECTION 6

1 A B C D E	11 A B C D E	21 A B C D E	31 A B C D E
2 A B C D E	12 A B C D E	22 A B C D E	32 A B C D E
3 A B C D E	13 A B C D E	23 A B C D E	33 A B C D E
4 A B C D E	14 A B C D E	24 A B C D E	34 A B C D E
5 A B C D E	15 A B C D E	25 A B C D E	35 A B C D E
6 A B C D E	16 A B C D E	26 A B C D E	36 A B C D E
7 A B C D E	17 A B C D E	27 A B C D E	37 A B C D E
8 A B C D E	18 A B C D E	28 A B C D E	38 A B C D E
9 A B C D E	19 A B C D E	29 A B C D E	39 A B C D E
10 A B C D E	20 A B C D E	30 A B C D E	40 A B C D E

SECTION 7

1 A B C D E	11 A B C D E	21 A B C D E	31 A B C D E
2 A B C D E	12 A B C D E	22 A B C D E	32 A B C D E
3 A B C D E	13 A B C D E	23 A B C D E	33 A B C D E
4 A B C D E	14 A B C D E	24 A B C D E	34 A B C D E
5 A B C D E	15 A B C D E	25 A B C D E	35 A B C D E
6 A B C D E	16 A B C D E	26 A B C D E	36 A B C D E
7 A B C D E	17 A B C D E	27 A B C D E	37 A B C D E
8 A B C D E	18 A B C D E	28 A B C D E	38 A B C D E
9 A B C D E	19 A B C D E	29 A B C D E	39 A B C D E
10 A B C D E	20 A B C D E	30 A B C D E	40 A B C D E

CONTINUE TO THE GRIDS FOR SECTION 7 ON THE NEXT PAGE WHEN INDICATED IN THE TEST BOOK.

Only answers entered in the ovals in each grid area will be scored.
No credit will be given for anything written in the boxes above the ovals.

SECTION 7 STUDENT-PRODUCED RESPONSES

6

7

8

9

10

11

12

13

14

15

16

Begin with number 1 for each section. If a section has fewer questions than answer spaces, leave the extra answer spaces blank. Be sure to erase any errors or stray marks completely.

SECTION 8

1 Ⓐ Ⓑ Ⓒ Ⓓ Ⓔ	11 Ⓐ Ⓑ Ⓒ Ⓓ Ⓔ	21 Ⓐ Ⓑ Ⓒ Ⓓ Ⓔ	31 Ⓐ Ⓑ Ⓒ Ⓓ Ⓔ
2 Ⓐ Ⓑ Ⓒ Ⓓ Ⓔ	12 Ⓐ Ⓑ Ⓒ Ⓓ Ⓔ	22 Ⓐ Ⓑ Ⓒ Ⓓ Ⓔ	32 Ⓐ Ⓑ Ⓒ Ⓓ Ⓔ
3 Ⓐ Ⓑ Ⓒ Ⓓ Ⓔ	13 Ⓐ Ⓑ Ⓒ Ⓓ Ⓔ	23 Ⓐ Ⓑ Ⓒ Ⓓ Ⓔ	33 Ⓐ Ⓑ Ⓒ Ⓓ Ⓔ
4 Ⓐ Ⓑ Ⓒ Ⓓ Ⓔ	14 Ⓐ Ⓑ Ⓒ Ⓓ Ⓔ	24 Ⓐ Ⓑ Ⓒ Ⓓ Ⓔ	34 Ⓐ Ⓑ Ⓒ Ⓓ Ⓔ
5 Ⓐ Ⓑ Ⓒ Ⓓ Ⓔ	15 Ⓐ Ⓑ Ⓒ Ⓓ Ⓔ	25 Ⓐ Ⓑ Ⓒ Ⓓ Ⓔ	35 Ⓐ Ⓑ Ⓒ Ⓓ Ⓔ
6 Ⓐ Ⓑ Ⓒ Ⓓ Ⓔ	16 Ⓐ Ⓑ Ⓒ Ⓓ Ⓔ	26 Ⓐ Ⓑ Ⓒ Ⓓ Ⓔ	36 Ⓐ Ⓑ Ⓒ Ⓓ Ⓔ
7 Ⓐ Ⓑ Ⓒ Ⓓ Ⓔ	17 Ⓐ Ⓑ Ⓒ Ⓓ Ⓔ	27 Ⓐ Ⓑ Ⓒ Ⓓ Ⓔ	37 Ⓐ Ⓑ Ⓒ Ⓓ Ⓔ
8 Ⓐ Ⓑ Ⓒ Ⓓ Ⓔ	18 Ⓐ Ⓑ Ⓒ Ⓓ Ⓔ	28 Ⓐ Ⓑ Ⓒ Ⓓ Ⓔ	38 Ⓐ Ⓑ Ⓒ Ⓓ Ⓔ
9 Ⓐ Ⓑ Ⓒ Ⓓ Ⓔ	19 Ⓐ Ⓑ Ⓒ Ⓓ Ⓔ	29 Ⓐ Ⓑ Ⓒ Ⓓ Ⓔ	39 Ⓐ Ⓑ Ⓒ Ⓓ Ⓔ
10 Ⓐ Ⓑ Ⓒ Ⓓ Ⓔ	20 Ⓐ Ⓑ Ⓒ Ⓓ Ⓔ	30 Ⓐ Ⓑ Ⓒ Ⓓ Ⓔ	40 Ⓐ Ⓑ Ⓒ Ⓓ Ⓔ

SECTION 9

1 Ⓐ Ⓑ Ⓒ Ⓓ Ⓔ	11 Ⓐ Ⓑ Ⓒ Ⓓ Ⓔ	21 Ⓐ Ⓑ Ⓒ Ⓓ Ⓔ	31 Ⓐ Ⓑ Ⓒ Ⓓ Ⓔ
2 Ⓐ Ⓑ Ⓒ Ⓓ Ⓔ	12 Ⓐ Ⓑ Ⓒ Ⓓ Ⓔ	22 Ⓐ Ⓑ Ⓒ Ⓓ Ⓔ	32 Ⓐ Ⓑ Ⓒ Ⓓ Ⓔ
3 Ⓐ Ⓑ Ⓒ Ⓓ Ⓔ	13 Ⓐ Ⓑ Ⓒ Ⓓ Ⓔ	23 Ⓐ Ⓑ Ⓒ Ⓓ Ⓔ	33 Ⓐ Ⓑ Ⓒ Ⓓ Ⓔ
4 Ⓐ Ⓑ Ⓒ Ⓓ Ⓔ	14 Ⓐ Ⓑ Ⓒ Ⓓ Ⓔ	24 Ⓐ Ⓑ Ⓒ Ⓓ Ⓔ	34 Ⓐ Ⓑ Ⓒ Ⓓ Ⓔ
5 Ⓐ Ⓑ Ⓒ Ⓓ Ⓔ	15 Ⓐ Ⓑ Ⓒ Ⓓ Ⓔ	25 Ⓐ Ⓑ Ⓒ Ⓓ Ⓔ	35 Ⓐ Ⓑ Ⓒ Ⓓ Ⓔ
6 Ⓐ Ⓑ Ⓒ Ⓓ Ⓔ	16 Ⓐ Ⓑ Ⓒ Ⓓ Ⓔ	26 Ⓐ Ⓑ Ⓒ Ⓓ Ⓔ	36 Ⓐ Ⓑ Ⓒ Ⓓ Ⓔ
7 Ⓐ Ⓑ Ⓒ Ⓓ Ⓔ	17 Ⓐ Ⓑ Ⓒ Ⓓ Ⓔ	27 Ⓐ Ⓑ Ⓒ Ⓓ Ⓔ	37 Ⓐ Ⓑ Ⓒ Ⓓ Ⓔ
8 Ⓐ Ⓑ Ⓒ Ⓓ Ⓔ	18 Ⓐ Ⓑ Ⓒ Ⓓ Ⓔ	28 Ⓐ Ⓑ Ⓒ Ⓓ Ⓔ	38 Ⓐ Ⓑ Ⓒ Ⓓ Ⓔ
9 Ⓐ Ⓑ Ⓒ Ⓓ Ⓔ	19 Ⓐ Ⓑ Ⓒ Ⓓ Ⓔ	29 Ⓐ Ⓑ Ⓒ Ⓓ Ⓔ	39 Ⓐ Ⓑ Ⓒ Ⓓ Ⓔ
10 Ⓐ Ⓑ Ⓒ Ⓓ Ⓔ	20 Ⓐ Ⓑ Ⓒ Ⓓ Ⓔ	30 Ⓐ Ⓑ Ⓒ Ⓓ Ⓔ	40 Ⓐ Ⓑ Ⓒ Ⓓ Ⓔ

ANSWER SHEETS - PRACTICE TEST 2

SECTION 1

Begin your essay on this page. If necessary, continue on the next page.

Continue on the next page if necessary.

Continuation of your essay from previous page, if necessary.

Begin with number 1 for each section. If a section has fewer questions than answer spaces, leave the extra answer spaces blank. Be sure to erase any errors or stray marks completely.

SECTION 2

1 A B C D E	11 A B C D E	21 A B C D E	31 A B C D E
2 A B C D E	12 A B C D E	22 A B C D E	32 A B C D E
3 A B C D E	13 A B C D E	23 A B C D E	33 A B C D E
4 A B C D E	14 A B C D E	24 A B C D E	34 A B C D E
5 A B C D E	15 A B C D E	25 A B C D E	35 A B C D E
6 A B C D E	16 A B C D E	26 A B C D E	36 A B C D E
7 A B C D E	17 A B C D E	27 A B C D E	37 A B C D E
8 A B C D E	18 A B C D E	28 A B C D E	38 A B C D E
9 A B C D E	19 A B C D E	29 A B C D E	39 A B C D E
10 A B C D E	20 A B C D E	30 A B C D E	40 A B C D E

SECTION 3

1 A B C D E	11 A B C D E	21 A B C D E	31 A B C D E
2 A B C D E	12 A B C D E	22 A B C D E	32 A B C D E
3 A B C D E	13 A B C D E	23 A B C D E	33 A B C D E
4 A B C D E	14 A B C D E	24 A B C D E	34 A B C D E
5 A B C D E	15 A B C D E	25 A B C D E	35 A B C D E
6 A B C D E	16 A B C D E	26 A B C D E	36 A B C D E
7 A B C D E	17 A B C D E	27 A B C D E	37 A B C D E
8 A B C D E	18 A B C D E	28 A B C D E	38 A B C D E
9 A B C D E	19 A B C D E	29 A B C D E	39 A B C D E
10 A B C D E	20 A B C D E	30 A B C D E	40 A B C D E

SECTION 4

1 A B C D E	11 A B C D E	21 A B C D E	31 A B C D E
2 A B C D E	12 A B C D E	22 A B C D E	32 A B C D E
3 A B C D E	13 A B C D E	23 A B C D E	33 A B C D E
4 A B C D E	14 A B C D E	24 A B C D E	34 A B C D E
5 A B C D E	15 A B C D E	25 A B C D E	35 A B C D E
6 A B C D E	16 A B C D E	26 A B C D E	36 A B C D E
7 A B C D E	17 A B C D E	27 A B C D E	37 A B C D E
8 A B C D E	18 A B C D E	28 A B C D E	38 A B C D E
9 A B C D E	19 A B C D E	29 A B C D E	39 A B C D E
10 A B C D E	20 A B C D E	30 A B C D E	40 A B C D E

Begin with number 1 for each section. If a section has fewer questions than answer spaces, leave the extra answer spaces blank. Be sure to erase any errors or stray marks completely.

SECTION 5

1 A B C D E	11 A B C D E	21 A B C D E	31 A B C D E
2 A B C D E	12 A B C D E	22 A B C D E	32 A B C D E
3 A B C D E	13 A B C D E	23 A B C D E	33 A B C D E
4 A B C D E	14 A B C D E	24 A B C D E	34 A B C D E
5 A B C D E	15 A B C D E	25 A B C D E	35 A B C D E
6 A B C D E	16 A B C D E	26 A B C D E	36 A B C D E
7 A B C D E	17 A B C D E	27 A B C D E	37 A B C D E
8 A B C D E	18 A B C D E	28 A B C D E	38 A B C D E
9 A B C D E	19 A B C D E	29 A B C D E	39 A B C D E
10 A B C D E	20 A B C D E	30 A B C D E	40 A B C D E

SECTION 6

1 A B C D E	11 A B C D E	21 A B C D E	31 A B C D E
2 A B C D E	12 A B C D E	22 A B C D E	32 A B C D E
3 A B C D E	13 A B C D E	23 A B C D E	33 A B C D E
4 A B C D E	14 A B C D E	24 A B C D E	34 A B C D E
5 A B C D E	15 A B C D E	25 A B C D E	35 A B C D E
6 A B C D E	16 A B C D E	26 A B C D E	36 A B C D E
7 A B C D E	17 A B C D E	27 A B C D E	37 A B C D E
8 A B C D E	18 A B C D E	28 A B C D E	38 A B C D E
9 A B C D E	19 A B C D E	29 A B C D E	39 A B C D E
10 A B C D E	20 A B C D E	30 A B C D E	40 A B C D E

SECTION 7

1 A B C D E	11 A B C D E	21 A B C D E	31 A B C D E
2 A B C D E	12 A B C D E	22 A B C D E	32 A B C D E
3 A B C D E	13 A B C D E	23 A B C D E	33 A B C D E
4 A B C D E	14 A B C D E	24 A B C D E	34 A B C D E
5 A B C D E	15 A B C D E	25 A B C D E	35 A B C D E
6 A B C D E	16 A B C D E	26 A B C D E	36 A B C D E
7 A B C D E	17 A B C D E	27 A B C D E	37 A B C D E
8 A B C D E	18 A B C D E	28 A B C D E	38 A B C D E
9 A B C D E	19 A B C D E	29 A B C D E	39 A B C D E
10 A B C D E	20 A B C D E	30 A B C D E	40 A B C D E

CONTINUE TO THE GRIDS FOR SECTION 7 ON THE NEXT PAGE WHEN INDICATED IN THE TEST BOOK.

Only answers entered in the ovals in each grid area will be scored.
No credit will be given for anything written in the boxes above the ovals.

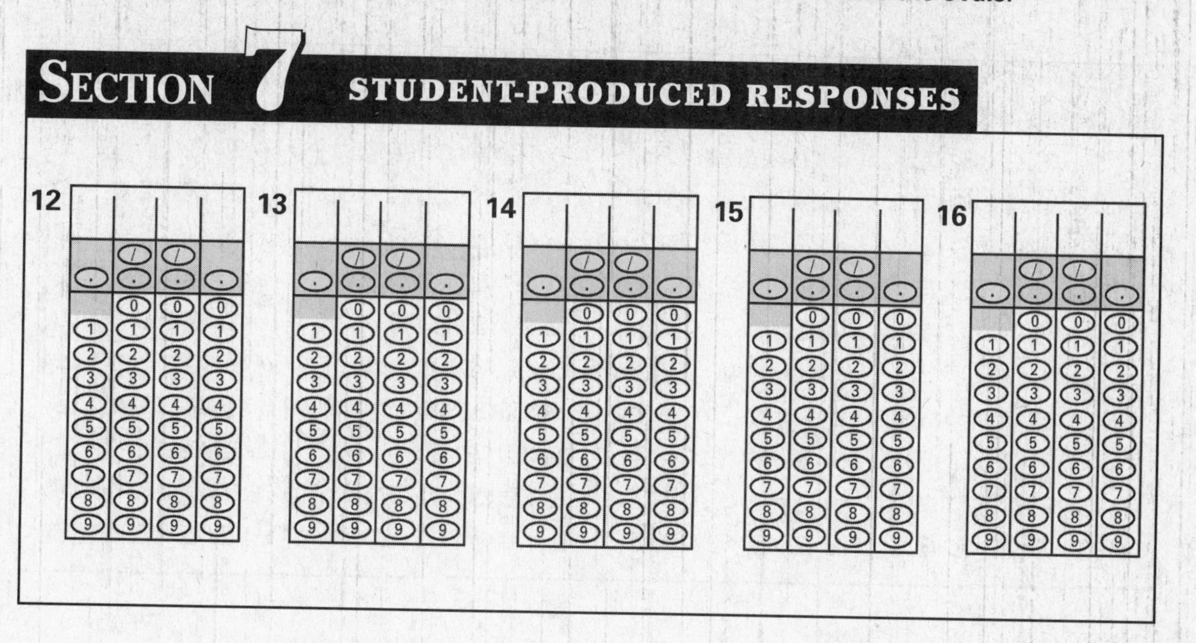

Begin with number 1 for each section. If a section has fewer questions than answer spaces, leave the extra answer spaces blank. Be sure to erase any errors or stray marks completely.

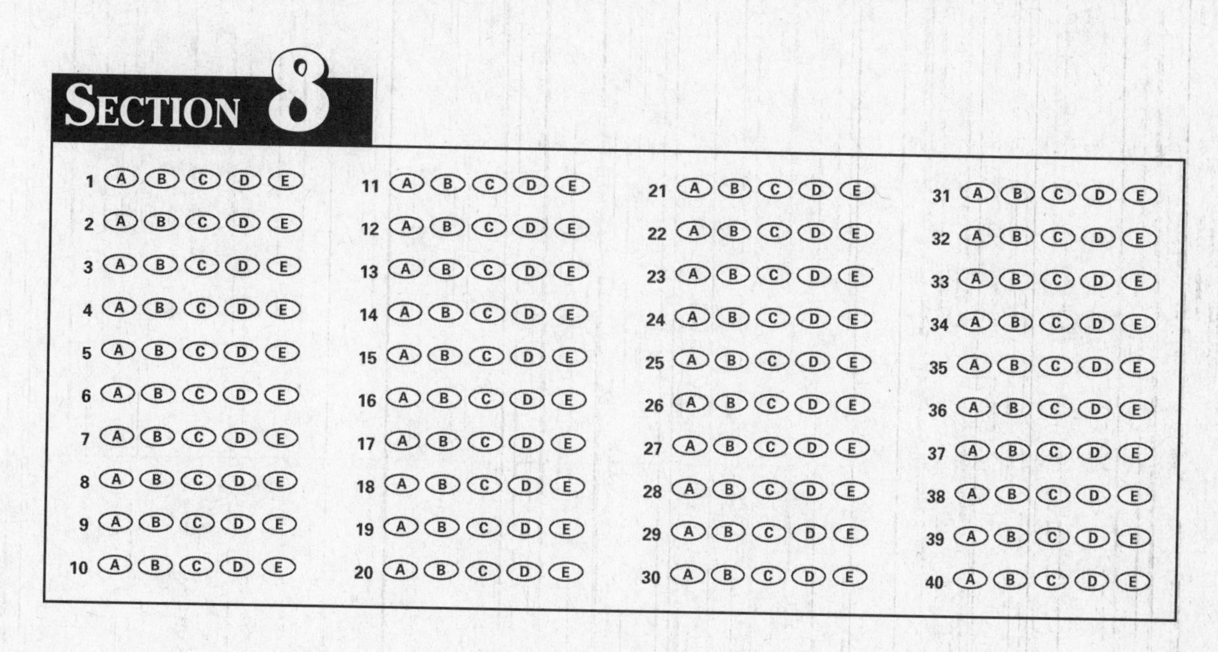

Begin with number 1 for each section. If a section has fewer questions than answer spaces, leave the extra answer spaces blank. Be sure to erase any errors or stray marks completely.

SECTION 9

1 Ⓐ Ⓑ Ⓒ Ⓓ Ⓔ	11 Ⓐ Ⓑ Ⓒ Ⓓ Ⓔ	21 Ⓐ Ⓑ Ⓒ Ⓓ Ⓔ	31 Ⓐ Ⓑ Ⓒ Ⓓ Ⓔ
2 Ⓐ Ⓑ Ⓒ Ⓓ Ⓔ	12 Ⓐ Ⓑ Ⓒ Ⓓ Ⓔ	22 Ⓐ Ⓑ Ⓒ Ⓓ Ⓔ	32 Ⓐ Ⓑ Ⓒ Ⓓ Ⓔ
3 Ⓐ Ⓑ Ⓒ Ⓓ Ⓔ	13 Ⓐ Ⓑ Ⓒ Ⓓ Ⓔ	23 Ⓐ Ⓑ Ⓒ Ⓓ Ⓔ	33 Ⓐ Ⓑ Ⓒ Ⓓ Ⓔ
4 Ⓐ Ⓑ Ⓒ Ⓓ Ⓔ	14 Ⓐ Ⓑ Ⓒ Ⓓ Ⓔ	24 Ⓐ Ⓑ Ⓒ Ⓓ Ⓔ	34 Ⓐ Ⓑ Ⓒ Ⓓ Ⓔ
5 Ⓐ Ⓑ Ⓒ Ⓓ Ⓔ	15 Ⓐ Ⓑ Ⓒ Ⓓ Ⓔ	25 Ⓐ Ⓑ Ⓒ Ⓓ Ⓔ	35 Ⓐ Ⓑ Ⓒ Ⓓ Ⓔ
6 Ⓐ Ⓑ Ⓒ Ⓓ Ⓔ	16 Ⓐ Ⓑ Ⓒ Ⓓ Ⓔ	26 Ⓐ Ⓑ Ⓒ Ⓓ Ⓔ	36 Ⓐ Ⓑ Ⓒ Ⓓ Ⓔ
7 Ⓐ Ⓑ Ⓒ Ⓓ Ⓔ	17 Ⓐ Ⓑ Ⓒ Ⓓ Ⓔ	27 Ⓐ Ⓑ Ⓒ Ⓓ Ⓔ	37 Ⓐ Ⓑ Ⓒ Ⓓ Ⓔ
8 Ⓐ Ⓑ Ⓒ Ⓓ Ⓔ	18 Ⓐ Ⓑ Ⓒ Ⓓ Ⓔ	28 Ⓐ Ⓑ Ⓒ Ⓓ Ⓔ	38 Ⓐ Ⓑ Ⓒ Ⓓ Ⓔ
9 Ⓐ Ⓑ Ⓒ Ⓓ Ⓔ	19 Ⓐ Ⓑ Ⓒ Ⓓ Ⓔ	29 Ⓐ Ⓑ Ⓒ Ⓓ Ⓔ	39 Ⓐ Ⓑ Ⓒ Ⓓ Ⓔ
10 Ⓐ Ⓑ Ⓒ Ⓓ Ⓔ	20 Ⓐ Ⓑ Ⓒ Ⓓ Ⓔ	30 Ⓐ Ⓑ Ⓒ Ⓓ Ⓔ	40 Ⓐ Ⓑ Ⓒ Ⓓ Ⓔ

ANSWER SHEETS - PRACTICE TEST 3

Begin your essay on this page. If necessary, continue on the next page.

Continue on the next page if necessary.

Continuation of your essay from previous page, if necessary.

Begin with number 1 for each section. If a section has fewer questions than answer spaces, leave the extra answer spaces blank. Be sure to erase any errors or stray marks completely.

SECTION 2

1	Ⓐ Ⓑ Ⓒ Ⓓ Ⓔ	11	Ⓐ Ⓑ Ⓒ Ⓓ Ⓔ	21	Ⓐ Ⓑ Ⓒ Ⓓ Ⓔ	31	Ⓐ Ⓑ Ⓒ Ⓓ Ⓔ
2	Ⓐ Ⓑ Ⓒ Ⓓ Ⓔ	12	Ⓐ Ⓑ Ⓒ Ⓓ Ⓔ	22	Ⓐ Ⓑ Ⓒ Ⓓ Ⓔ	32	Ⓐ Ⓑ Ⓒ Ⓓ Ⓔ
3	Ⓐ Ⓑ Ⓒ Ⓓ Ⓔ	13	Ⓐ Ⓑ Ⓒ Ⓓ Ⓔ	23	Ⓐ Ⓑ Ⓒ Ⓓ Ⓔ	33	Ⓐ Ⓑ Ⓒ Ⓓ Ⓔ
4	Ⓐ Ⓑ Ⓒ Ⓓ Ⓔ	14	Ⓐ Ⓑ Ⓒ Ⓓ Ⓔ	24	Ⓐ Ⓑ Ⓒ Ⓓ Ⓔ	34	Ⓐ Ⓑ Ⓒ Ⓓ Ⓔ
5	Ⓐ Ⓑ Ⓒ Ⓓ Ⓔ	15	Ⓐ Ⓑ Ⓒ Ⓓ Ⓔ	25	Ⓐ Ⓑ Ⓒ Ⓓ Ⓔ	35	Ⓐ Ⓑ Ⓒ Ⓓ Ⓔ
6	Ⓐ Ⓑ Ⓒ Ⓓ Ⓔ	16	Ⓐ Ⓑ Ⓒ Ⓓ Ⓔ	26	Ⓐ Ⓑ Ⓒ Ⓓ Ⓔ	36	Ⓐ Ⓑ Ⓒ Ⓓ Ⓔ
7	Ⓐ Ⓑ Ⓒ Ⓓ Ⓔ	17	Ⓐ Ⓑ Ⓒ Ⓓ Ⓔ	27	Ⓐ Ⓑ Ⓒ Ⓓ Ⓔ	37	Ⓐ Ⓑ Ⓒ Ⓓ Ⓔ
8	Ⓐ Ⓑ Ⓒ Ⓓ Ⓔ	18	Ⓐ Ⓑ Ⓒ Ⓓ Ⓔ	28	Ⓐ Ⓑ Ⓒ Ⓓ Ⓔ	38	Ⓐ Ⓑ Ⓒ Ⓓ Ⓔ
9	Ⓐ Ⓑ Ⓒ Ⓓ Ⓔ	19	Ⓐ Ⓑ Ⓒ Ⓓ Ⓔ	29	Ⓐ Ⓑ Ⓒ Ⓓ Ⓔ	39	Ⓐ Ⓑ Ⓒ Ⓓ Ⓔ
10	Ⓐ Ⓑ Ⓒ Ⓓ Ⓔ	20	Ⓐ Ⓑ Ⓒ Ⓓ Ⓔ	30	Ⓐ Ⓑ Ⓒ Ⓓ Ⓔ	40	Ⓐ Ⓑ Ⓒ Ⓓ Ⓔ

SECTION 3

1	Ⓐ Ⓑ Ⓒ Ⓓ Ⓔ	11	Ⓐ Ⓑ Ⓒ Ⓓ Ⓔ	21	Ⓐ Ⓑ Ⓒ Ⓓ Ⓔ	31	Ⓐ Ⓑ Ⓒ Ⓓ Ⓔ
2	Ⓐ Ⓑ Ⓒ Ⓓ Ⓔ	12	Ⓐ Ⓑ Ⓒ Ⓓ Ⓔ	22	Ⓐ Ⓑ Ⓒ Ⓓ Ⓔ	32	Ⓐ Ⓑ Ⓒ Ⓓ Ⓔ
3	Ⓐ Ⓑ Ⓒ Ⓓ Ⓔ	13	Ⓐ Ⓑ Ⓒ Ⓓ Ⓔ	23	Ⓐ Ⓑ Ⓒ Ⓓ Ⓔ	33	Ⓐ Ⓑ Ⓒ Ⓓ Ⓔ
4	Ⓐ Ⓑ Ⓒ Ⓓ Ⓔ	14	Ⓐ Ⓑ Ⓒ Ⓓ Ⓔ	24	Ⓐ Ⓑ Ⓒ Ⓓ Ⓔ	34	Ⓐ Ⓑ Ⓒ Ⓓ Ⓔ
5	Ⓐ Ⓑ Ⓒ Ⓓ Ⓔ	15	Ⓐ Ⓑ Ⓒ Ⓓ Ⓔ	25	Ⓐ Ⓑ Ⓒ Ⓓ Ⓔ	35	Ⓐ Ⓑ Ⓒ Ⓓ Ⓔ
6	Ⓐ Ⓑ Ⓒ Ⓓ Ⓔ	16	Ⓐ Ⓑ Ⓒ Ⓓ Ⓔ	26	Ⓐ Ⓑ Ⓒ Ⓓ Ⓔ	36	Ⓐ Ⓑ Ⓒ Ⓓ Ⓔ
7	Ⓐ Ⓑ Ⓒ Ⓓ Ⓔ	17	Ⓐ Ⓑ Ⓒ Ⓓ Ⓔ	27	Ⓐ Ⓑ Ⓒ Ⓓ Ⓔ	37	Ⓐ Ⓑ Ⓒ Ⓓ Ⓔ
8	Ⓐ Ⓑ Ⓒ Ⓓ Ⓔ	18	Ⓐ Ⓑ Ⓒ Ⓓ Ⓔ	28	Ⓐ Ⓑ Ⓒ Ⓓ Ⓔ	38	Ⓐ Ⓑ Ⓒ Ⓓ Ⓔ
9	Ⓐ Ⓑ Ⓒ Ⓓ Ⓔ	19	Ⓐ Ⓑ Ⓒ Ⓓ Ⓔ	29	Ⓐ Ⓑ Ⓒ Ⓓ Ⓔ	39	Ⓐ Ⓑ Ⓒ Ⓓ Ⓔ
10	Ⓐ Ⓑ Ⓒ Ⓓ Ⓔ	20	Ⓐ Ⓑ Ⓒ Ⓓ Ⓔ	30	Ⓐ Ⓑ Ⓒ Ⓓ Ⓔ	40	Ⓐ Ⓑ Ⓒ Ⓓ Ⓔ

SECTION 4

1	Ⓐ Ⓑ Ⓒ Ⓓ Ⓔ	11	Ⓐ Ⓑ Ⓒ Ⓓ Ⓔ	21	Ⓐ Ⓑ Ⓒ Ⓓ Ⓔ	31	Ⓐ Ⓑ Ⓒ Ⓓ Ⓔ
2	Ⓐ Ⓑ Ⓒ Ⓓ Ⓔ	12	Ⓐ Ⓑ Ⓒ Ⓓ Ⓔ	22	Ⓐ Ⓑ Ⓒ Ⓓ Ⓔ	32	Ⓐ Ⓑ Ⓒ Ⓓ Ⓔ
3	Ⓐ Ⓑ Ⓒ Ⓓ Ⓔ	13	Ⓐ Ⓑ Ⓒ Ⓓ Ⓔ	23	Ⓐ Ⓑ Ⓒ Ⓓ Ⓔ	33	Ⓐ Ⓑ Ⓒ Ⓓ Ⓔ
4	Ⓐ Ⓑ Ⓒ Ⓓ Ⓔ	14	Ⓐ Ⓑ Ⓒ Ⓓ Ⓔ	24	Ⓐ Ⓑ Ⓒ Ⓓ Ⓔ	34	Ⓐ Ⓑ Ⓒ Ⓓ Ⓔ
5	Ⓐ Ⓑ Ⓒ Ⓓ Ⓔ	15	Ⓐ Ⓑ Ⓒ Ⓓ Ⓔ	25	Ⓐ Ⓑ Ⓒ Ⓓ Ⓔ	35	Ⓐ Ⓑ Ⓒ Ⓓ Ⓔ
6	Ⓐ Ⓑ Ⓒ Ⓓ Ⓔ	16	Ⓐ Ⓑ Ⓒ Ⓓ Ⓔ	26	Ⓐ Ⓑ Ⓒ Ⓓ Ⓔ	36	Ⓐ Ⓑ Ⓒ Ⓓ Ⓔ
7	Ⓐ Ⓑ Ⓒ Ⓓ Ⓔ	17	Ⓐ Ⓑ Ⓒ Ⓓ Ⓔ	27	Ⓐ Ⓑ Ⓒ Ⓓ Ⓔ	37	Ⓐ Ⓑ Ⓒ Ⓓ Ⓔ
8	Ⓐ Ⓑ Ⓒ Ⓓ Ⓔ	18	Ⓐ Ⓑ Ⓒ Ⓓ Ⓔ	28	Ⓐ Ⓑ Ⓒ Ⓓ Ⓔ	38	Ⓐ Ⓑ Ⓒ Ⓓ Ⓔ
9	Ⓐ Ⓑ Ⓒ Ⓓ Ⓔ	19	Ⓐ Ⓑ Ⓒ Ⓓ Ⓔ	29	Ⓐ Ⓑ Ⓒ Ⓓ Ⓔ	39	Ⓐ Ⓑ Ⓒ Ⓓ Ⓔ
10	Ⓐ Ⓑ Ⓒ Ⓓ Ⓔ	20	Ⓐ Ⓑ Ⓒ Ⓓ Ⓔ	30	Ⓐ Ⓑ Ⓒ Ⓓ Ⓔ	40	Ⓐ Ⓑ Ⓒ Ⓓ Ⓔ

Begin with number 1 for each section. If a section has fewer questions than answer spaces, leave the extra answer spaces blank. Be sure to erase any errors or stray marks completely.

SECTION 5

1 (A) (B) (C) (D) (E)	11 (A) (B) (C) (D) (E)	21 (A) (B) (C) (D) (E)	31 (A) (B) (C) (D) (E)
2 (A) (B) (C) (D) (E)	12 (A) (B) (C) (D) (E)	22 (A) (B) (C) (D) (E)	32 (A) (B) (C) (D) (E)
3 (A) (B) (C) (D) (E)	13 (A) (B) (C) (D) (E)	23 (A) (B) (C) (D) (E)	33 (A) (B) (C) (D) (E)
4 (A) (B) (C) (D) (E)	14 (A) (B) (C) (D) (E)	24 (A) (B) (C) (D) (E)	34 (A) (B) (C) (D) (E)
5 (A) (B) (C) (D) (E)	15 (A) (B) (C) (D) (E)	25 (A) (B) (C) (D) (E)	35 (A) (B) (C) (D) (E)
6 (A) (B) (C) (D) (E)	16 (A) (B) (C) (D) (E)	26 (A) (B) (C) (D) (E)	36 (A) (B) (C) (D) (E)
7 (A) (B) (C) (D) (E)	17 (A) (B) (C) (D) (E)	27 (A) (B) (C) (D) (E)	37 (A) (B) (C) (D) (E)
8 (A) (B) (C) (D) (E)	18 (A) (B) (C) (D) (E)	28 (A) (B) (C) (D) (E)	38 (A) (B) (C) (D) (E)
9 (A) (B) (C) (D) (E)	19 (A) (B) (C) (D) (E)	29 (A) (B) (C) (D) (E)	39 (A) (B) (C) (D) (E)
10 (A) (B) (C) (D) (E)	20 (A) (B) (C) (D) (E)	30 (A) (B) (C) (D) (E)	40 (A) (B) (C) (D) (E)

SECTION 6

1 (A) (B) (C) (D) (E)	11 (A) (B) (C) (D) (E)	21 (A) (B) (C) (D) (E)	31 (A) (B) (C) (D) (E)
2 (A) (B) (C) (D) (E)	12 (A) (B) (C) (D) (E)	22 (A) (B) (C) (D) (E)	32 (A) (B) (C) (D) (E)
3 (A) (B) (C) (D) (E)	13 (A) (B) (C) (D) (E)	23 (A) (B) (C) (D) (E)	33 (A) (B) (C) (D) (E)
4 (A) (B) (C) (D) (E)	14 (A) (B) (C) (D) (E)	24 (A) (B) (C) (D) (E)	34 (A) (B) (C) (D) (E)
5 (A) (B) (C) (D) (E)	15 (A) (B) (C) (D) (E)	25 (A) (B) (C) (D) (E)	35 (A) (B) (C) (D) (E)
6 (A) (B) (C) (D) (E)	16 (A) (B) (C) (D) (E)	26 (A) (B) (C) (D) (E)	36 (A) (B) (C) (D) (E)
7 (A) (B) (C) (D) (E)	17 (A) (B) (C) (D) (E)	27 (A) (B) (C) (D) (E)	37 (A) (B) (C) (D) (E)
8 (A) (B) (C) (D) (E)	18 (A) (B) (C) (D) (E)	28 (A) (B) (C) (D) (E)	38 (A) (B) (C) (D) (E)
9 (A) (B) (C) (D) (E)	19 (A) (B) (C) (D) (E)	29 (A) (B) (C) (D) (E)	39 (A) (B) (C) (D) (E)
10 (A) (B) (C) (D) (E)	20 (A) (B) (C) (D) (E)	30 (A) (B) (C) (D) (E)	40 (A) (B) (C) (D) (E)

SECTION 7

1 (A) (B) (C) (D) (E)	11 (A) (B) (C) (D) (E)	21 (A) (B) (C) (D) (E)	31 (A) (B) (C) (D) (E)
2 (A) (B) (C) (D) (E)	12 (A) (B) (C) (D) (E)	22 (A) (B) (C) (D) (E)	32 (A) (B) (C) (D) (E)
3 (A) (B) (C) (D) (E)	13 (A) (B) (C) (D) (E)	23 (A) (B) (C) (D) (E)	33 (A) (B) (C) (D) (E)
4 (A) (B) (C) (D) (E)	14 (A) (B) (C) (D) (E)	24 (A) (B) (C) (D) (E)	34 (A) (B) (C) (D) (E)
5 (A) (B) (C) (D) (E)	15 (A) (B) (C) (D) (E)	25 (A) (B) (C) (D) (E)	35 (A) (B) (C) (D) (E)
6 (A) (B) (C) (D) (E)	16 (A) (B) (C) (D) (E)	26 (A) (B) (C) (D) (E)	36 (A) (B) (C) (D) (E)
7 (A) (B) (C) (D) (E)	17 (A) (B) (C) (D) (E)	27 (A) (B) (C) (D) (E)	37 (A) (B) (C) (D) (E)
8 (A) (B) (C) (D) (E)	18 (A) (B) (C) (D) (E)	28 (A) (B) (C) (D) (E)	38 (A) (B) (C) (D) (E)
9 (A) (B) (C) (D) (E)	19 (A) (B) (C) (D) (E)	29 (A) (B) (C) (D) (E)	39 (A) (B) (C) (D) (E)
10 (A) (B) (C) (D) (E)	20 (A) (B) (C) (D) (E)	30 (A) (B) (C) (D) (E)	40 (A) (B) (C) (D) (E)

CONTINUE TO THE GRIDS FOR SECTION 7 ON THE NEXT PAGE WHEN INDICATED IN THE TEST BOOK.

Only answers entered in the ovals in each grid area will be scored.
No credit will be given for anything written in the boxes above the ovals.

SECTION 7 STUDENT-PRODUCED RESPONSES

7 8 9 10 11

12 13 14 15 16

Begin with number 1 for each section. If a section has fewer questions than answer spaces, leave the extra answer spaces blank. Be sure to erase any errors or stray marks completely.

SECTION 8

1 Ⓐ Ⓑ Ⓒ Ⓓ Ⓔ	11 Ⓐ Ⓑ Ⓒ Ⓓ Ⓔ	21 Ⓐ Ⓑ Ⓒ Ⓓ Ⓔ	31 Ⓐ Ⓑ Ⓒ Ⓓ Ⓔ
2 Ⓐ Ⓑ Ⓒ Ⓓ Ⓔ	12 Ⓐ Ⓑ Ⓒ Ⓓ Ⓔ	22 Ⓐ Ⓑ Ⓒ Ⓓ Ⓔ	32 Ⓐ Ⓑ Ⓒ Ⓓ Ⓔ
3 Ⓐ Ⓑ Ⓒ Ⓓ Ⓔ	13 Ⓐ Ⓑ Ⓒ Ⓓ Ⓔ	23 Ⓐ Ⓑ Ⓒ Ⓓ Ⓔ	33 Ⓐ Ⓑ Ⓒ Ⓓ Ⓔ
4 Ⓐ Ⓑ Ⓒ Ⓓ Ⓔ	14 Ⓐ Ⓑ Ⓒ Ⓓ Ⓔ	24 Ⓐ Ⓑ Ⓒ Ⓓ Ⓔ	34 Ⓐ Ⓑ Ⓒ Ⓓ Ⓔ
5 Ⓐ Ⓑ Ⓒ Ⓓ Ⓔ	15 Ⓐ Ⓑ Ⓒ Ⓓ Ⓔ	25 Ⓐ Ⓑ Ⓒ Ⓓ Ⓔ	35 Ⓐ Ⓑ Ⓒ Ⓓ Ⓔ
6 Ⓐ Ⓑ Ⓒ Ⓓ Ⓔ	16 Ⓐ Ⓑ Ⓒ Ⓓ Ⓔ	26 Ⓐ Ⓑ Ⓒ Ⓓ Ⓔ	36 Ⓐ Ⓑ Ⓒ Ⓓ Ⓔ
7 Ⓐ Ⓑ Ⓒ Ⓓ Ⓔ	17 Ⓐ Ⓑ Ⓒ Ⓓ Ⓔ	27 Ⓐ Ⓑ Ⓒ Ⓓ Ⓔ	37 Ⓐ Ⓑ Ⓒ Ⓓ Ⓔ
8 Ⓐ Ⓑ Ⓒ Ⓓ Ⓔ	18 Ⓐ Ⓑ Ⓒ Ⓓ Ⓔ	28 Ⓐ Ⓑ Ⓒ Ⓓ Ⓔ	38 Ⓐ Ⓑ Ⓒ Ⓓ Ⓔ
9 Ⓐ Ⓑ Ⓒ Ⓓ Ⓔ	19 Ⓐ Ⓑ Ⓒ Ⓓ Ⓔ	29 Ⓐ Ⓑ Ⓒ Ⓓ Ⓔ	39 Ⓐ Ⓑ Ⓒ Ⓓ Ⓔ
10 Ⓐ Ⓑ Ⓒ Ⓓ Ⓔ	20 Ⓐ Ⓑ Ⓒ Ⓓ Ⓔ	30 Ⓐ Ⓑ Ⓒ Ⓓ Ⓔ	40 Ⓐ Ⓑ Ⓒ Ⓓ Ⓔ

Begin with number 1 for each section. If a section has fewer questions than answer spaces, leave the extra answer spaces blank. Be sure to erase any errors or stray marks completely.

SECTION 9

1 Ⓐ Ⓑ Ⓒ Ⓓ Ⓔ	11 Ⓐ Ⓑ Ⓒ Ⓓ Ⓔ	21 Ⓐ Ⓑ Ⓒ Ⓓ Ⓔ	31 Ⓐ Ⓑ Ⓒ Ⓓ Ⓔ
2 Ⓐ Ⓑ Ⓒ Ⓓ Ⓔ	12 Ⓐ Ⓑ Ⓒ Ⓓ Ⓔ	22 Ⓐ Ⓑ Ⓒ Ⓓ Ⓔ	32 Ⓐ Ⓑ Ⓒ Ⓓ Ⓔ
3 Ⓐ Ⓑ Ⓒ Ⓓ Ⓔ	13 Ⓐ Ⓑ Ⓒ Ⓓ Ⓔ	23 Ⓐ Ⓑ Ⓒ Ⓓ Ⓔ	33 Ⓐ Ⓑ Ⓒ Ⓓ Ⓔ
4 Ⓐ Ⓑ Ⓒ Ⓓ Ⓔ	14 Ⓐ Ⓑ Ⓒ Ⓓ Ⓔ	24 Ⓐ Ⓑ Ⓒ Ⓓ Ⓔ	34 Ⓐ Ⓑ Ⓒ Ⓓ Ⓔ
5 Ⓐ Ⓑ Ⓒ Ⓓ Ⓔ	15 Ⓐ Ⓑ Ⓒ Ⓓ Ⓔ	25 Ⓐ Ⓑ Ⓒ Ⓓ Ⓔ	35 Ⓐ Ⓑ Ⓒ Ⓓ Ⓔ
6 Ⓐ Ⓑ Ⓒ Ⓓ Ⓔ	16 Ⓐ Ⓑ Ⓒ Ⓓ Ⓔ	26 Ⓐ Ⓑ Ⓒ Ⓓ Ⓔ	36 Ⓐ Ⓑ Ⓒ Ⓓ Ⓔ
7 Ⓐ Ⓑ Ⓒ Ⓓ Ⓔ	17 Ⓐ Ⓑ Ⓒ Ⓓ Ⓔ	27 Ⓐ Ⓑ Ⓒ Ⓓ Ⓔ	37 Ⓐ Ⓑ Ⓒ Ⓓ Ⓔ
8 Ⓐ Ⓑ Ⓒ Ⓓ Ⓔ	18 Ⓐ Ⓑ Ⓒ Ⓓ Ⓔ	28 Ⓐ Ⓑ Ⓒ Ⓓ Ⓔ	38 Ⓐ Ⓑ Ⓒ Ⓓ Ⓔ
9 Ⓐ Ⓑ Ⓒ Ⓓ Ⓔ	19 Ⓐ Ⓑ Ⓒ Ⓓ Ⓔ	29 Ⓐ Ⓑ Ⓒ Ⓓ Ⓔ	39 Ⓐ Ⓑ Ⓒ Ⓓ Ⓔ
10 Ⓐ Ⓑ Ⓒ Ⓓ Ⓔ	20 Ⓐ Ⓑ Ⓒ Ⓓ Ⓔ	30 Ⓐ Ⓑ Ⓒ Ⓓ Ⓔ	40 Ⓐ Ⓑ Ⓒ Ⓓ Ⓔ

ANSWER SHEETS - PRACTICE TEST 4

Begin your essay on this page. If necessary, continue on the next page.

Continue on the next page if necessary.

Continuation of your essay from previous page, if necessary.

Begin with number 1 for each section. If a section has fewer questions than answer spaces, leave the extra answer spaces blank. Be sure to erase any errors or stray marks completely.

SECTION 2

1	A B C D E	11	A B C D E	21	A B C D E	31	A B C D E
2	A B C D E	12	A B C D E	22	A B C D E	32	A B C D E
3	A B C D E	13	A B C D E	23	A B C D E	33	A B C D E
4	A B C D E	14	A B C D E	24	A B C D E	34	A B C D E
5	A B C D E	15	A B C D E	25	A B C D E	35	A B C D E
6	A B C D E	16	A B C D E	26	A B C D E	36	A B C D E
7	A B C D E	17	A B C D E	27	A B C D E	37	A B C D E
8	A B C D E	18	A B C D E	28	A B C D E	38	A B C D E
9	A B C D E	19	A B C D E	29	A B C D E	39	A B C D E
10	A B C D E	20	A B C D E	30	A B C D E	40	A B C D E

SECTION 3

1	A B C D E	11	A B C D E	21	A B C D E	31	A B C D E
2	A B C D E	12	A B C D E	22	A B C D E	32	A B C D E
3	A B C D E	13	A B C D E	23	A B C D E	33	A B C D E
4	A B C D E	14	A B C D E	24	A B C D E	34	A B C D E
5	A B C D E	15	A B C D E	25	A B C D E	35	A B C D E
6	A B C D E	16	A B C D E	26	A B C D E	36	A B C D E
7	A B C D E	17	A B C D E	27	A B C D E	37	A B C D E
8	A B C D E	18	A B C D E	28	A B C D E	38	A B C D E
9	A B C D E	19	A B C D E	29	A B C D E	39	A B C D E
10	A B C D E	20	A B C D E	30	A B C D E	40	A B C D E

SECTION 4

1	A B C D E	11	A B C D E	21	A B C D E	31	A B C D E
2	A B C D E	12	A B C D E	22	A B C D E	32	A B C D E
3	A B C D E	13	A B C D E	23	A B C D E	33	A B C D E
4	A B C D E	14	A B C D E	24	A B C D E	34	A B C D E
5	A B C D E	15	A B C D E	25	A B C D E	35	A B C D E
6	A B C D E	16	A B C D E	26	A B C D E	36	A B C D E
7	A B C D E	17	A B C D E	27	A B C D E	37	A B C D E
8	A B C D E	18	A B C D E	28	A B C D E	38	A B C D E
9	A B C D E	19	A B C D E	29	A B C D E	39	A B C D E
10	A B C D E	20	A B C D E	30	A B C D E	40	A B C D E

Begin with number 1 for each section. If a section has fewer questions than answer spaces, leave the extra answer spaces blank. Be sure to erase any errors or stray marks completely.

SECTION 5

1	A B C D E	11	A B C D E	21	A B C D E	31	A B C D E
2	A B C D E	12	A B C D E	22	A B C D E	32	A B C D E
3	A B C D E	13	A B C D E	23	A B C D E	33	A B C D E
4	A B C D E	14	A B C D E	24	A B C D E	34	A B C D E
5	A B C D E	15	A B C D E	25	A B C D E	35	A B C D E
6	A B C D E	16	A B C D E	26	A B C D E	36	A B C D E
7	A B C D E	17	A B C D E	27	A B C D E	37	A B C D E
8	A B C D E	18	A B C D E	28	A B C D E	38	A B C D E
9	A B C D E	19	A B C D E	29	A B C D E	39	A B C D E
10	A B C D E	20	A B C D E	30	A B C D E	40	A B C D E

SECTION 6

1	A B C D E	11	A B C D E	21	A B C D E	31	A B C D E
2	A B C D E	12	A B C D E	22	A B C D E	32	A B C D E
3	A B C D E	13	A B C D E	23	A B C D E	33	A B C D E
4	A B C D E	14	A B C D E	24	A B C D E	34	A B C D E
5	A B C D E	15	A B C D E	25	A B C D E	35	A B C D E
6	A B C D E	16	A B C D E	26	A B C D E	36	A B C D E
7	A B C D E	17	A B C D E	27	A B C D E	37	A B C D E
8	A B C D E	18	A B C D E	28	A B C D E	38	A B C D E
9	A B C D E	19	A B C D E	29	A B C D E	39	A B C D E
10	A B C D E	20	A B C D E	30	A B C D E	40	A B C D E

SECTION 7

1	A B C D E	11	A B C D E	21	A B C D E	31	A B C D E
2	A B C D E	12	A B C D E	22	A B C D E	32	A B C D E
3	A B C D E	13	A B C D E	23	A B C D E	33	A B C D E
4	A B C D E	14	A B C D E	24	A B C D E	34	A B C D E
5	A B C D E	15	A B C D E	25	A B C D E	35	A B C D E
6	A B C D E	16	A B C D E	26	A B C D E	36	A B C D E
7	A B C D E	17	A B C D E	27	A B C D E	37	A B C D E
8	A B C D E	18	A B C D E	28	A B C D E	38	A B C D E
9	A B C D E	19	A B C D E	29	A B C D E	39	A B C D E
10	A B C D E	20	A B C D E	30	A B C D E	40	A B C D E

CONTINUE TO THE GRIDS FOR SECTION 7 ON THE NEXT PAGE WHEN INDICATED IN THE TEST BOOK.

Only answers entered in the ovals in each grid area will be scored.
No credit will be given for anything written in the boxes above the ovals.

SECTION 7 STUDENT-PRODUCED RESPONSES

6

7

8

9

10

11

12

13

14

15

16

Begin with number 1 for each section. If a section has fewer questions than answer spaces, leave the extra answer spaces blank. Be sure to erase any errors or stray marks completely.

SECTION 8

1 (A) (B) (C) (D) (E)	11 (A) (B) (C) (D) (E)	21 (A) (B) (C) (D) (E)	31 (A) (B) (C) (D) (E)
2 (A) (B) (C) (D) (E)	12 (A) (B) (C) (D) (E)	22 (A) (B) (C) (D) (E)	32 (A) (B) (C) (D) (E)
3 (A) (B) (C) (D) (E)	13 (A) (B) (C) (D) (E)	23 (A) (B) (C) (D) (E)	33 (A) (B) (C) (D) (E)
4 (A) (B) (C) (D) (E)	14 (A) (B) (C) (D) (E)	24 (A) (B) (C) (D) (E)	34 (A) (B) (C) (D) (E)
5 (A) (B) (C) (D) (E)	15 (A) (B) (C) (D) (E)	25 (A) (B) (C) (D) (E)	35 (A) (B) (C) (D) (E)
6 (A) (B) (C) (D) (E)	16 (A) (B) (C) (D) (E)	26 (A) (B) (C) (D) (E)	36 (A) (B) (C) (D) (E)
7 (A) (B) (C) (D) (E)	17 (A) (B) (C) (D) (E)	27 (A) (B) (C) (D) (E)	37 (A) (B) (C) (D) (E)
8 (A) (B) (C) (D) (E)	18 (A) (B) (C) (D) (E)	28 (A) (B) (C) (D) (E)	38 (A) (B) (C) (D) (E)
9 (A) (B) (C) (D) (E)	19 (A) (B) (C) (D) (E)	29 (A) (B) (C) (D) (E)	39 (A) (B) (C) (D) (E)
10 (A) (B) (C) (D) (E)	20 (A) (B) (C) (D) (E)	30 (A) (B) (C) (D) (E)	40 (A) (B) (C) (D) (E)

SECTION 9

1 (A) (B) (C) (D) (E)	11 (A) (B) (C) (D) (E)	21 (A) (B) (C) (D) (E)	31 (A) (B) (C) (D) (E)
2 (A) (B) (C) (D) (E)	12 (A) (B) (C) (D) (E)	22 (A) (B) (C) (D) (E)	32 (A) (B) (C) (D) (E)
3 (A) (B) (C) (D) (E)	13 (A) (B) (C) (D) (E)	23 (A) (B) (C) (D) (E)	33 (A) (B) (C) (D) (E)
4 (A) (B) (C) (D) (E)	14 (A) (B) (C) (D) (E)	24 (A) (B) (C) (D) (E)	34 (A) (B) (C) (D) (E)
5 (A) (B) (C) (D) (E)	15 (A) (B) (C) (D) (E)	25 (A) (B) (C) (D) (E)	35 (A) (B) (C) (D) (E)
6 (A) (B) (C) (D) (E)	16 (A) (B) (C) (D) (E)	26 (A) (B) (C) (D) (E)	36 (A) (B) (C) (D) (E)
7 (A) (B) (C) (D) (E)	17 (A) (B) (C) (D) (E)	27 (A) (B) (C) (D) (E)	37 (A) (B) (C) (D) (E)
8 (A) (B) (C) (D) (E)	18 (A) (B) (C) (D) (E)	28 (A) (B) (C) (D) (E)	38 (A) (B) (C) (D) (E)
9 (A) (B) (C) (D) (E)	19 (A) (B) (C) (D) (E)	29 (A) (B) (C) (D) (E)	39 (A) (B) (C) (D) (E)
10 (A) (B) (C) (D) (E)	20 (A) (B) (C) (D) (E)	30 (A) (B) (C) (D) (E)	40 (A) (B) (C) (D) (E)